Collective Excitations in Solids

NATO Advanced Science Institutes Series

A series of edited volumes comprising multifaceted studies of contemporary scientific issues by some of the best scientific minds in the world, assembled in cooperation with NATO Scientific Affairs Division.

This series is published by an international board of publishers in conjunction with NATO Scientific Affairs Division

A Life Sciences **B Physics**	Plenum Publishing Corporation New York and London
C Mathematical and Physical Sciences	D. Reidel Publishing Company Dordrecht, Boston, and London
D Behavioral and Social Sciences **E Applied Sciences**	Martinus Nijhoff Publishers The Hague, Boston, and London
F Computer and Systems Sciences **G Ecological Sciences**	Springer Verlag Heidelberg, Berlin, and New York

Recent Volumes in Series B: Physics

Volume 84 —Physical Processes in Laser–Materials Interactions
edited by M. Bertolotti

Volume 85 —Fundamental Interactions: *Cargèse 1981*
edited by Maurice Lévy, Jean-Louis Basdevant, David Speiser, Jacques Weyers, Maurice Jacob, and Raymond Gastmans

Volume 86 —Surface Mobilities on Solid Materials: Fundamental Concepts and Applications
edited by Vu Thien Binh

Volume 87 —Relativistic Effects in Atoms, Molecules, and Solids
edited by G. L. Malli

Volume 88 —Collective Excitations in Solids
edited by Baldassare Di Bartolo

Volume 89a—Electrical Breakdown and Discharges in Gases: Fundamental Processes and Breakdown
edited by Erich E. Kunhardt and Lawrence H. Luesen

Volume 89b—Electrical Breakdown and Discharges in Gases: Macroscopic Processes and Discharges
edited by Erich E. Kunhardt and Lawrence H. Luessen

Collective Excitations in Solids

Edited by

Baldassare Di Bartolo

Boston College
Chestnut Hill, Massachusetts

Assistant Editor

Joseph Danko

Boston College
Chestnut Hill, Massachusetts

Plenum Press
New York and London
Published in cooperation with NATO Scientific Affairs Division

Proceedings of the NATO Advanced Study Institute on
Collective Excitations in Solids,
held June 15–29, 1981,
in Erice, Italy

Library of Congress Cataloging in Publication Data

NATO Advanced Study Institute on Collective Excitations in Solids (1981: Erice, Italy)
Collective excitations in solids.

(NATO advanced science institutes series. Series B, Physics; v. 88)
"Proceedings of the NATO Advanced Study Institute on Collective Excitations in Solids, held June 15–29, 1981, in Erice, Italy"—P.
Includes bibliographical references and index.
1. Collective excitations—Congresses. 2. Exciton theory—Congresses. I. Di Bartolo, Baldassare. II. Danko, Joseph. III. North Atlantic Treaty Organization. Scientific Affairs Division. IV. Title. V. Series.
QC176.8.E9N36 1981 530.4′1 82-18928
ISBN 978-1-4684-8880-7 ISBN 978-1-4684-8878-4 (eBook)
DOI 10.1007/978-1-4684-8878-4

Softcover reprint of the hardcover 1st edition 1983
A Division of Plenum Publishing Corporation
233 Spring Street, New York, N.Y. 10013

"Friendship is the inexpressible comfort of feeling safe with a person, having neither to weigh thoughts nor measure words."

G. Eliot

PREFACE

This book presents an account of the NATO Advanced Study Institute on "Collective Excitations in Solids," held in Erice, Italy, from June 15 to June 29, 1981. This meeting was organized by the International School of Atomic and Molecular Spectroscopy of the "Ettore Majorana" Centre for Scientific Culture.

The objective of the Institute was to formulate a unified and coherent treatment of various collective excitation processes by drawing on the current advances in various branches of the physics of the solid state.

A total of 74 participants came from 54 laboratories and 20 nations (Australia, Belgium, Burma, Canada, China, France, F. R. Germany, Greece, Israel, Italy, Mexico, The Netherlands, Pakistan, Poland, Portugal, Romania, Switzerland, Turkey, The United Kingdom, and The United States). The secretaries of the course were: Joseph Danko for the scientific aspects and Nino La Francesca for the administrative aspects of the meeting.

Fourty-four lectures divided in eleven series were given. Nine "long" seminars and eight "short" seminars were also presented. In addition, two round-table discussions were held.

The Institute was concerned first with establishing the framework required for describing the physical processes resulting in the collective excitations of a solid; this task was the responsibility of the first three lecturers. Once the basic principles were established, the other lecturers developed in detail specific treatments of the various types of collective excitations. For example, one lecturer treated vibrations in solids (phonons), another the excitations resulting from the perturbation of a system of locked and aligned spins (magnons), etc. The subject of excitons included a general treatment and a more detailed examination of specific cases. Other excitations such as plasmons, polarons and polaritons were also treated.

Specific aspects of the theory were also presented in seminar form together with several applications to various areas of current interest.

Professor Williams led the attendees during the final session which considered a synthesis of the material presented in the lectures and identified directions of profitable efforts.

I would like to thank their help Dr. A. Gabriele, Ms. P. Savalli, and all the personnel of the "Ettore Majorana" organization in Erice, Prof. R. L. Carovillano, Chairman of the Department of Physics at Boston College, Prof. V. Adragna, Dr. G. Denaro, Rag. M. Strazzera, and Avv. G. Luppino.

I would like to thank for their help Prof. A. Zichichi, Director of the "Ettore Majorana" Centre for Scientific Culture, Dr. A. Gariele, Ms. P. Savalli, and all the personnel of the "Ettore Majorana" organization in Erice, Prof. R. L. Carovillano, Chairman of the Department of Physics at Boston College, Prof. V. Adragna, Dr. G. Denaro, Rag. M. Strazzera, and Avv. G. Luppino.

I would like to thank the members of the organizing committee (Profs. Williams and Scharmann), Prof. Knox, and Dr. Auzel for their valuable help and advice.

I am especially grateful to my friend and collaborator, A. La Francesca, who helped me tremendously during the two weeks of the meeting, and to J. Danko for his help during the meeting and for his patient and intelligent work as assistant editor of this book.

I am looking forward to the next meeting of our school in 1983 and to a new occasion to meet again in Erice many of the fine people with whom I shared the experience of this Institute.

Arrivederci, my friends!

B. Di Bartolo
Editor and Director of
the Institute

Erice, June 1981

CONTENTS

INELASTIC SCATTERING OF FAST PARTICLES BY PLASMONS
A. A. Lucas

FROM MAGNONS TO SOLITONS
D. C. Mattis

LONG SEMINARS

SURFACE COLLECTIVE EXCITATIONS
G. Benedek

COLLECTIVE EXCITATIONS IN CONCENTRATED Mn^{2+} SYSTEMS: SPECTRAL PROPERTIES
D. P. Pacheco

QUANTUM MECHANICAL DESCRIPTION OF SOLIDS*

F. Williams

Institut fur Festkorperphysik II
Technische Universität
D 1000 Berlin 12, Germany†

and

Physics Department
University of Delaware
Newark, Delaware 19711, U.S.A.‡

ABSTRACT

Condensed matter is a many-body problem in quantum mechanics. The many-electron and many-atom systems are separated from each other by the adiabatic approximation. These two many-body problems are then treated differently: the electron problem in the Hartree-Fock approximation leading to single electron states and specifically to band structure for crystals; the atomic dynamics of crystals as collective excitations which when quantized lead to phonons. Collective excitations of electrons are then considered and plasmons formulated. Some general characteristics of quasi-particle and collective excitations will be discussed and inter-related where possible. In addition to intrinsic properties of solids, some of their extrinsic properties are treated theoretically, including the effects of applied stress on electronic states and vibrational levels of ions and molecules in crystals.

*Supported in part by a grant from the Army Research Office-Durham and in part by a Humboldt Prize (Senior U.S. Scientist Award).

†Present address

‡Permanent address

I. INTRODUCTION

Crystalline solids, like polyatomic molecules, liquids and amorphous solids, consist of many electrons and many nuclei and thus constitute a many-body problem in quantum mechanics. Thus for all these states of matter there is the question of how to subdivide the problem, and this is done rather generally to obtain separate electronic and atomic many-body problems by the adiabatic approximation which is based on the relatively rapid motion of electrons compared to the slower motion of the nuclei. The separate solutions of these two problems constitute electronic states and vibrational levels, respectively.

Crystals differ from molecules, liquids and amorphous materials by translational symmetry, that is invariance by translations corresponding to multiples of the unit vectors $\underline{a}$, $\underline{b}$, $\underline{c}$ which define the unit cell. Thus, the potential $V_p(\underline{r})$ satisfies the following:

$$V_p(\underline{r} \pm \ell\underline{a} \pm m\underline{b} \pm n\underline{c}) = V_p(\underline{r}) \tag{1}$$

where ℓ, m, n, are positive integers or zero. Atoms are invariant under infinitesimal rotations, providing the basis for quantized angular momenta for atoms; molecules, under finite rotations, providing the basis for the symmetries of molecular wave functions. Liquids and amorphous solids are generally without long-range and in most cases without intermediate range order and thus are less amenable to quantum mechanical analyses. Returning to crystals it is the translational invariance, represented by Eq. (1), which facilitates solutions to both the many-electron problem leading to band structure and the atomic vibrations problem leading to phonons.

For real crystals the presence of impurities and lattice defects results in departures from perfect translational periodicity and the application of an electrical or a magnetic field or a mechanical stress results in the complete Hamiltonian having non-periodic components. Thus further approximations must be made to describe the properties of real crystals and the response of all materials to applied perturbations. The emphasis in the theory of the solid state is rather generally on the proper approximations to be used for particular types of materials subjected to particular perturbations to yield particular phenomena. This is a consequence of the complete many-body problem not being soluble exactly.

Some properties of matter such as cohesive energy and photoelectric emission depend on the total ground state energy, however, many properties such as specific heat and optical excitation and emission spectra depend on differences between excited and ground states. The latter class of properties leads to the concept of elementary excitations, the idea being that these differences or excitations can be determined theoretically when the determination of the total energies is almost impossible to calculate so as to be in good agreement with experiment. Elementary excitations divide into two classes: quasi-particle excitations which in some approximations go over to single real particles, e.g. electrons; and collective excitations which do not, for example, phonons which represent collective atomic motions and in no sense go over to single atoms in any approximation. In this introductory chapter we shall consider both types of elementary excitations, as well as the total energy problem, in order to introduce collective excitation within a general perspective of the quantum mechanics of solids.

II. ADIABATIC APPROXIMATION

According to quantum mechanics a system is describable as completely as is possible by a wavefunction which satisfies the appropriate Schrödinger equation. In this chapter we focus attention on stationary states and therefore our wavefunction $\psi(\underline{r},\underline{R})$ satisfies the time-independent Schrödinger equation:

$$H\psi(\underline{r},\underline{R}) = E\psi(\underline{r},\underline{R}) \quad (2)$$

where H is the Hamiltonian operator of the system with eigenvalues E. In subsequent chapters Di Bartolo [1] describes transitions between stationary states and therefore obtains solutions to the time-dependent Schrödinger equation. In our definition of $\psi(\underline{r},\underline{R})$ we are working in the position coordinate representation: for a system of n electrons and N nuclei, $\underline{r}$ specifies the position

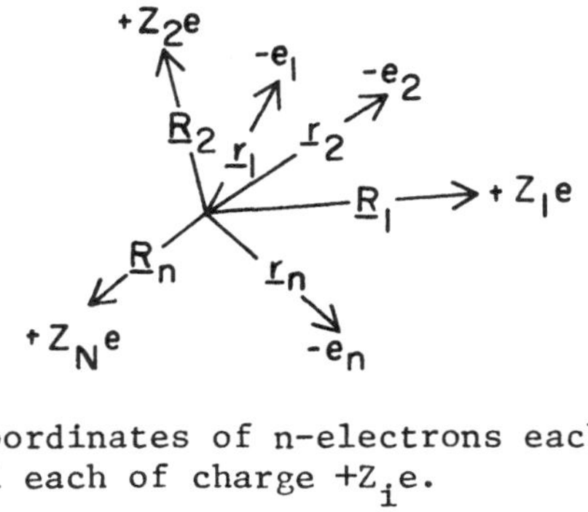

Fig. 1. Position coordinates of n-electrons each of charge -e and of N-nuclei each of charge $+Z_ie$.

coordinates of all electrons, e.g., $\underline{r}_1 \ldots \underline{r}_2$, and $\underline{R}_n$ specifies the position coordinates of all nuclei, i.e., $\underline{R}_1$, $\underline{R}_2 \ldots \underline{R}_N$, as illustrated in Fig. 1.

The basic idea for the separation of the electronic and nuclear dynamics is that the electrons move fast compared to the nuclei and thus the eigenstates of the electron system can be obtained by taking the coordinates $\underline{R}$ as fixed and therefore we have a time-independent Schrödinger equation for the electrons:

$$H_e \Phi(\underline{r};\underline{R}) = E_e(R)\Phi(\underline{r};\underline{R}) \tag{3}$$

where $H_e = H - T_N$ and T_N is the kinetic energy operator of the nuclei. The electronic eigenfunction $\Phi(\underline{r};\underline{R})$ and eigenvalue $E_e(\underline{R})$ both have parametric dependences on the nuclear coordinates $\underline{R}$. In other words, $\Phi(\underline{r};\underline{R})$ and $E(\underline{R})$ are smooth functions of R as long as changes in $\underline{R}$ occur adiabatically, that is, slowly so that no transitions occur between eigenstates of the electrons. Note that H_e in this formulation includes the potential energy operator for the interaction of nuclei with each other.

In the adiabatic approximation the total wavefunction is chosen as follows:

$$\psi(\underline{r},\underline{R}) = \chi(\underline{R})\Phi(\underline{r};\underline{R}) \tag{4}$$

which corresponds to one term in the complete Born-Oppenheimer expansion[2]. The $\chi(\underline{R})$ is the eigenfunction for the dynamics of the atoms. Substituting Eq. (4) into Eq. (2) and using Eq. (3) we obtain the following Schrödinger equation, if we neglect certain

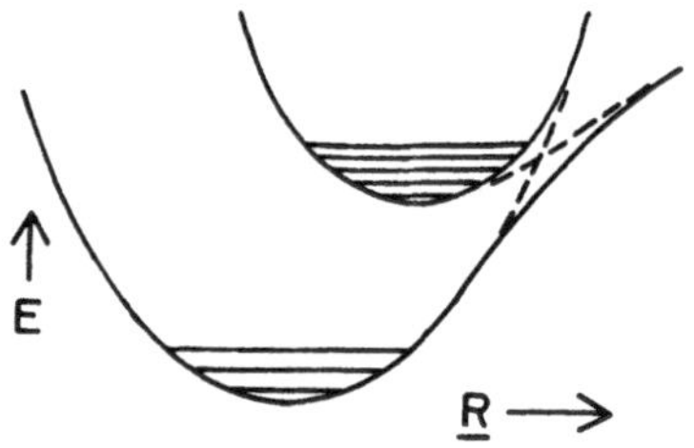

Fig. 2. Diabatic potentials, which maintain invariant characteristics but cross--; adiabatic potentials, which change characteristic but do not cross——; vibrational levels for each electronic state.

terms which will be discussed later in this section:

$$\{-\sum_j \frac{\hbar^2}{2M_j} \nabla_{\underline{R}}^2 + E_e(\underline{R})\}\chi(\underline{R}) = E\chi(R). \qquad (5)$$

The dynamics of atomic motion is thus governed by the effective potential $E_e(R)$, the eigenvalue of Eq. (3), which is identified as the adiabatic potential if the interaction between electronic states is included. Neglecting interaction between electronic states yields diabatic potentials [3]. Both are illustrated in Fig. 2 for different electronic states. Representative eigenvalues E of Eq. (2) and (5) are also shown.

The terms neglected in obtaining Eq. (5) are the following:

$$\sum \frac{h^{-2}}{2M_j} \{2\iint \chi_\alpha^*(\underline{R})\Phi_\zeta^*(\underline{r};\underline{R})\nabla_R\chi_\beta(\underline{R})\nabla_R\Phi_\mu(\underline{r};\underline{R})d\underline{r}d\underline{R}$$
$$+ \iint \chi_\alpha^*(\underline{R})\Phi_\zeta^*(\underline{r};\underline{R})\chi_\beta(\underline{R})\nabla_R^2\Phi_\mu(r;R)d\underline{r}d\underline{R}\}. \qquad (6)$$

The diagonal components $\alpha=\beta$, $\zeta=\mu$ of the first term are zero for a non-magnetic system whose particles are conserved; those of the second term are smaller than the kinetic energy operator for electrons by the factor m/M. The off-diagonal components of Eq. (6) are identified as electron-phonon interaction.

The many-body problem of n electrons and N nuclei has thus been transformed to two many-particle problems: one for electrons, the other for nuclei. Eqs. (3) and (5), are their respective many-particle time-independent Schrödinger equations, which are coupled. We next consider approximating the electron problem in terms of single particle wavefunctions first in general and then applied to crystals, and then analyze the atomic dynamics in terms of collective excitations.

III. HARTREE-FOCK APPROXIMATION

The many-electron wavefunction $\Phi(\underline{r}_1\underline{r}_2...\underline{r}_n)$ is separable into one-electron wavefunction $\phi_\ell(r_i)$ by means of the Hartree or Hartree-Fock approximation. Both make use of the variational principle to obtain minimum eigenvalues, and also of self-consistency of the wavefunction for each electron and its contribution to the effective potential of every other electron. The approximations differ in the form of $\Phi_\ell(r)$.

The Hartree approximation [4] uses a single product function: $\Phi(\underline{r}_1\underline{r}_2...\underline{r}_n) = \phi_1(\underline{r}_1)\phi_2(\underline{r}_2)...\phi_n(\underline{r}_n)$, which on substitution in Eq. (3) and subdividing H yield n coupled one-electron Schrödinger equations, each as follows:

$$H_i \phi_i(\underline{r}_i) = \left[-\frac{h^2}{2m}\nabla_i^2 + V_R(\underline{r}_i) + \sum_{j\neq i} e^2 \int \frac{|\phi_j|^2}{|r_i - r_j|} \underline{dr}_j\right]\phi_i(\underline{r}_i) = E_i\phi_i(\underline{r}_i) \qquad (7)$$

where $V_R(\underline{r}_i)$ is the potential energy operator for all nuclei. The parametric dependences of ϕ_i and E_i on R are implicit. The Hartree ϕ_i do not satisfy the Pauli principle.

The Hartree-Fock approximation [5] uses the Slater determinant for ϕ which is antisymmetric and therefore, satisfies the Pauli principle:

$$\Phi(\underline{r}_1\underline{r}_2\ldots\underline{r}_n) = (n!)^{-\frac{1}{2}} \begin{vmatrix} \phi_1(\underline{r}_1)\ldots\ldots\phi_1(\underline{r}_n) \\ \vdots \qquad\qquad \vdots \\ \phi_n(\underline{r}_1) \qquad \phi_n(\underline{r}_n) \end{vmatrix}, \qquad (8)$$

which on substitution in Eq. (3) leads to Eq. (7) but with additional terms:

$$e^2 \int \frac{\phi_i{}^*\phi_j}{|\underline{r}_i - \underline{r}_j|} \underline{dr}_j ;$$

in the effective potential for the ith electron. These are the exchange terms, which are spin-dependent since spin as well as orbital coordinates are included in the $\underline{r}_i$ of Eq. (8) and the antisymmetry can be in either the spin or orbital dependence. The Hartree-Fock one-electron wavefunctions are well established for free ions and provide the starting point for tight-binding, self-consistent wavefunctions for solids. We now consider one-electron functions for a period potential.

IV. ELECTRONIC BANDS IN CRYSTALS

The Schrödinger equation for an electron in a periodic potential $V_p(\underline{r})$ is:

$$\left[-\frac{\hbar^2}{2m}\nabla^2 + V_p(\underline{r})\right]\phi(\underline{r}) = E\phi(\underline{r}) \qquad (9)$$

whose solutions are Bloch functions:

$$\phi_{\eta k}(\underline{r}) = U_{\eta k}(\underline{r})e^{i\underline{k}\cdot\underline{r}} \qquad (10)$$

where $U_{\eta k}(\underline{r})$ is periodic in the same way as $V_p(\underline{r})$, as given in Eq. (1); η indexes allowed bands; and $\underline{k}$ is the wave number vector, or quasi-momentum, in units of $\hbar$.

If we define the units of the wave number vector space as follows, in terms of unit vectors of reciprocal lattice space:

$$\underline{a}^* = \underline{b} \times \underline{c}/v, \ \underline{b}^* = \underline{c} \times \underline{a}/v, \ \underline{c}^* = \underline{a} \times \underline{b}/v, \quad (11)$$

where v is the volume of the unit cell, and $\underline{a}$, $\underline{b}$ and $\underline{c}$ are given by Eq. (1), we can show that $\underline{k}$ is not uniquely determined in $\phi_{\eta k}$ since $\exp i(\underline{k}+2\pi\ell\underline{a}^*)\cdot\underline{a}=\exp i\underline{k}\cdot\underline{a}$, thus the infinite range of $\underline{k}$ in the extended zone can be reduced to $-\pi\underline{a}^*,-\pi\underline{b}^*,-\pi\underline{c}^* \leq \underline{k} \leq \pi\underline{a}^*,\pi\underline{b}^*,\pi\underline{c}^*$. The existence of forbidden energy gaps at these boundaries can be shown from $V_p(\underline{r})$ splitting the degenerate free particle states $\phi_o \sim \exp(i\underline{k}\cdot\underline{r})$. The extended and reduced zones and the gaps are illustrated in Fig. 3. The E(k) are eigenvalues of Eq. (9).

The Bloch functions $\phi_{\eta k}(\underline{r})$ are exact solutions to the periodic potential and make clear the plane wave nature of these solutions. If we take the Fourier transform of $\phi_{\eta k}(\underline{r})$ the Wannier functions $W_\eta(\underline{r}-\underline{R}_j)$ are obtained. The Bloch function is therefore:

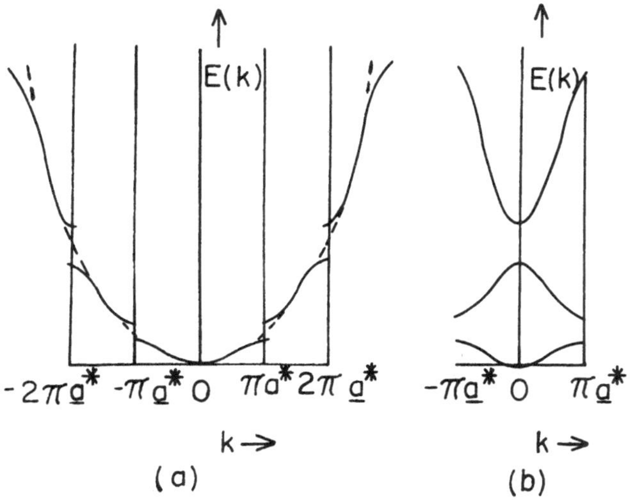

Fig. 3. Electronic states: (a) in extended zone of free particle-- and particle in periodic potential——, and (b) in reduced zone of particle in periodic potential.

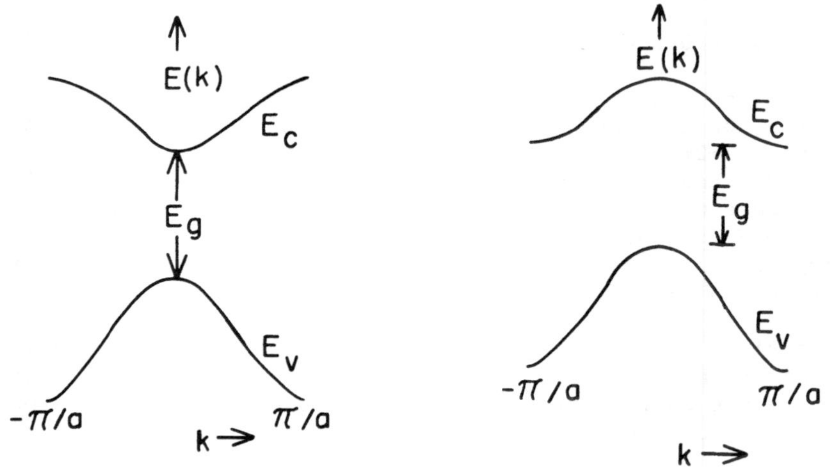

(a) (b)

Fig. 4. Band structure: (a) direct gap semiconductor; (b) indirect gap semiconductor. E_g is the forbidden gap betwee allowed bands.

$$\phi_{\eta k}(r) = N^{-\frac{1}{2}} \sum_j e^{ik\cdot R_j} W_\eta(\underline{r}-\underline{R}_j) \quad (12)$$

In the tight binding limit $W_\eta(\underline{r}-\underline{R}_j)$ approaches localized atomic wavefunctions. Thus the role of tight binding, as well as plane waves, in $\phi_{\eta k}(r)$ becomes evident.

The characteristics of band structure are determined by crystal symmetry and by the unit cell potential. For many materials $E(\underline{k})$ is more complex than simple theory predicts. In some cases the extrema of the valence and conduction band lie in the same region of k-space; in others, in different regions. Direct and indirect gap semiconductors are illustrated in Fig. 4. The theory gives the form and symmetry of band structure; from experiment quantitative features are obtained. For example, the response of an electron in a band to an applied field is in accordance with its effective mass m* given by:

$$m^* = \hbar^2 / \frac{\partial^2 E}{\partial k^2}, \tag{13}$$

which can be accurately measured, for example by cyclotron resonance experiments. The properties of shallow states of point charged dopants in semiconductors can be determined from effective mass theory. Many intrinsic and extrinsic electronic properties of semiconductors are in accordance with one-electron states. We now turn to the other part of the results of the adiabatic separation and analyze lattice vibrations as collective excitations. In section VI we shall return to electrons in solids and formulate their collective excitations.

V. LATTICE DYNAMICS AS COLLECTIVE EXCITATIONS: PHONONS

The $E_e(\underline{R})$ of Eq. (5) is the adiabatic potential $V_a(\underline{R})$ for atomic dynamics. We shall first analyze a linear chain of alternating atoms of mass M_1 and M_2, as shown in Fig. 5, and later generalize the formulation to three dimensions.

The $V_a(R)$ is expanded as Taylor series in the displacements u and v of the masses M_1 and M_2, respectively, from their equilibrium positions R_0:

$$V_a(R) = V_a(R_0) + \left(\frac{\partial V_a}{\partial u}\right)_{R_0} u + \left(\frac{\partial V_a}{\partial v}\right)_{R_0} v + + \tfrac{1}{2}\left(\frac{\partial^2 V_a}{\partial u^2}\right)_{R_0} u^2 + \tfrac{1}{2}\left(\frac{\partial^2 V_a}{\partial v^2}\right)_{R_0} v^2 + \cdots \tag{14}$$

The first term is independent of displacements; the second and third are zero from equilibrium; the fourth and fifth are harmonic terms. If we neglect the higher terms and assume nearest-neighbor interactions only, the equations of motion for M_1 and M_2 are:

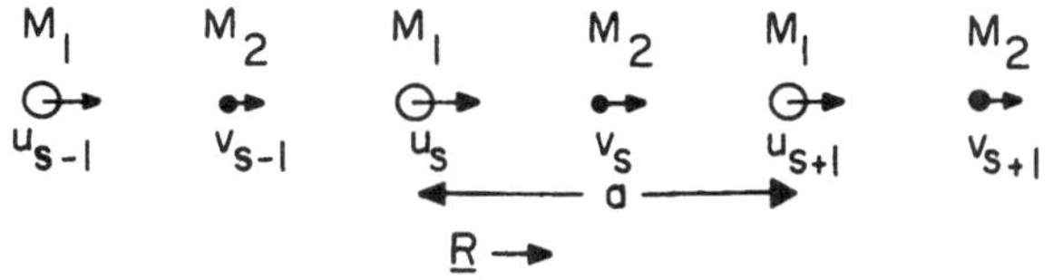

Fig. 5. Displacements from equilibrium for two atoms per link in linear chain.

$$M_1 \frac{d^2u_s}{dt^2} = K(v_s + v_{s-1} - 2u_s)$$

$$M_2 \frac{d^2v_s}{dt^2} = K(u_{s+1} + u_s - 2v_s) \tag{15}$$

where $K \equiv (\partial^2 V_a/\partial u^2)_{Ro}$, $(\partial^2 V_a/\partial v^2)_{Ro}$. We search for propagating wave solutions of the form $u_s = u \exp (i\, q\, s\, a - i\, \omega\, t)$ and $v_s = v \exp (i\, q\, s\, a - i\, \omega\, t)$, and obtain two solutions:

$$\omega_{\pm}^2 = K(\frac{1}{M_1} + \frac{1}{M_2}) \pm K\sqrt{(\frac{1}{M_1} + \frac{1}{M_2})^2 - \frac{4\sin^2 q\, a/2}{M_1 M_2}} \tag{16}$$

The $\omega_+(q)$ solution is the optical branch, and $\omega_-(q)$ is the acoustical. Because the force constants K are different for transverse and longitudinal displacements two sub-branches exist. The dispersion curves are shown in Fig. 6.

For the generalized three dimensional analysis the K is replaced by a force tensor:

$$G_{jj's\ell s'\ell'} \equiv [\partial^2 V_a / \partial u_{js\ell} \partial u_{j's'\ell'}]_{Ro}$$

where $u_{js\ell}$ is the j-component of displacement of the s-atom in the ℓ-cell, and the solutions are in terms of the Fourier transform of the force tensor:

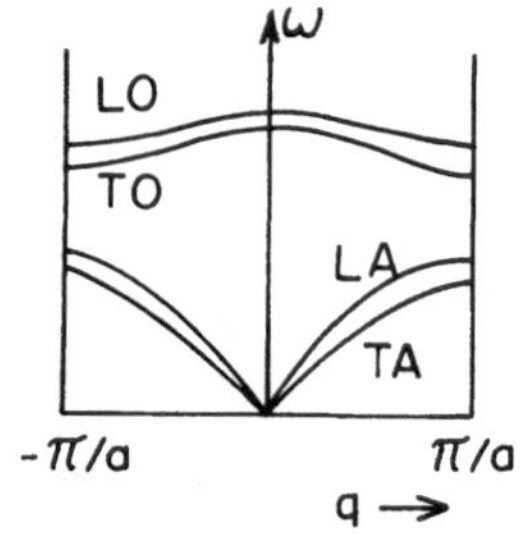

Fig. 6. Longitudinal optical LO, transverse optical TO, longitudinal acoustical LA, transverse acoustical TA, modes of lattice with spacing a versus wave number vector q.

$$\Sigma_{sj}\{G_{jj'ss'}(q) - \omega^2 M_{s'}\delta_{ss'}\delta_{jj'}\}u_{js} = 0 \tag{17}$$

The essential physics remain the same as shown in Fig. 6 except for complexity in k-space. The detailed structure in k-space can be determined experimentally, for example by neutron scattering experiments.

For each mode specified by branch ξ and wave number vector q quantized harmonic oscillator levels, such as shown in Fig. 2, exist:

$$E_{\xi,q}(b^\dagger b) = (b^\dagger b + \tfrac{1}{2})\hbar\omega_{\xi,q} \tag{18}$$

In the second quantized representation these states are describable in terms of occupation number, that is, number of phonons; $b^\dagger b$.

Finally points of similarity and of differences are to be noted between electronic band structure and lattice dynamics. As evident from comparison of Figs. 4 and 6 both have allowed bands separated by forbidden energy gaps. At small k and q, respectively, the dispersion relations are different: $E(\underline{k})$ is generally quadratic in $\underline{k}$; for the acoustical branch $E(\underline{q})$ is linear in $\underline{q}$.

VI. COLLECTIVE EXCITATION OF ELECTRONS: PLASMONS

In addition to their single particle states discussed in Sections III and IV, electrons in solids are capable of collective excitations. These when quantized are called plasmons. Conduction electrons in metals and in degenerate semiconductors exhibit plasma oscillation; valence electrons in semiconductors exhibit plasma can also exhibit plasmon effects. We shall make an elementary analysis of collective excitations of an electron gas and then establish some relationship between these states and single particle states.

We consider an electron gas of density n with a rigid fixed background of positive charge, and let the electron gas be displaced by x. The polarization P = nex leads to a depolarization field $F_d = -4\pi P = -4\pi nex$, from which we can write the equation of motion:

$$m\ddot{x} = eF_d = -4\pi ne^2 x, \tag{19}$$

and obtain the solution:

$$\omega_p^2 = 4\pi ne^2/m \tag{20}$$

the well-known plasma frequency. If we include the polarization of the ion cores of the background positive charge there is a factor of the very high frequency dielectric constant, specifically ε_∞^{-1} in ω_p^2. The dielectric constant including the response of the cores and the electron gas to an applied field is:

$$\varepsilon(\omega) = \varepsilon(\infty)\left[1 - \omega_p^2/\omega^2\right] \tag{21}$$

Thus a resonance is observed in optical absorption due to plasma oscillations. Also, energetic electrons traversing thin films are observed to lose energy in quanta $\hbar\omega_p$.

In the case of metals $n \sim 10^{22}cm^{-3}$ and $h\omega_p$ lies in the ultraviolet; for degenerate semiconductors $n \sim 10^{18}cm^{-3}$ and $h\omega_p$ lies in infrared. The semiconductors have another resonance due to collective displacements of all the valence band electrons, which lies in the ultraviolet. The collective oscillations of the valence band electrons in dielectrics also have their resonance in the ultraviolet because $n \sim 10^{22}cm^{-3}$.

For metals and dielectrics the inter-electron distances are comparable with unit cell dimensions a, b, c and an individual electron is correlated with specific ion cores during an oscillation so that the periodic potential does not enter into the equations of motion, therefore the m in Eq. (20) is to a good approximation the free electron mass. This is also the case for the collective oscillations of all the valence band electrons in semiconductors. On the other hand, for conduction electrons or valence band positive holes in degenerate semiconductors the inter-electron or inter-hole distances are large compared to a, b, c and any correlation is with charged dopants, so that the periodic potential does enter into the equations of motion, therefore, the m in Eq. (20) is to a good approximation the effective mass m* characteristic of the conduction or valence band edge, depending on whether the material is n- or p-type. The plasmons of degenerate semiconductors can thus be related to single particle electron states.

VII. EXTRINSIC STATES

In addition to the intrinsic properties of solids formulated theoretically in Sections IV, V and VI we now analyze selected extrinsic properties. We first present the effective mass theory of electron states of point charged defects in semiconductors [6] and then the Fröhlich Hamiltonian formulation [7] of states of point and associated charged defects including the effects of electron-phonon interaction [8].

The one-electron wavefunctions for bound states of point defects in elemental semiconductors are expanded in Bloch functions.

A continuum approximation is made for the perturbing potential energy operator: $V' = -e^2/\varepsilon r$. This leads to the eigenvalue equation:

$$[E_\eta(\underline{k}) - E]c_\eta(\underline{k}) - \sum_{\eta'\underline{k}'} \int \phi^*_{\eta\underline{k}} (e^2/\varepsilon r)\phi_{\eta'\underline{k}'} d\underline{r}\, c_{\eta'}(\underline{k}') = 0 \quad (22)$$

where $c_\eta(\underline{k})$ are the coefficient in the expansion and $E_\eta(\underline{k})$ and $\phi_{\eta\underline{k}}$ are given by Eqs. (9) and (10), respectively. For shallow states the c_η are negligible except for c_0 which is either c_c or c_v depending on whether the dopant is a donor or acceptor. After substituting $E_0(k) = \hbar^2 k^2/2m^*$ for that band and introducing the Fourier transform of c_0, that is:

$$F(r) = \sum_{\underline{k}} c_0(\underline{k}) \exp i\underline{k}\cdot\underline{r} \quad (23)$$

which is called the effective mass function. After several transformations we obtain the effective mass equation:

$$\left[-\frac{\hbar^2}{2m^*}\nabla^2 - \frac{e^2}{\varepsilon r}\right]F(r) = E\,F(r) \quad (24)$$

with hydrogenic solutions:

$$E_\xi = -\frac{m^* e^4}{2\hbar^2\varepsilon^2\xi^2} \quad (25)$$

These are one-electron states valid for elemental semiconductors, neglecting anisotropy, degeneracy in the band structure and the central cell potential of the dopant.

For compound semiconductor the effects of electron-phonon interaction must be included. These are represented by the off-diagonal components of Eq. (6). Frohlich et al [7]deduced the following effective Hamiltonian for this interaction:

$$H_i = \sum_q (V_q b_q e^{i\underline{q}\cdot\underline{r}} + V_q^* b_q^+ e^{-i\underline{q}\cdot\underline{r}}) \quad (26)$$

where:

$$V_q = -\frac{e}{q}\left(\frac{2\pi\hbar\omega}{\Omega}\right)^{\frac{1}{2}}\left(\frac{1}{\varepsilon_\infty} - \frac{1}{\varepsilon_0}\right)^{\frac{1}{2}} \quad (27)$$

where Ω is the volume of the crystal, the operator b^+ and b are respectively creation and annihilation operators for phonons, previously used in Eq. (18).

The application of Eq. (26) to point and associated defects in compound semiconductors leads to effective mass solutions which include electron-phonon interaction [8]. For example, the zero-phonon spectra of donor-acceptor (DA) pairs can be thus calculated theoretically. The electron-phonon interaction is found to be de-

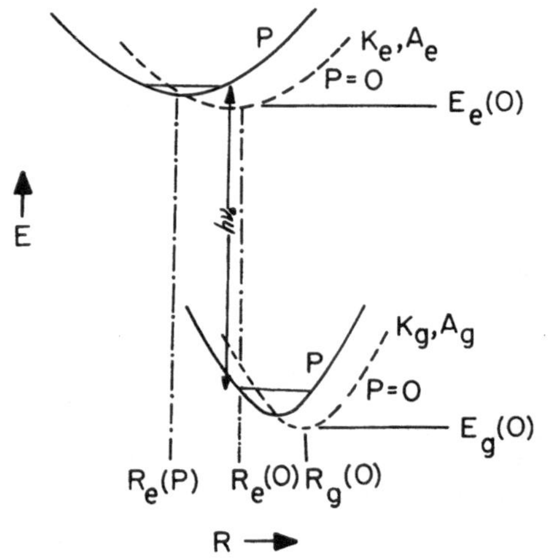

Fig. 7. Adiabatic potentials for non-degenerate states, without - - and with hydrostatic pressure —.

pendent on DA distance. Basically the single particle electronic excitation is clothed in virtual phonons, that is a collective excitation of lattice displacements. This illustrates coupling between single or few particle excitations and collective excitations.

VIII. SOME EFFECTS OF APPLIED STRESS

A great deal of the investigations of the solid state depends on perturbations by applied fields. We shall analyze the effects of mechanical stress on the adiabatic potentials, described in Section II, which are prerequisite to the determination of radiation transitions between states represented by these potentials [9].

We first consider hydrostatic pressure applied to a cubic material with dopants which interact with the lattice in the harmonic approximation, as shown in Fig. 7 for two states, each non-degenerate.

The application of pressure P via a coupling constant A displaces the equilibrium coordinate by ΔR: $PA = K\Delta R$, leaving the force constant K unchanged. The zero-phonon (ZP) transition $h\nu_0$ is thus the following:

$$h\nu_0(P) = h\nu_0(o) + \tfrac{1}{2}\left[\frac{Ae^2}{K_e} - \frac{Ag^2}{K_g}\right]P^2 \tag{28}$$

as shown in Fig. 7. In addition to the quadratic spectral shifts of ZP and vibronic transitions with P, the relative intensities change linearly with P because of the change in phase of the vibrational wavefunctions of the two states with respect to each other for $\Delta R_e \neq \Delta R_g$. In addition, linear shifts with P in $h\nu$ can readily be shown to originate from anharmonicity.

We next consider the effects of uniaxial stress. For non-orbitally degenerate electronic states of dopants or molecules in solids the theoretical analysis of the effects of uniaxial stress is essentially the same as for hydrostatic pressure. The deformation is divided among orthogonal components and the analysis proceeds as just outlined.

However, the situation is quite different for orbitally degenerate states exhibiting the Jahn-Teller (J-T) effect, for which the degeneracy is reduced by distortions due to electron-phonon interaction. In Fig. 8a the adiabatic potentials are shown for a system whose ground state is non-degenerate and whose excited state is degenerate (J-T).

The effect of uniaxial stress on a J-T system can be divided into two parts: first the stress is applied reversibly so that the applied force just balances the crystal force. In this way zero work is done moving the system from the outer well to the inner well, and then work is done on the system moving the inner minimum to the new equilibrium: $W = P^2A^2/2K$, where K characterizes small oscillations in the inner well. Second,

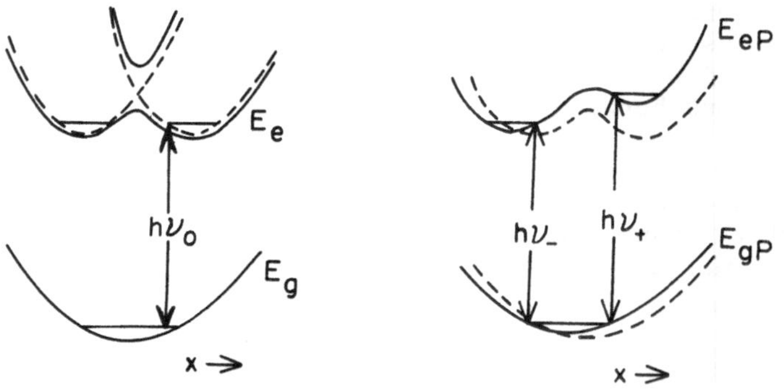

(a) (b)

Fig. 8. (a) Diabatic and adiabatic potentials for Jahn-Teller systems; (b) Adiabatic potentials including the effects of constant applied stress. The splitting of the zero-phonon transition is also shown.

dynamics at constant P are analyzed: in addition to work done against crystal forces there is now work done against constant pressure, specifically in going from the inner to outer minimum

$$W' = PA(x_o - x_i) \tag{29}$$

therefore a linear splitting of $h\nu_o$ occurs for J-T systems. This is shown in Fig. 8b. For the stresses possible in uniaxial experiments ΔR is small, however, $|x_o - x_i| >> \Delta R$ with these stresses.

Implicit in both the analysis of non-degenerate states under hydrostatic pressure and the analysis of J-T systems under uniaxial stress is a type of double adiabatic approximation: the inertia of the pressure apparatus is assumed large compared to that of all parts of the system under investigation, and therefore its time constant is long compared to that of the nuclei while of course the electrons have shorter time constants than both, in accordance with the usual adiabatic approximation.

These examples of effects of applied stress illustrate the perturbation of single particle states coupled to collective excitation, that is, of Hartree-Fock states coupled to phonons.

IX. CONCLUSIONS

The many-particle problem of electrons and nuclei in solids can be divided, with some rather generally valid approximations, into the problem of electronic states and that of phonons. In addition to the quasi-particle electronic band structure there are plasmon states of collective excitations of electrons. These two classes of electron states can in part be inter-related. The quantized collective modes for atomic motion divide into acoustical and optical phonon branches. The translational periodicity of crystals is a major factor in obtaining solutions for the quasi-particle electron states and separately for the phonon states. For both, forbidden energy gaps arise from this periodicity, however, the dispersion relations for electron and phonon states differ. The eigenstates for some classes of dopants and defects can be solved by perturbation methods applied to the electron and phonon states of perfect crystals, with attention to the effects of electron-phonon interaction. Similar perturbative methods can be used to determine the effects of applied fields on the intrinsic properties of solids, for example, the effects of hydrostatic and uniaxial stresses on the electronic states of dopants, which are coupled to collective lattice deformations. Other examples of collective excitations and of their interactions with quasi-particles and with other collective excitations will be evident in the subsequent chapters devoted to electron-hole droplets, excitons, magnons, polaritons, polarons and solitons.

REFERENCES

1. B. Di Bartolo, "Introduction to Collective Excitations in Solids," in this book.
2. M. Born and J. R. Oppenheimer, Ann. d. Phys. 84, 457 (1927).
3. F. Williams, D. E. Berry, and J. E. Bernard, in Radiationless Processes, edited by B. Di Bartolo (Plenum, 1979), p. 22.
4. D. R. Hartree, Proc. Cambridge Phil. Soc. 24, 89 (1928).
5. V. Fock, Zeits. f. Physik 61, 126 (1930).
6. W. Kohn, "Solid State Physics 5," (Academic Press, 1957).
7. H. Fröhloch, M. Pelzer, S. Zienau, Phil. Mag. 41, 221 (1950).
8. R. Evrard and F. Williams, in Luminescence of Inorganic Solids, edited by B. Di Bartolo (Plenum, 1977), p. 419.
9. D. Curie, D. E. Berry and F. Williams, Phys. Rev. B20, 2323 (1979).

INTRODUCTION TO COLLECTIVE EXCITATIONS IN SOLIDS*

B. Di Bartolo

Department of Physics
Boston College
Chestnut Hill, Massachusetts 02167, U.S.A.

ABSTRACT

The subject of collective excitations in solids is introduced by first considering the effects of time-dependent and time-independent perturbations on a two-level system. Subsequently, the physics of collective excitations is developed. After reviewing the process of absorption and emission of radiation, the interaction of radiation with collective excitations is dealt with. Final emphasis is accorded to examples of collective excitations.

I. INTERACTION IN A TWO-LEVEL SYSTEM

I.A. Quantum Mechanical Resonance

Let us consider a system with a time-independent Hamiltonian H_0. The time-dependent Schrödinger equation gives

$$H_0 \Psi = i\hbar \frac{\partial \Psi}{\partial t} \tag{1}$$

If the system is in a stationary state labeled i

$$\Psi(t) = \Psi_i(t) = e^{-i(E_i/\hbar)t}\, \Psi_i(0) \tag{2}$$

where the energy values are given by

*Supported in part by NATO Research Grant No. 1169.

$$H_o \Psi_i(0) = E_i \Psi_i(0) \tag{3}$$

We shall assume that the wavefunctions $\Psi_i(t)$ are orthonormal.

Let us now suppose that the system is subjected to a time-dependent perturbation represented by $H'(t)$. The system will be represented by a wavefunction $\Psi(t)$ such that

$$H\Psi(t) = (H_o + H')\,\Psi(t) = i\hbar \frac{\partial\Psi(t)}{\partial t} \tag{4}$$

We can expand $\Psi(t)$ in terms of the complete set $\Psi_i(t)$

$$\Psi(t) = \sum_i c_i(t)\,\Psi_i(t) \tag{5}$$

If $H' = 0$, the coefficients c_i's are time-independent. Replacing Eq. (5) in Eq. (4),

$$(H_o + H')\sum_i c_i(t)\,\Psi_i(t) = i\hbar\left[\sum_i c_i(t)\,\frac{\partial\Psi_i(t)}{\partial t} + \sum_i \frac{\partial c_i(t)}{\partial t}\,\Psi_i(t)\right] \tag{6}$$

Then

$$\sum_i c_i(t)\,H'\,\Psi_i(t) = i\hbar \sum_i \frac{\partial c_i(t)}{\partial t}\,\Psi_i(t) \tag{7}$$

where we have taken advantage of Eqs. (2) and (3). Multiplying by $\Psi_k^*(t)$ and integrating over all space we obtain

$$i\hbar \frac{\partial c_k(t)}{\partial t} = \sum_i c_i(t) \;\langle\Psi_k(t)|H'|\Psi_i(t)\rangle = \sum_i c_i(t)\,M_{ki}\; e^{i\omega_{ki}t} \tag{8}$$

where

$$\omega_{ki} = \frac{E_k - E_i}{\hbar} \quad ; \quad M_{ki} = \langle\Psi_i(0)|H'|\Psi_i(0)\rangle \tag{9}$$

We shall now make the following simplifying assumptions:

(a) The system has only two energy levels, say 0 and k;

(b) the diagonal matrix elements of H' are zero; and

(c) the perturbation H' is constant, but is turned on at time $t = 0$.

The coupled equations (8) become

$$\begin{cases} \dot{c}_k(t) = -\frac{i}{\hbar} c_o(t)\, e^{i\omega_{ko}t} M_{ko} \\ \dot{c}_o(t) = -\frac{i}{\hbar} c_k(t)\, e^{-i\omega_{ko}t} M_{ko}^* \end{cases} \tag{10}$$

Differentiating the first equation above and using the second equation, we obtain for $t > 0$

$$\ddot{c}_k(t) - i\omega_{ko}\,\dot{c}_k(t) = \frac{|M_{ko}|^2}{\hbar^2} c_k(t) = 0 \tag{11}$$

We expect $c_k(t)$ to be of the form

$$c_k(t) = Ae^{\alpha_1 t} + Be^{\alpha_2 t} \tag{12}$$

where α_1 and α_2 are solutions of the equation

$$\alpha^2 - i\omega_{ko}\alpha + \frac{|M_{ko}|^2}{\hbar^2} = 0 \tag{13}$$

or

$$\alpha_{1,2} = \frac{i}{2}\left[\omega_{ko} \pm \left(\omega_{ko}^2 + 4\frac{|M_{ko}|^2}{\hbar^2}\right)^{1/2}\right] = \frac{i\omega_{ko}}{2} \pm ia \tag{14}$$

where

$$a = \frac{1}{2}\left(\omega_{ko}^2 + 4\frac{|M_{ko}|^2}{\hbar^2}\right)^{1/2} \tag{15}$$

Let us study the time evolution of such a system starting from the following initial conditions

$$c_k(0) = 0 \quad , \quad c_o(1) = 1 \tag{16}$$

i.e., the system is in the state o at time $t = 0$. The initial condition on c_k gives $A = -B$; then Eq. (12) becomes

$$c_k(t) = A(e^{\alpha_1 t} - e^{\alpha_2 t})$$

$$= A\left(e^{i(\omega_{ko}/2)t+iat} - e^{i(\omega_{ko}/2)t-iat}\right)$$

$$= Ce^{i(\omega_{ko}/2)t} \sin at \tag{17}$$

where

$$C = 2iA \tag{18}$$

The expression for $c_o(t)$ can be derived from the first of Eq. (10) and the expression (17) for $c_k(t)$:

$$c_o(t) = \frac{i\hbar\, e^{-i\omega_{ko}t}}{M_{ko}} \dot{c}_k(t)$$

$$= \frac{i\hbar\, e^{-i(\omega_{ko}/2)t}\, C}{M_{ko}} \left(i \frac{\omega_{ko}}{2} \sin at + a \cos at\right) \tag{19}$$

The initial condition on $c_o(t)$ gives $C = M_{ko}/(i\hbar a)$. Therefore, c_k becomes

$$c_k(t) = Ce^{i(\omega_{ko}/2)t} \sin at = \frac{M_{ko}}{i\hbar a} e^{i(\omega_{ko}/2)t} \sin at \tag{20}$$

and

$$|c_k(t)|^2 = \frac{|M_{ko}|^2}{\hbar^2} \frac{1}{a^2} \sin^2 at \tag{21}$$

On the other hand, we can now write

$$c_o(t) = e^{-i(\omega_{ko}/2)t} \left(\cos at + i \frac{\omega_{ko}}{2a} \sin at\right) \tag{22}$$

and

$$|c_o(t)|^2 = \cos^2 at + \frac{\omega_{ko}^2}{4a^2} \sin^2 at \tag{23}$$

We verify that

$$|c_o(t)|^2 + |c_k(t)|^2 = 1 \tag{24}$$

Considering the expressions for $|c_o(t)|^2$ and $|c_k(t)|^2$ of the last section, we note that at time $t = 0$ $|c_o(0)|^2 = 1$ and $|c_k(0)|^2 = 0$, as expected. Following the time $t = 0$, the two probabilities have oscillatory behaviors; at time $t = \pi/2a$, $|c_k|^2$ and $|c_o|^2$ take their maximum and minimum values, respectively:

$$|c_k|^2_{max} = \frac{|M_{ko}|^2}{\hbar^2 a^2} = \frac{4|M_{ko}|^2}{\hbar^2\omega_{ko}^2 + 4|M_{ko}|^2} \xrightarrow[\omega_{ko}\to 0]{} 1 \tag{25}$$

$$|c_o|^2_{min} = \frac{\omega_{ko}^2}{4a^2} = \frac{\hbar^2\omega_{ko}^2}{\hbar^2\omega_{ko}^2 + 4|M_{ko}|^2} =$$

$$= 1 - \frac{4|M_{ko}|^2}{\hbar^2\omega_{ko}^2 + 4|M_{ko}|^2} = 1 - |c_k|^2_{max} \xrightarrow[\omega_{ko}\to 0]{} 0 \tag{26}$$

The condition $\omega_{ko} = 0$ (i.e., levels 0 and k degenerate) corresponds to a "quantum mechanical resonance." In this particular case

$$a = \frac{|M_{ko}|}{\hbar} \tag{27}$$

and

$$|c_k(t)|^2 = \sin^2 at = \sin^2 \frac{|M_{ko}|}{\hbar} t \tag{28}$$

$$|c_o(t)|^2 = \cos^2 at = \cos^2 \frac{|M_{ko}|}{\hbar} t \tag{29}$$

At the time $t = 0$

$$|c_k(0)|^2 = 0 \quad , \quad |c_o(0)|^2 = 1$$

After a time

$$t = \frac{\pi}{2a} = \frac{\hbar}{2|M_{ko}|} \tag{30}$$

the system shifts from the state 0 to the state k:

$$|c_k(\frac{\pi}{2a})|^2 = 1 \quad , \quad |c_o(\frac{\pi}{2a})|^2 = 0$$

I.B. Static Effects of Perturbation

Let us consider a system with the Hamiltonian

$$H = H_o + H' \tag{31}$$

where both H_o and H' are independent of time. Let us call ψ_ℓ the orthonormal eigenfunctions of H_o, and ψ'_ν the eigenfunctions of H:

$$H_o \psi_\ell = E_\ell \psi_\ell \tag{32}$$

$$H \psi'_\nu = E'_\nu \psi'_\nu \tag{33}$$

We can expand the eigenfunctions of H as follows:

$$\psi'_\nu = \sum_\ell a_{\nu\ell} \psi_\ell \tag{34}$$

where ℓ ranges over all the eigenfunctions of H_o. Replacing Eq. (34) in Eq. (33),

$$H_o \sum_\nu a_{\nu\ell} \psi_\ell + H' \sum_\ell a_{\nu\ell} \psi_\ell = E'_\nu \sum_\ell a_{\nu\ell} \psi_\ell \tag{35}$$

or, because of Eq. (32)

$$\sum_\ell a_{\nu\ell} (E'_\nu - E_\ell) \psi_\ell = \sum_\ell a_{\nu\ell} H' \psi_\ell \tag{36}$$

Multiplying by ψ_m^* and integrating over the space coordinates we find

$$a_{\nu\ell}(E'_\nu - E_\ell) = \sum_\ell a_{\nu\ell} H'_{m\ell} \tag{37}$$

where

$$H'_{m\ell} = \langle\psi_m|H'|\psi_\ell\rangle \tag{38}$$

We shall assume at this point that the energy level scheme of H_o consists of two degenerate levels; the equations (37) can then be written as follows:

$$\begin{cases} a_{\nu 1}(E'_\nu - E_1) = a_{\nu 1} H'_{11} + a_{\nu 2} H'_{12} \\ a_{\nu 2}(E'_\nu - E_2) = a_{\nu 1} H'_{21} + a_{\nu 2} H'_{22} \end{cases} \tag{39}$$

Setting $E_1 = E_2 = E$, we get

$$\begin{cases} (E'_\nu - E - H'_{11})\, a_{\nu 1} + (-H'_{12})\, a_{\nu 2} = 0 \\ (-H'_{21})\, a_{\nu 1} + (E'_\nu - E - H'_{22})\, a_{\nu 2} = 0 \end{cases} \tag{40}$$

which implies that

$$\begin{vmatrix} E'_\nu - E - H'_{11} & -H'_{12} \\ -H'_{21} & E'_\nu - E - H'_{22} \end{vmatrix} = 0 \tag{41}$$

If we assume for simplicity that $H'_{11} = H'_{22}$, we get

$$(E'_\nu - E - H'_{11})^2 - |H'_{12}|^2 = 0 \tag{42}$$

The equation above gives us the two eigenvalues

$$E'_1 = E + H'_{11} + |H'_{12}| \tag{43}$$

$$E'_2 = E + H'_{11} - |H'_{12}| \tag{44}$$

We have still to find the coefficients $a_{\nu m}$. If we take $\nu = 1$ and replace E'_1 by the expression above in the relations (40), we obtain $a_{11} = a_{22}$ and

$$\psi'_1 = \frac{1}{\sqrt{2}} (\psi_1 + \psi_2) \tag{45}$$

Taking $\nu = 2$ and replacing the expression for E'_2 in Eqs. (40), we obtain $a_{21} = -a_{22}$ and

$$\psi_2' = \frac{1}{\sqrt{2}} (\psi_1 - \psi_2) \tag{46}$$

It is easy to verify that

$$<\psi_1'|H|\psi_1'> = E + H_{11}' + |H_{12}'| \tag{47}$$

$$<\psi_2'|H|\psi_2'> = E + H_{11}' - |H_{12}'| \tag{48}$$

and

$$<\psi_1'|H|\psi_2'> = 0 \tag{49}$$

Let us relate now these findings to the results of the previous Section I.A:

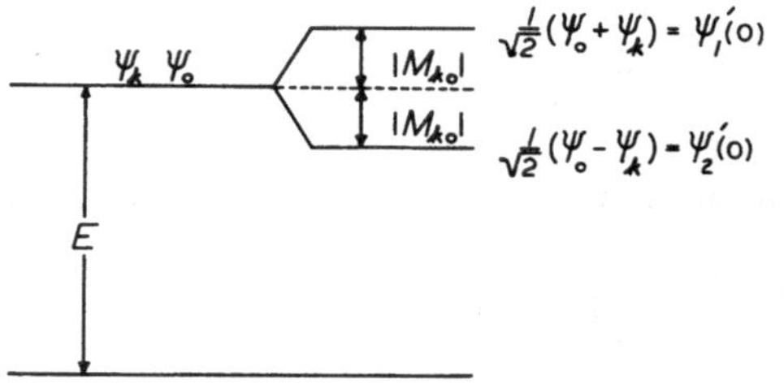

Fig. 1. Static Effect of Perturbation on a System Consisting of Two Degenerate States.

Since in the present case we consider two degenerate states we are clearly dealing with a "quantum mechanical resonance." If, in addition, we assume, as we did in Section I.A, that the diagonal matrix elements of H' are zero, we can represent the situation as in Figure 1.

We can now make the following observations:

(1) The same perturbation that is responsible for the change from the state 0 to state k is giving us a splitting of levels equal to $2|M_{ko}|$.

(2) Consider this splitting to be 1 cm^{-1} $(2|M| = hc)$. Then the time it takes the system to switch from state 0 to state k is

$$t = \frac{\pi}{2a} = \frac{\pi\hbar}{2|M|} = \frac{1}{2c} = 1.67 \times 10^{-11} \text{ sec}^{-1} \tag{50}$$

This is a very short time indeed. This case provides an illustrative example of how a perturbation that produces "negligible" (1 cm^{-1}) static effects may cause "dramatic" (1.67 x 10^{-11} sec) dynamical effects.

The time-dependent wavefunctions of the perturbed states are given by

$$\begin{cases} \psi_1'(t) = \frac{1}{\sqrt{2}} (\psi_o + \psi_k)\, e^{-(i/\hbar)(E_o+|M|)t} \\ \psi_2'(t) = \frac{1}{\sqrt{2}} (\psi_o - \psi_k)\, e^{-(i/\hbar)(E_o-|M|)t} \end{cases} \tag{51}$$

It is

$$\begin{cases} \psi_1'(0) = \frac{1}{\sqrt{2}} (\psi_o + \psi_k) \\ \psi_2'(0) = \frac{1}{\sqrt{2}} (\psi_o - \psi_k) \end{cases} \tag{52}$$

and conversely

$$\begin{cases} \psi_o = \frac{1}{\sqrt{2}} [\psi_1'(0) + \psi_2'(0)] \\ \psi_k = \frac{1}{\sqrt{2}} [\psi_1'(0) - \psi_2'(0)] \end{cases} \tag{53}$$

Let us assume that at time t = 0 the system is in the state ψ_o:

$$\Psi(t=0) = \psi_o = \frac{1}{\sqrt{2}} [\psi_1'(0) + \psi_2'(0)] \tag{54}$$

The wavefunction of the system at time t is then

$$\Psi(t) = \frac{1}{\sqrt{2}} \left[\frac{1}{\sqrt{2}} (\psi_o + \psi_k) e^{-i[(E+M)/\hbar]t} + \frac{1}{\sqrt{2}} (\psi_o - \psi_k) e^{-i[(E-M)/\hbar]t} \right]$$

$$= e^{-i(E_o/\hbar)t} \left(\psi_o \cos \frac{M}{\hbar} t - i \psi_k \sin \frac{M}{\hbar} t \right) \tag{55}$$

At the time t = 0, as we said before, the system is in the state ψ_o; at the time $t = \pi\hbar/(2|M|)$ the wavefunction of the system is

$$\psi(t = \frac{\pi\hbar}{2|M|}) = e^{-i(E_o/\hbar)(\pi\hbar/2M)} (-i \psi_k) \tag{56}$$

namely, the system is in the state ψ_k. After a time $t = 2\pi\hbar/(2|M|)$ the system goes back to the state ψ_o and so on.

We note here that the rate at which the energy is being exchanged between the two states is proportional to the "coupling energy" $|M|$.

We can make the following observations on these results:

(a) When the system is in a stationary state the wavefunction, as seen in Eq. (51), includes both ψ_o and ψ_k. There is an equal probability of finding the system in state ψ_o or in state ψ_k.

(b) In order to put the system in a state ψ_o, it is necessary to include the two wavefunctions ψ_1' and ψ_2' which have different energies (ψ_o is not a stationary state of $H = H_o + H'$).

The quantum mechanical system that we have examined is analogous to the mechanical system consisting of two equal pendula coupled by a weak spring. This system presents two normal modes whose patterns are presented in the following Figure 2.

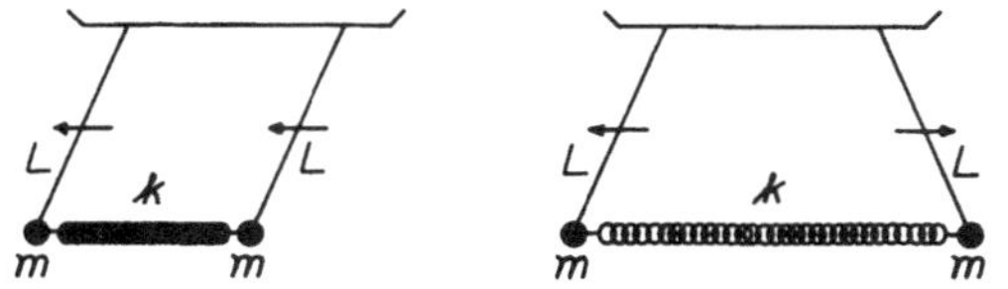

Fig. 2. Normal Modes of Two Coupled Pendula.

If the pendula oscillate according to one of the two normal modes the system is in a "stationary" state and the same amount of energy is associated to each pendulum. If, however, we start from the initial conditions in which one pendulum is in the rest position and the other pendulum is deflected, an exchange of energy between the two pendula takes place.

We note also that the closer the coupling, i.e., the stronger is the spring, the greater is the energy transfer rate.

II. COLLECTIVE EXCITATIONS

II.A. Setting of the Problem

Consider a linear crystal consisting of a chain of N identical atoms with a unit cell of length a.

Assume that the Hamiltonian of the system is the following

$$H = H_o + H' \tag{57}$$

where

$$H_o = \sum_{s=1}^{N} H_s \tag{58}$$

$$H' = \frac{1}{2} \sum_{\substack{s,s' \\ s \neq s'}} V_{ss'} \tag{59}$$

The H_s are one-atom Hamiltonians and the $H_{ss'}$ are terms representing interactions between different atoms. The ground state wavefunction of H_o is given by

$$\psi_g = u_1 u_2 \cdots u_{N-1} u_N \tag{60}$$

The first excited state of H_o corresponds to the following N degenerate wavefunctions

$$\begin{cases} \psi_1 = v_1 u_2 \cdots u_N \\ \psi_2 = u_1 v_2 \cdots u_N \\ \cdots\cdots\cdots\cdots\cdots\cdots \\ \psi_N = u_1 u_2 \cdots v_N \end{cases} \tag{61}$$

In these wavefunctions u_i represents a localized one-atom ground state and v_i a localized one-atom excited state. Both u_i and v_i are assumed to be nondegenerate.

The ground state of the Hamiltonian H_o is nondegenerate; the first excited state of H_o is N-fold degenerate, because there are N distinct ways to excite only one atom. For all these wavefunctions the energy is E_o:

$$H_o \psi_\ell = E_o \psi_\ell \tag{62}$$

The eigenfunctions of H are given by

$$H \psi'_\nu = (H_o + H') \psi'_\nu = E'_\nu \psi'_\nu \tag{63}$$

The excited-state eigenfunctions of H can be expressed as follows:

$$\psi'_\nu = \sum_{\ell=1}^{N} a_{\nu\ell} \psi_\ell \tag{64}$$

Replacing this in Eq. (63) we find

$$\sum_\ell a_{\nu\ell}(E_o - E'_\nu) \psi_\ell + \sum_\ell a_{\nu\ell} H' \psi_\ell = 0 \tag{65}$$

Multiplying by ψ_m^* and integrating

$$a_{\nu m}(E'_\nu - E_o - H'_{mm}) - \sum_{\ell\neq m} a_{\nu\ell} H'_{m\ell} = 0 \tag{66}$$

We note the following:

E'_ν = total energy of the system

$E_o + H'_{mm}$ = energy of the chain if the excitation cannot move

$E'_\nu - (E_o + H'_{mm}) = \varepsilon_\nu$ = energy associated with the motion of the excitation.

We see from Eq. (66) that the motion of the excitation is related to the off-diagonal terms of the Hamiltonian H!

II.B. Eigenfunctions

In order to look for functions that approximate well the wavefunctions ψ'_ν that satisfy the eigenvalue equation (63) we want to take advantage of the symmetry of the system. We shall assume periodic boundary conditions to take care of "surface" effects and translational symmetry for the chain

$$H(x + \ell a) = H(x) \tag{67}$$

Later in this treatment we shall generalize our results to three-dimensional crystals. It seems appropriate at this time to digress from the present sequence of derivations and obtain a general result for the three-dimensional case.

Let us assume that an electorn is in a periodic field of force and its Hamiltonian is such that

$$H(\vec{r} + \vec{R}_n) = H(\vec{r}) \tag{68}$$

The eigenfunctions of H are given by the equation

$$H\psi(\vec{r}) = E\psi(\vec{r}) \tag{69}$$

We shall assume these functions to be orthonormal. Let us introduce a translation operation $T_{\vec{n}}$ that displaces the origin by $-\vec{R}_n$, $\vec{R}_n$ being a <u>lattice vector</u>:

$$T_{\vec{n}}\psi(\vec{r}) = \psi(\vec{r} + \vec{R}_n) \tag{70}$$

Dropping for simplicity the subscript of the operation T we can write

$$TH\psi(\vec{r}) = ET\psi(\vec{r}) \tag{71}$$

or

$$H[T\psi(\vec{r})] = E[T\psi(\vec{r})] \tag{72}$$

Therefore, $T\psi(\vec{r})$ is an eigenfunction of H belonging, as $\psi(\vec{r})$, to the eigenvalue E.

If E is nondegenerate, then

$$T\psi(\vec{r}) = c\psi(\vec{r}) \tag{73}$$

with c = const. $c\psi(\vec{r})$ differs from $\psi(\vec{r})$ only because of a change in the origin. Therefore

$$\int |c\psi(r)|^2 \, d\tau = |c|^2 \int |\psi(r)|^2 d\tau = |c|^2 = 1 \tag{74}$$

If E is an m-degenerate eigenvalue of the energy, then

$$H[T\psi_\mu(\vec{r})] = E[T\psi_\mu(\vec{r})] \tag{75}$$

with

$$T\psi_\mu(\vec{r}) = \psi_\mu(\vec{r} + \vec{R}_n) = \sum_{\ell=1}^{m} c_{\mu\ell} \, \psi_\ell(\vec{r}) \tag{76}$$

with $\mu = 1,2, \ldots, m$. As said before, the functions ψ are orthonormal:

$$\int \psi_\mu^*(\vec{r}) \, \psi_\nu(\vec{r}) \, d\tau = \delta_{\mu\nu} \tag{77}$$

These integrals are not affected by a change of origin; therefore, the same is true of the new functions

$$\begin{aligned} \int \psi_\mu^*(\vec{r} + \vec{R}_n) \, \psi_\nu(\vec{r} + \vec{R}_n) \, d\tau &= \int \sum_{\ell=1}^{m} c_{\mu\ell}^* \, \psi_\ell^* \sum_{\ell'=1}^{m} c_{\nu\ell'} \, \psi_{\ell'} \, d\tau \\ &= \sum_{\ell=1}^{m} \sum_{\ell'=1}^{m} c_{\mu\ell}^* \, c_{\nu\ell'} \int \psi_\ell^* \, \psi_{\ell'} \, d\tau \\ &= \sum_{\ell=1}^{m} \sum_{\ell'=1}^{m} c_{\mu\ell}^* \, c_{\nu\ell'} \, \delta_{\ell\ell'} = \sum_{\ell} c_{\mu\ell}^* \, c_{\nu\ell} = \delta_{\mu\nu} \end{aligned} \tag{78}$$

We observe then the following:

(1) The m x m matrix of the coefficient c is unitary:

$$\underset{\sim}{C} \, \underset{\sim}{C}^+ = \underset{\sim}{1} \tag{79}$$

(2) Any unitary matrix can be brought to diagonal form by a unitary transformation.

(3) There exists an orthonormal set of linear combinations

of the functions ψ which are related to a diagonal matrix $\underset{\sim}{c}$.

$$T \begin{pmatrix} \psi_1 \\ \psi_2 \\ .. \\ .. \\ \psi_m \end{pmatrix} = \begin{pmatrix} c_{11} & 0 & ..... & 0 \\ 0 & c_{22} & ..... & 0 \\ 0 & 0 & ..... & 0 \\ 0 & 0 & ..... & 0 \\ 0 & 0 & ..... & c_{mm} \end{pmatrix} \begin{pmatrix} \psi_1 \\ \psi_2 \\ .. \\ .. \\ \psi_m \end{pmatrix} \tag{80}$$

This is equivalent to

$$\psi(\vec{r} + \vec{R}_n) = c\psi(\vec{r}) \tag{81}$$

being valid for each function ψ_μ.

(4) Having made the matrix $\underset{\sim}{C}$ diagonal for a particular translation $\vec{R}_n$, we can do likewise for any other translation because the translation operations commute. The matrices $\underset{\sim}{C}$ related to any two lattice vector translations commute, and a set of commuting matrices can always be diagonalized simultaneously.

On the basis of the above we can assume that all the eigenfunctions can be made to satisfy the condition

$$\psi(\vec{r} + \vec{R}_n) = c\psi(\vec{r}) \tag{82}$$

for any lattice vector displacement. Therefore, we have

$$\psi(\vec{r} + \vec{R}_n) = c(\vec{R}_n)\ \psi(\vec{r}) \tag{83}$$

$$\psi(\vec{r} + \vec{R}_n) = c(\vec{R}_m)\ \psi(\vec{r}) \tag{84}$$

and

$$\psi(\vec{r} + \vec{R}_n + \vec{R}_m) = c(\vec{R}_m)\ \psi(\vec{r} + \vec{R}_n)$$

$$= c(\vec{R}_n)\ c(\vec{R}_m)\ \psi(\vec{r}) = c(\vec{R}_n + \vec{R}_m)\ \psi(\vec{r}) \tag{85}$$

The coefficients c must respect two conditions

(1) $|c(\vec{R}_n)|^2 = 1$

(2) $c(\vec{R}_n + \vec{R}_m) = c(\vec{R}_n)\, c(\vec{R}_m)$

Therefore we set

$$c(\vec{R}_n) = e^{i\vec{k}\cdot\vec{R}_n} \tag{86}$$

and

$$\psi(\vec{r} + \vec{R}_n) = e^{i\vec{k}\cdot\vec{R}_n}\, \psi(\vec{r}) \tag{87}$$

The vector $\vec{k}$ is defined, apart from any reciprocal lattice vector $\vec{K}_s$, since for any such vector

$$e^{i\vec{K}_s\cdot\vec{R}_n} = 1 \tag{88}$$

for any $\vec{R}_n$. This gives us the possibility of keeping $\vec{k}$ within the <u>first Brillouin zone</u>.

The values of $\vec{k}$ are determined by the periodic boundary conditions that we impose on the wavefunctions.

We apply now these considerations to our linear crystal. As we noted in Eq. (64), the desired wavefunctions have the form

$$\psi'_k = \sum_{\ell=1}^{N} a_{k\ell}\, \psi_\ell \tag{89}$$

where we have changed the wavefunction index from ν to k. In the expressions above k runs over N possible values and we note that the wavefunctions ψ'_k may involve some degeneracy.

As we established in Eqs. (61), the first excited state of H_o is related to the wavefunctions

$$\begin{cases} \psi_1(x) = v_1(x)\, u_2(x-a)\, u_3(x-2a) \ldots u_N[x-(N-1)\,a] \\ \psi_2(x) = u_1(x)\, v_2(x-a)\, u_3(x-2a) \ldots u_N[x-(N-1)\,a] \\ \cdots\cdots\cdots\cdots\cdots\cdots\cdots\cdots\cdots\cdots \\ \psi_N(x) = u_1(x)\, u_2(x-a)\, u_3(x-2a) \ldots v_N[x-(N-1)\,a] \end{cases} \tag{90}$$

Let us consider the effect of a shift of the origin in the +x direction by an amount a:

$$\psi_i(x) \rightarrow \psi_i(x - a) \tag{91}$$

It is

$$\begin{aligned}\psi_1(x - a) &= v_1(x - a)\ u_2(x - 2a)\ u_3(x - 3a)\ \ldots\ u_N(x - Na) \\ &= v_1(x - a)\ u_2(x - 2a)\ u_3(x - 3a)\ \ldots\ u_N(x) \\ &= \psi_2(x)\end{aligned} \tag{92}$$

We have made use of the periodic boundary conditions in the intermediate step. In general it may be seen that

$$\psi_i(x - a) = \psi_{i+1}(x) \tag{93}$$

Let us now examine the effect of this shift of origin on the wavefunctions ψ'_k. First, according to the result (87)

$$\begin{aligned}\psi'_k(x - a) &= e^{-ika}\ \psi'_k(x) \\ &= e^{-ika}(a_{k1}\ \psi_1 + a_{k2}\ \psi_2 + \ldots\ a_{kN}\ \psi_N)\end{aligned} \tag{94}$$

On the other hand, we may also write

$$\psi_k(x - a) = a_{k1}\ \psi_2 + a_{k2}\ \psi_3 + \ldots\ a_{kN}\ \psi_1 \tag{95}$$

The comparison of the coefficients in Eqs. (94) and (95) gives us

$$a_{k\ell} = e^{i(\ell-1)ka}\ a_{k1} \tag{96}$$

The desired wavefunctions can now be written

$$\psi_k = a_{k1}(\psi_1 + e^{ika}\ \psi_2 + e^{i2ka}\ \psi_3 + \ldots) \tag{97}$$

The normalization of ψ'_k yields

$$|a_{k1}|^2 = \frac{1}{N} \tag{98}$$

and we may choose the phase of a_{k1} so that

$$a_{k1} = \frac{1}{\sqrt{N}}\ e^{ika} \tag{99}$$

resulting in

$$a_{k\ell} = \frac{1}{\sqrt{N}} e^{i\ell ka} \tag{100}$$

Therefore the approximate wavefunctions of H are

ground state

$$\psi_g = |u_1 \, u_2 \, \dots \, u_N\rangle \tag{101}$$

first excited state

$$\psi'_k = \frac{1}{\sqrt{N}} \sum_{\ell=1}^{N} e^{i\ell ka} |u_1 \, u_2 \, \dots \, v_\ell \, \dots \, u_N\rangle \tag{102}$$

The above wavefunctions are approximate for the following reasons:

(1) We treated the localized electronic wavefunctions u_j as non-overlapping. In fact, any functions used to approximate them do overlap.

(2) We did not take explicitly into account the interaction terms V_{ss} in generating the ψ_g and ψ'_k.

Finally, we consider the allowed values of k which are, as we said, determined by the boundary conditions that we impose. If we choose periodic boundary conditions, $\psi'_k(x + Na) = \psi'_k(x)$ and

$$a_{k,N+1} = a_{k1} \tag{103}$$

or

$$e^{ik(N+1)a} = e^{ika} \tag{104}$$

This implies

$$e^{ikNa} = 1 \tag{105}$$

with the solutions

$$k = \frac{2\pi n}{Na} \tag{106}$$

If we take

$$-\frac{N}{2} < n \leq \frac{N}{2}$$

the range for n is compatible with k being in the first Brillouin zone.

II.C. Dispersion Relations

We have already seen in the previous section that

$$a_{k\ell} = \frac{1}{\sqrt{N}} e^{i\ell ka} \tag{107}$$

These coefficients appear in Eq. (66) as follows

$$a_{km}\, \varepsilon_k - \sum_{\ell \neq m} a_{k\ell}\, H'_{m\ell} = 0 \tag{108}$$

where

$$\varepsilon_k = E'_k - E_o - H'_{mm} \tag{109}$$

Substituting Eq. (107) in Eq. (108), we obtain a relationship between ε_k and k

$$\frac{1}{\sqrt{N}} e^{imka}\, \varepsilon_k - \sum_{\ell \neq m} \frac{1}{\sqrt{N}} e^{i\ell ka}\, H'_{m\ell} = 0$$

and

$$\varepsilon_k = \sum_{\ell \neq m} e^{i(\ell-m)ka}\, H'_{m\ell} \tag{110}$$

This dispersion relation is independent of m; in addition, the N different values of k generate N values for ε_k. We may recall that

$$H'_{m\ell} = \langle \psi_m | H' | \psi_\ell \rangle \tag{111}$$

which in the present case reduces to

$$H'_{m\ell} = \langle v_m\, u_\ell | H' | u_m\, v_\ell \rangle \tag{112}$$

In many cases only nearest neighbor interactions are non-negligible; for such systems the matrix elements may be written

$$H'_{m\ell} = M\, \delta_{\ell, m\pm 1} \tag{113}$$

where M is the strength of the interaction. The dispersion relation represented by Eq. (110) then becomes

$$\varepsilon_k = Me^{ika} + Me^{-ika} = 2M \cos ka \tag{114}$$

The total energy of the system is then of the form

$$E'_k = E_o + H'_{mm} + \varepsilon_k = E_o + H'_{mm} + 2M \cos ka \tag{115}$$

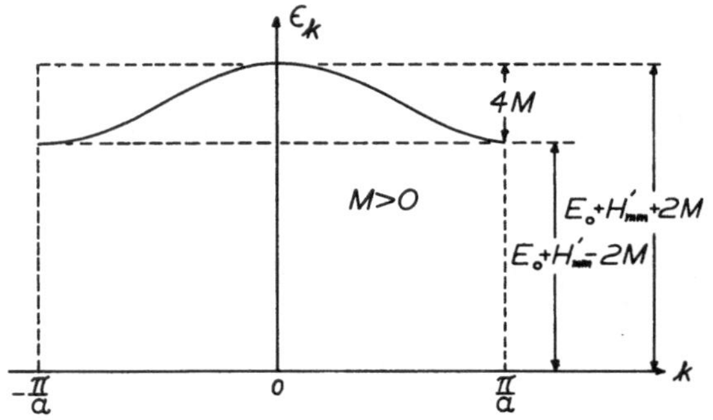

Fig. 3. Dispersion Relation for Collective Excitations of a Linear Chain.

where we have taken into account the expression (109). We observe that, as a result of the perturbation H', the N-fold degenerate state has become a band of N states (recall the N allowed values of k).

The dispersion relation is sketched in Figure 3.

II.D. Effective Mass

These collective excitations can be treated as quasi-particles with an effective mass. By forming wave packets with a spread $\Delta\vec{k}$ about $\vec{k}$ we can define a particle velocity as the group velocity

$$v_k = \frac{d\omega}{dk} = \frac{1}{\hbar}\frac{\partial\varepsilon_k}{\partial k} \tag{116}$$

We note that this formula is also valid for electromagnetic waves.

Since, for a free particle,

$$\varepsilon_k = \frac{\hbar^2 k^2}{m} \tag{117}$$

we can associate to our quasi-particle the mass

$$m^* = \frac{\hbar^2}{\left(\frac{\partial^2\varepsilon}{\partial k^2}\right)} \tag{118}$$

In the present case

$$\varepsilon_k = 2M \cos ka \tag{119}$$

and

$$\frac{d\varepsilon_k}{dk} = -2Ma \sin ka \tag{120}$$

$$\frac{d^2\varepsilon_k}{dk^2} = -2Ma^2 \cos ka \tag{121}$$

Therefore

$$v_k = \frac{1}{\hbar}\frac{\partial\varepsilon_k}{\partial k} = -\frac{2Ma \sin ka}{\hbar} \tag{122}$$

and

$$m^* = \frac{\hbar^2}{\left(\frac{\partial^2\varepsilon_k}{\partial k^2}\right)} = -\frac{\hbar^2}{2Ma^2 \cos ka} \tag{123}$$

ε_k, v_k and m^* are represented in Figure 4. We note that for small k

$$v_k = -\frac{2Mka^2}{\hbar} \tag{124}$$

$$m^* = -\frac{\hbar^2}{2Ma^2} \tag{125}$$

In order to establish some qualitative features of the effective mass approach, let us treat only the absolute values of v and m^*. We note:

(1) The velocity is linearly dependent on both M and k. As the strength of the interaction increases, the speed of excitation propagation also increases (all other factors being equal). The dependence of v on the interatomic distance a is more subtle, since a portion of this dependence is "hidden" in Mk. As a typical example, let us consider the electric dipole-electric dipole interaction between nearest neighbors. This interaction goes as a^{-3} [1]. In this case, we have $v \propto a^{-2}$, since $k \propto a^{-1}$. We therefore see that the velocity decreases (possibly dramatically) as a increases.

(2) The effective mass is inversely proportional to M. A strong interaction between nearest neighbors would then result in a relatively small effective mass (all other factors being equal). This reinforces our general notion that a strong interaction enhances the delocalization of excitation energy in the system. As for the velocity, the dependence of m^* on a requires some specification of the interaction M. For the electric dipole-electric dipole case given above, it is easy to verify that $m^* \propto a^1$. This results in the physically reasonable behavior that an increased separation distance hampers the movement of excitation energy.

II.E. Generalization to Three Dimensions

We generalize our treatment to an ordered solid in three dimensions. Let $\vec{a}_1$, $\vec{a}_2$ and $\vec{a}_3$ be the primitive basic lattice vectors and $N = N_1 \times N_2 \times N_3$ the total number of atoms. We introduce the notion of reciprocal lattice, a geometrical construction which consists of an array of points. The primitive basic vectors of the reciprocal lattice are given by

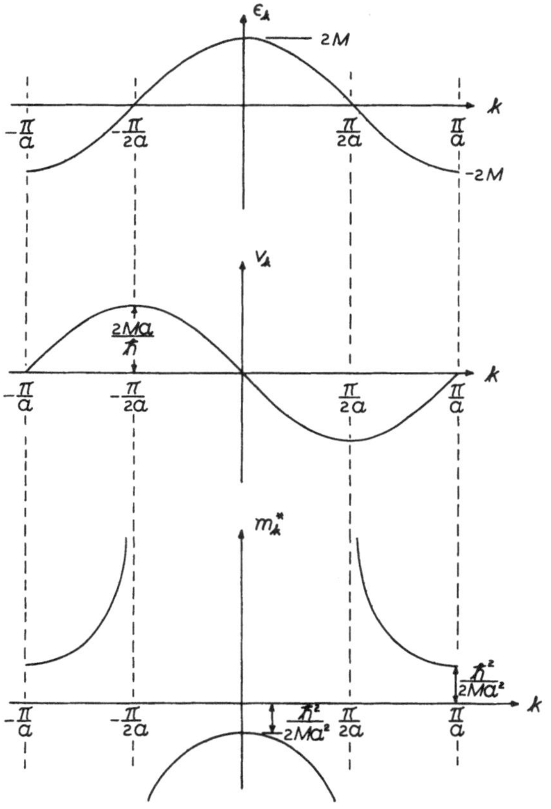

Fig. 4. Energy, Velocity and Equivalent Mass of Collective Excitations.

$$\begin{cases} \vec{b}_1 = 2\pi \dfrac{\vec{a}_2 \times \vec{a}_3}{\vec{a}_1 \cdot \vec{a}_2 \times \vec{a}_3} \\ \vec{b}_2 = 2\pi \dfrac{\vec{a}_3 \times \vec{a}_1}{\vec{a}_1 \cdot \vec{a}_2 \times \vec{a}_3} \qquad (\vec{b}_i\, \vec{a}_j = 2\pi\, \delta_{ij}) \\ \vec{b}_3 = 2\pi \dfrac{\vec{a}_1 \times \vec{a}_2}{\vec{a}_1 \cdot \vec{a}_2 \times \vec{a}_3} \end{cases} \tag{126}$$

In the reciprocal lattice we carve out the Brillouin Zone, another geometrical construction in $\vec{k}$ space. It is constructed by starting from a lattice point of the reciprocal lattice, drawing lines to the nearest neighbor points and cutting these lines halfway with perpendicular planes; the smallest volume enclosed by these planes is the (first) Brillouin Zone.

The wavefunction of the crystal corresponding to the first excited state is now designated as follows:

$$\psi_k = \frac{1}{\sqrt{N}} \sum_{\ell} e^{i\vec{k}\cdot\vec{R}_\ell} |u_1\, u_2\, \dots\, v_\ell\, \dots\, u_N> \tag{127}$$

The allowed values of $\vec{k}$ are determined by the periodic boundary conditions.

The dispersion relation is given by

$$\varepsilon_{\vec{k}} = \sum_{\substack{\ell \\ \ell\neq m}} e^{i\vec{k}\cdot(\vec{R}_\ell-\vec{R}_m)} H'_{m\ell} \tag{128}$$

The group velocity is given by

$$\vec{v}_{\vec{k}} = \frac{1}{\hbar} \vec{\nabla}_k\, \varepsilon_{\vec{k}} \tag{129}$$

and the mass tensor by

$$m^* = \frac{\hbar^2}{\vec{\nabla}_{\vec{k}}\, \vec{\nabla}_{\vec{k}}\, \varepsilon_{\vec{k}}} \tag{130}$$

II.F. Periodic Boundary Conditions and Density of States

Consider a crystal with

N_1 cells in the $\vec{a}_1$ direction,

N_2 cells in the $\vec{a}_2$ direction, and

N_3 cells in the $\vec{a}_3$ direction.

Call

$$\vec{R}_N = N_1 \vec{a}_1 + N_2 \vec{a}_2 + N_3 \vec{a}_3 \tag{131}$$

If we impose periodic boundary conditions (PBC), we obtain the allowed values of $\vec{k}$:

$$\psi_{\vec{k}}(\vec{r} + \vec{R}_N) = \psi_{\vec{k}}(\vec{r}) \tag{132}$$

namely,

$$e^{i\vec{k}\cdot\vec{R}_N} = 1 \tag{133}$$

This equation determines the allowed values of $\vec{k}$. Equation (133) implies that

$$\vec{k} \cdot \vec{R}_N = 2\pi s \qquad (s = \text{integer}) \tag{134}$$

The wave vector can be expressed as follows

$$\vec{k} = k_1 \hat{b}_1 + k_2 \hat{b}_2 + k_3 \hat{b}_3 \tag{135}$$

where $\hat{b}_i$ = unit vector in the $\vec{b}_i$ direction.

Therefore

$$\begin{aligned}\vec{k} \cdot \vec{R}_N &= (k_1 \hat{b}_1 + k_2 \hat{b}_2 + k_3 \hat{b}_3) \cdot (N_1 \vec{a}_1 + N_2 \vec{a}_2 + N_3 \vec{a}_3) \\ &= N_1 k_1 \vec{a}_1 \cdot \hat{b}_1 + N_2 k_2 \vec{a}_2 \cdot \hat{b}_2 + N_3 k_3 \vec{a}_3 \cdot \hat{b}_3 \\ &= 2\pi s \end{aligned} \tag{136}$$

This relation is satisfied if

$$\begin{cases} k_1 = \dfrac{2\pi s_1}{N_1 \vec{a}_1 \cdot \hat{b}_1} \\ \\ k_2 = \dfrac{2\pi s_2}{N_2 \vec{a}_2 \cdot \hat{b}_2} \\ \\ k_3 = \dfrac{2\pi s_3}{N_3 \vec{a}_3 \cdot \hat{b}_3} \end{cases} \tag{137}$$

with s_1, s_2 and s_3 positive or negative integers or zero.

Claim

The values of s_1 greater than N_1,

the values of s_2 greater than N_2, and

the values of s_3 greater than N_3

are redundant.

Proof

Let us assume that $s_i < N_i$ and let us replace s_i by $s_i + N_i$. Then

$$k_1 = \frac{2\pi s_1}{N_1 \vec{a}_1 \cdot \hat{b}_1} + \frac{2\pi N_1}{N_1 \vec{a}_1 \cdot \hat{b}_1} = \frac{2\pi s_1}{N_1 \vec{a}_1 \cdot \hat{b}_1} + \frac{2\pi|\vec{b}_1|}{\vec{a}_1 \cdot \vec{b}_1}$$

$$= \frac{2\pi s_1}{N_1 \vec{a}_1 \cdot \hat{b}_1} + |\vec{b}_1|$$

$$k_2 = \frac{2\pi s_2}{N_2 \vec{a}_2 \cdot \hat{b}_2} + |\vec{b}_2|$$

$$k_3 = \frac{2\pi s_3}{N_3 \vec{a}_3 \cdot \hat{b}_3} + |\vec{b}_3|$$

and

$$\vec{k} = \frac{2\pi s_1}{N_1 \vec{a}_1 \cdot \hat{b}_1} \hat{b}_1 + \frac{2\pi s_2}{N_2 \vec{a}_2 \cdot \hat{b}_2} \hat{b}_2 + \frac{2\pi s_3}{N_3 \vec{a}_3 \cdot \hat{b}_3}$$

$$+ \vec{b}_1 + \vec{b}_2 + \vec{b}_3$$

But

$$e^{i\vec{k}\cdot(\vec{b}_1+\vec{b}_2+\vec{b}_3)} = 1$$

Therefore the values of $s_1 > N_1$, $s_2 > N_2$ and $s_3 > N_3$ are redundant. Q.E.D.

We can then limit k_1, k_2 and k_3 to the following values:

$$k_1 = \frac{2\pi s_1}{N_1(\vec{a}_1 \cdot \hat{b}_1)} \qquad s_1 = 1, 2, \ldots, N_1 \quad \text{or } \pm 1, \pm 2, \ldots, \pm \frac{N_1}{2}$$

$$k_2 = \frac{2\pi s_2}{N_2(\vec{a}_2 \cdot \hat{b}_2)} \qquad s_2 = 1, 2, \ldots, N_2 \quad \text{or } \pm 1, \pm 2, \ldots, \pm \frac{N_2}{2}$$

$$k_3 = \frac{2\pi s_3}{N_3(\vec{a}_3 \cdot \hat{b}_3)} \qquad s_3 = 1, 2, \ldots, N_3 \quad \text{or } \pm 1, \pm 2, \ldots, \pm \frac{N_3}{2}$$

The number of states with k_1 in $(k_1 + dk_1)$, k_2 in $(k_2 + dk_2)$ and k_3 in $(k_3 + dk_3)$ is given by

$$ds_1\, ds_2\, ds_3 = \frac{N_1 N_2 N_3}{(2\pi)^3} [(\vec{a}_1 \cdot \hat{b}_1)(\vec{a}_2 \cdot \hat{b}_2)(\vec{a}_3 \cdot \hat{b}_3)] \times dk_1\, dk_2\, dk_3 \tag{138}$$

The infinitesimal volume element in $\vec{k}$ space is given by

$$d^3\vec{k} = dk_1\, dk_2\, dk_3(\hat{b}_1 \cdot \hat{b}_2 \times \hat{b}_3) \tag{139}$$

Therefore

$$ds_1\ ds_2\ ds_3 = \frac{N_1\ N_2\ N_3}{8\pi^3}\ d^3\vec{k}\ \frac{(\vec{a}_1 \cdot \hat{b}_1)(\vec{a}_2 \cdot \hat{b}_2)(\vec{a}_3 \cdot \hat{b}_3)}{\hat{b}_1 \cdot \hat{b}_2 \times \hat{b}_3} \tag{140}$$

But

$$\frac{(\vec{a}_1 \cdot \hat{b}_1)(\vec{a}_2 \cdot \hat{b}_2)(\vec{a}_3 \cdot \hat{b}_3)}{\hat{b}_1 \cdot \hat{b}_2 \times \hat{b}_3} = \frac{(\vec{a}_1 \cdot \vec{b}_1)(\vec{a}_2 \cdot \vec{b}_2)(\vec{a}_3 \cdot \vec{b}_3)}{\hat{b}_1 \cdot \hat{b}_2 \times \hat{b}_3} \tag{141}$$

We know that

$$\vec{a}_i \cdot \vec{b}_j = 2\pi\ \delta_{ij}$$

and

$$\vec{b}_1 \cdot \vec{b}_2 \times \vec{b}_3 = \text{volume of unit cell of reciprocal lattice} = 8\pi^3/\Omega_a$$

where

$$\Omega_a = \text{unit cell of "direct" lattice}$$

Therefore

$$\frac{(a_1 \cdot b_1)(a_2 \cdot b_2)(a_3 \cdot b_3)}{b_1 \cdot b_2 \cdot b_3} = \frac{8\pi^3}{\frac{8\pi^3}{\Omega_a}} = \Omega_a \tag{142}$$

and

$$ds_1\ ds_2\ ds_3 = \frac{N_1\ N_2\ N_3}{8\pi^3}\ d^3\vec{k}\ \Omega_a = \frac{V}{8\pi^3}\ d^3\vec{k} \tag{143}$$

because

$$N_1\ N_2\ N_3\ \Omega_a = V = \text{volume of the crystal}$$

The volume of the (first) Brillouin Zone is also $8\pi^3/\Omega_a$. The number of allowed $\vec{k}$ values in this zone is

$$\frac{8\pi^3}{\Omega_\alpha}\ \frac{V}{8\pi^3} = \frac{N\Omega_a}{\Omega_a} = N \tag{144}$$

where

$$N = N_1 \; N_2 \; N_3$$

III. INTERACTION OF RADIATION WITH COLLECTIVE EXCITATIONS

III.A. The Radiation Field

1. The Classical Radiation Field. In a region free of currents and charges, a radiation field is defined by a vector potential $\vec{A}(\vec{r},t)$ [2,3]

$$\nabla^2\vec{A}(\vec{r},t) - \frac{1}{c^2}\frac{\partial^2\vec{A}(\vec{r},t)}{\partial t^2} = 0 \tag{145}$$

$$\vec{\nabla} \cdot \vec{A}(\vec{r},t) = 0 \tag{146}$$

$$\vec{E}(\vec{r},t) = -\frac{1}{c}\frac{\partial\vec{A}(\vec{r},t)}{\partial r} \tag{147}$$

$$\vec{B}(\vec{r},t) = \vec{\nabla} \times \vec{A}(\vec{r},t) \tag{148}$$

Equation (145) is the so-called "field equation," Eq. (146) indicates that we have adopted the "Coulomb gauge," and Eqs. (147) and (148) give the electric field and the magnetic field in terms of $\vec{A}(\vec{r},t)$, respectively.

A typical solution of the field equation is given by

$$A(\vec{r})\; q(t) = \vec{\pi}\; [(\frac{4\pi c^2}{V})^{1/2}\; e^{i\vec{k}\cdot\vec{r}}][|q|e^{-i\omega t}] \tag{149}$$

where $\omega = kc$. We note that Eq. (146) implies

$$\vec{\pi} \cdot \vec{k} = 0 \tag{150}$$

namely, that the polarization of the wave is perpendicular to the direction of the wave vector. The allowed values of $\vec{k}$ are determined by the boundary conditions of the problem.

Summing over all $\vec{k}$ and all polarizations (σ), we obtain the general solution of the field equation

$$\vec{A}(\vec{r},t) = \sum_{\alpha}\sum_{\sigma}\; [q_\alpha^\sigma(t)\; \vec{A}_\alpha^\sigma(r) + q_\alpha^{\sigma *}(t)\; \vec{A}_\alpha^{\sigma *}(\vec{r})] \tag{151}$$

where

$$q_\alpha^\sigma(t) = |q_\alpha^\sigma| e^{-i\omega_\alpha t} \tag{152}$$

$$\vec{A}_\alpha^\sigma(\vec{r}) = \vec{\pi}_\alpha^\sigma \left(\frac{4\pi c^2}{V} \right)^{1/2} e^{i\vec{k}_\alpha \cdot \vec{r}} \tag{153}$$

The Hamiltonian of the radiation field can be derived from the expression for the energy of the field

$$\frac{1}{8\pi}\int [(\vec{E})^2 + (\vec{B})^2] \, d^3\vec{r} \tag{154}$$

and the relations (147) and (148) for $\vec{E}$ and $\vec{B}$, respectively. Because of the orthogonality of the various Fourier components of the field, cross terms with subscripts $\alpha\alpha'$ ($\alpha \neq \alpha'$) drop and we are left with (see [2]):

$$H = \sum_\alpha \sum_\sigma \omega_\alpha^2 (q_\alpha^\sigma q_\alpha^{\sigma *} + q_\alpha^{\sigma *} q_\alpha^\sigma) \tag{155}$$

We note the following:

(1) The Hamiltonian H of the radiation field is the sum of independent terms

$$H_\alpha^\sigma = \omega_\alpha^2 (q_\alpha^\sigma q_\alpha^{\sigma *} + q_\alpha^{\sigma *} q_\alpha^\sigma) \tag{156}$$

(2) The coordinates q_α^σ represent the "normal coordinates" of the field.

(3) No approximation has been made.

(4) We use the form $\omega_\alpha^2 (q_\alpha q_\alpha^* + q_\alpha^* q_\alpha)$ rather than the form $2\omega_\alpha^2 q_\alpha q_\alpha^*$ in preparation of our move into quantum mechanics: in the quantum mechanical treatment q and q^* become non-commuting operators.

The values of $\vec{k}$, as we said before, are determined by the boundary conditions. When, as it is the case here, the wavelength of the radiation is much smaller than the dimension of the spatial region under consideration, any sum over $\vec{k}$ is in effect an integral and the relevant information is the density of states, i.e., the number of states with $\vec{k}$ in $(\vec{k}, \vec{k} + d\vec{k})$. In order to find this quantity, periodic boundary conditions can be used; these conditions give for the possible values of the three components of $\vec{k}$

$$k_x = n_x \frac{2\pi}{L_x}$$

$$k_y = n_y \frac{2\pi}{L_y}$$

$$k_z = n_z \frac{2\pi}{L_z} \tag{157}$$

where the volume has been taken as a cube of sides L_x, L_y and L_z and where

$$n_x, n_y, n_z = 0, \pm 1, \pm 2, \pm 3, \ldots \tag{158}$$

The subscript α used before stands for a particular choice of n_x, n_y, and n_z.

The number of modes with $\vec{k}$ in $(\vec{k}, \vec{k} + d\vec{k})$ is given by

$$\frac{L_x dk_x}{2\pi} \frac{L_y dk_y}{2\pi} \frac{L_z dk_z}{2\pi} = \frac{L_x L_y L_z}{8\pi^3} dk_x\, dk_y\, dk_z$$

$$= \frac{V}{8\pi^3} k^2\, dk \sin\theta\, d\theta\, d\phi = \frac{V}{8\pi^3} \frac{\omega^2}{c^3} d\omega\, d\Omega \tag{159}$$

where

$$d\Omega = \sin\theta\, d\theta\, d\phi$$

2. The Quantum Radiation Field. Consider one term of the Hamiltonian (155)

$$H_\alpha = \omega_\alpha^2 (q_\alpha q_\alpha^* + q_\alpha^* q_\alpha) \tag{160}$$

where we have dropped for convenience the superscript σ. We introduce two new real variables for each α

$$Q_\alpha = q_\alpha + q_\alpha^*$$

$$P_\alpha = -i\omega_\alpha (q_\alpha - q_\alpha^*) = \dot{Q}_\alpha \tag{161}$$

The Hamiltonian H_α, when written in terms of Q_α and P_α, takes the form

$$H_\alpha = \frac{1}{2} \omega_\alpha^2 Q_\alpha^2 + \frac{1}{2} P_\alpha^2 \tag{162}$$

Q_α and P_α are real variables that satisfy Hamilton's equations:

$$\{Q_\alpha, P_{\alpha'}\} = \sum_i \left(\frac{\partial Q_\alpha}{\partial Q_i} \frac{\partial P_{\alpha'}}{\partial P_i} - \frac{\partial Q_\alpha}{\partial P_i} \frac{\partial P_{\alpha'}}{\partial Q_i} \right) = \delta_{\alpha\alpha'}$$

$$\{Q_\alpha, Q_{\alpha'}\} = \{P_\alpha, P_{\alpha'}\} = 0 \tag{163}$$

The prescription for moving over from a classical to a quantum mechanical treatment is simple. In the latter treatment Q_α and P_α become Hermitian operators and their commutator is obtained by replacing the Poisson brackets as follows:

$$\{Q_\alpha, P_{\alpha'}\} \rightarrow \frac{1}{i\hbar} [Q_\alpha, P_{\alpha'}] \tag{164}$$

Then

$$[Q_\alpha, P_{\alpha'}] = i\hbar \, \delta_{\alpha\alpha'}$$

$$[Q_\alpha, Q_{\alpha'}] = [P_\alpha, P_{\alpha'}] = 0 \tag{165}$$

q_α and q_α^*, which are related to Q_α and P_α by the relations (161), become two (non-Hermitian) operators which we shall call q_α and q_α^+, respectively. The commutation relation of these two operators are easily derived:

$$[q_\alpha, q_{\alpha'}^+] = \frac{\hbar}{2\omega_\alpha} \delta_{\alpha\alpha'}$$

$$[q_\alpha, q_{\alpha'}] = [q_\alpha^+, q_{\alpha'}^+] = 0 \tag{166}$$

We may replace q_α and q_α^+ by the dimensionless operators

$$a_\alpha = \left(\frac{2\omega_\alpha}{\hbar}\right)^{1/2} q_\alpha \quad , \quad a_\alpha^+ = \left(\frac{2\omega_\alpha}{\hbar}\right)^{1/2} q_\alpha^+ \tag{167}$$

It is

$$[a_\alpha, a_{\alpha'}^+] = \delta_{\alpha\alpha'}$$

$$[a_\alpha, a_{\alpha'}] = [a_\alpha^+, a_{\alpha'}^+] = 0 \tag{168}$$

The Hamiltonian of the radiation field can now be written

$$H = \sum_\alpha \omega^2 (q_\alpha q_\alpha^+ + q_\alpha^+ q_\alpha) = \sum_\alpha \hbar \, \omega_\alpha (a_\alpha^+ a_\alpha + \frac{1}{2}) \tag{169}$$

Reintroducing the polarization index σ

$$H = \sum_\alpha \sum_\sigma \hbar\,\omega_\alpha (a_\alpha^{\sigma+} a_\alpha^\sigma + \frac{1}{2}) \qquad (170)$$

The Hamiltonian

$$H_\alpha^\sigma = \hbar\,\omega_\alpha (a_\alpha^{\sigma+} a_\alpha^\sigma + \frac{1}{2}) \qquad (171)$$

has the energy eigenvalues

$$E_\alpha^\sigma = \hbar\,\omega_\alpha (n_\alpha^\sigma + \frac{1}{2}) \qquad (172)$$

where $n_\alpha^\sigma = 0, 1, 2, \ldots.$ The eigenfunctions of H_α^σ are simply given by the kets $|n_\alpha^\sigma\rangle$.

The Hamiltonian, eigenvalues and eigenfunctions of the radiation field are now listed:

$$H = \sum_\alpha \sum_\sigma H_\alpha^\sigma = \sum_\alpha \sum_\sigma \hbar\,\omega_\alpha (a_\alpha^{\sigma+} a_\alpha^\sigma + \frac{1}{2}) \qquad (173)$$

$$E_{n_1^{\sigma_1} n_1^{\sigma_2} n_2^{\sigma_1} \ldots} = \sum_\alpha \sum_\sigma \hbar\,\omega_\alpha (n_\alpha^\sigma + \frac{1}{2}) \qquad (174)$$

$$\psi_{n_1^{\sigma_1} n_1^{\sigma_2} n_2^{\sigma_1} \ldots} = \prod_\alpha \prod_\sigma |n_\alpha^\sigma\rangle \qquad (175)$$

One can see from the above relations that the radiation field may be thought of as a collection of an infinite number of harmonic oscillators, one for each (α,σ) component, with different degrees of excitation n_α^σ. Alternatively, the radiation field may be thought of as an ensemble of photons: n_α^σ is the number of photons present for each wave vector $\vec{k}_\alpha$ and polarization σ.

In the quantum-mechanical treatment the vector potential represents an operator which can be expressed as follows:

$$\vec{A} = \sum_\alpha \sum_\sigma [\vec{A}_\alpha^\sigma q_\alpha^\sigma + \vec{A}_\alpha^{\sigma *} a_\alpha^{\sigma+}]$$

$$= \sum_\alpha \sum_\sigma \left(\frac{4\pi c^2}{V}\right)^{1/2} \left(\frac{\hbar}{2\omega_\alpha}\right)^{1/2} \vec{\pi}_\alpha^\sigma (e^{i\vec{k}_\alpha\cdot\vec{r}} a_\alpha^\sigma + e^{-i\vec{k}_\alpha\cdot\vec{r}} a_\alpha^{\sigma+})$$

$$= \sum_\alpha \sum_\sigma \left(\frac{hc^2}{\omega_\alpha V}\right)^{1/2} \vec{\pi}_\alpha^\sigma (e^{i\vec{k}_\alpha\cdot\vec{r}} a_\alpha^\sigma + e^{-i\vec{k}_\alpha\cdot\vec{r}} a_\alpha^{\sigma+}) \qquad (176)$$

We note that the operators a_α^σ and $a_\alpha^{\sigma+}$ operate as follows:

$$a_\alpha^\sigma |n_\alpha^\sigma\rangle = \sqrt{n_\alpha^\sigma}\,|n_\alpha^\sigma - 1\rangle$$

$$a_\alpha^{\sigma+}|n_\alpha^\sigma> = \sqrt{n_\alpha^\sigma + 1}\,|n_\alpha^\sigma + 1> \tag{177}$$

III.B. The Form of the Interaction

Consider a particle of mass m and charge q under the action of a radiation field $\vec{A}(\vec{r},t)$ and of a potential field $\phi(r,t)$.

The equation of motion of such a particle is given by the Hamiltonian

$$H = \frac{(\vec{p} - \frac{q}{c}\vec{A})^2}{2m} + q\phi \tag{178}$$

where $\vec{p}$ = linear momentum of the particle. This can be easily justified by considering the Hamilton's equations, which give the correct expression for the (Lorentz) force acting on the particle.

The Hamiltonian H can be written as follows:

$$H = \frac{(\vec{p})^2}{2m} - \frac{q}{2mc}(\vec{p}\cdot\vec{A} + \vec{A}\cdot\vec{p}) + \frac{q^2}{2mc^2}(\vec{A})^2 + q\phi$$

$$= \frac{(\vec{p})^2}{2m} - \frac{q}{mc}(\vec{p}\cdot\vec{A}) + \frac{q^2}{2mc^2}(\vec{A})^2 + q\psi \tag{179}$$

since $[\vec{p},\vec{A}] = 0$, because of the Coulomb gauge. The interaction term, which is linear in the field, is relevant here. This term is

$$H_1 = -\frac{q}{mc}\vec{p}\cdot\vec{A}$$

$$= -\frac{q}{m}\sum_\alpha\sum_\sigma\left(\frac{h}{\omega_\alpha V}\right)^{1/2}\left(a_\alpha^\sigma e^{i\vec{k}_\alpha\cdot\vec{r}} + a_\alpha^{\sigma+} e^{-i\vec{k}_\alpha\cdot\vec{r}}\right)\vec{\pi}_\alpha^\sigma\cdot\vec{p} \tag{180}$$

In case of several particles

$$H_1 = -\left\{\sum_\alpha\sum_\sigma\left(\frac{h}{\omega_\alpha V}\right)^{1/2}\sum_i\left[\frac{q_i}{m_i}\left(a_\alpha^\sigma e^{i\vec{k}_\alpha\cdot\vec{r}_i} + a_\alpha^{\sigma+} e^{-i\vec{k}_\alpha\cdot\vec{r}_i}\right)\cdot(\vec{\pi}_\alpha^\sigma\cdot\vec{p}_i)\right]\right\} \tag{181}$$

III.C. Absorption and Emission Processes

Let us continue with the case of a charged particle under the action of a potential ϕ and of a radiation field $\vec{A}$. The Hamiltonian of the system which consists of the particle and the radiation field is given by

$$H = \frac{1}{2m} (\vec{p} - \frac{q}{c} \vec{A})^2 + q\phi + \frac{1}{8\pi} \int [(\vec{E})^2 + (\vec{B})^2] \, d^3\vec{r}$$

$$= \left[\frac{(\vec{p})^2}{2m} + q\phi\right] + \frac{1}{8\pi} \int [(\vec{E})^2 + (\vec{B})^2] \, d^3\vec{r}$$

$$- \frac{q}{2mc} (\vec{p} \cdot \vec{A}) + \frac{q^2}{2mc^2} (\vec{A})^2 \tag{182}$$

We can express H as follows:

$$H = H_o + H_1 + H_2 \tag{183}$$

where

$$H_o = \frac{(\vec{p})^2}{2m} + q\phi + \frac{1}{8\pi} \int [(\vec{E})^2 + (\vec{B})^2] \, d^3\vec{r}$$

$$= - \frac{\hbar^2}{2m} \nabla^2 + q\phi + \sum_\alpha \sum_\sigma \hbar \, \omega_\alpha (a_\alpha^{\sigma+} \, a_\alpha^\sigma + \frac{1}{2}) \tag{184}$$

and H_1 and H_2 are the terms linear and quadratic in the field, respectively.

The method to be applied here consists in considering H_o as the Hamiltonian of the "unperturbed" system, given simply by the sum of the Hamiltonian of the particle and the Hamiltonian of the radiation field and taking H_1 and H_2 as time-dependent perturbations of the system which may induce transitions between the different eigenstates of H_o. These eigenstates are given by

$$\psi_{e;n_1^{\sigma_1} \, n_1^{\sigma_2} \, \ldots} = \psi^e \prod_\alpha \prod_\sigma |n_\alpha^\sigma> \tag{185}$$

where ψ^e = eigenfunction of the particle and $|n_\alpha^\sigma>$ eigenfunction of the (α,σ) radiation oscillator. The energies of these states are given by

$$E_{e;n_1^{\sigma_1} \, n_1^{\sigma_2} \, \ldots} = E^e + \sum_\alpha \sum_\sigma \hbar \, \omega_\alpha \, n_\alpha^\sigma + \frac{1}{2}) \tag{186}$$

where E^e = energy of the particle and the sum over α and σ gives the energy of the radiation field.

In the case of one photon absorption, the initial and the final states are given by

$$\psi_i = |\psi_i^e\rangle|n_\alpha^\sigma\rangle|n_{\alpha'}^{\sigma'}\rangle \ldots \tag{187}$$

$$\psi_f = |\psi_f^e\rangle|n_\alpha^\sigma - 1\rangle|n_{\alpha'}^{\sigma'}\rangle \ldots \tag{188}$$

respectively. It is

$$\langle n_\alpha^\sigma - 1|a_\alpha^\sigma|n_\alpha^\sigma\rangle = \sqrt{n_\alpha^\sigma} \tag{189}$$

therefore, the relevant matrix element for the process of absorption of one photon is

$$\langle\psi_f^e;n_\alpha^\sigma - 1|H_1|\psi_i^e;n_\alpha^\sigma\rangle$$

$$= -\frac{q}{m}\left(\frac{h}{\omega_\alpha V}\right)^{1/2}\langle\psi_f^e|e^{i\vec{k}\cdot\vec{r}}\,\vec{\pi}_\alpha^\sigma\cdot\vec{p}|\psi_i^e\rangle\sqrt{n_\alpha^\sigma} \tag{190}$$

In the case of one photon emission, the initial and the final states are given by

$$\Psi_i = |\psi_i^e\rangle|n_\alpha^\sigma\rangle|n_{\alpha'}^{\sigma'}\rangle \ldots \tag{191}$$

$$\Psi_f = |\psi_f^e\rangle|n_\alpha^\sigma + 1\rangle|n_{\alpha'}^{\sigma'}\rangle \ldots \tag{192}$$

respectively. It is

$$\langle n_\alpha^\sigma + 1|a_\alpha^{\sigma+}|n_\alpha^\sigma\rangle = \sqrt{n_\alpha^\sigma + 1} \tag{193}$$

therefore the relevant matrix element for the process of emission of one photon is

$$\langle\psi_f^e;n_\alpha^\sigma + 1|H_1|\psi_i^e;n_\alpha^\sigma\rangle$$

$$= -\frac{q}{m}\left(\frac{h}{\omega_\alpha V}\right)^{1/2}\langle\psi_f^e|e^{-i\vec{k}\cdot\vec{r}}\,\vec{\pi}_\alpha^\sigma\cdot\vec{p}|\psi_i^e\rangle\sqrt{n_\alpha^\sigma + 1} \tag{194}$$

Since the radiation field has a continuous density of states (see for this Section III.A), both absorption and emission processes are associated with a probability per unit time. By applying the Fermi Golden Rule, we derive that the probability per unit time of finding the system (particle + radiation field) with one less or one more photon of energy $h\omega_\alpha$ and polarization $\vec{\pi}_\alpha^\sigma$ in the solid angle $(\Omega_\alpha, \Omega_\alpha + d\Omega_\alpha)$ is given by

$$P_\alpha^\sigma \, d\Omega_\alpha = \frac{2\pi}{\hbar^2} |M_\alpha^\sigma|^2 g(\omega_\alpha) \tag{195}$$

where

$$g(\omega_\alpha) = \frac{V \, \omega_\alpha^2}{8\pi^3 \, c^3} \, d\Omega_\alpha \tag{196}$$

and

$$|M_\alpha^\sigma|^2 = \begin{cases} \dfrac{q^2}{m^2} \dfrac{h}{\omega_\alpha V} \, |\langle\psi_f^e| e^{i\vec{k}_\alpha \cdot \vec{r}} \, \vec{\pi}_\alpha^\sigma \cdot \vec{p} |\psi_i^e\rangle|^2 \, n_\alpha^\sigma \\ \\ \dfrac{q^2}{m^2} \dfrac{h}{\omega_\alpha V} \, |\langle\psi_f^e| e^{-i\vec{k}_\alpha \cdot \vec{r}} \, \vec{\pi}_\alpha^\sigma \cdot \vec{p} \, \psi_i^e\rangle|^2 \, (n_\alpha^\sigma + 1) \end{cases} \tag{197}$$

where the upper (lower) row corresponds to the process of absorption (emission) of one photon. Replacing Eq. (197) in Eq. (195) and taking Eq. (196) into account, we find:

$$P_\alpha^\sigma \, d\Omega_\alpha = \frac{\omega_\alpha q^2}{hc^3 \, m^2} \, \left|\psi_f^e \left| \begin{matrix} e^{i\vec{k}_\alpha \cdot \vec{r}} \, \vec{\pi}_\alpha^\sigma \cdot \vec{p} \\ \\ e^{-\vec{k}_\alpha \cdot \vec{r}} \, \vec{\pi}_\alpha^\sigma \cdot \vec{p} \end{matrix} \right| \psi_i^e\right\rangle\right|^2 \begin{pmatrix} n_\alpha^\sigma \\ \\ n_\alpha^\sigma + 1 \end{pmatrix} d\Omega_\alpha \tag{198}$$

Let us consider two quantum states of the particle ψ_ℓ^e (ℓ stands for lower) and ψ_u^e (u stands for upper) with energies E_ℓ and E_u, respectively and $E_\ell < E_u$. It is possible to show that

$$|\langle\psi_\ell^e| e^{-i\vec{k}\cdot\vec{r}} \, (\vec{\pi} \cdot \vec{p}) |\psi_u^e\rangle|^2 = |\langle\psi_u^e| e^{i\vec{k}\cdot\vec{r}} \, (\vec{\pi} \cdot \vec{p}) |\psi_\ell^e\rangle|^2 \tag{199}$$

The last squared matrix element is the one that would enter the transition probability for an $\ell \to u$ (absorption) process.

On the basis of the above result, Eq. (198) becomes

$$P_\alpha^\sigma \, d\Omega_\alpha = \frac{\omega_\alpha q^2}{hc^3 m^2} |\langle\psi_u^e| e^{i\vec{k}\cdot\vec{r}} \vec{\pi}_\alpha^\sigma \cdot \vec{p} |\psi_\ell^e\rangle|^2 \begin{pmatrix} n_\alpha^\sigma \\ \\ n_\alpha^\sigma + 1 \end{pmatrix} d\Omega_\alpha \tag{200}$$

The transition probability for absorption is always proportional to the number of photons n_α^σ present; the transition probability for emission consists of one part, called "induced emission," which is proportional to n_α^σ and of another part, called "spontaneous emission," which is present even when $n_\alpha^\sigma = 0$. We note here that the transition probability of absorption and the transition probability for induced emission between two states are equal. If two or more charged particles are present

$$P_\alpha^\sigma \, d\Omega_\alpha = \frac{\omega_\alpha}{hc^3} |\langle \sum_t \frac{q_t}{m_t} (\vec{\pi}_\alpha^\sigma \cdot \vec{p}_t) \, e^{i\vec{k}_\alpha\cdot\vec{r}_t} \rangle u\ell|^2 \begin{pmatrix} n_\alpha^\sigma \\ \\ n_\alpha^\sigma + 1 \end{pmatrix} d\Omega_\alpha \tag{201}$$

III. D. Interaction of Photons with Collective Excitations

The relevant quantity in the creation or annihilation of a quantum of collective excitation via the absorption or emission of a photon is

$$\langle\psi_u \Big| \sum_i \frac{q_i}{m_i} e^{i\vec{k}_\alpha\cdot\vec{r}_i} (\vec{\pi}_\alpha^\sigma \cdot \vec{p}_i) \Big| \psi_\ell\rangle \tag{202}$$

where $\vec{k}_\alpha$ = wave vector of the photon.

In our case

$$\psi_\ell = |u_1 \, u_2 \, \ldots \, u_N\rangle \tag{203}$$

$$\psi_u = \frac{1}{\sqrt{N}} \sum_s e^{i\vec{k}\cdot\vec{R}_s} |u_1 \, u_2 \, \ldots \, v_s \, \ldots \, u_N\rangle \tag{204}$$

We can now write

$$<\psi_u\left|\sum_i \frac{q_i}{m_i} e^{i\vec{k}_\alpha\cdot\vec{r}_i}(\vec{\pi}_\alpha^\sigma \cdot \vec{p}_i)\right|\psi_\ell> = <\psi_u\left|\sum_i C_i\, e^{i\vec{k}_\alpha\cdot\vec{r}_i}\right|\psi_\ell>$$

$$= \frac{1}{\sqrt{N}}\sum_s e^{-i\vec{k}\cdot\vec{R}_s}<u_1\, u_2\, \ldots\, v_s\, \ldots\, u_N|\sum_i C_i\, e^{i\vec{k}_\alpha\cdot\vec{r}_i}|u_1\, u_2\, \ldots\, u_s\, \ldots\, u_N>$$

$$= \frac{1}{\sqrt{N}}\sum_i\sum_s e^{-i\vec{k}\cdot\vec{R}_s}<v_s|C_i\, e^{i\vec{k}_\alpha\cdot\vec{R}_i}\, e^{i\vec{k}_\alpha\cdot\vec{r}_i}|u_s>\,\delta_{is}$$

$$= \frac{1}{\sqrt{N}}\sum_s e^{i(\vec{k}_\alpha-\vec{k})\cdot\vec{R}_s}<v|C\, e^{i\vec{k}_\alpha\cdot\vec{r}\,'}|u>$$

$$= \sqrt{N}\, <v|C\, e^{i\vec{k}_\alpha\cdot\vec{r}\,'}|u>\, \delta_{\vec{k}_\alpha,\vec{k}+\vec{K}_s} \qquad (205)$$

where $C = \frac{q}{m}\vec{\pi}_\alpha^\sigma \cdot \vec{p}$ and $\vec{r}_i = \vec{R}_i + \vec{r}_i'$.

Setting $\vec{K}_s = 0$ (no umklapp processes)

$$<\psi_u|\sum_i C_i\, e^{i\vec{k}_\alpha\cdot\vec{r}_i}|\psi_\ell> = \sqrt{N}\, <v|C\, e^{i\vec{k}_\alpha\cdot\vec{r}\,'}|u>\, \delta_{\vec{k},\vec{k}_\alpha} \qquad (206)$$

This selection rule is illustrated in Fig. 5 which indicates that only excitations with $\vec{k} = \vec{k}_\alpha \simeq 0$ can be created in absorption and only excitations with $\vec{k} = \vec{k}_\alpha \simeq 0$ can produce the emission of a photon. We observe also that no dispersion effect can be seen because the dispersion curve of the collective excitations and that of the photons cross at <u>one</u> point.

The $\vec{k} = \vec{k}_\alpha$ rule is relaxed if more than one collective excitation is involved in the radiative process. For example, in absorption

$$\vec{k}_\alpha = \vec{k}_1 + \vec{k}_2 \qquad (207)$$

would correspond to the creation of a collective excitation of wave vector $\vec{k}_1$ and of a collective excitation of wave vector $\vec{k}_2$;

$$\vec{k}_\alpha = \vec{k}_1 - \vec{k}_2 \qquad (208)$$

would correspond to the creation of a collective excitation of wave

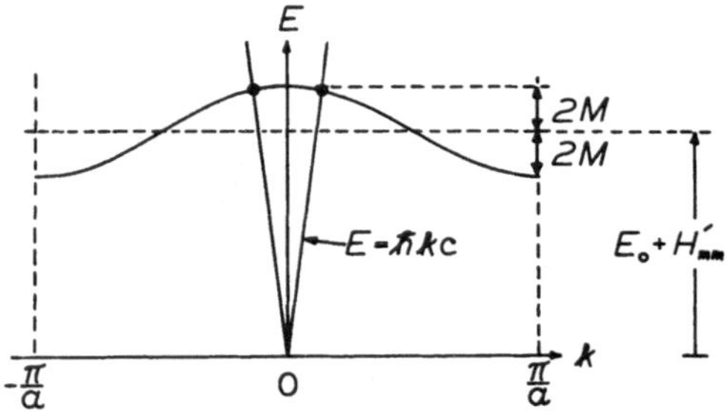

Fig. 5. Radiative Processes and Collective Excitations.

vector $\vec{k}_1$ and the annihilation of a collective excitation of wave vector $\vec{k}_2$.

IV. PROPAGATION OF RADIATION IN A DISPERSIVE MEDIUM

IV.A. Introduction

Let us consider an ensemble of N particles of mass m and negative charge e, each elastically bound to an equilibrium position where a positive charge $|e|$ resides. Let $n_o = N/V$ be the density of these particles, ω_o their natural angular frequency of oscillation and γ the damping constant. Let also

$$\vec{E}(\vec{r},t) = \vec{E}_o(\vec{r})\ e^{i\omega t} \tag{209}$$

represent an electromagnetic (plane) wave present in the medium.

The equation of motion of the individual particle is given by

$$\ddot{z} = -\frac{K}{m}\vec{z} + \frac{e}{m}\vec{E} - \gamma\dot{\vec{z}} = -\omega_o^2\vec{z} + \frac{e}{m}\vec{E} - \gamma\dot{\vec{z}} \tag{210}$$

where $\vec{z}(\vec{r},t)$ = displacement of the individual charged particle from its equilibrium position $\vec{r}$. We shall make the assumption that these displacements are always much smaller than the wavelength of the radiation.

The steady-state solution is

$$\vec{z}(\vec{r},t) = \vec{z}_o(\vec{r})\, e^{i\omega t} \tag{211}$$

where

$$\vec{z}_o(\vec{r}) = \frac{e/m}{\omega_o^2 - \omega^2 + i\gamma\omega}\, \vec{E}_o(\vec{r}) \tag{212}$$

The induced dipole moment is

$$e\vec{z}(\vec{r},t) = \frac{e^2/m}{\omega_o^2 - \omega^2 + i\gamma\omega}\, \vec{E}(\vec{r},t) = \alpha(\omega)\, \vec{E}(\vec{r},t) \tag{213}$$

where the <u>polarizability</u> $\alpha(\omega)$ is given by

$$\alpha(\omega) = \frac{e^2/m}{\omega_o^2 - \omega^2 + i\gamma\omega} = \mathrm{Re}\ \alpha(\omega) + i\ \mathrm{Im}\ \alpha(\omega) \tag{214}$$

and

$$\mathrm{Re}\ \alpha(\omega) = \frac{e^2}{m} \frac{\omega_o^2 - \omega^2}{(\omega_o^2 - \omega^2)^2 + \gamma^2\, \omega^2} \tag{215}$$

$$\mathrm{Im}\ \alpha(\omega) = -\frac{e^2}{m} \frac{\gamma\omega}{(\omega_o^2 - \omega^2)^2 + \gamma^2\, \omega^2} \tag{216}$$

The real part of the induced dipole moment is

$$\mathrm{Re}[e\vec{z}(\vec{r},t)] = \mathrm{Re}\{[\mathrm{Re}\ \alpha(\omega) + i\ \mathrm{Im}\ \alpha(\omega)][\vec{E}_o \cos \omega t + i\ \vec{E}_o \sin \omega t]\}$$

$$= \vec{E}_o\{[\mathrm{Re}\ \alpha(\omega)] \cos \omega t - [\mathrm{Im}\ \alpha(\omega)] \sin \omega t\}$$

or

$$\mathrm{Re}[e\vec{z}(\vec{r},t)] = \frac{e^2}{m} \frac{\vec{E}_o(\vec{r})}{\sqrt{(\omega_o^2 - \omega^2)^2 + \gamma^2 \omega^2}} \cos(\omega t - \phi) \tag{217}$$

where

$$\tan \phi = \frac{\gamma\omega}{\omega_o^2 - \omega^2} \tag{218}$$

The values of ϕ for $\omega = 0$, ω_o, and ∞ are $\phi = 0$, $\pi/2$, and π, respectively.

The magnetic force acting on the individual particle is equal to $\sim(v/c)$ x electric force. If we let $v = v(\text{oscillations}) \simeq \omega z_o$,

$$\frac{v}{c} = \frac{\omega z_o}{c} = \frac{2\pi z_o}{\lambda}$$

Taking $z_o = 10^{-8}$ cm and $\lambda = 1{,}000$ Å, v/c is on the order of 10^{-3} and the magnetic force is negligible.

IV.B. Dielectric Constant

We shall introduce at this point some important definitions.

Polarization = dipole/unit volume:

$$\vec{P} = n_o\ \alpha(\omega)\ \vec{E} \tag{219}$$

where

$\vec{E}$ = "average" applied field.

Dielectric susceptibility

$$\chi(\omega) = \frac{\vec{P}}{\vec{E}} = \frac{\text{dipole/unit volume}}{\text{applied field}} = n_o\ \alpha(\omega) \tag{220}$$

Electric displacement

$$\vec{D} = \vec{E} + 4\pi\ \vec{P} = \vec{E} + 4\pi\ \chi\ \vec{E}$$

$$= (1 + 4\pi\ \chi)\ \vec{E} = K\ \vec{E} \tag{221}$$

Dielectric constant

$$K(\omega) = 1 + 4\pi\ \chi(\omega) = n^2(\omega) \tag{222}$$

where $n(\omega)$ = index of refraction. Note that

$$\chi(\omega) = \frac{K(\omega) - 1}{4\pi} \tag{223}$$

For the system introduced in the previous section:

$$K(\omega) = 1 + 4\pi\ \chi(\omega) = 1 + 4\pi\ n_o\ \alpha(\omega)$$

$$= 1 + 4\pi\ n_o \frac{e^2/m}{\omega_o^2 - \omega^2 + i\gamma\omega} \tag{224}$$

If more than one type of oscillator is present in the system

$$K(\omega) = 1 + 4\pi \sum_s \frac{n_s\ e_s^2/m_s}{\omega_s^2 - \omega^2 + i\gamma_s\omega} = n^2(\omega) \tag{225}$$

For very low ω each oscillator type adds a constant contribution $[4\pi\ n_s\ e_s^2/(\omega_s^2\ m_s)]$ to the static dielectric constant. For $\omega >> \omega_s$ the contribution of the s-type oscillator becomes negligible.

For $\omega \simeq \omega_o$

$$K(\omega) = n^2(\omega) = A + \frac{4\pi\ n_o\ e^2/m}{\omega_o^2 - \omega^2 + i\gamma\omega} \tag{226}$$

where

$$A = 1 + \sum_{s>0} \frac{4\pi\ n_s\ e_s/m_s}{\omega_s^2} \tag{227}$$

A = 1 if there is only one oscillator type.

We can express the complex dielectric constant as follows

$$K(\omega) = K_r(\omega) + i\ K_i(\omega)$$

For $\omega \simeq \omega_o$ [4]

$$K_r = A + \frac{4\pi n_o e^2/m}{(\omega_o^2 - \omega^2)^2 + \gamma^2\omega^2}(\omega_o^2 - \omega^2) \simeq A + \frac{B\Delta\omega}{(\Delta\omega)^2 + (\gamma/2)^2} \tag{228}$$

$$K_i = -\frac{\omega\gamma(4\pi n_o e^2/m)}{(\omega_o^2 - \omega^2)^2 + \gamma^2\omega^2} \simeq -\frac{B\gamma/2}{(\Delta\omega)^2 + (\gamma/2)^2} \tag{229}$$

where $B = \dfrac{4\pi n_o e^2/m}{2\omega_o}$.

The approximate expressions for K_r and K_i are represented in Fig. 6.

The condition $K_r(\omega) \leq 0$ does not allow the propagation of radiation in the medium. This condition in the limit $\gamma \to 0$ corresponds to the "forbidden" frequency region

$$\omega_o < \omega \leq \omega_L \tag{230}$$

where

$$\omega_L = \omega_o\sqrt{1 + \frac{4\pi n_o e^2/m}{A\omega_o^2}} \tag{231}$$

Still in the limit $\gamma \to 0$:

$$K(\omega_L) = 0, \quad K(\omega_o) = \infty$$

$$K(\infty) = A$$

$$K(0) = A + \frac{4\pi n_o e^2/m}{\omega_o^2}$$

and

$$\frac{K(\infty)}{K(0)} = \frac{\omega_o^2}{\omega_L^2} \tag{232}$$

The last relation is called the Lyddane-Sachs-Teller relation.

IV.C. Propagation of Electromagnetic Waves in a Dispersive Medium

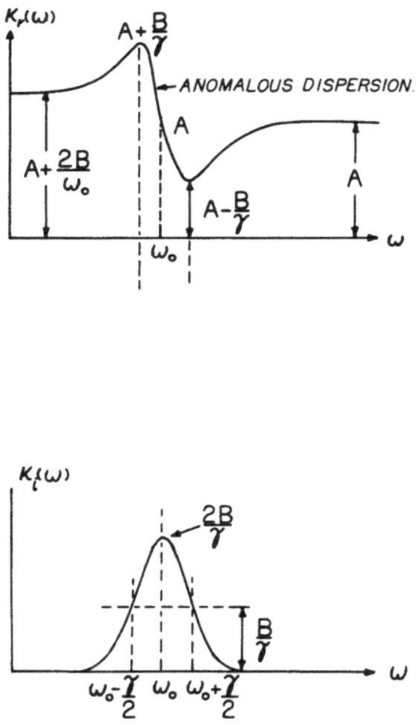

Fig. 6. Real and Imaginary Part of the Dielectric Constant of an Ensemble of Oscillating Charges.

The Maxwell equations can be expressed in general as follows:

$$\begin{cases} \vec{\nabla} \cdot \vec{B} = 0 \\ \vec{\nabla} \times \vec{E} + \frac{1}{c}\frac{\partial \vec{B}}{\partial t} = 0 \\ \vec{\nabla} \cdot \vec{E} = 4\pi \rho \\ \vec{\nabla} \times \vec{B} - \frac{1}{c}\frac{\partial \vec{E}}{\partial t} = \frac{4\pi}{c}\vec{j} \end{cases} \tag{233}$$

Consider now a medium which is uncharged ($\rho = 0$), polarizable (polarization = $\vec{P}$), and non-magnetic ($\mu = 1$; $\vec{B} = \vec{H}$). We have in this case

$$\rho = \rho_{pol} = -\vec{\nabla} \cdot \vec{P} \tag{234}$$

$$\vec{j} = \vec{j}_{pol} + \vec{j}_{true} = \frac{\partial \vec{P}}{\partial t} + \sigma \vec{E} \tag{235}$$

where σ = conductivity of the medium. Now we can write

$$\vec{\nabla} \cdot \vec{E} = 4\pi \rho_{pol} = -4\pi \vec{\nabla} \cdot \vec{P} \tag{236}$$

$$\vec{\nabla} \cdot (\vec{E} + 4\pi \vec{P}) = \vec{\nabla} \cdot K \vec{E} = 0 \tag{237}$$

Also

$$\vec{\nabla} \times \vec{B} = \frac{1}{c}\frac{\partial \vec{E}}{\partial t} + \frac{4\pi}{c}\vec{j} = \frac{1}{c}\frac{\partial \vec{E}}{\partial t} + \frac{4\pi}{c}(\vec{j}_{true} + \vec{j}_{pol})$$

$$= \frac{1}{c}\frac{\partial \vec{E}}{\partial t} + \frac{4\pi}{c}(\sigma \vec{E} + \frac{\partial \vec{P}}{\partial t})$$

$$= \frac{1}{c}\frac{\partial}{\partial t}(\vec{E} + 4\pi \vec{P}) + \frac{4\pi}{c}\sigma \vec{E}$$

$$= \frac{1}{c}(K \frac{\partial \vec{E}}{\partial t} + 4\pi \sigma \vec{E}) \tag{238}$$

Therefore the Maxwell equations for the medium can be expressed as follows:

$$\begin{cases} \vec{\nabla} \cdot \vec{H} = 0 \\ \vec{\nabla} \times \vec{E} + \frac{1}{c}\frac{\partial \vec{H}}{\partial t} = 0 \\ \vec{\nabla} \cdot K\vec{E} = 0 \\ \vec{\nabla} \times \vec{H} = \frac{1}{c}\left(K\frac{\partial \vec{E}}{\partial t} + 4\pi\sigma\vec{E}\right) \end{cases} \tag{239}$$

We look for wave-like solutions

$$\vec{E}(\vec{r},t) = \vec{E}_o\, e^{-i(\vec{\tau}\cdot\vec{r}-\omega t)}$$

$$\vec{H}(\vec{r},t) = \vec{H}_o\, e^{-i(\vec{\tau}\cdot\vec{r}-\omega t)} \tag{240}$$

Using these expressions in the Maxwell equations (239) we obtain

$$\begin{cases} \vec{\tau} \cdot \vec{H}_o = 0 \\ \vec{\tau} \times \vec{E}_o = \frac{\omega}{c}\vec{H}_o \\ K\,\vec{\tau} \cdot \vec{E}_o = 0 \\ -\vec{\tau} \times \vec{H}_o = \frac{1}{c}(\omega K - 4\pi\sigma i)\,\vec{E}_o \end{cases} \tag{241}$$

We shall consider first the case

$$K = 0;\ \vec{\tau} \cdot \vec{E} \neq 0 \tag{242}$$

It is

$$\vec{D} = K\vec{E} = \vec{E} + 4\pi\vec{P} = 0 \tag{243}$$

and

$$\vec{E} = -4\pi\vec{P} \tag{244}$$

If we neglect damping ($\sigma \simeq 0$)

$$\vec{H}_o = 0 \quad \text{and} \quad \vec{\tau} \times \vec{E} = 0 \tag{245}$$

Therefore in this case $\vec{E}$ represents a longitudinal wave with the wave vector $\vec{\tau}$ parallel to $\vec{E}$ and $\vec{P}$.

We consider next the case

$K \neq 0$

that gives

$$\begin{cases} \vec{\tau} \cdot \vec{H}_o = 0 \\ \vec{\tau} \times \vec{E}_o = \frac{\omega}{c} \vec{H}_o \\ \vec{\tau} \cdot \vec{E}_o = 0 \\ -\vec{\tau} \times \vec{H}_o = \left(\frac{\omega K}{c} - \frac{4\pi \sigma}{c} i \right) \vec{E}_o \end{cases} \tag{246}$$

In this case $\vec{\tau}$ is perpendicular to both $\vec{E}$ and $\vec{H}$. We can take the y-axis in the $\vec{E}$ direction, the z-axis in the $\vec{H}$ direction, and the x-axis in the $\vec{\tau}$ direction. We obtain from Eqs. (246)

$$\begin{cases} \tau E_o = \frac{\omega}{c} H_o \\ \tau H_o = \left(\frac{\omega K}{c} - \frac{4\pi \sigma i}{c} \right) E_o \end{cases} \tag{247}$$

or, combining the two equations above

$$\left(\frac{c\tau}{\omega} \right)^2 = K - \frac{4\pi \sigma}{\omega} i = K_r + i K_i \tag{248}$$

Therefore

$$\begin{cases} K_r = K \\ K_i = -\frac{4\pi \sigma}{\omega} \end{cases} \tag{249}$$

The index of refraction is also complex

$$\frac{c\tau}{\omega} = n = n_r + i n_i = \sqrt{K_r + i K_i} \tag{250}$$

The real and imaginary parts of the complex index of refraction and of the complex dielectric constant are related as follows:

$$\begin{cases} n_r^2 - n_i^2 = K_r \\ 2n_i n_r = K_i \end{cases} \tag{251}$$

Now we can write

$$\vec{E}(\vec{r},t) = \vec{j}\, E_o\, e^{-i\tau x + i\omega t}$$

$$= \vec{j}\, E_o\, e^{-i(\omega/c)nx + i\omega t} = \vec{j}\, E_o\, e^{-i(\omega/c)(n_r + in_i)x - i\omega t}$$

$$= \vec{j}\, E_o\, e^{-i[(\omega/c)n_r - \omega t]}\, e^{(\omega/c)n_i x} \tag{252}$$

$|\vec{E}|^2$ drop as

$$e^{2(\omega/c)n_i x} = e^{-\eta(\omega)x} \tag{253}$$

where $\eta(\omega)$ is defined as the absorption coefficient and is given by

$$\eta(\omega) = -2\,\frac{\omega}{c}\, n_i = -\frac{\omega K_i}{c n_r} \tag{254}$$

Note that if $\sigma = 0$, $K_i = 0$ and $\eta(\omega) = 0$ (no absorption).

Let us now relate these findings to the system of N-charged particles introduced in Section IV.A. We know that

$$K(\omega) = K_r(\omega) + i\, K_i(\omega) \tag{255}$$

where now, setting K = 1,

$$K_r(\omega) = 1 + \frac{4\pi\, n_o\, e^2/m}{(\omega_o^2 - \omega^2)^2 + \gamma^2\,\omega^2}\,(\omega_o^2 - \omega^2) \underset{\gamma\to 0}{\to} 1 + \frac{4\pi\, n_o\, e^2/m}{\omega_o^2 - \omega^2} \tag{256}$$

$$K_i(\omega) = \frac{-4\pi\, n_o\, e^2/m}{(\omega_o^2 - \omega^2)^2 + \gamma^2\,\omega^2}\,(\omega\gamma) = -\frac{4\pi\,\sigma}{\omega} \tag{257}$$

or

$$\sigma(\omega) = \frac{\omega^2\,\gamma(n_o\, e^2/m)}{(\omega_o^2 - \omega^2)^2 + \gamma^2\,\omega^2} \underset{\gamma\to 0}{\to} 0 \tag{258}$$

Consider now the case K = 0. Neglecting losses this corresponds to

$$K_r = 1 + \frac{4\pi\ n_o\ e^2/m}{\omega_o^2 - \omega^2} = 0 \tag{259}$$

The frequency for which $K_r = 0$ is

$$\omega_L = \omega_o \sqrt{1 + \frac{4\pi\ n_o\ e^2/m}{m\ \omega_o^2}} \tag{260}$$

Also, since

$$\vec{P} = n_o\ \alpha(\omega)\ \vec{E} = n_o\ \frac{e^2/m}{\omega_o^2 - \omega^2}\ \vec{E} \tag{261}$$

we have

$$\vec{D} = K\ \vec{E} = \vec{E} + 4\pi\ \vec{P} = (1 + 4\pi\ n_o\ \alpha)\ \vec{E}$$

$$= \left[1 + \frac{4\pi\ n_o\ e^2/m}{\omega_o^2 - \omega^2}\right] \vec{E} = 0 \tag{262}$$

as expected, in agreement with Eq. (243). $\vec{E}$ represents here a longitudinal wave with the wave vector $\vec{\tau}$ parallel to the induced dipole moments. Also, $\omega_L > \omega_o$, due to the fact that the longitudinal electric field has the effect of increasing the force constant of the oscillators.

If $K \neq 0$ we have from Eqs. (248) and (256), neglecting losses,

$$\left(\frac{c\tau}{\omega}\right)^2 = K_r = 1 + \frac{4\pi\ n_o\ e^2/m}{\omega_o^2 - \omega^2} = n_r^2 \tag{263}$$

and

$$\tau = \frac{\omega}{c} \sqrt{1 + \frac{4\pi\ n_o\ e^2/m}{\omega_o^2 - \omega^2}} \tag{264}$$

The dispersion curves are represented in Fig. 7.

We can now summarize our findings as follows [5,6]:

(1) For $\omega = \omega_L$, $K = 0$ and only longitudinal waves can propagate in the medium; the dispersion curve for longitudinal modes is

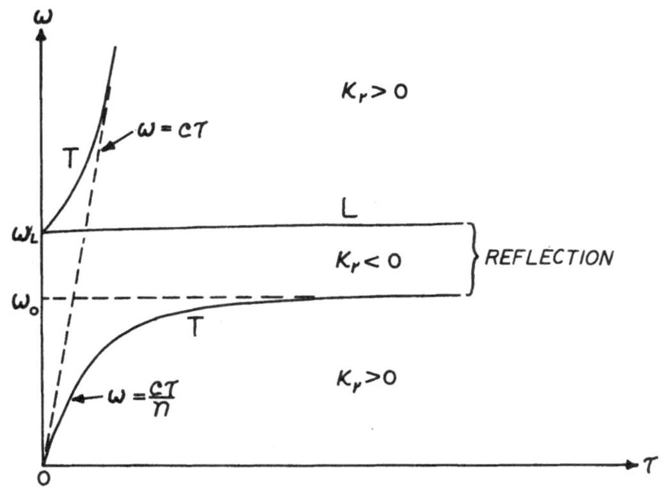

Fig. 7. Dispersion Relations for L (Longitudinal) and T (transverse) Modes.

a straight line with ω independent of τ.

(2) For values of ω such that $\omega_0 < \omega < \omega_L$, K(ω) is negative and the index of refraction is imaginary. No propagation is possible and the radiation is reflected. An incoming wave with a frequency in the above interval would be transmitted only through a thin slab of thickness $1/|\tau|$.

(3) When dealing with the transverse modes we cannot think of "photons" and "oscillators" independently; in effect, we have a completely new system of "modes" due to the coupling of the radiation with the oscillating charges. We can eliminate the effect of this coupling by letting $e \to 0$ or $m \to \infty$; in this

case $\omega = \tau c$ (photons have vacuum dispersion) and $\omega_L = \omega_o$ (oscillators oscillate at the frequency ω_o).

(4) For the lower branch of the T modes the closer is ω to ω_o, namely the larger is τ, less electromagnetic energy and more mechanical energy is present in the wave. The contrary is true for small τ and ω, but as $\tau, \omega \to 0$ some residual mechanical energy (which represents the energy of static polarization) is present [7].

(5) For the upper T branch, the larger is τ, less mechanical energy and more electromagnetic energy is present in the wave [7].

This general behavior has been looked at in the limiting case of $\gamma = \sigma = 0$ (no losses). The losses of the mechanical oscillators will make a transverse wave (which is in part mechanical and in part electromagnetic) lose energy. This may give us a mechanism or model for the losses experienced by an electromagnetic wave propagating through the system.

The incoming radiation couples with a certain number of transverse modes of the system. These modes are of mixed (mechanical and electromagnetic) nature; the mechanical part of these modes represents the sink for the energy. The excitation of these modes is restored by the incoming radiation. An electromagnetic wave of high ω (upper branch) will not lose any energy and will pass unattenuated through the system.

V. EXAMPLES OF COLLECTIVE EXCITATIONS

V.A. Phonons

1. Summary of properties.

(a) In a molecule or solid, the electrons provide the potential in which the atoms perform their vibrational motions. These motions in solids are often called lattice vibrations.

(b) These motions can be thought of as a superposition of normal modes of vibration which are 3N in number, if N = number of atoms in the solid. More precisely, since the solid has also three translational and three rotational degrees of freedom, the number of normal modes of vibration is 3N - 6.

(c) A normal mode of vibration is a pattern of motion in which all the atoms of the solid participate. There may, however, be localized modes which are related to a relatively small number of atoms.

(d) In treating the vibration of solids the harmonic approximation is used. This approximation implies that there is no exchange of energy between the normal modes.

(e) Each normal mode is equivalent to a harmonic oscillator. The vibrations of a solid are then equivalent to a collection of 3N harmonic oscillations. If an oscillator of frequency ω is in its nth excited state, this fact is also expressed by saying that n phonons, each of energy $h\omega$, are present in the solid.

(f) The different normal modes are not completely isolated, but are, rather, in speaking terms, due to their anharmonicity: this provides the mechanism by which the solid reaches thermal equilibrium.

(g) When considering the thermal vibrations of a solid, it may be interesting to have an idea of how many phonons may be present in a solid at, say, room temperature. The number of phonons in a frequency interval $(\omega, \omega + d\omega)$ is given by $\bar{n}(\omega)\ \rho(\omega)\ d\omega$ where:

$$\bar{n}(\omega) = (e^{\hbar\omega/kT} - 1)^{-1}$$

and

$$\rho(\omega) = \text{density of phonon states} \propto \omega^2/c_s^3$$

It is the value of c_s = velocity of sound in solids $\sim 5 \times 10^5$ cm/sec (versus c = velocity of light = 3×10^{10} cm/sec) that makes the number of phonons extremely large. For the sake of comparison, the total number of photons/cm^3 in black body radiation at T = 300°K is $\sim 6.4 \times 10^8$, the number of photons/cm^3 in a typical laser medium (λ = 6,300 Å, 1 W/cm^2) is 10^8, the number of phonons/cm^3 in a solid (with a Debye temperature T_D = 1,000 K) at T = 300°K is $\sim 3.5 \times 10^{23}$! The sheer number of phonons may give us an idea of their importance in affecting the spectral characteristics of ions in solids.

(h) If more than one atom is present in the unit cell, the dispersion curves of the vibrations include optical branches.

2. Infrared absorption by ionic solids. In this case the "oscillators" interacting with the electromagnetic radiation are the optical modes of lattice vibrations with small $\vec{k}$ (intra-molecular vibrations).

The splitting of optical phonon branches into longitudinal and transverse takes place in accord with the Lyddane-Sachs-Teller relation (232). This is experimentally confirmed when comparing ω_L/ω_T obtained by inelastic neutron scattering [8,9] with the experimental values of $[K(0)/K(\infty)]^{1/2}$ [10, p. 185]:

	NaI	KBr	GaAs
$\frac{\omega_L}{\omega_T}$	1.44 ± 0.05	1.39 ± 0.02	1.07 ± 0.02
$\left[\frac{K(0)}{K(\infty)}\right]^{1/2}$	1.45 ± 0.03	1.38 ± 0.03	1.08

The process of infrared absorption can be related to the diagram in Fig. 8. If we consider "the system" as encompassing the vibrating solid and the electromagnetic radiation, a state of the system with a photon of wave vector, say, $\vec{k}_A$ may be represented as follows:

$$\underbrace{|1_{\vec{k}_A}\rangle|0\rangle|0\rangle \;\ldots}_{\text{photon part}}\quad \underbrace{|0\rangle|1\rangle|n\rangle|n'\rangle\;\ldots}_{\text{phonon part}} \tag{265}$$

The two states

$|1_{\vec{k}_o}\rangle|0_{\vec{f}_o}\rangle$ = 1 photon of wave vector $\vec{k}_o$, no phonon of wave vector $\vec{f}_o = \vec{k}_o$

$|0_{\vec{k}_o}\rangle|1_{\vec{f}_o}\rangle$ = no photon of wave vector $\vec{k}_o$, 1 phonon of wave vector $\vec{f}_o = \vec{k}_o$

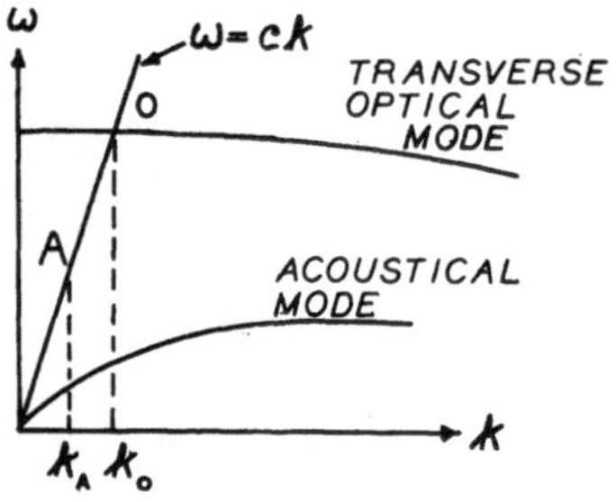

Fig. 8. The Coupling of Electromagnetic Radiation with Transverse Optical Phonon Modes.

are in resonance and may be split by the ion-photon interaction. The matrix element of this interaction taken among the two states above may be zero to first-order, but may be different from zero to second-order. In any case, a splitting of the two degenerate states will follow, and the situation around the crossing point 0 of Fig. 8 will be distorted as in Fig. 9. The real stationary states are a mixture of phonons and photons; they are what we call polariton states.

Polariton states occupy a small part of the Brillouin zone:

$$|\vec{k}_o| \simeq 10^4 \text{ cm}^{-1}$$

whereas the dimension of the Brillouin zone is $\sim 10^8 \text{ cm}^{-1}$.

The various points in Fig. 9 represent the following:

A = photon in the solid

P_o = pure photon

P_2 = phonon

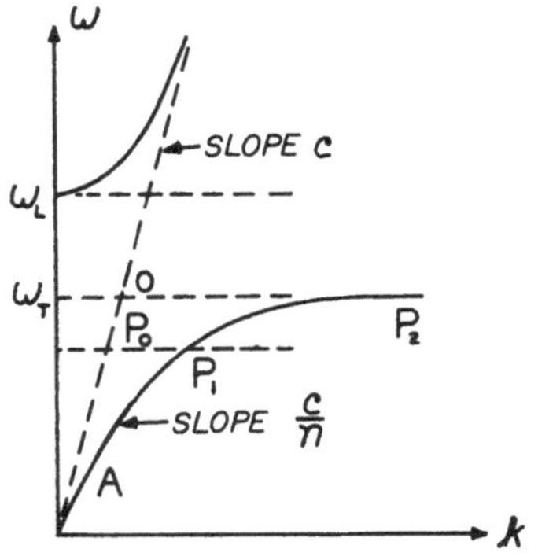

Fig. 9. Polariton States Due to the Interaction of Transverse Optical Phonons with Electromagnetic Radiation.

P_1 = polariton

For large $\vec{k}$ (small λ) we have pure photons and pure phonons.

The mechanism for absorption consists of the following:

(1) The incoming beam maintains a certain excitation of the T (transverse) normal modes.

(2) These T-modes decay because their mechanical part is damped.

(3) The excitation is continuously restored by the incoming beam.

V.B. Excitons

1. General theory. Consider a system of N atoms. The "electronic" Hamiltonian is given by

$$H = H_o + H' \tag{266}$$

where

$$H_o = -\sum_i \frac{\hbar^2 \nabla_i^2}{2m} + \sum_i V(\vec{r}_i) \tag{267}$$

$$H' = \sum_{i<j} \frac{e^2}{r_{ij}} \tag{268}$$

We shall assume the following:

(1) The nuclei are at rest in their equilibrium position, and

(2) The electrons are extra-core, one per atom.

The approximate ground state wavefunction of the system can be written as follows:

$$A|a_{01}\ a_{02} \cdots a_{ON}\rangle = \frac{1}{\sqrt{N!}} \begin{vmatrix} a_{01}(1) & a_{01}(2) & \cdots & a_{01}(N) \\ a_{02}(1) & a_{02}(2) & \cdots & a_{02}(N) \\ \cdot & \cdot & \cdot & \cdot \\ \cdot & \cdot & \cdot & \cdot \\ a_{ON}(1) & a_{ON}(2) & \cdots & a_{ON}(N) \end{vmatrix} . \tag{269}$$

where

$a_{0i}(j) = a_0(\vec{r}_j - \vec{R}_i)$ = ground state "atomic" wavefunction

A = antisymmetrizing operator

For each atom

$$h_i(\vec{r}) = -\frac{\hbar^2}{2m}\nabla^2 + V(\vec{r} - \vec{R}_i) \tag{270}$$

and

$$h_i(\vec{r})\, a_{ni}(\vec{r}) = \varepsilon_n\, a_{ni}(\vec{r}) \tag{271}$$

where

$V(\vec{r}-\vec{R}_i)$=potential due to the nucleus and to the core electrons

If the overlap of atomic functions is large more appropriate one-electron states are the eigensolutions of

$$h_o\, \psi_{n\vec{k}}(\vec{r}) = \varepsilon_{n\vec{k}}\, \psi_{n\vec{k}}(\vec{r}) \tag{272}$$

where

$$h_o = -\frac{\hbar^2}{2m}\nabla^2 + V(\vec{r}) \tag{273}$$

and

$V(\vec{r})$ = periodic crystal potential

A solution of Eq. (272) is a Block-type wavefunction

$$\psi_{n\vec{k}}(\vec{r}) = e^{i\vec{k}\cdot\vec{r}}\, u_{n\vec{k}}(\vec{r}) \tag{274}$$

These functions have the usual properties of orthonormality and closure

$$\left\{ \begin{array}{l} \int \psi^*_{n\vec{k}}(\vec{r})\, \psi_{n\vec{k}'}(\vec{r})\, d^3\vec{r} = \delta_{\vec{k}\vec{k}'} \\ \\ \sum_{\vec{k}} \psi^*_{n\vec{k}}(\vec{r})\, \psi_{n\vec{k}}(\vec{r}') = \delta(\vec{r} - \vec{r}') \end{array} \right. \tag{275}$$

The Block function $\psi_{n\vec{k}}(\vec{r})$ are completely delocalized. We can construct localized states known as Wannier functions, which we define as follows:

$$a_{ni}(\vec{r}) = a_n(\vec{r} - \vec{R}_i) = \frac{1}{\sqrt{N}} \sum_{\vec{k}} e^{-i\vec{k}\cdot\vec{R}_i} \psi_{n\vec{k}}(\vec{r})$$

$$= \frac{1}{\sqrt{N}} \sum_{\vec{k}} e^{i\vec{k}\cdot(\vec{r}-\vec{R}_i)} u_{n\vec{k}}(\vec{r}) \tag{276}$$

Unlike the Block functions which depend on both $\vec{k}$ and $\vec{r}$, the Wannier functions depend on $(\vec{r} - \vec{R}_i)$ only and are localized; in fact, the phases for the different terms are ~ zero when $\vec{r} \simeq \vec{R}_i$, and the functions $u_{n\vec{k}}$ add constructively.

The Wannier functions are orthonormal:

$$\int a^*_{ni}(\vec{r})\, a_{nj}(\vec{r})\, d^3\vec{r} = \frac{1}{N} \int \sum_{\vec{k}} e^{i\vec{k}\cdot\vec{R}_i} \psi^*_{n\vec{k}}(\vec{r}) \sum_{\vec{k}'} e^{-i\vec{k}'\cdot\vec{R}_j} \psi_{n\vec{k}'}(\vec{r})\, d^3\vec{r}$$

$$= \frac{1}{N} \sum_{\vec{k}} \sum_{\vec{k}'} e^{i\vec{k}\cdot\vec{R}_i} e^{-i\vec{k}'\cdot\vec{R}_j} \int \psi^*_{n\vec{k}}(\vec{r})\, \psi_{n\vec{k}'}(\vec{r})\, d^3\vec{r}$$

$$= \frac{1}{N} \sum_{\vec{k}} e^{i\vec{k}\cdot(\vec{R}_i-\vec{R}_j)} = \delta_{ij} \tag{277}$$

and have the closure property

$$\sum_i a^*_{ni}(\vec{r})\, a_{ni}(\vec{r}') = \frac{1}{N} \sum_{\vec{k}} \sum_{\vec{k}'} \psi^*_{n\vec{k}}(\vec{r})\, \psi_{n\vec{k}'}(\vec{r}) \sum_i e^{i(\vec{k}-\vec{k}')\cdot\vec{R}_i}$$

$$= \frac{1}{N} \sum_{\vec{k}} \sum_{\vec{k}'} \psi^*_{n\vec{k}}(\vec{r})\, \psi_{n\vec{k}'}(\vec{r}')\, N\, \delta_{\vec{k},\vec{k}'} = \sum_{\vec{k}} \psi_{n\vec{k}}(\vec{r})\, \psi_{n\vec{k}'}(\vec{r}') = \delta(\vec{r} - \vec{r}') \tag{278}$$

We note that for each atomic state n, there exists a set of N Block wavefunctions characterized by the N possible values of $\vec{k}$. Such being the case, if n = 0 (ground state) the zero-order ground state of the system is given by

$$\psi_G = A|\psi_{0\vec{k}_1}(\vec{r}_1)\ \psi_{0\vec{k}_2}(\vec{r}_2)\ \dots\ \psi_{0\vec{k}_N}(\vec{r}_N)> \tag{279}$$

It can be shown [6] that this function, apart sign, is the same as

$$\psi_G = A|a_{01}(\vec{r}_1)\ a_{02}(\vec{r}_2)\ \dots\ a_{0N}(\vec{r}_N)> \tag{280}$$

In general the determinantal functions

$$A|\psi_{n\vec{k}_1}(\vec{r}_1)\ \psi_{n\vec{k}_2}(\vec{r}_2)\ \dots\ \psi_{n\vec{k}_N}(\vec{r}_N)> \tag{281}$$

and

$$A|a_{n\vec{R}_1}(\vec{r}_1)\ a_{n\vec{R}_2}(\vec{r}_2)\ \dots\ a_{n\vec{R}_N}(\vec{r}_N)> \tag{282}$$

are identical if the state they refer to can be represented by a single determinant (closed-shell situation). The a-based function is equal to the ψ-based function times a unitary matrix whose elements are

$$\frac{1}{\sqrt{N}}\, e^{-i\vec{k}_s\cdot\vec{R}_t} \qquad (s,t = 1,2,\ \dots,\ N)$$

We want now to describe an excited state of the system. The simplest excitation is obtained by raising an electron from a ground state orbital a_{0h} to an orbital a_{1e} corresponding to a higher atomic state. This situation is represented by the wavefunction

$$\psi_1 = A|a_{01}\ a_{02}\ \dots\ a_{0,h-1}\ a_{1e}\ a_{0,h+1}\ \dots\ a_{0N}> \tag{283}$$

and corresponds to a "hole" in the "atom" at position $\vec{R}_h$ and an extra electron in the "atom" at position $\vec{R}_e$. States of the type ψ_1 above are called <u>electron transfer states</u>.

Let us call

$$\beta = \vec{R}_e - \vec{R}_h \tag{284}$$

For each $\vec{\beta}$, there will be several states of the type ψ_1. In the case of a linear crystal with four atoms we get the following excited states:

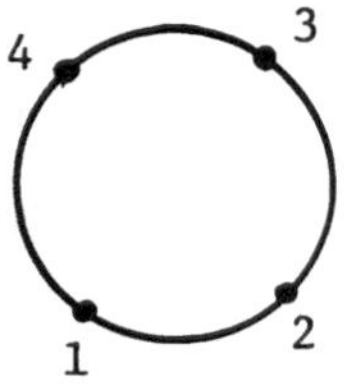

$\beta = 0$:	$\|1\ 0\rangle$	$\|2\ 0\rangle$	$\|3\ 0\rangle$	$\|4\ 0\rangle$
	$R_1\ R_1$	$R_2\ R_2$	$R_3\ R_3$	$R_4\ R_4$
$\beta = 1$:	$\|1\ 1\rangle$	$\|2\ 1\rangle$	$\|3\ 1\rangle$	$\|4\ 1\rangle$
	$R_1\ R_2$	$R_2\ R_3$	$R_3\ R_4$	$R_4\ R_1$
$\beta = -1$:	$\|1\,{-1}\rangle$	$\|2\,{-1}\rangle$	$\|3\,{-1}\rangle$	$\|4\,{-1}\rangle$
	$R_1\ R_4$	$R_2\ R_1$	$R_3\ R_2$	$R_4\ R_3$
$\beta = 2$:	$\|1\ 2\rangle$	$\|2\ 2\rangle$	$\|3\ 2\rangle$	$\|4\ 2\rangle$
	$R_1\ R_3$	$R_2\ R_4$	$R_3\ R_1$	$R_4\ R_2$

We have a total of 16 $\psi_1(\vec{R}_h,\vec{\beta})$ states and the matrix of the Hamiltonian will be 16 x 16. In the designation above $R_i\ R_j$ indicates that the hole is at position R_i and the electron at position R_j.

We note that in general the diagonal terms of the Hamiltonian $\langle\psi_1(\vec{R}_h,\vec{\beta})|H|\psi_1(\vec{R}_h,\vec{\beta})\rangle$ are the same for a certain value of $\vec{\beta}$ and for all $\vec{R}_h$, because of the translational symmetry of the system.

At this point we construct states in the <u>exciton representation</u>:

$$\psi_1(\vec{k},\vec{\beta}) = \frac{1}{\sqrt{N}} \sum_{\vec{R}_h} e^{i\vec{k}\cdot\vec{R}_h}\, \psi_1(\vec{R}_h,\vec{\beta}) \tag{283}$$

These states allow a diagonalization with respect to $\vec{R}_h$, but not $\vec{\beta}$:

$$\langle\psi_1(\vec{k},\vec{\beta})|H|\psi_1(\vec{k}',\vec{\beta})\rangle$$

$$= \frac{1}{N} \sum_{\vec{R}_h} \sum_{\vec{R}_{h'}} e^{-i\vec{k}\cdot\vec{R}_h}\, e^{i\vec{k}'\cdot\vec{R}_h} \langle\psi_1(\vec{R}_h,\vec{\beta})|H|\psi_1(\vec{R}_{h'},\vec{\beta}')\rangle\, \delta_{\vec{k}\vec{k}'}$$

$$= \frac{1}{N} \sum_{\vec{R}_h} \sum_{\vec{R}_{h'}} e^{-i\vec{k}\cdot(\vec{R}_h-\vec{R}_{h'})} \langle\psi_1(\vec{R}_h,\vec{\beta})|H|\psi_1(\vec{R}_{h'},\vec{\beta}')\rangle\, \delta_{\vec{k}\vec{k}'}$$

$$= \sum_{\vec{R}_{h'}} e^{-i\vec{k}\cdot(\vec{R}_h-\vec{R}_{h'})} \langle\psi_1(\vec{R}_h,\vec{\beta})|H|\psi_1(\vec{R}_{h'},\vec{\beta}')\rangle\, \delta_{\vec{k}\vec{k}'}$$

$$= \sum_{\vec{R}_{h'}} e^{i\vec{k}\cdot\vec{R}_{h'}} \langle\psi_1(0,\vec{\beta})|H|\psi_1(\vec{R}_{h'},\vec{\beta}')\rangle \,\delta_{\vec{k}\vec{k}'}$$

$$= H_{\vec{\beta}\vec{\beta}'} \,\delta_{\vec{k}\vec{k}'} \tag{284}$$

where

$$H_{\vec{\beta}\vec{\beta}'} = \sum_{\vec{R}_h=1}^{N} e^{i\vec{k}\cdot\vec{R}_h} \langle\psi_1(0,\vec{\beta})|H|\psi_1(\vec{R}_h,\vec{\beta}')\rangle \tag{285}$$

2. <u>The Frenkel Exciton</u>. In this case $\vec{\beta} = 0$ and the electron making the transition does not leave the "cell" where it resides. The <u>exciton</u> is in this case, called a <u>Frenkel exciton</u>, a packet of waves which are linear combinations of atomic excited wavefunctions [11]. No $\vec{\beta}$ diagonalization is necessary.

The ground state is given by

$$\psi_G = A|a_0(\vec{r} - \vec{R}_1)\; a_0(\vec{r} - \vec{R}_2)\; \ldots\; a_0(\vec{r} - \vec{R}_N)\rangle \tag{286}$$

The electron transfer state is given by

$$\psi_1(\vec{R}_h) = A|a_0(\vec{r} - \vec{R}_1)\; a_0(\vec{r} - \vec{R}_2)\; \ldots\; a_1(\vec{r} - \vec{R}_h)\; \ldots\; a_0(\vec{r} - \vec{R}_N)\rangle \tag{287}$$

and the generic matrix element of the Hamiltonian is

$$\langle\psi_1(\vec{R}_h)|H|\psi_1(\vec{R}_{h'})\rangle \tag{288}$$

We construct states in the <u>exciton representation</u> as follows:

$$\psi_1(\vec{k}) = \frac{1}{\sqrt{N}} \sum_{\vec{R}_h} e^{i\vec{k}\cdot\vec{R}_h}\, \psi_1(\vec{R}_h) \tag{289}$$

The ground state energy of the system is

$$\langle\psi_G|H|\psi_G\rangle = \langle\psi_G|H_0|\psi_G\rangle + \langle\psi_G|H'|\psi_G\rangle = N\,\varepsilon_0 + \Delta E_g \tag{290}$$

where

ε_0 = energy of the ground atomic state

$\Delta E_g = \langle\psi_G|H'|\psi_G\rangle$

The expectation value of the energy in the first excited state is

$$\langle\psi_1(\vec{k})|H|\psi_1(\vec{k})\rangle$$

$$= \frac{1}{N}\sum_{\vec{R}_h}\sum_{\vec{R}_{h'}} e^{-i\vec{k}\cdot(\vec{R}_h-\vec{R}_{h'})}\langle\psi_1(\vec{R}_h)|H|\psi_1(\vec{R}_{h'})\rangle$$

$$= \sum_{\vec{R}_{h'}} e^{-i\vec{k}\cdot(\vec{R}_h-\vec{R}_{h'})}\langle\psi_1(\vec{R}_h)|H|\psi_1(\vec{R}_{h'})\rangle$$

$$= \sum_{\vec{R}_h} e^{i\vec{k}\cdot\vec{R}_h}\langle\psi_1(0)|H|\psi_1(\vec{R}_h)\rangle$$

$$= \langle\psi_1(0)|H|\psi_1(0)\rangle + \sum_{\vec{R}_h\neq 0}\langle\psi_1(0)|H|\psi_1(\vec{R}_h)\rangle$$

$$= \varepsilon_1 + (N-1)\,\varepsilon_o + \langle\psi_1(0)|H'|\psi_1(0)\rangle + \sum_{\vec{R}_h\neq 0} e^{i\vec{k}\cdot\vec{R}_h}\langle\psi_1(0)|H'|\psi_1(\vec{R}_h)\rangle$$

$$= \varepsilon_1 + (N-1)\,\varepsilon_o + H'_{\vec{R}_h\vec{R}_h} + \sum_{\vec{R}_h\neq 0} e^{i\vec{k}\cdot\vec{R}_h} H'_{0\vec{R}_h} \qquad (291)$$

where ε_1 = energy of the excited atomic state.

The matrix elements of H taken between two ψ_1 states with different $\vec{k}$ are zero.

The total change in energy going from the ground state to an excited state is

$$\Delta E = [(N-1)\,\varepsilon_o + \varepsilon_1 + H'_{\vec{R}_h\vec{R}_h} + \sum_{\vec{R}_h\neq 0} e^{i\vec{k}\cdot\vec{R}_h} H'_{0\vec{R}_h}]$$

$$- [N\,\varepsilon_o + \Delta E_g]$$

$$= (\varepsilon_1 - \varepsilon_o) + (H'_{\vec{R}_h\vec{R}_h} - \Delta E_g) + \sum_{\vec{R}_h\neq 0} e^{i\vec{k}\cdot\vec{R}_h} H'_{0\vec{R}_h}$$

$$= \Delta E \text{ (atomic)} + \Delta V \text{ (= change in interaction energy)} + \varepsilon(\vec{k}) \tag{292}$$

where

$$\Delta E(\text{atomic}) = \varepsilon_1 - \varepsilon_0 \simeq 3 \text{ eV} \tag{293}$$

$$\Delta V = H'_{\vec{R}_h\vec{R}_h} - \Delta E_g \tag{294}$$

and finally

$$\varepsilon(\vec{k}) = \sum_{\vec{R}_h \neq 0} e^{i\vec{k}\cdot\vec{R}_h} H'_{0\vec{R}_h} \tag{295}$$

gives the dispersion relation.

We can make at this point the following observations:

(1) If $H' = 0$ the excited state is N-fold degenerate. As a result of the perturbation H' we have a band of states. (See Fig. 10).

(2) The matrix elements

$$H'_{\vec{R}_h\vec{R}_{h'}}$$

are responsible for the fact that the excitation energy does not stay in one atom, but moves along.

(3) Nuclear vibrations and other periodicity-destroying things such as defects, destroy the coherence of the exciton states and result in the scattering of excitons.

(4) If we consider only nearest neighbor interactions and set $H'_{0\vec{R}_h} = M$ we obtain for a linear lattice

$$\varepsilon(\vec{k}) = 2M \cos ka \tag{296}$$

and for a cubic lattice

$$\varepsilon(\vec{k}) = 2M \sum_i \cos k_i a \tag{297}$$

(5) Finally, if $\varepsilon(\vec{k})$ is known we can define

$$\vec{v}_g = \frac{1}{\hbar} \vec{\nabla}_{\vec{k}} \, \varepsilon(\vec{k}) \tag{298}$$

and

$$\frac{1}{m^*} = \frac{1}{\hbar^2} \vec{\nabla}_{\vec{k}} \vec{\nabla}_{\vec{k}} \, \varepsilon(\vec{k}) \tag{299}$$

We have now to direct some attention to the matrix elements responsible for the transfer of energy

$$H'_{\vec{R}_h \vec{R}_{h'}} = \langle \psi_1(\vec{R}_h) | H' | \psi_1(\vec{R}_{h'}) \rangle \tag{300}$$

We define the "product states" as follows:

$$\pi_h = | a_{01} \, a_{02} \cdots a_{1h} \cdots a_{0N} \rangle \tag{301}$$

$$\pi_{h'} = | a_{01} \, a_{02} \cdots a_{1h'} \cdots a_{0N} \rangle \tag{302}$$

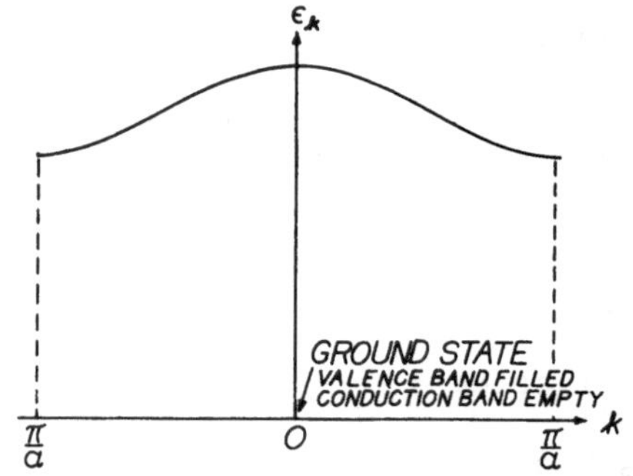

Fig. 10. Frenkel Exciton States.

Then

$$\psi_1(\vec{R}_h) = A\ \pi_h \tag{303}$$

$$\psi_1(\vec{R}_{h'}) = A\ \pi_{h'} \tag{304}$$

It is easy to show (see [12], p. 165) that

$$A(A\pi) = A^2\pi = \sqrt{N!}\ A\pi \tag{305}$$

and

$$<\psi_1(\vec{R}_h)|H'|\psi_1(\vec{R}_{h'})> = <A\ \pi_h|H'|A\ \pi_{h'}>$$

$$= <\pi_h|H'|A^2\ \pi_h> = \sqrt{N!}\ \pi_h|H'|A\ \pi_{h'}> \tag{306}$$

In order to relate the present notation with that used in Section II we shall rename the a functions as follows:

$$a_0 = \text{ground state function} = u$$

$$a_0(\vec{r}_i - \vec{R}_h) = u_h(i) \tag{307}$$

$$a_1 = \text{excited state function} = v$$

$$a_1(\vec{r}_i - \vec{R}_h) = v_h(i) \tag{308}$$

Then

$$\pi_o = v_o(1)\ u_h(2)\ \ldots \tag{309}$$

$$\pi_h = u_o(1)\ v_h(2)\ \ldots \tag{310}$$

and

$$<\psi_1(0)|H'|\psi_1(\vec{R}_h)> = \sqrt{N!}\ <\pi_o|H'|A\ \pi_h>$$

$$= <v_o(1)\ u_h(2)\left|\frac{e^2}{r_{12}}\right|[u_o(1)\ v_h(2) - u_o(2)\ v_1(1)]>$$

$$= <v_o(1)\ u_h(2)\left|\frac{e^2}{r_{12}}\right|u_o(1)\ v_h(2)>$$

$$- \langle v_o(1)\ u_h(2) \left| \frac{e^2}{r_{12}} \right| v_h(1)\ u_o(2) \rangle \tag{310}$$

The first term represents the interaction between two charge distributions $v_o^*(1)\ u_o(1)$ located at position 0 and $u_h^*(2)\ v_h(2)$ located at position $\vec{R}_h$; this direct interaction is coulombic and long-range in nature; this term is the same that appears in Eq. (112). The other term represents an exchange interaction; its range is short.

If each atom contains two electrons, rather than one, the formula (310) can be written as follows:

$$\langle v_o(1,2)\ u_h(3,4) | H' | u_o(1,2)\ v_h(3,4) \rangle$$

$$- \langle v_o(1,2)\ u_h(3,4) | H' | v_h(1,2)\ u_o(3,4) \rangle \tag{311}$$

where the $u(i,j)$ and $v(k,\ell)$ functions represent the ground and the excited atomic states, respectively. The ground state is generally a singlet, the excited state can be a singlet or a triplet. The direct interaction term vanishes if the excited state is a triplet; if the excited state is a singlet this term can be written as follows:

$$2\langle v_o(1)\ u_h(3) \left| \frac{e^2}{r_{13}} \right| u_o(1)\ v_h(3) \rangle \tag{312}$$

where $u(i)$ and $v(j)$ are ground and excited state electron orbitals, respectively.

Up to now we have considered only one atom or one molecule per unit cell. If there are several, say n, identical, but differently oriented molecules per unit cell, the interaction among these molecules produces a splitting into n levels and, in essence, n exciton states for each $\vec{k}$. The greater is the interaction among different molecules in the unit cell, the greater is the splitting of exciton states for a certain $\vec{k}$. The splitting for the case n = 2 is reported in Fig. 11 [13]. The additional splitting that is present even if the excited state of the molecule is non-degenerate is called the Davydov splitting. It is obvious that this splitting is restricted to excited states and it cannot be referred to one molecule in the crystal. It is also clear that such splitting does not exist for molecules in solution.

The Davydov splitting has been observed in the spectra of organic molecules. It may range from a few hundred to several

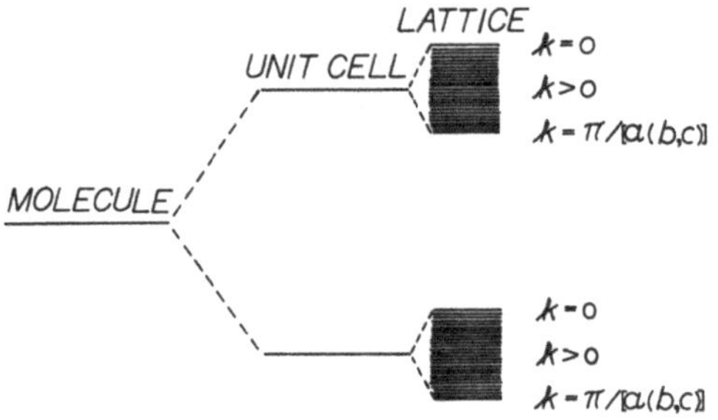

Fig. 11. Davydov Splitting Due to the Presence of Two Interacting Molecules in the Unit Cell.

thousand cm^{-1} [13]. It may also occur in inorganic cyrstals such as MnF_2; for this system there are two Mn^{2+} ions per unit cell; they are distinguished from each other by the orientation of the F^- ions. The interaction among the Mn^{2+} ions is of antiferromagnetic nature. However, since this interaction is small, the Davydov splitting is not experimentally measurable [14].

3. <u>The Wannier Exciton</u>. The general matrix element $<\psi_1(\vec{k},\vec{\beta})|H|\psi_1(\vec{k},\vec{\beta}')>$ was used by Wannier to develop his effective mass theory [15]. We can arrive at the same conclusions by making the following observations. Certain semiconductors such as germanium and silicon have large dielectric constant, large bands and small gaps. In these materials the interaction between electron and hole is weak and in fact electron and hole maintain much of their "free" character.

The Hamiltonian of the Wannier exciton can be written as follows:

$$H = -\frac{\hbar^2}{2m_e}\nabla_e^2 - \frac{\hbar^2}{2m_h}\nabla_h^2 - \frac{e^2}{K|\vec{r}_e - \vec{r}_h|} \tag{313}$$

where m_e and m_h are the effective masses of the electron and of the hole, respectively, $\vec{r}_e$ and $\vec{r}_h$ are the coordinates of the electron and of the hole, respectively, and K is the dielectric constant of the medium. The following equivalence can be established.

Hydrogen atom	Wannier exciton
e^2	e^2/K
m = mass of the electron	m_e
M = mass of the nucleus	m_h
$\mu = \frac{mM}{m+M} \simeq m$	$\mu' = \frac{m_e m_h}{m_e + m_h}$
$\text{Energy} = \frac{\hbar^2 K^2}{2(m+M)} + E_n$ (314)	$\text{Energy} = \varepsilon_c + \frac{\hbar^2 k^2}{2(m_e+m_h)} + E_n$ (315)
	(ε_c = energy of bottom level of conduction band)
$E_n = -\frac{me^4}{2\hbar^2}\frac{1}{n^2} = -\frac{e^2}{2a_0}\frac{1}{n^2} = -\frac{13.6}{n^2}\ \text{eV}$ (316)	$E_n = -\frac{\mu' e^4}{2\hbar^2 K^2}\frac{1}{n^2} = -\frac{e^2}{2a_0'}\frac{1}{Kn^2} = -\frac{13.6}{n^2}\left(\frac{\mu'}{m}\frac{1}{K^2}\right)\ \text{eV}$ (317)
$a_0 = \frac{\hbar^2}{me^2} \simeq 0.5\ \text{Å}$ (318)	$a_0' = \frac{\hbar^2 K}{\mu' e^2} = a_0\left(\frac{m}{\mu'}K\right)$ (319)
$a_n = n^2 a_0$ (320)	$a_n' = n^2 a_0' = n^2 a_0\left(\frac{m}{\mu'}K\right)$ (321)

Example

Energy gap between the valence and the conduction bands = 0.3 eV

$m_e = 0.02\ m$

$m_h = 0.4\ m$

$K = 12$

We find

$$E_n = -\frac{13.6}{n^2}\left(\frac{\mu'}{m}\frac{1}{K^2}\right) = -\frac{0.0018}{n^2}\ \text{eV}$$

$$a'_0 = a_0\left(\frac{m}{\mu'}K\right) = 0.5\,\frac{12}{0.019} \simeq 316\ \text{Å}$$

$$a'_n = 316\ n^2\ \text{Å}$$

The binding energy of the exciton is 0.0018 eV, versus $kT \simeq (1/40)$ eV at room temperature; an exciton created in this system at room temperature would be thermally unstable and would quickly decay into an electron and a hole.

We note the following:

(1) In general an effective mass may not be defined. Rather, because of the directionality of the dependence of the one-electron energy in $\vec{k}$ space, only an "effective mass tensor" has meaning. In this case it is not possible to make the center of mass transformation which is necessary to define the simple exciton model. The valence energy band (for example, the 2p band) may also be degenerate at the top along a certain $\vec{k}$ direction and not along another $\vec{k}$ direction; likewise for the conduction band at its bottom. This could contribute to the anisotropy of the problem and to additional degeneracy of the exciton levels.

(2) The 1s eigenfunction of the H atom is given by

$$\psi_{100} = \frac{1}{\sqrt{\pi}}\,a_0^{-3/2}\,e^{-(r/a_0)}$$

and is different than zero for $r = 0$, namely at the position of the nucleus. This fact leads to the so-called quantum-defects in the spectra of hydrogen-like atoms.

A similar situation arises when an exciton is in the 1s state in that the probability of finding both electron and hole in the same cell is not negligible. This may produce deviation from the Wannier-type situation and, consequently, a correction to the "ground state" energy of the exciton may be necessary.

(3) Consider an exciton of reduced mass μ and radius a. Its angular momentum being, say, h is given by

$$\hbar = \mu\, va = \mu\, a^2\, \omega$$

where $v = a\omega$ and ω is the angular frequency of the rotational motion of the exciton. It is

$$\omega = \frac{\hbar}{\mu a^2}$$

$$a = \sqrt{\frac{\hbar}{\mu\omega}}$$

We call <u>critical radius</u> the one that we obtain by putting $\omega = \omega_0$ = lowest resonance frequency of the system = frequency of "optical" lattice vibration branch:

$$a_{crit} = \sqrt{\frac{\hbar}{\mu\, \omega_o}}$$

By setting

$$\omega_o \simeq 3 \times 10^{-13} \sec^{-1}$$

$$\mu = 0.5\ m$$

we obtain

$$a_{crit} = \sqrt{\frac{\hbar}{m\, \omega_o}} = \sqrt{\frac{10^{-27}}{0.5 \times 9.1 \times 10^{-28} \times 3 \times 10^{13}}} \simeq 27\ \text{Å}$$

If $a < a_{crit}$ the electron-hole pair revolves at frequencies greater than ω_o and the high-frequency dielectric constant has to be used. If $a >> a_{crit}$ the electron-hole pair revolves at frequencies smaller than ω_o and the low-frequency dielectric constant is the one that has to be used.

For many semiconductors the low-frequency dielectric constant may be used for explaining the experimental spectra data.

4. The Intermediate Case. We have examined the two limiting cases: (i) exciton with zero radius (Frenkel exciton) and (ii) exciton with large radius (Wannier exciton). By "large" radius we mean a radius large in comparison to the lattice constant.

This can be seen by considering the following argument. When electron and hole are close to each other the limitation of the spatial dimension of the pair makes it necessary, as dictated by the uncertainty principle, the use of a large number of waves (namely a large portion of the $\vec{k}$ space) in order to describe the exciton. The range of $\vec{k}$ values necessary may actually involve regions of the $\vec{k}$ space where the $\varepsilon(\vec{k})$ dependence for both the valence and the conduction band is not parabolic. For small radii the Wannier model breaks down.

On the other hand, the Frenkel model breaks down for large radii, namely for a solid in which the electrons are not tightly bound to the parent atoms or molecules. As a consequence of the overlap of atomic or molecular orbital, a Frenkel-type of diagonalization of the matrix of the Hamiltonian is not possible and consequently waves of the Frenkel-type do not represent adequately the situation.

Even if the two extreme simplifications are not applicable in the intermediate case, we note that the "general theory" presented in Section V.B.1 is adequately set to handle in principle any case. In particular, Frenkel (atomic-like) and Wannier (non-atomic-like) excitons could occur in the same solid. We may cite the case of solid xenon where the $n = 1$ is a Frenkel exciton and the $n = 2,3$ are Wannier excitons [5].

An interesting "intermediate" case is the one presented by simple ionic crystals like the alkali halides. The basic excitation due to the absorption of a photon is the transfer of an electron from a halogen atom to a neighboring alkali atom. In this case it is clear that the Frenkel model cannot be used because of the electron transfer and the Wannier model cannot be used because of the small exciton radius. For more details on this subject we refer the reader to [6].

5. The Photon-Exciton System. In analogy with the phonon-photon case we may say the following:

(1) Photons couple with long-wavelength exciton states.

(2) Exciton states are "modes," namely express "resonances" of the system and we associate "oscillators" to them and apply to them the general considerations of Section IV.

(3) Are there longitudinal and transverse exciton states? Yes! This is the case when the ground atomic state is of the 1s type and the excited atomic state is of the 2p type. The relevant quantity, according to Eq. (206) is

$$\vec{p}(\vec{k}_\alpha) = \langle v|\vec{p}\, e^{i\vec{k}_\alpha\cdot\vec{r}'}|u\rangle \simeq \langle v|\vec{p}|u\rangle \tag{322}$$

which has three components proportional to x, y and z. If the polarization vector of the radiation field is in the z-direction only the z-component of $\vec{p}(\vec{k}_\alpha)$ couple with the field.

(4) Finally the phonon-transverse exciton interaction creates <u>polaritons</u>, as depicted in Fig. 12, where

A = photon in the solid

P_o = pure photon

P_2 = exciton

P_1 = polariton

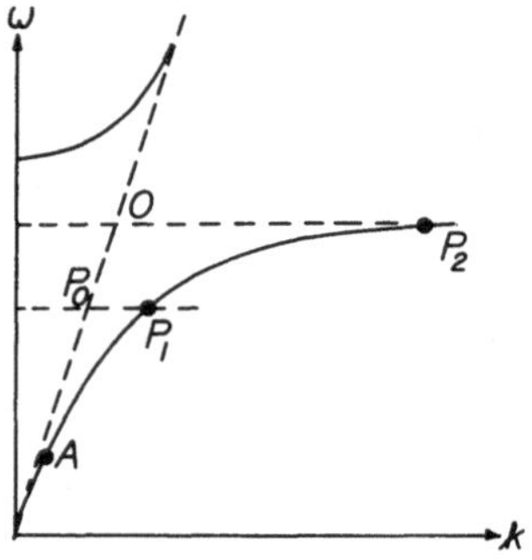

Fig. 12. Polariton States Due to the Interaction of Transverse Exciton States with Electromagnetic Radiation.

6. The Photon-Exciton-Phonon System. In order to have a more complete model of the interactions of the radiation field with a solid we have to include in our treatment both the exciton system and the phonon system. Figure 13 illustrates the situation. We shall assume that the photon energies are such that the switch "infrared absorption" is open and the switch "absorption" is closed. The process of absorption takes place as follows:

(1) An incoming photon excites the solid to a polariton state (or creates a polariton wave packet).

(2) The polariton travels undisturbed and recreates the photon at the surface of the solid, or the polariton is scattered by a phonon or a point defect into a nonradiative state. (See Fig. 14.)

(3) In the words of Pekar [16]:

"Absorption becomes possible only as a result of exciton decay accompanied by a thermal transition of the system to excited states of another type."

7. Indirect Transitions. The presence of lattice vibrations makes it possible to observe phonon-assisted transitions. The selection rules are simply determined by the conservation of linear momentum and of energy. We shall call

$\vec{k}_e$ = wave vector of the exciton

$\vec{q}$ = wave vector of the phonon

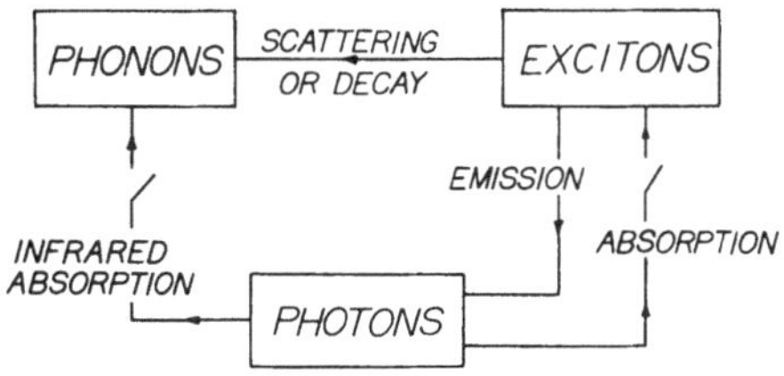

Fig. 13. The Photon-Exciton-Phonon System.

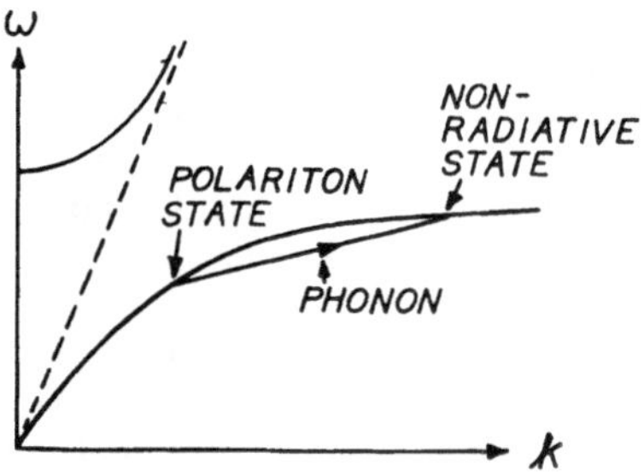

Fig. 14. Scattering of Polariton into a Non-radiative State.

$\vec{k}_\alpha$ = wave vector of the photon

E_e = energy of the exciton

$\hbar\omega_q$ = energy of the phonon

and here are the possible cases:

(a) Creation of an exciton, accompanied by the creation of a phonon

$$\begin{cases} \vec{k}_e = -\vec{q} \\ \hbar c k_\alpha = E_e + \hbar\omega_q \end{cases} \tag{323}$$

(b) Creation of an exciton, accompanied by the annihilation of a phonon

$$\begin{cases} \vec{k}_e = \vec{q} \\ \hbar c k_\alpha = E_e = \hbar\omega_q \end{cases} \tag{324}$$

This process does not take place at very low temperature (when there are no phonons around)

(c) Annihilation of an exciton, accompanied by the creation of a phonon

$$\begin{cases} \vec{k}_e = \vec{q} \\ \hbar c k_\alpha = E_e - \hbar\omega_q \end{cases} \tag{325}$$

(d) Annihilation of an exciton, accompanied by the annihilation of a phonon

$$\begin{cases} \vec{k}_e = -\vec{q} \\ \hbar c k_\alpha = E_e + \hbar\omega_q \end{cases} \tag{326}$$

This process does not take place at very low temperature.

The cases (a) and (b) are illustrated in Fig. 15. Since we are dealing with phonons near the zone boundary, the phonon energy $\hbar\omega_q$ is nearly independent of $\vec{q}$.

Near T = 0, only phonon creation is relevant. By considering exciton states with energy near E_e, it is evident that a band exists in absorption for these temperatures. The absorption edge starts at $E_e + \hbar\omega_q$ and extends to higher energies. As the temperature is raised, a new absorption edge appears at $E_e - \hbar\omega_q$ corresponding to phonon absorption. As a result, the absorption spectrum is now comprised of two absorption edges separated by $2\hbar\omega_q$.

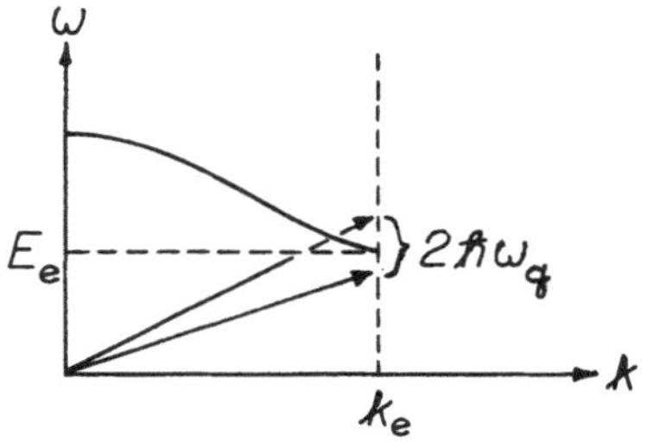

Fig. 15. Indirect Transitions.

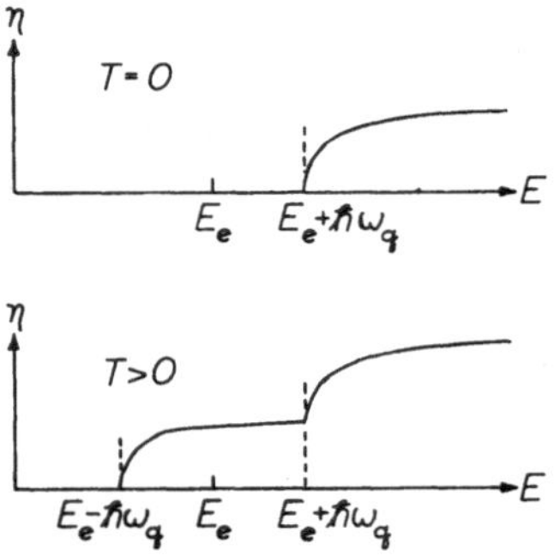

Fig. 16. Absorption Coefficient for Indirect Transitions at Different Temperatures.

One can therefore discern the presence of such indirect transitions by examining the temperature dependences and separation of the absoprtion edges. A thorough review of both the theoretical and experimental aspects of indirect transitions is given by McLean [17]. The effects of temperature on the absorption coefficient are illustrated in Fig. 16.

V.C. Magnons

1. Setting of the Problems. Consider a system consisting of N spins coupled ferromagnetically. This system may be realized by N atoms with orbital angular momentum zero and spin angular momentum S = 1/2. The ground state of this system consists of all spins lining in the same direction, say z.

An excitation of this system may be produced by raising the temperature by a small amount so that one spin is flipped. Since each spin has equal probability of being flipped, a spin wave perturbation will travel throughout the system.

If the temperature is increased further, two spins may be flipped and two spin waves will travel with different velocities, will meet and will scatter. If the flipped spins are adjacent we have what is called a spin complex and a situation in which the energy is lower than that when the flipped spins are separated. If more than two spins are flipped, more scattering may be present and larger spin complexes may be formed.

The basic approximation that is made in spin wave theory is that spin waves are independent of each other; this approximation is clearly valid for $T \ll T_f$. T_f, called the ferromagnetic Curie temperature, is the temperature above which all ferromagnetic order is destroyed, and the system behaves as a paramagnet.

Within the limits of this approximation the total energy of the spin waves is equal to the sum of the energies of the individual spin waves. Dyson has shown that the error in the calculation of the magnetization when the above approximation is used is less than 5% for $T = 0.5\ T_f$ [18].

2. Hamiltonian and Eigenstates. The relevant Hamiltonian of the coupled spin system is given by

$$H = -2J \sum_{\ell,m}{}' \vec{S}_\ell \cdot \vec{S}_m \tag{327}$$

where the prime in the sum excludes the term with $\ell = m$, and where $J > 0$. $J = 230k$ for Ni and $205k$ for Fe (k = Boltzmann constant). $\vec{S}$ are operators represented by the Pauli matrices:

$$S^x = \frac{1}{2}\begin{pmatrix} 0 & 1 \\ 1 & 0 \end{pmatrix}$$

$$S^y = \frac{1}{2}\begin{pmatrix} 0 & -i \\ i & 0 \end{pmatrix}$$

$$S^z = \frac{1}{2}\begin{pmatrix} 1 & 0 \\ 0 & -1 \end{pmatrix} \tag{328}$$

The eigenfunctions of S^2 and S^z are

$$\alpha = \begin{pmatrix} 1 \\ 0 \end{pmatrix} \quad ; \quad S^2\alpha = \frac{3}{4}\alpha \quad , \quad S_z\alpha = \frac{1}{2}\alpha \tag{329}$$

$$\beta = \begin{pmatrix} 0 \\ 1 \end{pmatrix} \quad ; \quad S^2\beta = \frac{3}{4}\beta \quad , \quad S_z\beta = -\frac{1}{2}\beta \tag{330}$$

We define

$$\begin{cases} S^+ = S^x + i\, S^y = \begin{pmatrix} 0 & 1 \\ 0 & 0 \end{pmatrix} \\ \\ S^- = S^x - i\, S^y = \begin{pmatrix} 0 & 0 \\ 1 & 0 \end{pmatrix} \end{cases} \tag{331}$$

It is

$$\begin{cases} S^+\beta = \alpha \\ S^-\alpha = \beta \end{cases} \tag{332}$$

Note also that

$$S_\ell^x\, S_m^x + S_\ell^y\, S_m^y = \frac{1}{2}\,(S_\ell^+\, S_m^- + S_\ell^-\, S_m^+) \tag{333}$$

Therefore we can write

$$\begin{aligned} H &= -2J \sum_{\ell,m}{}' \vec{S}_\ell \cdot \vec{S}_m \\ &= -2J \sum_{\ell,m}{}' [S_\ell^x\, S_m^x + S_\ell^y\, S_m^y + S_\ell^z\, S_m^z] \\ &= -J \sum_{\ell,m}{}' [S_\ell^+\, S_m^- + S_\ell^-\, S_m^+ + 2S_\ell^z\, S_m^z] \end{aligned} \tag{334}$$

The ground of the system is

$$\psi_G = |\alpha(1)\, \alpha(2)\, \alpha(3)\, \dots\, \alpha(N)> \tag{335}$$

When we operate on this state with H we obtain

$$\begin{aligned} H\,\psi_G &= -J \sum_{\ell,m}{}' [S_\ell^+\, S_m^- + S_\ell^-\, S_m^+ + 2S_\ell^z\, S_m^z]\,|\alpha(1)\, \alpha(2)\, \dots\, \alpha(N)> \\ &= -J\,\frac{1}{4}\, n\, N\, \psi_G = E_o\, \psi_G \end{aligned} \tag{336}$$

$E_o = -J(1/4)\, nN$ is the energy of the ground state, n is the number of nearest neighbors. Also, we have counted only once the interaction between two atoms; from this the extra factor of 1/2. The state of the system when one spin is flipped is described by the wavefunction

$$\psi_1(p) = |\alpha(1)\ \alpha(2)\ \ldots\ \beta(p)\ \ldots\ \alpha(N)> \tag{337}$$

When we operate on this state with H we obtain

$$\begin{aligned} H\,\psi_1(p) &= -J \sum_{\ell,m}{}' \,[S_\ell^+ S_m^- + S_\ell^- S_m^+ + 2S_\ell^z S_m^z] \\ &\quad \cdot |\alpha(1)\ \alpha(2)\ \ldots\ \beta(p)\ \ldots\ \alpha(N)> \\ &= -J[\sum_{\substack{\ell,m \\ (\ell\neq p,m\neq p)}}{}' S_\ell^z S_m^z + \sum_{r=1}^{n} S_p^+ S_{p+r}^- + \sum_{r=1}^{n} S_p^z S_{p+r}^z] \\ &\quad \cdot |\alpha(1)\ \alpha(2)\ \ldots\ \beta(p)\ \ldots\ \alpha(N)> \\ &= [-\frac{1}{4}\,n(N-2)\,J + \frac{1}{2}\,nJ]\,\psi_1(p) - J \sum_{r=1}^{n} \psi_1(p+r) \\ &= (-\frac{1}{4}\,nN + n)\,J\,\psi_1(p) - J \sum_{r=1}^{n} \psi_1(p+r) \end{aligned} \tag{338}$$

or

$$H\,\psi_1(p) = A\,\psi_1(p) - J \sum_{r=1}^{n} \psi_1(p+r) \tag{339}$$

where

$$A = nJ(1 - \frac{N}{4}) \tag{340}$$

The N functions $\psi_1(p)$ $(p = 1,2,\ \ldots,\ N)$ are <u>not</u> eigenfunctions of H.

We can form linear combinations of $\psi_1(p)$ functions as follows:

$$\psi_1(\vec{k}) = \sum_{p=1}^{N} e^{i\vec{k}\cdot\vec{R}_p}\,\psi_1(p) \tag{341}$$

where $\vec{k}$ is determined by the usual periodic boundary conditions.

Now we obtain

$$H\,\psi_1(\vec{k}) = \sum_{p=1}^{N} e^{i\vec{k}\cdot\vec{R}_p} H\,\psi_1(p)$$

$$= \sum_{p=1}^{N} e^{i\vec{k}\cdot\vec{R}_p}[A\,\psi_1(p) - J\sum_{r=1}^{n}\psi_1(p+r)]$$

$$= A\,\psi_1(\vec{k}) - J\sum_{r=1}^{n}\sum_{p=1}^{N} e^{i\vec{k}\cdot\vec{R}_p}\psi_i(p+r) \qquad (342)$$

But

$$\sum_{p=1}^{N} e^{i\vec{k}\cdot\vec{R}_p}\psi_1(p+r) = \sum_{p=1}^{N} e^{i\vec{k}\cdot\vec{R}_{p+r}}\psi_1(p+r)\,e^{i\vec{k}\cdot(\vec{R}_p-\vec{R}_{p+r})}$$

$$= \sum_{p=1}^{N} e^{i\vec{k}\cdot\vec{R}_p}\psi_1(p)\,e^{i\vec{k}\cdot(\vec{R}_{p+r}-\vec{R}_p)} = \psi_1(\vec{k})\,e^{i\vec{k}\cdot\vec{r}} \qquad (343)$$

where $\vec{r}$ = vectorial distance from given atom to nearest neighbor. Therefore

$$H\,\psi_1(\vec{k}) = A\,\psi_1(\vec{k}) - J(\sum_r e^{i\vec{k}\cdot\vec{r}})\,\psi_1(\vec{k})$$

$$= (-n\frac{NJ}{4} + nJ - J\sum_r e^{i\vec{k}\cdot\vec{r}})\,\psi_1(\vec{k})$$

$$= [E_o + (n - \sum_r e^{i\vec{k}\cdot\vec{r}})\,J]\,\psi_1(\vec{k}) = E_{\vec{k}}\,\psi_1(\vec{k}) \qquad (344)$$

where

$$E_{\vec{k}} = E_o + (n - \sum_r e^{i\vec{k}\cdot\vec{r}})\,J \qquad (345)$$

is the dispersion relation. For small values of $\vec{k}$

$$e^{i\vec{k}\cdot\vec{r}} = 1 + i\vec{k}\cdot\vec{r} - \frac{(\vec{k}\cdot\vec{r})^2}{2} \qquad (346)$$

$$\sum_r e^{i\vec{k}\cdot\vec{r}} = n - \frac{1}{2}\sum_r (\vec{k}\cdot\vec{r})^2 \tag{347}$$

and

$$E_{\vec{k}} - E_o = \frac{1}{2} J \sum_r (\vec{k}\cdot\vec{r})^2 \tag{348}$$

where $|\vec{k}\cdot\vec{r}| << 1$. For cubic symmetry

$$\sum_r (\vec{k}\cdot\vec{r})^2 = \frac{1}{3} k^2 \sum_r r^2 \tag{349}$$

and

$$E_{\vec{k}} - E_o = \frac{1}{6} J k^2 \sum_r r^2 = J k^2 a^2 \tag{350}$$

where a = lattice constant.

The system of spin waves can be thought of as a collection of harmonic oscillators. Second quantization techniques can be applied and can allow us to express the dispersion relation (345) as follows:

$$\hbar\omega = (n - \sum_r e^{i\vec{k}\cdot\vec{r}}) J \tag{351}$$

The z-component of the total spin is easily obtained by operating on the wavefunction of the spin system by means of the operator

$$S_z = \sum_{i=1}^{N} S_i^z$$

By operating with S_z on ψ_G of Eq. (335) we find that the z-component of the total spin when the system is in the ground state is

$$M_S = N\frac{1}{2} = NS \tag{352}$$

The z-component of the total spin when a spin wave of wave vector $\vec{k}$ is excited is obtained as follows:

$$S_z\,\psi_1(\vec{k}) = \sum_i S_i^z \sum_p e^{i\vec{k}\cdot\vec{R}_p}\,\psi_1(p) = \sum_p e^{i\vec{k}\cdot\vec{R}_p} \sum_i S_i^z\,\psi_1(p)$$

$$= \sum_p e^{i\vec{k}\cdot\vec{R}_p} (N - 2) \frac{1}{2} \psi_1(p) = (N \frac{1}{2} - 1) \psi_1(\vec{k}) \qquad (352)$$

Therefore when a spin wave is excited

$$M_S = NS - 1 \qquad (353)$$

and the z-component of each spin will be $(S - 1/N)$.

If $n_{\vec{k}}$ spin waves of wave vector $\vec{k}$ are excited the z-component of the total spin is

$$M_S = NS - n_{\vec{k}} \qquad (354)$$

and the z-component of each spin will be

$$\frac{M_S}{N} = S - \frac{n_{\vec{k}}}{N} \qquad (355)$$

3. Semiclassical Treatment [10,19]. In this approximation each spin is a vector of length $[S(S + 1)]^{1/2}$ precessing about the z-axis. If a number of spin waves $n_{\vec{k}}$ of wave vector $\vec{k}$ are excited then each spin precesses about the z-axis describing a circle of amplitude u as in Fig. 17.

By approximating $[S(S + 1)]^{1/2}$ with S the z-component of a spin for $u << S$ is given by

$$\sqrt{S - u^2} \simeq S - \frac{u^2}{2S} \qquad (356)$$

We have to set this equal to the expression (354):

$$S - \frac{u^2}{2S} = S - \frac{n_{\vec{k}}}{N} \qquad (357)$$

and we find, now calling u by a new name, $u_{\vec{k}}$,

$$u_{\vec{k}}^2 = \frac{2S\, n_{\vec{k}}}{N} \qquad (358)$$

The radius $u_{\vec{k}}$ is the same for all spins. The precession of a linear string of spins is now illustrated in Fig. 18.

4. Thermodynamics of Magnons. The energy of the spin-wave system is given by

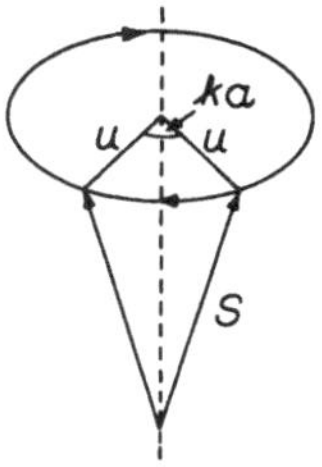

Fig. 17. Precession of a Spin About the z-Axis.

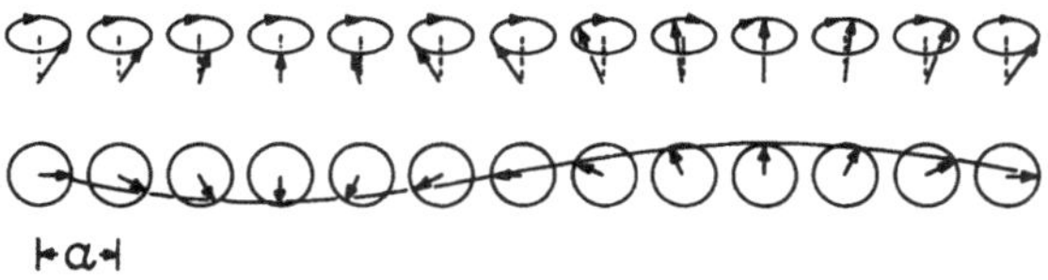

Fig. 18. Precession of a Linear String of Spins [10,19].

$$E_{n_{\vec{k}_1} n_{\vec{k}_2} \cdots n_{\vec{k}_N}} = \sum_{\vec{k}} n_{\vec{k}} E_{\vec{k}} \tag{359}$$

We simplify the notation as follows:

$$n_{\vec{k}_i} = n_i \tag{360}$$

$$E_{\vec{k}_i} = E_i \tag{361}$$

Then we can write the partition sum of the system as follows:

$$Z = \sum_r e^{-\beta E_r} = \sum_{n_1 n_2 \ldots n_N} e^{-\beta(n_1 E_1 + n_2 E_2 + \ldots n_N E_N)}$$

$$= \sum_{n_1=0}^{\infty} e^{-\beta n_1 E_1} \sum_{n_2=0}^{\infty} e^{-\beta n_2 E_2} \ldots \sum_{n_N=0}^{\infty} e^{-\beta n_N E_N}$$

$$= \frac{1}{1 - e^{-\beta E_1}} \frac{1}{1 - e^{-\beta E_2}} \cdots \frac{1}{1 - e^{-\beta E_N}} \tag{362}$$

and

$$\ell n\ Z = - \sum_{i=1}^{N} \ell n(1 - e^{-\beta E_i}) \tag{363}$$

where $\beta = 1/kT$. Therefore the internal energy of the system is given by

$$\overline{E} = - \frac{\partial\ \ell n\ z}{\partial \beta} = \sum_{\vec{k}} \frac{E_{\vec{k}}}{e^{\beta E_{\vec{k}}} - 1} = \sum_{\vec{k}} \overline{n}(E_{\vec{k}})\ E_{\vec{k}} \tag{364}$$

where

$$\overline{n}(E_{\vec{k}}) = \frac{1}{e^{E_{\vec{k}}/kT} - 1} \tag{365}$$

represents the average number of spin waves. We can also think of a spin wave of wave vector $\vec{k}$ as a particle called magnon of moment

$\hbar\vec{k}$ and energy $E_{\vec{k}}$; $n_{\vec{k}}$ is then the average number of magnons of energy $E_{\vec{k}}$ present at temperature T.

Magnons are clearly bosons since the interchange of two reversed spins does not change the state of the system. Indeed, Eq. (365) expresses the statistics one would expect for bosons whose total number is not fixed.

The number of allowed $\vec{k}$ values in $(\vec{k}, \vec{k} + d\vec{k})$ is given by

$$g(\vec{k})\ d^3\vec{k} = \frac{V}{8\pi^3}\ d^3\ \vec{k} \tag{366}$$

The number of allowed $\vec{k}$ values with $k = |\vec{k}|$ in $(k, k + dk)$ is given by

$$g(k)\ dk = 4\pi\ k^2\ \frac{V}{8\pi^3}\ dk = \frac{Vk^2}{2\pi^2}\ dk \tag{367}$$

The number of magnons with energy in $(E, E + dE)$ is then given by

$$g(E)\ dE = g(k)\ dk = g(k)\ \frac{dk}{dE}\ dE$$

$$= \frac{Vk^2}{2\pi^2}\ \frac{dk}{d(Ja^2\ k^2)}\ dE = \frac{V}{2\pi^2}\ \frac{k}{2Ja^2}\ dE$$

$$= \frac{V}{4\pi^2}\ \left(\frac{1}{Ja^2} \right)^{3/2}\ E^{1/2}\ dE \tag{368}$$

where we have used the dispersion relation (350).

The total number of magnons is given by

$$\sum_{\vec{k}} \bar{n}(E_{\vec{k}}) = \int_0^{E_{max}} \bar{n}(E)\ g(E)\ dE$$

$$= \frac{V}{4\pi^2}\ \left(\frac{1}{Ja^2} \right)^{3/2} \int_0^{E_{max}} \frac{E^{1/2}}{e^{E/kT} - 1}\ dE$$

$$= \frac{V}{4\pi^2} \left(\frac{1}{Ja^2} \right)^{3/2} (kT)^{3/2} \int_0^{E_{max}} \frac{x^{1/2}}{e^x - 1} dx \tag{369}$$

For $T << E_{max}/k$ we have

$$\sum_k \bar{n}(E_{\vec{k}}) \simeq \frac{V}{4\pi^2} \left(\frac{1}{Ja^2} \right)^{3/2} (kT)^{3/2} \int_0^{\infty} \frac{x^{1/2}}{e^x - 1} dx$$

$$= 0.1174 \, V \left(\frac{kT}{Ja^2} \right)^{3/2} \tag{370}$$

Example

$J = 205\ k \quad ; \quad a = 2 \times 10^{-8}$ cm; $\quad T = 20°K$

$$\frac{\Sigma \bar{n}}{V} = 0.01174 \left(\frac{kT}{Ja^2} \right)^{3/2} = \left(0.1174 \frac{k \times 20}{205k \times (2 \times 10^{-8})^2} \right)^{3/2}$$

$$\simeq 4.5 \times 10^{20} \ cm^{-3}$$

The change of the spontaneous magnetization due to the excitation of spin waves is

$$\frac{\Delta M}{M(0)} = \frac{M(0) - M(T)}{M(0)} = \frac{\sum_{\vec{k}} n_k}{V} = 0.1174 \left(\frac{kT}{Ja^2} \right)^{3/2} \frac{V}{N} \tag{371}$$

But

$$N = V \frac{Q}{a^3} \tag{372}$$

where Q = 1, 2, and 4 for simple cubic, body-centered cubic and face-centered cubic lattice, respectively. Then

$$\frac{\Delta M}{M(0)} = 0.1174 \left(\frac{kT}{Ja^2} \right)^{3/2} \frac{a^3}{Q} = \frac{0.1174}{Q} \left(\frac{kT}{J} \right)^{3/2} = AT^{3/2} \tag{373}$$

where $A = [(0.1174/Q)(k/J)^{3/2}]$. The above formula expresses the so-called Block $T^{3/2}$ law.

The internal energy of the system is given by

$$\sum_{\vec{k}} \bar{n}(E_{\vec{k}})\, E_{\vec{k}} = \int_0^{E_{max}} \bar{n}(E)\, E g(E)\, dE$$

$$= \frac{V}{4\pi^2} \left(\frac{1}{Ja^2} \right)^{3/2} \int_0^{E_{max}} \frac{E^{3/2}}{e^{E/kT} - 1}\, dE$$

$$= \frac{V}{4\pi^2} \left(\frac{1}{Ja^2} \right)^{3/2} (kT)^{5/2} \int_0^{E_{max}/(kT)} \frac{x^{3/2}}{e^x - 1}\, dx \tag{374}$$

For $T \ll E_{max}/k$ we have

$$\sum_{\vec{k}} \bar{n}(E_{\vec{k}})\, E_{\vec{k}} \simeq \frac{V}{4\pi^2} \left(\frac{1}{Ja^2} \right)^{3/2} (kT)^{5/2} \int_0^{\infty} \frac{x^{3/2}}{e^x - 1}\, dx$$

$$= \frac{1.7844}{4\pi^2} \frac{V}{Ja^2} k^{5/2}\, T^{5/2} \tag{375}$$

and the specific heat per unit volume of the magnon system is

$$\frac{1.7844}{4\pi^2} \left(\frac{1}{Ja^2} \right)^{3/2} k^{5/2} \frac{5}{2} T^{3/2} = 0.113 \frac{k^{5/2}}{(Ja^2)^{3/2}} T^{3/2} \tag{376}$$

V.D. Plasmons

1. Dielectric Response of an Ensemble of Oscillators. It is appropriate at this point to summarize some of the results obtained in Section IV. We dealt then with a system of N particles of mass m and negative charge e, each elastically bound to an equilibrium position where a positive charge $|e|$ resides; the angular frequency of oscillation is ω_0 and γ is the damping constant. If the system is under the action a plane wave

$$\vec{E}(\vec{r},t) = \vec{E}_o(\vec{r})\, e^{i\omega t} \tag{377}$$

then the particles oscillate displacing themselves from their equilibrium positions by the amount

$$\vec{z}(\vec{r},t) = \frac{e/m}{\omega_o^2 - \omega^2 + i\gamma\omega} \vec{E}_o(\vec{r}) \, e^{i\omega t} \tag{378}$$

The induced dipole moment is

$$e\vec{z}(\vec{r},t) = \alpha(\omega) \, \vec{E}(\vec{r},t) \tag{379}$$

where

$$\alpha(\omega) = \text{polarizability} = \frac{e^2/m}{\omega_o^2 - \omega^2 + i\gamma\omega} \tag{380}$$

The real part of the induced dipole moment is

$$\text{Re}[e\vec{z}(\vec{r},t)] = \frac{e^2}{m} \frac{\vec{E}_o(\vec{r})}{\sqrt{(\omega_o^2 - \omega^2)^2 + \gamma^2 \omega^2}} \cos(\omega t - \phi) \tag{381}$$

where

$$\tan\phi = \frac{\gamma\omega}{\omega_o^2 - \omega^2} \tag{382}$$

$\phi = 0$, $\pi/2$ and π for $\omega = 0$, ω_o and ∞, respectively.

The polarization is given by

$$\vec{P}(\vec{r},t) = n_o \, \alpha(\omega) \, \vec{E}(\vec{r},t) = \chi(\omega) \, \vec{E}(\vec{r},t) \tag{383}$$

where

$$\chi(\omega) = \text{dielectric susceptibility} = n_o \, \alpha(\omega) \tag{384}$$

$$n_o = N/V \tag{385}$$

The electric displacement is given by

$$\vec{D}(\vec{r},t) = \vec{E}(\vec{r},t) + 4\pi \, \vec{P}(\vec{r},t)$$

$$= [1 + 4\pi \, \chi(\omega)] \, \vec{E}(\vec{r},t) = K(\omega) \, \vec{E}(\vec{r},t) \tag{386}$$

where the dielectric constant $K(\omega)$ is given by

$$K(\omega) = 1 + \frac{4\pi\, n_o\, e^2/m}{\omega_o^2 - \omega^2 + i\gamma\omega} \tag{387}$$

In the limit of $\gamma = 0$

$$K(\omega) = 1 + \frac{4\pi\, n_o\, e^2/m}{\omega_o - \omega^2} \tag{388}$$

This function is represented in Fig. 19. We note that

$$K(\omega) < 0 \qquad \text{for} \qquad \omega_o < \omega < \omega_L \tag{389}$$

where

$$\omega_L = \omega_o \sqrt{1 + \frac{4\pi\, n_o\, e^2/m}{\omega_o^2}} \tag{390}$$

and

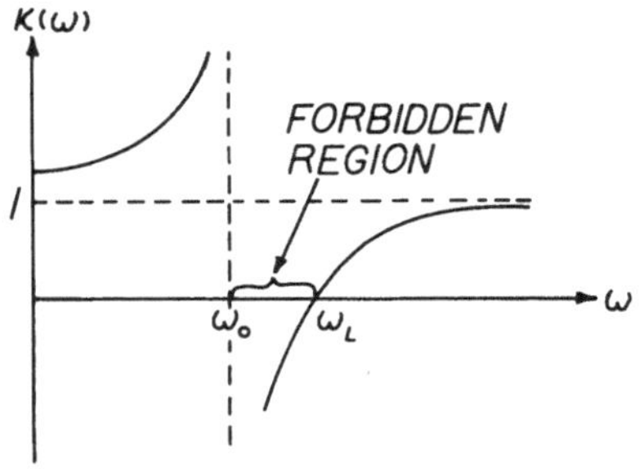

Fig. 19. Dielectric Constant of an Ensemble of Oscillators.

$$K(\omega_L) = 0 \tag{391}$$

$$K(\infty) = 1 \tag{392}$$

$$K(0) - K(\infty) = \frac{4\pi\, n_o\, e^2/m}{\omega_o^2} \tag{393}$$

2. Dielectric Response of a Free Electron Gas. In order to handle this case we set in the expressions we reported

$$\omega_o = 0 \quad , \qquad \gamma = 0 \tag{394}$$

and find that, in this case, under the action of the field $\vec{E}(\vec{r},t)$ given in Eq. (377)

$$e\vec{z}(\vec{r},t) = \alpha(\omega)\ \vec{E}(\vec{r},t) \tag{395}$$

where

$$\alpha(\omega) = -\frac{e^2}{m\omega} \qquad\qquad (\phi = 0) \tag{396}$$

The polarization is given by

$$\vec{P}(\vec{r},t) = n_o\ \alpha(\omega)\ \vec{E}(\vec{r},t) = \chi(\omega)\ \vec{E}(\vec{r},t) \tag{397}$$

where

$$\chi(\omega) = n_o\ \alpha(\omega) = -\frac{n_o\ e^2}{m\omega^2} \tag{398}$$

The electric displacement is given by

$$\vec{D}(\vec{r},t) = K(\omega)\ \vec{E}(\vec{r},t) = [1 + 4\pi\ \chi(\omega)]\ \vec{E}(\vec{r},t) \tag{399}$$

where the dielectric constant $K(\omega)$ is given by

$$K(\omega) = 1 - \frac{4\pi\, n_o\, e^2/m}{\omega^2} \tag{400}$$

This function is represented in Fig. 20. We note that

$$K(\omega_L) = 0 \tag{401}$$

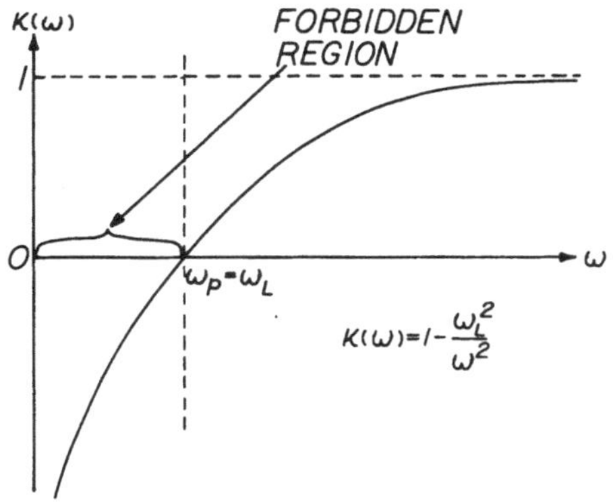

Fig. 20. Dielectric Constant of a Free Electron Gas.

where

$$\omega_L = \sqrt{\frac{4\pi n_o e^2}{m}} = \omega_p = \underline{\text{plasma frequency}} \tag{402}$$

and K(ω) can be written

$$K(\omega) = 1 - \frac{\omega_L^2}{\omega^2} \tag{403}$$

3. Transverse Optical Modes in a Plasma. A plasma can be defined as a medium with equal concentration of positive and negative charges, of which at least one charge type is mobile [10]. In solids such a plasma may be realized by the conduction electrons

whose negative charges are balanced by the positive charges of the ion cores.

The dispersion relation for transverse electromagnetic waves in a plasma is given by

$$\omega = \frac{ck}{\sqrt{K}} = \frac{ck}{\sqrt{1 - \frac{\omega_p^2}{\omega^2}}} \tag{404}$$

or

$$\omega^2 = \omega_p^2 + c^2 k^2 \tag{405}$$

or

$$\frac{\omega}{\omega_p} = \sqrt{1 + \left(\frac{ck}{\omega_p}\right)^2} \tag{406}$$

The last expression is represented in Fig. 21. ω cannot be smaller than ω_p for transverse electromagnetic waves; when such waves have $\omega < \omega_p$ and are incident on the plasma they are reflected. If $\omega < \omega_p$, $K(\omega) < 0$, as in Fig. 20.

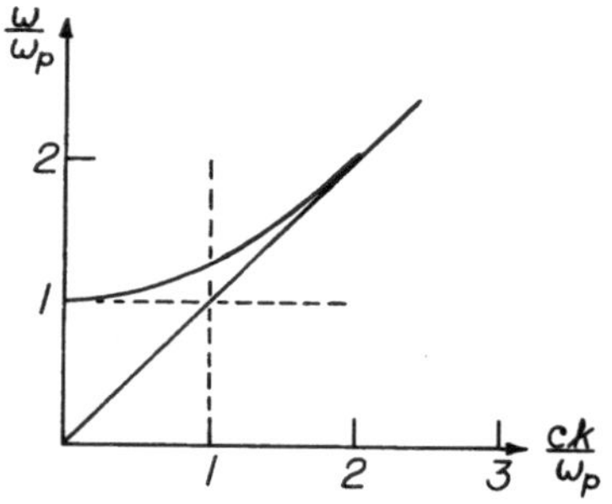

Fig. 21. Dispersion Relation for Transverse Electromagnetic Waves in a Plasma [10].

If we use for e and m the values of the charge and mass of the electron, respectively, we find

$$\omega_p = \omega_L = \sqrt{\frac{4\pi\, n_o\, e^2}{m}} \simeq 5.7 \times 10^4 \sqrt{n_o} \tag{407}$$

and

$$\omega_p = \frac{2\pi c}{\omega_p} \simeq \frac{3.32 \times 10^6}{\sqrt{n_o}} \tag{408}$$

Examples of ω_p and λ_p are given in the following table:

$n_o(cm^{-3})$	10^{22}	10^{18}	10^{14}	10^{10}
$\omega_p(sec^{-1})$	$5.7x10^{15}$	$5.7x10^{13}$	$5.7x10^{11}$	$5.7x10^{9}$
$\lambda_p(cm)$	$3.3x10^{-5}$	$3.3x10^{-3}$	0.33	33

λ_p is the cut-off wavelength; no radiation with $\lambda > \lambda_p$ can be transmitted through the plasma.

This result is in accord with the experimental data; for alkali metals it is found that a metal film is transparent only if $\lambda < \lambda_p$. We report in the following table, taken from [10], the relevant information.

	Li	Na	K	Rb	Cs
λ_p (Å) (calculated)	1550	2090	2870	3220	3620
λ_p (Å) (observed)	1550	2100	3150	3400	-

The optical properties of metals are reviewed in [20].

We note also that the reflection of light from metals is similar to the reflection of radio waves from the ionosphere.

4. Longitudinal Optical Modes in a Plasma. The electron gas undergoes a free (longitudinal) oscillation at $\omega = \omega_p$, the cut-off frequency of the transverse electromagnetic waves.

The electrons move with respect to the positive background:

$$\vec{P} = -n_o e\vec{z} \tag{409}$$

But $K(\omega_p) = 0$, then

$$\vec{D} = K\vec{E} = \vec{E} + 4\pi\vec{P} = 0 \tag{410}$$

and

$$\vec{E} = -4\pi\vec{P} = 4\pi n_o e\vec{z} \tag{411}$$

The equation of motion of the electrons in the unit volume is

$$n_o m \frac{d^2\vec{z}}{dt^2} = -n_o e\vec{E} = -4\pi n_o^2 e^2\vec{z} \tag{412}$$

or

$$\frac{d^2\vec{z}}{dt^2} + \frac{4\pi n_o e^2}{m}\vec{z} = 0 \tag{413}$$

or

$$\frac{d^2\vec{z}}{dt^2} + \omega_p^2 z = 0 \tag{414}$$

where ω_p is given by Eq. (402).

A plasma oscillation of small wavevector has $\omega \simeq \omega_p$. The dispersion relation for longitudinal oscillators will not be derived here; the reader is referred to Pines [21] who gives

$$\omega = \omega_p \left(1 + \frac{3v_F^2}{10\omega_p^2} k^2 + \ldots\right) \tag{415}$$

where v_F = Fermi velocity.

Plasma oscillations in metals are collective longitudinal excitations of the gas which consists of the conduction electrons.

A plasmon is a quantum of plasma oscillation. The typical energy of a plasmon can be obtained by using for n_0 the typical metallic density of $n_0 \simeq 10^{23}$. We obtain then

$$\omega_p = 5.7 \times 10^4 \sqrt{n} = 5.7 \times 10^4 \sqrt{10^{23}} \simeq 2 \times 10^{16} \text{ sec}^{-1}$$

$$\hbar\omega_p = 10^{-27} \times 2 \times 10^{16} = 2 \times 10^{11} \text{ ergs} \simeq 12 \text{ eV}$$

Therefore no thermal excitation of plasmons takes place.

Plasmons can be produced by passing an electron through a thin metallic film or reflecting an electron from a film. The charge of the electron couples with the fluctuations of the $\vec{E}$ field. The transmitted or reflected electron shows an energy loss equal to $n(\hbar\omega_p)$ where n = number of plasmons created.

We note also that there may be "surface" plasmons, different in ω_p from "volume" plasmons.

Finally, plasma oscillations can also be excited in dielectric films of, say, Si, Ge, InSb. In a dielectric all the valence band electrons oscillate with respect to the ion cores.

The following table presents some typical values of plasmon energies.

Material	Observed $\hbar\omega_p$ (in eV)	Calculated $\hbar\omega_p$ (in eV)
Li	7.12	7.96
Na	5.71	5.58
K	3.72	3.86
Mg	10.6	10.9
Al	15.3	15.8
Dielectrics		
Si	16.4 - 16.9	16
Ge	16 - 16.4	16
InSb	12 - 13	12

REFERENCES

1. D. L. Dexter, J. Chem. Phys. 21, 836 (1953).
2. B. Di Bartolo, in Spectroscopy of the Excited State, B. Di Bartolo, ed., Plenum Press, New York and London, p. 17 (1976).
3. B. Di Bartolo, in Luminescence of Inorganic Solids, B. Di Bartolo, ed., Plenum Press, New York and London, p. 1 (1978).
4. A. von Hippel, Dielectrics and Waves, Wiley, New York (1954).
5. D. L. Dexter and R. S. Knox, Excitons, Wiley Interscience, New York (1965).
6. R. S. Knox, Theory of Excitons, Academic Press, New York and London (1963).
7. M. Born and K. Huang, Dynamical Theory of Crystal Lattices, Oxford, London (1954).
8. A. D. B. Woods et al, Phys. Rev. 131, 1025 (1963).
9. J. L. T. Waugh and G. Dolling, Phys. Rev. 132, 2410 (1963).
10. C. Kittel, Introduction to Solid State Physics, Wiley, New York (1971).
11. J. Frenkel, Phys. Rev. 37, 17 (1931); Phys. Rev. 37, 1276 (1931).
12. B. Di Bartolo, Optical Interactions in Solids, Wiley, New York (1968).
13. H. C. Wolf, Solid State Physics 9, 1 (1959).
14. J. Hegarty, Ph.D. Thesis, University College, Galway, Ireland (1976), unpublished.
15. G. H. Wannier, Phys. Rev. 52, 191 (1937).
16. S. I. Pekar, Soviet Physics JETP 11, 1286 (1960).
17. T. P. McLean, in Progress in Semiconductors, Vol. 5, A. F. Gibson et al, eds., Wiley, New York (1961).
18. F. J. Dyson, Phys. Rev. 102, 1217 (1956); Phys. Rev. 102, 1230 (1956).
19. A. H. Morrish, The Physical Principles of Magnetism, Wiley, New York (1965).
20. M. P. Givens, Solid State Physics 6, 313 (1958).
21. D. Pines, Elementary Excitations in Solids, Benjamin, New York and Amsterdam (1963).

QUASI-PARTICLES AND EXCITONS: MODELS OF STRUCTURE AND CORRELATION

G. Mahler

Institut für theoretische Physik
Universität Stuttgart
D-7000 Stuttgart 80

ABSTRACT

The electronic excitation spectrum of a material depends on its structure, range and type of electronic correlations and the nature of the reference state considered. We discuss these relations by means of simple models, which include disordered and externally driven systems.

I. INTRODUCTION

In real material systems including solids the basic constituents are fermions (electrons, nucleons). Bosons may result from a grouping of an even number of fermions into certain complexes (e.g. He^4). Bosons may also occur from pairing of two quasi-particles (e.g. Cooper pairs) or of a fermion with a fermion hole, which both play a crucial role in the theory of electronic excitations. In any case Bose- is obeyed only approximately, though deviations may be negligible for some applications.

In this lecture we do not discuss collective properties of those elementary excitations (excitons). We are rather interested in the interplay of structure and correlations in producing the respective excitation spectrum. Here, structure is taken as a given property to be defined independently from the electronic subsystem (though a consistent theory for the structure would certainly have to include the electrons). Correlations - besides those induced from quantum statistics - derive from the interaction of electrons as charges elementary particles; this interaction may be modified by other fields due to the solid background.

A derivation of the electronic (low) excitation spectrum poses in general a very difficult problem, and a vast literature exists on most topics included here [1]. However, instead of discussing the various approximations involved in the study of a given material, we focus on a number of basic models (the selection of which is at present rather arbitrary). Such an approach has proven to be very fruitful in the understanding of phase transitions and critical phenomena as they depend on structure, dimension, range and type of interaction [2]. Here we propose to follow the same line with respect to the type of excitation spectra. In this context it is interesting to recall that there is even an intimate relationship between the kind of excitation spectrum and phase transitions.

The most transparent way to discuss elementary excitation is based on quantum-field theoretical methods [3,4] originally developed in elementary particle physics. It therefore seems worthwhile to briefly summarize the concepts of quantum-field theory before discussing models of structure and correlation in the main part.

II. BASIC CONCEPTS OF QUANTUM-FIELD THEORY

II. A. Quantization of Classical Fields

Besides attempts on axiomatic foundations, quantum-field theory is commonly introduced in the frame-work of the Lagrange-formalism, which, despite its shortcomings, is still the simpliest and most intuitive way.

Suppose we consider a classical scalar field $F(\underline{r},t)$ in real space $\underline{r} = (x,y,z)$ with the Lagrangian

$$L = \int d^3r\, \mathcal{L}(F,\nabla F, \dot{F},t) \quad . \tag{1}$$

Generalizing the ideas of classical mechanics of mass points, the equation of motion for F (field equation) is given by the Lagrange equation.

$$\frac{\delta\mathcal{L}}{\delta F} - \frac{d}{dt}\frac{\delta\mathcal{L}}{\delta\dot{F}} = 0 \tag{2}$$

where functional derivatives are formally to be carried out like the respective partial derivatives

$$\frac{\delta}{\delta \dot{F}} = \frac{\partial}{\partial \dot{F}} \quad ,$$

$$\frac{\delta}{\delta F} = \frac{\partial}{\partial F} - \nabla \frac{\partial}{\partial (\nabla F)} \quad . \quad * \tag{3}$$

The generalized momentum,

$$\pi = \frac{\delta \mathcal{L}}{\delta \dot{F}} \quad , \tag{4}$$

allows to define the Hamiltonian

$$H = \int d^3r \; \mathcal{H} \tag{5}$$

by means of the well-known Legendre transformation for the Hamiltonian density

$$\mathcal{H}(F, \nabla F, \pi, t) = \pi \dot{F} - \mathcal{L} \quad . \tag{6}$$

Field quantization means to interpret $\mathcal{H}$ as an operator. This requires to know the basic commutation rules. If F was taken as one component X of $\underline{R}(\underline{r},t)$ describing the motion of a continuous mechanical system we would relate

$$\pi \leftrightarrow p_i \quad \text{(momentum of particle i)} \quad ,$$

$$X \leftrightarrow q_i \quad \text{(position of particle i)} \quad , \tag{7}$$

and thus expect the following generalized commutation relations to hold for the corresponding (conjugated) operators:

$$[\hat{\pi}(\underline{r},t),\ \hat{F}(\underline{r}',t)] = \frac{h}{i}\,\delta(\underline{r} - \underline{r}') \quad ,$$

$$[\hat{\pi}(\underline{r},t),\ \hat{\pi}(\underline{r}',t)] = [\hat{F}(\underline{r}',t),\ \hat{F}(\underline{r},t)] = 0 \quad . \tag{8}$$

$$* \quad \nabla \frac{\partial A}{\partial (\nabla F)} \equiv \sum_{i=1}^{3} \frac{\partial}{\partial x_i} \left[\frac{\partial A}{\partial \left(\frac{\partial F}{\partial x_i}\right)} \right]$$

As the ordering of π and F becomes important after quantization, we must in addition observe symmetrization rules in

Eq. (6) to make the procedure unique. The basic assumption of quantum field theory is that Eq. (8) should hold for <u>any</u> quantized field which has a classical counterpart. There is, in principle, a very simple recipe: If you have a classical field equation, then find the Lagrangian, derive the generalized momentum according to Eq. (4) and quantize the resulting Hamiltonian H according to

$$\hat{H} = \int d^3r\, \mathcal{H}(\hat{F}, \nabla\hat{F}, \hat{\pi}, t) \tag{9}$$

observing the rules, Eq. (8). For a vector field, $\underline{F}$, Eq. (8) has to be generalized according to

$$[\pi_x(\underline{r},t),\ \hat{F}_y(\underline{r}',t)] = \frac{\hbar}{i}\,\delta_{xy}\ \delta(\underline{r} - \underline{r}') \quad . \tag{10}$$

As a simple example let us consider the linear distortion $u(\underline{r},t)$ of an elastic continuum described by the wave equation

$$\rho\frac{\partial^2 u}{\partial t^2} - d\nabla^2 u = 0, \tag{11}$$

where ρ is the mass-density and d characterizes the elastic property of the system. By Eq. (2) we easily convince ourselves that

$$\mathcal{L}(\dot{u}, \nabla u) = \frac{1}{2}\rho\dot{u}^2 - \frac{1}{2}d(\nabla u)^2 \tag{12}$$

and

$$\pi = \rho\dot{u}.$$

We thus get

$$\mathcal{H} = \frac{1}{2\rho}\ \pi^2 + \frac{1}{2}d(\nabla u)^2,$$

which in quantized form leads to

$$\hat{H} = \frac{1}{2\rho}\int\hat{\pi}^2\ d^3r + \frac{d}{2}\int(\nabla\hat{u})^2\ d^3r. \tag{13}$$

Of course, we are, in principle, free to choose any field equation for subsequent quantization, i.e. any equation relating F, ∇F, $\dot{F}$, which has a Lagrangian. In practice, however, only very few types of field equations are physically relevant. While we leave the Dirac-field to the elementary particle physicists, we may focus for the study of the solid state on the Schrödinger-field, the electromagnetic field and the phonon field. In the continuum limit the phonon field can be regarded as a modified Klein-Gordon-field, a fact which is exploited especially with respect to its non-linear versions (e.g. the Sine-Gordon-equation). In the following we will restrict ourselves to the Schrödinger-field.

II.B. Schrödinger-Field

We take the time-dependent Schrödinger equation for a single particle as a classical scalar field equation (Lagrange equation):

$$i h \dot{F}(\underline{r},t) + \frac{h^2}{2m} \Delta F(\underline{r},t) + V(\underline{r},t)\, F(\underline{r},t) = 0 \tag{14}$$

Contrary to the simple example considered above, this field is in general complex and we have to consider F and F* as two independent fields, where F* obeys the equation conjugated to Eq. (14). We may use the (non-hermitian) Lagrange-density

$$\mathcal{L}(F^*, \nabla F^*, \dot{F}^*, t) =$$

$$= i h F^* \dot{F} - \frac{h^2}{2m} \nabla F^* \nabla F - V(\underline{r},t)\, F^* F, \tag{15}$$

from which we recover Eq. (14) by means of Eq. (2). By the general rule the momentum conjugated to F is given by

$$\pi = \frac{\delta \mathcal{L}}{\delta F} = i\hbar F^*, \tag{16}$$

and

$$\mathcal{H} = \frac{h^2}{2m} \nabla F^* \nabla F + V(\underline{r},t)\, F^* F. \tag{17}$$

Quantization leads to

$$\hat{H} = \int [\frac{h^2}{2m} \nabla \hat{F}^+ \nabla \hat{F} + V(\underline{r},t)\, \hat{F}^+ \hat{F}] d^3 r \tag{18}$$

with

$$[\hat{F}(\underline{r},t),\ \hat{F}^+(\underline{r},t)] = \partial(\underline{r}-\underline{r}'). \tag{19}$$

While in the usual (first) quantization we would require

$$1 = \int F^*\, F\, d^3 r \tag{20}$$

and interprete F as the wave function Ψ, the expression on the right-hand side of Eq. (20) is now an operator, which has no classical analogue:

$$\hat{N} = \int \hat{F}^+ \hat{F}\, d^3 r. \tag{21}$$

We convince ourselves that

$$[\hat{H}, \hat{N}] = 0, \tag{22}$$

so that we can find eigenstates of $\hat{H}$ which are also eigenstates of N. A state with the property

$$\hat{N}|0> = 0 \tag{23}$$

i.e., with the eigenvalue N = 0 is called vacuum state, while for the state

$$|1> = \hat{F}^+(\underline{r},t)|0> \tag{24}$$

one finds the eigenvalue N = 1. A general N-particle state can be written as

$$N> = \int d^3r_1..d^3r_N\ \Psi(\underline{r}_1..\underline{r}_N)\hat{F}^+(\underline{r},t)..\hat{F}^+(\underline{r}_N,t)|0> \quad . \tag{25}$$

One shows that if this state is to be eigenstate also to $\hat{H}$ with eigenvalue E we must require

$$\sum_{i=1}^{N} [-\frac{\hbar^2}{2m}\Delta_i + V(\underline{r}_i)]\ \Psi(\underline{r}_1..\underline{r}_N) = E\Psi(\underline{r}_1..\underline{r}_N) \quad . \tag{26}$$

This shows that our field-theoretical approach is but an alternative formulation of standard quantum mechanics of non-interacting particles where, however, the particle number is not fixed. Furthermore, as Fermion-fields have no classical limit, their quantum-field-theoretical treatment requires a slight modification of our recipe: In the commutation relations like Eq. (8) the minus-commutator has to be substituted by the plus-commutator.

Finally, if one wants to refer to a Fermion ground state of fixed particle number, one may study the modified Hamiltonian

$$\hat{H}^N = \hat{H} - \mu\,\hat{N}, \tag{27}$$

where $\mu \geq 0$ serves here as a Lagrangian multiplier, which selects a state of given N as that of lowest energy. The parameter μ is identical with the chemical potential of the system at T = 0, $\mu = (\partial E/\partial N)|_{V,T}$, indicating that we are talking about a state in thermal equilibrium.

II.C. Occupation Number Representation

The classical field $F(\underline{r},t)$ may be expanded in any complete orthogonal set of functions $u_\ell(\underline{r})$ with

$$\langle u_\ell(\underline{r},t)\ |u_{\ell'}\ (\underline{r},t)\rangle = \delta_{\ell\ell'}. \tag{28}$$

This means for the operators

$$\hat{F}(\underline{r},t) = \sum_\ell \hat{a}_\ell\ (t)\ u_\ell\ (\underline{r}),$$
$$F^+(\underline{r},t) = \sum_\ell \hat{a}^+_\ell\ (t)\ u_\ell^*\ (\underline{r}). \tag{29}$$

When put into Eq. (8), one sees that the $\hat{a}_\ell$ are to obey the commutation rules

$$[\hat{a}^+_\ell\ (t),\ \hat{a}_{\ell'}(t)]_\pm = \delta_{\ell\ell'} \tag{30}$$

(upper sign: Fermion field).
$\hat{H}$ and $\hat{N}$ expressed in the new operators read

$$\hat{N} = \sum_\ell \hat{a}^+_\ell\ \hat{a}_\ell\ , \tag{31}$$

$$\hat{H} = \sum_{\ell\ell'} H_{\ell\ell'}\ \hat{a}^+_\ell\ \hat{a}_{\ell'}\ , \tag{32}$$

with

$$H_{\ell\ell'} = \langle u_\ell|(-\frac{\hbar^2}{2m}\ \Delta + V(\underline{r},t)\,|u_{\ell'}\rangle. \tag{33}$$

Obviously, $H_{\ell\ell'}$ becomes diagonal in ℓ, if u_ℓ happen to be the eigenfunctions of

$$\hat{h} = -\frac{\hbar^2}{2m}\ \Delta + V(\underline{r}t). \tag{34}$$

The corresponding N-particle states can now be written as

$$N\rangle = \pi\ (\hat{a}^+_i\ )^{n_i}\ |0\rangle, \tag{35}$$

where for Fermion fields $n_i = 0,1$; for Boson-fields $n_i = 0,1,2,\ldots\infty$, with

$$N = \sum_i n_i \tag{36}$$

The vacuum state $\hat{N}|0\rangle = 0$ now requires

$$\hat{a}_i \ |0\rangle = 0 \tag{37}$$

for all i.

The choice of a certain representation u_i is mainly a matter of convenience. One would certainly like to use the quantum numbers of the representation for a classification of states, at least approximately.

II.D. Self-Interaction

An obvious deficiency of the present status of our approach is that it seems to exclude - in the language of first quantization - any particle-particle-interactions. To overcome this restriction we have to supplement $V(\underline{r})$ in the field equation (14) by a potential $\tilde{V}(\underline{r},t)$ produced by the field itself (self-interaction). For a charged field we may write

$$\Delta\tilde{V} = -4\pi eF^*F \quad ,$$

with the solution

$$\tilde{V}(\underline{r},t) = \int d^3r' F^*(r,t)V(\underline{r},\underline{r}')F(\underline{r}',t) \quad , \tag{38}$$

$$V(r,r') = \frac{e}{|\underline{r}-\underline{r}'|} \quad .$$

With this $\tilde{V}(\underline{r},t)$ the Schrödinger equation has become nonlinear. (Simplified versions are investigated for soliton solutions.)

One verifies that the Lagrangian density then gets an additional

$$\mathcal{L}_w(F^*) = -\frac{1}{2}\int F^*(\underline{r}',t)F^*(\underline{r},t)V(\underline{r},\underline{r}')F(\underline{r},t)F(\underline{r}',t)d^3\underline{r} \tag{39}$$

while the conjugated momentum remains unchanged. As a consequence with Hamiltonian in quantized form also gets an additional term

$$\hat{H}_w = \frac{1}{2}\int\int\hat{F}^+(\underline{r}',t)\hat{F}^+(\underline{r},t)V(\underline{r},\underline{r}')\hat{F}(\underline{r},t)\hat{F}(\underline{r}',t)d^3r d^3r' \tag{40}$$

In occupation number formalism this leads to

$$\hat{H}_w = \frac{1}{2} \sum_{\substack{k,l\\m,n}} V_{klmn} \hat{a}_k^+ \hat{a}_\ell^+ \hat{a}_m \hat{a}_n \quad , \tag{41}$$

where

$$V_{klmn} = \int\int u_k^*(\underline{r}')u_\ell^*(\underline{r})V(\underline{r},\underline{r}')u_m(\underline{r})u_n(\underline{r}')d^3rd^3r' \tag{42}$$

II.E. Sources

Another and very important extension of the original field equation (14) is to include external sources. In the simple mechanical example of Eq. (11) such a source may be visualized as a driving force creating, e.g., forced oscillations. A simple extension of Schrödinger-field reads

$$i\hbar\dot{F}(\underline{r},t) + \frac{\hbar^2}{2m} \Delta F + V(\underline{r},t)F = g(\underline{r},t) \quad , \tag{43}$$

where F* should obey the conjugated equation. As a result we get the correction to the Lagrangian density

$$\mathcal{L}_s(F^*) = g(\underline{r},t)F^*(\underline{r},t) \quad , \tag{44}$$

and the hermitian correction to $\hat{H}$ as

$$\hat{H}_s = \int d^3r' [\hat{F}^+(\underline{r}',t)g(\underline{r},t) + h.c.] \quad . \tag{45}$$

We immediately see that now, in general, $[\hat{H},\hat{N}]=0$, as $\hat{H}_s$ acts as a particle source. If g is interpreted as

$$g(\underline{r},t) = \gamma(\underline{r})\ \phi(\underline{r},t) \quad , \tag{46}$$

where γ denotes the coupling to another field ϕ, we end up with

$$\hat{H}_s = \int d^3r\ \gamma(\underline{r})\hat{F}^+(\underline{r}',t)\ \phi(\underline{r}',t) + h.c. \tag{47}$$

$\hat{H}_s$ describes the destruction of one type of particle and the simultaneous creation of a particle of different type. We may also consider the source term to be proportional to the Schrödinger-field considered

$$g(\underline{r},t) = \lambda(\underline{r})F(\underline{r},t) \quad , \tag{48}$$

so that

$$\hat{H}_s = \int d^3r \hat{F}^+ \lambda(\underline{r}) \hat{F} \quad . \tag{49}$$

Many other possibilities could be discussed. Again, physically relevant are only a few types, from which we will encounter some examples.

II.F. Effective Vacuum

For many applications especially for electronic excitations, the true vacuum state of a Fermion system (no particle present) does not serve as a convenient starting point for investigations.

Here, the particle-hole-concept allows to use a reference state with a fixed number of particles, N, as an effective vacuum, e.g.

$$|N> = \prod_i \hat{a}_i^+ |0> \quad ; \quad E_i \leq \mu \tag{50}$$

for a non-interacting Fermi-field. The unitary transformation,

$$\begin{aligned} \hat{c}_i^+ &= \hat{a}_i^+ \qquad E_i > \mu \\ \hat{b}_i^+ &= a_i \qquad E_i < \mu \end{aligned} \tag{51}$$

which conserves the commutation rules, then leads to

$$\hat{H} = E_o + \sum_{\ell,\ell'} \hat{c}_\ell^+ \hat{c}_{\ell'} H_{\ell\ell'} - \sum_{\ell,\ell'} \hat{b}_\ell^+ \hat{b}_\ell H_{\ell\ell'} +$$

$$+ \sum_{\ell,\ell'} (\hat{c}_\ell^+ \hat{b}_{\ell'}^+ + \hat{b}_{\ell'} \hat{c}_\ell) H_{\ell\ell'} \tag{52}$$

In regard to this equation: $E(\ell) > \mu$ and $E(\ell') > \mu$ for the first summation, $E(\ell) < \mu$ and $E(\ell') < \mu$ for the second summation, and $E(\ell) > \mu$ and $E(\ell') < \mu$ for the third summation. E_o is such that

$$E_o^N = <N|\hat{H}|N> = \sum_\ell H_{\ell\ell'} \tag{53}$$

where $E(\ell) < \mu$. We see that

$$c_i|N> = \hat{b}_j|N> = 0 \quad . \tag{54}$$

Defining

$$\hat{N}_e = \sum_{\ell} \hat{c}^+_{\ell} \hat{c}_{\ell}$$

where $E(\ell) > \mu$ and

$$\hat{N}_h = \sum_{\ell} \hat{b}^+_{\ell} \hat{b}_{\ell} \quad , \tag{55}$$

where $E(\ell) < \mu$,

we conclude that in the reference state $|N\rangle$ no particles (electrons) or holes are present:

$$\hat{N}_e|N\rangle = \hat{N}_h|N\rangle = 0 \quad , \tag{56}$$

which justifies to call $|N\rangle$ the effective vacuum state $|\tilde{0}\rangle$ for the particle-hole system. From

$$\hat{N} - \sum_{\ell} \hat{a}^+_{\ell} \hat{a}_{\ell} = \hat{N}_e - \hat{N}_h = 0 \quad . \tag{57}$$

We infer that the number of particles and holes should be equal, but is in general not conserved (though the number of real particles N is fixed). On the other hand, the particle-hole picture is a useful concept only as long as $N_e = N_h << N$, i.e., for states not too far from the reference state, so that the particles and holes can, at least in a first approximation, be treated independently.

The particle-hole scheme can also be allied in the case of interacting Fermions. Then, however, the reference state is known only approximately, and so are the particles and holes based on it. As a consequence of the interaction of the real particles, the particle and hole states represent collective properties of the whole system and usually gain a finite lifetime.

Finally, if the ground state correlations are too large, the distinction between particle- and hole-states may become artificial. In such a case, a two-quasi-particle approximation based on a quasi-particle vacuum can be more appropriate. This is the case for the superconductor; a further example will be discussed in Section III.F.

We are now ready to apply this field theoretical scheme to the solid state, in particular to the elementary excitations of the electronic subsystem. This requires to specify

a, the structure of the solid by the one-particle (effective) potential $V(\underline{r})$, b, the electron correlations by the self-interaction $V(\underline{r},\underline{r}')$, and

c, the reference state by the definition of the effective vacuum $|\hat{0}\rangle$.

III. MODELS

III.A. Empty Band Model

In a first simplified model we may neglect the self-interaction of the electronic Schrödinger field and focus on the consequences of the discrete translational invariance of $V(\underline{r})$ in Eq. (14):

$$V(\underline{r}) = V(\underline{r} + \ell \cdot \underline{R}); \quad \ell = 0, \pm 1 \ldots \tag{58}$$

As is well-known from elementary quantum theory, momentum will no longer be conserved (i.e. will no longer be a good quantum number) if the particle is in a spatially varying potential. However, if this variation is strictly periodic, quasi-momentum is conserved, which is defined only up to multiples of $h\underline{g}$, where $\underline{g}$ is a reciprocal lattice vector. As a consequence, single particle states can now be classified by their quasi-momentum, which, to avoid ambiguities, is deliberately restricted to the first Brillouin-zone. The eigenfunctions are modulated plane waves (Bloch-functions)

$$\Psi_{\underline{k}}(\underline{r}) = \Psi_{\underline{k}}(\rho,R) = \chi_{\underline{k}}(\rho)\, e^{i\underline{k}\cdot\underline{r}} \tag{59}$$

where $\underline{r} = \underline{R} + \underline{k} \cdot \underline{\rho}$ is the position within the Wigner-Seitz-cell $\underline{R}$, and $\underline{k}$ is the wave vector. If Eq. (59) is to solve the 1-particle-time-independent Schrödinger equation, $\chi_{\underline{k}}(\rho)$ must satisfy the condition

$$\left[-\frac{\hbar^2}{2m}(\nabla + i\underline{k})^2 + V(\rho)\right]\chi_{\underline{k}}(\rho) = E(\underline{k})\chi_{\underline{k}}(\underline{\rho}), \tag{60}$$

where the boundary condition that $\chi_{\underline{k}}(\rho)$ should match the solution in the adjacent Wigner-Seitz-cell. For a given $\underline{k}$ these constraints lead to a discrete spectrum with band index λ. The single-particle states can thus be classified as

$$\psi_{\lambda\underline{k}}(\underline{r}) = \chi_{\lambda\underline{k}}(\underline{r})\, e^{i\underline{k}\cdot\underline{r}} \tag{61}$$

$$\langle \chi_{\lambda\underline{k}}(\underline{r}) \mid \chi_{\lambda'k'}(\underline{r})\rangle = \delta_{\lambda\lambda'}\,\delta_{kk'}, \tag{62}$$

with the corresponding dispersion relations $\underline{E}_\lambda(k)$. Though being a clearly defined property of the system (i.e. exclusively a result of the underlying structure) such states are of only very restricted value, e.g. for dilute systems, as their reference state is the true vacuum. However, for qualitative purposes one can construct N-particle reference states to get a first qualitative idea of the typical excitation spectra of metals and insulators, respectively. This is discussed next.

III.B. Fermi Gas Model

For a non-interacting Fermi-system, the exact N-particle ground state is

$$|N\rangle = \prod_{\underline{k},\lambda} \hat{a}^+_{\underline{k}\lambda} |0\rangle \equiv |\tilde{0}\rangle \tag{63}$$

where $\hat{F}(\underline{r},t)$ is expanded in terms of the Bloch-functions according to Eq. (35) and $E_\lambda(\underline{k}) < \mu$. If the equation

$$E_\lambda(\underline{k}) - \mu = 0 \tag{64}$$

has a solution for a specific λ_M, it is usually sufficient to consider this single band, and the properties of the ground state are best characterized by the Fermi-surface, Eq. (64). In this single band model, the corresponding Hamiltonian reads

$$\hat{H} = \sum_{k<k_F} [\, E(\underline{k})\, \hat{a}^+_{\underline{k}}\, \hat{a}_{\underline{k}} - \mu\hat{N}], \tag{65}$$

where $E(k_F) = \mu.$

Taking the ground state as an effective vacuum state, we get in the particle-hole picture (compare to Section II.F.)

$$\hat{H} = E_o^N + \sum_{k<k_F} [\mu - E(\underline{k})]\, \hat{b}^+_{\underline{k}}\, \hat{b}_{\underline{k}} + $$

$$+ \sum_{k>k_F} [E(\underline{k}) - \mu]\, \hat{c}^+_{\underline{k}}\, \hat{c}_{\underline{k}} \quad , \tag{66}$$

with

$$E_o^N = \sum_{k<k_F} E(\underline{k}). \tag{67}$$

The corresponding particle-hole-spectrum may look already quite complicated. However, if $E(\underline{k})$ has the simple form

$$E(\underline{k}) = \hbar^2 k^2/2m^*, \tag{68}$$

where m* is some isotropic effect mass, the spectrum has the shape shown in Fig. 1a. As we saw in Section II.F. particles and holes can be created only in pairs. The simpliest excitation of total momentum $\underline{K}$ would therefore consist of an electron-hole-pair,

$$|\underline{K}\rangle = \hat{c}^+_{\underline{k}} \hat{b}^+_{\underline{K}-\underline{k}} |\tilde{0}\rangle, \tag{69}$$

with energy

$$\begin{aligned} E(\underline{k},K) &= E_e(k) + E_h(\underline{K} - \underline{k}) \\ &= \frac{\hbar^2}{2m} (K^2 + 2\underline{K}\cdot\underline{k});\ 0\leq\underline{k}\leq k_F. \end{aligned} \tag{70}$$

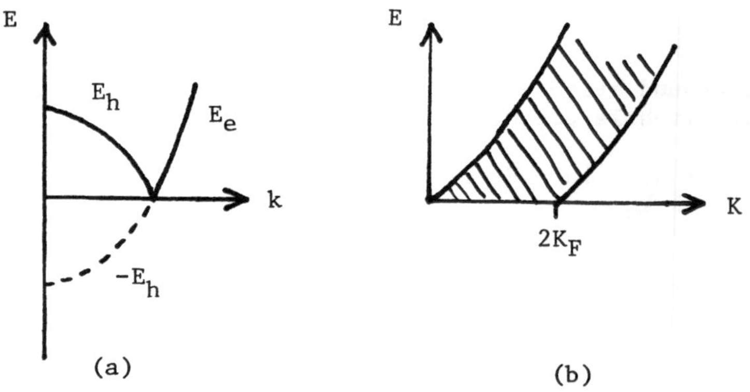

Fig. 1. Quasi-particle spectrum (a) and pair spectrum (b) of a non-interacting 1-band model; reference state is the filled Fermi sphere E_F.

This spectrum as sketched in Fig. 1b may be taken as zero order approximation for metals.

We encounter a completely different situation, if Eq. (64) has no solution. This is the case when μ lies within a band gap E_g, and we have to include at least the 2 bands closest to μ (two-band model). For small k the corresponding dispersion relations can be expanded near the band extrema, i.e. for simple isotropic bands (λ= c for conduction band, v for valence band)

$$E_c(k) = E_g + \frac{\hbar^2 k^2}{2m_c} + \ldots,$$

$$E_\nu(k) = -\frac{\hbar^2 k^2}{2m_\nu} - \ldots . \tag{71}$$

Using the state with filled valence band as our reference state, we may apply the particle-hole scheme and again get a Hamiltonian of the type shown in Eq. (66),

$$\hat{H} = E_0^N + \sum_k [\mu - E_v(\underline{k})]\hat{b}^+_{\underline{k}}\hat{b}_{\underline{k}} + \sum_k [E_c(\underline{k}) - \mu]\hat{c}^+_{\underline{k}}\hat{c}_{\underline{k}} \tag{72}$$

with $\mu = E_g/2$.

The simpliest excitation of total momentum $\underline{K}$ would now consist of an electron-hole pair of the type

$$|\underline{K}\rangle = \hat{c}^+_{\underline{k}}\,\hat{b}_{\underline{K}-\underline{k}}\,|\tilde{0}\rangle \quad , \tag{73}$$

with energy

$$E(\underline{k},\underline{K}) = E_e(\underline{k}) + E_h(\underline{K}-\underline{k}). \tag{74}$$

For given $\underline{K}$ the minimal energy is

$$E_{min}(\underline{K}) = E_g + \frac{\hbar^2 K^2}{2(m_\nu + m_c)}$$

with no upper limit (under assumption of Eq. (71), i.e., the effective mass approximation). For later reference we may consider the T = 0 electron-hole-system with $k_F^e = k_F^h$ = const. as our reference state. The resulting quasi-particle spectrum doubles the features of Fig. 1a and is shown in Fig. 2b (note that all energies are counted positive). This is but a somewhat unusual way of representing various kinds of pair spectra; intra-band transitions (particle-hole excitations with same band index), recombination (hole-hole-excitation with different index) and absorption (particle-particle excitation). Such a situation may be encountered under stationary high excitation conditions.

III.C. Tight Binding Model (TBM)

Let the one-particle potential $V(\underline{r})$ of Eq. (14) be given by

$$V(\underline{r}) = \sum_{\ell} V_{\ell}(\underline{r}-\underline{R}_{\ell}) \quad , \tag{75}$$

where V_1 is a local contribution at position R_1, e.g., by an atom. Then we may define local eigenfunctions with local quantum number ν by

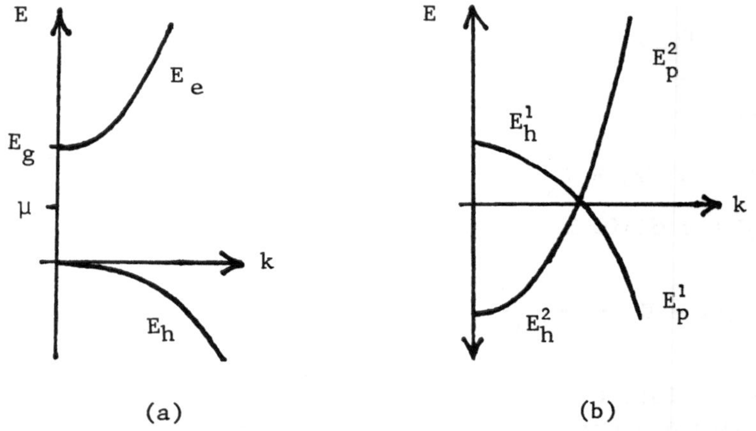

Fig. 2. Quasi-particle-spectrum of a non-interacting 2-band model (a) reference state is the filled lower band; (b) reference state is a T = 0 electron-hole system with $k_F^e = k_F^h$.

$$\hat{H}_\ell \, \mu_{\ell\nu}(\underline{r}) = E_{\ell\nu}(\underline{r}) \, u_{\ell\nu}(\underline{r}) \tag{76}$$

with

$$\hat{H}_\ell = - \frac{h^2}{2m} \Delta + V_\ell \tag{77}$$

Assuming

$$\langle \mu_{\ell'\nu'} | \mu_{\ell\nu} \rangle \simeq \delta_{\nu\nu'} \, \delta_{\ell\ell'} \tag{78}$$

we may expand the Schrödinger-field-operator as

$$\hat{F}(\underline{r},t) = \sum_{\ell,\nu} \hat{a}_{\ell\nu}(t) \, \mu_{\ell\nu}(\underline{r}-\underline{R}_\ell) \tag{79}$$

to get the Hamiltonian

$$\hat{H} = \sum_{\substack{\ell_1\ell_2 \\ \nu_1\nu_2}} H^{\nu_1\nu_2}_{\ell_1\ell_2} \, \hat{a}^+_{\ell_1\nu_1} \, \hat{a}_{\ell_2\nu_2} \tag{80}$$

with

$$H^{\nu_1\nu_2}_{\ell_1\ell_2} = E^{\nu_1}_{\ell_1} \, \delta_{\ell_1\ell_2} \, \delta_{\nu_1\nu_2}$$

$$+ \sum_{\ell \neq \ell_2} \langle \mu_{\ell_1\nu_1} | V_\ell(\underline{r}-\underline{R}_\ell) | \mu_{\ell_2\nu_2} \rangle \; . \tag{81}$$

Equation (80) presents an example of the often-used tight-binding Hamiltonian, where particle-interactions are neglected [7]. In the simplest case one restricts the Hamiltonian H in Eq. (80) to $\nu = 1$ (single orbit) and finds

$$\hat{H} = \sum_\ell \varepsilon^o_\ell \, \hat{a}^+_\ell \, \hat{a}_\ell + \sum_{\substack{\ell,\ell' \\ \ell\neq\ell'}} V_{\ell\ell'} \, \hat{a}^+_\ell \, \hat{a}_{\ell'} \quad , \tag{82}$$

with

$$\varepsilon^o_\ell = E_\ell + \sum_{i\neq\ell} \langle \mu_\ell | V_i(\underline{r}-\underline{R}_i) | \mu_\ell \rangle$$

$$V_{\ell\ell'} = \sum_{i\neq\ell'} \langle\mu_\ell|V_i(\underline{r}-\underline{R}_i)|\,\mu_{\ell'}\rangle \tag{83}$$

and the eigenvalues can be obtained from a diagonalization procedure

This Hamiltonian becomes interesting also if the system under consideration has no long range order so that the positions $\underline{R}_l$ are no longer equivalent. Then the varying local environments produce locally different energies E^0 and in general, fluctuating transfer integrals $V_{\ell\ell'}$.

Such a Hamiltonian of a disordered system has exciting new properties, a study of which, however, is unexpectedly difficult in spite of the highly approximate Hamiltonian. In a first step, one may therefore avoid to follow up the microscopic evaluation of ε_1^o and $V_{\ell\ell'}$ and, instead, simply use ad hoc distributions for them. This is the line followed in the Anderson model [6] and subsequent modifications. The important result is that states become delocalized provided the disorder-"strength" exceeds a critical value [7].

Instead of considering local excited states, $\nu > 1$, we may study singly and doubly occupied local ground states. However, the doubly occupied state ("upper band") only exists if the "lower band" is occupied, i.e. by virtue of particle correlations, which so far have not been discussed. We come back to this problem in Section III.E.

III.D. Models with Particle Interactions

Including the influence of structure and correlations generally poses an untractable problem. The Jellium model[5], to start with, considers Coulomb correlations, which in plane wave representation lead to (see Eq. (42)).

$$V_{\underline{k}_1\underline{k}_2\underline{k}_3\underline{k}_4} = \sum_{\underline{q}} V(q)\; \delta_{\underline{q},\underline{k}_1-\underline{k}_4}\; \delta_{\underline{q},\underline{k}_3-\underline{k}_2}, \tag{84}$$

where $$V(q) = \frac{4\Pi e^2}{\Omega\, q^2},$$

while the structure is reduced to a positive background charge to compensate the negative charge of the electron system (electron-liquid). In a first approximation one may still use the product-wave function, Eq. (63), as the reference state, which then leads to the HF-ground state; improved approximations are also at hand, which, however, all reproduce qualitatively the result of Fig. 1.

Modified Coulomb-correlations,

$$V(q) = \frac{4\Pi e^2}{\Omega(q^2+\lambda^2)}, \tag{85}$$

(λ = screening vector) together with an (attractive) Fröhlich-interaction will lead to a ground state of different nature, the BCS- or superconducting state. Then the single particle spectrum changes qualitatively as a gap opens, which makes the distinction between particle- and hole branch obsolete.

Finally, short-range repulsive interactions,

$$V_{ijmn} = U\,\delta_{ij}\,\delta_{k\ell}\,\delta_{ik}; \quad U>0, \tag{86}$$

where the indices refer here to positions in a local representation, are considered in the Hubbard model [8] which thus combines structure and correlation, and is able to describe narrow bands.

For an insulator one may also use a HF-scheme to calculate the particle-hole spectra with reference to the filled valance band, where, however, a plane wave representation is no longer adequate. Though the results are qualitatively satisfactroy, one may prefer for many purposes to start from an "empirical" band-parameters.

Then the particle-hole scheme leads to an Hamiltonian

$$\hat{H} = E_o + \hat{H}_e + \hat{H}_h + \hat{H}_{ee} + \hat{H}_{hh} + \hat{H}_{eh}, \tag{87}$$

where $\hat{H}_{eh}$ comprises the direct Coulomb-interaction between electrons and holes as well as the electron-hole-exchange, a true many-body effect. The exciton

$$|\phi\ (\underline{K})> = \sum_{\underline{k}} Z(\underline{k},\ \underline{K}-\underline{k})\ \hat{c}^+_{\underline{k}}\ \hat{b}^+_{\underline{K}-\underline{k}}\ |\mathcal{O}> \tag{88}$$

is then found to be an approximate eigenstate of $\hat{H}$. If Z is localized in k-space near $k \simeq 0$ (which means that the Fourier-transform in real space is delocalized), we get the well-known dispersion-relation for Wannier excitons:

$$E(\underline{K},n) = E_g + \frac{\hbar^2 K^2}{2(m_\nu+m_c)} - \frac{\mu e^4}{2h^2\varepsilon_\infty n^2}, \tag{89}$$

where the exchange splitting has been neglected. If Z is delocalized we approach the limit of Frenkel excitations [1].

Unfortunately, the situation for non-metallic systems without long-range order is far less understood, even with respect to the mere existence of excitons. Mostly two lines of approach have been adopted: One is to consider Wannier-excitons in a random field taken as a perturbation [10]. The other line is to start from bound excitons, a local description, which should be valid for dilute disordered systems [11]. Up to now, however, there is no systematic approach to cover part of the range in between the two limits. We want to propose one model in the next section.

III.E. Random Cell Model (RCM) [12]

While the classification of crystal order can unambiguously be carried out by reference to symmetry, this is not the case for disordered systems. Besides their lack of long-range order, there is, to our knowledge, no general positive definition. To avoid ambiguities one may therefore refer to structure models, which can clearly be defined in mathematical terms [7]. The structure model with the "highest degree of disorder" is that for which the positions of atoms etc. are completely uncorrelated. The corresponding distribution is the Poisson-distribution

$$P(\nu, x) = x^{\nu} e^{-x}/\nu! \quad , \tag{90}$$

which gives the probability to find ν atoms in an arbitrary volume V_c, if $n = x/V_c$ is the average density. Starting from this extreme case, other models could be introduced by taking in account specific correlations (e.g. hard core repulsion etc.). Presently we will restrict ourselves to the Poisson-model.

We again start from the Schrödinger-field $F(\underline{r},t)$ with a one-particle potential as given in Eq. (75)

$$V(\underline{r}) = \sum_{\ell} V_{\ell}(\underline{r}-\underline{R}_{\ell}) \tag{91}$$

and the self-interaction $V(\underline{r},\underline{r}')$.

We suppose that artificially we can decompose the space into a regular net of cells of volume V_c which, due to the random nature of the system, now contain fluctuating arrangements of potentials and particles. We further suppose that the ground state ϕ^o of the system can be approximated by a simple product of cell wave functions ϕ^o (1) which describe the respective neutral cell 1 in its local equilibrium ground state. $\phi^o(1)$, in general, is a few-body function with "(1)" denoting all electron coordinates in 1:

$$\phi^o = \prod_{\ell} \phi^o\ (\ell) = |\gamma> \tag{92}$$

Let $\hat{H}_1$ denote the cell Hamiltonian,

$$\hat{H}_{\ell} = \sum_{n=1}^{n_{\ell}} [-\frac{\hbar^2}{2m} \nabla_n^2 + \sum_{j=1}^{n_{\ell}} V(\underline{r}_n-\underline{R}_j) + \frac{1}{2}\sum_{n,n'}^{n_{\ell}} V(\underline{r}_n,\underline{r}_{n'}) , \tag{93}$$

which in short-hand notation reads,

$$\hat{H}_{\ell} = \hat{H}^o_{\ell}\ (\ell) + V_{\ell}(\ell) + V\ (\ell,\ell). \tag{94}$$

We suppose to have solved the eigenequations

$$\hat{H}_{\ell}\ \phi^o(\ell) = E^o_{\ell}\ \phi^o(\ell) \tag{95}$$

By first-order perturbation theory we thus get

$$E^o = <\phi^o\ |\hat{H}\ |\phi^o> = \sum_j E^o_j + \\ + \frac{1}{2}\sum_{i,j}<\phi^o(i)\phi^o(j)\ |W(^{oo}_{ij})\ |\phi^o(j)\phi^o(i)> \tag{96}$$

with

$$W(^{oo}_{ij}) = V_i(j) + V_j(i) + V(i,j). \tag{97}$$

Particle-hole-states can then be described by creation of an additional electron in cell k (one hole in cell k'):

$$\hat{c}^+_k\ |\gamma> = \phi^o(1)\ldots\phi^-(k)\ldots >, \\ \hat{b}^+_k\ |\gamma> = \phi^o(1)\ldots\phi^+(k')\ldots>, \tag{98}$$

where $\phi^{\pm}$ (1) is the ground state eigenfunction of cell 1 with one additional electron and one electron missing, respectively, and we assume

$$<\phi^{\pm}(k)\ |\ \phi^{\pm}(\ell)> = \delta_{k\ell}. \tag{99}$$

The corresponding Hamiltonian can be written as

$$\hat{H}_{eff} = \sum_{k,\ell} (H^e_{k\ell}\ \hat{c}^+_k\ \hat{c}_{\ell} + H^h_{k\ell}\ \hat{b}^+_k\ \hat{b}_{\ell}) + E_o, \tag{100}$$

where

$$H^e_{k\ell} \simeq (E^-_\ell - E^o_\ell)\,\delta_{k\ell} +$$

$$+ \langle\phi^-(k)\phi^o(\ell)\mid W(^{o-}_{k\ell})\mid\phi^o(k)\,\phi^-(\ell)\rangle \quad , \tag{101}$$

$$H^h_{k\ell} \simeq (E^+{}_\ell - E^o_\ell)\,\delta_{k\ell} +$$

$$+ \langle\phi^+(k)\phi^o(\ell)\mid W(^{o+}_{k\ell})\mid\phi^o(k)\,\phi^+(\ell)\rangle \quad . \tag{102}$$

$W(^{o-}_{ij})$ is of the type given in Eq. (97) but with one additional electron in cell j, $W(^{o+}_{ij})$ with one electron missing in cell j.

Let us consider the special ground state with just one electron in each cell and $V_1(\underline{r}-R_1)$ being a Coulomb-potential centered at R_1. Then $\phi^o(1) = \phi_{1s}(\underline{r}-\underline{R})$, where ϕ_{1s} is the Hydrogen-ground state wave-function, and we immediately obtain

$$H^h_{kk} \simeq |E_D| = 1 \text{ Ryd.}, \tag{103}$$

$$H^e_{kk} \simeq |E_D| - |E^-_D| \simeq -0.05 \text{ Ryd.}, \tag{104}$$

where $-E_D$ is the effective Rydberg-energy (Ryd.). This hole-band is the simpliest version of a tight binding model, derived from a microscopic structure model. Its off-diagonal matrix-elements are (neglecting 3-center integrals):

$$H^h_{k\ell} = -2|E_D|\left(1 + \frac{|\underline{R}_k-\underline{R}_\ell|}{a_B}\right)\exp\{-|\underline{R}_k-\underline{R}_\ell|/a_B\} \quad , \tag{105}$$

i.e. depend on the relative positions of the ion-centers in cell k and ℓ.

In general, however, one would not expect to find in each cell just one donor: The actual distribution is given by Eq. (90), but, unfortunately, the reference volume V_c is not defined yet. But suppose that any cluster in a spherical cell of Radius R_c, in which there are <u>no</u> centers in the outer skin of fixed thickness R_d, would approximately <u>not</u> interact with any other cluster (cell state). (This condition gives a lower bound for R_d; we use R_d = 3 Bohr radii.) Then one way to choose V_c is to maximize the statistical weight of these quasi-isolated cells, a procedure, which may be considered to be a random analogy to the usual diagonalization. It turns out that the total statistical weight for quasi-isolated cells is

$$\phi_\infty(x_d,u) = \frac{\exp[x_d u \cdot f(u)]-1}{\exp[x_d u]-1} \quad , \tag{106}$$

where

$$x_d = n\,V_d,\quad u = V_c/V_d,$$
$$f(u) = (1 - u^{-1/3})^3 \,, \qquad (107)$$

and $V_d = \frac{4}{3}R_d^3$ is taken as a convenient unit of volume. $\partial\phi_\infty/\partial u=0$ gives the optimal choice $u_m(x_d)$, i.e. gives V_c for given n and R_d. At this V_c the corresponding cell-particle-hole-states show a sharp maximum of their respective average life-times, $\tau \simeq |\bar{H}_{k\ell}|^{-2}$, provided $x_d \lesssim 0.1$. Figure 3a shows this function $u_m(x_d)$ and Fig. 3b the fluctuation in the number of donors per cell, $\delta\nu = (x_d u_m)^{-1/2}$. We conclude that especially in the dilute limit the system is expected to be strongly random.

For practical reasons only clusters of finite number of centers ν can be considered. Therefore one may ask about the convergence of cluster expansions. The answer can be found in Fig. 3c, where the statistical weights of expansions including clusters of up to ν_{max} donors are shown. We see that a 3-center cluster expansion is sufficient for densities up to $x_d \simeq 10^{-1}$ (the Mott-density corresponds to $x_d \simeq 2$).

We thus conclude that this scheme should be applicable in a rather broad range of densities, where microscopic structure models including correlation have so far not been available. The model may thus be used as a microscopic foundation of commonly used tight-binding models, where now all matrix elements can be (approximately) calculated from an underlying structure model.

Similarly one can obtain an effective pair Hamiltonian

$$\hat{H} = \sum_{k,\ell} H^e_{k\ell}\,\hat{c}_k\hat{c}_\ell + \sum_{k,\ell} H_k\,\hat{b}_k\,\hat{b}_\ell + \sum_{\substack{k\ell\\ mn}} \chi_{km\ell n}\,\hat{c}_k\,\hat{b}_m\,\hat{c}\,\hat{b}_n \,, \qquad (108)$$

where e.g.

$$\chi_{k\ell k\ell} \simeq \langle\phi^+(k)\ \phi^-(\ell)|W(^{+-}_{k\ell})|\ \phi^+(k)\ \phi^-(\ell)\rangle - \langle\phi^0(k)\ \phi^0(\ell)|W(^{oo}_{k\ell})|\ \phi^0(k)\ \phi^0(\ell)\rangle \,. \qquad (109)$$

The pair state

$$|P\rangle = \hat{c}^+_\ell,\hat{b}^+_\ell|\ \mathcal{O}\rangle \qquad (110)$$

would therefore have the approximate energy

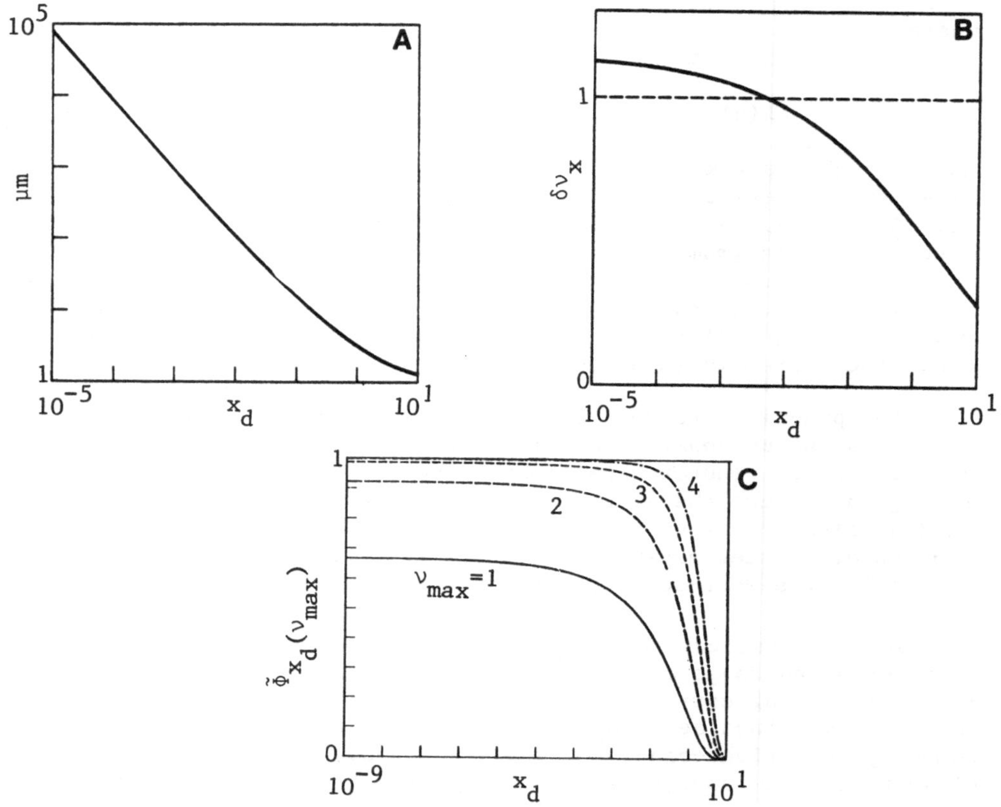

Fig. 3. (a) Optimized cell size u_m and (b) the relative fluctuation in number of centers per cell $\delta\nu$. (c) Total statistical weights for cluster sizes up to ν_{max}.

$$E^p_{\ell\ell} = H^e_{\ell'\ell'} + H^h_{\ell\ell} + \chi_{\ell'\ell\ell'\ell} \quad , \tag{111}$$

Contrary to the exciton-state in an ordered system all degeneracies are lifted, i.e. the pair excitation energy depends on ℓ' and the position vector to cell ℓ containing the hole. We expect that this pair state is localized. However, as the particle-hole states become delocalized, this should be mirrored also in the behaviour of excitons. So far there seems to be no work published on the localization and delocalization of excitonic states.

III.F. Driven Systems

Non-equilibrium states are investigated in many branches of physics [13]. One important example are states produced in matter by interaction with strong light sources. One comparatively simple class of such states are stationary states, which are controlled by a periodic driving force. In any case, the model of a driven system does not consider any feed-back by the system, an assumption, which clearly limits its application.

As a starting point we consider the Schrödinger equation

$$[i\hbar \frac{\partial}{\partial t} - \frac{1}{m} (\frac{\hbar}{i} \nabla - \frac{e}{c} \underline{A})^2 + V(\underline{r})]F(\underline{r},t) = 0 \tag{112}$$

of a charged particle in an external electromagnetic field specified by a vector-potential

$$\underline{A}(\underline{r},t) = \underline{A}_o \cos(wt - \underline{q} \cdot \underline{r}) \quad , \tag{113}$$

with

$$\text{div } \underline{A} = 0.$$

Neglecting the A^2-term, this equation can be written as

$$[i\hbar \frac{\partial}{\partial t} + \frac{\hbar^2}{2m} \nabla^2 + V(\underline{r})]F(\underline{r},t) = \frac{i\hbar e}{mc} \underline{A}\nabla F(\underline{r},t). \tag{114}$$

Now, this equation is taken as a classical field equation where the right-hand side can be interpreted as a source-term for the Schrödinger-field considered before (compare Sect. II.E.) As already discussed, this source-term leads to the additional term in the Hamiltonian

$$\hat{H}_s = \int d^3r \, \hat{F} + [\frac{i\hbar e}{mc} \underline{A}\cdot\underline{\nabla}] \, \hat{F}. \tag{115}$$

Again we expand $\hat{F}$ in terms of Block-functions,

$$\hat{F}(\underline{r},t) = \sum_{\underline{k},\lambda} \hat{a}_{\underline{k},\lambda} (t) \, \chi_{\underline{k}_\lambda} (\underline{r}) e^{i\underline{k}\cdot\underline{r}} \tag{116}$$

For a 2-band insulator-model $\lambda = v,c$, we immediately go over to the particle-hole-picture with respect to the filled valence band,

$$\hat{F}(\underline{r},t) = \sum_{\underline{k}} [\hat{b}^+_{-k}(t) \, \chi_{\underline{k}\nu} \, e^{i\underline{k}\cdot\underline{r}} + \hat{c}_k(t) \, \chi_{\underline{k}c} \, e^{i\underline{k}\cdot\underline{r}}] \tag{117}$$

so that, neglecting momentum $\underline{q}$,

$$\hat{H}(t) = \hat{H}_o + \hat{H}_s(t),$$

$$\hat{H}_o = E_o + \sum_{\underline{k}} E_c(k)\, \hat{c}^+_k \hat{c}_k - \sum_{\underline{k}} E_\nu(k)\, \hat{b}_k + \hat{b}_k ,$$

$$E_o = \sum_k E_\nu(k) , \tag{118}$$

$$\hat{H}_s = \sum_k \lambda(\underline{k})\, [\hat{c}^+_k \hat{b}^+_{-k}\, e^{i\omega t} + \text{h.c.}],$$

with the dipole-matrix element

$$\lambda(\underline{k}) = \frac{1}{2m} \underline{A}_o \int \chi_{\underline{k}c}\, \underline{\hat{P}}\, \chi_{\underline{k}\nu}\, d^3_r \simeq \text{const.} \tag{119}$$

Intra-band processes have been neglected. $\hat{H}_s(t)$ describes processes where one electron-hole pair is created or destroyed.

For resonant interactions $\hat{H}_s$ cannot be treated by perturbation-theory. In a different line of approach, then, the coupling to the light field is no longer treated as a source for the otherwise free Schrödinger field. As the electromagnetic field does appear here only as a driving force without considering its respective internal degrees of freedom, it is tempting to view $\hat{H}$ as describing an electron system in a potential periodic in space <u>and</u> time.

For simplicity let us consider first a Hamiltonian periodic only in time with

$$\hat{H}(t + \tau) = \hat{H}(t), \tag{120}$$

$$\omega = \frac{2\pi}{\tau} . \tag{121}$$

Usually the periodic part derives from an external perturbation which must have been switched on inducing, in general, a rather complicated transient behaviour. We focus on the asymptotic behaviour, which, under suitable conditions, will be the regime of stationary states.

These stationary states [14] are defined as solutions of the time-dependent Schrödinger-equation of the Bloch-like form

$$\psi(\underline{r},t) = u(\underline{r},t)e^{-iEt/\hbar}, \tag{122}$$

with

$$u(\underline{r},\ t+\tau) = u(\underline{r},t) \tag{123}$$

and

$$[\hat{H}(t) - i\hbar \frac{\partial}{\partial t}]\ u_E(\underline{r},t) = u_E(\underline{r}t). \tag{124}$$

The real number E is called quasi-energy. It follows that

$$\psi(\underline{r},\ t+\tau)\ e^{-iEt/\hbar}\ \psi(\underline{r},t), \tag{125}$$

i.e. after a period of time the wave function has changed only by a phase factor. If $u_E(\underline{r},t)$ is solution of Eq. (124), so is $e^{i\omega t}$ $u_E(\underline{r},t)$, however, with quasi-energy $E + \hbar\omega$:

$$e^{i\omega t}\ u_E(\underline{r},t) = u_{E+\hbar\omega}\ (\underline{r},t) \quad , \tag{126}$$

while the total wave function remains unchanged. So, as expected by analogy to the quasi-momentum (see Section III.A), the quasi-energy is only defined modulo $h\omega$:

$$E' = E \pm n\hbar\omega; \quad n = 0,1,2,\ldots \tag{127}$$

and we may restrict ourselves to a reduced energy range

$$\varepsilon \leq E \leq \varepsilon + \hbar\omega \tag{128}$$

with a convenient choice of the energy zone ε. One easily shows that states with $E - E' \neq n\hbar\omega$ are orthogonal, and a general stationary state (at least in a finite dimensional Hilbert-space) may be expanded as

$$\psi(t) = \sum_n c_n\ \psi_{E_n}\ (t), \tag{129}$$

where the coefficients are independent of time. Expectation values of time-independent operators $\hat{L}$ have period τ. Observations on a time scale $T \gg \tau$, however, will be characterized by the averaged values

$$\ll \hat{L} \gg\ = \frac{1}{\tau} \int_0^\tau <\psi_E\ (t')\ |\ \hat{L}\ |\ \psi_E(t')>dt'. \tag{130}$$

There are various methods to calculate the quasi-energy spectrum. One way is to Fourier-analyze

$$\hat{H}(t) = \sum_p e^{-ip\omega t} H_p,$$
$$u(t) = \sum_s e^{-is\omega t} u_s, \tag{131}$$

put these expressions into the Schrödinger-equation, Eq. (124),

$$\sum_p (H_p u_{s-p} - s\omega u_s) = E u_s, \tag{132}$$

and try to solve this coupled system of equations approximately. This method can easily be generalized to include the effect of spatial periodicity [15[. The eigenvalues E then describe the "empty" hyperband-structure in the sense discussed in Section III.A. In this scheme, however, the inclusion of particle-interactions and the introduction of a reference state for particle-hole excitations faces severe problems and has not been attempted as yet.

$$U^+ [\hat{H}(t) - i\hbar \frac{\partial}{\partial t}] \hat{U} = \hat{H}_u - i\hbar \frac{\partial}{\partial t}, \tag{133}$$

with

$$\frac{\partial \hat{H}_u}{\partial t} = 0.$$

The transformed wave function

$$\phi = U^+ u \tag{134}$$

then obeys the equation

$$[\hat{H}_u - i\hbar \frac{\partial}{\partial t}] \phi = E \phi . \tag{135}$$

As by definition $\hat{H}_u$ is time-independent, one can use the Ansatz

$$\frac{\partial}{\partial t} \phi = 0 \tag{136}$$

and the quasi-energy spectrum E can be found from the effective Schrödinger-equatior

$$\hat{H}_u \phi = E \phi .$$

For our present problem, Eq. (118), such a unitary transformation can be found:

$$\hat{U} = \exp\left[-\frac{i\hbar\omega}{2}t \sum_{\underline{k}'} (\hat{c}^{+}_{k'}, \hat{c}_{k'} + \hat{b}^{+}_{k'}, \hat{b}_{k'},)\right] \quad , \tag{137}$$

so that

$$\hat{H}_u = E_o + \sum_{\underline{k}} \varepsilon_e(\underline{k})\, \hat{c}^{+}_{\underline{k}} \hat{c}_{\underline{k}} + \sum_{\underline{k}} \varepsilon_h\,(k)\, \hat{b}^{+}_{k} \hat{b}_{k} +$$

$$+ \sum_{\underline{k}} \lambda\,(\underline{k})\,[\hat{c}^{+}_{\underline{k}}\, \hat{b}^{+}_{-\underline{k}} + \text{h.c.}], \tag{138}$$

with

$$\varepsilon_e\,(\underline{k}) = E_e\,(k) - \frac{\hbar\omega}{2},$$
$$\varepsilon_n\,(\underline{k}) = -E_\nu(k) - \frac{\hbar\omega}{2}$$

This Hamiltonian is diagonalized by means of a Bogoliobov transformation

$$\hat{c}_{\underline{k}} = u_{\underline{k}}\,\hat{\alpha}_{\underline{k}} + \nu_{\underline{k}}\,\hat{\beta}^{+}_{\underline{k}},$$
$$\hat{b}_{\underline{k}} = u_k\,\hat{\beta}_{\underline{k}} - \nu_{\underline{k}}\,\hat{\alpha}^{+}_{\underline{k}}, \tag{139}$$
$$u^2_{\underline{k}} + \nu^2_{\underline{k}} = 1; \quad u_{\underline{k}}, \nu_{\underline{k}} \text{ real},$$

$$\hat{H}'_u = E_o + \Delta E_o\,(\lambda,\omega) +$$

$$+ \sum_{\underline{k}} [\varepsilon_\alpha(k)\, \hat{\alpha}^{+}_{\underline{k}}\hat{\alpha}_{\underline{k}} + \varepsilon_\beta(k)\, \beta^{+}_{\underline{k}}\beta_{\underline{k}}] \quad ; \tag{140}$$

where

$$\varepsilon_{\alpha,\beta}\, [\lambda^2 + \xi^2_{+}]^{1/2} \pm \xi^{-} , \tag{141}$$

(the upper sign refers to α),

$$\xi \pm (\underline{k}) = \frac{1}{2}\,[\varepsilon_e(\underline{k}) \pm \varepsilon_h\,(k)], \tag{142}$$

and

$$\nu_k^2 = \frac{1}{2} [1 - \frac{\xi(k)}{[\lambda^2+\xi_+^2]^{1/2}}]. \tag{143}$$

The constant energy shift is

$$\Delta E_o(\lambda,\omega) = \sum_{\underline{k}} [\varepsilon_e(\underline{k}) + \varepsilon_h(\underline{k})] \nu_k^2 + \tag{144}$$

$$+ 2 \sum_{\underline{k}} \lambda(\underline{k}) u_k \nu_k$$

Defining the new effective vacuum state (at zero temperature by

$$\hat{\alpha}_k | \tilde{0} >_u = \hat{\beta}_k | \tilde{0} >_u = 0 \tag{145}$$

we get

$$_u< \tilde{0} | \hat{H}_u | 0 >_u = E_o + \Delta E_o \tag{146}$$

as the reference energy. This reference state is not a vacuum state for the original electron-hole-system: The probability to find an electron or a hole with a wave-vector $\underline{k}$ in the state $\tilde{0}>_u$ is

$$n_{\underline{k}} = {}_u< \tilde{0} | \hat{a}^+_{\underline{k}} \hat{a}_{\underline{k}} | \tilde{0} >_u = \nu_{\underline{k}}^2$$

Obviously, in the limit $\lambda(k) \to 0$ we get a free electron-hole gas with the Fermi-level $n_{\underline{k}} = \frac{1}{2}$ at $\xi_+ = 0$ i.e.

$$E_c(\underline{k}_F) - E_\nu(\underline{k}_F) = \hbar\omega . \tag{148}$$

In this limit we get with Eq. (71)

$$\Delta E_o = \sum_{k<k_F} [\frac{\hbar^2 k^2}{2m_\nu} + \frac{\hbar^2 k^2}{2m_c} + Eg - \mu]$$

$$= \frac{3}{5} (E_F^e + E_F^h) N - (\mu - Eg)N \tag{149}$$

with $\mu = \hbar\omega$, and $E_F^{e,h}$ the respective Fermi-energies. Though being a clearly defined consequence of our model, this is not what one would expect on physical grounds, as this state should decay via spontaneous emission of photons, a process not considered here. Either one has to include this effect or to restrict this kind of model to large coupling strength λ Finally, the quasi-particle-spectrum is sketched in Fig. 4. For $\lambda = 0$ we recover the result of Fig. 2b. For finite λ it resembles the spectrum of a superconductor except that here we have two branches.

The same procedure can be carried out for a Schrödinger-field with self-interaction $V(\underline{r},\underline{r}')$. As the interaction term is not affected by the transformation U, we get now

$$\hat{H}_u^W = \hat{H}_u + \hat{H}_w \tag{150}$$

where H_w is for an electron-hole liquid

$$\hat{H}_w = \frac{1}{2} \sum_{\underline{kk}'} \sum_{q\neq 0} V(q)\,[\hat{c}^+_{\underline{k+q}}\, \hat{c}_{\underline{k-q}}\, \hat{c}_{\underline{k}'}\, \hat{c}_{\underline{k}}$$

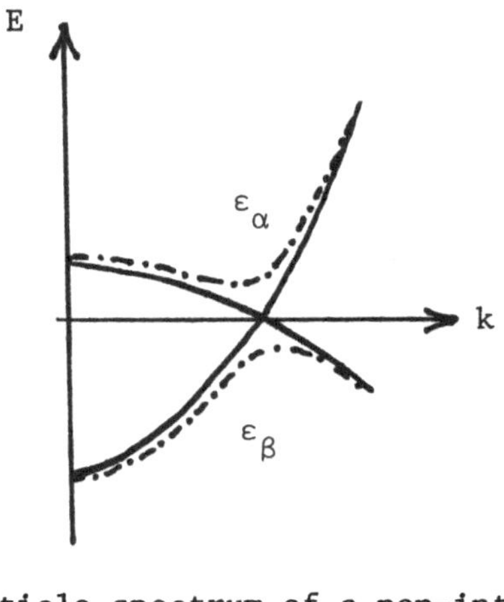

Fig. 4. Quasi-particle spectrum of a non-interacting driven 2-band model; reference state is the light deformed electron-hole gas. Solid line: $\lambda = 0$. Dash-dotted line: finite λ.

$$+ \hat{b}^{+}_{\underline{k}+\underline{q}} \hat{b}^{+}_{\underline{k}'-\underline{q}} \hat{b}_{\underline{k}'} \hat{b}_{\underline{k}}] + \hat{H}_{eh} + \text{const.}$$

We treat the influence of $\hat{H}_W$ by first-order perturbation theory, i.e. we calculate the "HF-ground state" with a variational wave function equivalent to the Bogoliubov transformation used before,

$$|\psi> = \prod_{\underline{k}'} (u_{\underline{k}'} + \nu_{\underline{k}'} \hat{a}^{+}_{\underline{k}'} \hat{b}^{+}_{\underline{k}'}) \mid \tilde{0} >_{u}, \tag{152}$$

so that

$$\Delta E^{W} = <\psi|\hat{H}_W|\psi>$$

$$= \sum_{\underline{k}} [2\nu^2_{\underline{k}} \xi_{+}(k) - 2\lambda\nu_{\underline{k}}(1 - \nu^2_{\underline{k}})^{1/2} - \nu^2_{\underline{k}} \sum_{\underline{q}} V(\underline{q}) \nu^2_{\underline{k}+\underline{q}}] \tag{153}$$

The T = 0 ground state can be found from the condition

$$\frac{\delta\Delta E^{W}}{\delta\nu^2_k} = 0 \quad , \tag{154}$$

from which we get

$$\nu^2_{\underline{k}} = \frac{1}{2}\left\{1 - \frac{\xi'_{+}(\underline{k})}{[\xi'_{+}(k) + \lambda^2]^{1/2}}\right\} \quad , \tag{155}$$

$$\xi'_{\underline{+}} (\underline{k}) = \xi_{\underline{+}} - \sum_{\underline{q}} V(\underline{q}) \nu^2_{\underline{k}+\underline{q}}. \tag{156}$$

For $V_q = 0$ we immediately recover the result of Eq. (143).

Equation (155) can be solved only approximately. It is interesting to note that in a certain parameter region (ω,λ) there is no longer a unique solution, but a cust behaviour Fig. 5). One shows that the middle brach is unstable, so that one should observe a behaviour of optical bistability.

The origin of this bistability is easily understood: If the laser frequency ω approaches Eg from above, the exchange energy lowers the effective gap, so that a high-density electron-hole plasma is still maintained below Eg, where the light source approaching Eg from below would not yet be able to produce a sizeable population of electron-hole pairs.

Finally, the corresponding HF-quasi-particle-energies are now

$$\varepsilon'_{\alpha,\beta} = [\lambda^2 + \xi'^{\,2}_{+}]^{1/2} \pm \xi'_{-} , \qquad (157)$$

which may be used also as the constituents of pair excitations.

On the other hand, true excitonic effects should only be expected if the electron-hole density of the reference state is small compared with the Mott density. Then, however, one would have to go beyond the HF-scheme and, in particular, study the influence of the strong field on the effective particle-hole interaction: As a function of the external control-parameters λ and ω the nature of the resulting pair spectrum would change even qualitatively. A similar situation is encountered if one neglects the photon field and uses a non-equilibrium model with N (electron-hole-number) as a control parameter [17]. This leads to the well-known effect of electron-hole-plasma-drops coexisting with a dilute exciton gas.

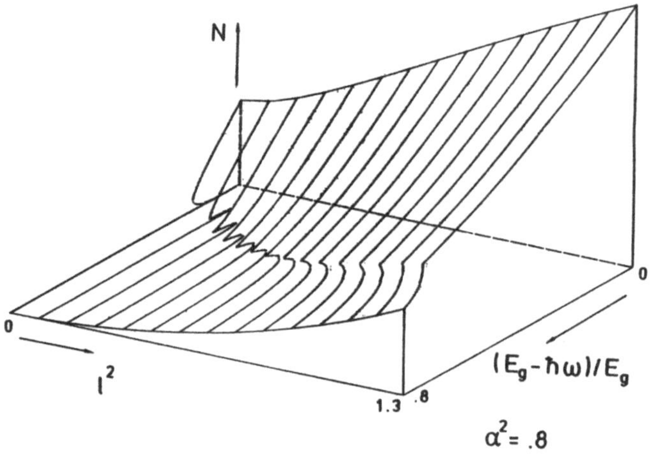

Fig. 5. HF-reference state of a driven 2-band model as a function of the external control-parameters $h\omega$ and λ. $l = 2/Eg$ and λ specifies the zero-field-band structure.

IV. CONCLUSIONS

We tried to outline some basic features of excitation spectra as they are influenced by models of structure and correlation. We restricted ourselves to a very limited number of examples, which, however, should show the delicate relationship on various model features. Categories, which may be used to classify the spectra, comprise the nature of quasi-particles, existence of gaps, the type of quasi-particle interactions (excitons) and the relation to phase transition (Anderson-type, superconducting/normal, metal/insulator).

I thank Mr. Th. Harbich and Dr. Jan Golka for valuable discussions.

REFERENCES

1. S. Nakajima, Y. Toyozawa, and R. Abe, The Physics of Elementary Excitations, Springer, New York (1980), and D. Pines, Elementary Excitations in Solids.
2. S. Ma, Modern Theory of Critical Phenomena, Benjamin, Reading (1976).
3. H. Haken, Quantenfeldtheorie des Festkörpers, Teubner, Stuttgart (1973).
4. E. M. Henley and W. Thirring, Elementary Quantum Field Theory, McGraw Hill, New York (1963).
5. D. Pines and P. Nozières, The Theory of Quantum Liquids, Benjamin, New York (1966).
6. P.W. Anderson, Phys. Rev. 109, 1492 (1952).
7. J.M. Ziman, Models of Disorder, Cambridge University Press, London (1979).
8. E.J. Yoffa, W.A. Rodriques, and D. Adler, Phys. Rev. B 19, 1203 (1979).
9. W. Hanke and L.J. Sham, Phys. Rev. B 21, 4656 (1980).
10. V.L. Bonch-Bruevich and V.D. Iskra, phys.stat.sol. (b) 68, 369 (1975).
11. M. Capizzi, G.A. Thomas, F. De Rosa, R.N. Bhatt, and T.M. Rice, Solid State Commun. 31, 611 (1979).
12. J. Golka and G. Mahler, J. Phys. C14, 21, 1445 (1981).
13. H. Haken, Synergetics, Springer, Berlin (1977).
14. H. Sambe, Phys. Rev. A7, 2203 (1973).
15. N. Tzoar and S.I. Gersten, Phys. Rev. B 12, 1132 (1975).
16. Th. Harbich and G. Mahler, phys. stat.sol. (b) 104, 185 (1981).
17. R.M. Rice, J.C. Hensel, T.G. Philips, and G.A. Thomas, Solid State Phys. 32 (1977).

COHERENT WAVEPACKETS OF PHONONS

N. Terzi

Istituto di Fisica dell'Università and
Gruppo Nazionale di Struttura della Materia del CNR
Via Celoria 16, Milano, Italy

ABSTRACT

The propagation of an elastic pulse in a one-dimensional elastic medium is investigated. First, normal modes and frequencies are introduced to characterize the elastic medium, consisting of a line of atoms harmonically vibrating around a stable equilibrium configuration. Second, the equation of motion of an elastic disturbance along the chain is obtained in the classical limit (d'Alembert equation) and the solution corresponding to a single elastic pulse is commented. Third, the quantum mechanical treatment is formulated and the pulse propagation is written in terms of phonons: the pulse is seen to correspond to a coherent wavepacket of phonons propagating at the "sound" velocity of the chain. Different coherent wavepackets are considered and their change during the motion is related to their composition in energy and wavevectors. Finally, "birth and death" of the pulse are discussed.

Two appendices are added at the end to collect in a compact way some properties of the coherent states (Appendix A) and useful relations dealing with the Poisson distribution (Appendix B).

I. ONE-DIMENSIONAL ELASTIC LINE [1,2]

Although one-dimensional (1-dim) systems are considered at every introductory course on lattice dynamics (and are probably well known by all the students), yet their usefulness to point out in an elementary way the physical meaning of every step done during the exposition is undeniable. Also for that reason, these lectures are dealing with the vibrations of a chain. Indeed, starting from

1-dim, extension to 2-dim and 3-dim cases can be done for some properties, such as the character of the solutions of the secular equation with its symmetries and the validity of the models adopted to describe lattice dynamics.

However, there are peculiar properties of 1-dim systems which are specifically related to the structure of the equation of motion in one spatial dimension. The chief example (not discussed here) is that of solitons, exact solutions of a non linear equation of motion in 1-dim. The example here discussed belongs instead to the class of solutions of the linear equation of motion, which describe the propagation of an elastic pulse in harmonic approximation. It is shown that the pulse proceeds without spreading when constituted by elementary waves whose frequency belongs to the region $\omega_k = v\, k$.

I.A. Monoatomic Array of Rigid Atoms

Consider a discrete chain, whose dynamics is first described by means of a model with the least number possible of parameters. To this end the chain is assumed to consist of a regular array of N equal, pointlike, undeformable (or rigid)atoms of mass M, linked by massless springs of stiffness K, which couple only nearest neighbour atoms (n.n). Modification of this oversimplified model are discussed in par.II.

In order to study the chain dynamics, let the atoms (labelled by $\ell=1,\dots,N$) free to displace longitudinally from their equilibrium position $x^o_\ell = \ell a$ to new position x_ℓ , and indicate by $u_\ell = x_\ell - x^o_\ell$ their small displacements ($u_\ell \ll a$). Since each atom is elastically coupled to its n.n., the equations of motion of the atoms form a coupled system (atoms are coupled oscillators). Therefore the first problem arising in the study of lattice dynamics is that to find a coordinate transformation $u_\ell \to q_k$ such that the equations of motion become decoupled, both in classical and quantum limits. This is achieved in the classical limit, by writing the net force f_ℓ acting on each atom as a sum of uncoupled contributions depending only on ℓ . And in quantum limit, when potential and kinetic energies can be written in the same way.

In this section, we show how, by taking advantage of the periodicity of the atomic array and of the boundary conditions, such coordinates (the socalled normal modes) can be easily found for our simple chain. It is worth stressing that the search of normal modes with their relative frequencies, is essentially a kinematical problem which consists in finding a decoupling procedure for forces and energies. It is therefore the same in the classical and quantum limits. It is clear that the secular equation which the decoupling condition leads to, depends necessarily on the physical characteristics of the lattice, such as its geometry and its atomic (or ionic or molecular) composition via forces or interactions. But it does not depend on the structure of the equations of motion.

I.B. Forces

The physical considerations leading to the transformation $u_\ell \to q_k$, are the following.

1. The chain is in a stable equilibrium configuration. Then the net force f_ℓ, applied on the atom ℓ ($\ell=1,\ldots N$) owing to the coupling between n.n., must be a restoring force. In the elastic Hooke's limit, f_ℓ is

$$f_\ell = K(x_{\ell+1}-x_\ell-a) - K(x_\ell-x_{\ell-1}-a) \tag{1a}$$

$$= K(u_{\ell+1}+u_{\ell-1}-2u_\ell) \tag{1b}$$

$$= M^{\frac{1}{2}}\,\Omega^2(q_{\ell+1}+q_{\ell-1}-2q_\ell)\ . \tag{1c}$$

$q_\ell = M^{\frac{1}{2}} u_\ell$ are the socalled reduced displacements and $\Omega^2 \equiv K/M$. Recall that K is the stiffness constant of the springs linking n.n.

2. A homogeneous displacement δ of all the atoms $x_\ell \to x_\ell+\delta$, causes no physical changes on the chain $f_{\ell+\delta} \to f_\ell$ in Eq. (1a). In other words, the forces (and the interactions) are obviously independent from the position of the observer.

3. The chain has finite length L=Na. Then Eq. (1b) for the first ($\ell=1$) and last ($\ell=N$) atoms does not contain the terms proportional to $u_{\ell-1}$ and to $u_{\ell+1}$, respectively. It could instead contain terms of coupling between the chain and its outer world. However, if the chain is long enough and, mainly, if we are concerned only with bulk properties, the asymmetry of the terminal atoms with respect to the bulk atoms can be removed by introducing periodic Born-von Karman boundary conditions: $u_1 = u_{N+1}$. That corresponds to fold the chain over itself and connect first and last atoms with a spring of stiffness K. It follows that

$$u_{N+\ell} = u_\ell\ . \tag{2}$$

4. Owing to point 2, the atoms are disposed in a regular periodic array. Then the Bloch's theorem holds and wavevectors $\underline{k}$ can be introduced such that q_ℓ is given by a superposition of partial terms $q_\ell^{(k)}$, where

$$q_\ell^{(k)} \sim e^{ik\ell a}\, q_o^{(k)} \tag{3a}$$

By using Eq.(2), one obtains for the transformation $q_\ell \rightarrow q_k$

$$q_\ell = \frac{1}{\sqrt{N}} \sum_k e^{ik\ell a} q_k, \tag{3b}$$

with their inverse $q_k \rightarrow q_\ell$

$$q_k = \frac{1}{\sqrt{N}} \sum_\ell e^{-ik\ell a} q_\ell \tag{3c}$$

where k belongs to the Brillouin zone ($k \in$ B.Z.), or

$$k = \frac{2\pi}{a} \frac{n}{N} = \frac{2\pi}{\lambda} \qquad (-\frac{N}{2} < n \leqslant \frac{N}{2}) \tag{4}$$

$\lambda = a\, N/n$ is the wavelength associated to the (quasi) wavevector k. The transformations (3b) and (3c) are consistent because

$$\sum_k e^{ik(\ell-\ell')a} = N\delta_{\ell\ell'} \tag{5a}$$

and

$$\sum_\ell e^{i(k-k')\ell a} = N\delta_{kk'} \ . \tag{5b}$$

Notice that

$$q_{-k} = q_k^* \tag{6}$$

because q_ℓ are real. q_k and q_k^* are complex conjugated.

In terms of the new coordinates q_k, the forces f_ℓ are

$$f_\ell = (\frac{M}{N})^{\frac{1}{2}} \Omega^2 \sum_k e^{ik\ell a}(e^{ika} + e^{-ika} - 2)\, q_k$$

$$= (\frac{M}{N})^{\frac{1}{2}} \Omega^2 \sum_k (-4\sin^2 \tfrac{1}{2}ka) e^{ik\ell a} q_k$$

$$= -(\frac{M}{N})^{\frac{1}{2}} \sum_k \omega_k^{\ 2}\, e^{ik\ell a}\, q_k \tag{7a}$$

$$= \sum_k f_\ell^{(k)} \tag{7b}$$

where ω_k^2 are the solutions of the so-called *secular equation*

$$\omega_k^2 - 4\Omega^2 \sin^2 \tfrac{1}{2}ka = 0. \tag{8}$$

From that said before about the requested decoupled form of forces, from Eq.(7) it is clear that q_k are the required normal coordinates and ω_k^2 their squared frequencies. In fact any coupling between adjacent atoms has disappeared from Eq.(7) at expenses of a sum over N terms.

ω_k as function of k gives the so called dispersion relations for frequency ω_k versus wavevector k. Comments on ω_k are given in Sect.II. Here we notice that the condition of elastic stability considered before, imposes on ω_k a very peculiar behaviour for small k, which is indeed a general feature of ω_k versus k for all the elastic materials. In fact: i) ω_k^2 must be positive and ω_k is real. Only one root ω_k (the positive one, because ω_k is a frequency) is considered for each k, when k goes over the entire B.Z.: $\omega_k = |\omega_k|$ so that $\omega_{-k} = \omega_k$.

ii) If ever among the solutions of the secular equation (8) one is found for a given $\bar{k}$ such as $\omega^2_{\bar{k}} < 0$, the corresponding $f_\ell^{(\bar{k})}$ in Eq.(7a) becomes positive. The chain is then unstable for the mode $q_{\bar{k}}$ because $f_\ell^{(k)}$ is a driving and not a restoring force.

iii) $\omega_k \to 0$ when $k \to 0$: whatever small the displacements q_ℓ (or q_k) might be, the chain is able to react with an adequately small restoring force (7a). Conditions of stability lead therefore to the existence of an acoustical branch such as: $\omega_{k=0} = 0$ and

$$\omega_k = vk \quad \text{when} \quad k \to 0. \tag{9}$$

$v = \left.\frac{d\omega_k}{dk}\right|_0$ is the sound velocity of the elastic medium.

For the simple chain here considered, from Eq.(8) one has:

$$v = \Omega a = a\sqrt{K/M} \; . \tag{10}$$

Eq.(10) supplies a relationship between a macroscopic measurable quantity (the sound velocity) and the microscopic elastic parameter (stiffness constant K) of the chain, provided the boundary conditions such that the chain is in a stable equilibrium when n.n. interatomic distance is a. The meaning of K in terms of microscopic interactions is commented in Sec.II.

I.C. Energy

1. Potential energy. A procedure analogous to that followed in the previous section is here used to deduce the transformation $u_\ell \to q_k$ which transform the total potential energy $V(\{x_\ell\})$ of the oscillating coupled atoms in a sum of suitable decoupled terms. Comments about the microscopic interatomic interactions are posponed to Sec.II. $\{x_\ell\}$ is a shortwriting for $(x_1, \ldots, x_N)$.

Expand $V(\{x_\ell\})$ around x_ℓ° in power series of the (small) displacements u_ℓ and take into account terms to the second order (harmonic approximations):

$$V(\{x_\ell\}) = V(\{x_\ell^\circ\}) + \sum_\ell F_\ell u_\ell + \tfrac{1}{2} \sum_{\ell\ell'} A_{\ell\ell'} u_\ell u_{\ell'} \; . \tag{11}$$

Here

$$F_\ell = \frac{\partial V}{\partial x_\ell}\bigg|_{\{x_\ell^o\}} \tag{12}$$

is the net force acting on the atom ℓ when the chain is at rest ($x_\ell = x_\ell^o$) and

$$A_{\ell\ell'} = \frac{\partial^2 V}{\partial x_\ell \partial x_{\ell'}}\bigg|_{\{x_\ell^o\}}, \tag{13}$$

where $A_{\ell\ell'} = A_{\ell'\ell}$ are the elements defining the harmonic *force constant* matrix A. As it is shown presently, $A_{\ell(\ell+1)}$ and $A_{(\ell-1)\ell}$ are related to the stiffness constant K of the springs which couple n.n. Then, unless all the off-diagonal terms of A are explicitly put equal to zero (Einstein model), $V(\{x_\ell\})$ contains terms relating displacements of different atoms. As before, a transformation to normal modes $q_\ell \to q_k$ is needed. If one uses the transformation (3), through which forces have been disantangled, one obtains the remarkable result that also $V(\{q_\ell \to q_k\}) = V(\{q_k\})$ is given by a sum of uncoupled terms in q_k variables: the same transformation (3) "decouples" forces and potential energy (and the kinetic energy, as it will be shown later).

So, let us apply again points 1-4 of the previous section to deduce the transformation $q_\ell \to q_k$ from the properties of $V(\{q_\ell \to q_k\})$.

1. The chain is in a stable equilibrium configuration: no net forces are present when $u_\ell = 0$. Then

$$F_\ell = 0 \qquad (\forall\, \ell) \tag{14}$$

in Eq. (11). In other words, the elastic equilibrium configuration $x_\ell^o = a\ell$ corresponds to a minimum of the potential energy.

Actually, this condition holds only at T=0 K, because at a finite temperature T the equilibrium is determined by the minimum of an appropriate thermodynamic potential at that temperature and not by the minimum of the simple mechanical potential (11). Depending on the mechanical constraints applied while the temperature is changed, the thermodynamic potential to be minimized is: the Helmoltz free energy F=V-TS (S is the entropy) if the chain length L=Na is kept fixed; the Gibbs free energy G=V-TS+pL, if the pressure applied on the ends of the chain is instead kept fixed. In this last case L and a are temperature dependent: a=a(T). Their value is determined at every temperature by the thermodynamic condition of stability,

$$\frac{\partial G}{\partial x_\ell}\bigg|_{\ell a(T)} = F_\ell(T) = 0 \qquad (\forall\, \ell) \tag{15}$$

(for fixed external pressure), instead of the mechanical condition (14). By further expanding G around $x_\ell^o(T)=\ell a(T)$, we obtain from the second-order terms the thermodynamic harmonic force constants $\tilde{A}_{\ell\ell'}$ and from the successive terms the anharmonic constants. This is the correct, but very cumbersome procedure. So, simplifying approximations are often used. The most simple is the *so-called quasi-harmonic approximation*, which in short says the following: i) use Gibbs potential to evaluate the equilibrium configuration $x_\ell^o(T)$ at every temperature by means of Eq. (15). ii) Use the mechanical potential expanded around $x_\ell^o(T)$, to evaluate by means of Eqs. (11) and (13) the harmonic and anharmonic force constants.

To avoid unnecessary complications, it will be hereafter assumed that T=0 K.
Inserting Eq. (14) in Eq. (11) one obtains

$$\Delta V = V(\{x_\ell\}) - V(\{x_\ell^o\}) = \tfrac{1}{2} \sum_{\ell\ell'} A_{\ell\ell'} u_\ell u_{\ell'} \tag{16a}$$

$$= \tfrac{1}{2} \sum_\ell (A_{\ell(\ell-1)} u_{\ell-1} + A_{\ell\ell}\, u_\ell + A_{\ell(\ell+1)}\, u_{\ell+1}) u_\ell \tag{16b}$$

since only n.n. interaction is considered.

2. (point 4 of Sec. I.B). The atoms form a regular array. Then

$$A_{\ell\ell'} = A(|\ell-\ell'|), \tag{17}$$

and in particular

$$A_{\ell(\ell-1)} = A_{\ell(\ell+1)} = A_1$$

$$A_{\ell\ell} = A_o. \tag{18}$$

3. (point 2 of Sec. I.B). ΔV must be invariant for a homogeneous displacement δ : $\Delta V(\{x_\ell\})=\Delta V(\{x_\ell+\delta\})$ in Eq.(16a). Since by Eq.(17) $A_{\ell\ell'}=A(\ell-\ell')=A(\ell+\delta-\ell'-\delta)$ it must be

$$\sum_{\ell\ell'} A_{\ell\ell'} u_\ell u_{\ell'} = \sum_{\ell\ell'} A_{\ell\ell'} (u_\ell+\delta)(u_{\ell'}+\delta).$$

It follows that

$$\delta \sum_\ell (\sum_{\ell'} A_{\ell\ell'}) u_\ell = 0 \quad \text{and} \quad \delta^2 \sum_{\ell\ell'} A_{\ell\ell'} = 0.$$

Since δ is arbitrary, the strongest condition is that

$$\sum_{\ell'} A_{\ell\ell'} = 0. \tag{19}$$

Then

$$A_o = -2A_1, \text{ where } A_1 = K \text{ (stiffness const.)}, \quad (20)$$

in our case. Defining the squared frequency as before

$$\Omega^2 = A_1/M = K/M ,$$

one obtains ($q_\ell = M^{\frac{1}{2}}u_\ell$)

$$\Delta V = \tfrac{1}{2}\Omega^2\sum_\ell q_\ell(2q_\ell - q_{\ell+1} - q_{\ell-1})$$
$$= \tfrac{1}{2}\Omega^2\sum_\ell(q_{\ell+1} - q_\ell)^2. \quad (21)$$

4. Periodic boundary conditions and periodicity of the atoms inside the array, lead as before to the transformations (3) with (4). Putting Eq. (3b) in Eq. (21) and recalling Eqs. (5b) and (8), one obtains:

$$V = \tfrac{1}{2}\frac{\Omega^2}{N}\sum_\ell\sum_{kk'}(2 - e^{ika} - e^{-ika})e^{i(k+k')\ell a}q_k q_{k'}$$
$$= \tfrac{1}{2}\sum_k \omega_k^2 q_k q_k^* , \quad (22)$$

where ω_k^2 is the solution of the secular equation (8). ΔV is the sum of potential energies of uncoupled oscillators, defined in the fictiticus displacement-space $\{q_k\}$.

2. Kinetic Energy. Indicating $\dot{u}_\ell = dx_\ell/dt = du_\ell/dt$ the velocity of the atom ℓ, by $M\dot{u}_\ell$ its momentum and by $p_\ell = M^{-\frac{1}{2}}(M\dot{u}_\ell) = \dot{q}$ its reduced momentum, the transformation $p_\ell \to p_k$ is obtained by deriving with respect to the time the transformation $q_\ell \to q_k$:

$$p_\ell = \frac{1}{\sqrt{N}}\sum_k e^{ika\ell}p_k \quad (23a)$$

where

$$p_k = \frac{1}{\sqrt{N}}\sum_k e^{-ika\ell}p_\ell, \quad (23b)$$

and where the complex p_k are such that (p_k^* is the conjugated value of p_k)

$$p_{-k} = p_k^*, \quad (24)$$

because p_ℓ are real quantities. The total kinetic energy K, where $K(\{p_\ell\}) = K(\{p_\ell \to p_k\})$, is

$$K = \tfrac{1}{2}\sum_\ell p_\ell^2 = \tfrac{1}{2}\Sigma_k p_k p_k^* = \tfrac{1}{2}\sum_k p_k p_{-k}. \quad (25)$$

K is a sum of uncoupled terms either in $\{q_\ell\}$-space or in $\{q_k\}$ space. This remarkable result follows from Eq.(5b). In the present case it is a consequence of the periodic arrangement of the atoms in the chain. More generally, normal coordinates q_k (with momenta p_k) are such that both potential and kinetic energies are at the same time sum of uncoupled terms.

When non periodic array of atoms are considered, the transformation $q_\ell \to q_\lambda$ and $p_\ell \to p_\lambda$ (λ labels the normal coordinates) is far from being so easily found and simple as that of Eqs.(3) and (23) with the property (5b). Such a transformation, however, always exists, because of the algebrical properties of quadratic forms with positive coefficients like $V(\{q_\ell\})$ and $K(\{p_\ell\})$. *The characteristic equation* of such a transformation is just the *secular equation* (a generalization of Eq.(8)) whose roots, when reported in function of λ, give the *frequency dispersion* relations (or curves if plotted) of the *normal frequencies* ω_λ.

3. Total Energy. The total energy in harmonic approximation is

$$H = T+V=\tfrac{1}{2}\Sigma_\ell\left(\frac{1}{M}\dot{u}_\ell^2 + \Sigma_{\ell'}A_{\ell\ell'}u_\ell u_{\ell'}\right)$$

$$=\tfrac{1}{2}\Sigma_\ell\left[p_\ell^2+\Omega^2(q_{\ell+1}-q_\ell)^2\right]$$

$$=\tfrac{1}{2}\Sigma_k(p_k p_k^* + \omega_k^2 q_k q_k^*) = \textstyle\sum_k H_k. \qquad (26)$$

I.D. Continuous Vibrating Line

The microscopic interatomic features become irrelevant when attention is focussed on oscillations extending over several interatomic distances ($\lambda \gg a$, or $k \ll \pi/a$ from Eq.(4)). With respect to such oscillations, the atomic array behave like an elastic string, whose dynamics is determined only by the macroscopic constants v (sound velocity) and L=Na. The microscopic constants Ω, a and N have to disappear. On the other hand, the continuous limit means also that only the acoustical region $\omega_k = vk$ in the frequency dispersion is of physical relevance. This fact greatly simplifies the form of the equations and of the results, as it will be shown later.

Consider therefore the limit $a\to 0$ and $N\to\infty$ of the dynamical quantities previously evaluated under the condition that L=Na remains finite. Introduce the continuous variable x

$$\ell a \to x$$

and the continuous displacement field q(x)

$$q_\ell \to a^{\frac{1}{2}}\ q(x) \tag{27}$$

Since $a \to dx$, differences become differentials

$$a^{-\frac{1}{2}}(q_{\ell+1}-q_\ell) \to q(x+a)-q(x) = \frac{dq(x)}{dx}\bigg|_a a + \frac{1}{2}\frac{d^2q(x)}{dx^2}\bigg|_a a^2 + o(a^2) \tag{28}$$

$$a^{-\frac{1}{2}}\ (q_{\ell+1}-q_\ell)-(q_\ell-q_{\ell-1}) \to \frac{d^2q(x)}{dx^2}\bigg|_a a^2 + o(a^2), \tag{29}$$

and summations become integrals

$$\sum_{\ell=1}^{N} \to \frac{1}{a}\int_0^L dx \tag{30}$$

so that Eqs. (5b) and (5a) become

$$\int_0^L dx\ e^{i(k-k')x} = L\ \delta_{k,k'} \tag{31}$$

$$\sum_k e^{ikx} = L\delta(x) \tag{32a}$$

$$\sum_k (ik)\ e^{ikx} = L\delta'(x) \tag{32b}$$

where $\delta(x)$ and $\delta'(x)$ are the Dirac delta function and its derivative, respectively. The transformations $q(x) \to q_k$ and $p(x) \to p_k$ become

$$q(x) = \frac{1}{\sqrt{L}} \sum_k e^{ikx} q_k \tag{33a}$$

$$p(x) = \frac{1}{\sqrt{L}} \sum_k e^{ikx}\ p_k. \tag{33b}$$

Notice that k remains numerable, in accordance with the periodic boundary conditions, which still hold while taking the continuous limit $q_\ell \to q(x)$.

By using Eqs. (7a) and (30) and recalling Eqs.(8)-(10) one obtains for the forces $f_\ell \to f(x)$

$$(aM)^{-\frac{1}{2}} f(x) = \Omega^2 a^2 \frac{d^2q(x)}{dx^2} = v^2 \frac{d^2q(x)}{dx^2} \tag{34a}$$

$$= -\frac{1}{\sqrt{L}} \sum_k e^{ikx} \omega_k^2 q_k \tag{34b}$$

$$= -\frac{1}{\sqrt{L}} \sum_k e^{ikx} v^2 k^2 q_k \text{ when } k \ll \frac{\pi}{a}\ . \tag{34c}$$

and for the potential energy

$$\Delta V = \frac{1}{2}\int_0^L dx a^2\Omega^2\left(\frac{dq(x)}{dx}\right)^2 = \frac{1}{2}v^2\int_0^L dx\left(\frac{dq(x)}{dx}\right)^2 \tag{35a}$$

$$= \frac{1}{2}\sum_k \omega_k^2 |q_k|^2 \tag{35b}$$

$$= \frac{1}{2}\sum_k v^2k^2|q_k|^2, \text{ when } k<<\frac{\pi}{a} \tag{35c}$$

II. BEYOND THE MODEL OF A CHAIN OF MASSES AND SPRINGS [3-7]

Before discussing the equation of motion, it must be recalled that the 1-dim chain is the simplest possible model by which lattice dynamics may be described. Since we are here concerned with the acoustic (low frequency and long wavelengths) region of the frequency spectrum, such a model seems reasonable. However, in other cases a more realistic microscopic description must be used, where due attention is given to:

i) Geometry. Geometrical factors make the structure of the secular equation more complicated than that considered in Eq.(8). New features arises when: 1) the atomic array is not periodic. 2) 2-dim and 3-dim elastic media are studied.

ii) Models. Credible models cannot consider atoms and springs as unvariable, uncorrelated local elementary units. In particular: 1) the elementary masses must be allowed to have an internal structure (nucleus plus electrons) and be deformable; 2) the elementary springs must be allowed to consist of electrons, dynamically coupled to the nuclear motion and delocalized if necessary (as for metals). Several models have been proposed in literature to take into account both the changes.

iii) Microscopic interactions. The interatomic and intermolecular potentials must provide physically reliable values to the parameters entering the models, and vicaversa.

iv) Finite temperature.

Any of these points cannot be ignored when information on microscopic properties are deduced from the analysis of the frequency spectrum or from the dispersion curves. All the possible and existing complications at the microscopic level must be however such as to give the correct macroscopic behaviour in the elastic limit. It is interesting at this point to examine the $k\to 0$ region of different dispersion curves, either obtained from neutron scattering experiments or evaluated by means of models. A surprising result of most of non conducting crystals is that the linear behaviour of ω_k versus k goes far beyond the $k \to 0$ region, sometimes as far as $k \lesssim \pi/2a$. So, the condition $k \ll \pi/a$, given for instance in Eqs.(34c) and (35c), may be safely replaced by the less restrictive $k \lesssim \pi/2a$ condition in several cases.

III. MOTION OF A PULSE IN THE CLASSICAL LIMIT [2,8]

III.A. General Features

Consider a string (continuous elastic line) characterized by a frequency dispersion $\omega_k = vk$. v is the constant sound velocity along the line. At t=0 distort the string locally but still over dimensions D>>a. Be the deformation process so quick to be instantaneous on the time-scale of the sound propagation. The plucked string is then left alone.

Indicate by $\phi(x)$ the shape given to the pulse at t=0 and analyze it in terms of normal mode components

$$\phi(x) = q(x,0) = \frac{1}{\sqrt{L}} \sum_k e^{ikx} q_k(0) \tag{36a}$$

$$= \frac{1}{L} \sum_k e^{ikx} \int_{-\infty}^{+\infty} d\xi \; \phi(\xi) \; e^{-ik\xi}. \tag{36b}$$

At a further time t>0, q(x,t) is obviously determined by the equation of motion.

Consider first the discrete chain and the Newton's form of laws. If $\ddot{u}_\ell$ is the acceleration of the atom ℓ , one has

$$M \ddot{u}_\ell = f_\ell , \qquad (\ell=1,\dots,N) \tag{37a}$$

which becomes, by using Eqs.(1), the system of N coupled differential equations:

$$\ddot{q}_\ell = \Omega^2(q_{\ell+1} + q_{\ell-1} - 2q_\ell). \tag{37b}$$

Consider then the continuous limit of Eq.(37) with Eq. (34). The famous d'Alembert equation for wave propagation

$$\frac{\partial^2 q(x,t)}{\partial t^2} = v^2 \frac{\partial^2 q(x,t)}{\partial x^2} \tag{38}$$

is obtained.

By introducing normal coordinates either in Eq. (37) or in Eq. (38), a system of decoupled equations

$$\ddot{q}_k = -\omega_k^2 q_k, \qquad (k \in \text{B.Z.}) \qquad (39)$$

is found, where ω_k^2 is given in Eqs.(8)-(10). The solutions of the system (39) are

$$q_k(t)=q_k(0)\cos\omega_k t+\frac{1}{\omega_k}p_k(0)\sin\omega_k t. \qquad (40)$$

Take now into account the initial conditions (36) together with the condition $p_k(0)=0$, so that

$$q(x,t)=\frac{1}{2\sqrt{L}}\sum_k(e^{i\omega_k t}+e^{-i\omega_k t})\,q_k(0)e^{ikx} \qquad (41a)$$

$$=\frac{1}{2L}\sum_k e^{ikx}\int_{-\infty}^{\infty}d\xi\;\phi(\xi)\,(e^{-i(k\xi+\omega_k t)}+e^{-i(k\xi-\omega_k t)}). \qquad (41b)$$

Equation (41b) can be integrated once the dispersion ω_k versus k of the string is known. In the long wave limit here considered (ω_k=vk), Eq. (41) may be exactly evaluated, by using Eqs. (31) and (32a).

$$q(x,t)=\tfrac{1}{2}\int_{-\infty}^{+\infty}d\xi\phi(\xi)\left[\delta(x-\xi-vt)+\delta(x-\xi+vt)\right]$$

$$=\tfrac{1}{2}\left[\phi(x-vt)+\phi(x+vt)\right]=\phi_1(x,t)+\phi_2(x,t): \qquad (42)$$

the initial pulse splits into two pulses ϕ_1 and ϕ_2, each propagating at velocity v, $\phi_1(x,t)$ to the right (x>0) and $\phi_2(x,t)$ to the left (x<0). With respect to the initial pulse, the travelling pulses are reduced but undistorted. This is a consequence of the assumed form (ω_k=vk) of the frequency dispersion in Eq. (41) and not of the harmonic approximation.

III.B. Pulse Shape and Dispersion

Whether or not the pulses $\phi_1(x,t)$ and $\phi_2(x,t)$ change their initial shape while travelling, depends on their recipe (Eq. (36)) in terms of normal modes $q_k(0)$. Consider for instance the extreme case when $q_k(0)\neq0$ for k belonging to a neighbourhood of the B.Z. boundary ($k\simeq\pi/a$); expand ω_k around $k=\pi/a$ and put it into Eq.(41b). The pulse would not travel at all, because there the linear term in k should be zero. Indeed, as it is well known in waves dynamics, the propagation of a wave packet (the present elastic pulse) is characterized by its group velocity and at the zone boundary the group velocity is zero.

Notice that the propagation of linear polarized *sound* in a perfect compressible fluid, obeys to the same wave equation (38) of the pulse along the elastic string. The pulse until now considered corresponds to a noise in acoustics. Monochromatic sound of frequency $\omega_{\bar{k}} = v\bar{k}$ corresponds instead to

$$\phi(x) = \frac{1}{\sqrt{L}} \Phi_o \cos(\bar{k}x + \delta). \tag{43}$$

In music a *note* of pitch $\nu_o=\omega_o/2\pi$ emitted by a musical instrument or by the voice, is a superposition of several monocromatic waves of frequency $\omega_n=n\omega_o$ (n=1,2,...), when end effects can be neglected. ν_o is called fundamental and ν_n its overtones. The recipe they enter in Eq. (36a), gives the quality or timbre or color of the note:

$$q(x,t) = \frac{1}{\sqrt{L}} \sum_n \phi_n \cos n(k_o x - \omega_o t). \tag{44}$$

It may be useful to comment at this point on the meaning of the expression *no dispersive region* in dealing with dispersion curves ω_k versus k. If one is looking at the propagation of pulses in the *real space*, the no dispersive region is the region where $\omega_k=v\,k$. In fact, owing to Eq.(42), linearly polarized sound propagates without dispersion and deformation when made up of modes, whose frequency belongs to that (acoustic) region. If on the contrary, one looks directly to *dispersion curves* ω_k versus k, a frequency branch is said no-dispersive in a region of the *inverse k-space*, when it maintains there a fairly constant value. In this latter context, the acoustic branch (Eq.(8)) is dispersive over the whole B.Z.

IV. MOTION OF A PULSE IN QUANTUM MECHANICS [2,9]

In this section the propagation of the pulse in 1-dim is studied by using a quantum mechanical approach. Obviously, the quantum description must coincide with Eq.(42) in the classical limit. The temperature of the vibrating string is assumed T=0 K, as before.

We proceed as follows. First, the useful quantum variables are identified. Then the state vector $|\psi(x,0)\rangle$, corresponding to the initial conditions (36) is written. Finally, the time evolution of $|\psi(x,t)\rangle$ is studied. The average value of the displacement $\langle\hat{q}(x)\rangle_t = q(x,t)$ in such a state is obtained and compared with Eq.(42). Moment $\langle\hat{p}(x)\rangle_t$ and energy $\langle\hat{H}_o\rangle$ are also evaluated and discussed. All the mathematical details are written in the Appendices.

IV.A. Variables

As in the classical limit, the natural variables are displacements $\hat{u}_\ell$ ($\hat{q}_\ell = M^{\frac{1}{2}}\hat{u}_\ell$ are reduced displacements) and momenta

$M\hat{u}_\ell^{\frac{1}{2}}(\hat{p}_\ell = M\,\hat{u}_\ell)$. A transformation to the normal modes is again needed in order to write the energy $\hat{H}_0$ as a summation of uncoupled terms. The transformation $\hat{q}_\ell \to \hat{q}_k$ is still given by Eq. (33).

Notice that q_ℓ and p_ℓ as classical variables are canonically conjugated, because $\dot{q}_\ell = \partial H_0/\partial p_\ell$ and $\dot{p}_\ell = -\partial H_0/\partial q_\ell$, where H_0 is given by Eq.(26). As quantum variables, they are hermitian ($\hat{q}_\ell^+ = \hat{q}_\ell$ and $\hat{p}_\ell^+ = \hat{p}_\ell$) non commuting operators, with commutation rules

$$[\hat{q}_\ell, \hat{p}_{\ell'}] = i\hbar\delta_{\ell,\ell'} \quad , \quad [\hat{q}_\ell, \hat{q}_{\ell'}] = [\hat{p}_\ell, \hat{p}_{\ell'}] = 0. \tag{45}$$

From definition (33), the variables $\hat{q}_k$ and $\hat{p}_k$ turn out to be not hermitian ($\hat{q}_k^+$ is the hermitian cojugate of $\hat{q}_k$)

$$\hat{q}_k^+ = \hat{q}_{-k} \quad \text{and} \quad \hat{p}_k^+ = \hat{p}_{-k}, \tag{46}$$

whose commutation rules, as obtained from Eq (45) with Eq. (33) are

$$[\hat{q}_k, \hat{p}_{k'}] = i\hbar\delta_{k,k'} \, , \, [\hat{q}_k, \hat{q}_{k'}] = [\hat{p}_k, \hat{p}_{k'}] = 0. \tag{47}$$

From Eqs.(26) and (46) the harmonic hamiltonian turns out to be

$$\hat{H}_0 = \tfrac{1}{2} \textstyle\sum_k (\hat{p}_k \hat{p}_k^+ + \omega_k^2 \hat{q}_k \hat{q}_k^+). \tag{48}$$

From (46) and (48), one can realise that the price to pay in quantum mechanics to write $\hat{H}_0 = \Sigma_k \hat{H}_k$ is that to work with "bad" variables: normal modes are in fact non hermitian variables. Once accepted this inevitable complication the best attitude is to work by using not $\hat{q}_k$ and $\hat{p}_k$ themselves, but their linear combinations simplifying as far as possible the study of the stationary states. Introduce to this end the annihilation ($\hat{a}_k$) and creation ($\hat{a}_k^+$) operators, so defined:

$$\hat{a}_k = \frac{1}{\sqrt{2\hbar\omega_k}} (\omega\hat{q}_k + i\hat{p}_k^+) \quad \text{and} \quad \hat{a}_k^+ = \frac{1}{\sqrt{2\hbar\omega_k}} (\omega_k \hat{q}_k^+ - i\hat{p}_k), \tag{49}$$

or

$$\hat{q}_k = \frac{1}{\sqrt{2\hbar\omega_k}} (\hat{a}_k + \hat{a}_{-k}^+) \quad \text{and} \quad \hat{p}_k = i\frac{\omega_k}{\sqrt{2\hbar\omega_k}} (\hat{a}_k^+ - \hat{a}_{-k}). \tag{50}$$

$\hat{H}_0$ takes a very simple form

$$\begin{aligned} \hat{H}_0 &= \tfrac{1}{2} \textstyle\sum_k \hbar\omega_k (\hat{a}_k^+ \hat{a}_k + \hat{a}_{-k}^+ \hat{a}_{-k} + 1) \\ &= \textstyle\sum_k \hbar\omega_k (\hat{a}_k^+ \hat{a}_k + \tfrac{1}{2}). \end{aligned} \tag{51}$$

The commutation rules of $\hat{a}_k$ and $\hat{a}_k^+$ are:

$$[\hat{a}_k,\hat{a}_{k'}^+] = \delta_{k,k'} \quad \text{and} \quad [\hat{a}_k,\hat{a}_{k'}] = [\hat{a}_k^+,\hat{a}_{k'}^+] = 0 \tag{52}$$

IV.B. Phonons and Phonon Wavepackets

The stationary states of the string are given by the eigenvalues and eigenvectors of $\hat{H}_o$ in Eq. (51):

$$E(\{n_k\}) = \sum_k E(n_k) = \sum_k \hbar\omega_k(n_k + \tfrac{1}{2}), \tag{53}$$

$$|\{n_k\}> = \Pi_k|n_k> = (n_1!n_2!\ldots)^{-\frac{1}{2}}(\hat{a}_1^+)^n\ (\hat{a}_2^+)^n \ldots|0>. \tag{54}$$

Here the states $|n_k>$ are eigenvectors of the socalled number operator $\hat{n}_k=\hat{a}_k^+\hat{a}_k$

$$\hat{a}_k^+\hat{a}_k|n_k> = n_k|n_k> \tag{55}$$

and n_k are positive integers.

$|0>$ is the vacuum state; its energy is the zero point energy $E_{zp}=\frac{1}{2}\sum_k\hbar\omega_k$ of the string (see Appendix A for further details).

Since $\hat{H}_o$ is the sum of uncoupled hamiltonians, $|\{n_k\}>$ is the product of numerable independent terms, one for each k. The energy for each k is an integer multiple of the fundamental frequency $\hbar\omega$ This leads to the interpretation of the stationary states $E(\{n_k\})$ in terms of independent particles. The state $|\{n_k\}>$ is said to have n_k *phonons* on the mode k, $n_{k'}$ phonons on the mode k', and so on. The operator $\hat{n} = \Sigma_k\hat{a}_k^+\hat{a}_k$ counts the total number n of phonons excited in the stationary state $|\{n_k\}>$ because

$$n = \sum_k n_k = <\{n_k\}|\hat{n}|\{n_k\}>. \tag{56}$$

n is a positive integer. Viceversa, a state with a fixed integer number of phonons is a stationalry state of $\hat{H}_o$.

Consider now *wavepackets of phonons*. We are particularly interested in the combination of stationary states (or of phonons) which gives the vector state of the string, when the elastic pulse travels at the sound velocity as in Eq.(42). More generally, as to correspond to a running wave. For what said, such a state cannot be a wavepacket made of a definite number of phonons.

In order to understand where to find the solution, focus attention on the process producing on the string the initial deformation

$\phi(x)$ of Eq.(36). A local deformation is generated when an external force, applied for a short but finite time to a piece of string, pushes it out of its initial equilibrium configuration (described by the vacuum state $|0\rangle$ in the present case, because T=0 K). But this is just a way to produce what is called a *coherent wavepacket* of phonons:[(o)] in fact the vacuum state, if driven by an external force, develops into a coherent state. Call t=0 the time at which the driving force ends acting and the deformation is let free to move. The initial state vector of the deformed string is therefore $|\psi(x,0)\rangle=|\{\alpha_k\}\rangle$, where $|\{\alpha_k\}\rangle$, given by Eq.(A-53), indicates a coherent wavepacket of phonons. $|\{\alpha_k\}\rangle$ is such to have a mean energy (but not a fixed energy) and to be composed by a mean (not fixed) number of phonons (Eqs.(A-7) and (A-8)).

The values of α_k ($k\varepsilon$B.Z.), distinguishing a coherent wavepacket from the other, turn out to be related to the initial conditions by Eq.(A-15a). Initial conditions are in turn related to the characteristics of the driving force (pulsed, periodic, etc.) producing the initial deformation.

In the particular case of the pulse here studied, consider $\phi(x)$ of Eq.(36) together with the condition $\langle\hat{p}_k\rangle=0$, as done in Sect.IIIA in the classical limit. From Eq.(A-15a)

$$\alpha_k=(\omega_k\langle\hat{q}_k\rangle+ i\langle\hat{p}_k\rangle)/\sqrt{2\hbar\omega_k} = \left(\frac{\omega_k}{2\hbar}\right)^{\frac{1}{2}}\langle\hat{q}_k\rangle, \tag{57}$$

where $\langle\hat{q}_k\rangle$ means $\langle\{\alpha_k\}|\hat{q}_k|\{\alpha_k\}\rangle$, etc., one obtains

$$\phi(x) = \langle\hat{q}(x,0)\rangle = \frac{1}{\sqrt{L}}\sum_k e^{ikx}\langle\hat{q}_k(0)\rangle$$

$$= \frac{1}{\sqrt{L}}\sum_k e^{ikx}\left(\frac{2\hbar}{\omega_k}\right)^{\frac{1}{2}}\alpha_k. \tag{58}$$

By inverting Eq. (58), the complex values of α_k to be put in $|\psi(x,0)\rangle$ are so obtained.

IV.C. Propagation of a Pulse: Moment and Energy

In Appendix A it is shown that a quantum harmonic oscillator, put initially into a coherent state, remains in a coherent state in

(°) Definition and mathematical properties of coherent or semiclassical states, if introduced at this point, should interrupt the physical discourse. Therefore, they are collected together in Appendix A, which we refer to hereafter. Equations written in the Appendix A are marked by the letter A.

the course of time. This property still holds in the present case, because the N normal modes are decoupled harmonic oscillators. Since Eq.(A-20) is identical to Eq. (40), the splitting of the pulse into two divergent pulses and their propagation are still described by Eq.(42), provided that q(x,t) is substituted by $<\hat{q}(x,t)>$ in the quantum limit. Also the same considerations about sound, reported in Sects.III.A and III.B, can be drawn in the quantum case.

In conclusion the following very important result may be stated: all *the running waves are represented in quantum mechanics by coherent wavepackets of phonons.* In particular, also monochromatic sound (Eq.(43)) corresponds to a Poisson distribution of phonons (Appendix B).

Moment and energy of the pulse can now be evaluated. The result is the following.

1. Moment. From Eq. (33b), with Eqs. (32b) and (A-21), one obtains

$$p(x,t) = <\hat{p}(x)>_t = \tfrac{1}{2}\, v[\dot{\phi}_x(x-vt) - \dot{\phi}_x(x+vt)] \qquad (59a)$$

$$= p^+(x,t) + p^-(x,t), \qquad (59b)$$

where $\dot{\phi}_x(x,t)$ is the x-derivative of $\phi(x,t)$ of Eq.(36). $p^+(x,t)$ and $p^-(x,t)$ are the momenta of the right side and left side travelling pulses, respectively. Since at t=0 p(x)=0, also p(x,t)=0 as it is clear from Eq.(59): *total moment is conserved.*

2. Energy. From Eqs. (A-56a), (57) and (9), one obtains

$$\bar{E} = <\hat{H}_o> = \Sigma_k \hbar\omega_k \left(\frac{\omega_k}{2\hbar} <\hat{q}_k(0)>^2 + \tfrac{1}{2}\right)$$

$$= \tfrac{1}{2}\, (\Sigma_k <\hat{q}_k(0)>^2 k^2) v^2 + E_{zp} \qquad (60)$$

Half of this energy goes at velocity v to the right side of the string and half to the left. When $\omega_k <\hat{q}_k(0)>^2 >> \hbar$, as in the classical limit, one obtains the interesting result

$$\bar{E} = \tfrac{1}{2}(M^+ + M^-)v^2 \qquad (61)$$

which looks like the kinetic energy of two particles of masses $M^+ = M^- = \tfrac{1}{2}\Sigma_k <\hat{q}_k(0)>^2 k^2$ propagating at constant velocity v along opposite directions. M^+ and M^- are related to the composition of the pulse in normal modes through the derivatives of $\phi(x)$.

V. BIRTH AND DEATH OF A PULSE [9,10]

Coherent states of phonons have been studied in the previous sections by using the continuous limit of the elastic vibrating chain and by considering the region $\omega_k = v\,k$ of the frequency dispersion. The conclusions are however more general. On microscopic scale, as when considering molecular and solid state physics, coherent states are in fact created every time an external force acts on a group of ions for a finite time, like during mechanical shocks, chemical reactions and even as a consequence of an electronic excitation. These coherent wavepackets then propagate according to their recipe (A-15) in terms of phonons and to the elastic and geometrical characteristics of the medium.

The 1-dim harmonic chain is indeed a very idealized system. Actually, the pulse dissolves also in 1-dim media, the sooner the more dispersive the medium and the stronger the anharmonicity, unless the non linear terms are so peculiar as to build a soliton.

In 2-dim and 3-dim cases the pulse always decreases with distance, owing to geometrical factors, unless it corresponds to a plane wave.

In both cases, the elastic medium reaches a stationary thermal equilibrium, when the initial coherence among phonons has been destroyed and independent incoherent phonons are present with a weight given by the thermal distributions. Therefore, in evaluating average values, the Poisson distribution over the phonons of the coherent wavepacket must be substituted by an appropriate exponential distribution.

In terms of energy, death of the pulse means that the initial free energy (61) (the entropy of a coherent wavepacket is zero!) has been completely and irreversibly transformed into heat.

REFERENCES

1. M. Born and K. Huang, Dynamical Theory of Crystal Lattices (Oxford University Press, London, 1954).
2. E. M. Henley and W. Thirring, Elementary Quantum Field Theory (McGraw-Hill, New York, 1962).
3. A. A. Maradudin, E. W. Montroll, G. H. Weiss, and I. P. Ipatova, in Solid State Physics (Academic, New York, 1972), Suppl. 3.
4. W. Jones and N. H. March, Theoretical Solid State Physics (Wiley Interscience, London, 1973).
5. J. A. Reissland, The Physics of Phonons (Wiley, London, 1973).
6. H. Bilz and W. Kress, Phonon Dispersion Relations in Insulators (Springer, Berlin, 1979).

7. G. Venkataraman, L. A. Feldkamp, and V. C. Sahni, Dynamics of Perfect Crystals (The M.I.T. Press, Cambridge, Mass., 1975).
8. A. Sommerfeld, Mechanics of Deformable Bodies (Academic, New York, 1964).
9. W. H. Louisell, Quantum Statistical Properties of Radiation (Wiley, New York, 1973).
10. A. Giorgetti and N. Terzi, Solid State Comm. 39, 635 (1981).

APPENDICES

Appendix A. The Coherent States [1,4]

The hamiltonian $\hat{H}_o$ of a linear harmonic oscillator, as function of its position $\hat{q}$ and its momentum $\hat{p}$, is

$$\hat{H}_o = \tfrac{1}{2}(\omega^2\hat{q}^2 + \hat{p}^2), \quad \text{(A-1a)}$$

where the conjugated (hermitian) operators $\hat{q}$ and $\hat{p}$ do not commute: $[\hat{q},\hat{p}] = i\hbar$.

$\hat{H}_o$ can also be written

$$\hat{H}_o = \hbar\omega(\hat{a}^+\hat{a}+\tfrac{1}{2}) \quad \text{(A-1b)}$$

in terms of the (non-hermitian) annihilation and creation operators $\hat{a}$ and $\hat{a}^+$, where

$$\hat{a} = \frac{1}{\sqrt{2\hbar\omega}}(\omega\hat{q} + i\hat{p}) \text{ and } \hat{a}^+ = \frac{1}{\sqrt{2\hbar\omega}}(\omega\hat{q} - i\hat{p}) \quad \text{(A-2a)}$$

(with their inverse transformations

$$\hat{q} = \frac{\hat{a} + \hat{a}^+}{\sqrt{2\omega/\hbar}} \quad\text{and}\quad \hat{p} = i\omega\,\frac{\hat{a}^+ - \hat{a}}{\sqrt{2\omega/\hbar}}) \quad \text{(A-2b)}$$

and where $[\hat{a},\hat{a}^+] = 1$.

In the socalled (phonon) number basis $|n>$

$$\hat{a}^+\hat{a}|n> = n|n> \qquad (n\in \mathbb{N}^+) \quad \text{(A-3)}$$

$$|n> = \frac{(\hat{a}^+)^n}{\sqrt{n!}}|0>,$$

eigenvalues and eigenfunctions of $\hat{H}_o$ are

$$\hat{H}_o|n> = \hbar\omega\,(n+\tfrac{1}{2})|n> . \quad \text{(A-1c)}$$

1. Definition of the Coherent State $|\alpha>$. A coherent state $|\alpha>$ is the eigenvector, solution of the equation

$$\hat{a}|\alpha\rangle = \alpha|\alpha\rangle \tag{A-4a}$$

where (α^* is the complex conjugated value of α)

$$\langle\alpha|\hat{a}^+ = \alpha^*\langle\alpha| . \tag{A-4b}$$

By using the basis $|n\rangle$ of Eq.(A-3), $|\alpha\rangle$ is given to a phase factor by

$$|\alpha\rangle = e^{-\frac{1}{2}|\alpha|^2} \sum_n \frac{\alpha^n}{\sqrt{n!}} |n\rangle \tag{A-5a}$$

$$= e^{-\frac{1}{2}|\alpha|^2} \sum_n \frac{1}{\sqrt{n!}} |\alpha|^n e^{in\phi}|n\rangle \tag{A-5b}$$

$$= e^{-\frac{1}{2}|\alpha|^2} \sum_n \frac{1}{n!} (\alpha \ \hat{a}^+)^n|0\rangle \tag{A-5c}$$

$$= e^{-\frac{1}{2}|\alpha^2|} e^{\alpha\hat{a}^+}|0\rangle . \tag{A-5d}$$

because

$$\langle n|\alpha\rangle = e^{-\frac{1}{2}|\alpha|^2} \frac{\alpha^n}{\sqrt{n!}} = \langle 0|\alpha\rangle \frac{\alpha^n}{\sqrt{n!}} . \tag{A.6a}$$

2. Some Properties.

I. $\langle\alpha|\alpha\rangle = 1$: $|\alpha\rangle$ are normalized.

II. $|\alpha\rangle$ do *not* form an *orthogonal* set, because $\hat{a}$ is not hermitian:

$$\langle\beta|\alpha\rangle = e^{-\frac{1}{2}|\alpha|^2 - \frac{1}{2}|\beta|^2 + \alpha\beta^*}$$

But

$$|\langle\beta|\alpha\rangle|^2 = e^{-|\alpha-\beta|^2} \to 0 \text{ when } |\alpha - \beta| \to \infty$$

III. For the same reason, $|\alpha\rangle$ form an *overcomplete* set: there are infinitely many different ways to expand an arbitrary state in terms of $|\alpha\rangle$. However, there is a resolution of the identity through $|\alpha\rangle$ states, which resembles in the complex α-plane to a "closure" property:

$$\frac{1}{\pi} \int |\alpha\rangle d^2\alpha \langle\alpha| = \frac{1}{\pi} \int |\alpha\rangle d\mathrm{Re}\alpha \ d\mathrm{Im}\alpha \langle\alpha| = 1.$$

IV. $$|\langle n|\alpha\rangle|^2 = e^{-|\alpha|^2}(|\alpha|^2)^n/n! \tag{A.6b}$$

is a Poisson distribution (See Appendix B) with respect to the (phonon) number states.

V. $|\alpha=0\rangle = |n=0\rangle$: the vacuum state $|n=0\rangle$ is the coherent state with eigenvalue $\alpha=0$.

3. Significant Averages: $\langle\hat{O}\rangle \equiv \langle\alpha|\hat{O}|\alpha\rangle$.

I. $$\langle\hat{n}\rangle \equiv \langle\alpha|\hat{a}^{+}\hat{a}|\alpha\rangle = |\alpha|^2 \tag{A.7}$$

II. $$\langle\hat{H}_o\rangle = \langle\hbar\omega(\hat{a}^{+}\hat{a} + \tfrac{1}{2})\rangle = \hbar\omega(|\alpha|^2 + \tfrac{1}{2}) \tag{A.8}$$

III. $$\langle\hat{q}\rangle = \sqrt{\frac{\hbar}{2\omega}}\,(\alpha+\alpha^*) \tag{A.9}$$

$$\langle\hat{q}^2\rangle = \frac{\hbar}{2\omega}\,(\alpha^{*2}+\alpha^2+2|\alpha|^2+1) \tag{A.10}$$

$$(\Delta q)^2 = \langle\hat{q}^2\rangle-\langle\hat{q}\rangle^2 = \hbar/2\omega \tag{A.11}$$

IV. $$\langle\hat{p}\rangle = i\sqrt{\frac{\hbar\omega}{2}}(\alpha^* - \alpha\ ) \tag{A-12}$$

$$\langle\hat{p}^2\rangle = -\frac{\hbar\omega}{2}\,(\alpha^{*2}+\alpha^2-2|\alpha|^2-1) \tag{A-13}$$

$$(\Delta p)^2 = \langle\hat{p}^2\rangle-\langle\hat{p}\rangle^2 = \hbar\omega/2. \tag{A-14}$$

V. Then, from Eqs. (A-9) and (A-12), it follows:

$$\alpha = \frac{1}{\sqrt{2\hbar\omega}}\,(\omega\langle\hat{q}\rangle + i\langle\hat{p}\rangle) = |\alpha|e^{i\phi} \tag{A-15a}$$

where

$$|\alpha|^2 = \frac{1}{2\hbar\omega}\,(\omega^2\langle\hat{q}\rangle^2+\langle\hat{p}\rangle^2) = E_{cl}/\hbar\omega \tag{A-15b}$$

$$\mathrm{tg}\phi = \langle\hat{p}\rangle/\omega\langle\hat{q}\rangle\ . \tag{A-15c}$$

VI. $$\Delta p\Delta q = \hbar/2 \tag{A-16}$$

VII. It is worth stressing that

$$\langle\hat{H}_o\rangle = \tfrac{1}{2}(\omega^2\langle\hat{q}\rangle^2+\langle\hat{p}\rangle^2) + \tfrac{1}{2}\hbar\omega = E_{cl} + E_{zp}, \tag{A-17}$$

as it follows from Eqs. (A-8) and (A-15b).

4. Time Evolution

I. Operators (Heisenberg picture). In the harmonic approximation, from

$$i\hbar \frac{d\hat{a}}{dt} = [\hat{a}, \hat{H}_o] = \hbar\omega\hat{a}$$
$$i\hbar \frac{d\hat{a}^+}{dt} = [\hat{a}^+, \hat{H}_o] = -\hbar\omega\hat{a}^+ \quad \text{(A-18a)}$$

one has

$$\hat{a}(t) = \hat{a}_o e^{-i\omega t}$$
$$\hat{a}^+(t) = \hat{a}_o^+ e^{i\omega t} \quad \text{(A-18b)}$$

Since $|\alpha\rangle = |\alpha(0)\rangle$, from Eq. (A-1) one obtains

$$\langle\hat{a}\rangle_t = \langle\alpha|\hat{a}(t)|\alpha\rangle = \alpha e^{-i\omega t} \text{ and } \langle\hat{a}^+\rangle_t = \alpha^* e^{i\omega t} \quad \text{(A-19)}$$

and from Eqs. (A-9) and (A-12)

$$\langle\hat{q}\rangle_t = \langle\alpha|\sqrt{\frac{\hbar}{2\omega}}\,(\hat{a}(t)+\hat{a}^+(t))|\alpha\rangle = \langle\hat{q}\rangle_o \cos\omega t + \frac{1}{\omega}\langle\hat{p}\rangle_o \sin\omega t, \quad \text{(A-20)}$$

$$\langle\hat{p}\rangle_t = \langle\alpha|i\sqrt{\frac{\hbar\omega}{2}}\,(\hat{a}^+(t)-\hat{a}(t))|\alpha\rangle = \langle\hat{p}\rangle_o \cos\omega t - \omega\langle\hat{q}\rangle_o \sin\omega t: \quad \text{(A-21)}$$

the mean values of the position and momentum operators behave like the corresponding variables of the classical harmonic oscillator.

II. State vectors (Schrödinger picture). The Schrödinger equation of motion of the state vector $|\psi(t)\rangle$

$$i\hbar \frac{\partial|\psi(t)\rangle}{\partial t} = \hat{H}|\psi(t)\rangle \quad \text{(A-22a)}$$

has the formal solution

$$|\psi(t)\rangle = u(t,0)|\psi(0)\rangle\,, \quad \text{(A-22b)}$$

where

$$i\hbar \frac{\partial u}{\partial t} = \hat{H}u, \text{ where } u(0,0) = 1 \quad \text{(A-23)}$$

Consider first the *conservative system* formed by a free oscillator in harmonic approximation. The Hamiltonian $\hat{H} = \hat{H}_o = \hbar\omega(\hat{a}^+\hat{a}+\frac{1}{2})$ is time independent, so that $u(t,0) = \exp(-i\hat{H}_o t/\hbar)$. $|\psi(t)\rangle$ turns out to be the wave packet

$$|\psi(t)\rangle = e^{-i\omega(\hat{a}^+\hat{a}+\frac{1}{2})t}|\psi(0)\rangle$$
$$= \sum_n \langle n|\psi(0)\rangle\, e^{-i\omega(n+\frac{1}{2})t}|n\rangle. \qquad \text{(A-24)}$$

Assume that at $t=0$ the oscillator is in the coherent state $|\psi(0)\rangle = |\alpha\rangle$. At $t>0$ the harmonic oscillator is still in a coherent state because using Eq. (A-3) in Eq. (A-24) one finds

$$|\psi(t)\rangle = e^{-\frac{1}{2}(|\alpha|^2+i\omega t)}\sum_n \frac{1}{\sqrt{n!}}(\alpha e^{-i\omega t})^n|n\rangle \ : \qquad \text{(A-25)}$$

a free harmonic oscillator put initially into a coherent state, remains in a coherent state.

Notice that, since $|\psi(t)\rangle = \exp(-i\hat{H}_o t/\hbar)|\alpha\rangle$ and

$$\langle\psi(0)|\psi(t)\rangle = e^{-\frac{1}{2}|\alpha|^2}\sum_n \frac{(\alpha^*)^n}{\sqrt{n!}}\, e^{-\frac{1}{2}(|\alpha|^2+i\omega t)}\sum_{n'}\frac{(\alpha e^{-i\omega t})^{n'}}{\sqrt{n'!}}\,\delta_{nn'}$$

$$= e^{-\frac{1}{2}i\omega t}\exp\{|\alpha|^2(e^{-i\omega t}-1)\}, \qquad \text{(A-26)}$$

we obtain the useful relation

$$\langle\alpha|e^{-i\omega a^+ at}|\alpha\rangle = \exp|\alpha|^2(e^{-i\omega t}-1) \ . \qquad \text{(A-27)}$$

The study of the non*conservative system* formed by an oscillator driven by an external force involve some mathematics and will be not treated here. However, the result is still simple in a very important case, when at $t=0$ the oscillator is in a coherent state and the time-dependent hamiltonian can be written ($f(t)$ is a complex-valued function) [1]:

$$\hat{H} = \hbar\omega\hat{a}^+\hat{a} + \hbar\{f(t)\hat{a}+f^*(t)\hat{a}^+\}. \qquad \text{(A-28)}$$

Then, at $t>0$ $|\psi(t)\rangle$ is still a coherent wave packet. In other words, the forcing term in Eq.(A-28) while changing amplitude and phase of $|\psi(t)\rangle$, does not destroy the coherence between the $|n\rangle$ terms in $|\psi(t)\rangle$. In particular, if $|\psi(0)\rangle = |0\rangle$, the phonon vacuum state, the coherent state $|\psi(t)\rangle$ into which $|0\rangle$ develops is

$$|\psi(t)\rangle = \exp\{A(t) + C(t)\hat{a}^{+}\}|0\rangle \tag{A-29}$$

$$A(t) = -\int_0^t dt'' \, f(t'') \int_0^{t''} dt' \, f(t') \, e^{i\omega(t'-t'')} \tag{A-30}$$

$$C(t) = -i \int_0^t dt' \, f^{*}(t') \, e^{i\omega(t'-t)} \quad . \tag{A-31}$$

5. Coordinate Representation (q-rep). Consider now the coordinate presentation $\psi(q,t)$ of the state vector $|\psi(t)\rangle$ (Schrödinger picture) of the free harmonic oscillator:

$$\psi(q,t) \equiv \langle q|\psi(t)\rangle = \sum_n \langle q|n\rangle\langle n|\psi(0)\rangle e^{-i\omega(n+\frac{1}{2})t} \tag{A-32}$$

where

$$\langle q|n\rangle = (\omega/\hbar\pi(2^n n!)^2)^{1/4} \, H_n(q\sqrt{\frac{\omega}{\hbar}}) \, e^{-\frac{1}{2}(q\sqrt{\frac{\omega}{\hbar}})^2} \tag{A-33}$$

and $H_n(x)$ is the Hermite polynomial of order n.

Assume as before that at t=0 $|\psi(0)\rangle=|\alpha\rangle$: $\psi(q,t)$ can then be evaluated directly from relation (A-32), by using the definition of $|\alpha\rangle$.

It is however simpler to work in the Heisenberg picture and express $|\psi(0)\rangle$ in terms of the eigenstates of $\hat{q}(t)$: $\psi(q,t)=\langle q(t)|\psi(0)\rangle$. By using the definition of $\hat{a}$ in terms of $\hat{q}$ and $\hat{p}$ and (A-18b), one obtains

$$\frac{1}{\sqrt{2\hbar\omega}} (\omega\hat{q} + i\hat{p}) = \hat{a}(0)e^{-i\omega t}. \tag{A-34}$$

By projecting Eq.(A-1a) with Eq.(A-34) on $|q(t)\rangle$ and recalling that $p =(\hbar/i)d/dq$, one obtains the differential equation ($\alpha=\alpha(0)$)

$$(\omega q+\hbar d/dq)\psi(q,t) = \sqrt{2\hbar\omega} \, \alpha \, e^{-i\omega t} \, \psi(q,t), \tag{A-35}$$

whose solution is

$$\psi(q,t) = N_1 \exp(- \frac{\omega}{2\hbar} q^2+\sqrt{\frac{2\omega}{\hbar}} \alpha \, e^{-i\omega t} q) \tag{A-36a}$$

$$= N_2 \exp\{- \frac{\omega}{2\hbar} (q-\sqrt{\frac{2\hbar}{\omega}} \alpha e^{-i\omega t})^2\} \tag{A-36b}$$

$$= N_3 \exp\{- \frac{\omega}{2\hbar} q^2+\sqrt{\frac{2\omega}{\hbar}} \alpha q - \tfrac{1}{2}|\alpha|^2(1+e^{2i\phi})\} \tag{A-36c}$$

where N_i are normalizing constants. On normalizing one finds

$$\psi(q,t)=\left(\frac{\omega}{\hbar\pi}\right)^{1/4}\exp\left\{-\frac{\omega}{2\hbar}\left[q-\sqrt{\frac{2\hbar}{\omega}}\,\alpha(\cos\omega t+2i\sin\omega t)\right]\times\left(q-\sqrt{\frac{2\hbar}{\omega}}\,\alpha\cos\omega t\right)\right\} \quad \text{(A-36d)}$$

and

$$|\psi(q,t)|^2=\left(\frac{\omega}{\hbar\pi}\right)^{\frac{1}{2}}e^{-(\omega/\hbar)\left(q-\alpha\sqrt{\frac{2\hbar}{\omega}}\cos\omega t\right)^2}. \quad \text{(A-37)}$$

If one introduces Eqs. (A-15)-(A-20), $\psi(q,t)$ can be also written (to a phase factor) in terms of the averages $\langle\hat{q}\rangle_t$ and $\langle\hat{p}\rangle_t$ as follows

$$\psi(q,t)=\langle q(t)|\alpha\rangle=\left(\frac{\omega}{\hbar\pi}\right)^{1/4}\exp\left\{-\frac{\omega}{2\hbar}(q-\langle\hat{q}\rangle_t)^2+\frac{1}{\hbar}\langle\hat{p}\rangle_t\, q\right\}. \quad \text{(A-38)}$$

In the form (A-38), it is easy to see that the coherent w.p. $|\alpha\rangle$ is represented in q-rep. by a plane wave modulated by a gaussian,

$$\exp\left\{-\frac{(q-\langle\hat{q}\rangle_t)^2}{2\sigma^2}\right\}$$

of variance $\sigma^2=\hbar/\omega=2(\Delta q)^2$.

6. Why Coherent. Be a state vector $|\psi(t)\rangle$ of the harmonic oscillator represented at t=0 by the most general wave packet of number states $|n\rangle$

$$|\psi(0)\rangle=\sum_n\langle n|\psi(0)\rangle\,|n\rangle=\sum_n c_n|n\rangle, \quad \text{(A-39a)}$$

where the coefficients $c_n=\langle n|\psi(0)\rangle=|c_n|e^{i\phi_n}$ take arbitrary complex values but for two conditions

$$\sum_n|c_n|^2=1 \quad \text{because } \langle\psi(0)|\psi(0)\rangle=1,$$

and

$$\langle\psi(0)|\hat{H}_0|\psi(0)\rangle=\sum_n E_n|c_n|^2 \text{ where } E_n=\hbar\omega(n+\tfrac{1}{2}).$$

$|c_n|$ is the amplitude and ϕ_n the phase with which each stationary state $|n\rangle$ enters into the composition of the wave packet.

Deduce now how ϕ_n and $|c_n|$ depend on n when $|\phi(0)\rangle$ is a coherent wave packet. First notice that all c_n can be written in terms of only two complex parameters c_0 and α, because from Eq. (A-4)

$$c_n = \langle n|\alpha\rangle = \frac{\alpha^n}{\sqrt{n!}} c_0 .$$

Since

$$\langle\psi(0)|\psi(0)\rangle = |c_0|^2 \sum_n \frac{(|\alpha|^2)^n}{n!} = |c_0|^2 e^{|\alpha|^2} = 1,$$

c_0 can be chosen to have the real value

$$c_0 = \exp - \tfrac{1}{2}|\alpha|^2 .$$

Then, from Eq. (A-39), $|c_n|$ and ϕ_n are found to depend only on α:

$$|\alpha\rangle = \sum_n |c_n| e^{+i\phi_n}|n\rangle = \sum_n e^{-\frac{1}{2}|\alpha|^2} \frac{1}{\sqrt{n!}} |\alpha|^n e^{in\phi}|n\rangle : \qquad \text{(A-39b)}$$

the *states* $|n\rangle$ *sum up coherently* in $|\alpha\rangle$ in the sense that their dephasing $\phi=\phi_n-\phi_{n-1}$ is constant and their relative amplitude decreases constantly with n: $|c_n| = \alpha|c_{n-1}|/\sqrt{n}$, or $(|c_{n-1}|^2-|c_n|^2)/|c_n|^2 = (n-\langle n\rangle)/\langle n\rangle$.
$|\alpha|$ and ϕ are called the amplitude and the phase, respectively, of the whole coherent state $|\alpha\rangle$.

7. Why Minimum Uncertainty. Given two operators $\hat{A}$ and $\hat{B}$ which do not commute

$$\hat{A}\hat{B} - \hat{B}\hat{A} = i\hat{C}, \qquad \text{(A-40)}$$

a minimum uncertainty wave packet is such to make the product of the uncertainties $\Delta A = \sqrt{\langle\hat{A}^2\rangle-\langle\hat{A}\rangle^2}$ and $\Delta B = \sqrt{\langle\hat{B}^2\rangle-\langle\hat{B}\rangle^2}$ as small as the noncommutativity:

$$\Delta A \Delta B = \tfrac{1}{2}|\langle\hat{C}\rangle| . \qquad \text{(A-41)}$$

It can be shown that such wave packet is the solution of the equation [3] :

$$(\hat{B}-\langle\hat{B}\rangle)|\psi\rangle = i \frac{\langle\hat{C}\rangle}{2(\Delta A)^2}(\hat{A}-\langle\hat{A}\rangle)|\psi\rangle . \qquad \text{(A-42)}$$

We are interested in the case $\hat{A} = \hat{q}$ and $\hat{B} = \hat{p}$, whose commutation is $[\hat{q},\hat{p}] = i\hbar$. When Eq. (A-42) is evaluated in q-rep, one finds

$$\left(\frac{\hbar}{i}\frac{d}{dq} - \langle\hat{p}\rangle\right)\psi(q) = \frac{i\hbar}{2(\Delta q)^2}(q-\langle\hat{q}\rangle)\psi(q), \qquad \text{(A-43)}$$

whose normalized solution is

$$\psi(q) = \left(\frac{1}{2\pi(\Delta q)^2}\right)^{1/4} \exp\left\{-\frac{(q-\langle\hat{q}\rangle)^2}{4(\Delta q)^2} + i\frac{\langle\hat{p}\rangle}{\hbar} q\right\} \qquad \text{(A-44)}$$

$\psi(q)$ is a plane wave ($\sim\exp i \frac{\langle\hat{p}\rangle q}{\hbar}$) modulated by a gaussian. Such a wave packet has by construction (Eq.(A-41))

$$\Delta q \Delta p = \tfrac{1}{2}\hbar, \tag{A-45}$$

but still depends on the arbitrary value of Δq. However, among all the possible minimum uncertainty wavepackets (Eq. (A-44)), only one that with the constant uncertainty $(\Delta q)^2=\hbar/2\omega$, remains minimum in the course of time. This is just the coherent state given by Eq. (A-38), because of Eq. (A-11).

8. Why Semiclassical. Ehrenfest's equations

$$\frac{d}{dt}\langle\hat{q}\rangle = \frac{\langle\hat{p}\rangle}{m} \text{ and } \frac{d}{dt}\langle\hat{p}\rangle = -\langle\frac{d\hat{H}}{d\hat{q}}\rangle, \tag{A-46}$$

describing the motion of the average values $\langle\hat{q}\rangle$ and $\langle\hat{p}\rangle$ over $|\psi(t)\rangle$ give the correspondence between quantum and classical mechanics.

They reduce formally to the classical equations of motion for the average values of a harmonic oscillator in a coherent state, because then (see Eqs.(A-19) and (A-8)).

$$\langle\hat{H}\rangle = \tfrac{1}{2}(\omega^2\langle\hat{q}\rangle^2 + \langle\hat{p}\rangle^2) + \tfrac{1}{2}\hbar\omega = E_{cl} + \tfrac{1}{2}\hbar\omega \tag{A-47}$$

and $$\langle\frac{d\hat{H}}{d\hat{q}}\rangle = \omega^2\langle\hat{q}\rangle . \tag{A-48}$$

Since a coherent w.p. does not spread in time (Eqs.(A-11) and (A-14)), the Ehrenfest's eqs. maintain a physical meaning at all times for the harmonic oscillator : *the coherent state is likened to a harmonic classical particle*. When $|\alpha|^2 \gg \frac{1}{2}$, $\langle\hat{H}\rangle \sim E_{cl}$: the mean energy over the coherent state coincides with the classical energy of the harmonic oscillator, when the zero-point energy $E_{z.p}=\frac{1}{2}\hbar\omega$ becomes negligible part of the total energy.

9. Why Displaced. Compare the probability distribution in q-rep of the vacuum state and of $|\alpha\rangle$, given by Eqs. (A-33) and (A-37) respectively:

$$|\langle q|0\rangle|^2 = \left(\frac{\omega}{\hbar\pi}\right)^{\frac{1}{2}} e^{-\omega q^2/\hbar} \tag{A-49}$$

$$|\langle q|\alpha\rangle|^2 = \left(\frac{\omega}{\hbar\pi}\right)^{\frac{1}{2}} \exp\{-\frac{\omega}{\hbar}(q-\sqrt{\frac{2\hbar}{\omega}}\,\alpha\cos\omega t)^2\}. \tag{A-50}$$

The gaussian equation (A-50) associated to the coherent state corresponds to that of the vacuum state rigidly displaced of the value $q_o=\sqrt{2\hbar/\omega}\ \alpha\cos\omega t$: the coherent wavepacket probability oscillates in time around q=0, withoug spreading and whatever be $<n>=|\alpha|^2$, like a classical harmonic oscillator shaped as the vacuum state probability.

10. Multimode Coherent State. In the text it is deduced the hamiltonian $\hat{H}_o$ of a system of elastically coupled oscillators. In terms of decoupled normal modes $\hat{q}_k$, H_o is

$$\hat{H}_o = \tfrac{1}{2} \sum_k (\hat{p}_k\hat{p}_k^+ + \omega_k^2\hat{q}_k\hat{q}_k^+) \tag{A-51a}$$

$$= \tfrac{1}{2} \sum_k \hbar\omega_k(\hat{a}_k^+\hat{a}_k+\tfrac{1}{2}) = \sum_k\hat{H}_k, \tag{A-51b}$$

where the relations between the operators $\hat{q}_k$ and $\hat{p}_k$ and the operators $\hat{a}_k$ $\hat{a}_k^+$ are the same as those in Eqs.(A-2) and (49)-(50) of the text.

The eigenvalues $E(\{n_k\})$ and the eigenvectors $|\{n_k\}>$ of Eq. (A-49) are

$$E(\{n_k\}) = \sum_k E(n_k) = \sum_k\hbar\omega_k(n_k + \tfrac{1}{2}), \tag{A-52a}$$

$$|\{n_k\}> = \Pi_k|n_k> = (n_1!n_2!\ldots)^{-\frac{1}{2}}(\hat{a}_1^+)^{n_1}(\hat{a}_2^+)^{n_2}\ldots|0> . \tag{A-52b}$$

By using Eqs.(A-4) for each normal mode k, a multimode coherent state $|\{\alpha_k\}>$ can be defined :

$$|\{\alpha_k\}> \equiv \Pi_k|\alpha_k> \tag{A-53a}$$

$$= \Pi_k e^{-\frac{1}{2}|\alpha_k|^2}\sum_k \frac{(\alpha_k)^{n_k}}{\sqrt{n_k!}}\ |n_k> \tag{A-53b}$$

$$= e^{-\frac{1}{2}|\alpha|^2}(\exp \sum_k\alpha_k\hat{a}_k^+)|0>, \tag{A-53c}$$

where

$$|\alpha|^2 = \sum_k|\alpha_k|^2. \tag{A-54}$$

$|\{\alpha_k\}>$ has the same properties of the one-mode coherent state. In particular:

$$<\hat{q}_k> = <\{\alpha_k\}|\hat{q}_k|\{\alpha_k\}> =<\alpha_k|\hat{q}_k|\alpha_k>=\sqrt{\hbar/2\omega_k}(\alpha_k+\alpha_k^*) \tag{A-55}$$

$$<\hat{H}_o> = <\{\alpha_k\}|\hat{H}_o|\{\alpha_k\}> = \sum_k \hbar\omega_k(|\alpha_k|^2+\tfrac{1}{2}) \tag{A-56a}$$

$$= \tfrac{1}{2}\sum_k(\omega_k^2<\hat{q}_k>^2 + <\hat{p}_k>^2)+\tfrac{1}{2}\sum_k \hbar\omega_k \tag{A-56b}$$

$$= E_{cl} + E_{zp}.$$

All the considerations done for $|\alpha>$ also hold for $|\{\alpha_k\}>$.

Appendix B. The Poisson Distribution [5]

Given a (one dimensional) probability distribution P(n) of the discrete variable $n(n\in \mathbb{N}^+)$, and by indicating

$$\overline{f(n)} \equiv \sum_n f(n)P(n), \tag{B-1}$$

one can define the following *functions*

generating function: $\gamma(\beta) = \overline{\beta^n}$ $(\beta\in\mathbb{C})$ (B-2)

moments generating f: $M(\beta) = \overline{e^{\beta n}}=\gamma(e^\beta)$ $(\beta\in\mathbb{C})$ (B-3)

characteristic function: $\chi(k) = \overline{e^{ikn}} = \gamma(e^{ik})$ $(k\in\mathbb{R})$, (B-4)

and the following *parameters* $(\ell\in\mathbb{N}^+)$

moments: $m_\ell = \overline{n^\ell} = i^{-\ell}\frac{d^\ell}{dk^\ell}\chi(k)\Big|_{k=0} =$

$$= \frac{d^\ell}{d\beta^\ell}M(\beta)\Big|_{\beta=0} \tag{B-5}$$

central moments: $\mu_\ell=\overline{(n-\overline{n})^\ell}$ (B-6)

factorial moments: $\nu_\ell=\overline{n(n-1)\dots(n-\ell+1)}= \frac{d^\ell}{d\beta^\ell}\gamma(\beta)\Big|_{\beta=1}$ (B-7)

semiinvariants or cumulants:

$$c_\ell = i^{-\ell}\frac{d^\ell}{dk^\ell}\ln\chi(k)\Big|_{k=0} =$$

$$= \frac{d^\ell}{d\beta^\ell}\ln M(\beta)\Big|_{\beta=0}. \tag{B-8}$$

The following relations hold:

$m_0 = 1$:normalization of P(n) (B-9)

$m_1 = \bar{n} = \nu_1 = c_1$: average of P(n) (B-10)

$\mu_2 = \sigma^2 = m_2 - \bar{n}^2 = \nu_2 - \bar{n}(\bar{n}-1) = c_2$: variance σ^2 of P(n) (B-11)

$\mu_3 = c_3$ and $\mu_4 = c_4 + 3c_2^2$ (B-12)

$m_2 = \nu_2 + \nu_1 = c_2 + c_1^2$ (B-13)

$m_3 = \nu_3 + 3\nu_2 + \nu_1 = c_3 + 3c_1c_2 + c_1^3$ (B-14)

Provided $M(\beta)$ and $\gamma(\beta+1)$ are analytic at $\beta=0$, functions and parameters are related by the series ($\ell \in \mathbb{N}^+$):

$$M(\beta) = \sum_\ell m_\ell \beta^\ell/\ell! \quad \text{(B-15)}$$

$$\ln M(\beta) = \sum_\ell c_\ell \beta^\ell/\ell! \quad \text{(B-16)}$$

$$\gamma(\beta+1) = \sum_\ell \nu_\ell \beta^\ell/\ell!. \quad \text{(B-17)}$$

1. Poisson Distribution. The Poisson distribution P(n) is

$$P(n) = e^{-\bar{n}} \frac{\bar{n}^n}{n!} \quad (n \in \mathbb{N}^+ \text{ and } \bar{n}>0) \quad \text{(B-18)}$$

and has the following properties

I. The generating function is

$$\gamma(\beta) = e^{\bar{n}(\beta-1)}, \text{ or } \gamma(\beta+1) = e^{\bar{n}\beta} \quad \text{(B-19)}$$

so that

$$\nu_\ell = \bar{n}^\ell. \quad \text{(B-20)}$$

II. All the *moments* m_ℓ are different from zero.

III. *Cumulants:* $c_0=0$ and $c_\ell=\bar{n}$ ($\ell=1,2,\ldots$). This follows from Eq. (B-16) when use is made of Eqs. (B-19) and (B-3).

IV. $m_1 = \bar{n}$ and $\sigma^2 = \bar{n}^2$,

$$m_2 = \bar{n}(\bar{n}+1), \quad m_3 = \bar{n}(\bar{n} + 3\bar{n}+1), \quad \text{(B-21)}$$

$\mu_3 = \bar{n}$, $\mu_4 = 3\bar{n}^2 + \bar{n}$.

V. $\mu_3/\sigma^3 = \bar{n}^{-\frac{1}{2}}$ in the so called coefficient of skewness and

$(\mu_4/\sigma^4)-3 = \bar{n}^{-1}$ in the so called coefficient of excess.

2. Perfect Gas in a Container. Consider a gas of N_0 non interacting particles in a container of volume V_0. Be such a system in equilibrium and isolated. Focus attention on an ideal subvolume v of the container and denote by n ($n=0,\ldots,N_0$) the number of particles that are found to be located within v.

Since the particles are non-interacting, they are in average uniformely distributed within V_0. The probability p that a given particle be found in v is given by $p=v/V_0$, so that the average number $\bar{n}$ of particles in v is

$$\bar{n} = p\, N_0 = v\,\frac{N_0}{V_0} = v\rho_0. \qquad \text{(B-1)}$$

In other words, since the gas is uniformly distributed, the average density in v is equal to the number density in V_0: $\bar{n}/v = N_0 V_0 = \rho_0$.

We ask now for the probability P(n) that during a measurement involving v, a given number n of particles is found inside v. To this end assume that $v \ll V_0$, so that $p \ll 1$. This is the situation when the container is made infinitely large ($V_0 \to \infty$) while the particle density ρ_0 remains constant. Then, also $N_0 = \rho_0 V_0 \to \infty$ and $n \ll N_0$, since n is finite. The probability P(n) of finding n particles in v is equal to the probability W(n) that an event, characterized by a probability $p=v/V_0$, occurs n times out of N_0 independent trials. W(n) is given by the well known binomial distribution

$$W(n) = \frac{N_0!}{n!(N_0-n)!}\, p^n (1-p)^{N_0-n} \qquad (0 \le p \le 1), \qquad \text{(B-22)}$$

which becomes the Poisson distribution P(n), when $p \ll 1$ and $n \ll N_0$ but $pN_0 = \bar{n}$ is finite:

$$P(n) = e^{-\bar{n}}\,\frac{\bar{n}^n}{n!}\,.$$

The scattering Δn of the values of n around $\bar{n}$ is measured by the standard deviation σ (or dispersion), where σ^2 is the variance (B-21)

$$\Delta n = \sigma = \bar{n}^{\frac{1}{2}}, \qquad \text{(B-23)}$$

while the fluctuations are measured by the relative dispersion:

$$\Delta n/\bar{n} = \bar{n}^{-\frac{1}{2}}: \tag{B-24}$$

the larger $\bar{n}$ the larger Δn, but the smaller the fluctuations $\Delta n/\bar{n}$.

3. Harmonic Oscillator in a Coherent State. Be now a harmonic oscillator, in a state $|\psi(t)\rangle$, with a mean energy $E = \langle\hat{H}_o\rangle$ Be $|\psi(t)\rangle$ a coherent wave packet $|\alpha\rangle$ of states $|n\rangle$; $|\alpha\rangle$ is Poisson distributed among $|n\rangle$, because of Eqs. (A-6) and (B-4).

The previous consideration on the uniform gas of non interacting particles can be used, by stating the following simile: 1) The non-interacting particles are the phonons, excited over the normal mode of the harmonic oscillator. 2) The normal mode, like the volume V_o, can accomodate any number N_o of particles with arbitrary density ρ_o (phonons are bosons). 3) The volumes v and V_o are kept fixed. Since the simile is suggested by the Poisson distribution, which in turn depends on the product $pN_o=v\rho_o=\bar{n}$, only one parameter (the particle density ρ_o) characterizes the gas. 4) ρ_o is determined by the value of the energy $\langle H_o\rangle=\frac{1}{2}(|\alpha|^2+\frac{1}{2})\hbar\omega$ (see Eqs. (A-8) and (A-6)) because

$$\rho_o = N_o/V_o = \bar{n}/v = |\alpha|^2/v: \tag{B-25}$$

the higher the mean energy $\langle\hat{H}_o\rangle$ the larger the particle density.

Therefore, one can imagine the normal mode of the oscillator put in a coherent state of mean energy $\langle H_o\rangle$ like a container of a fixed volume V_o filled uniformly with a phonon gas of density ρ_o.

Do now measurements involving a fixed volume v of gas, as when in a sampling process the same volume of gas is always removed. The probability that, doing in that way we capture n particles, is P(n), the Poisson distribution. So P(n) is the probability that n phonons with the quantized energy $E_n=\hbar\omega(n+\frac{1}{2})$, are involved in a measurement process of the harmonic oscillator in a coherent state. Dispersion and fluctuation are given by Eqs. (B-23) and (B-24).

The previous simile can be generalized to the case when the system is a multi-mode harmonic oscillator, so that several independent normal modes q_k contribute to the state vector $|\psi(t)\rangle$. We assume again that the system is in a coherent state, represented by the vector $|\{\alpha_k\}\rangle=\Pi_k|\alpha_k\rangle$ where each $|\alpha_k\rangle$ is given by Eq. (A-7).

Following the previous simile, the state $|\{\alpha_k\}\rangle$ corresponds to an isolated container of volume V_o, filled with a uniformely distributed gas of phonons of several components k. Partial (ρ_k) and total (ρ_o) densities are related by $\Sigma_k\rho_k=\Sigma_k N_k/V_o=N_o/V_o=\rho_o$, because $N_o=\Sigma_k N_k$.

As before, the mean energy $\langle\hat{H}_o\rangle = \Sigma_k \hbar\omega_k(|\alpha_k|^2+\frac{1}{2})$ fixes the total density of the phonon gas $\rho_o = n/v = \Sigma_k |\alpha_k|^2/v$, while the partial densities $\rho_k = n_k/v = |\alpha_k|^2/v$ are related to the energy distribution among the normal modes.

The probability that drawing away from V_o a fixed volume v of gas we find n_1 particles of type 1 (n_1 phonons on the normal mode Q_1), etc. is given by the multiple Poisson distribution ($n_k \in \mathbb{N}^+$, $\bar{n}>0$, $\bar{n}_k>0$)

$$P(\{n_k\}) \equiv P(n_1,\ldots,n_k,\ldots) = \Pi_k P(n_k) =$$

$$= \Pi_k \, e^{-\bar{n}_k} \frac{\bar{n}_k^{\,n_k}}{n_k!} = e^{-\bar{n}} \, \Pi_k \frac{\bar{n}_k^{\,n_k}}{n_k!} . \qquad \text{(B-26)}$$

$P(\{n_k\})$ is therefore the probability that the sequence $\{n_k\}$ of phonons, with its quantized energy

$$E(\{n_k\}) = \sum_k \hbar\omega_k(n_k + \tfrac{1}{2}),$$

enters in a physical process involving the multimode oscillator in the coherent state $|\{\alpha_k\}\rangle$.

REFERENCES

1. W. H. Louisell, Quantum Statistical Properties of Radiation (Wiley, New York, 1973).
2. E. M. Henley and W. Thirring, Elementary Quantum Field Theory (McGraw-Hill, New York, 1962).
3. A. Messiah, Quantum Mechanics (Wiley, New York, 1970).
4. E. Merzbacher, Quantum Mechanics (Wiley, New York, 1970).
5. G. A. Korn and T. M. Korn, Mathematical Handbook (McGraw-Hill, New York, 1978).

INTRODUCTION TO EXCITON PHYSICS

R. S. Knox

Department of Physics and Astronomy
University of Rochester
Rochester, New York 14627, USA

ABSTRACT

Nonmetallic solids have a band gap in the one-electron picture. When particle interactions are considered, a large number of bound electron-hole states are found within the gap. These states, called excitons, play a large role in determining the optical properties of nonmetals. We discuss the electronic structure of excitons, their interactions with phonons and photons, and their further role of transporting excitation energy.

I. INTRODUCTION

Conventionally the term exciton is used to describe a delocalized electronic excitation whose energy is just below that of the one-electron band gap of a condensed system. However, free electron-hole pairs can be considered to be ionized excitons, excited states of defects are equally well thought of as trapped excitons, and the exciton and plasmon can be related formally. Therefore, in a certain sense exciton physics is concerned with all optical properties of all condensed matter. In these lectures we concentrate on the conventional exciton with emphasis on its universal properties and its role in energy transport. The author's old review [1] will be called upon to supply certain details. Some updating of that article will be attempted.

The term exciton is frequently liberated from its strict role of describing the completely delocalized eigenstate of a perfect crystal. Thus a wave packet made up of delocalized excitons may also be called an "exciton," and even a stationary wave packet which is localized can be called an "exciton." The context should always make the meaning clear.

In part II the electronic structure or basic framework of the exciton will be described. Although this structure is an indispensable starting point, the relatively strong and widely variable interactions of excitons with phonons and photons dictate most of their observable manifestations. Part III introduces the exciton-phonon interaction and surveys its magnitude in various systems. Part IV introduces the exciton-photon interaction, whose special attributes make exciton spectroscopy a vastly richer field than ordinary gas-phase spectroscopy. The process of exciton transport of energy is surveyed in part V.

II. ELECTRONIC STRUCTURE

II.A. Excitons in a Static Crystal

1. Atomic and Molecular Approach. We will not present a historical development, for which readers may consult our earlier work [1]. Nonetheless it is worth mentioning that Frenkel's [2] theory was the first and it was based on a model which is quite valid today in an important limiting case. Consider N identical atoms or molecules, one of which considered alone and situated at the origin would have stationary states obeying a Schrodinger equation

$$H_0(\vec{r})\phi_n(\vec{r}) = \varepsilon_n\phi_n(\vec{r}) \ . \tag{1}$$

Suppose, further, that in the presence of all other unexcited members of the system, each atom or molecule situated at a point $\vec{R}_I$ contributes a set of states obeying

$$H_{0I}(\vec{r}-\vec{R}_I)\phi_{In}(\vec{r}-\vec{R}_I) = \varepsilon_{In}(\vec{r}-\vec{R}_I)\phi_{In}(\vec{r}-\vec{R}_I). \tag{2}$$

In both Eqs. (1) and (2) the index n may stand for a set of quantum numbers, including those for spin and vibrational states. Also, the vector $\vec{r}$ stands for all electronic degrees of freedom associated with the atom or molecule. In Eq. (2), the site index I when used on the Hamiltonian as well as on the energies and wave functions implies that all sites may not be equivalent. Equation (2) represents a "crystal field" approach to localized states, in that an effective potential due to the surrounding ground-state atoms is included in H_{0I}. It may vary from site to site because of vacancies, dislocations, or a general disordering as in an amorphous system. More generally, the situation in the case of Eq. (2) can involve non-identical substituents and the one at some $\vec{R}_I$ may be an impurity.

Gas-phase spectroscopy generally can rely on Eq. (1) for a good first-order description, and the spectroscopy of crystal impurities generally can rely on Eq. (2), with much assistance from site symmetry properties. In each case the central assumption is that the N members of the system are acting independently and that observed spectra are simple superpositions. Some broadening might occur on the observed levels of Eq. (1) as a result of collisions or Doppler effect. Some might occur on the levels of Eq. (2) as a result of phonon interactions (homogeneous broadening) or physical variations among sites (inhomogeneous broadening). Exciton physics enters when interactions between excitations on different sites cannot be ignored.

The crystal-field nature of H_{0I} in Eq. (2) assumes that neighboring sites are in their ground states. For a complete description we must consider the problem as a many-body problem and find the eigenstates of the complete Hamiltonian. In general, for fixed nuclei of charge $Z_I e$ (e = positive by convention),

$$H = \Sigma_i \frac{\vec{p}_i^{\,2}}{2m} - \Sigma_I \Sigma_i \frac{Z_I e^2}{|\vec{R}_I - \vec{r}_i|} + \Sigma_{I<J} \frac{Z_I Z_J e^2}{|\vec{R}_I - \vec{R}_J|} + \Sigma_{i<j} \frac{e^2}{r_{ij}} \tag{3}$$

where capital letters and subscripts refer to nuclei and lower case to electrons. Sums run over all corresponding particles in the system. By standard methods H must be broken into a zero-order part and a perturbation. Details may be found in [1]; similar decompositions are discussed elsewhere in this volume. For most purposes H may be written in the form

$$H = \Sigma_I H_{0I} + \Sigma_{I<J} V_{IJ}, \tag{4}$$

where H_{0I} is the local crystal-field Hamiltonian and V_{IJ} provides the remaining site-site interaction. The passage from Eq. (3) to Eq. (4) is not simply a matter of sorting out terms. The operators H_{0I} include part of the intersite interaction because of the crystal field. Thus V_{IJ} in Eq. (4) must have built-in corrections to avoid duplications of interactions. Moreover, the simplified form Eq. (4) conceals the indistinguishability of electrons, with the result that exchange interactions must be handled with care. These problems can be minimized by use of the standard creation-annihilation operator formalism to be introduced later.

The atomic/molecular approach to excitons, due to Frenkel, regards the zero-order ground state as a product of ground-state crystal field functions (in practice, antisymmetrized)

$$\Phi_0 = \phi_{10}\phi_{20}\phi_{30} \cdots \phi_{I0} \cdots \quad (5)$$

and a zero-order excited state as a similar product

$$\Phi_{In} = \phi_{10}\phi_{20} \cdots \phi_{I-1,0}\phi_{In}\phi_{I+1,0} \cdots \quad (6)$$

There are of course many possible states at each site and many will be degenerate, so n refers to a set of quantum numbers as before.

The matrix element $(\Phi_0,H\Phi_0)$ is the first-order ground state energy, and its second-order correction is the usual van der Waals energy. We are concerned only with first-order excitation energies, which will be simplified by adjusting $(\Phi_0,H\Phi_0)$ to zero.

In the case of isolated, non-interacting impurities one needs only diagonalize the matrices $(\Phi_{In},H\Phi_{In'})$. For the exciton, by definition, there are nonvanishing matrix elements of the form $(\Phi_{In},H\Phi_{Jn'})$ where $I \neq J$. The site-site correlation created by this coupling has a profound effect on spectra. In order to examine the essentials, we consider the simplified case in which there is one important nondegenerate local excited state which we label with n = 1. Other local excited states will be assumed to perturb this state negligibly. The many states of the system built from local state 1 are connected by the matrix $(\Phi_{I1},H\Phi_{J1})$, whose size is that of the number of sites in the system. We denote its elements by H_{IJ} and temporarily drop the label "1." Its eigenvalues and eigenvectors, labelled in appropriate way by $\alpha,\beta,\ldots$, are given by

$$\Psi_\alpha = \Sigma_I S_{I\alpha}\Phi_I \quad (7)$$

and

$$E(\alpha) = \Sigma_{I,J} S^{-1}_{\alpha I} H_{IJ} S_{J\alpha} \quad (8)$$

where S is a unitary matrix.

In the case of a perfect crystal with one atom or molecule per unit cell, the matrix S has a well-known form. The set of labels α is the set of crystal $\vec{K}$ vectors, running over the first Brillouin zone, and under the usual assumption of periodic boundary conditions we have $S_{I\vec{K}} = N^{-\frac{1}{2}}e^{i\vec{K}\cdot\vec{R}_I}$. Therefore

$$\Psi_{\vec{K}} = N^{-\frac{1}{2}}\Sigma_I e^{i\vec{K}\cdot\vec{R}_I}\Phi_I \quad (9)$$

and

$$E(\vec{K}) = \Sigma_J e^{i\vec{K}\cdot(\vec{R}_J-\vec{R}_I)} H_{IJ} . \tag{10}$$

In Eq. (10) the index I is free and irrelevant because all sites are equivalent under crystal symmetry conditions. We now reintroduce "1," write $H_{II} = \varepsilon_1$, make the assumption that H_{IJ} is nonvanishing only for nearest neighbors, with value J_{11}, and apply Eq. (10) to the case of a triclinic lattice with nearest-neighbor distance a, taken to lie along the x direction:

$$E_1(\vec{K}) = \varepsilon_1 + 2J_{11} \cos K_x a . \tag{11}$$

The effect of the intersite coupling J_{11} is to broaden the site excitation energies into a band. Indeed, Eq. (11) is the dispersion relation of a particle (exciton) of effective mass

$$\begin{aligned} M^* &= \hbar^2(\partial^2 E_1/\partial K_x{}^2)^{-1} \\ &= - (\hbar^2/2a^2 J_{11}) \cos K_x a \end{aligned} \tag{12}$$

Thus, in this elementary case, for motion in the x direction one has positive-mass excitons at $\vec{K} = 0$ (for negative J_{11}) or at $K_x = \pi/a$ (for positive J_{11}). Because of the assumed vanishingly small J's in other directions, the effective mass for motion in the y and z directions is infinite. The effect of J_{11} on a spectrum, in this simple model, is to shift a strong absorption band either upward or downward in energy (the choice depending on J's sign). This follows from the rule, to be developed in part III, that only $\vec{K} \approx 0$ states are involved in a strongly allowed transition.

Should there be several closely degenerate states n = 1, 2, ..., d associated with each cell of the lattice, by virtue of degeneracy in the molecules, or the presence of more than one molecule in the cell, or both, the transformation (9) is insufficient to diagonalize the Hamiltonian matrix. As in the case of elementary tight-binding theory of electron bands, there will remain d^2 matrix elements for each $\vec{K}$, and diagonalization of the $d \times d$ matrix results in d sub-bands. In molecular crystals the splittings between these sub-bands are well known as Davydov splittings, and are primarily due to the presence of nonequivalent molecules in the unit cell (see part IV.C).

Another complication which exists even in the case of high symmetry is the anisotropy of the interactions J_{nn}. For example, if we consider an exciton built from an atomic p state, the matrix element H_{IJ} will depend on the angle between $\vec{R}_{IJ}$ and the axes of quantization of the localized wave functions. In particular, there will be both a $J_{\sigma\sigma}$ and a $J_{\pi\pi}$ for any pair of neighbors, where σ and π are p wave functions parallel and perpendicular to $\vec{R}_{IJ}$. They may have different signs and a complete analysis is necessary for even the simplest crystals. We note that in general Eq. (12) must be replaced by a more complete effective mass tensor

$$(M^*)^{-1}_{\mu\nu} = \hbar^{-2}(\partial^2 E_n(\vec{K})/\partial K_\mu \partial K_\nu), \tag{13}$$

where μ and ν refer to any pair of Cartesian labels. All of the complications arising from degeneracies, unit cell composition, and J anisotropy must be built into the computation of $E_n(\vec{K})$ before exciton dynamics may be predicted. Only in special cases will the generalized tensor (13) be diagonal.

Examples of two fairly complicated applications of the Frenkel formalism may be found in an early work from Rochester on solid argon by the author [3] and a more recent one on tetrachlorobenzene by Davidovich [4].

2. Effective Mass Approach. The tight binding method, applied in the previous section, emphasizes and requires knowledge of both the detailed structure of the system and its translational symmetry. As is well known, the electrical properties of many solids, particularly metals and semiconductors, may be explained remarkably well by a model which ignores the detailed structure and assumes an even greater translational symmetry. Electrons and holes are taken to move in potential regions of macroscopic size, their dynamics determined by assumed effective masses and assumed rates of scattering with a background provided by phonons and impurities. In the mid 1930's the foundation for adding excitons to this picture was laid by Slater and Shockley [5], Wannier [6], and Mott [7]. In the 1950's Elliott [8] developed the theory to include line strengths.

Effective mass (EM) theory essentially gives each electron, when residing in a region whose average potential energy is constant, a Hamiltonian $\vec{p}_e^{\,2}/2m_e^*$. Similarly, each hole gets a Hamiltonian $\vec{p}_h^{\,2}/2m_h^*$. This drastic simplification was long regarded as a fortunate empirical fact, since the entire rapidly varying crystal potential appears to be ignored successfully by these model particles. About twenty years ago the nature of the EM approximation was clarified by the discovery that the carriers, in keeping their wave functions orthogonal to the core electrons, employ an effective potential energy to do so which cancels much of the true core potential. The net pseudopotential is small [9].

At reasonable carrier densities the Coulomb repulsion keeps the electrons away from each other and the holes away from each other. Collisions between electrons and holes result in recombination. Let us, however, examine the situation before recombination of one particular electron-hole pair occurs. The Hamiltonian of the pair, following the EM philosophy, will be*

$$H_{eh} = \frac{\vec{p}_e^{\,2}}{2m_e^*} + \frac{\vec{p}_h^{\,2}}{2m_h^*} - \frac{e^2}{\varepsilon|\vec{r}_e - \vec{r}_h|} , \tag{14}$$

where $\vec{p}$ and $\vec{r}$ represent the momentum and position coordinates, respectively, of the particles, and ε is a dielectric constant of the medium between the electron and the hole. Equation (14) is properly constructed in the context of the EM approximation, because, from the viewpoint of either particle, the other is producing a slowly-varying "external" potential not accounted for by its spatially-invariant effective mass parameter.

Since Eq. (14) is identical to the hydrogen atom Hamiltonian with adjusted parameters, its energy eigenvalues and eigenfunctions are completely known. Choosing the zero of energy at the ground state of the crystal with no free carriers, we see that the unbound electron-hole pair has an energy equal to the band gap E_G and that the eigenvalues of Eq. (14) are

$$E_n(\vec{K}) = E_G - \frac{G}{n^2} + \frac{\hbar^2 K^2}{2M^*} , \tag{15}$$

where $G = \mu e^4/2\hbar^2\varepsilon^2$, $\mu^{-1} = (m_e^*)^{-1} + (m_h^*)^{-1}$, $M^* = m_e^* + m_h^*$, $\hbar\vec{K}$ is the total momentum of the electron and hole, and n is a hydrogenic quantum number, i.e., any positive integer. G is called the exciton Rydberg, which, because of the weakening of the Coulomb attraction represented by ε, is usually 10 to 1000 times smaller than the hydrogenic Rydberg of 13.6 eV.

As we will note in part II.B, the spectra of many crystals are explained by the first two terms on the right side of Eq. (15). The $\vec{K}$-dependent term corresponds to the K^2 term in the expansion of the cosine in Eq. (11), and all such terms implying motion will be considered later.

Like the tight-binding result Eq. (11), Eq. (15) is correct in principle but limited when applied to real systems. Degeneracy is again an important complicating factor. Electrons and holes in different branches of their respective bands will have different effective masses, and this complicates the effective Hamiltonian considerably (Dresselhaus [10]; [1], pp. 42-43). A problem frequently encountered in the applicaiton of Eq. (15) is that the

---

*Gaussian units are used throughout this article.

observed lowest level (n = 1) does not fit the hydrogenic sequence while the others do. This can be traced to the small radius of the n = 1 exciton. For arbitrary n, the radius is $n^2(\varepsilon/\mu)a_0$, where a_0 is the Bohr radius (0.53×10^{-8} cm). Thus, when n = 1, the motions of the electron and hole are very closely correlated in space and any effects on the energy due to crystal potential variation are magnified. Donors and acceptors having hydrogenic series also evidence this problem, known generally as the "central cell correction" problem.

3. Unified Approach. The Frenkel and Wannier exciton models are limiting cases of a more general formalism which need not be detailed here (see [1], Chapter II). It may be useful, however, to consider their relationship in qualitative terms (Fig. 1). In section 1 we developed the Frenkel exciton from localized functions, i.e., from the lower left box in Fig. 1. In section 2 we developed the Wannier exciton from free carrier functions, i.e., from the upper right boxes in Fig. 1. The general formalism results from the following scenario: a crystal Hamiltonian is given; we may choose any convenient representation as long as it spans the interesting states of the system; and then with more or less hard work, we diagonalize the Hamiltonian and arrive at the "general delocalized exciton state," i.e., one of the true eigenstates. The choice of starting representation is made easier by some knowledge of the spectrum, the dielectric constant, and the wave functions of the constituent atoms or molecules. The decision to stop at the Frenkel or Wannier box or to proceed to the generalized state is based on optimizing the information contained in the calculation, considering the cost of getting it.

The horizontal transitions in Fig. 1 are purely mathematical, but the vertical transitions are purely physical. It is instructive to consider some "wrong" choices in starting an analysis.

Suppose we want to describe excitons in a molecular crystal (CO_2). We are attracted to the fact that approximate outer-orbital electron (and hole) Wannier functions constructed from symmetrical plane wave states, or any approximate Bloch functions, are beautifully orthogonal. That is, the functions constructed at any lattice site are orthogonal to all others, even nearest neighbors. We set out to diagonalize the Hamiltonian, succeed, and then find that predicted excitation energies are hopelessly wrong, probably ten times too small. What went wrong? The Coulomb interaction within each CO_2 molecule was badly approximated by wave functions which were designed to satisfy nothing but crystal symmetry. Many higher sets of Wannier functions would be needed to be mixed in to do a job which a better choice of zero-order states would have done.

At another extreme, suppose we have a covalent solid (Si) and

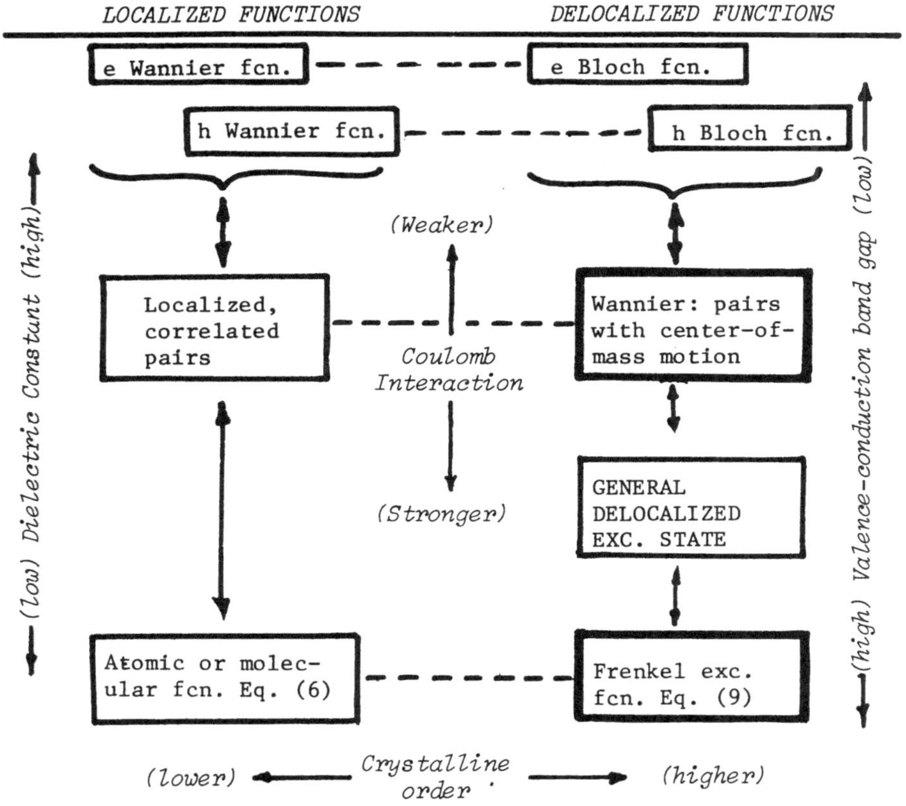

Fig. 1. Interrelations among one- and two-electron representations of states of solids. Horizontal dashed lines represent connections between representations which can be made by use of the coordinate-momentum transformation [typically $v_K = N^{-\frac{1}{2}}\Sigma_J \exp(iK \cdot R_J)u_J$]. Various boxes and the vertical transformations are discussed in the text. Boxes outlined heavily are the traditional Frenkel and Wannier limits discussed in parts II.A.1 and 2, respectively.

choose to compute with atomic wave functions. To justify this we might argue that central cell corrections would be small in such a representation. Again we compute, and find no agreement with experiment. What went wrong this time? We did not notice that the wave functions we used were, in fact, badly non-orthogonal at different lattice sites. Any attempt to orthogonalize them would have great convergence difficulties and again large numbers of higher states would need to be mixed in to do the job which ordinary Bloch functions would have done better in the first place.

On the vertical scale of Fig. 1 we have noted the trends of dielectric constant and band gap which are generally consistent with the representations shown. However, these are trends only, and other considerations such as lattice structure and ionicity will help dictate the best model.

4. Excitons in Disordered Materials. One must always pay close attention to the true eigenstates in any given situation. In the discussion just concluded, the "general exciton state" appeared in the column of delocalized representations because the true energy eigenstates of crystals are indeed delocalized. The localized functions listed in the left column are of interest in ordered systems only as components of the delocalized states or as short-lived non-stationary states which might be produced by radiation.

In recent years disordered or amorphous materials have become important technologically and the study of their properties has become intense. We can appreciate the problems involved immediately by returning to the general Hamiltonian

$$H = \Sigma_I H_{0I} + \Sigma_{I<J} V_{IJ} \, . \qquad (4)$$

Nothing in this Hamiltonian requires crystal symmetry. We may again set up states (5) in zero order and calculate the matrix elements of Eq. (4) in this representation. We again obtain a matrix H_{IJ} when we make the nondegenerate excited state assumption. We may even assert that Eqs. (7) and (8) are the eigenstates and eigenvalues. However, a standard analysis must stop at this point. The matrix H_{IJ} is random. Its diagonal elements have some average value ε_1, but centered in a distribution of width Δ_D. Its off-diagonal elements H_{IJ} have some average value J centered in a distribution of width Δ_{OD}.

In 1958 Anderson treated this disorder problem in the context of spin diffusion and electrical conduction. In the tight-binding approximation, the exciton problem is mathematically the same (cf. top and bottom of the left column of Fig. 1, with disorder). When Δ_{OD} is zero, it turns out [11] that all states are localized if

$$\Delta_D > czJ, \qquad (16)$$

where J is a nearest-neighbor intersite coupling energy, z the coordination number of the lattice, and c is 0, 3/2, or 2 in 1, 2, or 3 dimensions. In the theory a restriction to nearest-neighbor interactions is not essential but the interaction must drop off faster than r^{-3}.

Recall that J is associated with kinetic energy of particle

motion. In a sense, the Anderson criterion specifies how much kinetic energy is needed to overcome pockets of diagonal disorder. If $J = 0$, clearly most states will be localized. If both J and Δ_D are zero, the situation is ambiguous because any linear combination of localized states is then an eigenstate.

Excitation transfer in mixed and random systems is a very active field of research. It is possible here only to mention some papers which can bring the reader into contact with the field... Monberg and Kopelman [12], who argue against an Anderson-like interpretation of exciton transport in mixed organic crystals, Lyo et al. [13], who treat spectral and spatial diffusion, and Klafter and Silbey [14], who discuss the disorder problem in terms of the generalized master equation.

5. Effects of External Fields. Excitons, having their origin in atoms and molecules, respond to electric and magnetic fields in a variety of ways. One of the more interesting effects, observed in semiconductors, is the magnetostark effect, through which the $\vec{K}$ vector of the exciton was first observed (by Thomas and Hopfield [15]; reviewed in [1], pp. 85-86).

Excitons of rather definite spin multiplicity can exist in organic crystals. Magnetic fields are useful in triplet exciton spectroscopy (see, e.g., [16]) and they strongly affect triplet exciton annihilation rates (see [17]).

Strain fields are unique to solids and affect excitons in a way which can be understood readily on the basis of the EM approximation. Inhomogeneous strains cause a variation in lattice constant which is reflected in the local band gap. If the exciton's energy changes throughout a crystal, diffusing excitons will be driven toward lower energy by the strain field. Wolfe and colleagues have used this phenomenon, along with very high excitation levels, to observe exciton flow and localized strain fields in semiconductor crystals [18].

6. Excitons as Quasiparticles. For detailed calculations of exciton bands and wave functions a complete and tedious formulation in terms of determinantal wave functions is generally necessary. However, the dynamics of excitons can be handled much more easily with standard creation and annihilation operators. If we consider the vacuum state $|0\rangle$ to be that of a nonconducting solid with no excitations, i.e., with the valence bands full and conduction bands empty, then the various states in Fig. 1 can be described as follows:

State with one electron in band n with wave vector $\vec{k}$: $c^{\dagger}_{m\vec{k}}|0\rangle$

Same, with one hole in ℓ, $\vec{k}$: $c_{\ell\vec{k}}|0\rangle$

Pair state, unmixed: $c^{\dagger}_{m\vec{k}} c_{\ell\vec{k}'} |0\rangle$

Localized Frenkel state: $\beta^{\dagger}_{In} |0\rangle$ (molecular state n, site $\vec{R}_I$)

Exciton state: $\beta^{\dagger}_{n\vec{K}} |0\rangle = N^{-\frac{1}{2}} \Sigma_I e^{i\vec{K}\cdot\vec{R}_I} \beta^{\dagger}_{In} |0\rangle$ (Frenkel)

$$\beta^{\dagger}_{n\vec{K}} |0\rangle = \Sigma_{m,\vec{k},\ell,\vec{k}'} A_{n\vec{K};m\vec{k},\ell\vec{k}'} c^{\dagger}_{m\vec{k}} c_{\ell\vec{k}'} |0\rangle \quad \text{(Wannier)}.$$

The coefficients A in the last equation are to be determined in principle by a direct calculation which is circumvented in practice by using the Wannier model. The notation will be explained in more detail as necessary. We follow the general second-quantization rule that any operator written in terms of creation and annihilation operators takes on whatever form is necessary to ensure that matrix elements, when calculated in the number representation, are the same as those which would be calculated in the wave-mechanics representation. We will reserve the letters a and b for photons and phonons, respectively. In the above equations, the letters ℓ, m, n may include any number of band, polarization, and spin indexes. An operator of the form o destroys a particle, while $o^{\dagger}$, the hermitean conjugate of o, creates one.

Although the exciton is constructed from two Fermions, it is not strictly a boson. The commutation relations for the β's are those of bosons plus a correction term proportional to the number of excitations per unit cell in the system ([1], pp. 88-89). The reason may be seen most readily in the localized Frenkel representation. Two excitations cannot reside on one molecule unless, by fortuitous circumstance, precisely the right electronic states exist. This may be contrasted with the case of phonons, where the second excitation of an oscillator is truly indistinguishable from the first.

In summary, the Hamiltonian for pure exciton states may be written

$$H = \Sigma_{n\vec{K}} E_n(\vec{K}) \beta^{\dagger}_{n\vec{K}} \beta_{n\vec{K}} \quad . \tag{17}$$

Here $E_n(\vec{K})$ is an energy eigenvalue computed in an appropriate model, and the β's are the operators introduced above. The eigenstates of H are $|\{n_{\beta\vec{K}}\}\rangle$ in a standard number-operator formalism.

7. Many-Body Considerations. Several important collective effects are too extensive to be dealt with in these lectures, but they deserve mention. First, the insertion of the dielectric constant in the effective Hamiltonian (14) need not be considered entirely phenomenological. Haken and Schottky [19] used model Hamiltonians to show how it follows from polarization of the medium and developed

an effective r-dependence to account for the response of the medium to the relative speed of the electron and hole in their orbit. Abe et al. [20] and Sham and Rice [21] have shown from first principles how the electronic polarization contributes $1/\varepsilon$ to the Wannier exciton Coulomb interaction.

The possibility that a phase transition might take place if the exciton binding energy were to exceed the band gap ([1], p. 100) has been examined extensively, particularly by des Cloizeaux [22] and Jerome et al. [23]. The author does not know of any experimental confirmation of this phase, known as the "excitonic insulator." Such a phase may also be approached from the metallic phase through the formation of charge density waves [24].

The exciton has been proposed as an intermediate for coupling Cooper pairs in schemes for high-temperature superconductors [25,26]. The author is not aware of experimental confirmation of exciton involvement in superconductivity.

II.B. Excitons in Real Materials

1. Observation. From its earliest days the observed exciton has had essentially a "dual existence." First, as an excited state of a solid, it can be said to be "observed" if the following conditions are met: an optical absorption or electron energy loss appears in the right place in the spectrum; it has the expected strength and width; and, in the optical case, the absorption activates appropriately little conduction. Two examples of this "spectral observation" of the exciton are shown in Fig. 2 [27,28].

One might argue that spectral observation is enormously indirect, that what has really been observed is merely the attenuation of a beam, and that the measurement of some dynamical feature of the exciton should constitute a true observation. In Frenkel's 1936 paper the concept of the diffusing exciton wave packet was introduced [29] and in 1948 Förster [30] wrote the appropriate diffusion equation along with a theory for computation of the diffusion constant. In 1956 the first experiments on exciton diffusion were performed by Simpson [31]. The idea was that excitons created in one region of a crystal could be observed in another region if the latter region contained detectors. Figure 3 and its caption explain this method. Successful mounting of such an experiment might be called "exciton observation by capture."

In molecular crystals (e.g., tetrachlorobenzene, Fig. 2a) the accomplishment of spectral observation is nontrivial. Molecular absorption lines remain fairly sharp in the condensed state, especially at low temperatures, and their shifts from vapor-phase

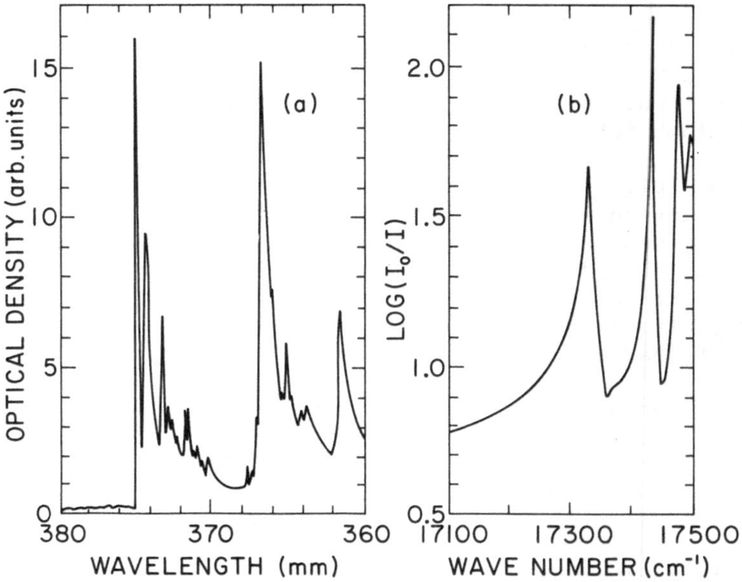

Fig. 2. Spectral observation of excitons. (a) Singlet-triplet exciton absorption of a single crystal 1, 2, 4, 5-tetrachlorobenzene at 4.2 K (unpolarized; ab face of crystal) [27]. (b) The "yellow" series of Cu_2O, measured in absorption at 4.2 K [28].

positions are not large. Therefore, spectral observation by our criterion relies on the observation of Davydov splittings caused by intermolecular interactions or of more subtle $\vec{K}$-dependent effects. In the case of semiconductors the appearance of a hydrogen-like series accomplishes spectral observation, provided the exciton Rydberg is consistent with the material parameters and the line strengths generally follow theoretical predictions (part IV).

Observations of excitons by capture requires great care in design and interpretation of the experiment. The detecting centers which capture the exciton may be fluorescing impurities in the case of large-band-gap materials or in the case of semiconductors, shallow traps which detect by ionization. In both cases the distribution of initial excitation must be known, solutions of the appropriate diffusion equation must be used in the analysis, and direct photo-excitation of the traps must be avoided or considered as a systematic correction. We will return to these issues in part V.

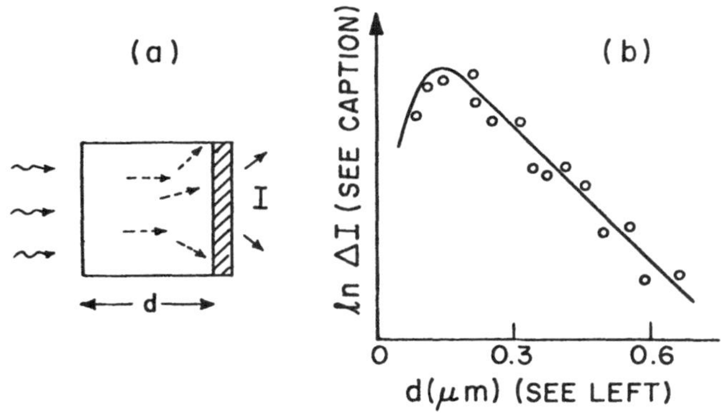

Fig. 3. (a) The geometry of an experiment for observing excitons by capture. Light (wavy arrows) is absorbed nonuniformly in a crystal, creating diffusing excitons (dashed arrows). On arriving at the shaded region, the excitons are captured by traps consisting of different molecules whose fluorescence indicates the capture event. (b) Typical analysis: emission from the detector layer contains a component not accounted for by direct excitation by transmitted photons. This excess (ΔI) may be fit to a predicted flux of excitons at the detector. The figure is adapted from Simpson [31]; the data points are here sketched, not authentic.

Spectral and capture observations have been the primary sources of information on the exciton. Modern optical techniques have made several other kinds of observation possible, but one may generally associate these methods with either of the basic types or a combination of the two. Some will be discussed later. We may mention direct emission from the exciton as a method which is not so readily categorized.

2. Spectral Characteristics. There is a rough correspondence between crystal binding types and the nature of exciton spectra. We have already discussed the case of *molecular crystals*, wherein the vapor-phase lines are shifted, broadened, and split. The broadening may be resolved into very detailed sidebands related to intramolecular vibrations and the phonons in the crystal [32-34]. The appropriate exciton model to be used is Frenkel's, and the lower section of Fig. 1 applied, because the dielectric constant is low, qualitatively speaking.

Covalent crystals, including most small-to-medium band gap semiconductors, display hydrogenic exciton series. This is exemplified in Fig. 2(b). However, the details of the band structure of these crystals are of great importance. If the smallest band gap is indirect, i.e., if the valence band maxima and conduction band minima do not occur at the same point in $\vec{k}$ space, the spectrum in the region of the indirect gap energy will contain broad bands. The bands are a result of phonon-assisted transitions and their shapes may be analyzed in such a way as to accomplish spectral observation of "indirect excitons" ([1], pp. 158-162).

Ionic crystals have a very large binding energy but have exciton spectra resembling more those of molecular crystals than those of covalent crystals. (While each binding type has a characteristic spectrum, the correspondence is not with the binding energy but rather with the dielectric constant.) Consider the alkali halides, typified by NaI [35] (Fig. 4a). The two strong lines at 5.8 and 6.8 eV are essentially molecular transitions within the I^- ions to excited states bound by the Coulomb field of the surrounding Na^+ ions. In free space I^- has no bound excited states. Here again the Frenkel model applies crudely, but the involvement of the localized excitation with anions and nearest cations forces a detailed analysis on a model intermediate between those of Frenkel and Wannier. In fact, hydrogenic series are observed in many alkali halides. The cation lattices are isoelectronic with the *rare gas crystals*, our final example (Fig. 4b). Here one may view the strongest lines as pure Frenkel excitons and the weaker ones as highly mixed Frenkel excitons [36]. But since the latter fall into a hydrogenic series, an equally valid view is that all the lines are part of a hydrogenic series but the central cell correction is such that the lowest lie accidentally near the position of atomic lines. The rare gas exciton and band structures were first considered about 20 years ago [3,37] and are still of interest [38,39] because of the increasing spectral detail available through synchrotron radiation sources (see, e.g., [40]).

3. Surface Excitons. The surface of a crystal is one of the canonical crystal imperfections whose effect must always be considered. As with electronic band structures, exciton band structures are augmented by surface states which are generally localized in the direction normal to the surface but delocalized in its plane. The subject of surface excitons is complicated by two factors: since coupling to the electromagnetic field is involved, and this field itself is being modified as it begins to interact with the bulk upon entering the surface, a difficult boundary value problem results. To observe the k-dependence of surface excitons grazing incidence of light is necessary. We will return to this topic in part III. A second complicating factor, existing even for normal incidence of light at the surface, is the "dead layer" effect, which we now discuss.

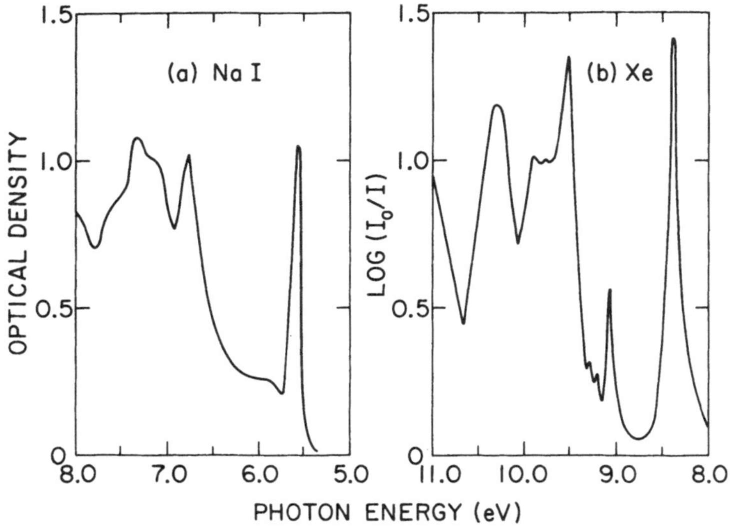

Fig. 4. (a) NaI absorption at 80 K (after Eby et al. [35]).
(b) Xenon absorption at 21 K (after Baldini [36]).

Phenomenologically, the surface constitutes a region into which bulk excitons cannot penetrate. The thickness of this region has not been evaluated theoretically, but one may expect it to be equal to the depth of the region of band-bending in a given situation or the diameter of a bulk excitons, whichever is larger. The surface region, having no bulk excitons, cannot contribute to the exciton population during optical excitation, and therefore constitutes a "dead layer" which was probably first observed by Apker and Taft [41]. Their experiments were an early attempt to observe excitons by capture in alkali halides, using F centers as detectors. The dead layer was seen more definitively by Hopfield and Thomas [42], who had to assume a potential barrier for excitons 10-15 nm inside the surface of a CdS crystal in order to explain the details of the reflectivity spectrum. In this case, the effect of the dead layer was to adjust the boundary conditions for the incident and reflected electromagnetic wave and the waves inside the crystal. In a recent theoretical treatment of surface excitons, Altarelli et al. [43] have predicted the existence of a "dead layer" from first principles for certain model parameters. However, the author does not know of any numerical applications of this dead-layer theory as yet.

4. <u>Trapped and Localized Excitons</u>. Since trapped excitons

are essential to the interpretation of effects observed in most real crystals, we include this qualitative discussion. Clearly surface excitons may be regarded as trapped on the surface, but "trapping" generally refers to localization at a point or a line. We will consider point trapping.

Formally, a trapped exciton is an occupied localized excited eigenstate of a crystal. The word "trapped" implies that a dynamic process has occurred before this state becomes occupied. In the prevailing picture an exciton wave packet collides with the special region containing the available localized eigenstate, and an irreversible transition takes place. The irreversibility is a result of the emission of phonons, the localized state typically being lower in energy than the exciton band by at least k_BT. It is worth remarking that if precisely the same state is occupied as a result of direct excitation, it is hardly ever called a trapped exciton, but rather an excited state of a trapping center.

In the preceding paragraph we have not specified the nature of the trap. It could be any point imperfection... a region of high disorder, for example. A collection of such traps would have an inhomogeneous energy distribution and would be best considered in connection with the Anderson problem. More typical traps are impurities with reasonably homogeneous energy levels such as tetracene in an anthracene crystal or thallium ions in an alkali halide crystal. Shallow traps can be introduced into organic crystals by selective protonation or deuteration (see, e.g., [44]). Finally, the special region may appear spontaneously as a result of vibrations, in which case one has "self-trapping." This will be discussed in part III.

The "mini-exciton" [45] has been studied in organic crystals. It is a special case of a trapped exciton in which the trap is a pair of impurity molecules at nearest-neighbor distance. When trapped at the pair, the exciton is delocalized over the two, which constitute a mini-crystal (N = 2). Theoretical and experimental work on these pairs, more frequently regarded as dimers, is reviewed by Burland and Zewail [46]. Recently we have discussed the role of off-diagonal density matrix elements in determining the polarization of fluorescence from molecular pairs [47].

III. INTERACTIONS WITH PHONONS

III.A. Formalism

1. Phonons. The lattice, heretofore assumed completely motionless, must now be treated realistically. As in the theory of small molecules it is generally feasible here to make the Born-Oppenheimer

or adiabatic approximation [49]. We assume that each nuclear coordinate $\vec{R}_I$ appearing in the Hamiltonian (3) differs by at most a small displacement $\vec{u}_I$ from its average position $\vec{R}_I^0$ in the perfect lattice. Since the nuclei will now be in motion their kinetic energy

$$T = \Sigma_I (p_I^2/2M_I) \tag{18}$$

must be reintroduced to produce a total Hamiltonian

$$\mathcal{H} = T + H(\{\vec{R}_I^0 + \vec{u}_I\}). \tag{19}$$

By making a Taylor expansion in the $\vec{u}_I$ we introduce the assumption about their smallness explicitly:

$$\mathcal{H} = T + H(\{R_I^0\}) + \Sigma_I (\vec{\nabla}_I H)_0 \cdot \vec{u}_I + \tfrac{1}{2} \Sigma_I \Sigma_J (\vec{\nabla}_I \vec{\nabla}_J H)_0 : \vec{u}_I \vec{u}_J + \ldots \tag{20}$$

The Born-Oppenheimer method proceeds, in what is subsequently found to be a self-consistent manner, by taking the displacements $\vec{u}_I$ as parameters as far as the electronic states are concerned. This is put into practice by assuming that the eigenstates of $\mathcal{H}$ are linear combinations of product states $|e,v\rangle = |e\rangle|v)$, where $|e\rangle$ are electronic eigenstates and $|v)$ are vibrational eigenstates.

For nonmetals the ground electronic state is well separated from the excited states and it is possible to work usefully with the subset of states $|0,v\rangle$. This allows us to write

$$\langle 0,v|\mathcal{H}|0,v'\rangle = (v|\mathcal{H}_{eff}|v'), \tag{21}$$

where $\mathcal{H}_{eff}$ is the expectation value of $\mathcal{H}$ in the electronic ground state and acts as an effective Hamiltonian for the displacement coordinates $\vec{u}_I$. For the ground electronic state $|0\rangle$ we could choose a particular model state such as Eq. (5) in the Frenkel model or continue in some generality by assuming a state $|0\rangle$ which is the best ground state available for the rigid lattice. Then

$$\mathcal{H}_{eff} = \langle 0|\mathcal{H}|0\rangle = T + \varepsilon_0 + \Sigma_I \langle 0|(\vec{\nabla}_I H)_o|0\rangle \cdot \vec{u}_I + \tfrac{1}{2}\Sigma_I \Sigma_J \vec{\vec{k}}_{IJ} : \vec{u}_I \vec{u}_J + \ldots \tag{22a}$$

or

$$H_{eff} = T + \tfrac{1}{2}\,\Sigma_I\Sigma_J\vec{\vec{k}}_{IJ}:\vec{u}_I\vec{u}_J + \ldots \, . \tag{22b}$$

The form (22b) results from the arbitrary choice $\varepsilon_o = 0$ and the observation that $<0|(\vec{\nabla}_I H)|0> = \vec{\nabla}_I E_o$ must be zero since the lattice is in mechanical equilibrium in the configuration described electronically by $|0>$.

It should be clear that Eq. (22b) is just the standard lattice Hamiltonian which leads to phonons. (We have included the derivation here, although phonons are covered elsewhere in this volume, to provide a close connection with the exciton problem.) The force-constant tensor k_{IJ} is calculable in principle and can be related to the dynamical matrix, the displacements $\vec{u}_I$ can be related to the normal coordinates Q, and these in turn finally to phonon creation and annihilation operators ($b^{\dagger}_{s\vec{q}}$ and $b_{s\vec{q}}$). The effective Hamiltonian becomes

$$H_{eff} = \Sigma_{s\vec{q}}\hbar\omega_{s\vec{q}}(b^{\dagger}_{s\vec{q}}b_{s\vec{q}} + \tfrac{1}{2}), \tag{23}$$

where $\omega_{s\vec{q}}$ is the frequency of the vibrational mode of wave vector $\vec{q}$ lying on one of the three branches s of the vibrational spectrum. Had our lattice contained a basis rather than being simple, the number of branches would be larger than three but the form of H_{eff} would be the same. The vibrational states $|v)$ are specified by the numbers of phonons in each mode: $|v) = |\{n_{s\vec{q}}\})$.

2. Linear Exciton-Phonon Coupling. We continue to work in a "separable" representation $|e,v> = |e>|v)$, where we have just given an approximate treatment of the states $|0,v>$. We now consider states in which one exciton is present, abbreviated $|n\vec{K},v>$, where "$n\vec{K}$" is in the notation of the previous section. Exciton-phonon coupling is important in two respects. It may appear in the matrix element diagonal in the excitons, thus contributing a first-order correction to the energy, and it may couple two different excitons, leading to non-zero scattering probabilities. Both of these will be seen to emerge directly from H in what follows.

Consider the matrix element $<n\vec{K},v|H|n'\vec{K}',v'>$ computed from Eq. (20). The first and fourth terms are completely diagonal in the approximation we adopt. Indeed, we shall assume that they completely reproduce the ground-state problem. To the extent that they do not, their contribution will be taken up below (part 5). The second term of Eq. (20) is also completely diagonal but more securely so: it contains no phonon operators by definition and is part of the presumably solved exciton problem, giving the unperturned exciton energy $E_n(\vec{k})$. Finally, the third term of Eq. (20), which vanished in the ground electronic state, contributes a term

$$<n\vec{K},v|\Sigma_I(\vec{\nabla}_I H)_o \cdot \vec{u}_I|n'\vec{K}',v'>. \tag{24}$$

The fact that $\vec{u}_I$ is a linear combination of phonon operators means that $|v)$ and $|v')$ must differ by exactly one phonon. If they do not, Eq. (24) is zero. If they do, the resulting combination of terms like $<n\vec{K}|(\vec{\nabla}_I H)_o|n'\vec{K}'>$ will not necessarily vanish. Consider the diagonal term: this would represent $\vec{\nabla}_I E_n(\vec{K})$. Whereas the ground state was at an overall extremum in the space of the $\vec{u}_I$, by definition, the excited state presents a different set of force constants to the nuclei and they might well shift. $\vec{\nabla}_I E_n(\vec{K})$ would be related to the net force on nucleus I in the state $|n\vec{K}>$. While there is no corresponding argument for the off-diagonal terms, there is nonetheless no argument that they should be zero. Generally they are not.

The matrix element (24) is the leading term of the exciton-phonon interaction and can be expressed concisely in particle operators as follows:

$$H_{ep} = \Sigma\Sigma\Sigma g(n\vec{K},n'\vec{K}',s\vec{q})\hbar\omega_{s\vec{q}}\beta^{\dagger}_{n\vec{K}}\beta_{n'\vec{K}'}(b_{s\vec{q}} + b^{\dagger}_{s,-\vec{q}}), \tag{25}$$

where the sums run over all exciton bands, exciton wave vectors, phonon branches, and phonon wave vectors which may be involved.

The coupling constant g is dimensionless, as defined by Eq. (25). It is zero when the interactive particles do not have momenta satisfying the conservation of crystal momentum: $\vec{k} - \vec{k}' - \vec{q}$ must be zero or 2π times some finite reciprocal lattice vector. We will assume it is zero in subsequent equations.

Up to this point, the approximate Hamiltonian of the system is the sum of three parts developed above. We assign the symbol $\mathcal{H}$ to it:

$$\mathcal{H} = H_{ex}[\text{Eq. (17)}] + H_{ph}[\text{Eq. (23)}] + H_{ep}[\text{Eq. (25)}]. \tag{26}$$

3. Clothing of Excitons. The term H_{ep} is the first in a series of phonon-dependent interaction terms, and this series usually converges or is assumed to do so without serious problems. When eigenstates of Eq. (26) are sought, yet another series will develop if H_{ep} is treated by perturbation theory. Experience has shown that this series is not generally convergent because the phonon coupling between excitons in the same band can be quite strong. (An analogous coupling leads to the Stokes shift in luminescent systems.) For simplicity, we consider a case in which only one exciton band and one phonon branch interact appreciably. Then in simplified notation $\mathcal{H}$ becomes

$$H = \Sigma_K E(K)\beta_K^\dagger\beta_K + \Sigma_q \hbar\omega_q(b_q^\dagger b_q + \tfrac{1}{2})$$

$$+ \Sigma g(K,q)\hbar\omega_q(b_q^\dagger + b_{-q})\beta_K^\dagger\beta_{K+q} \quad (27)$$

The linear coupling term can be removed by the use of a well-known unitary transformation. Unfortunately, the resulting coupling constants and eigenvalues are rather complicated in the momentum-space exciton representation. A clearer physical picture is obtained by starting in the exciton site representation for which the Hamiltonian becomes, approximately,

$$H \approx \Sigma_I \varepsilon_1 \beta_I^\dagger\beta_I + \Sigma_{I\neq I'} J_{II'}\beta_I^\dagger\beta_{I'}$$

$$+ \Sigma\hbar\omega_q(b_q^\dagger b_q + \tfrac{1}{2})$$

$$+ \Sigma_{I,q} g(I,q)\hbar\omega_q(b_q^\dagger + b_{-q})\beta_I^\dagger\beta_I \quad (28)$$

This form is approximate because the linear phonon interaction does not modulate the intermolecular coupling, i.e., in the last term of Eq. (28) we have omitted a $\beta_I^\dagger\beta_{I'}$ term with $I \neq I'$. This approximation may be removed when modulation of J appears to be appreciable.

The first, third, and fourth terms of Eq. (28) can be visualized semiclassically (Fig. 5). For each site I the electronic states are taken to depend on the coordinate Q corresponding to one of the phonons, say the one which interacts most strongly. The coupling $g\hbar\omega_q$ is proportional to $(\partial\varepsilon_1(Q)/\partial Q)_0$. As is well known from the theory of localized impurity states, there occurs in the excited state a shift in the equilibrium position to $Q = Q_0$ with a concomitant shifts of the energy from $\varepsilon_1(0)$ to $\varepsilon_1(Q_0)$. The ratio of this shift to $\hbar\omega_q$ is a convenient measure of the strength of the coupling and is essentially equal to $|g|^2$. The ratio of this shift to J, a typical intersite matrix element from the second term of Eq. (28), is an important parameter in exciton transport as we will see below.

A quantum-mechanical treatment which parallels the semiclassical treatment makes use of a unitary transformation (see, e.g., [49])

$$\tilde{H} = e^{iS}He^{-iS}$$

$$= \Sigma_I[\varepsilon_1 - \Sigma_q|g(I,q)|^2\hbar\omega_q]\beta_I^\dagger\beta_I + \Sigma_q\hbar\omega_q(b_q^\dagger b_q + \tfrac{1}{2})$$

$$+ \Sigma_{I<I'}(J_{II'}e^{-i(\alpha_I-\alpha_{I'})}e^{i\theta_{II'}}\beta_I^\dagger\beta_{I'} + \text{h.c.}), \quad (29)$$

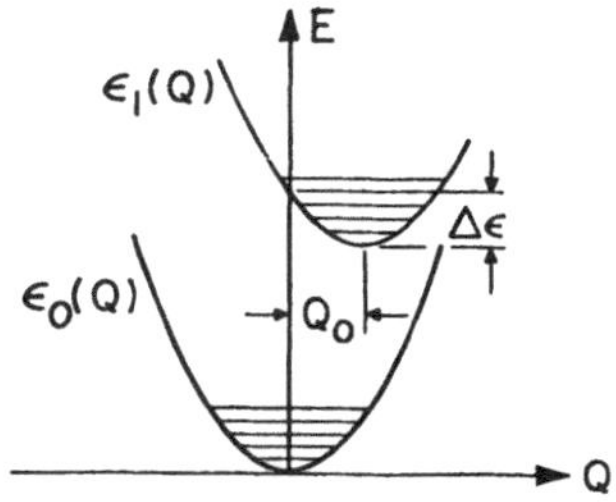

Fig. 5. Configuration coordinate diagram, illustrating semiclassically the effect of linear electron-phonon coupling. The coordinate Q is associated with a normal mode for which in the excited electronic state, $(\partial\varepsilon_1(q)/\partial Q)_0 \neq 0$. The shift of the upper-state minimum is a direct result of the coupling.

where

$$S = i\Sigma_I \alpha_I \beta_I^\dagger \beta_I, \tag{30}$$

$$\alpha_I = \Sigma_q g(I,q)(b_q - b_{-q}^\dagger), \tag{31}$$

and

$$\theta_{II'} = 2\text{Im}\Sigma_q g(I,q)g(I',q)^*. \tag{32}$$

Note that the linear coupling term no longer appears in $\tilde{H}$, that the "bare" exciton energy ε_1 has been shifted downward, and that a new effective interaction between sites I and I' has appeared. The operators $\beta_I{}^\dagger$ create "clothed" excitons in this representation. They have a lower energy because of the cloud of phonons surrounding them, in analogy with the polaron.

Extensive use of the clothed exciton representation has been made by Silbey and colleagues [50,51] in connection with excitation transport in molecular crystals. Duke and Soules [49], whose notation we have followed, applied it to the problem of excitation transfer between localized impurities, generalizing Förster's theory [30] to include Orbach's phonon-assisted transfer mechanism [52].

4. Effects of Higher-Order Coupling. Earlier we made the approximation that the quadratic terms of Eq. (20) had exciton-state matrix elements which were identical to those of the ground electronic state. When this approximation fails, its principal effect is to change the vibrational frequencies in the excited state (the curve ε_1 in Fig. 5 becomes flatter, for example). Although canonical transformations which eliminate $(b_q + b_{-q}^{\dagger})^2$-like interactions have been found [51], the resulting formalism is very complex.

III.B. Applications

1. Coupling Strengths. Estimations or actual computations of the coupling parameter g in Eq. (25) differ widely in techniques and accuracy. As with most exciton properties, the results can be classified usefully according to solid types. One generalization seems to be possible: the value of g is a linear combination of the corresponding coupling constants for an electron and a hole in the same material, the binding of the pair seeming to have little qualitative influence. Naturally, in some cases, the linear combination is such as to cause cancellation. An upper limit on $|g|$ is safely assumed to be $|g_e| + |g_h|$. A general formulation is given in [1], pp. 137-144, for the Frenkel and Wannier limits.

In *semiconductors* the coupling of the exciton to acoustic phonons may be estimated from deformation potentials. Application of [1], Eq. (10.22) to a typical case in which $|\vec{k}'-\vec{k}| \sim 10^6$ cm^{-1} yields $g \lesssim 0.1$. This is a relatively weak coupling.

Molecular crystals contain a large variety of vibrational and librational modes, and the calculation of g is very complicated. We note that much work must first be done in sorting out the interactions with intramolecular vibrations; these modes may be viewed as optical phonons of zero dispersion. Duke et al. [53] have made considerable quantitative progress in the case of benzene using a model based explicitly on modulation of electron orbital energies. The coupling constant connecting the B_{2u} molecular excitation and the breathing vibrational mode is $g = 1.4$; this is representative of the many non-zero coupling constants in benzene. Duke et al. have correlated their computations with x-ray photoelectron and ultraviolet absorption spectra. In anthracene, calculations based on modulation of dipole-dipole interactions [54] show that g for $\vec{K} = 0$ excitons is of the order of 0.5 to 1 for librational modes, transverse optical phonons, and others. Molecular crystals are thus in the intermediate coupling range.

Alkali halides represent an extreme case of strong exciton-phonon coupling, as a result of the tendency for a hole to coalsce on two halide ions to form an X^- covalent bond (the V_K center [55]). The corresponding coupling may be estimated from an observed Stokes shift of 2 or 3 eV, which must equal $|g|^2\hbar\omega$ (next section); for a typical phonon with $\hbar\omega \sim 0.02$ eV, we therefore have $g \sim 10$.

2. Scattering. Normally, particle interactions are manifest in scattering experiments. Unlike the situation for charge carriers, information on exciton scattering by phonons is not readily extracted from transport data. Therefore, although exciton-phonon scattering cross sections can be calculated ([1], pp. 170-174), there is no appropriate large body of temperature- and dopant-concentration-dependent transport data with which to compare them. The exciton-phonon matrix elements are, however, prominently involved in the theory of optical properties, which will occupy most of our attention until part V, where transport will be discussed.

Scattering processes involving both phonons and photons simultaneously are much more specific than transport processes. Much information on exciton and defect states has been obtained from Raman, resonant Raman, Brillouin, and spin-flip scattering. These processes generally depend on the exciton-phonon interaction for their existence. Two recent reviews may be consulted for details [56,57].

3. Self-Trapping. Strong electron-phonon coupling leads to a phenomenon known as self-trapping, in which the exciton "localizes." Before discussing the dynamics of this process, we would like to make an attempt to clarify the concept of "localization."

There are two convenient representations in which we may express the state ψ of a system containing an exciton. An expansion in terms of local (site) states is

$$\psi(t) = \Sigma_{In} a_{In}(t)\Phi_{In}, \tag{33}$$

where Φ_{In} is the Frenkel state (6). On the other hand, in terms of crystal momentum states, we have the expansion

$$\psi(t) = \Sigma_{\nu\vec{K}} b_{\nu\vec{K}}(t)\Psi_{\nu\vec{K}}, \tag{34}$$

where $\Psi_{\nu\vec{K}}$ is the Frenkel state (9) generalized to the case of many bands, many molecules per unit cell, etc. The indices n and ν are here assumed to include vibrational quantum numbers as well as other kinds. When we are considering a perfect crystal, Eq. (34) is an expansion in delocalized eigenstates. Obviously, the unitary transformation connecting Φ and Ψ can be used to find a relation between the a's and the b's.

There are two broad classes of "localized" excitons. If one of the eigenstates is local, as in the case of an impurity state or an Anderson-localized state, then that state can be occupied indefinitely, in principle, subject to decay through explicit channels such as photon emission or radiationless processes.

The second class of "localized" excitons contains those which might exist in the perfect, translationally invariant crystal. We may ask whether states in the site representation (Φ_{In} or linear combinations $\Sigma_n c_n \Phi_{In}$) are ever valid descriptions of the exciton. In general, such states cannot persist, since the coefficients $a_{In}(t)$ in Eq. (33) evolve in time. However, it is possible in principle to set up excitation conditions such that $\psi(0)$ is completely localized or, at least, less spatially homogeneous than the time eigenstates. "Initial localization" of this sort occurs in all site-selection spectroscopy.

The exciton site representation may provide a useful description even at long times after an exciton packet of <u>any</u> kind is set up. Consider the density matrix

$$\rho_{In,I'n'}(t) = \langle a_{In}(t) a_{I'n'}(t)^* \rangle \ , \tag{35}$$

where the brackets represent an ensemble average. Any complete description of an exciton state must specify this matrix or its counterpart in some other representation. Absorption of a photon creates an exciton packet whose density matrix is more nearly diagonal in the $\vec{K}$ representation, meaning that in this case Eq. (35) will initially have many large off-diagonal elements. Under the influence of scattering processes these off-diagonal elements may decay rapidly to zero* even though the diagonal elements do not; then the exciton dynamics can be described adequately in terms of the probabilities of localized excitation

$$P_{In}(t) = \rho_{In,In}(t) = \langle |a_{In}(t)|^2 \rangle. \tag{36}$$

---

*Only in a true eigenstate representation will off-diagonal elements of ρ decay rigorously to zero. Here, we ignore the residual correlations between sites which must remain even under severe vibrational dephasing. If the off-diagonal elements of Eq. (35) do <u>not</u> decay during the lifetime of the exciton, the localized description is surely inappropriate and a conventional scattering picture should be used.

Summed over n, this quantity is $P_I(t)$, the probability that excitation is on site I. It is interesting to note that the motion of an exciton can be described exactly in terms of $P_I(t)$ if the initial density matrix happens to be diagonal in the site representation. This requires the use of the generalized master equation (GME) [58]. Kenkre has shown, in addition, that the site-representation GME gives an exact description when the initial density matrix is diagonal in the delocalized representation [59].

After scattering has minimized all of the off-diagonal elements and other coherence effects have died out, the quantities P_I obey the Pauli Master Equation in which rates of transfer of excitation between sites appear. Under certain conditions these rates become virtually zero. The exciton is then said to be self-trapped.

The phenomenon of self-trapping has been studied theoretically for many years by Toyozawa, who has recently prepared a concise review of the physics involved [60]. Figure 6 shows how the various parameters of Eq. (28) influence the behavior of the exciton. Figure 6(a), a conventional cc diagram like Fig. 5, shows the effect of coupling of one vibrational mode to a localized exciton state (one which would persist if J were zero). The excited-state curve is lowered by $E_{LR} = |g|^2\hbar\omega$ and shifted by $Q_0 \propto |g|$. D is the width of an absorption band resulting from upward transitions centered at $\varepsilon_1(0)$, and the hollow arrow shows the resulting Stokes-shifted emission process. If E_{LR} and Q_0 were zero the upper parabola would be directly above the lower one and all transitions would occur vertically at energy $\varepsilon_1(0)$. In (c), a conventional static-lattice band picture, if J were zero the allowed transition would also occur at $\varepsilon_1(0)$.

In Fig. 6(b) the simultaneous effects of J and E_{LR} are sketched. When E_{LR} is much larger than J (a case not shown), the point S falls below the point F and the exciton becomes self-trapped. In another case not shown, when E_{LR} is much smaller than J (in which case Q_0, which is proportional to $|g|\sim\sqrt{E_{LR}}$, may also be considered smaller for a given J), the point S falls among the many free states in the diagram and is never significantly occupied. The case shown is an intermediate one in which the free and trapped exciton may coexist through tunneling or thermal activation. The "momentarily self-trapped exciton" [61] is part of one explanation of a characteristic optical absorption edge shape called Urbach's rule, which we discuss in part IV. . Mott and Stoneham [62] have discussed the free-to-self-trapped transition in terms of a delay time for the process.

As discussed by Toyozawa, the ratio E_{LR}/J determines the nature of the relaxed exciton. If self-trapped, its emission will display the large Stokes shift characteristic of the localized state, and if free, it will display a sharp emission characteristic of momentum states. Because of the short range of the exciton-phonon interaction [60,63] the transition between these regimes is a sharply discontinuous function of E_{LR}/J. On the other hand, the absorption spectrum and the "unrelaxed" behavior of the exciton after absorption depend on the relative values of the parameters D (temperature dependent) and J. There is no abrupt transition as D/J varies.

The most intensively studied self-trapped excitons are those in the alkali haldies. It was recognized long ago by Kabler [64] and Toyozawa [65] that the exciton's relaxed configuration could be that of a trapped hole (V_K center) binding an electron. This view has been confirmed by extensive studies of emission polarization, spin resonance of the triplet state, and transient absorption spectroscopy.

All the rare gases except helium condense into face-centered cubic lattices. They are therefore isomorphic and isoelectronic to the halide sublattice of the face-centered alkali halide crystals. It is not surprising, therefore, that their exciton states are found to be similar in all respects. The self-trapped exciton [40] is almost certain to consist of a self-trapped hole (Ar_2^+) binding an electron, but the identification is not complete since the self-trapped hole has not been observed directly in the rare gases.

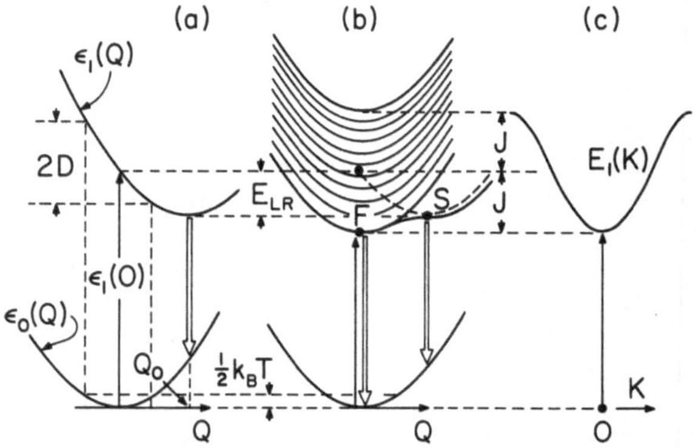

Fig. 6. Relationships among energy parameters influencing self-trapping and band motion. (a) Configuration coordinate diagram for localized excitations; (c) Band structure for delocalized excitons in a rigid lattice; (b) Combination of the two resulting in a free (F) or self-trapped (S) exciton. Adapted from Y. Toyozawa [60]. See text for explanation of symbols.

In two particular ionic crystals [66] the self-trapped exciton has a hole polarization along the <100> crystal axes instead of <110>, as in the simple alkali halides. The case of CsI is most easily understood, since CsI has a simple cubic lattice and the self-trapped hole forms an I_2^- molecule as in other iodides. The states of this self-trapped triplet were identified by optically detected magnetic resonance and phosphorescence polarization. In AgCl, the hole orientation is <100> apparently because the d shell of Ag^+ becomes involved. This is a result of its rather shallow energy compared with that of the highest filled alkali ion. The configuration of the self-trapped exciton, according to Hayes and Owen, is $(Cl\cdot Ag\cdot Cl)^-$. They have also observed the species $(Cl\cdot Ag\cdot Br)^-$.

Cho [67] has given a concise review of the subject of self-trapped excitons.

IV. INTERACTIONS WITH PHOTONS

IV.A. Semiclassical Radiation Theory

Excitons can be generated in a condensed system at least four different ways: by absorption of photons, recombination of carriers, thermal release of trapped excitations, or bombardment of electrons or other charged particles. For study of exciton properties, photon absorption is the most useful method. The probability of absorption due to excitons is sensitive to both energy and polarization and these are easily controlled.

In this section we will apply standard semiclassical radiation theory to the exciton production problem. We will see that the exciton-photon interaction is frequently very strong and that the semiclassical method is adequate only for overall line strength calculations. It fails to account easily for line broadening and ignores completely the dynamics of the coupled exciton-photon system. Nonetheless, a semiclassical treatment is worthwhile. The matrix elements calculated have wider application and a variety of spectra can be interpreted in terms of line strengths alone. In part B we will examine the dynamics associated with the absorption.

1. Transition Rates. The general nonrelativistic interaction between an electromagnetic field and N electrons is

$$H' = -\frac{e}{mc}\Sigma_i\vec{A}(\vec{r}_i,t)\cdot\vec{p}_i + \frac{e^2}{2mc^2}\Sigma_i\vec{A}(\vec{r}_i,t)^2 \tag{36}$$

where the sum runs over all electrons, $\vec{A}(\vec{r},t)$ is the vector potential of the field, $\vec{p}_i$ is the momentum of electron i, and the other symbols have their standard meanings. The quadratic term contributes to higher-order processes which are negligible in ordinary spectroscopy (see part 4, below). A semiclassical treatment of absorption runs as follows: the interaction (36) is computed for a plane wave represented by

$$\vec{A}(\vec{r},t) = \vec{A}_o e^{i(\vec{\kappa}\cdot\vec{r}-\omega t)} + \text{c.c.} \tag{37}$$

and is "turned on" at t = 0. The electronic system is taken to be in its ground state at $t \leq 0$. The amplitudes of the excited states of the system are calculated at $t > 0$ by first-order perturbation theory, using Eq. (36) as a perturbation. The resulting expressions are linear in time when averaged over a distribution of final states which is relatively smooth and continuous in energy. In atomic systems, or wherever the amplitude of a particular precisely located excited state is being considered, such an average cannot be performed and a rate of production of the excited state cannot be defined. However, if the radiation field has a range of frequencies ω, an average may be performed over this variable, leading to a calculation of an absorption rate. This procedure is described in various quantum theory textbooks. For our purposes, it will be sufficient to observe that the predicted rate of absorption contains the square of the matrix element of H' connecting the ground state and the excited state.

Consider now a transition between the ground state of a crystal and the exciton state (9). General symmetry considerations predict that the matrix element of H' connecting these states will be zero unless the field $\vec{A}(\vec{r})$ and the state (9) have the same symmetry; in particular, their translational symmetry must be the same, i.e., their wave length and therefore wave vector must be the same.* The calculation ([1], pp. 114-115) yields

$$\langle \nu K|H'|0\rangle = N^{\frac{1}{2}}\delta_{\vec{\kappa},\vec{K}}(-e/mc)A_o(\xi,\vec{\kappa})\vec{e}_\xi(\vec{\kappa})\cdot(e^{i\vec{\kappa}\cdot\vec{r}}\vec{p})_{\nu 0}\ , \tag{38}$$

where we have added a label ν to specify the exciton band, A_o is the magnitude of $\vec{A}_o$, $\vec{e}_{\xi\vec{\kappa}}$ is a unit vector along $\vec{A}_o$ specifying the polarization ($\xi = 1,2$) and $(\ldots)_{\nu 0}$ is a one-electron matrix element between the hole and electron in the exciton.

From the definition of absorption coefficient α as the relative decrease in energy flow per unit distance in the direction of propagation, we can use Eq. (38) to obtain

*Strictly speaking, $\vec{\kappa}$ and $\vec{K}$ may differ by 2π times any reciprocal lattice vector, but the energy conservation requirement forces $\vec{\kappa} = \vec{K}$ in practice.

$$\alpha = \frac{4\pi^2 e^2}{m^2 nc\hbar\omega} n_o \delta_{\vec{\kappa},\vec{K}} f(\hbar\omega - E_\nu) |\vec{e}_\xi(\vec{\kappa}) \cdot (e^{i\vec{\kappa}\cdot\vec{r}} \vec{p})_{\nu o}|^2. \qquad (39)$$

The new symbols are n, refractive index of the crystal; n_o, density in lattice points per unit volume; f(E), a normalized shape function; and E_ν, the energy of the state with $\vec{\kappa} = \vec{K}$. There are two approaches leading to f: If sharp levels and a broad excitation beam are assumed, one finds that f is a delta function. If a monochromatic excitation beam and a level broadened by vibrational interactions are assumed, then f reflects the assumed density of states. The latter approach does not really improve on the former. To predict the shape function f, phonons must be introduced explicitly.

Along with the energy conservation requirement represented by f, the wave vector conservation requirement greatly restricts the possible final states in an absorption process. As shown in Fig. 7, only the state at the crossing point of the photon and exciton dispersion curves is allowed to absorb. Since $\kappa \ll 1/a$, where a is a lattice constant, the transition is usually considered "vertical" (see inset in Fig. 7) or is said to produce excitons "at K = 0."

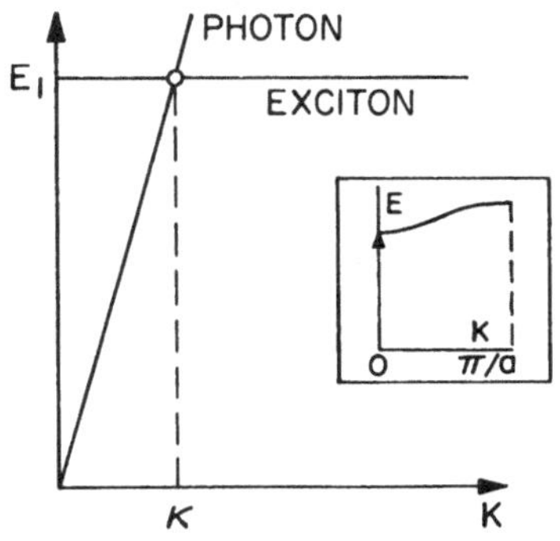

Fig. 7. The photon dispersion curve is $\hbar\omega = c\kappa$, and an exciton may be created when $\kappa = K(\text{exc.})$ and $\hbar\omega = E_1(K)$. Inset: same curves on a contracted scale (see text).

Under certain circumstances, Eq. (39) can be the basis for a theory of line shape. If $\nu\vec{K}$ were a state containing one exciton plus one or more phonons, the state having total energy E_ν and total wave vector $\vec{K}$, then one could superimpose the absorption due to all such states. Such an approach works in the case of indirect transitions (part 3) but for the broadening of strong lines, it is cumbersome and ignores phase relations among neighboring states in the continuum. We comment later (part C.3) on theories of the line shape.

2. Direct Transitions. In a "direct transition" no phonons are required to satisfy the momentum selection rule. The transition involves final states centered around a pure exciton state with $\vec{\kappa} = \vec{K}$. While the absorption band may be broadened by participation of low-energy and low-momentum phonons, they do not affect the total strength of the transition. Setting $\vec{\kappa} = \vec{K} \approx 0$ in Eq. (39), one regains an absorption expression identical to that for an atomic or molecular system of density n_0. All of the interest focuses on the matrix element $(\vec{p})_{\nu 0}$ and therefore on the localized hole and electron functions in the system. Direct transitions are most easily observed in systems with a direct gap, as described in Fig. 8. Four categories of spectra can be distinguished.

(a) Pure Frenkel case. The hole and electron are within the same molecule. Matrix elements of $\vec{p}$ are essentially identical to those in a molecule. The spectrum appears essentially identical to that of a dense gas of molecules. Exception: symmetry of the unit cell introduces certain interactions causing splitting (see part C.1).

(b) Charge-transfer case. Occurs in ionic crystals or in higher transitions in molecules. In the state $|\nu\rangle$ the hole and electron are on different, but neighboring, molecules. Transitions weaker than in the pure Frenkel case but higher multiplicity of levels.

(c) Wannier allowed case. Applies to semiconductors and to monatomic insulators (rare gas crystals). Hole and electron widely separated but in a relative orbit of s-like symmetry so that there exists a reasonable probability of both being at the same site. Characteristic n-dependence, where n is the hydrogenic level: $\alpha \propto n^{-3}$

(d) Wannier forbidden case. Applies mainly to semiconductors. Same as (c) but the relative orbit is p-like symmetry, so there exists no chance of the hole and electron visiting the same site. Characteristic n-dependence: $\alpha \propto (n^2-1)/n^5$.

The theory of cases (c) and (d) was worked out by Dresselhaus [10] and by Elliott [8]. Cases (a) and (b) are limiting cases of (c) and (d), respectively, for very small radius excitons. When an

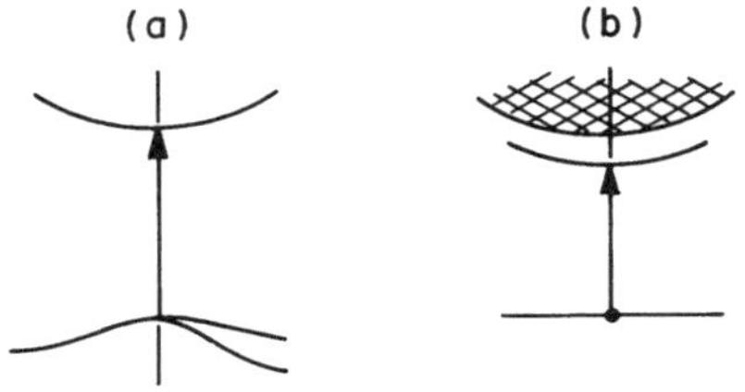

Fig. 8. (a) Schematic band structure of a crystal with a direct gap. The arrow shows a transition with $\vec{K} \simeq 0$ between valence and conduction bands creating an electron-hole pair. (b) Many-body picture of the same crystal. Here the arrow shows a transition creating an exciton. The cross-hatched region corresponds to all the free-pair states of (a).

exciton has a very small radius, regularities such as the n-dependence of the transition probabilities are overwhelmed by the atomic or molecular details of the unit cell.

3. Indirect Transitions. Many important semiconductors are "indirect," as illustrated in Fig. 9 (a). Electrons can be excited thermally at energies lower than that of the optical band gap. The electron-hole states of the indirect-gap system are found in the shaded area of Fig. 9 (b), which shows all the low-lying pair states in terms of their total energy E and total momentum $\vec{K} = \vec{k}_e - \vec{k}_h$. At any given $\vec{K}$ the Coulomb interaction induces a set of exciton states just below the lowest energy of the dissociated pairs of that $\vec{K}$. At $\vec{K} = 0$ these are called "direct excitons" (DE) and at other points "indirect excitons" (IE).

Since free electron-hole pairs are essentially continuum states of the exciton, the $\vec{K} = 0$ selection rule applies here entirely. In Fig. 9 (a) it requires that $\vec{k}_e = \vec{k}_h$. All vertical transitions are allowed as far as wave vector constraints are concerned, and "O" is the one at lowest energy in this case. Despite the fact that transitions "T" are disallowed, there is a weak absorption edge in the vicinity of the thermal-gap energy. It results from breakdown of the momentum rule by emission or absorption of phonons. Transitions "IE" are made possible by the same mechanism. Producing an

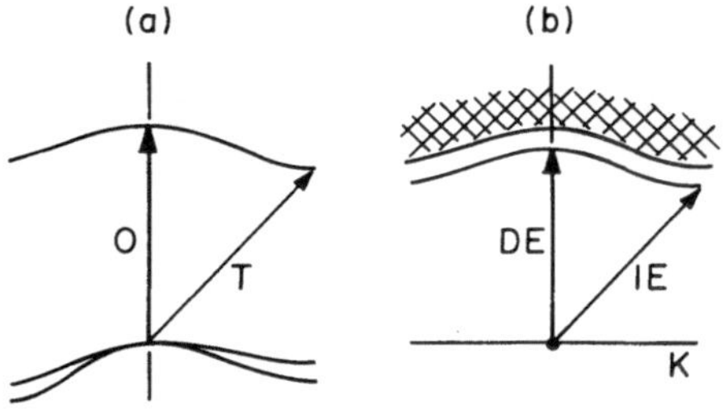

Fig. 9. (a) Schematic band structure of a crystal with an indirect gap. T shows a transition across the "thermal" gap and O shows a transition across the "optical" gap. (b) Many-body picture of the indirect-gap solid (see text).

exciton of wave vector K_{IE} optically requires emission of a phonon with $q = -K_{IE}$ or the absorption of a phonon with $q = +K_{IE}$. The detailed shape of the absorption edge will depend on temperature and the selection rules operating in the particular crystal (see, e.g. [68,69]). Generally speaking there will be a monotonic increase in absorption with an increase in photon energy, because the density of exciton and pair states rises rapidly from zero at the edge.

If the DE transition happens to be allowed, it will be broader than it would be if DE were at the minimum of the band. It can be seen by comparing Figs. 8 and 9 that in the latter case the number of final states available at equal energy is much larger. A resonance occurs between DE and the states in the upper range of the IE transitions. A typical example of this situation is found in AgCl. In Fig. 10 we show spectra measured by Brown et al. [69] and Tutihasi [70] at comparable temperatures (77 K and 111 K, respectively). In (a) the broadened direct peak is seen at enormously higher absorption coefficient than that of the indirect edge seen in (b).

In many cases the absorption edge has a particular exponential dependence first described by Urbach [71]. In part C we discuss this edge in some detail.

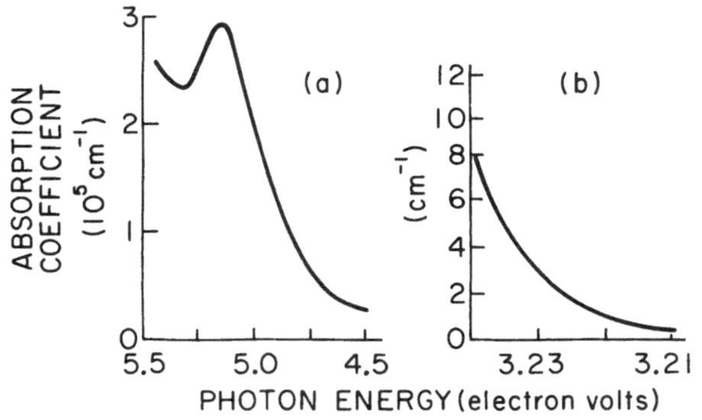

Fig. 10. Absorption of AgCl in the direct region (a) at 111 K and the indirect region (b) at 77 K (after references [70] and [69], respectively). The peak in (a) shifts to longer wavelengths and broadens at higher temperatures; the smooth curve in (b) shifts to shorter wavelengths and develops edge features at lower temperatures.

4. Giant Two-Photon Effects. Simultaneous absorption of two photons is made possible by a first-order effect of the A^2 term in (36) or by a second-order effect of the term linear in $\vec{A}$. This normally results in a very small absorption probability seen near photon energies $\hbar\omega + \hbar\omega' \gtrless E_G$. One photon creates a virtual excited state and the second one takes the system to an energy-conserving real state (or even parity). Loudon [72] and Hopfield and Worlock [73] did the earliest theoretical and experimental work on two-photon creation of excitons; see the review by Reynolds and Collins ([57], pp. 212-220).

The probability of any multiphoton process involving intermediate states is greatly enhanced when a resonance occurs, i.e., when an intermediate state energy is close to the photon energy. The second-order two-photon matrix element is a sum of terms containing the factor

$$\frac{\langle f|\vec{A}\cdot\vec{p}|m\rangle\langle m|\vec{A}\cdot\vec{p}|i\rangle}{E_m - E_i - \hbar\omega} \tag{40}$$

where i, m, and f refer to initial, intermediate, and final states, respectively; $\hbar\omega$ refers to the energy of one photon in the beam from which two will make the transition: $E_f - E_i = 2\hbar\omega$. Hanamura [74] proposed that two-photon processes should be particularly efficient in creating bound exciton pairs (biexcitons) by virtue of the fact that $\hbar\omega$ would be near an important intermediate state, which in this case is the single exciton state itself. Thus, with intense irradiation, a band with giant strength (compared with normal two-photon strengths) should appear at laser frequencies such that

$$2\hbar\omega = 2E_1 - E_b, \tag{41}$$

where E_1 is the single exciton energy and E_b the binding energy of the pair. This prediction has been confirmed in detail in CuCl [75-77]. Furthermore, differences in emission rates among biexcitons created with different total momenta have been observed in CuCl [77].

Two-photon creation of exciton complexes may be accomplished with the cooperation of other kinds of intermediate states, such as impurity-bound excitons, which have appropriate energies [78].

IV.B. The Polariton

1. Background. Two strongly interacting states whose energy levels attempt to cross when a parameter is varied "repel" each other. This is seen in a typical coupled two-state system whose Hamiltonian matrix is

$$\begin{pmatrix} E_1 & V \\ V^* & E_2 \end{pmatrix} \tag{42}$$

where we take $E_1 = E_o$ and $E_2 = E_o + c\xi$, in which E_o and c are constants and ξ is a parameter. The eigenvalues of Eq. (42) are

$$W_\pm = E_o + (c\xi/2) \pm [(c\xi/2)^2 + |V|^2]^{\frac{1}{2}} , \tag{43}$$

As Fig. 11 shows, the levels E_1 and E_2 are "repelled" and never come closer together than $2|V|$. At large values of $|\xi|$ the eigenstates are almost purely $|1\rangle$ and $|2\rangle$, the unperturbed eigenstates. If ξ varies slowly from negative to positive values, the system goes adiabatically from one $|1\rangle$ to $|2\rangle$ or vice versa. If ξ varies

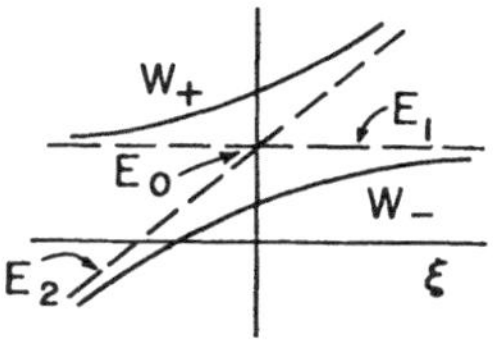

Fig. 11. Repelling of two coupled states whose unperturbed energies cross. ξ is a parameter chosen to vary E_2 linearly. $W_\pm$ are the eigenvalues, Eq. (43).

rapidly, crossing may occur. This problem was treated long ago by von Neumann and Wigner [79].

The parameter ξ in the above example is entirely arbitrary. It might be a physical quantity such as an electric field strength or it could characterize a sequence of different coupled matrices. Looking back at Fig. 7 and thinking about the matrix element (38), we find such a sequence. The two-state system is the pair of states $|\text{photon } \kappa = K\rangle$ and $|\text{exciton } K\rangle$ which are coupled by V [Eq. (38)].* An entire sequence of 2 × 2 problems is obtained as we let K vary, and so the parameter ξ is in this case equal to K. The sequence 2 × 2 problems arises because of the momentum selection rule, which ensures coupling only between one particular exciton and one particular photon. Figure 12 is the result (for details, see [1], p. 108).

The analogy between Figs. 11 and 12 is correct in principle, but certain details must be supplied to obtain a true picture of the mixed states, called "polaritons," whose eigenvalues are $W_\pm$ in Fig. 12. For example, at K = 0 the lower branch merges with the photon branch (E_2) because there can be no state lower in energy than the vacuum level. The upper branch converges to $E_1 + \Delta$, the energy of a longitudinal exciton state. This is ultimately attributable to long-range effects omitted from the simplified Hamiltonian

*In Eq. (38) the one-photon state is shown as $|0\rangle$ because photons were not included explicitly in the states; $|0\rangle$ refers to the electron in ground state.

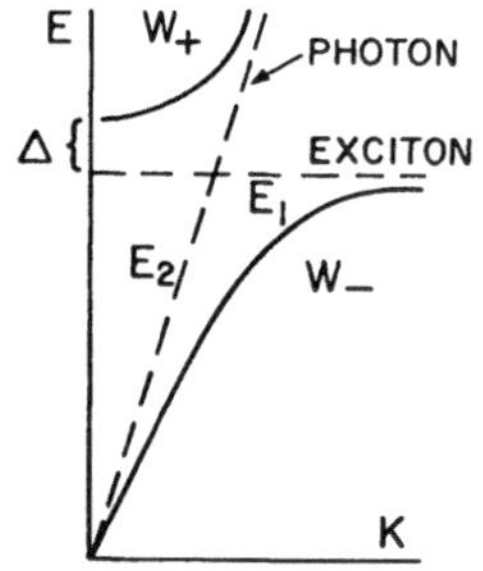

Fig. 12. Repelling of photon and exciton states whose zero-order energies cross. States $|1\rangle$ and $|2\rangle$ are coupled at each K by the matrix element Eq. (38). The mixed states whose energies are $W_\pm$ are known as "polaritons."

matrix. That is to say, many more than two states must be considered in an isotropic crystal near $\vec{K} = 0$.

The importance of the polariton in the interpretation of optical phenomena in the exciton region was recognized by Fano [80] in a general context and brought into sharp focus by Hopfield [81]. The general idea existed previously in the analogous case of photon-optical-phonon coupling [82], and one may find curves like Fig. 12 can be found in Born and Huang's book ([48], p. 91). We have previously discussed the Lorentz-model aspects of the polariton ([1], pp. 109-110 and 163-168).

2. Polariton Picture of Optical Properties. Because photons are explicitly included in the polariton picture, the notion of vertical transitions (Figs. 7, 8) must be modified. In fact, the transition becomes "horizontal" in the following sense (Fig. 13). The photon outside the crystal is on the pure photon curve (say at A). On entering the crystal its energy is maintained but its momentum need not be conserved at the surface; it is therefore transferred to B, on the polariton curve.

The transition A → B is not sufficient for absorption, because B → A is also possible. A → B → A would correspond to transmission,

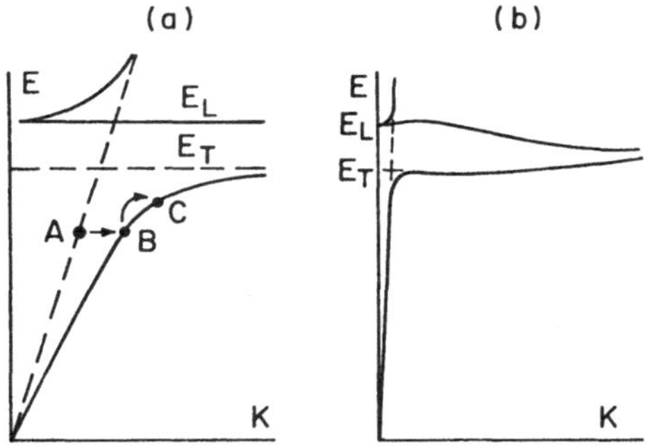

Fig. 13. (a) Photon absorption as a horizontal transition, photon (A) to polariton (B), followed by scattering to polariton (C). E_T and E_L are the transverse and longitudinal exciton branch energies at K = 0. (b) Same as (a) with a compressed K scale. The K-dependence of the exciton energies, often referred to as spatial dispersion, fills in the density of polariton states between E_L and E_T.

the final state also being outside the crystal.* For absorption, it is required that scattering occur from state B into other states C from which the probability of emergence from the crystal is low.

Horizontal transitions of low probability will be found in the vicinity of E_T and E_L where, as model calculations show, reflectivity is high; a photon will be more likely to stay outside the crystal than enter it. Indeed, between E_T and E_L reflection is perfect because there are no exciton states at all in resonance with the photon state (see Born and Huang [48], p. 118). Conditions

*Clearly the second state "A" referred to here would be a wave packet constructed from plane waves appropriately phased to be on the other side of the crystal.

for perfect reflectance do not occur in most real crystals, because there is curvature in the exciton bands which cannot show on the expanded scale of Fig. 13 (a). As shown schematically in Fig. 13 (b) there are generally states in the $E_L - E_T$ gap. Hopfield and Thomas [42] first showed how this band curvature effect quantitatively explains reflectivity in the CdS exciton region.

Returning to the absorption process: we see that transitions upward along the polariton curve, driven by phonons and the increasing density of states toward larger K, remove excitation which might otherwise emerge as transmitted energy. Not only is the C → A rate lower, but also the number of competing processes increases with the density of states. Excitations which emerge after many real scattering events would be regarded as luminescent emission. Sell et al. [83] have identified emission from both the upper (W_+) and lower (W_-) branches in GaAs at 2°K.

3. Direct Observations of Polariton Dispersion. The $\vec{K}$ dependence of polariton energies was first measured directly for optical-phonon polaritons by Henry and Hopfield [84] using Raman scattering. They varied $\vec{K}$ by varying the observed direction of the scattered light with respect to that of the incident light from a He-Ne laser; the energy difference then provided $\hbar\omega(\vec{K})$ for the phonons. Recently some very powerful optical methods have been used to obtain full dispersion curves for exciton polaritons. Following a method proposed by Brenig et al. [85], Ulbrich and Weisbuch [86] used resonant Brillouin scattering to obtain $W_\pm(\vec{K})$ curves in GaAs whose parameters, such as $E_L - E_T$ and the bare exciton band masses, agreed with independently measured values. Itoh and Suzuki [87] and Mita et al. [88] have used two-photon Raman scattering with the excitonic molecule as an intermediate state to determine many points of $W_\pm(\vec{K})$ in CuCl.

Masumoto et al. [89] have used picosecond pulses to obtain polariton group velocities directly by time-of-flight measurements. In CuCl crystals of thickness on the order of 10-100 μm, they found readily observable delays in the transmitted light intensity (the speed of light corresponds to a delay 0.0033 ps/μm; correspondingly an exciton with the same momentum and effective mass $2m_e$ experiences a delay of 1250 ps/μm). Polariton dispersion curves were verified and the effective mass of the CuCl exciton was determined directly as $(2.0 \pm 0.1)\ m_e$; somewhat different from an earlier indirect determination of $(2.5 \pm 0.3)\ m_e$ by Hönerlage et al. [90]. Time-of-flight experiments were also done by Ulbrich and Fehrenbach [91] on GaAs crystals.

4. Surface Polaritons. In addition to breaking the momentum selection rule for excitations at normal incidence, a surface provides additional eigenstates. These states exist within the $E_L - E_T$ gap and most of them lie close to the curve $E_L(\vec{K})$. They do not couple with ordinary photons and are closely related to the

evanescent waves which are solutions of Maxwell's equations at an interface. Indeed, one of the methods of exciting the surface polariton is to create evanescent waves in the adjacent material by total internal reflection. When the wave vector parallel to the surface is matched between the evanescent wave and the surface polariton, the internal reflection is attenuated [92]. This and other techniques are examined in a recent review by Fischer and Lagois [93].

Lagois [94] has proposed that the surface polariton is an excellent probe of the "dead layer," since its interaction with bulk excitons is localized near the surface. It is found that predicted attenuated total reflection spectra which agree with experiment are appreciably changed by only a 10% change in the assumed dead layer thickness in ZnO. The fit is best for a thickness 70 A, consistent with those discussed above.

IV.C. Special Topics

1. Davydov Splitting. Organic molecular crystals usually have several molecules per unit cell; however, the molecules are identical. This leads to an interesting situation in which exciton states can be built on each of several sublattices, and they will all have the same first-order energies. But then they interact with each other and repel each other, causing a multiplicity in the spectrum known as Davydov splitting [95]. Consider the model of Fig. 14. Each molecule, represented by "A," sees the same relative environment. All the upright A's can be coupled to form $\psi_A(\vec{K})$ and all the upside-down A's can be coupled to form $\psi_{\forall}(\vec{K})$. Both will have first-order eigenvalues with the same $\vec{K}$ dependence. (There may be some differences along the most general direction in $\vec{K}$ space, but the eigenvalues can be considered equal for our purposes. See [1], pp. 29 However, the Hamiltonian matrix element $V_{A\forall}$, where A and ∀ are nearest neighbors, does not necessarily vanish and may be the largest coupling present. At $\vec{K} = 0$, the nominally degenerate A and ∀ bands will be split by twice the sum of the four nearest-neighbor interactions. When there are n identical molecules per unit cell, there are at most n Davydov components.

A ∀ A ∀

∀ A ∀ A

A ∀ A ∀

Fig. 14. Part of a model molecular crystal with two molecules per unit cell.

Observation of Davydov splitting depends on a particular aspect of the photon-exciton interaction: polarization. In most cases, especially for small n, all the Davydov components are not simultaneously resolved and must be selected by varying the polarization of the exciting light with respect to the crystal axes (see, e.g. [96]).

2. Excitons in Mixed Crystals. Exciton spectra can be a useful probe of the degree of order in a mixed crystal. Consider two components X and Y, such that the pure phases have distinct absorption peaks (Fig. 15). What happens as Y is mixed into X? In

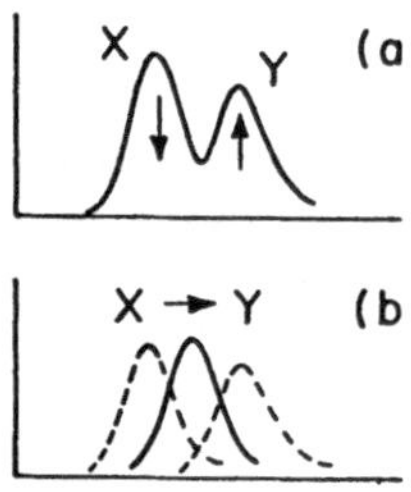

Fig. 15. Two possible effects on mixed-crystal spectra as [Y] increases and [X] decreases. (a) The persistence limit, (b) the amalgamation limit.

two extreme views, as the concentration of Y increases, (a) a new absorption band arises as that of X falls, or (b) a continuous wavelength shift of the original X band occurs, ending at the position of Y when [Y] = 100%. Case (a), which has been called the "persistence" case, is essentially a superposition of two spectra. This might be expected in practice if the excited states of X and Y were highly localized, or if X and Y were not mixing uniformly. Case (b), the "amalgamation" case, might be expected if the excited states were delocalized and X and Y were uniformly mixed. In this case, the presence of Y is felt as a perturbation on the energy of X's absorption peak, increasingly as [Y] increases.

In practice, spectra of mixed crystals do not act precisely as they should in these extreme limits. The case of the alkali halides is rather interesting because both types of behavior can be distinguished. In the first work (by Mahr [97]; see also Murata and Nakai [98]) the spectra of KCl-KBr mixtures were found to be essentially of the persistence type. One may imagine regions of Cl^- and Br^- ions (even as few as one) maintaining their original excited states. On the other hand, when the cations are mixed as in KCl-RbCl and KCl-NaCl (Nakai et al. [99]), amalgamation behavior is observed. In this case, one may imagine that the mix of cations forms a smoothly varying perturbation on the environment of the Cl^- ions. Central to these arguments, of course, is the fact that the anion contains the highest filled orbitals and is more heavily involved than the cation in determining the spectrum of the pure crystal.

Onodera and Toyozawa [100] have developed a theory and applied it to a variety of inorganic mixed crystals. Kopelman [44] has reviewed theories applicable primarily to organic mixed crystals.

3. Line Shapes and Widths. One may take either of two extreme approaches to the problem of line shapes. In the simple approach, we can regard the exciton line as essentially collision-broadened. The collisions are with phonons, and if the exciton-phonon coupling constant is small, a quasi-Lorentzian line shape results. The line width is linear in temperature except at low temperature, where phonon emission leaves a residual width. If the exciton-phonon coupling constant is large, multiphonon processes dominate and the line becomes more Gaussian in the manner of localized impurity bands. To a certain extent this may be understood in terms of the exciton having been slowed down to a nearly localized condition when the coupling is strong. The line width is again constant at low temperatures and proportional to T at high temperatures. It goes as $A[\coth(\hbar\omega/2k_BT)]^{\frac{1}{2}}$ on this model, where A is proportional to the square of the electron-phonon coupling constant and ω is the angular frequency of the one vibrational mode assumed to be involved.

The alternative approach to line shapes is the formal one. Recognizing that the photon, exciton, and phonon must be treated as a fully interacting system, we are forced to abandon semiclassical radiation theory. One of the earliest formal treatments was Toyozawa's [101]. By treating the exciton-photon coupling in first order but the exciton-phonon coupling exactly, he was able to develop a formalism which would reproduce both the weak and strong coupling limits discussed above as special cases (see [1], pp. 144-152). An entire lecture series would be required to cover the general problem of line shapes. We will limit our discussion to the question of Urbach's Rule, which has intrigued theorists for many years.

4. Urbach's Rule. In the silver haldies [71], alkali halides

[102], other crystals both pure and impure, the absorption coefficient α on the long wavelength side of the absorption peak is given by

$$\alpha = \alpha_o \exp[\sigma(\hbar\omega_o - \hbar\omega)/k_B T], \quad (\omega < \omega_o) \tag{44}$$

where α_o and ω_o are empirical constants. Generally $\hbar\omega_o$ is of the order of the absorption peak energy and is known as the "focal point energy" (from a manner of data fitting). The factor σ is of the order of unity at high temperatures but is temperature dependent The fact that Eq. (44) is not Lorentzian, not Gaussian, and does not even conform to a Boltzmann law puzzled exciton researchers for many years. Developments may be followed through our review ([1], pp. 153-158) and Kurik's [103].

There have been three distinct approaches to explaining Urbach's rule. In the first category are very general arguments such as Dexter's [104] in which thermal fluctuations shift the band edge. This theory did not predict a sufficiently wide range of validity for Eq. (44), which is indeed usually valid over three or more decades in α. Another general argument was given by Hopfield [105], who adopted the polariton picture of absorption and pointed out that a polariton would thermally ionize at a rate proportional to the Boltzmann factor. This very attractive picture unfortunately fails to predict any value of σ other than unity, and σ can range from near zero to 2.5. Recently Skettrup [106] has put Dexter's fluctuation model into completely quantitative form, predicting the temperature dependence of σ and several other quantitative relations involving the parameters, applying it successfully to ZnO.

In the second category is the electric-field modulation argument initiated by Redfield [107] and given in detail by Dow and Redfield [108]. Redfield noted that random impurities present a random electric field to the electronic states of a solid, and he showed that an exponential edge would result, i.e., $\alpha \sim \exp[a(\hbar\omega_o - \hbar\omega)]$. In a pure crystal a similar dependence would result from the electric fields due to phonons, whence would come a temperature dependence of $\underline{a}$. Mohler and Thomas [109] find the Dow-Redfield picture consistent with their measurements of the absorption edges of CuCl and TlCl in applied fields.

The third approach is the application of general optical response theory directly to the problem of edge absorption. Sumi and Toyozawa [61] found that Eq. (44) could be derived along with a specific temperature dependence of σ, apparently first noticed by Mahr [110] in connection with impurity spectra:

$$\sigma(T) = \sigma_o (2k_B T/\hbar\omega_p) \tanh(\hbar\omega_p/2k_B T), \tag{45}$$

where ω_p is the angular frequency of a phonon mode or modes interacting sufficiently strongly with the exciton that it momentarily self-traps.

In most instances where data follow Urbach's rule, each of the three theories are likely to be able to provide parameters for an explanation. However, Wiley and colleagues [111] claim to have found a case for which only the Sumi-Toyozawa theory holds. In GeS, in one mode of polarization Urbach's rule holds in detail: σ is given by Eq. (45) and a phonon frequency is thereby selected. In another mode of polarzation Urbach's rule does not hold. The argument of Wiley et al. runs as follows: the phonon selected by $\sigma(T)$ has no electric field associated with it, thus eliminating the Dow-Redfield explanation. Moreover, the generality of Skettrup's theory would predict that both polarizations should have an Urbach edge. They do not, eliminating the Skettrup explanation. Wiley et al. choose not to eliminate Sumi-Toyozawa even though they have also found an inconsistency there: the self-trapping criterion necessary to validate its applicability is not satisfied by the resulting parameters.

Exponential absorption edges may well be the effect of many causes, each with its own temperature dependence. Thus many "Urbach Rules" probably exist. It is clearly too much to expect the same theory to handle the "piezo-Urbach rule" in GaAs [112] and the "Urbach rule" in amorphous organic solids [113].

V. KINETICS AND DYNAMICS AT LOW AND INTERMEDIATE DENSITIES

V.A. Introduction

The processes by which excitons decay have attracted the attention of condensed-matter researchers for nearly fifty years. In 1963 it still made sense to ask whether excitons transport energy in any significant amount over some distance before decay takes place [1]. In 1981, largely thanks to the development of high-intensity and short-pulse spectroscopy, any doubt has been removed. This question has been replaced by several which inquire about the mode of transport (diffusive or coherent), about quantitative agreement with theory, about the competition between free and self-trapped states, and about literally dozens of effects attributable to true exciton motion.

Every type of solid has been involved in exciton energy transport studies, except for metals, where the exciton is either a virtual state or is Xray-excited and immobile. In rare gases: exciton diffusion lengths have been measured by photoelectron yields in Kr-Au "sandwiches" [114]. In semiconductors: Wolfe and co-workers have made exciton transport completely visible by guiding clouds of luminescing excitons through silicon with the use of strain fields [18]. In ionic crystals: acoustic-mode scattering of free excitons in KBr [115] and NaCl [116] over a wide temperature range has been inferred from data on emission of impurities as detectors. In organic crystals: a large number of trapping experiments have been done, some of which we will mention in part B, but in addition quite a few important conclusions about transport have been reached indirectly through linewidth and line shape determinations [46,117]. In 1, 2, 4, 5-tetrachlorobenzene, for example, Burland et al. [118] see the transition from Lorentzian to Gaussian shapes predicted by Toyozawa's theory [101].

In this lecture it will be possible to cover only some of the essential features of modern excitation transport research. Part B concentrates on the process of diffusion, the most prevalent characterization of excitation transport. In part C we comment on the general problem of coherence and examine the ways it can be sorted out of data dominated by diffusion effects. In part D we touch on annihilation and pair formation, phenomena introduced by high-intensity pulse excitation.

V.B. Exciton Diffusion

1. Historical. Frenkel alluded to the exciton as a diffusing particle [29], but an actual diffusion equation for excitons was not written until twelve years later when Förster [30] approximated energy transfer in high-concentration solutions by exciton flow on a lattice. In 1956 the first diffusion lengths for singlet excitons were obtained, in anthracene, experimentally (46 nm [31]) and theoretically (35 nm [119]). In 1964 the first measured diffusion length for triplet excitons appeared (20 μm [120]).

Exciton motion is involved quantitatively in the competition between host-crystal fluorescence and excitation trapping. Therefore the trap concentration dependence of fluorescence yield provides an alternative to Simpson's method (see part II.B, esp. Fig. 3). When Wolf reviewed the field in 1967 [121] there seemed to be general harmony between measurements of singlet diffusion lengths and diffusion constants (about $5 \times 10^{-3} cm^2 s^{-1}$) and elementary theories of the Stern-Vollmer type based on application of the Smoluchowski theory of coagulation [122]. In these theories the trapping rate is given by $4\pi D r_0 n_0$, where D is the exciton diffusion constant, r_0 an arbitrarily chosen trap radius, and n_0 the density of traps. Such a theory appears to have been first adopted for excitons by Galanin and Chizikova [119].

Powell and Kepler's pioneering time-resolved fluorescence work [123] completely obliterated the calm which had prevailed. To interpret their results on anthracene-tetracene mixed crystal fluorescence, these authors used the time dependence of the Smoluchowski theory... the theory whose steady-state predictions had appeared to work well before. They found it impossible to fit their data with reasonable parameters and introduced the alternative of direct long-range trapping with an extremely small diffusion constant. Over the next few years, many alternative interpretations were introduced, often with vigor and conviction. Soos and Powell dealt with several of them in 1972 [124]. By 1975 everyone seemed to be worn out and calm again returned, with no clear resolution to the issues, in our opinion. The situation at that point was reviewed by Powell and Soos [125].

During the past six years there have been several attempts to clarify the time-dependent fluorescence situation. Unfortunately most of the important details in the signals occur at very short times, where instrumental resolution is just beginning to become truly adequate.

2. Förster's Theories. There are many things each called "Förster's theory" and each very important to exciton physics. It is useful to distinguish among them. First, note that this entire discussion is within the context of the site representation. It is assumed that somehow all observables of interest can be related to $\rho_I(t)$, the probability that excitation is on site I. In particular, it is often assumed that an initial state consists of excitation localized on a particular site.

Pairwise rate of excitation transfer. Förster's theory [30] of this rate is based on time-dependent perturbation theory. Consider excitation initially on molecule D whose normalized emission spectrum is $I_D(\omega)$ ($\int I_D(\omega)d\omega = 1$), whose natural radiative lifetime is τ_{D0}, and whose fluorescence yield is ϕ_D (observed lifetime $\tau_D = \phi_D\tau_{D0}$). If a molecule A having an optical absorption cross section $\sigma_A(\omega)$ is located at a distance R from molecule D, then the rate constant for flow of excitation from D to A will be

$$F_{A\leftarrow D} = F_{AD} = \frac{\phi_D}{\tau_D} \cdot \frac{1}{R^6} \cdot \frac{3\kappa^2}{4\pi} \int \lambda\!\!\!^{-}(\omega)^4 I_D(\omega)\sigma_A(\omega)\,d\omega \tag{46}$$

where $\lambda\!\!\!^{-}(\omega) = c/\omega n(\omega)$, $n(\omega)$ is the index of refraction of the medium in which D and A are imbedded, and κ^2 is a geometrical factor due to the relative orientations of the transition dipole moments on the molecules, which are assumed to have electronically nondegenerate excited states. The average of κ^2 when no restrictions are placed on the geometry is 2/3. Förster's rate is usually written as $F_{A\leftarrow D} = \tau_D^{-1}(\bar{R}_o/R)^6 = \tau_{D0}^{-1}(R_o/R)^6$, where the parameters $\bar{R}_o$ and R_o are defined implicitly via Eq. (46). The Förster theory represented by Eq. (46) is based on the assumption of dipole-allowed transitions in both molecules and has been applied widely to excitation transfer in solution and in crystals. Dexter [126] generalized the theory to non-dipole-allowed transitions.

Concentration quenching of fluorescence polarization. The central issue confronted in Förster's paper [30] was the fact that fluorescence depolarization increases with concentration in a solution of similar molecules. The Förster theory of polarization quenching is based on the pairwise rate (46), depending for its success particularly on the R^{-6} dependence, and certain assumptions about probabilities of transfer to neighbors more distant than the nearest. This particular "Förster theory" is distinct from the others and is relevant to excitons only in the context of the next paragraph.

Diffusion of excitons. The assumptions about distant neighbors mentioned above were not valid at high solute concentrations, so Förster assumed the extreme case of excitation transfer in a close-packed lattice of molecules. In so doing, he wrote the first exciton diffusion equation, basing it on the Pauli master equation:

$$\frac{d\rho_I}{dt} = \sum_{J\neq I}\left[F_{IJ}\rho_J - F_{JI}\rho_I\right] - \frac{\rho_I}{\tau_I} \qquad (47)$$

By a well-known procedure, the probabilities ρ_I can be smoothed into a continuum excitation probability density $\rho(\vec{r})$ while the sum on the right side of Eq. (47) becomes a diffusion term [30, 127, 128]:

$$\frac{\partial\rho(\vec{r},t)}{\partial t} = D\nabla^2\rho(\vec{r},t) - \frac{\rho(\vec{r},t)}{\tau} \qquad (48)$$

The lattice and continuum are assumed perfect and homogeneous here. The relation between D and the value of the Förster rate when only nearest-neighbor pairwise transfer rates F are assumed nonzero is $D = cFa^2$, where c is a numerical constant of order unity depending on the lattice and the dimensionality ([128], appendix 5) and a is the lattice constant.

Fluorescence quenching by excitation transfer. The fourth "Förster theory" is another application of the rate (46), this time to the problem of an excitation donor surrounded by a random distribution of excitation acceptors. If these acceptors were to be located at specific positions $\{\vec{R}_A\}$ then the kinetic equation and its solution would be trivial:

$$\frac{d\rho_D}{dt} = -\sum_A F_{AD}\rho_D - \frac{\rho_D}{\tau_D} \qquad (49)$$

$$\rho_D(t) = \exp\left(-\frac{t}{\tau_D} - \sum_A F_{AD} t\right) \tag{50}$$

Here ρ_D is the probability that the donor is excited at time t and back-transfer is assumed absent ($F_{DA} \equiv 0$). The non-trivial part of the problem is averaging Eq. (50) over a random acceptor configuration recalling that $F_{AD} \sim R_{AD}^{-6}$. Förster [129] showed that the appropriate average of Eq. (50) is

$$\langle\rho_D(t)\rangle = \exp\left(-\frac{t}{\tau_D} - n_A \bar{V}_o\sqrt{\frac{\pi t}{\tau_D}}\right), \tag{51}$$

where n_A is the acceptor density and $\bar{V}_o = 4\pi\bar{R}_o^3/3$ where $\bar{R}_o$ is defined as in the discussion following Eq. (46). This theory is valid for those cases in which there is no donor-donor transfer during the lifetime τ_D. The theory must be modified for high donor and low acceptor concentrations, and is specifically inapplicable to the typical crystalline host-guest experiment.

3. Time-Resolved Studies of Singlets Assumed to Diffuse. Many studies of exciton motion by fluorescence quenching are based on the following picture: light is absorbed, creating a known or assumed distribution of excitations. They "localize," in the sense that the probability distribution is taken to be well described by the density matrix in the site representation (see sec. III.B.3). The probabilities of site excitation are thought to obey an equation like (47) or (48), but there are special "guest" or "trap" sites at which F(in) > F(out). Excitation still on the lattice is detected by the host fluorescence and trapped excitation is detected by guest fluorescence if there is any. When this picture is adequate, two Förster theories apply but two do not. The rate of hopping between sites is given by Eq. (46) where (D,A) may be (host,host), (guest, host) or (host, guest). In this picture Eqs. (47), (48) apply. But concentration quenching of fluorescence polarization is not explicitly involved, and, because there is host-host transfer, the formalism of Eqs. (49)-(51) does not apply.

Making quantitative predictions based on the picture just described has been found extremely tricky. Equation (47) is the best starting point, in principle, but guest concentrations are usually very small ($\sim 10^{-2}$ to 10^{-4} mol %). A massive computation would be required since 10^4 to 10^6 host sites per trap need to be included. Therefore the continuum form (48) is usually used. Another problem is that the guest sites are randomly located. The usual approach is to consider a single guest surrounded by a number of hosts sufficient to represent the average composition of the system. Then this assembly is treated as a continuum and the guest and the outer boundaries provide boundary conditions for Eq. (48), as sketched in Fig. 16. Next, usually without much discussion, Eq.

(48) is integrated over a volume which excludes the trap and the space outside the assembly,

$$\int_V \frac{d\rho}{dt}\, d\vec{r} = D\left(\int_1 d\vec{A} + \int_2 d\vec{A}\right) \cdot \vec{\nabla}\rho - \int_V \frac{\rho}{\tau}\, d\vec{r} \ , \tag{52}$$

where Green's theorem has been used on the diffusion term. The numbers 1 and 2 refer to the inner and outer surfaces of the volume V (see Fig. 16). The effect of this integration is to create an equation for the observed quantity $n(t) = \int_V \rho d\vec{r}$, the probability that the excitation is anywhere in the region V,

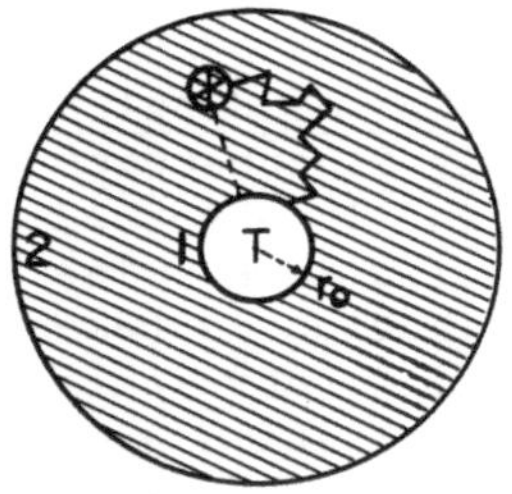

Fig. 16. A two-dimensional version of the scheme or model upon which the use of Smoluchowski's theory of coagulation for exciton trapping is based. The volume between surfaces 1 and 2 is V.

$$\frac{dn}{dt} = -D \cdot 4\pi r_o^{\,2}\left(\frac{\partial\rho}{\partial r}\right)_{r_o} - \frac{n}{\tau} \tag{53}$$

The outer surface has been forgotten and the inner surface integral written in terms of the solution of Eq. (48) under these boundary conditions. The outer surface integral might be neglected without harm if the gradient of ρ is indeed zero there, as it would be in the absence of excitation flow. However, "forgotten" appears to be a better description in most treatments, which use the original Smoluchowski boundary condition of ρ = constant at r = infinity.

The attempt to produce an equation for n must now deal with the diffusion term of Eq. (53). It is assumed (see, e.g., [125]) that $(\partial\rho/\partial r)_{r_0}$ is proportional to n and to the time-dependent part of the solution of the Smoluchowski problem at $r = r_0$. The result is

$$\frac{dn}{dt} = -\frac{n}{\tau} - \left[4\pi D r_0\left(1 + \frac{r_0}{\sqrt{\pi D t}}\right)\right] n. \tag{54}$$

The quantity in square brackets is a rate "constant" which must be used with much discretion because it arises from the following transient effect. At t = 0 the Smoluchowski solution assumes ρ = constant everywhere. The excitons near the trap are quickly swept in until the conflicting boundary condition $\rho = 0$ [or $(\partial\rho/\partial r)_{r_0}$ = constant] is satisifed. This sweep results in the $t^{-\frac{1}{2}}$ term in the rate. At long times, the trapping rate per trap relaxes to $4\pi D r_0$, the one used by Wolf [121], as mentioned above.

The solution of Eq. (54) has the form $\exp(-at-b\sqrt{t})$, unfortunately because it is in the same form as Eq. (51) and the two formalisms are very different. We hope to have made it clear that the basis of Eq. (54) as a tool for exciton fluorescence quenching analysis is extremely shaky. Many authors have tried to fit the short-time dependence of n(t) with a function $\exp(-at-b\sqrt{t})$. In most cases some other function probably would have fit as well. In any finite system we expect the true solution to be a linear combination of exponentials. Other fits are probably fortuitous and certainly hard to interpret. Wong and Kenkre have shown theoretically that trapping from momentum states does not have a $t^{-\frac{1}{2}}$ term in the rate [130].

The competition between diffusion and direct transfer to the trap, shown respectively as solid and dashed lines in Fig. 16, can be included in the diffusion equation but only with considerable mathematical difficulty or approximations. A term proportional to r^{-6} must be included in the diffusion equation (48). If in this case diffusion is then totally negligible, the conditions of the fourth Förster theory are effectively met: the donors (host crystal) are initially excited at random distances from the acceptor (guest). However, a combined diffusion and transfer theory is generally required. To the best of our knowledge this was first done in a paper by Tunitskii and Bogdasarian [131]. It was given a highly useful form by Yokota and Tanimoto [132], and has recently been treated in great detail by a group in Stuttgart [133]. Yokota and Tanimoto, using an approximate solution to the augmented Smoluchowski problem, find a steady-state trapping rate

$$k = 4\pi\cdot(0.676)[\tau^{-1}(\bar{R}_0)^6]^{1/4}\, D^{3/4}\ , \tag{55}$$

where $\bar{R}_0$ is the Förster parameter relating to the long-range transfer.

This rate reduces essentially to $4\pi Dr_0$ when the Förster theory of D is used, r_0 is taken to be the nearest neighbor distance, and the long-range $\bar{R}_0$ is taken to be the same as the one appearing in D.

The effect of trap randomness has received increasing attention. Under certain conditions the solutions are the same as those for regular trap distributions [134]. We cannot afford the time to discuss the wealth of recent treatments including those of Wieting et al. [135], Blumen and Manz [136], Huber [137], and Chow and Powell [138].

4. Triplet Exciton Diffusion. The longer lifetime of triplet excitations, on the order of microseconds and greater, allow them to diffuse further than singlets of comparable diffusion constant. Unfortunately the long lifetime also means very low emission. The first detection of triplet diffusion was accomplished by making use of collisions between triplets. In part V.D. this process will be discussed quantitatively. Here we merely note that the energy of two triplets is greater than that of one singlet, and at high triplet concentrations fluorescing singlets are produced. Avakian and Merrifield [120] used this fact to monitor triplet exciton concentrations in anthracene.

Generally speaking, the theory of triplet exciton diffusion is similar to that of singlet diffusion, but certain differences are noteworthy: there is no long-range trapping complication because the Förster dipole-dipole interaction is absent. Only the Dexter exchange term remains to accomplish excitation transfer, and since this mechanism depends heavily on wave-function overlap there is much more anisotropy in the diffusion constant (more properly diffusion tensor). In the extreme, some crystals support virtually one-dimensional triplet exciton diffusion, a prime example being 1, 2, 4, 5-tetrachlorobenzene(TCB). Three examples of the large literature on this crystal are the observation of trapping of triplets by trap phosphorescence (Güttler et al. [139]), the previously mentioned line-shape analysis [118], which has implications for triplet exciton motion, and the work of Dlott et al. [135,140] in which the effects of motion in directions other than the high-diffusion direction are found necessary to explain the details of exciton trapping.

A further qualitative difference between triplet and singlet exciton migration has been found in the context of exciton percolation (see, e.g., [141]).

V.C. Coherence

1. General Remarks. There is only a broad consensus on the

meaning of "coherence" in exciton physics. From a quantum mechanical viewpoint, coherence is the persistence or recurrence of an initial state, especially the persistence of the phase of a state. Therefore we might not expect an exciton wave packet which has been scattered several times to be called coherent. Nevertheless, the word "coherent" is frequently used as a synonym for "being in a wave packet of fairly definite wave vector." We prefer to characterize such excitons as "nearly wave-like."

The problem Förster solved in his classic 1948 paper [30] was closely related to the subject of coherence. Perrin [142] had assumed that excitation transfer between molecules took place at a rate proportional to J (the same intermolecular interaction used elsewhere in these lectures). His reasoning: excitation placed on one molecule in resonance with another will flow back and forth between them with a period h/2J, according to an elementary quantum mechanical calculation. Thus in a time h/4J, all excitation will have moved from the first to the second molecule. But the use of 4J/h as a rate of transfer led to very poor agreement with experiments on fluorescence depolarization. Förster pointed out the importance of the large density of nearly-degenerate states on the two molecules in removing the coherence of Perrin's process, and showed that a time-dependent perturbation calculation leading to a rate proportional to J^2 was the correct approach to a rate calculation. A complete discussion of this problem as it stood in 1965 was given by Förster [143].

2. Coherence as Described by a Memory Function. As of 1965, and for several years thereafter, there were believed to be two approaches to the calculation of intermolecular energy transfer rates. Förster's mechanism was accepted for most applications, but there remained an undercurrent of opinion that the Perrin process was "faster" or represented "stronger coupling." Much of this was fueled by Förster's own review [143]. Kenkre and the present author showed [144,145] that Perrin's and Förster's mechanisms were two limiting cases of a single mechanism. This was done in two stages. First, we pointed out that a more general form of Eq. (47) could be introduced [144]:

$$\frac{d\rho_I(t)}{dt} = \Sigma_{J\neq I}\left[\int W_{IJ}(t-t')\rho_J(t')dt' - \int W_{JI}(t-t')\rho_I(t')dt'\right] \quad (56)$$

Here the decay term of Eq. (47) has been omitted for convenience In Eq. (56) the functions are kernels providing "memory" of previous states, thus introducing the possibility of coherence in the general sense. Equation (56) is a form of the generalized master equation (GME) and is exact in cases where the density matrix at t = 0 is diagonal [58]. We showed [144,146] that $\mathcal{W}$ can be evaluated in an approximation whose level is the same as that of Förster's theory, with the result

$$\mathcal{W}_{AD} = \text{const. } \mathcal{R}e\ I_D(t)\sigma_A(t)^*, \tag{57}$$

where $I_D(t)$ and $\sigma_A(t)$ are the Fourier transforms of $\omega^{-3}I_D(\omega)$ and $\omega^{-1}\sigma_A(\omega)$, respectively [cf. Eq. (46)]. The constant is known, and may be found by equating the integral of $\mathcal{W}(t)$ over all time to the Förster rate (46). Thus we provided a way to deduce information on coherence, embodied in the time dependence of $\mathcal{W}$, from independently measured quantities. For broad spectra the coherence time is very short ($\sim$ 20 fs) [146].

In the second stage [145] we observed that for pairwise transfer a rate could be defined in terms of the buildup of excitation to a certain level on the second molecule. The critical parameter is α/J, where α is a measure of the rate of decay of the memory. If the memory is assumed to decay as $e^{-\alpha t}$, then the transfer rate k is the solution of

$$(\alpha/k) + e^{-\alpha/k} - 1 = \alpha^2\hbar^2/2|J|^2 , \tag{58}$$

which approaches $2\pi J/h$ at small α/J and $8\pi^2J^2/h^2\alpha$ at large α/J. Thus the Förster and Perrin theories are unified. And it should be noted that the Perrin "rate" is delusory. Whereas near the Perrin limit excitation does move from one molecule to another, it immediately moves back again reversibly and transfer is not really accomplished. Only after Förster's irreversible processes have time to operate can true transfer be said to have occurred.

In a proper description of exciton motion in a system containing more than two sites, the memory function is very complicated. Kenkre has treated more general cases and has developed an extensive theory of exciton transport based on the GME [147,148].

It should be emphasized that coherence effects on the site probability densities ρ_I are completely accounted for by the GME under localized or delocalized initial conditions [59]. However, if a physical quantity depends on off-diagonal density matrix elements, and one needs to know them, one must use a second GME dealing with these matrix elements or use the Stochastic Liouville Equation [50,149,150]. Kenkre has shown the connection between the GME and SLE approaches in detail [151]. We have used the SLE to add coherence effects to Förster's depolarization theory [47].

3. Observation of Coherence. In part II we classified observation of excitons into spectral observation and observation by capture. A similar classification can be made for exciton

coherence. Some experiments claim to have demonstrated coherence by spectral features, others by particular aspects of diffusion. We will not be able to cover the entire range of possibilities in this rapidly developing field.

Spectral observation of coherence can be claimed in a number of experiments. Provided that by "coherence" we mean "behavior indicating wave-like states," the first observation was that of Thomas and Hopfield [15], who could essentially tune a transition energy by using the state's $\vec{K}$ dependence. Another case is the optical line study by Burland et al. [118], in which one must interpret "coherent" in the Toyozawa sense of "non-trapped," i.e., free, and thus again "wavelike." Another case is the triplet exciton spin resonance line width, whose narrowing is taken to indicate rapid exciton scattering and therefore loss of coherence [16,152]. A similar argument has been applied to optical transitions [153].

Observation of coherence through its effect on the diffusion of excitons is rather indirect. The SLE theory of the diffusion constant is usually formulated in such a way that it breaks into two parts, D = D(incoh) + D(coh). This two-term diffusion constant is essentially a result of choosing to approximate the exciton dynamics in such a way as to introduce a two-term memory function, the first term being a delta-function [151]. As may be seen from Eq. (56), replacing $\mathcal{W}(t)$ by $F\delta(t)$ reproduces the incoherent transfer terms of the Förster equation (47). The theory must therefore predict some feature of D(coh) which would make it observable. Generally this would be the temperature dependence. Unfortunately, a rather accurate and reliable way of subtracting out the correct temperature dependence of D(incoh) would also be required. An application of the SLE equations to excitons in antiferromagnets has been made by Ueda and Tanabe [154].

A very interesting method of observing exciton motion is the transient grating, used by Salcedo et al. [155] in experiments on pentacene in p-terphenyl. In principle, excitons are created in a geometric pattern resembling a grating by combining two simultaneously pulsed intense beams at an appropriate angle to produce the required excitation pattern by interference. A third beam is then used to look at the amplitude of the "grating," which decays as the excitons diffuse into the spacing. Kenkre [156] has shown that the signal from the transient grating contains, in principle, all of the information needed to reconstruct the exciton memory function. Thus far the grating method appears not to have been applied to pure crystals.

Kenkre and Wong [157] have reexamined the entire theoretical problem of exciton flow in the Simpson geometry (Fig. 3). They find considerable sensitivity of the inferred diffusion constant to the assumed detrapping or reflection rates at the trapping region and to the extent of coherence. In both instances, it appears that diffusion constants have been underestimated in the past by neglecting these effects.

V.D. Phenomena at Intermediate Densities

1. General Remarks. The cross section per unit cell for one-photon absorption in an exciton band ranges from arbitrarily small values to a few times 10^{-16} cm^2. Typical photon fluxes currently attainable from short-pulse lasers range from 10^{14} to 10^{18} per cm^2 per pulse. It is therefore easy to obtain a very high density of excitons whenever an appreciable cross section exists. When such a high density is produced, the outcome is different for different crystals: in organics annihilation takes place. The site representation seems to be highly appropriate to describe the situation. When two excitons are nearest neighbors, an autoionization takes place leading to the destruction of at least one of them. In some semiconductors, on the other hand, the collision of two excitons leads to a bound state that lives as long as a radiative lifetime, and a shifted emission line occurs. Other than guessing that matrix elements for annihilation are widely different in the two cases, we seem to have no explanation of this difference in behavior.

2. Annihilation. The fusion of two excitations to form ionized states or a higher-energy fluorescing excitation was discovered soon after the introduction of the laser ([1], pp. 170 and 178). Kepler et al. [158] wrote the appropriate kinetic equation in a form which has survived,

$$dn/dt = \alpha I - \beta n - \gamma n^2, \tag{59}$$

where n is the density of excitons, in their case triplets in anthracene, αI the volume rate of exciton production with an excitation intensity I, a monomolecular decay constant $\beta(=1/\tau)$, and a bimolecular rate constant γ. This equation, along with a corresponding singlet equation and generalizations to include singlet-triplet annihilation, is well established. Typical values of γ range from 10^{-8} cm^3 s^{-1} to 10^{-11} cm^3 s^{-1}. Under steady excitation, the constant solution of Eq. (59) is

$$n = \frac{\sqrt{\beta^2 + \gamma\alpha I} - \beta}{2\gamma} . \tag{60}$$

Since the fluorescence yield ϕ is proportional to $\beta n/I$, one sees that the effect of γ is to decrease ϕ from a constant value to zero at high intensities. Under pulsed excitation, where αI is zero except during the time it takes to produce an initial density n(0), the solution of Eq. (59) is given by

$$n(t) = \frac{1}{\frac{\gamma}{\beta} + \left[\frac{1}{n(0)} - \frac{\gamma}{\beta}\right] e^{\beta t}} \quad . \tag{61}$$

This form has been fit satisfactorily to fluorescence data in numerous experiments, e.g., in the case of tetracene crystals [159]. The process of singlet-singlet annihilation has been remarkably useful in estimating the size of the domains of exciton migration in the chlorophyll aggregates of chloroplasts [160].

In the case of singlet-triplet annihilation, which is a well established mechanism in anthracene crystals [161] and photosynthetic systems [162,163], Förster's theory, Eq. (46), can be used to estimate the annihilation rate. It is found [164] that chlorophyll triplets carry with themselves a "sphere of destruction" of radius $\bar{R}_0 \sim 35$ Å. The Förster theory applies here because a triplet-to-higher-triplet transition is highly allowed and is induced by transfer from the singlet. The triplet therefore acts as the excitation acceptor. A similar calculation has been done for the singlet-singlet annihilation rate in chlorophyll [165], with similar results.

The magnetic field dependence of triplet-triplet fusion rates, as well as those of the inverse process fission, can be used to obtain a large amount of information about the triplets themselves, their dynamics, and the structure of a system. For reviews, the articles by Geacintov and Swenberg should be consulted [17,166].

Recently Kenkre has constructed a theory of annihilation which incorporates exciton coherence [167]. It is likely that such a theory will ultimately be essential in the interpretation of all high-intensity-pulsed-laser induced fluorescence data at short times.

3. Biexciton Formation. In certain semiconductors such as silicon, the attempt to induce high exciton densities leads to the electron-hole droplet, as described elsewhere in these lectures. In CuCl it is possible to produce biexcitons whose presence is detected by emission lines shifted by the biexciton binding energy [168].

At present the main interest in biexcitons is the possibility that they might form a Bose-Einstein condensate. If this were possible, it would be a unique system on which to test the laws of statistical mechanics. A comprehensive review of the theory of biexcitons is given by Hanamura and Haug [169]. Representative experimental papers which give the flavor of the current situation are those of Nagasawa et al. [170] and Chase et al. [171].

ACKNOWLEDGEMENTS

The author thanks Prof. Y. Nakai of Kyoto University for discussions and encouragement which, in part, led to this review. He also thanks the National Science Foundation for partial support of the work through grant PCM-80-11819.

REFERENCES

1. R. S. Knox, Theory of Excitons (Supplement 5 to Solid State Physics, ed. by F. Seitz and D. Turnbull, Academic Press, N. Y., 1963).
2. J. Frenkel, Phys. Rev. 37, 17, 1276 (1931).
3. R. S. Knox, J. Phys. Chem. Solids 9, 238, 265 (1959).
4. M. A. Davidovich, Phys. Stat. Solidi (b)98, 735 (1980).
5. J. C. Slater and W. Shockley, Phys. Rev. 50, 705 (1936).
6. G. H. Wannier, Phys. Rev. 52, 191 (1937).
7. N. F. Mott, Trans. Faraday Soc. 34, 500 (1938).
8. R. J. Elliott, Phys. Rev. 108, 1384 (1957).
9. F. Bassani and V. Celli, Nuovo Cim. 11, 805 (1959), J. Phys. Chem. Solids 20, 64 (1961); E. Antoncik, J. Phys. Chem. Solids 10, 314 (1959); J. C. Phillips and L. Kleinman, Phys. Rev. 116, 287, 880 (1959).
10. G. Dresselhaus, J. Phys. Chem. Solids 1, 14 (1956).
11. Anderson's original paper [P. W. Anderson, Phys. Rev. 109, 1492 (1958)] does not contain explicit results of this kind. See, however, the discussion and references in C. B. Duke, Mol. Cryst. Liq. Cryst. 50, 63 (1979).
12. E. M. Monberg and R. Kopelman, Chem. Phys. Lett. 58, 497 (1978).
13. S. K. Lyo, T. Holstein, and R. Orbach, Phys. Rev. B18, 1637 (1978).
14. J. Klafter and R. Silbey, J. Chem. Phys. 72, 843 (1980).
15. D. G. Thomas and J. J. Hopfield, Phys. Rev. 124, 657 (1961).
16. J. Zieger and H. C. Wolf, Chem. Phys. 29, 209 (1978).
17. N. E. Geacintov and C. E. Swenberg, in Luminescence Spectroscopy (M. D. Lumb, ed., Academic Press, N. Y., 1978), p. 239.
18. R. S. Markiewicz, J. P. Wolfe, and C. D. Jeffries, Phys. Rev. B15, 1988 (1977); P. L. Bourley and J. P. Wolfe, Phys. Rev. Lett. 40, 526 (1978); M. A. Tamor and J. P. Wolfe, Phys. Rev. Lett. 44, 1703 (1980).
19. H. Haken, Nuovo Cim. [10] 3, 1230 (1956); H. Haken and W. Schottky, Z. Physik Chem. 16, 218 (1958).

20. Y. Abe, Y. Osaha, and A. Morita, J. Phys. Soc. Japan 17, 1576 (1962).
21. L. J. Sham and T. M. Rice, Phys. Rev. 144, 708 (1966).
22. J. DesCloiseaux, J. Phys. Chem. Solids 26, 259 (1965).
23. D. Jerome, T. M. Rice, and W. Kohn, Phys. Rev. 158, 462 (1967).
24. W. F. Brinkman and T. M. Rice, Phys. Rev. B2, 4302 (1970).
25. W. A. Little, Phys. Rev. 134, A1416 (1964); D. Davis, H. Gutfreund, and W. A. Little, Phys. Rev. B13, 4766 (1976).
26. D. Allender, J. W. Bray, and J. Bardeen, Phys. Rev. B7, 1020 (1973); B8, 4433 (1973).
27. G. A. George and G. C. Morris, Molec. Cryst. Liq. Cryst. 11, 61 (1970).
28. S. Nikitine, J. B. Grun, and M. Sieskind, J. Phys. Chem. Solids 17, 292 (1961).
29. J. Frenkel, Phys. Z. Sowjetunion 9, 158 (1936).
30. Th. Förster, Ann. Physik [6] 2, 55 (1948).
31. O. Simpson, Proc. Roy. Soc. (London) A238, 402 (1956).
32. E. I. Rashba, Fiz. Tverd. Tela 5, 1040 (1963) [= Sov. Phys. Solid State 5, 757 (1963)].
33. V. L. Broude and V. K. Dolganov, Fiz. Tverd. Tela 14, 274 (1972) [= Sov. Phys. Solid State 14, 225 (1972)].
34. E. F. Sheka, Mol. Cryst. Liq. Cryst. 29, 323 (1975).
35. J. E. Eby, K. J. Teegarden, and D. B. Dutton, Phys. Rev. 116, 1099 (1959).
36. G. Baldini, Phys. Rev. 128, 1562 (1962).
37. R. S. Knox and F. Bassani, Phys. Rev. 124, 652 (1961).
38. L. Resca, R. Resta, and S. Rodriguez, Phys. Rev. B18, 696, 702 (1978).
39. S. Baroni, G. Grosso, L. Martinelli, and G. Pastori Parravicini, Phys. Rev. B20, 1713 (1979).
40. N. Schwentner, Appl. Optics 19, 4104 (1980).
41. L. Apker and E. A. Taft, Phys. Rev. 81, 678 (1951).
42. J. J. Hopfield and D. G. Thomas, Phys. Rev. 132, 563 (1963).
43. M. Altarelli, G. Bachelet, and R. Del Sole, J. Vac. Sci. Tech. 16, 1370 (1979).
44. R. Kopelman, in Excited States (ed. by E. C. Lim. Academic Press, N. Y., 1975) vol. 2, p. 33.
45. B. J. Botter, C. J. Nonhof, J. Schmidt, and J. H. van der Waals, Chem. Phys. Lett. 43, 210 (1976).
46. D. M. Burland and A. H. Zewail, Adv. Chem. Phys. 40, 369 (1979).
47. T. S. Rahman, R. S. Knox, and V. M. Kenkre, Chem. Phys. 44, 197 (1979).
48. See, e.g. M. Born and K. Huang, Dynamical Theory of Crystal Lattices (Oxford U. Press, Oxford, 1954), Chap. IV and Appendices VII and VIII.
49. C. B. Duke and T. F. Soules, Phys. Lett. 29A, 117 (1969); T. F. Soules and C. B. Duke, Phys. Rev. B3, 262 (1971).
50. M. Grover and R. Silbey, J. Chem. Phys. 54, 4843 (1971); S. Rackovsky and R. Silbey, Molec. Phys. 25, 61 (1973).

51. R. W. Munn and R. Silbey, J. Chem. Phys. 68, 2439 (1978).
52. R. Orbach, in Optical Properties of Ions in Crystals (ed. by H. M. Crosswhite and H. W. Moos, Interscience Publishers, N. Y., 1967), p. 445.
53. C. B. Duke, N. O. Lipari, and L. Pietronero, Chem. Phys. Lett. 30, 415 (1975).
54. D. P. Craig and L. A. Dissado, Proc. Roy. Soc. (London) A363, 153 (1978).
55. W. Känzig, Phys. Rev. 99, 1890 (1955); T. G. Castner and W. Känzig, J. Phys. Chem. Solids 3, 178 (1957).
56. P. Y. Yu, in Excitons (ed. by K. Cho, Springer-Verlag, Berlin, 1979), Chap. 5, p. 211.
57. D. C. Reynolds and T. C. Collins, Excitons: Their Properties and Uses (Academic Press, Inc., New York, 1981), Chap. 7.
58. See, e.g., R. W. Zwanzig, Physica 30, 1109 (1964).
59. V. M. Kenkre, J. Stat. Phys. 19, 333 (1978).
60. Y. Toyozawa, in Relaxation of Elementary Excitations (ed. by R. Kubo and E. Hanamura, Springer-Verlag, Berlin, 1980), p. 3.
61. H. Sumi and Y. Toyozawa, J. Phys. Soc. Japan 31, 342 (1971).
62. N. F. Mott and A. M. Stoneham, J. Phys. C 10, 3391 (1977).
63. Y. Toyozawa, Prog. Theor. Phys. 26, 29 (1961).
64. M. N. Kabler, Phys. Rev. 136, A1296 (1964).
65. Y. Toyozawa, Tech. Report A-119 of the Institute of Solid State Physics, Tokyo (1964), unpublished.
66. CsI: T. Iida, Y. Nakaoka, J. P. von der Weid, and M. A. Aegerter, J. Phys. C 13, 983 (1980); L. Falco, J. P. von der Weid, M. A. Aegerter, T. Iida, and Y. Nakaoka, J. Phys. C 13, 993 (1980); J. P. Pellaux, T. Iida, J. P. von der Weid, and M. A. Aegerter, J. Phys. C 13, 1009 (1980); AgCl: W. Hayes and I. B. Owen, J. Phys. C 11, L607 (1978).
67. K. Cho, in Excitons (ed. by K. Cho, Springer-Verlag, Berlin, 1979), p. 1.
68. G. G. MacFarlane, T. P. MacLean, J. E. Quarrington, and V. Roberts, J. Phys. Chem. Solids 8, 388 (1959).
69. F. C. Brown, T. Masumi, and H. H. Tippins, J. Phys. Chem. Solids 22, 101 (1962).
70. S. Tutihasi, Phys. Rev. 105, 882 (1957).
71. F. Urbach, Phys. Rev. 92, 1324 (1953).
72. R. Loudon, Proc. Phys. Soc. (London) 80, 952 (1962).
73. J. J. Hopfield and J. M. Worlock, Phys. Rev. 137, A1455 (1965).
74. E. Hanamura, Solid State Comm. 12, 951 (1973).
75. A. Souma, T. Goto, T. Ohta, and M. Ueta, J. Phys. Soc. Japan 29, 697 (1970).
76. R. W. Svorec and L. L. Chase, Solid State Comm. 20, 353 (1976).
77. L. L. Chase, N. Peyghambarian, G. Grynberg, and A. Mysyrowicz, Opt. Comm. 28, 189 (1979).
78. E. I. Rashba, Fiz. Tekh. Poluprovodn. (USSR) 8, 1241 (1974) [= Sov. Phys. Semiconductors 8, 807 (1975)].

79. J. von Neumann and E. Wigner, Physik. Z. 30, 467 (1929) [English translation: R. S. Knox and A. Gold, in Symmetry in the Solid State (Benjamin, N. Y., 1964), p. 167].
80. U. Fano, Phys. Rev. 103, 1202 (1956).
81. J. J. Hopfield, Phys. Rev. 112, 1555 (1958).
82. K. B. Tolpygo, Z. Eksp. Teor. Fiz. 20, 497 (1950); K. Huang, Proc. Roy. Soc. (London) A208, 352 (1951).
83. D. D. Sell, S. E. Stokowski, and J. V. DiLorenzo, Phys. Rev. Lett. 27, 1644 (1971).
84. C. H. Henry and J. J. Hopfield, Phys. Rev. Lett. 15, 964 (1965).
85. W. Brenig, R. Zeyher, and J. L. Birman, Phys. Rev. B6, 4617 (1972).
86. R. G. Ulbrich and C. Weisbuch, Phys. Rev. Lett. 38, 865 (1977).
87. T. Itoh and T. Suzuki, J. Phys. Soc. Japan 45, 1939 (1978).
88. T. Mita, K. Sotome, and M. Ueta, Solid State Comm. 33, 1135 (1980).
89. Y. Masumoto, Y. Unuma, Y. Tanaka, and S. Shionoya, J. Phys. Soc. Japan 47, 1844 (1979).
90. B. Hönerlage, A. Bivas, and V. D. Phach, Phys. Rev. Lett. 41, 49 (1978).
91. R. G. Ulbrich and G. W. Fehrenbach, Phys. Rev. Lett. 43, 963 (1979).
92. A. Otto, Z. Physik 216, 398 (1968).
93. B. Fischer and J. Lagois, in Excitons (K. Cho, ed. Springer-Verlag, Berlin, 1979), Chap. 4, p. 183.
94. J. Lagois, Ph.D. Thesis, Stuttgart (1976). See [93].
95. A. S. Davydov, Zh. Eksp. Teor. Fiz. 18, 210 (1948).
96. H. C. Wolf, in Solid State Physics (F. Seitz and D. Turnbull, eds., Academic Press, Inc., N. Y., vol. 9, 1959), p. 1.
97. H. Mahr, Phys. Rev. 122, 1464 (1961).
98. T. Murata and Y. Nakai, J. Phys. Soc. Japan 23, 904 (1967).
99. Y. Nakai, T. Murata, and K. Nakamura, J. Phys. Soc. Japan 18, 1481 (1963).
100. Y. Onodera and Y. Toyozawa, J. Phys. Soc. Japan 24, 341 (1968).
101. Y. Toyozawa, Prog. Theor. Phys. 20, 53 (1958); 27, 89 (1962).
102. W. Martienssen, J. Phys. Chem. Solids 2, 257 (1957).
103. M. V. Kurik, Phys. Stat. Sol. (a)8, 9 (1971).
104. D. L. Dexter, Nuovo Cim. Suppl. 7, 245 (1958).
105. J. J. Hopfield, J. Phys. Chem. Solids 22, 63 (1961).
106. T. Skettrup, Phys. Rev. B18, 2622 (1978).
107. D. Redfield, Phys. Rev. 130, 916 (1963).
108. J. D. Dow and D. Redfield, Phys. Rev. B5, 594 (1972).
109. E. Mohler and B. Thomas, Phys. Rev. Lett. 44, 543 (1980).
110. H. Mahr, Phys. Rev. 125, 1510 (1962).
111. J. D. Wiley, E. Schonheer, and A. Breitschwerdt, Solid State Comm. 34, 891 (1980); J. D. Wiley, D. Thomas, E. Schonheer, and A. Breitschwerdt, J. Phys. Chem. Solids 41, 801 (1980).
112. J. D. Dow, M. Bowen, R. Bray, D. L. Spears, and K. Hess, Phys. Rev. B10, 4305 (1974).
113. J. Klafter and J. Jortner, Chem. Phys. 26, 421 (1977).

114. N. Schwentner, G. Martens, and H. W. Rudolf, Phys. Stat. Sol. (b) 106, 183 (1981).
115. M. Itoh and Y. Nakai, Solid State Comm. 27, 1155 (1978).
116. M. Itoh and Y. Nakai, J. Phys. Soc. Japan 46, 546 (1979).
117. C. B. Harris and D. A. Zwemer, Ann. Rev. Phys. Chem. 29, 473 (1978).
118. D. M. Burland, D. E. Cooper, M. D. Fayer, and C. R. Gochanour, Chem. Phys. Lett. 52, 279 (1977).
119. M. D. Galanin and Z. A. Chizikova, Opt. i. Spektrosk. 1, 175 (1956).
120. P. Avakian and R. E. Merrifield, Phys. Rev. Lett. 13, 541 (1964).
121. H. C. Wolf, Adv. Atomic and Molec. Phys. 3, 119 (1967).
122. M. von Smoluchowski, Z. Physik. Chem. (Leipzig) 92, 129 (1917); B. Sveshnikoff, Acta Physicochimica URSS 3, 257 (1935); S. Chandrasekhar, Rev. Mod. Phys. 15, 1 (1943). An especially lucid discussion is given by F. C. Collins and G. E. Kimball, J. Colloid Sci. 4, 425 (1949).
123. R. C. Powell and R. G. Kepler, Phys. Rev. Lett. 22, 636, 1232 (1969); J. Luminescence 1, 254 (1970).
124. R. C. Powell and Z. G. Soos, Phys. Rev. B5, 1547 (1972); Z. G. Soos and R. C. Powell, Phys. Rev. B6, 4035 (1972).
125. R. C. Powell and Z. G. Soos, J. Luminescence 11, 1 (1975).
126. D. L. Dexter, J. Chem. Phys. 21, 836 (1953).
127. Z. Bay and R. Pearlstein, Proc. Nat. Acad. Sci. (US) 50, 962, 1071 (1963).
128. R. M. Pearlstein, Ph.D. thesis, University of Maryland (1966).
129. Th. Förster, Z. Naturf. 4a, 321 (1949).
130. Y. M. Wong and V. M. Kenkre, Phys. Rev. B20, 2438 (1979).
131. N. N. Tunitskii and Kh. S. Bagdasar'yan, Opt. i Spek. 15, 100 (1963) [= Opt. and Spectr. 15, 50 (1963).
132. M. Yokota and O. Tanimoto, J. Phys. Soc. Japan 22, 779 (1967).
133. U. K. A. Klein, R. Frey, M. Hauser, and U. Gösele, Chem. Phys. Lett. 41, 139 (1976).
134. R. P. Hemenger, R. M. Pearlstein, and K. Lakatos-Lindenberg, J. Math. Phys. 13, 1056 (1972).
135. R. D. Wieting, M. D. Fayer, and D. D. Dlott, J. Chem. Phys. 69, 1996 (1978).
136. A. Blumen and J. Manz, J. Chem. Phys. 71, 4694 (1979).
137. D. L. Huber, Phys. Rev. B20, 2307 (1979).
138. H. C. Chow and R. C. Powell, Phys. Rev. B21, 3785 (1980).
139. W. Güttler, J. U. von Schutz, and H. C. Wolf, Chem. Phys. 24, 159 (1977).
140. D. D. Dlott, M. D. Fayer, and R. D. Wieting, J. Chem. Phys. 69, 2752 (1978).
141. D. C. Ahlgren, E. M. Monberg, and R. Kopelman, Chem. Phys. Lett. 64, 122 (1979).
142. F. Perrin, Ann. Physique 17, 283 (1932).
143. Th. Förster, in Modern Quantum Chemistry, part III (O.

Sinanoglu, ed., Academic Press, Inc., N. Y., 1965), p. 93.
144. V. M. Kenkre and R. S. Knox, Phys. Rev. B9, 5279 (1974).
145. V. M. Kenkre and R. S. Knox, Phys. Rev. Lett. 33, 803 (1974).
146. V. M. Kenkre and R. S. Knox, J. Luminescence 12, 197 (1976).
147. V. M. Kenkre, in Statistical Mechanics and Statistical Methods and Application (U. Landman, ed., Plenum Publ. Corp., N. Y., 1977), p. 441.
148. V. M. Kenkre, in Exciton Dynamics in Molecular Crystals and Aggregates (H. Haken, ed., Springer-Verlag, in preparation).
149. H. Haken and P. Reineker, Z. Physik 249, 253 (1972); H. Haken and G. Strobl, Z. Physik 262, 135 (1973).
150. P. Reineker, Phys. Rev. B19, 1999 (1979).
151. V. M. Kenkre, Phys. Rev. B12, 2150 (1975).
152. R. M. Shelby, A. H. Zewail, and C. B. Harris, J. Chem. Phys. 64, 3192 (1976).
153. C. B. Harris, Chem. Phys. Lett. 52, 5 (1977).
154. K. Ueda and Y. Tanabe, J. Phys. Soc. Japan 48, 1137 (1980).
155. J. R. Salcedo, A. E. Siegman, D. D. Dlott, and M. D. Fayer, Phys. Rev. Lett. 41, 131 (1978).
156. V. M. Kenkre, Phys. Lett. 82A, 100 (1981).
157. V. M. Kenkre and Y. M. Wong, Phys. Rev. B22, 5716 (1980), and to be published.
158. R. G. Kepler, J. C. Caris, P. Avakian, and E. Abramson, Phys. Rev. Lett. 10, 400 (1963).
159. A. J. Campillo, S. L. Shapiro, and C. E. Swenberg, Chem. Phys. Lett. 52, 11 (1977).
160. G. Paillotin, C. E. Swenberg, J. Breton, and N. E. Geacintov, Biophys. J. 25, 513 (1979).
161. S. D. Babenko, V. A. Benderskii, V. L. Goldanskii, A. G. Lavrushko, and V. P. Tychinskii, Phys. Stat. Solidi (b)45, 91 (1971).
162. N. E. Geacintov and J. Breton, Biophys. J. 17, 1 (1977).
163. T. G. Monger and W. W. Parson, Biochim. Biophys. Acta 460, 393 (1977).
164. T. S. Rahman and R. S. Knox, Phys. Stat. Sol. (b)58, 715 (1973).
165. B. P. Wittmershaus and R. S. Knox, unpublished.
166. C. E. Swenberg and N. E. Geacintov, in Organic Molecular Photophysics (J. B. Birks, ed., John Wiley and Sons, N. Y., 1973), vol. 1, p. 489.
167. V. M. Kenkre, Phys. Rev. B22, 2089 (1980).
168. A. Mysyrowicz, J. B. Grun, R. Levy, A. Bivas, and S. Nikitine, Phys. Lett. 26A, 615 (1968).
169. E. Hanamura and H. Haug, Phys. Lett. 33C, 209 (1977).
170. N. Nagasawa, T. Mita, and M. Ueta, J. Phys. Soc. Japan 47, 909 (1979).
171. L. L. Chase, N. Peyghambarian, G. Grynberg, and A. Mysyrowicz, Phys. Rev. Lett. 42, 1231 (1979).

EXCITONS IN SEMICONDUCTORS

P. J. Dean

Royal Signals and Radar Establishment
St Andrews Road, Malvern
Worcs, U.K.

ABSTRACT

These lectures are intended to provide a coherent account of the fundamental properties of both free and bound excitons in semiconducting crystals. The emphasis is upon a description of those properties most relevant to the exploitation of exciton-related properties in the derivation of parameters of the semiconductor, both intrinsic and the extrinsic properties related to impurity and defect states, rather than the more formal treatment which is already provided in other chapters of this book. The discussions naturally centre on the optical properties, from which most aspects of the behaviour of excitons in semiconductors have been discovered and interpreted. Section I concerns the properties of free excitons, Section II contains the main characteristics of weakly bound excitons other than those related to satellite optical transitions, caused by phonon coupling and excited electronic states of the centre binding the exciton which are the subject of Section III. Special effects related to high excitation intensity are briefly reviewed in Section IV, other than electron-hole drops. The latter do not involve exciton states are are treated separately in this book.

I. BASIC PROPERTIES OF FREE EXCITONS

I.A. Concept of the Free Exciton

We are interested in these lectures in the semiconducting class of crystalline solids. The key property of these materials in their perfect state is the existence of a finite energy gap between the upper region of the highest band containing electrons at low temperatures and the next band of energies allowed on the usual Bloch description of the energy band structure (Fig 1a). This energy gap defines the upper edge of the window of optical transmission allowed by the perfect crystal. The lower limit in the infrared spectral region is usually set by strong lattice-Reststrahl-absorption for the majority of crystals which possess a significant ionic component in their interatomic bonding. This class of crystals includes all the compound semiconductors. However, some of the most technologically significant semiconductors, such as Ge and particularly Si, crystallise into a lattice with completely covalent bonding. Such lattices do not exhibit first

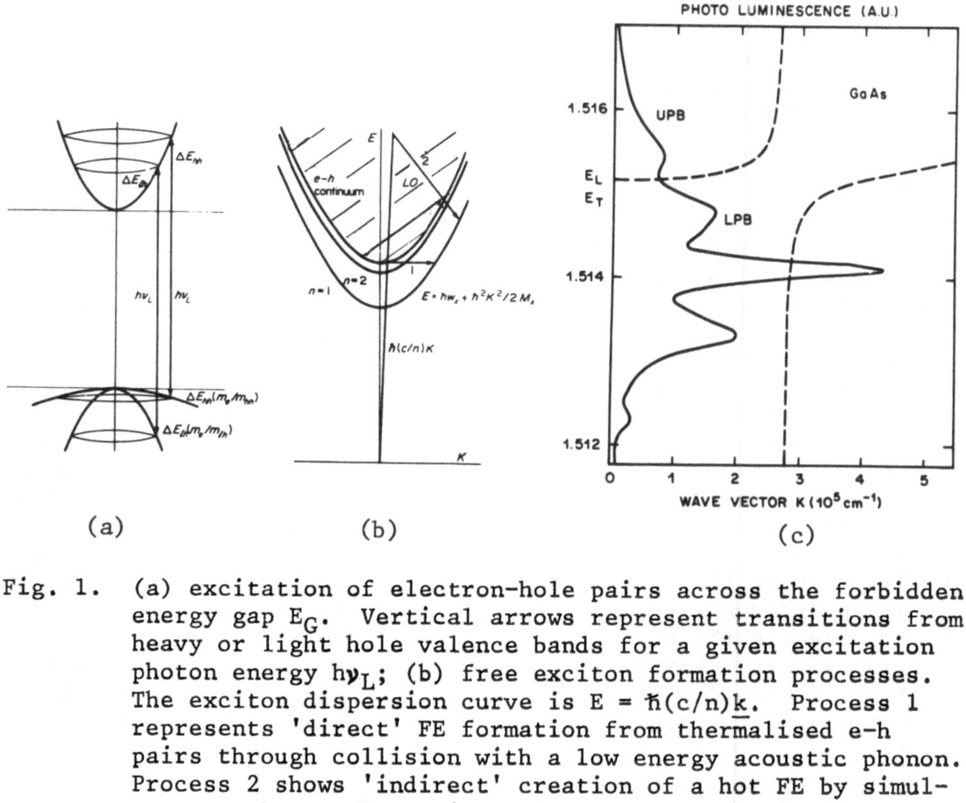

Fig. 1. (a) excitation of electron-hole pairs across the forbidden energy gap E_G. Vertical arrows represent transitions from heavy or light hole valence bands for a given excitation photon energy $h\nu_L$; (b) free exciton formation processes. The exciton dispersion curve is $E = \hbar(c/n)\underline{k}$. Process 1 represents 'direct' FE formation from thermalised e-h pairs through collision with a low energy acoustic phonon. Process 2 shows 'indirect' creation of a hot FE by simultaneous photon absorption and LO phonon emission (Nakamura and Weisbuch, ref 115). (c) luminescence in relatively pure single crystal of GaAs at 2^oK (full curve) interpreted by the polariton dispersion curve (dashed). The lower,unlabelled features involve impurity related luminescence, mainly BE, $E_L(E_T)$ is the longitudinal (transverse) FE energy (Weisbuch, ref 146).

order, single-phonon infrared absorption in the perfect state.

Semiconductors differ from insulators primarily in the magnitude of the fundamental energy gap, E_G. This clearly has a direct influence upon the possibility that significant conductivity may occur as a result of thermal excitations of electrons across this energy gap. The equilibrium carrier density n_i resulting from such a process is strongly temperature dependent

$$n_i = 2(m_e^* kT/2\pi\hbar^2)^{3/2} \exp [E_F/kT] \quad (1)$$

where $\hbar$ and k are Planck's and Boltzmann's constants and T is the absolute temperature. The Fermi level E_F is defined relative to the conduction band edge, so that

$$E_F = -E_G/2 + \frac{3kT}{4} \ln(m_h^*/m_e^*) \quad (2)$$

The densities of thermally generated free electrons and holes n_e and n_h are equal to n_i for this intrinsic thermal excitation process, while the Fermi level lies exactly at mid-gap if the electron and hole effective masses m_e^* and m_h^* are equal. Here m_h^* defines the response to an applied electric field of the vacant electron (hole) state in an otherwise completely full valence band. We shall see that m_e^* is generally $\ll m_h^*$ for direct gap semiconductors, but these masses are much more similar for indirect gap semiconductors such as Si or GaP.

This temperature dependence of carrier density, to which the bulk electrical conductivity σ_{el} is proportional, gives a semiconductor its most characteristic property. If E_G is very large, then n and $\sigma = ne\mu$ are clearly both very small. However, more subtle effects conspire to further increase the quantitative difference between the electronic properties of semiconductors and insulators. The large band gap automatically gives relatively large effective masses m_e^* and m_h^* for free electrons and holes, as described for esample through the k.p formalism [1] for the band structure of a zincblende structure semiconductor in the vicinity of a symmetry point such as Γ, where the minimum forbidden gap occurs for a large number of direct gap semiconductors. This formalism takes account of spin-orbit coupling. The electronic wavefunctions for small wave-vector $\underline{k}$ and in the presence of spin-orbit coupling are expanded in terms of the spin-free electronic states exactly at Γ. Only the states derived from the lowest Γ_1 conduction band and highest Γ_{15} (zero spin) valence bands are considered in the first order treatment. This leads to 4 energy bands at k = 0, 5 at finite k. Only the upper pair of valence bands, degenerate at k = 0, are shown in Fig 1a. The neglected higher and lower lying bands influence the non-parabolicity,also omitted in Fig 1a and the mass of the heavy holes in valence band V1. The light hole band $m^*_{1h}(v_2)$ is coupled to the conduction band c and to the band v3. The electron mass at the band edge is given by

$$\frac{1}{m_e^*} = 1 + \frac{2m_o P^2}{3\hbar^2}\left[\frac{3E_G + 2\Delta}{E_G(E_G+\Delta)}\right] \quad (3)$$

and is therefore a strong function of E_G, being very small when E_G is small. The light hole mass m^*_{1h} is influenced in a similar manner. Large values of these masses imply small values of μ and therefore even smaller σ, since $\mu = e\tau/m^*$. The mean time τ between inelastic scattering events also tends to be short in very wide gap materials, which usually possess strongly ionic lattices and therefore very strong scattering by optical phonons. Sufficiently strong electron-phonon interaction may produce a further increase in m^* due to polaron effects. These effects lead in the limit to strong local lattice distortion and the formation of a small polaron or self-trapped hole with only a very restricted, thermally activated, hopping type of mobility [2]. In addition, the wider gap materials tend to possess many deep trapping states which may overwhelm the electrical effects of any shallow conductivity-enhancing donor or acceptor states that may be present, and so produce a semi-insulating behaviour over a wide range of experimental conditions such as temperature and density of the shallow centres [3]. We will not be much concerned with true insulating crystals such as the alkali halides or wider gap oxides in these lectures.

The elementary interband excitation process in Fig. 1a automatically creates a positive hole in the vicinity of the excited electron at the bottom of the conduction band. The long range electromagnetic interaction between these oppositely charged electronic particles results in the formation of mutually bound electron-hole pair states (Fig 1b). This is the free exciton, FE, free in the sense that the periodicity invested in the electron-hole states by the Bloch theorem is not removed by this electron-hole interaction. The centre of mass of the exciton can move through the crystal by diffusional or drift processes, just like the individual electronic particles. However, this exciton migration does not of itself produce electrical conductivity, since each exciton contains a pair of charges of opposite sign. Creation of free or bound exciton, BE states often contributes a positive response in photoconductivity excitation spectra. In these cases, some type of dissociation process must occur after FE creation, for example due to impact ionization of the FE or BE during lattice scattering or even from processes internal to the BE, such as the Auger effect discussed in Section 2f. This translational character of the FE was a subject of considerable research interest about 20 years ago [4]. However, after a number of proofs of this property which were not universally accepted, very direct evidence for the translational motion of the FE was demonstrated from a magneto-Stark effect [5] in CdS (Fig 2). More indirect evidence of exciton mobility is now commonplace, obtained from their recombination luminescence spectra which contain evidence of energy distributions which have the width and form of free electronic particles at equilibrium with the lattice [6] (Fig 3a). Measurements at very low temperatures and high optical excitation rates indicate a distribution with a well-defined electronic particle temperature appreciably in excess of that of the lattice [7] (Fig 3b). This excess temperature is a consequence of one of the most basic properties of excitons, namely the fact that they are not equilibrium states of the system. Recombination involving annihilation of the electron by the hole is always an important equilibrium-restoring process, since the exciton states are invariably studied under conditions where the density of electron states far exceeds the thermal equilibrium value defined from the carrier density in Eq. 1.

I.B. Energy States of the Free Exciton

The existence of the FE state is revealed through the characteristic form it introduces in the optical absorption edge due to inter-band electronic excitations (Fig 4). The general effect is a considerable increase in the oscillator strength for excitations near E_G and the appearance of additional absorption over an energy range below E_G set by the internal bound states of the FE. We call the lowest energy bound state the ground state of the FE.

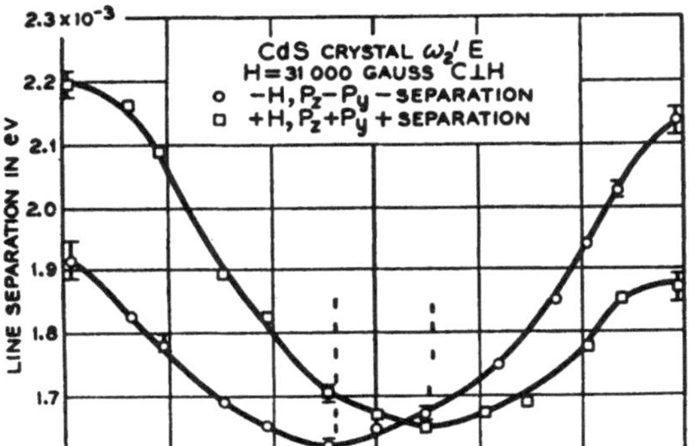

Fig. 2. The separation between two pairs of Zeeman-split subcomponents of the n = 2 FE in CdS as a function of the applied electric field F at 1.6°K. The notation is like Fig 14. The displacement of the two minima consequent upon reversal of magnetic field H is the quasi-electric field E_q due to the centre of mass velocity $\underline{v}$ of the FE in H; $E_q = 1/c\ \underline{v} \times \underline{H}$ (Thomas and Hopfield, ref 5).

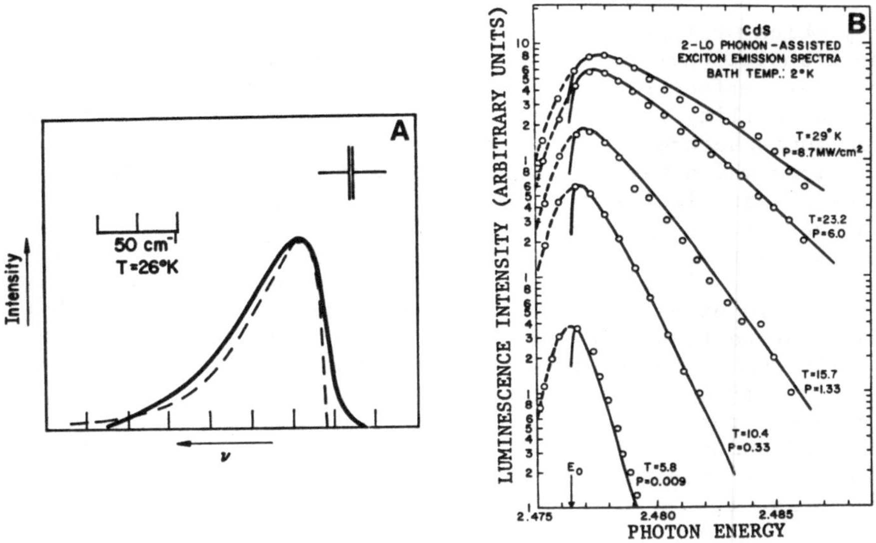

Fig. 3. (a) solid curve is the experimental n = 1 Γ_9-Γ_7 FE luminescence assisted by the emission of two LO phonons in single crystal CdS at 26°K. The dashed curve is the Maxwell-Boltzmann (MB) theoretical form for T = 26°K. The low energy threshold is near 2.51 eV. The no-phonon luminescence is narrower and more complex, like Fig 1c. (Gross et al, ref 6). (b) lineshapes of the 2 LO phonon-assisted luminescence of FE in CdS, with a lattice (bath) temperature of 2°K, measured for the indicated powers P of laser excitation. The points are experimental, the solid curves represents MB with the indicated temperatures of the FE distribution. (Leheny et al, ref 7).

binding energy E_X, and define the exciton energy gap E_{GX} such that $E_G = E_{GX} + E_X$. The exciton energy states are each spread into energy bands as a consequence of their translational energy (Fig 1b), according to the equation

$$E_X = \frac{\hbar^2 \underline{k}^2}{2M_X} \tag{4}$$

where $M_X = m_e^* + m_h^*$.

This equation is valid for small k. The free exciton energies are subject to the influence of lattice periodicity at large $\underline{k}$ in just the same way as the free electron and hole states. Processes involving the optical creation of FE, or their annihilation by interactions with single quanta of electromagnetic radiation in the absence of phonon interactions, involve FE states of very low

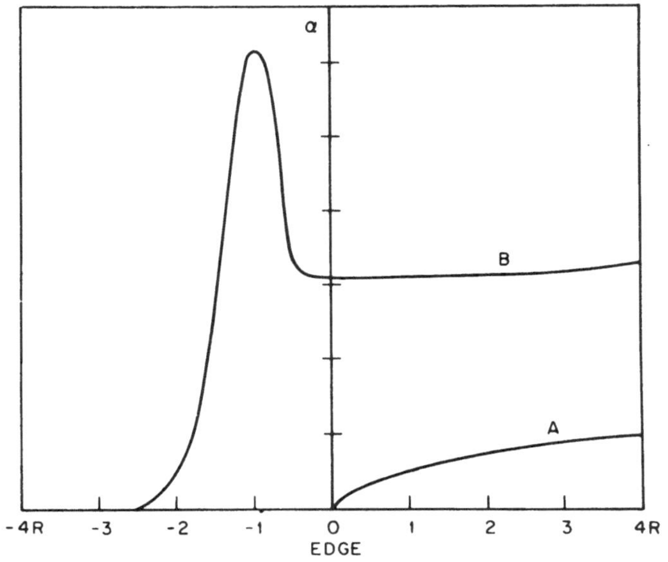

Fig. 4. The form of the optical absorption edge in a direct gap semiconductor (A) without and (B) with account of FE creation. The horizontal energy scale is given in terms of the FE Rhydberg R = E_X (Eq. (4)).(Bergh and Dean, ref 81).

k which match the very small photon wave-vectors. In this case, appropriate for transitions in direct gap semiconductors, the FE spectra exhibit the well-defined k ≈ 0 energy states given in the simplest approximation by the hydrogenic effective mass theory

$$E_X^n = - \frac{\mu R_o}{\varepsilon^2 n^2} \tag{5}$$

where μ is the reduced electron-hole mass, $\mu = m_e^* m_h^*/M_X$, ε is the static dielectric constant for energy states $E_X^n < h\omega_{LO}$ and n is the principal (orbital) quantum number of the energy state. There is an analogy with the quantum mechanical treatment of the problem of the hydrogen atom which leads to the Rydberg R_0. Transitions between these well-defined energy states produce corresponding sharp line absorption (Fig 5) and luminescence spectra.

A more rigorous treatment of FE theory using Wannier functions to represent the bound states leads to formulae for the transition oscillator strength and a treatment of selection rules. The optical transition integrals are usually expanded as a function of the FE wave-vector k for small k of order the photon wave-vector. Two important cases are distinguishable by whether the leading

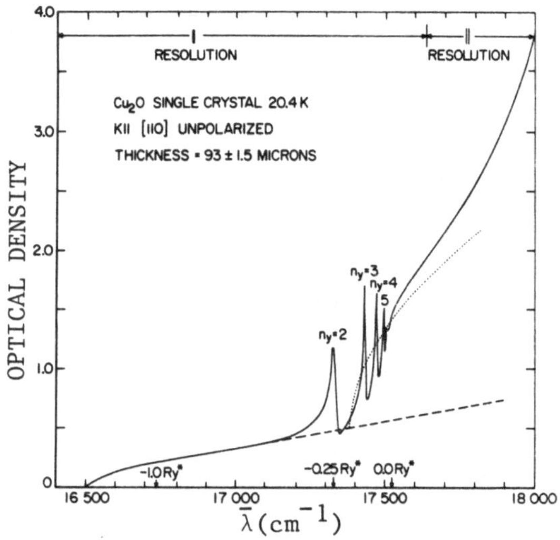

Fig. 5. The optical absorption spectrum of a melt-grown single crystal of Cu_2O in the region of the yellow FE series. Note that the dashed background of phonon-assisted absorption is depressed by an interference effect just above the $n_y = 2$ line (Forman et al, ref 28).

term in this expansion is zero or not [8]. If not, the first class FE is observed, with very strong optical absorption at the n = 1 FE state, perhaps $> 10^5$ cm^{-1} for a high quality crystal in which the linewidths are small. Electric dipole-allowed optical transitions occur to a series of s-like FE states with oscillator strength α n^{-3}. When the first term in the integral expansion is zero, the second is usually finite and allowed transitions occur only to p-like FE states, giving the second class FE spectrum. In this case, the predicted oscillator strength α n^{-5}. The leading transition to the n = 2 FE state has maximum absorption coefficient of $\sim 10^2$ cm^{-1} and is much more readily studied at convenient crystal thicknesses. We shall see that GaAs provides an example of a first class spectrum, while Cu_2O provides an outstanding example of a second class spectrum (Fig 5). The greater experimental accessibility of the latter class is undoubtedly responsible for the fact that the hydrogenic-type absorption series characteristic of the formation of a series of FE excited states was first discovered for Cu_2O in 1952 [9].

I.C. Frenkel and Wannier-Mott Free Excitons

Usually, FE are classified into two extremes, the Frenkel exciton and the Wannier-Mott exciton. These are distinguished by the radius a_{FE} of the FE ground state and therefore by the magnitude of E_X relative to other energies of the system, particularly the long wavelength longitudinal optical phonon energy $\hbar\omega$ which couples strongly to the electronic particles. The Frenkel exciton has a_{FE} comparable with the interatomic separation. The exciton can then be quite well approximated by an excitation within the anion which dominates the upper valence band states in a strongly ionic crystal such as the alkali halides. The exciton migrates within the equivalent sites of the anion sublattice with limited, thermally activated mobility. Considerable lattice distortion, which strongly lowers the energy of the FE and completely removes its translational mobility, frequently occurs before the FE state can be studied through its recombination radiation. The branching ratio of luminescence between the unrelaxed and relaxed FE states is a sensitive function of the host lattice when the relaxation is of the molecular polaron or self-trapped exciton type, involving a motion of the lattice ions governed by short range interactions rather than a purely dielectric effect. The theoretical description of this problem [2] is analogous to that for the isoelectronic trap (Sections 2a, 2e) [10], where no bound state exists for a carrier or FE unless the magnitude of the local potential well exceeds a well-defined threshold value. The existence of a potential barrier before self-trapping can occur in particular lattices leads to a delay in the formation of the relaxed state and the chance of observing luminescence from the unrelaxed as well as the relaxed FE.

Although the ground state of the Frenkel FE may have a large binding energy, perhaps ~ 1 eV and not amenable to description in terms of effective mass theory, the same difficult problem that exists for any deep state in a semiconductor, the very much shallower excited states may well be accounted for by the simple form in Eqn (5). This is best seen in intermediate cases, where the n = 1 FE radius is only ~ 10 lattice spacings and the excess binding of the observed ground state FE binding energy over the theoretical value derived from the Wannier-Mott excited states is only ~ 10%, as for the first class FE in CdS [11] or ~ 12% for the second class FE in SnO [12] and ~ 43% for the second class FE in Cu_2O [13]. The FE excited states are hard to see in the transmission spectra of the very thin film samples necessary to observe the first class spectra in the alkali halides, where the FE is very Frenkel-like. Some evidence of the n = 2 FE can be seen more readily in low temperature reflectivity of uncontaminated single crystal surfaces [15] (Fig 6). Deviations from effective mass theory are also smaller for FE states with binding energies $\ll \hbar\omega_{LO}$, so that polaron effects are reduced to the usual simple mass-enhancement derived from the Frohlich electron-phonon interaction. The coupling to polar LO phonons resulting from electron scattering in the lattice distortion associated with the vibrational mode is represented by the electron-phonon coupling

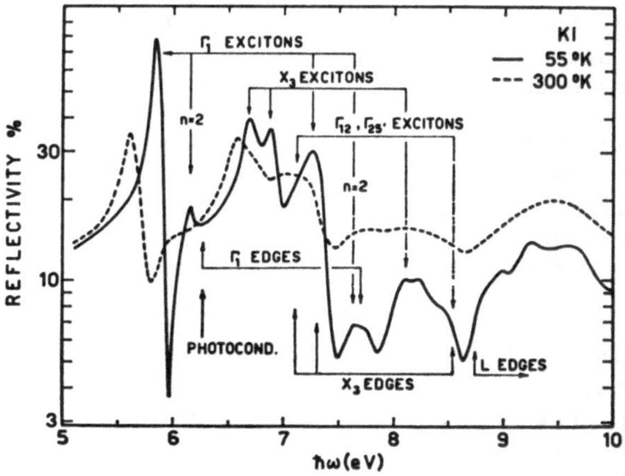

Fig. 6. Optical reflection spectrum from a clean cleaved surface of a pure KI,single crystal at 300°K and 55°K. The Γ_1 and Γ_{12}, $\Gamma_{25'}$ FE are formed from the various conduction band states at the centre of the Brillouin zone, whereas the X_3 FE involve conduction band states at the <100> zone boundary (Baldini and Bosacchi, ref 15).

constant [16]

$$A(q) = -\frac{i\hbar\omega}{q}\left(\frac{\hbar}{2m^*\omega}\right)^{\frac{1}{4}}\left(\frac{4\pi\alpha}{V}\right)^{\frac{1}{2}} \tag{6}$$

where the dimensionless Frohlich constant α is given by

$$\alpha = \frac{e^2}{\hbar}\left(\frac{1}{\varepsilon_\infty} - \frac{1}{\varepsilon_0}\right)\left(\frac{m^*}{2\hbar\omega}\right)^{\frac{1}{2}} \tag{7}$$

where ω is the LO phonon frequency, V is the crystal volume and ε_∞, ε_0 are respectively the high and low frequency dielectric constants of the lattice.

The ground state FE radius a_X is given in terms of the ground state radius a_0 of the H atom from effective mass theory simply as

$$a_x = \varepsilon a_0/\mu \tag{8}$$

The value of a_X may be $\gg a_0$ for a semiconductor with a relatively small value of m_e^* or m_h^*, such as GaAs, where $m_e^* = 0.067\ m_0 \approx \mu$ and $\varepsilon = 12.6$ so that $a_X \sim 100$ Å. It is particularly important for the accurate applicability of simple effective mass theory that a_X should be large compared with the unit cell diameter, ~ 5.65 Å for GaAs. The appropriate value of m_h^* to be used in μ

is a complex problem in view of the non-spherical, non-parabolic form of the valence band structure. There is no simple analytic solution to this problem. However, the numerical data of Baldereschi and Lipari [17] give E_X = 4.2 meV in GaAs, in good agreement with experiment [18] (Fig 7). The FE in GaAs is a classic example of the opposite limit to the Frenkel exciton, namely the Wannier-Mott exciton. This opposite extreme is a characteristic of semiconductors possessing properties quite different from the alkali halides; relatively high degree of covalent bonding and therefore high dielectric constant, high mobility and low effective mass. As we have seen from Eqn 3 this latter feature is particularly well demonstrated by the lowest energy electron states in direct gap semiconductor with moderate to small E_G, as occurs in GaAs. Then, the fact that m_h^* may be relatively large, dominated by the contribution from m_{hh} in the averaging of the valence states near $\underline{k}$ = 0 (Fig 1a), has only a minor effect on the reduced mass μ and therefore on E_X.

I.D. Direct and Indirect Gap Semiconductors

There are two main differences between the properties of FE in indirect gap semiconductors and those of direct gap type shown in Figs 5-7. These are both associated with the very different form of the electron states. We emphasise electron states, since

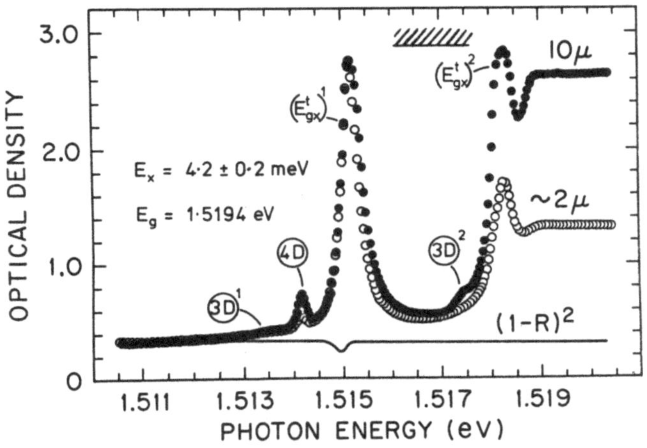

Fig. 7. Optical density of a pure GaAs single crystal at 2°K for two sample thicknesses; 10μm solid circles, about 2μm open circles. The solid curve is the structure due to the $(1-R)^2$ term, where R is the single surface reflectance. The structure labelled (4D) and (3D) involves DBE states. (Sell, ref, 18).

the indirect nature of many semiconductores, including the most technologically important such as Ge, Si, AℓAs and GaP, comes from a displacement of the lowest conduction band minimum far from the centre of the Brillouin zone where the valence band maxima is still situated. Usually, the conduction band minima lie along major symmetry axes, often at or relatively close to the boundary of the Brillouin zone. The case of GaP, (Fig. 8) is particularly complex, since the conduction band surface exhibits a saddle point (camel's back) structure about the symmetry point X, the <100> boundary of the reduced zone [19]. Neglecting this extra complication for the moment, the salient feature of these indirect band structures is a strong deviation of the E,$\underline{k}$ contours from the simple spherical forms found in zone centre minima toward strongly eccentric prolate ellipsoids whose major axes are aligned along the principal crystallographic axes on which these minima may fall. Thus, in Ge there are 4 equivalent ellipsoidal minima centred at L, the <111> boundary points of the Brillouin zone, whereas in Si there are 6 minima centred at points Δ, some 82% of the distance between Γ and X (Fig 8b). The free and weakly bound electron states, such as are involved in the shallow excited states of donors or in the FE, are described in terms of the longitudinal m_ℓ^* and transverse

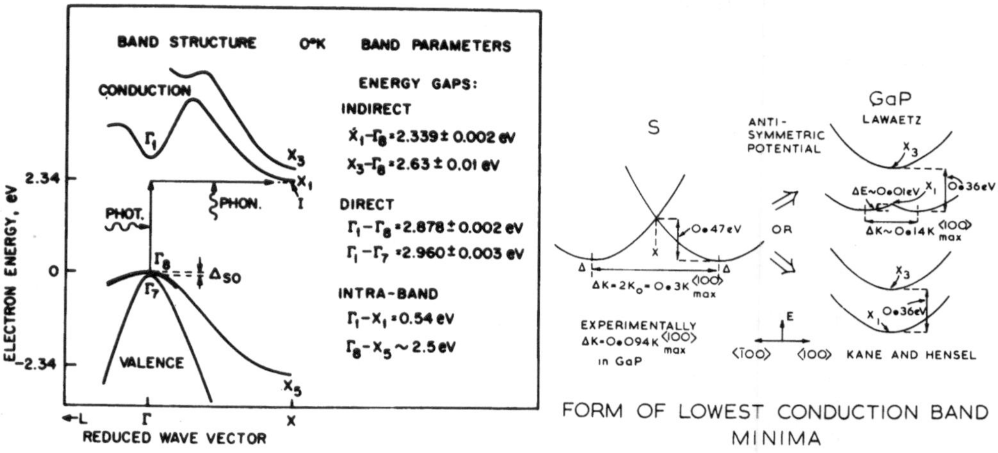

(a) (b)

Fig. 8. (a) the band structure of GaP in the vicinity of the lowest indirect Γ_8-X_1 and direct Γ_8-Γ_1 energy gaps. The indirect transition shown involves momentum conservation through phonon scattering of an electron from a virtual state derived from Γ_1 to X_3. The indirect E_G quoted is ~ 11 meV too small, the direct about 6 meV too small. (b) sections along the Γ-X direction in the Brillouin zone showing how the conduction band structure of GaP near X may evolve from that of Si. It is now known that the 'camel's back' form predicted by Lawaetz is appropriate (Dean and Herbert, ref 50).

m_t^* effective masses which describe the major and minor axes of these prolate ellipsoids [20]. The effective mass Hamiltonian for a donor bound state then takes the form

$$H_o = \frac{\hbar^2}{2m_t^*}\left(\frac{\partial^2}{\partial x^2} + \frac{\partial^2}{\partial y^2}\right) + \frac{\hbar^2}{2m_\ell^*}\frac{\partial^2}{\partial z^2} - \frac{e^2}{\varepsilon r} \tag{9}$$

In contrast to the spherically symmetric case of the H atom or an electron state at a zone centre spherical conduction band minimum, Eq. (9) cannot be solved analytically. Usually, a variational approach is used, such as that applied by Faulkner to shallow donor states [21] and the problem is solved numerically across a net of appropriately selected variational parameters, two in this particular case. This procedure ensures that the analysis yields an upper bound to the energy states of the real system. The much more complicated problem of the FE states in a semiconductor such as GaP, with a strongly non-parabolic energy surface along the major axis of the conduction band minima as a consequence of the relatively small value of the X_3^c-X_1^c splitting, δ introduced by the heteropolar component of the bonding in a compound semiconductor, has received a satisfactory treatment only relatively recently [19]. The exciton Hamiltonian is written as

$$H_{ex}(\underline{p}_e,\underline{p}_h) = H_e(\underline{p}_e) + H_h(\underline{p}_h) - \frac{e^2}{\varepsilon|\underline{r}_e-\underline{r}_h|} \tag{10}$$

where the electron kinetic energy operator is written

$$H_e(\underline{p}_e) = \frac{\underline{p}_{e_t}^{\,2}}{2m_t^*} + \frac{\underline{p}_{e_\ell}^{\,2}}{2m_\ell^*} + \frac{\delta}{2} - \left(\frac{\delta^2}{4} + \frac{D^2}{h^2}\,\underline{p}_{e_\ell}^{\,2}\right)^{\frac{1}{2}} \tag{11}$$

and D is the X_1^c-X_3^c interband momentum matrix element

$$D = \frac{i\hbar}{m_o} < X_1^c|p_{e\ell}|\, X_3^c> \tag{12}$$

The hole kinetic energy operator was described through Kane's spherical zero-order approximation

$$H_h(\underline{p}_h) = \frac{p_h^{\,2}}{2m_h^*} + H_h^{(1)}(\underline{p}_h) \tag{13}$$

where $m_h^* = m_o/\gamma_1$, γ_1 is the leading Luttinger spherical valence band parameter and the non-spherical operator $H_h^{(1)}(\underline{p}_h)$ is treated as a perturbation in the form $H_h^{(1)}(\underline{K}, \underline{p})$ after transformation to centre of mass co-ordinates.

Typical values of m_ℓ^* and m_t^* are much more comparable with m_h^* in these indirect gap transitions than in direct gap semiconductors. For example, in Si, $m_\ell^* = 0.916\ m_0$, $m_t^* = 0.191\ m_0$ and $m_h^* \simeq 0.24\ m_0$, taking a simple average of light and heavy hole bands. This is the first major difference, with the consequence that the theoretical values of E_X are much more equally sensitive to detailed properties of hole as well as electron states than in small to moderate gap direct semiconductors. The second difference is intimately linked with the large difference in electron wave-vector between the hole states $\underline{k}_h^v$ and electron states $\underline{k}_e^c$. This means that the optical transitions through which the exciton states are probed are of fundamentally different character. Only phonon-assisted creation or annihilation of FE can occur in a perfect crystal, where the phonons of energy $\hbar\omega$ and wave-vectors $\underline{q}$ are selected to conserve momentum according to

$$\underline{q} = \underline{k}_e^c - \underline{k}_h^v \tag{14}$$

where the comparatively negligible photon wave-vector $\underline{k}_p$ has been neglected. The Hamiltonian for this three-body interaction therefore contains essential terms due to the process by which momentum is conserved due to electron-phonon scattering processes. The theoretical forms for phonon-assisted optical absorption by (first class) and forbidden (second-class) FE transitions then take the form [8]

$$\left.\begin{aligned} \alpha &= C\,(h\nu - E_{GX} \pm \hbar\omega)^{\frac{1}{2}} \\ &\text{and} \\ \alpha &= D\,(h\nu - E_{GX} \pm \hbar\omega)^{3/2} \end{aligned}\right\} \tag{15}$$

where $\hbar\omega$ is the energy of a phonon which satisfies Eqn (14) and C, D are approximately constant.

The effect of the electron-phonon interactions on the optical spectra therefore involves the replacement of the sharp line absorption structure due to inter-bound state excitations of the FE (Figs 5-7) by stepped structure (Fig 9). The major features of these steps are the successive thresholds caused by the ability of phonons from different branches of the lattice vibrational frequency spectrum to satisfy Eq. (14) with quite different values of $h\omega$. The individual components appear as absorption edges rather than lines because of the ability to balance any particular value of $\underline{k}_e^c - \underline{k}_h^v$ by a phonon of appropriate wave-vector $\underline{q}$, so that the optical

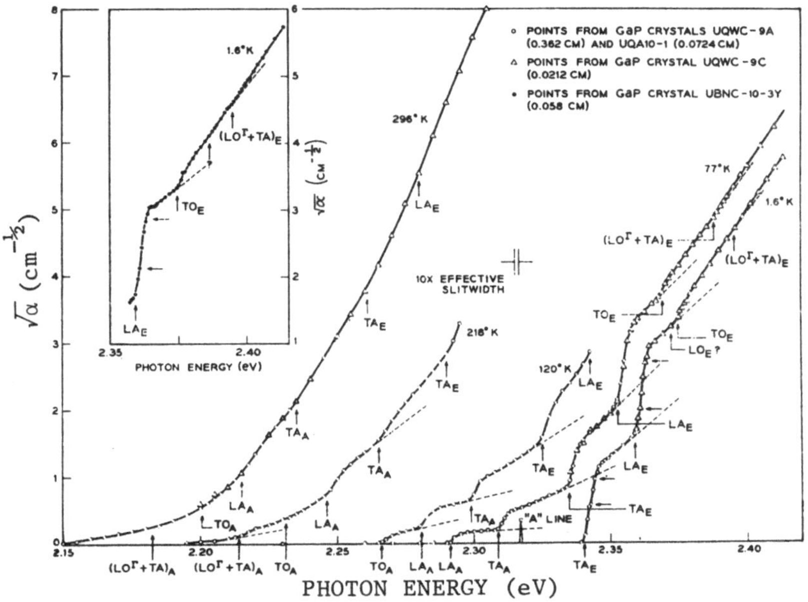

Fig. 9. The threshold region of the intrinsic interband optical absorption spectrum of a pure single crystal of GaP at the indicated temperatures. Transmission spectra of samples with three different thicknesses were used. Vertical arrows indicate thresholds for different momentum-conserving phonons emitted (E) or absorbed (A) in the electronic transition. Horizontal arrows denote electronic structure (Fig 10). (Dean and Thomas, ref 24).

transitions are not restricted to narrow, well defined ranges of electron and hole energies very close to the band extrema as in the case of direct transitions. In this way, FE involving electron-hole states extending far above the minimum energy difference $E_e^c - E_h^v - \frac{e^2}{\varepsilon|r_e - r_h|}$ can be photo-created (or annihilated), since the excess centre of mass momentum of the FE can be balanced by the appropriate small change in $\underline{q}$. It is important to remember that the necessity to match the essentially zero value of the photon wave-vector $\underline{k}_p$ also imposes stringent limitations on the form of no-phonon FE luminescence in direct gap semiconductors. The form of this luminescence must be described with the aid of exciton-photon polariton theory, through which the energy regions of the exciton-photon states of strongly mixed character from which luminescence can occur may be recognized [22] (Fig. 1c). These restrictions also become relaxed through phonon interactions, so only the LO phonon replicas of the FE luminescence show the simple Boltzmann form anticipated from their kinetic energy distribution (Fig 3a). A similar factor must be applied to the absorption component for each individual phonon-assisted transition in Eq. (15), to obtain the form of the intrinsic luminescence for an indirect gap semiconductor.

Allowed transitions to orbital internal excited states of the FE are expected to be much weaker than those to the n=1 ground state, approximately according to the variation of the oscillator strength as n^{-3}. We have seen that the same variation holds for allowed direct FE. In practice, this means that these excited states are difficult to resolve in absorption spectra of indirect gap semiconductors, since they become lost against the strong background from the n=1 FE. They may perhaps be best seen in luminescence, where contributions from states of higher centre of mass kinetic energy are reduced by the form of the Boltzmann distribution fixed at some appropriate temperature. Not much attention has been paid to this possibility since the appearance of an ~8 meV excited FE state in the intrinsic luminescence of Si [23].

The absorption edge spectrum of GaP [24] (Fig 9), contains some fine structure near the threshold of the absorption component associated with each phonon branch. The origin and curious form of this structure was an unsolved mystery when these data were taken in 1966, since only a single splitting of approximate order $E_X/10$ was expected due to the influence of valley anisotropy of the electron on the valence band states of the hole. Physically, this means that FE states of different angular momentum projection, ½ or 3/2 on the <100> symmetry axis imposed by the ellipsoidal conduction band minima have significantly different binding energies. Analysis of this structure for GaP [19] (Fig. 10) in terms of the full description of Eq. (11), including the effect of the 'camel's back' on the effective value of $m^*_{e\ell}$, has been very successful in accounting for the extra splitting and the curious shape deriving from the existence of P_1 as well as the simple P_0 type critical points (Van Hove singularities) in the conduction band and therefore FE energy band structure. The derived camel's back parameter $E_{X_1^c} - E_{\Delta min_1^c}$ is 3.5 ± 0.3 meV, very similar for both FE states and free electron states. The minima fall at about 92% of the Γ-X_1 wave-vector separation. The $E^{\frac{1}{2}}{}_{1s}(0)-E^{3/2}{}_{1s}(0)$ valley anisotropy splitting is 1.9_6meV, while the ground state binding energy $E^{3/2}_{1s}(0)$ = E_X is 21.4 meV, much larger than the original estimate [24]. This revision has led to a new value of E_G=2.350 ± 0.001 meV in GaP (Fig 8). This, and re-adjustments in the values of E_A for acceptors in GaP has produced a much more settled and self-consistent set of values of E_D and E_A for the whole range of donor and acceptor impurity states in GaP. For example, values of E_D for the remarkably deep donor O obtained from independent optical techniques now agree at 0.898 ± 0.001 eV to better than 1 part in 900! [26].

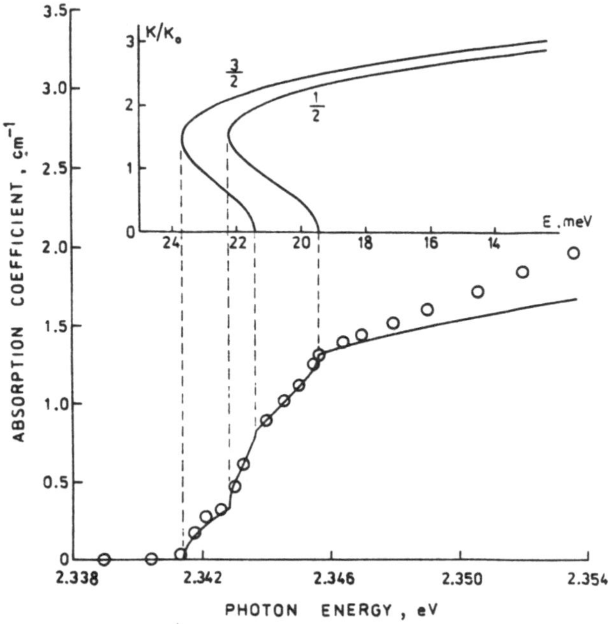

Fig. 10. Detail in the indirect FE absorption for GaP in the region of the TA_E component of Fig 9. Circles are experiment, solid line is the theoretical FE density of states allowing for the 'camel's back' form of the exciton dispersion shown in the inset (Glinskii et al, ref 19).

These results are in fair agreement with a recent study of the behaviour of these absorption edges under uniaxial stress, although a rather smaller value of the camel's back parameter was obtained, ~ 2.4-2.8 meV [27]. This stress work, like many similar investigations of bound exciton, BE as well as FE states in GaP, fully confirmed that the conduction band minima in GaP lie along the <100> axes of the Brillouin zone. This result is also firmly established from the fit of the phonon energies derived from the zero-stress intrinsic FE-related absorption and luminescence spectra to the phonon dispersion curves [147].

I.E. Phonon-Assisted Free Exciton Absorption in Direct Transitions

The remarkable absorption spectra of Cu_2O in Fig 5 [28] contains a prominent absorption step near 16,500 cm^{-1} (~ 2.046 eV) and 17,380 cm^{-1} (2.155 eV). These steps have similar form to those in Fig 9, and are also attributed to phonon-assisted FE creation. The necessity for such absorption processes arises for Cu_2O in a quite different way than for Si or GaP. The energy gaps all occur at Γ and the interband transitions are all direct-type (Fig 11),

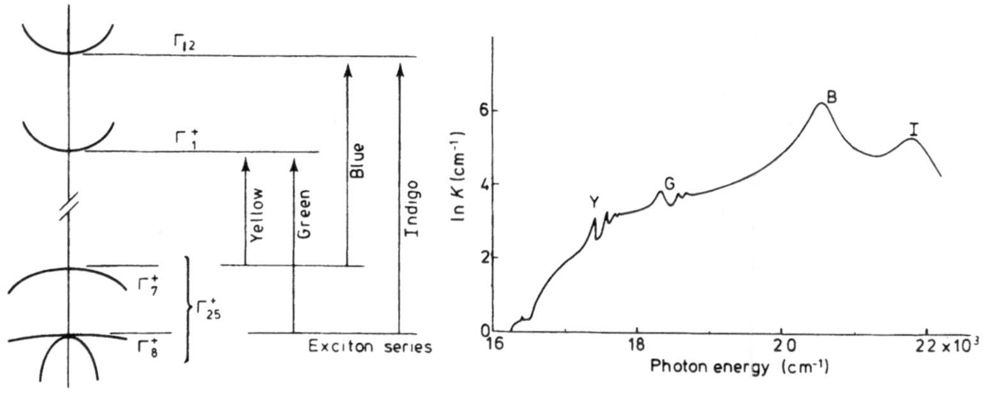

(a) (b)

Fig. 11. (a) the valence and conduction bands of Cu_2O near the centre of the Brillouin zone (Elliott, ref 30). (b) a schematic representation of the inter-band optical absorption in Cu_2O at 4.2°K. K is the absorption coefficient in cm^{-1} and the different absorption contributions from the individual FE series in (a) are clearly separated (Grossmann, ref 37).

so there is no problem of momentum conservation. The Γ_{25}^+ valence band from the Cu 3d states is split into a Γ_7^+ level and a Γ_8^+ level [30] effectively with the opposite sign from the usual valence band splittings in semiconductors made from main group elements which do not have open 3d (or 3f) shells (Fig 11a). Conduction bands are predominantly Cu4s-like, the lowest having symmetry Γ_1^+ (or Γ_6^+ with spin). Photo-creation of 1s FE in the lowest energy series (yellow, Y in Fig 11b and Fig 5) is forbidden for electric dipole transitions. Thus, only transitions to 2p FE states are strong, and are responsible for the sharp lines in Fig 5. Creation of the n = 1 Y FE state is significant only in the presence of a strong electric field $\gtrsim 10^5 V\ cm^{-1}$, sufficient to break the electric dipole selection rule [31], or with the emission (also absorption at higher lattice temperatures) of a phonon of appropriate symmetry. The threshold of the step near 2.046 eV is attributed to the creation of the 1s Y FE plus emission of an 13.6 meV Γ_{12}^- symmetry phonon. Corresponding processes have been observed in luminescence at low temperatures [32]. Broadening of the phonon-assisted component but not the no-phonon component was observed as the temperature was increased, so increasing the kinetic energy of the FE system, as expected from the discussion in Section 1b. The absorption lines due to creation of np FE states, where $n \gtrsim 2$, are then superposed upon the continuum due to 1s + phonon. Quantum mechanical interference is possible involving the wave-functions for different optical processes contributing through different pathways to the same final states in the vicinity of these lines [33].

These effects, similar to interference involving BE and FE, are responsible for the asymmetric lineshapes of the n=2 and n=3 Y FE states, and the excursion of the full lines below the dotted continuum in Fig 5. The 2p energy states of the yellow and green FE series in Cu_2O accurately follow the simple form of Eq. (5) [13]. The predicted p-like character of the n=3 FE state, which is less broadened by the interference effects than the n=2 state and less subject to diamagnetic shifts than the FE states of $n \geqslant 4$, has been established from Zeeman measurements at ~ 10T [34]. The value of μ was obtained independently from oscillatory structure in the interband magneto-absorption.

I.F. Higher Energy Free Exciton States

So far, we have only considered FE states arising from the extremal regions of the conduction and valence band structure. These states dominate the intrinsic recombination luminescence, because of strong energy relaxation to these lowest states before luminescence. This occurs especially for studies at the lowest temperatures, which are required for observation of full details in the optical spectra, uncluttered by anti-Stokes phonon-assisted transitions and other, phonon-broadening effects inherent in high temperature spectra.

However, sharp structure due to exciton states associated with higher-lying regions of the conduction band and lower-lying regions of the valence band structure are readily seen. Two such features appear for GaP in Fig. 12. The original data of Dean et al [35] has been re-analysed [36] in terms of bound to bound transitions, in

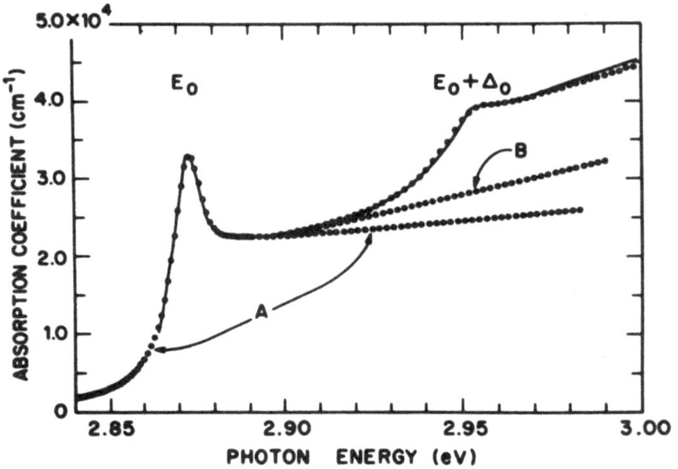

Fig. 12. Direct E_0 and $E_0+\Delta$ FE absorption of a pure single crystal of GaP recorded at 25°K. The solid curve is experimental, the dotted curves are various theoretical fits (Sell and Lawaetz, ref 36).

which that to the 1s state responsible for the broadened peak predominates, together with FE continuum described by the expression [8]

$$\alpha(h\nu) = \frac{2(2\mu)^{3/2}e^2 f_{cv}}{n\, c\hbar^2 m_o}\left[\sum_n 4\pi R_D E_X^{3/2}\,\delta\left(\frac{h\nu - h\nu_n}{n^3}\right) + \frac{2\pi E_X^{\frac{1}{2}}\, u(h\nu - E_G)}{1-\exp(-2\pi Z)}\right] \qquad (16)$$

where f_{cv} is the oscillator strength, n is the refractive index at photon energy $h\nu$, $Z = E_X/(h\nu - E_G)$ and u(x) is the unit step function. Curve A in Fig 12 is for E_X = 11 meV and Γ = 6 meV, where Γ is the Lorentzian broadening parameter which represents the effects of rapid scattering out of the electron states derived from the Γ_1 minimum, which are iso-energetic with energy states of large $\underline{k}_e$ in the Brillouin zone (Fig 8). Curve B in Fig 12 attempts to allow for the influence of strong non-parabolicty in the valence band, while the remaining broad absorption band near 2.95 eV is attributed to direct FE involving hole states from the spin-orbit split-off valence band, with $E_X \sim 10$ meV and Γ = 11 meV. In this case, the holes are also subject to strong intra (valence) band scattering processes. The value of the spin-orbit splitting parameter at k = 0 is $\Delta_o = 78 \pm 2$ meV [36], slightly smaller than the original value obtained from the more elementary analytical procedure [35].

The remarkable absorption spectrum of Cu_2O in Fig 11b also contains well-defined structure for the green series, involving transitions from the spin-orbit split-off valence band, splitting $\Delta_o \sim 0.12$ eV [37]. Very broad absorption components also appear involving transitions to the much higher energy Γ_{12}^- conduction band states, where the $\Gamma_{12}^- - \Gamma_1^+$ conduction band splitting is $\sim 4.5 \times \Delta_o$, or ~ 0.56 eV. The latter absorption bands involve allowed photo-creation of 1s FE, however the strong broadening prevents direct observation of the discrete FE structure, rather as in Fig 12.

I.G. Zeeman Splittings and Diamagnetic Shifts

The transitions to the n = 1 FE state in a first class spectrum are usually so strong that direct transmission measurements are impracticable. The transition can be conveniently studied in reflectance. However, the peaks are usually broad even at 4.2°K, and Zeeman splittings can only be resolved at very large magnetic fields. Such fields have only become widely available in recent years. In addition, the Zeeman effects are then observable only in an intermediate field limit, where the dimensionless parameter $\gamma = \hbar\omega_c/2R_o$ is not sufficiently small for analysis in terms of the linear Zeeman effect [38]. Here, ω_c is the cyclotron resonance frequency $eH/\mu c$, usually dominated by the magnetic effects on the low mass electron for transitions in a direct gap semiconductor. Fortunately, the strong transitions associated with the FE A in CdS, involving the Γ_9 valence band split by both spin-orbit interaction and the local crystal field in this wurtzite semiconductor (Fig 13), are not active for $E||C$. A great deal of detailed information was therefore particularly readily obtainable in this polarization, and on the weaker absorption at the n = 2 FE for $E\perp C$ as well as $E||C$ [11]. The longitudinal-transverse exciton splitting E_{LT} (Fig 1c) was clearly observed, since the longitudinal exciton state $A_L(1s_L)$ is seen for $E||C$ and the transverse $A(1s_T)$ for $E \perp C$. This splitting is a general effect observed for any FE states created by allowed electric dipole transitions, the first class spectra of Section 1b [39]. The splitting arises from the long range part of the electron-hole exchange interaction within orbital singlet FE states and is formally equivalent to the well-known Lyddane-Sachs-Teller splitting of the longitudinal and transverse optical phonon branches for small $\underline{k}_p$, which is due to the macroscopic field associated with long wavelength LO ionic vibrations in a

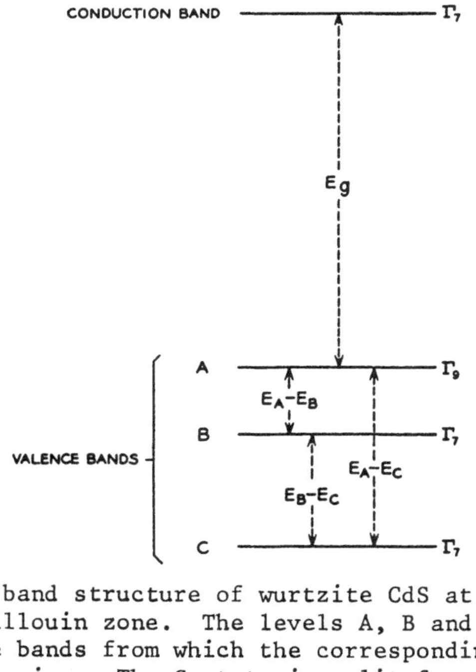

Fig. 13. Energy band structure of wurtzite CdS at the centre of the Brillouin zone. The levels A, B and C indicate the valence bands from which the corresponding FE and BE states arise. The C state is split from A and B by spin-orbit coupling, while A and B are split in the axial crystal field (Thomas and Hopfield, ref 148).

partially polar lattice. Optical absorption generally produces transverse FE in cubic semiconductors, because the FE momentum $\underline{k}_{FE}$ must match that of the absorbed photon which has a transverse electric field. E_{LT} is 0.85 meV for the n = 1 FE in CdS, and about 0.4 meV for the n = 2 FE. The short range part of the e-h exchange, equivalent to the exchange interactions in BE states discussed in Sections 2b → 2e was unmeasurably small for FE in CdS. It can often only be given an upper limit in analyses of FE magneto-optical properties even at high magnetic fields, as in ZnSe where E_{LT} is ~ 1.1 meV, but the short range exchange splitting is $\lesssim$ 0.1 meV [38].

The n = 2 state of the A FE in CdS has two resolved components at H = 0 (Fig 14). The higher energy splits into 4 components and is the P_xP_y state formally analogous to the $P_\pm$ states in the theory of donors in a semiconductor with ellipsoidal conduction band minima (Section 1d). In this case, the electron and hole masses and the dielectric constant are defined relative to the c-axis of the wurtzite lattice. The lower lying magnetic doublet P_Z in Fig 14 is analogous to P_0 for the donor. The reader is referred to the original paper [11] for the detailed analysis of the very complex

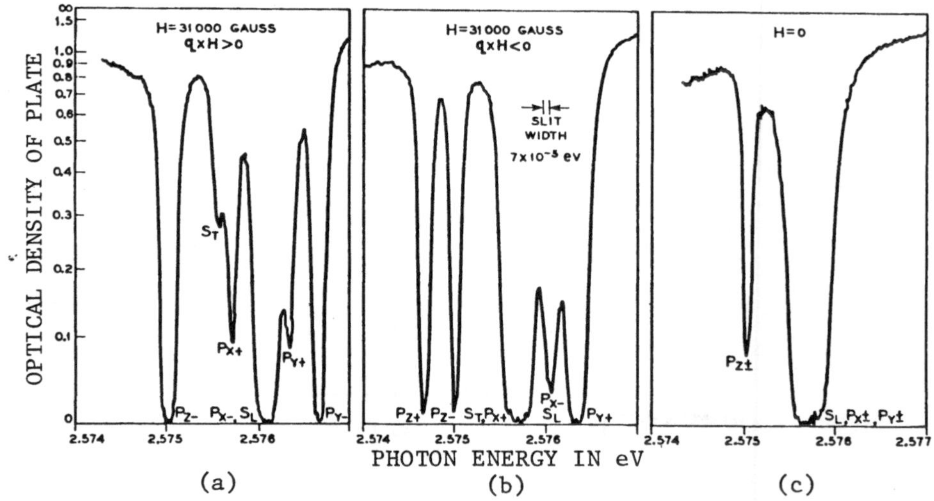

Fig. 14. Microphotometer trace of optical absorption, increasing downwards in these transmission spectra for the n = 2 FE states of single crystal CdS recorded at 1.6°K. The orientations are c⊥H, H||x, q⊥H and q⊥c, where q is the photon wave-vector and c the symmetry axis of the wurtzite crystal. The magnetic field is reversed between (a) and (b), while H = 0 in (c). (Thomas and Hopfield, ref 11).

properties of this system, which include substantial diamagnetic shifts for the large radius FE states of higher orbital quantum number. Very good agreement was obtained between theory and experiment for $n = 2 \rightarrow n = 4$ states, and an accurate set of electron and anisotropic hole masses and g vlues were derived. The importance of contributions from optical transitions whose matrix elements are proportional to the photon wave-vector, the second class transitions of Section 1b, for the large number of exciton lines observed and the sensitivity of the optical spectra to reversal of the magnetic field (Fig 14a and b) was emphasised. The effect can be quantitatively analysed in terms of interference between these second class transitions and first class transitions in which the optical matrix elements are independent of the photon wave-vector. This paper set the scene for subsequent magneto-optical investigations of FE in many direct gap semiconductors.

II. INTRODUCTION TO BOUND EXCITONS

II.A. History of Bound Excitons

The existence of BE states was predicted in 1958, when the analogy between excitons weakly bound to donors and acceptors and various types of atomic and molecular hydrogenic systems become recognised [40]. These BE states may be classified as states of localised electronic excitation containing an excited electron-hole pair (BE) whose characteristics derive closely from the extremal energy bands of the electron and hole states adjacent to the band gap of the semiconductor. We define the binding energy E_{BX} of the BE simply as the difference between the transition energy of the absorption (or luminescence) line by which the BE state is usually identified and the FE energy gap E_{GX}. It is possible to observe BE associated with excited FE states of internal orbital quantum number $n \gtrsim 2$ (Section 3c) as well as with FE states involving higher energy gaps of the host semiconductor, when these higher gaps lie relatively closely above the minimal gaps. However, most BE reported in the literature involve states derived from the lowest FE state at E_{GX}. The values of E_{BX} are usually quite small. Indeed, the clear identification of a BE state through the definition just given implies that E_{BX} should be small compared with the internal energies of the system which binds the exciton. When these are the shallow neutral donors and acceptors familiarly used in the control of electrical conductivity of semiconductors, we are inevitably dealing with BE states for which effective mass theory is well-suited. However, the advantage this should bring to the theoretical description of the BE states is frequently outweighed by the complicated many-body aspects of the problem, which usually remove the possibility of simple analytical solutions. The limiting case of this complexity is the multiple bound exciton complex discussed in Section 4c.

The theoretical descriptions become relatively simple again only when the number of electronic particles is strictly limited and when their individual binding energies within the BE are appreciably different from one another. A good example is the BE at an isoelectronic trap [41]. Calculation of the E_{BX} value of such a BE is very difficult [42]. The difficulties are closely related to the reasons that first principles calculations of the binding energy of any truly deep impurity state are difficult, connected with the multi-band character of the wave-function of a tightly bound, highly localised state. The first electronic particle binds at an isoelectronic trap entirely through the short range interactions which define the binding at any deep state. This is true even though the actual magnitude of the single particle binding energies are often not very large, as in the case of the technologically important N isoelectronic trap in GaP [41]. However, the binding of the second particle, which always interacts weakly through the long range Coulomb field of the first, can be calculated quite accurately, especially when the first particle is bound tightly enough to fully justify the isoelectronic donor or isoelectronic acceptor description [43]. A good example is the NN_1 associate in GaP, where the electron is bound by ~ 124 meV and the hole is bound to the electron by about 40 meV.

Many of the fundamental properties of BE were first demonstrated and classified for shallow donors and acceptors in CdS [44]. This semiconductor is relatively easily prepared in quite pure single crystal form by vapour transport, has readily accessible near gap optical properties in the blue-green spectral region and quite conveniently resolvable energy states due to the different BE and FE species. However, the chemical nature of the donors and acceptors, particularly the latter, was not established until nearly a decade after the first detailed optical studies of the near-gap (edge) absorption and luminescence [45]. The work on BE complemented earlier FE studies in CdS [11] (Section 1g), which provided a sound basis for identification of the parentage of the FE states. These early studies on CdS illuminated many aspects of behaviour, including transition oscillator strengths, electron-hole j-j coupling, phonon coupling, and the use of optical polarization, magneto-optics and optical bias to establish the generic indentity of the BE states as well as the qualitative sizes of their binding energies. Later work, involving the same authors, also led to the discovery of the isoelectronic trap in GaP [41].

II.B. Typical Near Gap Bound Exciton Luminescence

The main donor and acceptor-related BE states [44] appear in the semi-schematic luminescence spectrum of CdS in Fig 15. The main lines were labelled $I_1 \rightarrow I_3$ in order of increasing transition

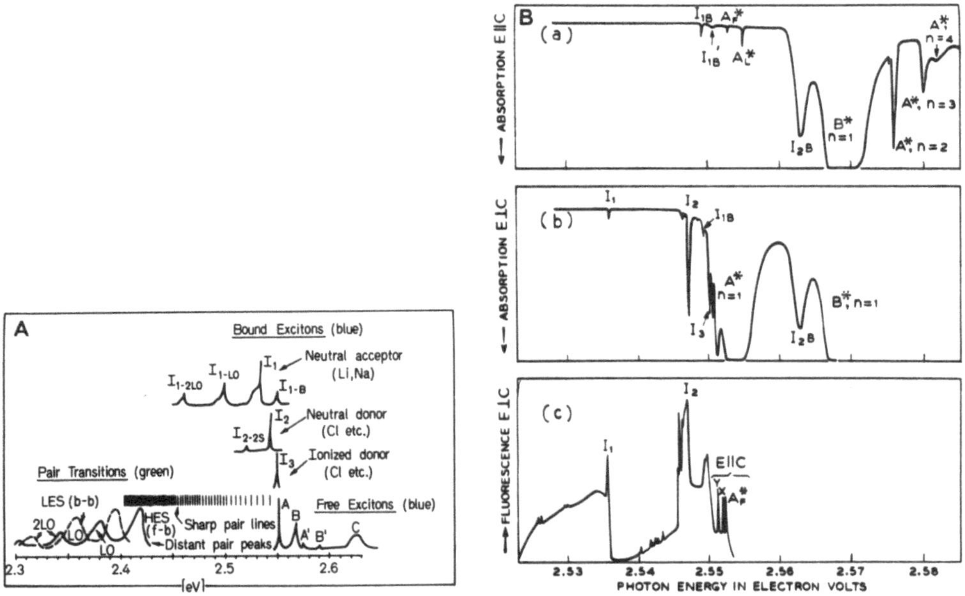

Fig. 15. (A) semi-schematic diagram of the edge luminescence of wurtzite CdS, with FE and BE contributions to the right and donor acceptor pair (-) and free electron to bound hole (f-b) at left. The LO(Γ) phonon dominates the phonon replicas. (B) detail in the strongly polarized near gap luminescence (c) and optical absorption obtained from a 5-50μm set of CdS single crystal plates (a) for $E \parallel c$ and (b) for $E \perp c$. A^* and B^* are FE, lines I are BE derived from FE B where indicated. (Thomas and Hopfield, ref 44).

energy and fall between ~ 18 and ~ 5 meV below the lowest n = 1 FE state labelled A (A^* in Fig 15B). We have seen in Fig 13 that this FE involves a Γ_9 hole state, with $P_{x,y}$ (m_i = ± 3/2) orbital character. The properties of the main BE lines shown in Fig 15 are listed in Table 1, together with their identifications. These identifications stem partly from the polarization properties and linewidths, also shown. The BE states involving holes from the intermediate Γ_7 (B) valence band not only have the expected, different polarization properties but also much greater linewidths, due to rapid Γ_7-Γ_9 hole relaxation by phonon emission and consequent lifetime broadening. The relaxation processes are responsible for the absence of these lines in the recombination luminescence. The fact that this excess linewidth is not present for I_{1B} can be understood if this BE state involves two holes with antiparallel angular momenta, the lowest lying energy state arising from hole-hole j-j coupling. Comparisons with atomic spin-orbit coupling confirms that a BE state containing two holes, one from each of the top two valence bands in Fig 13, will exhibit a j-j coupling reduced from ~ 0.4 eV, the atomic value for neutral S, by the probability of finding the two holes in the same unit cell, an overlap reduction factor for these extended crystalline states. The state of highest angular momentum lies uppermost. The observed splitting of I_{1B}-$I_{1B'}$ of ~ 1.4 meV is considered reasonable [44]. The electron-hole j-j splitting of 0.27 meV, resolvable only for I_{1B}, also seems of reasonable magnitude based upon the energy levels of its atomic analogue, the neutral Ar atom, with atomic splitting of only 0.04 eV.

The general assignment of the BE lines given in Table 1 stems from further evidence. Firstly, it was known even in 1962 that donors are much shallower than acceptors in CdS, $E_D \sim$ 32 meV, $E_A \sim$ 150 meV. It is therefore natural to assign the lower-lying I_1 BE to acceptors, since E_{BX} for the lowest BE states is generally expected to exhibit some proportionality with E_D or E_A, and to be in general 5-20% of these quantities, as discussed in Section 2c. This identification was extended from experiments with infrared bias light [44], which quenched the I_1 and I_2 lines but enhanced the I_3 lines for luminescence recorded under minimum intensity levels of above-gap exciting light. This is just as expected from the attributions in Table 1, since the infra-red light should favour BE recombinations transitions at ionized donor states on the simple view that it photoionizes these shallow centres. These experiments also showed that I_1 exhibited larger quenching effects than I_2, consistent with the fact that CdS is an n-type semiconductor and it is necessary to photoneutralise acceptors by the above gap excitation before the neutral acceptor BE, (A^o,X) can be formed. In addition, I_B and $I_{B'}$ quenched together with I_1, confirming the close relationship suggested in Table 1.

Table 1
A summary of the properties of the lines described in the text.

Line	Active for	Energy (eV) (λ (Å))	Approx. apparent width (10^3 eV)	Energy below exciton A, eV (2.5537 eV)	Energy below exciton B, eV (2.5687 eV)	Ground state
I_1	$E \perp c$	2.53595 (4888.5)	0.1	0.0177		Neutral acceptor
I_2 (Many lines)	$E \perp c$	e.g., 2.5471 (4867.15)	0.1	0.0066		Neutral donor
I_{1B}	$E \parallel c$	2.54887 (4863.7)	0.1		0.0198 }	Neutral acceptor (trapped exciton from Band B)
I_{1B}	$E \perp c$	2.54914	0.1		0.0196 }	
I_3	$E \perp c$	2.5499 (4861.7)	0.1	0.0038		Ionized donor
I_{1B}'	$E \parallel c$	2.5504 (4860.8)	0.5		0.0183	Neutral acceptor (trapped exciton from Band B)
Y	$E \parallel c$ $E \perp c$?	2.55127 (4859.1)	0.1	0.0024		
X	$E \parallel c$ $E \perp c$	2.55206 (4857.6)	0.1	0.0016		
I_{2B}	$E \parallel c$ $E \perp c$	2.5626 (4837.7)	1.7		0.0061	Neutral donor (trapped exciton from Band B)

The third, generally very important proof of the assignments came from the zero-field and Zeeman splittings of these states (Table 2). Only the I_3 lines exhibit zero-field splittings [44], because the D^0,X and A^0,X complexes can only exist with the like particles in antiparallel spin states (considering here only I_1 and not I_{1B}) a result of the exclusion principle. The 0.31 meV zero-field splitting of the I_3 line is attributed to electron-hole j-j coupling in the excited state. The final state of this transition is just the ground state of the crystal with no unpaired electronic particles, and so exhibits no zero-field or magnetic splittings. The exciton bound to an isoelectronic trap (Section 2a) has identical properties in this respect (Section 2e). However, it is not constrained to lie at approximately $E_G - E_D + \Delta E$ as the I_3 (D^+X) transition must, where ΔE represents the weak binding of the hole to the neutral donor. Since ΔE is so small, only ~ 1 meV in CdS, the I_3 state is a particularly evanescent BE state with respect to thermal ionization. It can be seen clearly only at the lowest temperatures in all semiconductors. The complementary BE state A^-,X is not bound in CdS nor in any semiconductor where $m_h^*/m_e^* \gtrsim 1$ (Section 2c).

The magnetic effects summarized in Table 2 [44] are interpreted in Fig. 16, where the characteristic magnetic properties of a Γ_9 hole state, $m_j = \pm 3/2$, are shown relative to the crystal c axis of wurtzite CdS. The corresponding magnetic g value, g_h goes to zero for magnetic field H⊥c. Parts (a) and (b) of Fig 16 show that D^0,X and A^0,X are distinguished in wurtzite CdS by an opposite ordering of the character of the ground and excited states of their

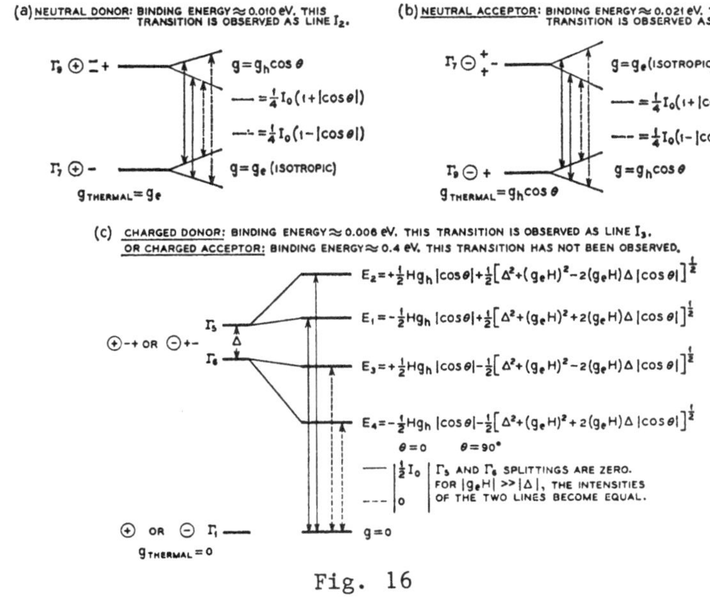

Fig. 16

transitions. The prediction that only the hole has an anisotropic g value [144], together with the observation of thermalisation between the magnetic subcomponents in absorption, allows a firm distinction to be drawn between I_1 (A^o,X) and $I_2(D^o,X)$. The Zeeman splittings of the I_1 and I_2 lines are accurately linear in H (Fig. 17). The I_3 lines have more complicated, nonlinear magnetic splittings, but their Zeeman properties are equally characteristic.

We shall see that I_1 and I_2-like lines can also be distinguished through the energy distributions of certain electronic satellites, which reveal internal excitations of the neutral acceptors and donors which bind these excitons (Section 3b). The existence of these satellites for I_2 was briefly mentioned by Thomas and Hopfield [44], though no details were presented. However, Thomas and Hopfield [44] noted that the electron-hole j-j splitting and the oscillator strength of the BE (or FE) transition are both proportional to the probability that the electron and hole are in the same unit cell in the crystal, and are therefore proportional to one another. Clear experimental evidence for this effect has not yet been obtained. The j-j splitting of the I_3 line is surprisingly nearly independent of E_D and E_{BX} for a series of D^+,X states for ZnSe [46] and CdS [47]. We also note that although the I_1 lines lie below I_2 in direct gap semiconductor, for reasons already given, the I_3 lines sometimes lie below I_2, as in ZnSe [46] and ZnTe [48] rather than above I_2, as in CdS where they were first identified [44]. Quite different behaviour may be expected in indirect gap semiconductors with $m_h^*/m_e^* \sim 1$, such as Si and GaP. The I_3 lines do not exist for either of these semiconductors, neither D^+,X nor A^-,X are stable (Section 2c). The I_2 and I_1-like lines cover quite a wide range with extensive overlap, as we now discuss, though the average trend is for I_1 to lie above I_2, an effect particularly marked in GaP (Fig 18)! [49].

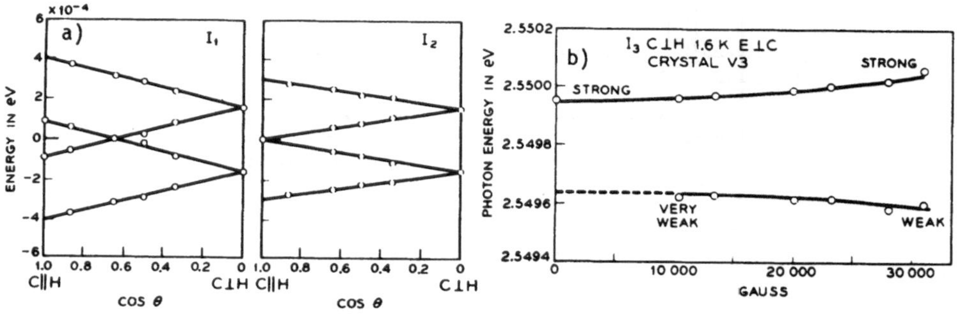

Fig. 17. Zeeman splittings of the I_1, I_2 (part (a)) and I_3 (part (b)) BE no-phonon transitions for wurtzite CdS at 1.6°K. The incidence of a single component for I_2, c||H is an accident of the electron and hole g values (Thomas and Hopfield, ref 44).

II.C. Trends of E_{BX} - Haynes' Rule

The main obstacle to a thorough theoretical description of shallow BE states, for which the effective mass approximation is valid, is a proper treatment of inter-particle correlations. The problem is usually treated with screened inter-particle and impurity or defect potentials, perhaps using a simple static $q = 0$ dielectric function [39,50]. Central cell effects arising from impurity-host core potential differences and associated local strain field and valence charge redistributions are significant and notoriously difficult to treat with accuracy. Polaron effects also introduce difficulties other than for very shallow bound states, $E_B \ll \hbar\omega_{LO}$, where the free polaron effective mass may be used with quite small mass corrections, particularly to electron states in most low mass direct gap semiconductors when α is then small (Eq. 7). Bound state polaron theory is still a contentious subject for electrons, and is quite undeveloped for holes [50]. There are problems for electron states in indirect gap semiconductors in handling the large intervally potential and kinetic energy terms, which tend to cancel and require detailed treatment [51].

The simplest case is the I_3-type centre, exemplified as we have seen by D^+,X for direct gap semiconductors. The basic Hamiltonian, neglecting band degeneracy and mass anisotropy, may be written

$$H = -\frac{\hbar^2}{2m_h^*}\nabla_h^2 - \frac{\hbar^2}{2m_e^*}\nabla_e^2 + \frac{e^2}{\varepsilon_o r_h} - \frac{e^2}{\varepsilon r_e} - \frac{e^2}{\varepsilon r_{eh}} \qquad (17)$$

It can be shown that the total energy $E(\sigma)$ is a monotically increasing function of the mass ratio $\sigma = m_e^*/m_h^*$. [50] There is a critical value of σ above which the D^+,X is unbound. The physical principle involved is that for large σ, the kinetic energy $p^2/2m_h^*$ required to localise the hole at the donor eventually exceeds the corresponding energy gain from any realistic potential energy for hole binding at a neutral centre (neutral donor). The critical mass ratio is inverted for binding at ionized acceptors. The extent of the difficulty of accurate calculations even for this simple system, with only two electronic particles, may be judged from the variety of values of σ_c derived [50]. However, the most realistic limit may be ~ 0.45 [52], that is no bound state exists for $\sigma > 0.45$. This is consistent with the absence of I_3-type states in Si and GaP, where σ is close to 1. It is certain that such BE states do not exist in these indirect gap semiconductors, since they would not be subject to the Auger effect unlike D^o,X and A^o,X (Section 2f), and therefore would be strongly radiative and hard to overlook.

Binding in D^o,X and A^o,X is an even more complicated problem, again with a wide variety of published theoretical predictions.

Generally, E_{BX} defined in units of E_D for D^oX increases from ~5.5% in the H^- ion limit ($m_h^* \ll m_e^*$) to ~ 33% in the H_2 molecule limit ($m_h^* \gg m_e$). There is no guarantee in this case that E_{BX} will increase monotically, however [50]. Given the prediction that a bound state should exist for D^o,X and A^o,X for all values of σ, the practically interesting question is how E_{BX} may vary between impurity centres of a given type in a given semiconductor. This problem automatically lies beyond the scope of effective mass theory, which gives only a single binding energy for impurities of a given type, say donors. The general monotonic trend of E_{BX} with E_D observed experimentally in several semiconductors has been explained in terms of an expression derived from first order perturbation theory, assuming relatively weak central cell enhancements of the effective mass binding potentials and energies, as holds in much experimental data. Then, it can be shown that [49,53]

$$E_{BX} = \left[(E_{BX})_{EM} - (E_i)_{EM} \frac{\delta \rho_c}{\rho_c} \right] + \frac{\delta \rho_c}{\rho_c} E_i \qquad (18)$$

where ρ_c is the electronic charge in the impurity central cell and $\delta\rho_c$ is the increased charge when the BE is present, E_i is the binding energy of an actual donor and $(E_i)_{EM}$ is the corresponding effective mass value. This equation has the form of much experimental data in Si [54], GaP [49] (Fig. 18), CdS [47] and ZnSe [46]. Since $\delta\rho_c/\rho_c$ and $(E_{BX}/E_i)_{EM}$ are both f(σ), the constant term a in the experimentally determined function, Eq. (19) below, may be positive (A^o,X in GaP), negative (D^o,X in GaP) or ~0 (both A^o,X and D^o,X in Si)

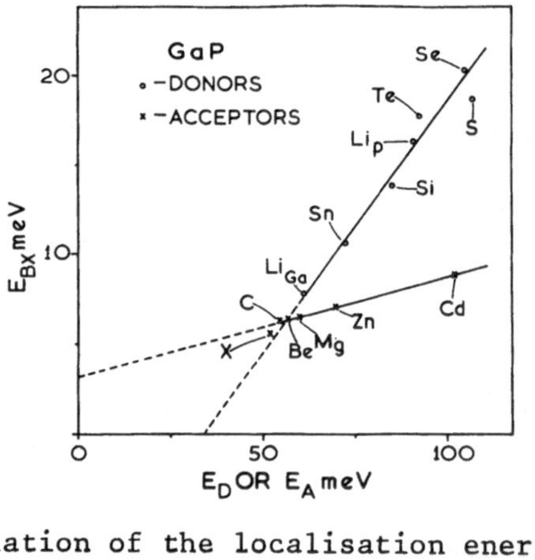

Fig. 18. The variation of the localisation energy E_{BX} for the lowest energy BE component involving the indicated neutral donors (steeper trend) and neutral acceptors in zinc-blende GaP. The axial acceptor X may be H_{Ga}-O_P (revised from Dean, ref 49).

$$E_{BX} = a + b\,E_i \tag{19}$$

Given that $a \sim 0$ in Si, then for σ significantly far away from this condition of intersection, a should have opposite sign for D^o,X and A^o,X as observed in GaP (Fig 18).

It is important to recognise that the value of b in Fig 18 is much smaller for excited DBE and ABE states [46,55]. This is intimately related to the strongly aberrant behaviour of ABE states in direct gap semiconductors. The Haynes' rule form of Eq. (19) is not obeyed for A^o,X complexes when $\sigma \ll 1$, as in direct gap semiconductors like GaAs [56], InP [57], ZnTe [48] and CdS [45]. Experimental data for some of these materials is limited to two or three shallow acceptors, but rather more comprehensive information is available for GaAs [56] and ZnTe [48] (Fig 19). The general effect is quite clear. The constant b in Eq. (19) is close to zero for $E_A \lesssim 3 \times (E_A)_{EM}$, and only becomes significant for very deep acceptors such as Sn in GaAs [58] or Au in ZnTe [59], where $E_A \sim$ 4-5 X $(E_A)_{EM}$. It is important to recognise the influence of the redistribution in the ordering of the j-j split states of the

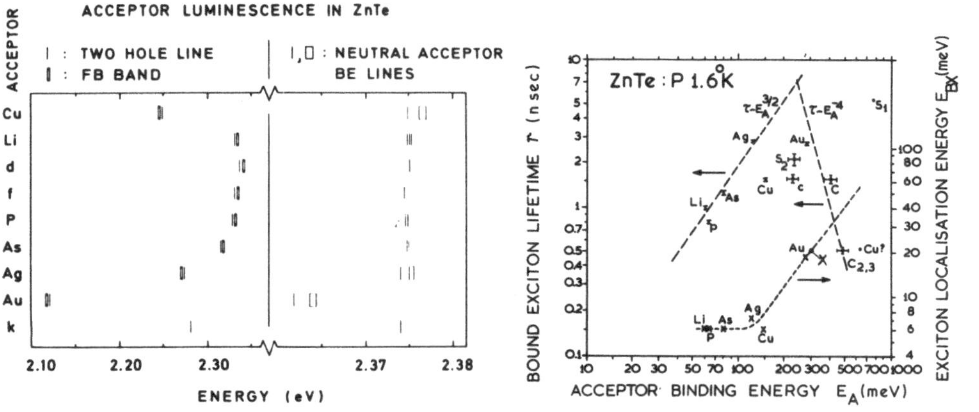

(a) (b)

Fig. 19. (a) a schematic diagram showing the F-B bands and 'two-hole' ABE satellites at the left and at the right the principal ABE lines for the indicated acceptors in zinc-blende ZnTe. The broader ABE transitions are lifetime broadened (Vengahus and Dean, ref 48). (b) upper points are variations of BE lifetimes for ABE involving the indicated acceptors in ZnTe as $f(E_A)$, with dashed theoretical forms (Section 2f), ordinate at left. Lower points drawn without error bars show variation of E_{BX} (ordinate at right) with E_A. Point Cu? is a Cu-related complex of less certain E_A (Schmid and Dean, ref 63).

A^o,X BE which occurs over this range of the functional dependence of E_{BX} on E_A, within which b is initially ~ 0. The J = ½ state formed by e-h coupling to the J = 0 hole-hole state lies highest for small E_A. The only other combination allowed by the exclusion principle for the j-j coupling of two j = 3/2 holes from the uppermost Γ_8 valence band of a zincblende structure semiconductor, J = 2 lies slightly lower in energy. The J = 2 state is slightly split into J = 3/2 and J = 5/2 by the e-h exchange in the BE (Fig 20). As E_A increases, the J = 0 hole-hole state moves down through the J = 2 state, apparently because the J = 0 state has an s-like envelope wave-function and is much more strongly influenced by the acceptor central cell potential than the d-like J = 2 states [60]. The J = ½ state eventually becomes lowest for sufficiently large E_A, and replaces the J = 5/2 state in the determination of the Haynes' rule behaviour, which has always been applied to the lowest energy state of the BE system. Although this model of combined h-h and e-h j-j coupling appears to give a good account of the zero-field splittings of A^o,X in GaAs and InP [66], (Fig 20), the measured intensity ratios for A^o,X BE states in Si (Section 4c) are in rather poor agreement with the prediction of 1:4:1 for the J = 5/2, 3/2, 1/2 states just described. For example these ratios are ~ $1:1.5_0:1.1_6$ for Si:Aℓ [67], which agrees better with expectation 1:3:2 obtained under the assumption that the crystal field splitting of the J = 2 state predominates over the e-h exchange splitting [68]. If the lowest energy component arises from the J = ½ state even for the shallowest A^o,X in Si, as suggested by

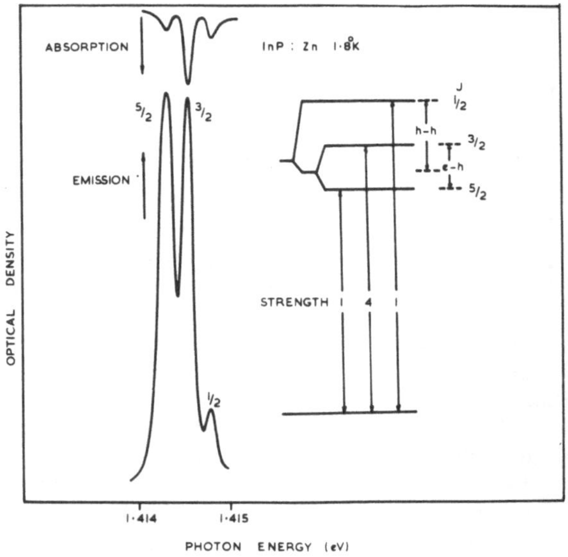

Fig. 20. Luminescence and optical transmission of the Zn ABE in zincblende InP, showing thermalisation in former at 1.8°K (left). The splittings are interpreted in terms of h-h and e-h exchange at right. (White et al, ref 66).

Thewalt [69], opposite to both the direct gap semiconductors and to comparable atomic systems, the increase in observed no-phonon oscillator strength from this state compared with expectation may be understood. Thus, it may be quite misleading to use Eq. (19) with b set to 0.1 → 0.2 from the observed behaviour of the shallowest A^o,X BE in order to estimate E_A from the E_{BX} measured from a much deeper A^o,X BE, as has frequently been attempted [61,62]. The detailed form of this nonlinear E_{BX}, E_A relation for A^o,X is not yet accurately established for any direct gap semiconductor, largely because of the scarcity of information for the deeper acceptors. However, Fig 25 shows the probable trend in ZnTe [63].

II.D. Additional Aspects of j-j Coupling and Other Zero-Field Bound Exciton States

The behaviour of D^o,X BE complexes is complicated in direct gap semiconductors mainly by the presence of a variety of excited states, lying closely above the lowest state which exhibits the strongest dependence of E_{BX} on E_D. These states have been studied recently by selective excitation spectroscopy, using tuneable dye lasers, in ZnSe [64] and ZnTe [65]. Comparison of the 'two-electron' transition satellites of these excited BE states with that of the ground state, discussed further in Section 3b, has confirmed the prediction that these arise from excited angular momentum states of the hole. Three excited states of total angular momentum $2p_{\frac{1}{2}}$, $2p_{3/2}$ and $2p_{5/2}$ are expected to appear closely above the $1s_{3/2}$ ground state [60].

Many of the D^o,X states in an indirect gap semiconductor like GaP and Si may be treated on a simplified conduction band model first introduced by Hopfield [70]. This model recognises that the complications stemming from the additional degeneracy associated with the multi-equivalent conduction band minima (Section 1d) can be removed for the ground state by the same feature which simplifies the donor ground state. These lowest D^o (Fig 21) and D^o,X states involve electrons whose wave-functions contain a symmetric contribution from all the energy equivalent conduction band minima, giving a $1s(A_1)$ state of Γ_1 symmetry. These electrons therefore behave as simple $S = \frac{1}{2}$ particles, and D^o,X contains an $S = 0$ antiparallel combination of these spin states. The overall wave-function of the DBE may therefore be written as $|\psi_h, \psi_e{\downarrow}\psi_e{\uparrow}\rangle$ and has purely the magnetic character of the hole, just as in a direct gap semiconductor. This simple result only applies for those DBE involving donors on the host sublattice of largest electronegativity, usually the anions [71]. The valley-orbit interaction, which splits the $1s(A_1)$ state from the orbital doublet 1s(E) state in a three-valley semiconductor, to which GaP approximates in first order neglecting the camel's back (Fig 8), is zero for Ga-site donors in GaP [71] (Fig 21) and for electrons in A^o,X BE complexes.

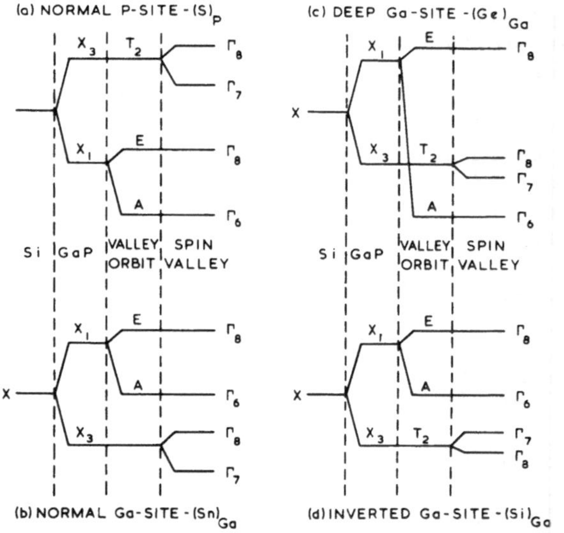

Fig. 21. Four possible schematic configurations for the electronic ground states of substitutional donors in zincblende GaP. All have T_d site symmetry. The different configurations arise from variations in the character of the Bloch term in the electron wave-function (Dean at al, ref 73).

Instead, the resulting T_2 state is subject to a form of spin-orbit 'spin-valley' splitting [72]. This produces a pseudo-p state with non-magnetic orbital properties, since the components of the p states originate from the conduction band valleys from different (inequivalent) points in the Brillouin zone [72]. The spin-valley splittings of the donor electrons are typically small, ~ few meV. Their ordering depends upon the relative magnitudes of the atomic numbers Z of the host atom and chemical substituent donor, since the atomic spin orbit splittings increase strongly and monotonically with Z. Thus, this splitting is inverted for the Sn_{Ga} donor compared with Si_{Ga} in GaP [73] (Figs. 21b and c). Corresponding spin-valley splittings have not yet been resolved for the DBE. These spin-valley split donor states also have peculiar effective spin g values [72].

The contribution of the valley-orbit splitting in the spectra of DBE complexes for P site donors in GaP has recently been considered and has also been extensively studied in Si in connection with the description of the multiple bound exciton (MBE) states (Section 4c). The situation in GaP is rather more speculative. The excited states of the S DBE in Fig. 22 contain a component, $S^o_o{}'$, with about 20% of the strength of the principal DBE line S^o_o. This

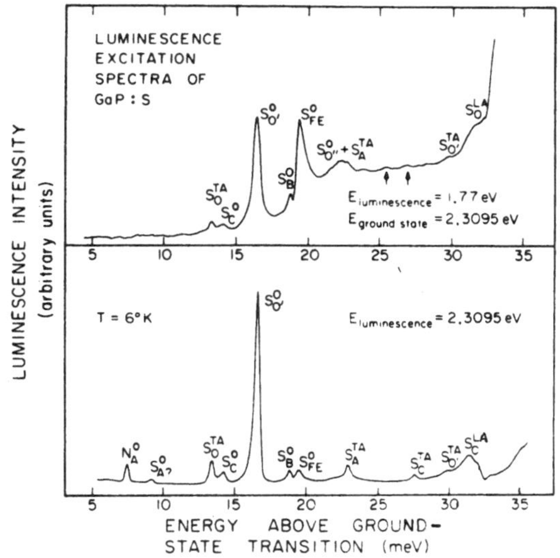

Fig. 22. Luminescence excitation spectra in GaP, upper for broad deep-centre luminescence at 1.77 eV and lower for the principal S DBE no-phonon line S_0^0 at 2.3095 eV, the energy zero in these spectra. The S_0^0 contribution is about four times $S_{0'}^0$. (Elliott and McGill, ref 74).

strength indicates that it cannot arise from the recombination of a DBE containing a Γ_1 and Γ_3 electron within the 1s envelope state. A Γ_1 (1s(A_1)) electron must remain after the transition,since the transition energies show that this recombination must occur to the 1s(A_1) donor ground state. However, the no-phonon recombination of a 1s(E) (Γ_3) electron is expected to be very weak, because the d-like 1s(E) state has very small wave-function overlap with the donor core which provides the oscillator strength for BE recombination in indirect gap semiconductors (Section IV.C). Elliott and McGill [74] have suggested that $S_{0'}^0$ arises from the recombination of a DBE containing two Γ_1 1s electrons and a $2s_{3/2}$ excited state of the hole. However, Mathieu et al [75] support their contention that this state arises from a DBE state in which only one electron is in an excited 2s Γ_1 state by detailed measurements of uniaxial stress splittings. Mathieu et al [75] also suggest that a further state 3.2 meV above $S_{0'}^0$, quite strong in total absorption [76], involves the same set of electron-hole states but split from the first mainly by an e-e exchange interaction. This splitting seems unduly large for such an effect, however. Moreover, Matthieu et al [75] did not recognise that this further excited state near 2.330 eV contains a major contribution from the no-phonon creation of FE whose oscillator strength is derived from exciton scattering in the potential of the neutral donor. Very large cross-sections are exhibited for FE scattering at isoelectronic traps or isoelectronic substituents which nearly form a bound state, like Sb in GaP [77].

Merle et al [78] have noticed anomalies in the line-shape and intensity of the upper energy sub-component of the S_0^o transition, which splits into two under a <111> uniaxial stress. Only the degeneracy of the Γ_8 valence band states is lifted for this stress direction, because the conduction band minima near X remain energetically equivalent. The anomalies occur near 3.25 k bar, which implies that the $1s(A_1)$,$1s(E)$ BE state lies ~ 12 meV above the $1s(A_1)$, $1s(A_1)$ ground state (S_0^o) if the stress splitting rates of both states are identical. Stress sub-components from these two zero-stress states exchange oscillator strength in an anti-crossing. However, there is some reason to believe that components S_A^o and S_c^o, observed with much weaker no phonon strengths relative to their momentum conserving phonon replicas than observed for S_0^o (Section III.A) at zero stress (Fig 22), may involve the $1s(A_1)1s(E)$ BE state [74]. Such an identification would imply a valley orbit splitting energy of the DBE of ~ 9.2 or ~ 14.2 meV. As already noted, we probably cannot use Fig 23 to predict an E_{BX} value for a DBE state involving two electrons of different symmetries, as Merle et al [78] have tried to do.

The j-j coupling of the (A^o,X) states is expected and observed to be very complicated for indirect gap semiconductors [55]. The identity of the lowest energy BE state, which exhibits a b value (Eq. (19)) much greater than for any other zero-field states, and which is used in Fig 18, has not yet been established, although magneto-optical studies show that it has simple Γ_6 (pure spin) character [55].

II.E. Magneto-Optical Effects

We describe here only the Zeeman effects, which we have already seen may be vital to the identification of BE states (Section 2b). Diamgnetic shifts, proportional to H^2 are also sometimes very useful, for example in the identification of BE states associated with n = 2 rather than n = 1 FE states (Section 3b).

The external magnetic field couples to both spin and orbital angular momenta, so the observed g values of electronic particles in solids may deviate considerably from the pure spin value of 2 [50]. The magnetic splitting involves the expectation value of the operator $2\beta\underline{H}.(\underline{L} + 2\underline{S})$ where L is the orbital angular momentum, $2\underline{S}$ the Pauli spin matrix, β is the Bohr magneton and $\underline{H}$ is the applied magnetic field. The vector sum $\underline{L} + 2\underline{S}$ often has different values parallel and perpendicular to a principal symmetry axis, for example the major axis of the prolate ellipsoidal conduction band minima of an indirect gap semiconductor (Section 1d) or at the valence band edge in a wurtzite semiconductor (Figs 16 and 17). Large values for the expectation value of L can occur for compound semiconductors with strong spin orbit coupling and narrow direct energy gaps. The Roth formula [79] describes the g value resulting from simple k.p theory, with a form similar to that for m^* in Eq. (3).

$$g \sim 2 - \frac{m_o}{m_e^*} \left[\frac{2\Delta}{3E_g + 2\Delta} \right] \tag{20}$$

Probably the most useful results of Zeeman data are the information they give on the nature of the centre to which the exciton is bound. This information includes the number of electronic particles involved in the transition and particularly whether any electronic particles remain after the BE has recombined, distinguishing the I_1 or I_2 lines from the I_3 line (Fig 17). Information on the symmetry properties of the centre binding the exciton is also very significant. Many details have been given in the review of Dean and Herbert [50]. We restrict discussion to two examples here. Both involve isoelectronic traps (Section 2a), which produce a BE state containing only the two electronic particles of the exciton.

The technologically important N_P isoelectronic trap is a tetrahedral substituent on the electron-attractive sublattice and introduces BE transitions of large oscillator strength due to the binding of the electron by short range forces to the strongly electronegative N core [41]. The electron therefore behaves like an $s=\frac{1}{2}$ particle (Fig 23), whereas the hole exhibits the much more complicated magnetic characteristics of the $j = 3/2$ states at the p-like valence band maximum. The linear Zeeman properties of the hole states are described by the energy equation

$$E^h = E_o^h - 2\beta \left[\tilde{k}\, J.H + \tilde{q}\, (J_X^3 H_X + J_Y^3 H_Y + J_Z^3 H_Z) \right] \tag{21}$$

where E_o^h is the hole binding energy in zero field, $\tilde{k}$ and $\tilde{q}$ are the isotropic and anisotropic hole g values and J are angular momentum operators for $j = 3/2$. The e-h exchange gives rise to two states of $J = 1$ and $J = 2$. Transitions from the lower-lying $J = 2$ state are a electric dipole-forbidden in zero magnetic field and zero strain. However, states of the same magnetic quantum number m_j mix together in the magnetic field. The linear Zeeman effect (Fig 23, lower) confirms the expected magnetic degeneracies 2J+1 for transitions from these single BE states to the J=0 ground state. The relative intensities of the transitions from the $J = 2$ state are observed to vary as H^2, as anticipated for a magnetic-induced oscillator strength f_B. These Zeeman splittings are essentially isotropic, confirming the T_d site symmetry of N_P, and there is no evidence of the crystal field splitting of the $J = 2$ state which

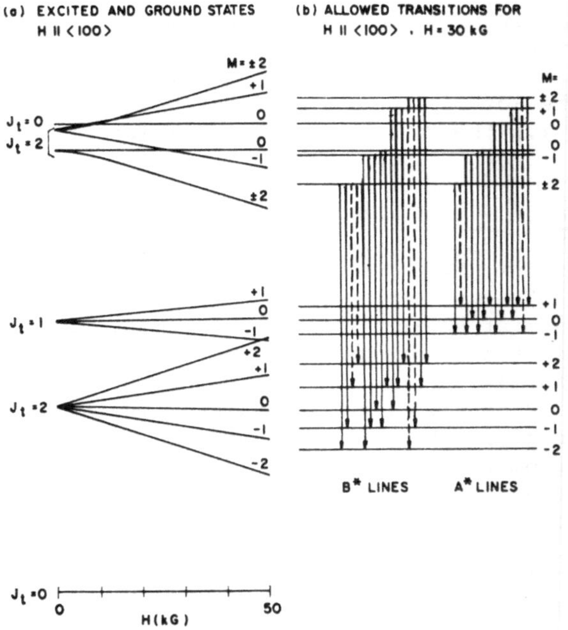

Fig. 23. Zeeman splittings of the bound excitonic molecule and single BE states at the N_P isoelectronic trap in GaP for magnetic field $H||\langle 100\rangle$. The excitonic molecule substates $M = \pm 2$ (upper) become pure $M = +2$ and $M = -2$ only at high fields $\gtrsim 50$ kG. The solid arrows indicate electric dipole excitonic molecule transitions allowed at all H, broken arrows are transitions allowed at weak H. (Merz et al, ref 131).

is predicted by group theory, and which is easily observed for the much more tightly bound Bi BE in GaP [80]. The linear magnetic splittings of the J = 1 state are about 3/5 of those of the J = 2 state, as expected from the theoretical ratio $(5g_h-g_e)/(3g_h + g_e)$ if $g_e \sim 2$ and $g_h \sim 1$. These magnetic properties gave very important support to the first recognition of the character of isoelectronic traps in semiconductors [41]. Non-linear magnetic effects have been investigated in more detail up to 150 kG [50].

The second example is the nearest-neighbor Cd_{Ga}-O_P associate, also present in GaP and a technologically important activator for light emitting diodes, this time for red luminescence [81]. Prior to the observation of the structured luminescence spectrum in Fig 24a, which has the form expected for BE recombinations at a centre with a moderate degree of phonon coupling (Section 3a), this luminescence had been attributed to distant DAP recombinations involving an O_P donor with $E_D \sim 0.4$ eV [82]. However, the distant pair-like kinetic properties of the red luminescence turned out [83]

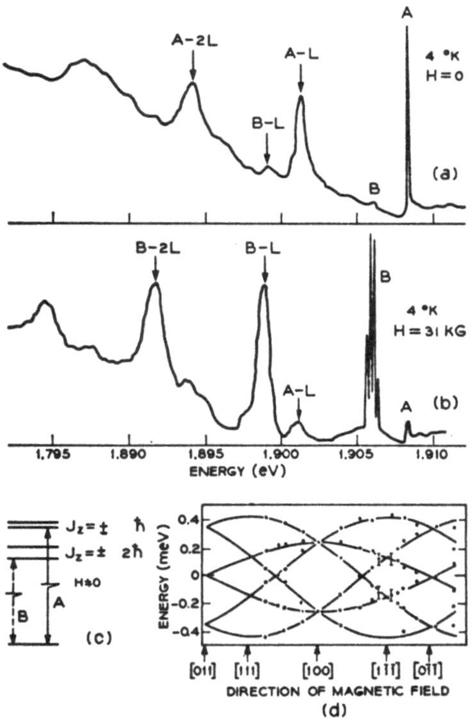

Fig. 24. (a) a small portion of the Cd_{Ga}-O_P BE luminescence in GaP at 4.2°K showing the A and B no-phonon lines and their ~ 7 meV acoustic phonon replicas L, for magnetic field H = 0. (b) is for H = 31 kG and H||<111> and shows the very great increase in the B/A intensity ratio. (c) is a level diagram for each centre with given <111>-type axis. (d) shows the variation of Zeeman splittings as H is turned in a plane perpendicular to the [$0\bar{1}1$] direction in the single crystal. (Henry et al, ref 83).

to originate from transitions between electrons bound to the Cd_{Ga}-O_P 'molecular' isoelectronic trap and holes at distant Zn acceptors, rather than normal distant donor acceptor pair recombinations. This type of distant pair process does not exhibit a spectral shift during time decay, because the members of this distant impurity pair have no first order mutual Coulomb interaction.

The Zeeman properties of the no-phonon line A in Fig 24 and a second line B split below A by ~ 2.2 meV due to e-h exchange interaction in the BE yield very important support to the Cd_{Ga}-O_P associate model. The A and B lines both split with strong anisotropy whose form (Fig 24d) indicates that the centre has the expected <111> local symmetry axis [83]. The Zeeman behaviour can be quantitatively understood if the hole states are split in the local crystal field to place the $m_j = \pm 3/2$ valence bands uppermost. This is analogous to the $\Gamma_9 - \Gamma_7$ splitting due to the host crystal field in wurtzite CdS (Fig 13), and gives rise to the type of hole magneto-optical behaviour shown in Figs 16 and 17. Thus, the splitting of the BE state becomes zero for H⊥ <111>. However, the intensity of the B line falls to zero in a strain-free crystal when H||<111>, since the intensity is due to a magnetic field-induced mixing of A and B magnetic sub-states and therefore varies as $H^2 \sin^2\theta$. The B line intensity is weak at zero field, since it involves a transition from the $|2, \pm 2\rangle$ BE state to the J = 0 ground state, whereas the A line comes from the $|2, \pm 1\rangle$ BE state and is allowed by Δm_j selection rules.

The attribution of the red luminescence to Cd_{Ga}-O_P associate recombinations immediately led to a drastic revision in $(E_D)_O$, since the electron can be regarded as trapped in this associate with an energy approximately $(E_D)_O - e^2/\varepsilon r$ where $-e^2/\varepsilon r$ represents the energy lowering due to the neighbouring ionized acceptor. Since $e^2/\varepsilon r \sim 0.5$ eV and $E_e \sim 0.4$ eV for the Cd_{Ga}-O_P associate, then $(E_D)_O \sim 0.9$ eV. This initial rough estimate has been verified with surprising accuracy through various infrared spectra associated with isolated O_P in GaP [84,85].

Quite different magneto-optical behaviour results if the Γ_7, $m_j = \pm \frac{1}{2}$ hole states lie uppermost on binding to an axial centre with the opposite sign of local crystal field (local strain [86]), or if the crystal field is so strong that the hole is reduced to a pure spin ½ particle by field-induced mixing between the $|3/2, 1/2\rangle$ and spin-orbit split-off $|1/2,1/2\rangle$ hole states. Then, the BE exhibits isotropic magnetic behaviour in a semiconductor like GaP where g_e is isotropic. The resulting J = 0, J = 1 states only reveal information about the symmetry axis of the centre to which the exciton is bound through the anisotropy of the D parameter which describes the splitting and interaction of the $m_j = 0$ and $m_j = \pm 1$ magnetic subcomponents of the J = 1 triplet state. One example in GaP is the Ge-Ge D-A associate [73]. Usually, Zeeman measurements are not sufficiently sensitive to detect the small residual anisotropy in the effective D parameter. However, optically detected magnetic resonance can still frequently do so [87]. A pair of J=0,1 states can also arise from exciton localisation at a centre which produces a local spin-orbit splitting of opposite sign to that exhibited by compound semiconductors containing main-group atoms (Fig 8). Examples of this occur for excitons bound to axial centres containing Cu in GaP [149], where the inverted spin-orbit splitting results from interactions with the open 3d electron shell. This is the same effect responsible for the inverted valence band structure in Cu_2O (Fig 11). In this extreme, the local field splitting of the axial impurity centre must be small compared with the inverted spin-orbit splitting if the hole is to retain the isotropic $s = \frac{1}{2}$ character of the $|\frac{1}{2}, \frac{1}{2}>$ valence states.

Other perturbation techniques can provide information on the centres which bind excitons, such as uniaxial stress [88] and Stark splittings in strong electric fields [89].

II.F. Auger Recombinations

Bound excitons may exhibit interbound state Auger recombinations if the BE complex contains more than two weakly bound electronic particles [90]. Then, two may recombine giving their annihilation energy to the third which is ejected deep into the bands (Fig 25). The excess kinetic energy of the ejected particle then appears as heat through phonon emission relaxation processes within the allowed energy bands of the host. The final stage in the cycle is recapture by the impurity. For example, the shallow donor which is ionized in the Auger recombination process of Fig 25 has a large capture cross-section for the ejected electrons at the low temperature necessary for the observation of the DBE states [91]. This process provides one explanation for the appearance of such DBE related absorption as positive features in low temperature photoconductivity spectra [92]. Other responsible effects include thermal dissociation and inter-site tunneling which becomes important for shallow BE states at relatively low impurity concentrations [41,99]. The existence of the Auger effect is frequently

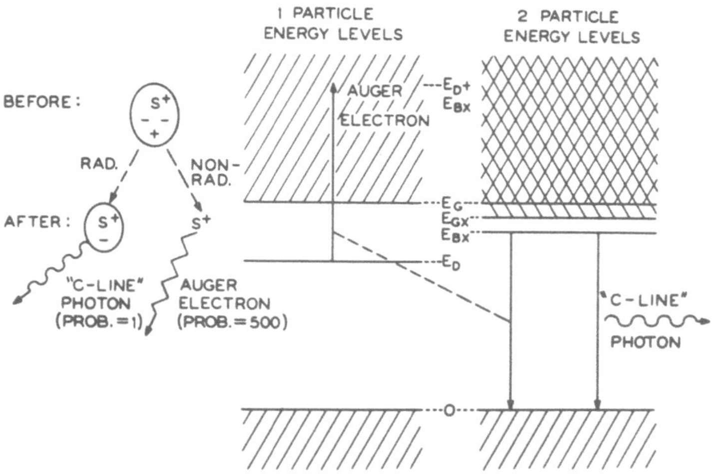

Fig. 25 A schematic representation of Auger recombinations within D^o,X BE states in GaP. The C line is caused by the no-phonon transition which dominates the luminescence. The parallel radiative and non-radiative (Auger) decay channels have the indicated relative probabilities (Nelson et al, ref 90).

recognised from a large discrepancy between the observed BE recombination rate R and that calculated from the optical absorption cross-section $\int \sigma d\nu$ using the expression from detailed balance

$$R = \frac{8\pi n^2 g_1}{\lambda^2 g_2} \int \sigma d\nu \qquad (22)$$

where n is the refractive index at wavelength λ and g_1, g_2 are the degeneracies in the initial and final states of the absorption process. The value of R calculated from Eq. (22) is ~ 500 times smaller than observed for S DBE in GaP [90]. Even larger discrepancies occur for ABE [55] or Ga site DBE [72], where the GaP band structure is less favourable for the no-phonon radiative recombination channel.

Very simple arguments show that the R_{Auger} should vary like E_A^4 within a series of ABE. This has been confirmed both in GaP [55] and in Si [93], where a more detailed theory leads to a very similar prediction. The Auger process is less important for shallow BE in direct gap semiconductors. There, the radiative rates are very high, both as an inherent consequence of the direct gap in which momentum conservation is easily achieved without strong coupling to an impurity and because of the giant oscillator strength of the very shallow BE states which accompany the small m_e^* (Section 2g). The Auger rates are very similar in direct and indirect gap semiconductors. They become enhanced only by aspects of the band structure independent of the classification direct-indirect, for example when $E_G \sim \Delta$ (Eq. 3), as in GaSb [94]. However, there is some indication that the Auger process begins to be significant for the more tightly bound BE even in direct gap semiconductors, such as ZnTe (Fig 19) [63].

II.G. Giant Oscillator Strength

Rashba and Gurgenishvili [95] were the first to point out that the oscillator strength f of a BE ($f = 8\pi/4.5\ n \int \sigma d\nu$ from Eq. 22) is given by the oscillator strength per molecule of the FE, f_{EX} multiplied by the number of host molecules covered in the e-h overlap. The latter factor can give an enhancement of many orders of magnitude. Henry and Nassau [96] showed that a calculation for CdS, using realistic parameters $f_{EX} = 2.6 \times 10^{-3}$ [97] and an enhancement of nearly 1000, yields $f_{BE} \sim 1$ and DBE lifetimes ~ 0.5 n sec. These are essentially identical to the measured values. The calculated shallow ABE lifetimes were nearly twice the observed values [96], suggesting that the Auger process might just become significant for the more tightly bound I_1 states (Fig 15). Of course, the observed and calculated lifetimes of BE at isoelectronic traps always agree whenever the data permit accurate comparison and tunneling is not serious [98]. This confirms that intra-centre Auger processes cannot occur. However, intercentre processes are always possible [85].

III. BOUND EXCITON SATELLITE STRUCTURE

We will consider only processes which occur widely in all semiconductors originating in phonon interactions and various types of electronic excitations rather than more exotic effects such as spin waves (magnons) which are a feature of restricted classes of material.

III.A. Phonon Replicas

This is a very wide subject, and can only be treated very briefly here. The localised electronic states in a BE, like the FE states, couple to the lattice through the electron-phonon interaction. This can cause local lattice distortion, which generally leads to a contribution to the binding energy. When this contribution is very significant, the system is usually described through a configurational co-ordinate diagram [100] (Fig 26). The effect of

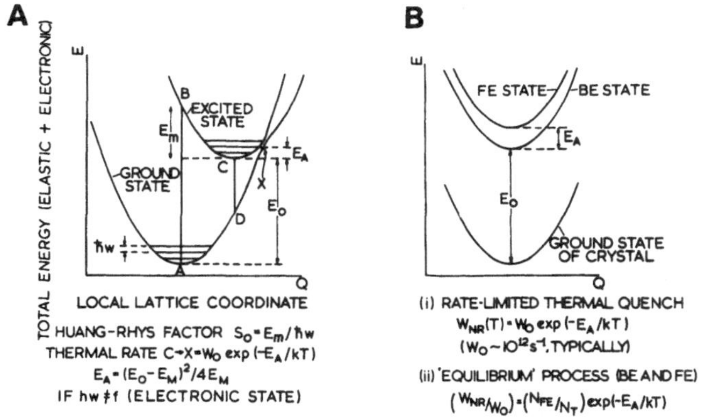

Fig. 26. (a) schematic representation of an electronic transition for strong phonon coupling where considerable change in the mean value of a representative local lattice co-ordinate Q occurs between the ground electronic state near A and the excited state near C. In (b) lattice relaxation is small and the parabolae which represent the lattice vibrational states on the configurational co-ordinate diagrams in the approximation of harmonic, linear electron-phonon coupling are nearly parallel to one another. The equations discuss the different types of thermal quenching of luminescence (Dean and Herbert, ref 50).

the lattice relaxation is approximated by means of a single, appropriately chosen local lattice parameter Q and interactions with a single lattice vibrational mode energy $\hbar\omega$. In such cases, the optical spectra are strongly broadened, and there is a large Stokes shift between the peaks of the emission and absorption spectra. This Stokes shift is 2 $\hbar\omega S_o$, where S_o is the most probable number of phonons emitted in the optical absorption or luminescence process. The Huang-Rhys factor S_o is a constant of the system if the electron-phonon coupling is linear, that is if the vibrational frequencies are independent of the electronic state of the system. This is frequently not true for moderate to strong phonon coupling strength. Such coupling frequently results in nearly featureless very broad optical spectra, as in the Zn_{Ga}-O_P BE [101].

One important consequence of significant non-linearity in the electron-phonon coupling is the opportunity this provides for isotope shifts to occur even in the no-phonon spectral lines, which represent transitions between zero-point vibrational states of the ground and excited electronic states of the system [101]. The sense of these isotope shifts suggests that bond softening occurs when extra electronic charge is localised at the centre. The effect is related to the electronic contribution to the temperature dependence of the energy gap of the host lattice [102]. Such isotope shifts have been of particular consequence for spectra related to the deep O donor in GaP [84,85,103]. Remarkably, the no-phonon lines of many of these spectra show a substantial (~ 0.7 meV) isotope shift when $O^{16} \rightarrow O^{18}$, for example the A,B lines in Fig 24a, although no shifts appear for those phonons which are clearly resolved in the richly structured sideband spectra. Observation of the no-phonon isotope shifts provided greatly needed proof of the involvement of the spectra with the elusive impurity O.

When the electron-phonon interaction is small or moderate, transitions in which phonons are involved give rise to clearly resolved sideband optical transitions (Fig 24a, b). The structure is best studied at very low temperatures, when the phonon occupation number $n = 1/[\exp(\hbar\omega/kT)-1] \rightarrow 0$, where $\hbar\omega$ is the phonon energy and kT represents the lattice kinetic energy at absolute temperature T. Only phonon emission (Stokes) phonon-assisted processes can then occur. The direct electron-phonon interaction is often dominated by long wavelength LO modes, particularly for compound semiconductors. These phonons have Γ_1 symmetry and like the acoustic phonon L in Figs 29a, b, do not alter the selection rules for the electronic transitions at a centre of symmetry T_d or C_{3v}. The associated phonon replicas may be recognised from their identical form as satellites from both the allowed A and forbidden B no-phonon lines (Fig 29b), and do not mix these electronic states [103]. They are receiving modes [104] which collectively represent the lattice relaxation. The high frequency LO modes themselves degenerate into pairs of acoustic phonons of lower energy and large wave-vector near maxima in the phonon density of states. The fractional oscillator strength within the multiple series often exhibited by receiving modes is represented by the Poisson distribution for small S_o

$$F(h\nu_o + nh\omega) = \frac{e^{-S_o} S_o^n}{n!} \qquad (23)$$

The envelope of Eq. 23 quickly approaches the Gaussian form familiar for strong phonon coupling when $S > 3$. The anti-Stokes processes which occur at finite T have a most noticeable effect on the spectral form nearest to the no-phonon transition.

Quite different considerations obtain when the phonon coupling is required to enhance the probability of the electronic transition. Two forms of this effect are common. The first has already been discussed in Section 1d, and involves phonons selected to satisfy momentum conservation in an indirect transition according to Eq. (14) [55] (Fig 27). The second involves coupling to phonons of symmetry appropriate to the removal of a transition selection rule, for example Γ_3 phonons in the case of the Cd_{Ga}-O_P associate in GaP [103], not shown in Fig 24a. These processes become significant

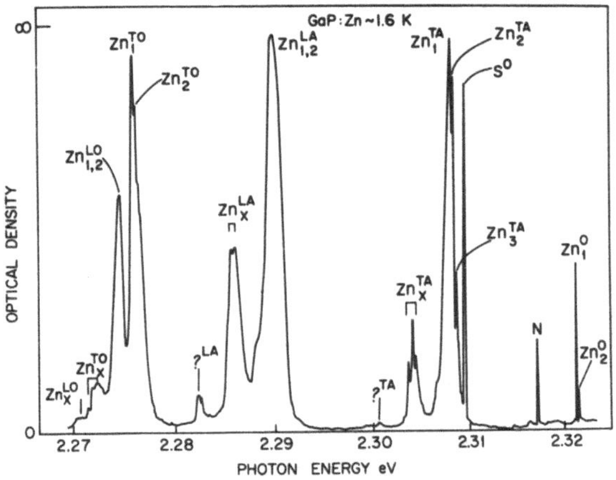

Fig. 27. The luminescence of ABE at neutral Zn acceptors in GaP, recorded photographically, together with weaker luminescence from neutral S donors and N isoelectronic traps. Superscripts 0 and TA, LA, TO, LO indicate, respectively, no-phonon transitions and indirect transitions in which momentum is conserved by the emission of appropriate phonons (Eqn 14) from the different branches of the lattice vibrational spectrum of GaP. Components subscript X and ? are due to MBE recombination (Section 4c) (Dean et al, ref 55).

when the basic electronic transition is forbidden, yet the corresponding state is strongly favoured by thermalisation, as is true for transitions from the B|2, ± 2> BE state in Fig 24c at low temperatures. Then, the form of the phonon sidebands of the allowed (A) and forbidden (B) transitions observed separately at sufficiently high and low temperatures are quite different (Fig 28c,d), with the Γ_3 promoting modes superimposed upon the receiving modes in the latter case. These modes mix together the four spin states of the hole, which are degenerate at zero local lattice distortion, and thereby contribute directly to the transition matrix element [103].

The promoting modes have less significance when the electron is the most tightly bound electronic particle in a zincblende semiconductor, as for ZnTe:O [105] (Fig 28), since the electron states

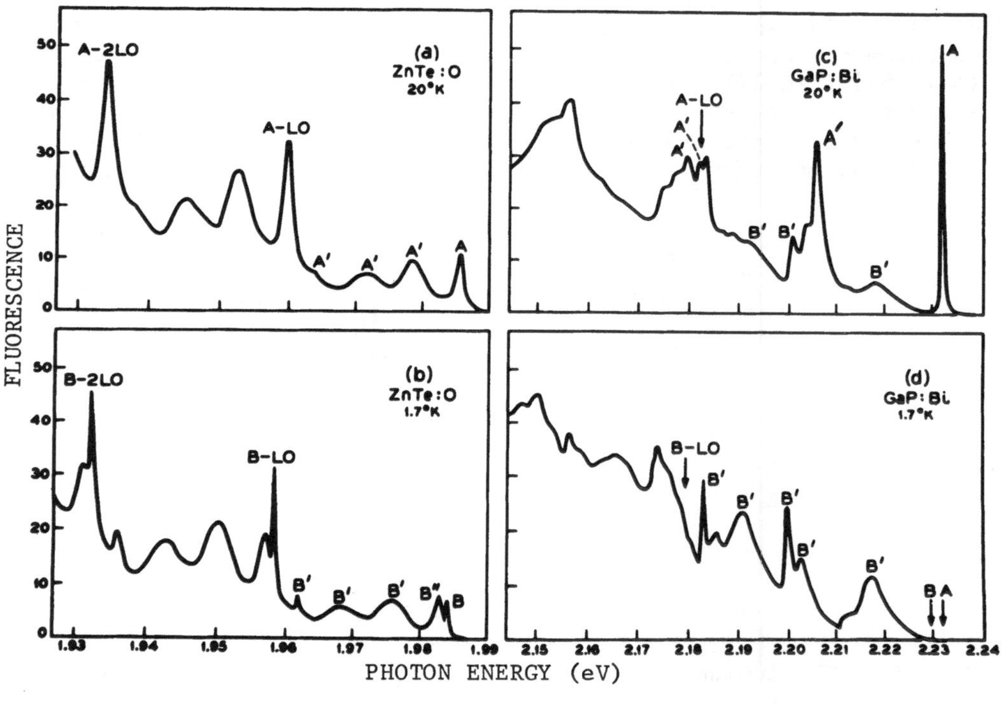

Fig. 28. Small portions of the photoluminescence of BE at the isoelectronic traps O_{TE} in ZnTe [(a) and (b)] and Bi_P in GaP [(c) and (d)]. The spectra are compared at high and low temperature to favour electric dipole allowed and forbidden transitions from exchange split no-phonon states A and B, respectively. Component B" for O_{TE} involves a special phonon-assisted transition which mixes B into A. (Hopfield et al, ref 43).

are spherically symmetric. Then, the oscillator strength ratio f_A/f_B tends to be much larger than when the hole is tightly bound, as for GaP:Bi, [98] consistent with the greater tendency of the phonon sidebands of the A and B transitions to be similar (Fig 28). One significant difference is the special phonon satellite B'' in Fig 28b, which results from a small degree of cross-coupling between the A and B states via the weakly bound hole. Satellites due to promoting modes are usually themselves replicated by the receiving modes when transition probability and lattice relaxation effects are both significant, as for the unmarked satellites for GaP:Bi in Fig 28.

When a spherically symmetric state is bound by short range forces yet with moderate lattice relaxation, as for GaP:N, the N BE sidebands provide quite an undistorted representation of the one-phonon density of states [41]. The main intrusive feature is the appearance of a local mode satellite. This is observed only as an overtone for isolated N, because the symmetric electron distribution couples mainly to Γ_1 phonons in which the surrounding lattice moves in a breathing mode, with the N_P atom at rest [105]. The fundamental mode is readily observed when the local symmetry is lowered, as for N_P-N_P associates [41] or the Li_I-Li_{Ga}-O_P centre, where the rich phonon sideband spectrum was crucial to the verification that the centre involved O_P and contained two inequivalent Li atoms [106]. The energies of these local modes are often difficult to predict because of significant changes in local force constants on introduction of the impurity. For example, although local mode energy of the N_P isoelectronic trap is very close to expectation from a very simple model with unaltered force-constants [41], this calculation gives much higher vibrational energies than are observed for Li_I-Li_{Ga}-O_P in GaP and also for the Li_{Zn} acceptor in ZnSe [107] (Fig 29). However, the isotope shift $\Delta h\omega_{LOC} \sim +\ 3.2$ meV when $Li^7 \rightarrow Li^6$ can be estimated rather accurately from the observed $h\omega_{LOC}$ for Li^7 and the fractional mass change $(\Delta M/M)$ through the simple relation

$$\Delta h\omega_{LOC} = \frac{1}{2}\left(\frac{\Delta M}{M}\right)h\omega_{LOC} \tag{24}$$

since the local force constants are insensitive to the change in isotopic mass. Observation of these weak local mode satellites of the I_1^X ABE of ZnSe provide important confirmation of the role of Li as a dominant shallow acceptor [108]. Since the phonon coupling is dominated by interactions with the asymmetric bound hole in this case, the symmetry argument given above for electron-attractive GaP:N trap does not apply, and appreciable strength appears in the fundamental of the local mode (Fig 29). No-phonon isotope shifts are too small to detect in this BE luminescence, since they depend essentially upon the change in local charge during the electronic transitions. This is very small for the principal ABE transitions,

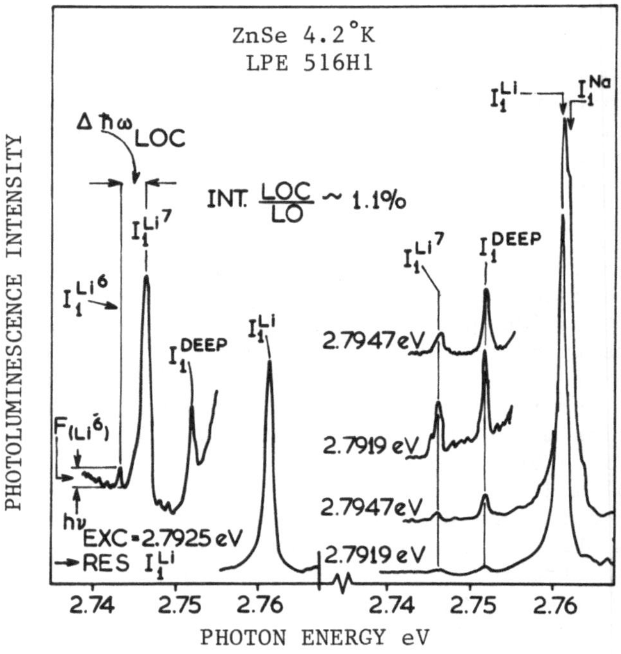

Fig. 29. One-LO phonon satellites of the Li, Na and Cu? (I_1^{DEEP}) A^o,X transitions in ZnSe. At the right, the weak local mode replica $I_1^{Li^7}$ does not shift as much as the LO replica when the energy of the exciting laser $h\nu_{EXC}$ is increased because the unresolved NP components of the Na ABE do not contribute to $I_1^{Li^7}$. The high gain spectra on the left show the weak additional satellite from the ~ 7% abundant Li^6 (Dean, ref 107).

since one hole remains closely bound to the acceptor core throughout the transition. However, no-phonon isotope shifts have been seen for distant DAP recombinations involving the Li_{Cd} acceptor in CdS, with a magnitude that decreases dramatically with decreasing DAP separation as the excited electronic state becomes more BE-like [45].

III.B. Electronic Satellites

The electronic satellites of main interest are those observed in BE complexes which contain more than two electronic particles. The D^o,X and A^o,X systems are very important members of this class of BE. We will first consider the D^o,X system, since the energy states of the donor are usually simpler and the principles are unaltered for the A^o,X systems. We may decompose the D^o,X into a BE bound to a neutral donor in its ground state. The recombination of the BE leaves the neutral donor unaltered, and is responsible for the principal BE luminescence lines discussed earlier. However, the wave-functions of electron states in the BE contain a superposition of contributions from excited states of the same parity as the ground state. These are particularly significant when the donor is relatively shallow, effective mass-like. The effect of a strong attractive central cell potential is primarily to lower the energy of the 1s ground state relative to all excited states, and this makes the wave-functions of electrons in the BE more 1s-like. However, there is always a finite probability that the donor will be left in an ns excited state following BE recombination. These transitions produce satellites displaced below the principal DBE by the excitation energy to the donor state involved. Thus, in Fig 30

$$h\nu_{D_1} - h\nu_{D_{2p}} = L_1 - L_{2p} = E_{D_{1s}} - E_{D_{2p}} \tag{25}$$

with a similar expression for excitations to the 2s states, which produce the satellites D_{2s}, L_{2s}. Such transitions were first recognised for shallow donors in GaP and were named as 'two-electron' transition satellites [109]. However, these satellites were not analysed correctly until relatively recently [110]. This delay arose partly from misconceptions about the parity of the excited states involved [109] and also from problems in the analysis of donor states in GaP caused by the camel's back effect in the conduction band, which was recognised only recently (Section II.D). It is now clear that transitions to s-like donor states predominate in the satellites of the ground state of the DBE, particularly in semiconductors like GaP and Si [111] which have relatively tightly bound donor states.

Transitions to 2p states are quite prominent in Fig 30, even for satellites of the non selectively excited principal DBE ground state D_1, for example D_{2p} which is about 20% as strong as D_{2s}. Henry and Nassau [112] noted that the presence of the satellites due to transitions to np states arises from an effect similar to the origin of the second class FE spectra discussed in Section I.B. The transition probability for BE decay is proportional to the square of an overlap integral of the form

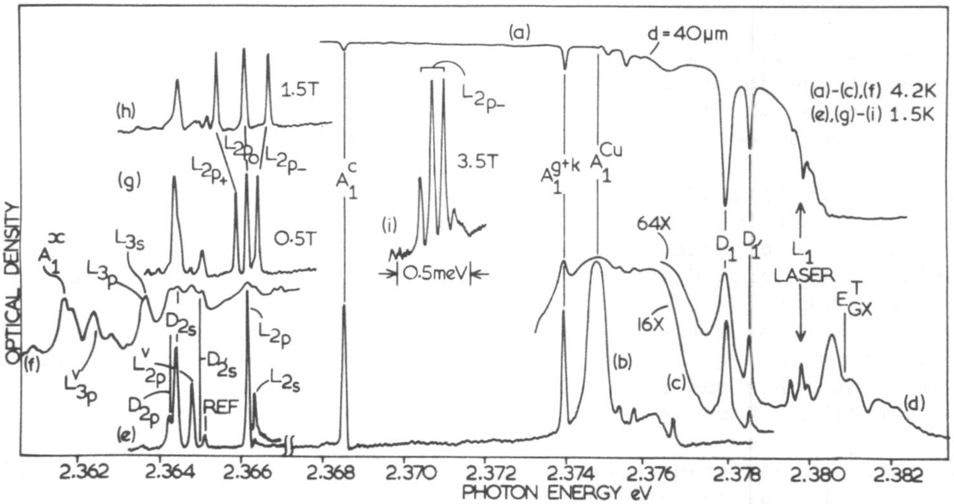

Fig. 30. Small portions of the near-gap optical spectra of ZnTe recorded photographically at 4.2 and 1.5°K. (a) is optical transmission, (b) - (i) photoluminescence, excited by above gap light for (b) - (d) and by light of energy at L_1 for (e) - (i). Spectra (g) - (i) were taken under the indicated magnetic fields. 'Two-electron' satellites from the resonantly pumped DBE excited state are $L_{ns,np}$ while corresponding satellites of the DBE ground state are $D_{ns,np}$. L^V_{np} originates from DBE excited state D_1'. The light near 2.375 eV is heavily overexposed on some recordings. (Dean et al, ref 114).

$$\left.\begin{aligned} &\iint \phi_D(\underline{r}_1)^* \, \phi(\underline{r}_1, \underline{r}_h, \underline{r}_h) \, e^{i\underline{q}\cdot\underline{r}_h} \, d\underline{r}_1 \, d\underline{r}_h \\ \approx &\iint \phi_D(\underline{r}_1)^* \, \phi(\underline{r}_1,\underline{r}_h,\underline{r}_h) \, d\underline{r}_1 \, d\underline{r}_h \\ &\quad + i\underline{q} \iint \phi_D(\underline{r}_1)^* \, \phi(\underline{r}_1,\underline{r}_h,\underline{r}_h) \, \underline{r}_h \, d\underline{r}_1 \, d\underline{r}_h \end{aligned}\right\} \qquad (26)$$

where $\phi_D(\underline{r}_1)$ is the donor state which remains after the transition. The BE wave-function $\phi(\underline{r}_1, \underline{r}_2, \underline{r}_h)$ has been written $\phi(\underline{r}_1,\underline{r}_h,\underline{r}_h)$ on the sudden approximation, in which it is realised that if the hole $\underline{r}_h$ recombines with electron $\underline{r}_2$, they must have the same spatial co-ordinate $\underline{r}_h$ at the instant of recombination. The small but finite wave-vector of the emitted photon is $\underline{q}$ in Eq. (26). This sudden approximation has also been used to account for the fact that 'two-electron' or 'two-hole' transitions are much more probable as satellites of no-phonon transitions than of momentum-conserving phonon replicas in the luminescence of indirect gap semiconductors [113]. The donor wave-function $\phi_D(\underline{r}_1)$ in Eq. 26 is much more 1s-like for momentum-conserving phonon-assisted transitions, since these occur with very different values of $\underline{r}_1$ and $\underline{r}_2 = \underline{r}_h$. It is interaction of the wavefunctions of the two electrons in the BE which mixes the higher donor excited states into $\phi_D(\underline{r}_1)$.

The first term on the right in Eq. 26 represents transitions to the donor ground state and to ns excited states. The second term in this expansion of the electric dipole operator for small $\underline{q}$ represents parity forbidden transitions in which the transition probability is proportional to q^2, just like the second class transitions in FE spectra (Section I.B), and involves the creation of the p-like BE states in Fig. 30. The importance of the np 'two-electron' satellites is much greater for DBE states in which $r_h \sim q^{-1}$, so they are much more significant in relatively narrow gap semiconductors where m_e^* and E_{BX} are both small.

Selective excitation with a narrow laser line promotes very sharp 'two-electron' satellites when the broadening is largely inhomogeneous [114], as is true for the very shallow DBE excited states in Fig 30. The laser line L, is resonant with the third DBE excited state which has E_{BX} only ~ 1.1 meV. However, the width of the sharpest two-electron satellites approaches the resolution limit of the detection system employed ~ 0.01_5 meV and is not far beyond the likely limiting homogeneous linewidth imposed by life-time broadening, ~ 0.01 meV. The lines are sufficiently narrow that the spin-splittings of transitions to each orbital sub-com ponent of the 2p states are fully resolved in a magnetic field of only ~ 3T (Fig 30i). Also, transitions to 3p states are clearly seen. The p excited states are clearly identified from their characteristic orbital magnetic splittings, shown in Fig 30 (e)-(h), from which the effective mass donor energy $E_D = 17.7_8$ meV and the central cell correction of 0.3_4 meV can be derived for the single donor observed. The value of m_e^* is 0.128 ± 0.004 m_o and $|g_e|$ is 0.40 ± 0.03. Selective excitation considerably simplifies the Zeeman spectrum of the 'two-electron' satellites. The resonance enhancement condition with the narrow laser line can be maintained only if the hole is treated as a magnetically decoupled particle. Thus, its complex Zeeman behavior (Section II.E) which has not yet been thoroughly analyzed for DBE states in ZnTe, is effectively removed form these spectra.

It is particularly helpful to use selective excitation of DBE luminescence in ZnTe, since this semiconductor is normally strongly p-type and DBE transitions are relatively weak even for strong excitation by above band gap light. Resonant excitation into the DBE gives dramatic increases in the intensities of their satellites. The same technique was used to enhance the relative intensities of the very weak Li ABE satellites in Fig 29 for a similar reason [107]. Here the problem is that ZnSe is a dominantly n-type semiconductor.

A key feature of Fig 30 is the dramatic reversal in the relative intensities of 'two-electron' satellites to 2p and 2s donor excited states for satellites of the DBE excited states compared with those of the ground DBE state D_1. These transitions are illustrated in Fig 31, where it is shown that the relative intensity of transitions to the 2p state have increased ~ 100-fold for excitation into third DBE excited state compared with the ground state. Large enhancements are also observed for the two other main DBE excited states shown in Fig 30, independent of whether they were directly excited by the laser line L_1. An 'H_2 molecule' analogue of the DBE excited states has been used [114] in a configuration which reproduces the p-like envelope of the hole which is believed to account for these DBE excited states (Section II.D)[60]. These 'two-electron' replicas then originate from hole recombinations

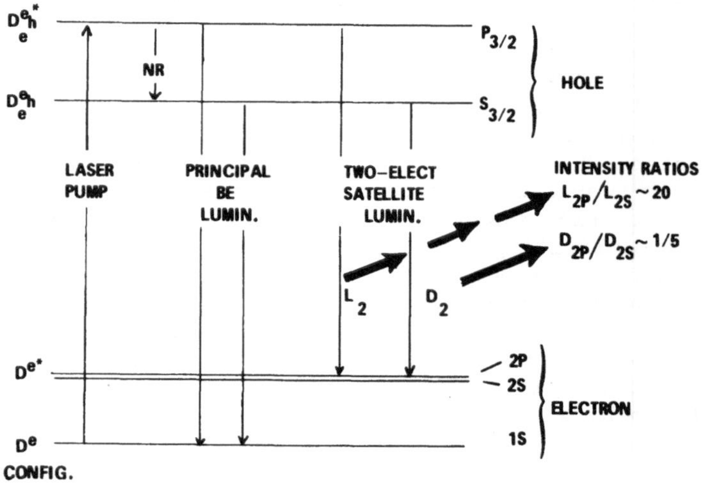

Fig. 31. A schematic representation of the DBE states in a direct gap semiconductor like ZnTe including the lowest state D^{e}_{eh} and an excited BE state $D^{e}_{e}h^{*}$, the donor ground state De and the n=2 excited donor states. 'Two-electron' transitions occur from the resonantly excited $D^{e}_{e}h^{*}$ state and from $D^{e}_{e}h$ after nonradiative relaxation (NR) (Dean et al, ref 114).

with the s-like electron around the donor, which guarantees strong no-phonon recombinations. The remaining electron is left in the p-like state, in which it correlated with the orbital motion of the hole around the neutral donor. These results give strong support to the predicted nature of these DBE excited states [60].

The resonance excitation effects are of quite general validity and have been reproduced in ZnTe [65] and ZnSe [64] using tuneable dye laser sources. Similar electronic characteristics were derived from a magneto-optical study for ZnSe [64]. A particular feature of the latter study was the observation of lines just below the n=2 FE, which gave particularly strong positive contributions in the photoluminescence excitation (PLE) of DBE components (Fig 32). These lines are quite broad, ~ 0.5 meV. However, selective excitation within them revealed satellite lines of the same half-width as the dye laser, ~ 0.09 meV, and displaced by the same 1s-2s, 2p excitation energies (Eq (25)) as observed for the usual DBE lines

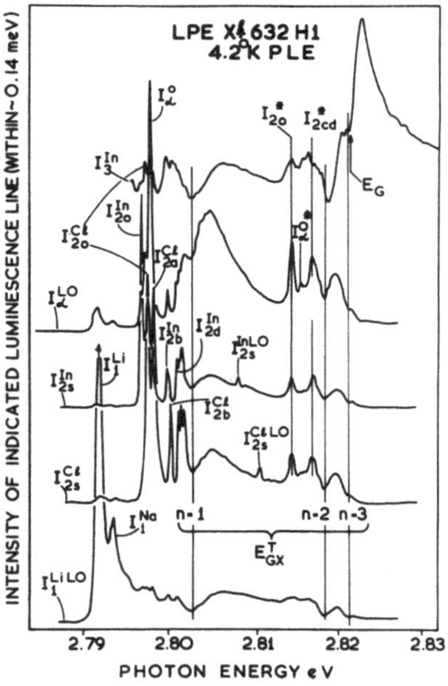

Fig. 32. Low temperature PLE spectra from a relatively pure single crystal of zincblende ZnSe. The different detected luminescence components are indicated at the left. α is an unidentified neutral (isoelectronic ?) centre. In and Cℓ are donors, Li is an acceptor. Other notation follows Fig 17. The transverse FE gap (Fig 1c) is noted for various internal excited states. Excited DBE states associated with the n=2 FE have superscript *. Lower DBE excited states carry subscripts a,b,c to distinguish the lowest DBE state, subscript o. (Dean et al, ref 64).

below the n=1 FE (Fig 33). Observation of these satellites clearly proves previous conjectures that corresponding lines observed for ZnTe involve high excited states of the DBE [48]. Different donors contribute to a different extent to excitations within different regions of these broadened lines, as shown from the lowest two spectra in Fig 33 and transitions to 2s donor states generally predominate to a larger extent than even satellites of the principal DBE. The large diagmagnetic shift rates of these broad lines, comparable with the n=2 FE, as well as their position suggest that they involve excitations within the BE rather than of the BE relative to the neutral donor responsible for the DBE states just below the n=1 FE.

Studies of corresponding 'two-hole' satellites have provided detailed information on acceptor excited states for several semiconductors, including GaAs [56], InP [57] and ZnTe [48]. In view of the failure of Haynes' rule for very shallow acceptors in direct gap semiconductors, these satellites tend to be displayed at spectral positions which relate to nearly constant transition energies of the principle ABE lines. Thus, the positions of the satellites given an immediate indication of the binding energies of the acceptors involved (Fig 19a). Transitions to 10 or 11 excited acceptor

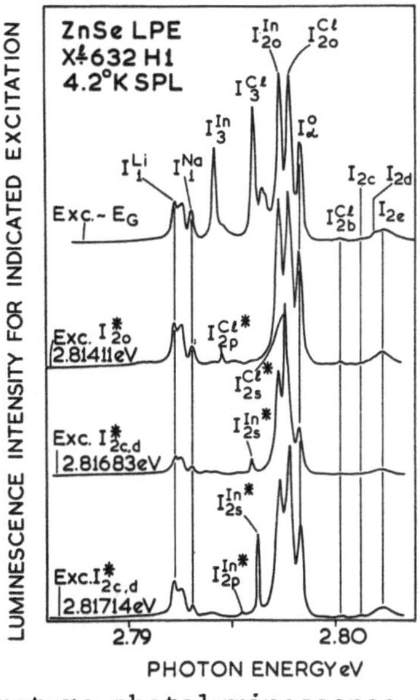

Fig. 33. Low temperature photoluminescence as in Fig 15, for selective excitation at the positions indicated on the left. The excitation is adjusted to optimise the I_3 DBE lines at the top and in lower spectra emphasises the * 'two-electron' satellites of the high D^0,X excited state components I^*_{2o}, and $I^*_{2c,d}$ (Dean et al, ref 64).

states can be resolved in 'two-hole' satellite spectra from the highest quality ZnTe [50]. 'Two-hole' transitions to the dominant 2s state have been clearly observed for acceptors with E_A as large as 0.43eV in ZnTe [105], despite the spectral broadening due to strong phonon coupling associated with such a tightly bound hole. Clear resolution of shallow acceptors with closely similar E_A is possible much more readily using the 'two-hole' ABE satellites (Fig 34a), than from the much broader free electron to bound hole luminescence bands also indicated schematically in Fig 19a.

The average slope of the empirical line through the experimental points for transitions to 2s states in Fig 34b [56] is about twice the value ($1/n^2 \sim 1/8$) expected for the effective mass approximation using the hydrogenic model. Corresponding data for shallow donors in CdS [47] and ZnSe [46] give more closely linear trends, but also with a larger slope than this simple theoretical prediction, respectively 0.22 and 0.20. These discrepancies remain unexplained at present.

III.C. The Use of Luminescence Excitation Spectra

We have already seen in Section 3b that very useful information can be obtained from selectively excited luminescence spectra, in

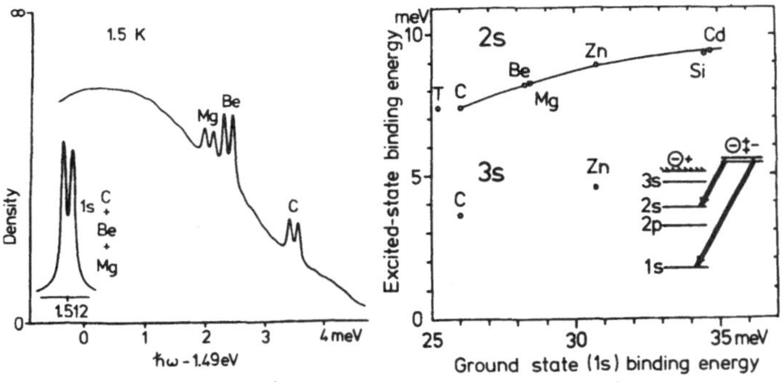

Fig. 34. (a) 'two-hole' photoluminescence satellites of a GaAs single crystal containing neutral C, BE and Mg acceptors superposed on the broad ~ 1.49 eV donor acceptor pair luminescence. The principal ABE transitions of these acceptors (inset) are not mutually resolved. The highest exchange split ABE subcomponent,(Fig 20) is thermalised away here. (b) the variation of the 2s and 3s acceptor excited state splittings with the 1s ground state for the indicated acceptors in GaAs, measured as in (a). T is the effective mass theoretical value for 2s. The 'two-hole' transition mechanism is indicated schematically at the right (Ashen et al, ref 56).

which the spectral contributions of different donors or acceptors, or indeed of any other optically active centres, can be selectively enhanced in turn. An additional important feature of the upper spectrum in Fig 33 is the strong selective enhancement of the I_3 (D^+,X) DBE luminescence for excitation just above the free particle energy gap, E_G = 2.821 eV in ZnSe [64]. The I_3 doublets fall below the I_2 lines in ZnSe, in contrast to CdS. It is therefore clear that it can be very useful to compare the photoexcitation spectra for various types of BE and other luminescence systems, since these spectra contain important evidence about the major pathways leading to the formation of the luminescent states. In the present example, the strong contribution from the structure near E_G and weak response near E_{GX} (Fig 32 upper spectrum) shows that photocreation of free electron-hole pairs is important. This indicates that the D^+,X centre is formed less by direct capture of FE than by the reaction

$$D^o + h \rightarrow D^+,X \tag{27}$$

Direct FE capture processes are evidently much more important for the creation of D^o,X and A^o,X (I_2 and I_1) luminescence. It is often convenient to measure these excitation spectra through the detection of phonon replicas or, in order to more specifically isolate a given BE system, using the 'two-electron' or 'two-hole' satellites In this way, the excitation spectrum may include the principal no-phonon transition itself (Fig 32).

Further information is provided when these spectra are measured over a range several optical phonon energies $\hbar\omega_{LO}$ above E_{GX} and E_G. Such spectra as Fig 35 provide important information about energy relaxation processes within the bands, and the nature of the capture processes leading to localised luminescence. The oscillations of periodicity $\sim \hbar\omega_{LO}$ indicate that the electronic relaxation (recombination) times τ_o, the electron-electron or exciton-electron scattering times τ_{ee} which control the equilibrium within the photoexcited carrier distributions, and the lifetime for inelastic energy loss to LO phonons τ_{LO} stand in the order

$$\tau_{LO} \ll \tau_{ee} \ll \tau_o \tag{28}$$

Capture of electrons by ionized donors enhances the D^o-h luminescence but decreases the photoconductivity [116]. The cross-sections of these capture processes decrease rapidly with the kinetic energy of the electronic particles, again since capture proceeds by cascade through very extended excited states at very low temperatures [91]. These conditions are sufficient to ensure oscillatory structure in antiphase between photoluminescence and photoconductivity excitation spectra [117] and with periodicity $\sim \hbar\omega_{LO}$ since τ_{LO} is $\sim 10^{-12}$ sec, much shorter than the capture times. Because of the very different curvatures of the conduction and valence

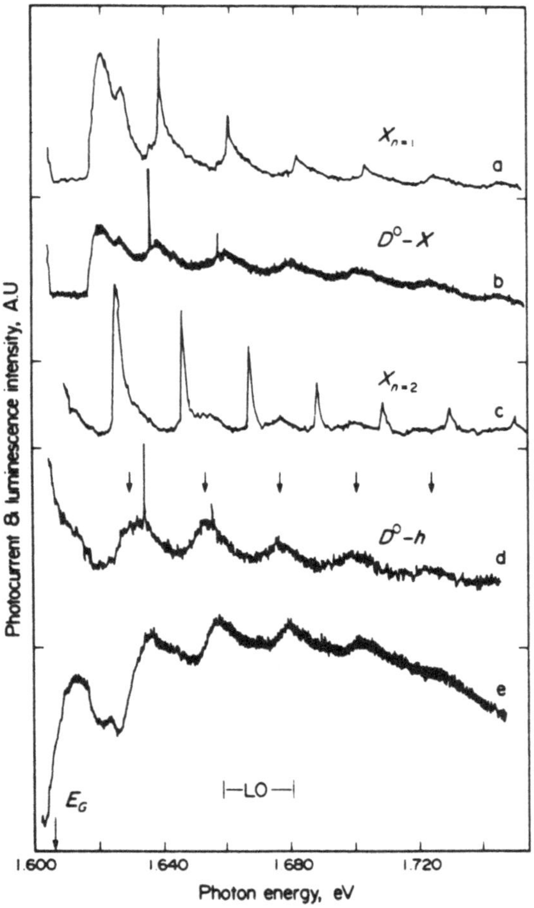

Fig. 35. Optical excitation spectra for luminescence and photoconductivity in pure zincblende CdTe measured at constant incident power at 1.8°K. The selectively detected luminescence components are indicated for spectra (a) - (d) where X represents FE luminescence. The arrows indicate the energies $E_G + 1.1\ n\hbar\omega_{LO}$ from Eqn 29 (Fig 1a). (Nakamura and Weisbuch, ref 115).

bands (Fig 1a), the exact periodicity of the low energy thresholds in these free carrier-related optical spectra (Fig 35d) is given by

$$h\nu = E_G + n\hbar\omega_{LO}\ (1 + m_e^*/m_h^*) \tag{29}$$

Corresponding processes occur in exciton-related luminescence. The cross-section for formation of shallow BE states may also be a rapidly decreasing function of the kinetic energy of the FE. Photo-creation of FE by indirect processes involving the emission of n LO phonons seems to be frequently more important for the eventual creation of low energy FE than formation from capture of free e-h pairs. In this case, the periodicity is different (Figs.35(a)-(c)). The low energy thresholds of the luminescence excitation peaks are simply given by

$$h\nu = E_{GX} + n\hbar\omega_{LO} \tag{30}$$

Another reason for the existence of a peak response close above these thresholds is that the matrix element for FE creation is expected to peak at $[\hbar^2/(2m_e+m_h)](2/a_{1B})^2$, where a_{1B} is the n=1 FE Bohr radius [115]. However, the importance of the concept of cold FE in the creation of shallow BE states is clearly seen by the observation of a strong peak near the direct FE gap (Fig 12) in the excitation spectra of the Bi isoelectronic trap luminescence in GaP [118]. This behaviour stands in striking contrast to the strong minima observed for distant shallow DAP luminescence, where the influence of enhanced surface recombination due to the onset of strong absorption near $(E_{GX})_{DIR}$ predominates. The overall form of these excitation spectra can be dramatically transformed by changes in conditions of surface charge, most probably due to their control of the bending of the semiconductor energy surfaces adjacent to the crystal surface and in the region where the photoexcitation is absorbed for a direct gap semiconductor. Surface recombination effects are responsible for the minima present near the centres of the strong n=1 and n=2 FE absorption lines even where I_2 BE luminescence is detected (Fig 32), although the weaker absorption in the wings of these transitions contributes a positive response. Such 'reversal' effects are extremely common in excitation spectra, and also sometimes even appear in the principal BE no-phonon lines of the centre whose luminescence is detected.

Resonant Raman scattering can also be very important [115]. The efficiency of these scattering processes is resonantly enhanced whenever the energy of the ingoing or outgoing photon, or of some intermediate state, is resonant with some energy level of the crystal, perhaps the n=1 FE or some BE state. The very sharp peaks in spectra a and b of Fig 35 represent these processes. Resonant Raman scattering is distinguished from resonant LO phonon assisted population of a given exciton energy state (hot excitons) by the sharpness of the peaks involved, the absence of any delay between the incoming and outgoing photons and the detailed optical selection rules exhibited in the scattering process. The resonant Raman process is favoured in very pure crystals at low temperatures, where FE capture times are relatively long. Raman scattering processes involving real excitations within the BE to excited states of the I_2 (D^o,X) and $I_1(A^o,X)$ systems have been reported recently in CdS [119].

IV. HIGH EXCITATION INTENSITY EFFECTS

IV.A. Stimulated Emission

There have been many effects observed as a consequence of creation of very high densities of free e-h and FE states in pure semiconductors at low temperatures. These include the onset of super-radiance. Given a suitably shaped piece of semiconductor relatively free from processes which provoke optical losses, true laser action can occur. Laser action can be obtained very readily in FE-related states at low temperatures. However, since FE distributions are governed by Bose statistics, rather than Fermi-Dirac statistics like electrons, inversion of the FE population to obtain laser action directly at the n=1 FE no-phonon state cannot occur [120]. Various types of electron-exciton and exciton-exciton inelastic scattering processes usually produce recombination leading to the optical gain necessary for laser action [121,122]. The general idea is that one exciton recombines radiatively, while the second exciton or electron conserves energy and momentum by being scattered into higher free particle energy states of the crystal. In this way, FE and electrons initially in these higher energy states can stimulate luminescence from a considerable range of the photon-like region of the polariton dispersion curve (Fig 1e). These processes are more important than FE LO phonon scattering processes [120]. Stimulated emission may also occur from BE states, particularly in the special circumstances of isoelectronic traps in indirect gap semiconductors like GaP [123], or especially $GaAs_{1-x}P_x$ at alloy compositions near conduction band cross-over [124]. There are difficulties associated with stimulated emission processes in BE states for direct gap semiconductors [125]. It is often rather difficult to give convincing proof that a given recombination process, out of the many which may occur within a very limited energy range close to the band gap of a semiconductor when such scattering processes are considered, is responsible for the optical gain observed in a particular case.

IV.B. Excitonic Molecules and Bose-Einstein Condensation

The excitonic molecule has been observed in the luminescence of a number of semiconductors, including Si [126], in addition to the luminescence from electron-hole liquid condensates, the electron-hole drops described in another chapter of this book. A considerable amount of work has been performed recently on CuCℓ. The predicted preferential formation of excitonic molecules by two photon excitation processes has been clearly demonstrated to occur by a giant 'two-photon' absorption process [127]. The Bose-like FE are theoretically predicted to condense into a completely mono-energetic high density distribution under certain conditions at sufficiently high excitation densities [128]. An excitonic molecule gas has favourable characteristics for the observation of the phase transition to the Bose-Einstein condensed state, including the small effective mass which increases the critical temperature for this phase transition. This condensation should be observable through sharp line luminescence. Attempts to observe this state experimentally have been frustrated by a number of effects, including strong competition from BE and other luminescence processes. Recent positive claims centre on certain observations on the free excitonic molecule luminescence in CuCℓ [129]. However, counter claims exist, and is too early to be certain whether this elusive condition has really been achieved.

IV.C. Multiple Bound Excitons

This is a BE state which for several years was clearly demonstrated in only a single case, that of the isoelectronic trap N in GaP [131]. Extensive and detailed magneto-optical studies clearly confirmed the symmetries of the states of the single BE and double or molecular BE respectively responsible for the AB and A^*,B^* sharp luminescence lines (Fig 23). The j-j coupling in the $N^{e\,e}_{h\,h}$ state is obtained simply from the two states formed by coupling two j = 3/2 holes, analogous to Fig 20. Once again, the two s = 1/2 electrons form a singlet S = 0 state according to the exclusion principle. The A^*,B^* luminescence is favoured over A, B at very low temperatures, when the single BE system thermalises into the J = 2 state from which transitions are forbidden and effective luminescence decay time becomes very long, $\gtrsim 1\mu$ sec [98]. Thus, the m=1 single exciton state is relatively easily optically saturated to favour luminescence from the m=2 bound excitonic molecule. The decay time of the A^*, B^* luminescence is $\sim 7.5 \pm 1$ nsec [31], compared with ~ 40 nsec for the allowed single BE transition [98]. This difference is attributed to Auger decay within $N^{e\,e}_{h\,h}$, not possible within N^e_h. This was confirmed since the total luminescence efficiency is observed to decrease by a similar proportion between 4.2°K, when A^*, B^* are weak, and 1.5°K, where they can be dominant.

The fact that the binding energy of the second exciton is almost as large as the first excited considerable interest [131]. Since no additional phonon replicas can be attributed to the A^*,B^* transitions, the large lattice relaxation, thought to be responsible for most of the reduction of the E_{BX} for the single BE compared with a simple pseudopotential calculation [42], is not much influenced by the binding of the second exciton [131]. Because $E_{BX} \ll h\omega_{LO}$, the lattice can re-adjust to the electronic motion. It was conjectured that lattice relaxation ensures that the two electrons in the m=2 state do not occupy the central cell at the same time, reducing the electron-electron repulsion. We shall see that tight binding of many excitons is a general feature of multiple bound exciton (MBE) systems involving binding to centres in which the single particle potential is long ranged, such as D^o and A^o rather than short range, such as the N isoelectronic trap. N-associated MBE states involving m > 2 have not been clearly identified. Initial reports of a radiative N_h^{ee} state were finally reduced to a debate over the properties of a single weak line [132].

Some experimental evidence suggestive of MBE effects in Si had been obtained both before the report of the N m=2 MBE luminescence in GaP in 1969, in unpublished work of J.R. Haynes, and soon after in 1970 when Kaminskii and Pokrovskii first announced a series of moderately well resolved lines which they attributed to MBE decay at neutral B acceptors [133]. The new luminescence lines, which occurred just below the normal m=1 ABE luminescence, were favoured at high excitation densities and low temperatures. Kaminskii and Pokrovskii [133] were also the first to suggest that these states might provide nucleation centres for electron-hole drops.

The subsequent history of development of this topic is quite fascinating [134]. Rapid advances were made in the quality of the optical data, with better laser excitation sources and eventually also special photodetectors. However, a combination of theoretical prejudice and uninspired physical insights caused some authors to remain highly skeptical of the validity of the basic MBE description for these line series, while other initial protagonists entered an extended period of revision during the middle 1970s! The regularly spaced luminescence lines, converging only slowly with decreasing energy below the m=1 BE state (Fig 36) [140], encouraged the initial veiw that E_{BX} increased continuously and strongly with m![135]. This apparent effect seemed much more surprising than the behaviour of the N MBE in GaP. The E_{BX} for the m=2 N MBE was 'only' 84% of the first, even for this special case where more reason could be offered for $E_{BX}^{m=2}$ to be large! [131]

However, it soon became clear that many of the detailed properties of these line series, such as their behaviour in magnetic fields and under uniaxial stress, initially held to be inconsistent with the basic MBE description [136,138], in fact gave it strong support [137,139,140]. Various alternative models proved unsatisfactory and had to be abandoned when measurements on Si:Aℓ and Si:Ga [141] (Fig 37) showed very clearly that the splittings in the m=1 states of these ABE due to J-J coupling (Fig 20) appeared in the final state of the m=2 transition and in reverse spectral energy order. This is just as expected from the MBE model, if the hole states are considered on a shell model (SM), in which the three holes in the m=2 MBE excited state are recognised as behaving like a single J = 3/2 particle in a shell of hole states which can accept up to four particles [137]. The two electrons should once

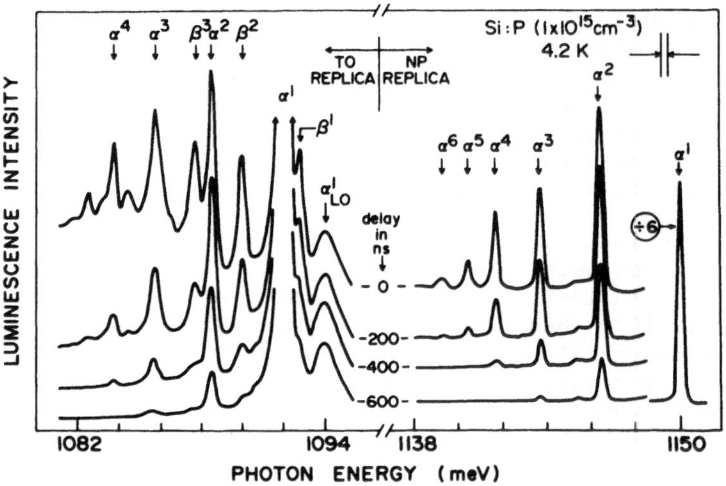

Fig. 36. The no-phonon (right) TO and LO momentum-conserving phonon-assisted luminescence (left) of neutral P DBE in single crystal Si, recorded with the indicated delays after excitation by 50-100 nsec optical pulses. All spectra are normalised to the same intensity of the principal D BE α^1, scaled as indicated. The α and β series are labelled according to the SM in Fig 38. The superscript indicates the number of excitons in each luminescing state. (Thewalt, ref 140).

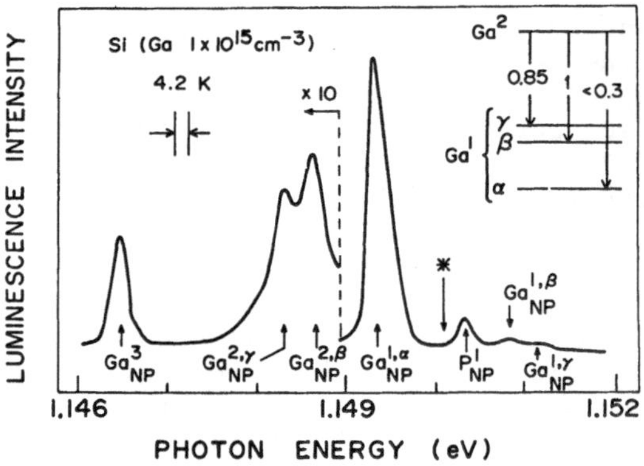

Fig. 37. The no-phonon photoluminescence of A^o,X at Ga acceptors in single crystal Si, with a small contribution from residual P donors. The level diagram interprets the sub-components arising from j-j coupling in the final state of the m=2 MBE luminescence, identical to that in the initial state of m=1 BE luminescence (Thewalt, ref 141). The weak $Ga^2_{NP}\alpha$ sub-component predicted near * has been confirmed by Lyon et al, ref 150.

again pair off, leaving an m=2 A BE state with total J = 3/2, unsplit in zero external field.

The key to a more realistic interpretation of the E_{BX} sequence in the MBE spectra was immediately provided from the much more complete SM Description given in Fig 38, derived by Kirczenow [142]. The electrons are placed in the states strongly split by valley-orbit coupling (Section II.D) rather than in a single shell with full degeneracy of the conduction band as assumed by Dean et al in their incomplete SM . The valley-orbit splittings are much less important for acceptor complexes. The symmetries of the electron and hole states given in Fig 38 define the nature of the transitions between the BE states. In particular, only BE decay involving the recombination of a Γ_1 electron, which has large overlap with the impurity core, can produce a luminescence series with strong no-phonon component in an indirect gap semiconductor such as Si (Section II.D).

Considering the influence of the exclusion principle, the SM predicts that the experimentally dominant α no-phonon line series (Fig 36) must involve transitions to excited states for all m > 1 (Fig 38). Analysis of the data in Fig 36 in these terms yields spectroscopic binding energies which generally decline monotonically with m for the DBE series [141,142], rather than increasing dramatically as was supposed at first [135].

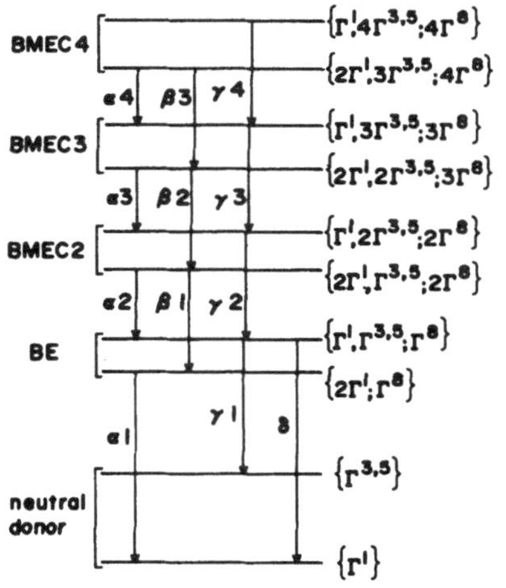

Fig. 38. Electronic configurations of the donor-related MBE states in Si on the SM and the transitions between them, neglecting the small splittings between the Γ_3 and Γ_5 valley-orbit excited states. The α and γ transition series have strong no-phonon members, unlike β and δ. Detection of γ and δ at low temperatures is made difficult by thermalisation, and δ is best studied in absorption. (Kirczenow, ref 142).

Very different types of MBE series are seen for substitutional donors such as P compared with interstitial donors like Li or substitutional acceptors [134]. This is explained on the SM through the very different types of valley-orbit splitting expected, depending upon the symmetry of the Bloch contribution to the electron wave-function at the impurity site (Section 2d). However, our theoretical understanding of the E_{BX} trends as f(m) remains far from satisfactory, and there are discrepancies between optically and thermodynamically-determined experimental values, which may indicate difficulties with the necessary assumptions of thermodynamic equilibrium [134]. Nevertheless, the SM has given a remarkably detailed account of the majority of the complex properties of MBE in Si and SiC. These include the presence of weak line series, the intensity effects attributed to shell closure also the lifetime, uniaxial stress and magneto-optical data. The many additional splittings possible in principle from interparticle interactions appear to be negligible in Si. Strangely they seem more significant in Ge, despite the generally smaller E_{BX} values fundamentally associated with the smaller m_e^* in Ge [143]. Similar line series have recently been reported for acceptor-doped GaP and interpreted in terms of MBE annihilation [144], displacing an earlier identification in terms of phonon-assisted intervalley scattering [145]. Once again, the detailed attribution of the electronic substructure for transitions involving a given m is less satisfactory than in Si. It is possible that both mechanisms contribute to the observed structure in this case, since the m=3 MBE satellites [144] close to the energies predicted for g-type intervalley scattering according to the revised understanding of the form of the 'camel's back' conduction band structure in GaP [19].

REFERENCES

1. E. O. Kane, J. Phys. Chem. Solids 1, 249 (1957).
2. N. F. Mott and A. M. Stoneham, J. Phys. C10, 3391 (1977).
3. M. G. Clark. J. Phys. C13, 2311 (1980).
4. See the spirited discussion in Session G of the 1958 International Conference on the Physics of Semiconductors, J. Phys. Chem. Solids 8, 193 (1959). This topic is also addressed by E. F. Gross Usp. Fiz. Nauk. 76, 433 (1962). [Engl. Transl. Sov. Phys. Uspekhi 5, 195 (1962)].
5. D. G. Thomas and J. J. Hopfield, Phys. Rev. 124, 657 (1961).
6. E. F. Gross, S. Permogorov and B. Razbirin, J. Phys. Chem. Solids 27, 1647 (1966).
7. R. F. Leheny, R. E. Nahory and K. L. Shaklee, Phys. Rev. Lett. 28, 437 (1972).
8. R. J. Elliott, Phys. Rev. 108, 1384 (1957).
9. E. F. Gross and N. A. Karryev, Dokl. Akad. Nauk. SSSR 84, 261 and 471 (1952).
10. R. A. Faulkner, Phys. Rev. 175, 991 (1968).

11. J. J. Hopfield and D. G. Thomas, Phys. Rev. 122, 35 (1961).
12. M. Nagasawa and S. Shionoya, Phys. Lett. 22, 409 (1966).
13. E. F. Gross and A. A. Kaplyanski, Dokl. Akad. Nauk. SSSR 132, 98 (1960).
14. K. J. Teegarden and G. Baldini, Phys. Rev. 155, 896 (1967).
15. G. Baldini and B. Bosacchi, Phys. Rev. 166, 863 (1968).
16. J. T. Devreese (ed), Polarons in Ionic Crystals and Polar Semiconductors (North Holland, Amsterdam, 1972).
17. A. Baldereschi and N. O. Lipari, Phys. Rev. B3, 439 (1971); B8, 2697 (1973).
18. D. D. Sell, Phys. Rev. B6, 3750 (1972).
19. G. F. Glinskii, A. A. Kopylov and A. N. Pikhtin, Solid State Commun. 30, 631 (1979).
20. W. Kohn, Solid State Physics, ed. by F. Seitz and D. Turnbull 5 (Academic Press, New York, 1955).
21. R. A. Faulkner, Phys. Rev. 184, 713 (1969).
22. D. D. Sell, S. E. Stokowski, R. Dingle and J. V. DiLorenzo, Phys. Rev. B7, 4568 (1973).
23. J. R. Haynes, M. Lax and W. F. Flood, Proc. Int. Conf. Phys. Semicond., Prague, 1960 (Academic Press, New York, 1961), p. 423. The ~ 8 meV energy was believed by Haynes et al to represent E_X, but is now known to be considerably smaller, K. L. Shaklee and R. E. Nahory, Phys. Rev. Lett. 24, 942 (1970).
24. P. J. Dean and D. G. Thomas, Phys. Rev. 150, 690 (1966).
25. A. A. Kopylov and A. N. Pikhtin, Solid State Commun. 26, 735 (1978); also, M. D. Sturge, A. T. Vink and F. P. J. Kuijpers, Appl. Phys. Lett. 32, 49 (1978).
26. P. J. Dean, to be published.
27. R. G. Humphreys, U. Rossler and M. Cardona, Phys. Rev. B18, 5590 (1978).
28. R. A. Foreman, W. S. Brower, Jr. and H. S. Parker, Phys. Lett. 36A, 395 (1971).
29. K. Shindo, T. Goto and T. Anzai, J. Phys. Soc. Japan 36, 753 (1974).
30. R. J. Elliott, Phys. Rev. 124, 340 (1961).
31. J. L. Deiss, A. Daunois and S. Nikitine, Solid State Commun. 8, 521 (1970).
32. Y. Petroff, P. Y. Yu and Y. R. Shen, Phys. Rev. B12, 2488 (1975).
33. J. J. Hopfield, J. Phys. Chem. Solids 22, 63 (1961).
34. A. G. Zhilich, J. Halpern and B. P. Zakharchenya, Phys. Rev. 188, 1294 (1969).
35. P. J. Dean, G. Kaminsky and R. B. Zetterstrom, J. Appl. Phys. 38, 3551 (1967).
36. D. D. Sell and P. Lawaetz, Phys. Rev. Lett. 26, 311 (1971).
37. M. Grossman, Polarons and Excitons, ed. by C. G. Kuper and G. D. Whitfield (Oliver and Boyd, Edinburgh, 1963), p. 373.
38. For example, H. Venghaus, Phys. Rev. B19, 3071 (1979).
39. R. Knox, Theory of Excitons, Solid State Physics, ed. by F. Seitz and D. Burnbull, Suppl. 5 (Academic Press, New York, 1963).

40. M. Lampert, Phys. Rev. Lett. 1, 450 (1958).
41. D. G. Thomas and J. J. Hopfield, Phys. Rev. 150, 680 (1966).
42. R. A. Faulkner, Phys. Rev. 175, 991 (1968).
43. J. J. Hopfield, D. G. Thomas and R. T. Lynch, Phys. Rev. Lett. 17, 312 (1966).
44. D. G. Thomas and J. J. Hopfield, Phys. Rev. 128, 2135 (1962).
45. C. H. Henry, K. Nassau and J. W. Shiever, Phys. Rev. B4, 2453 (1971).
46. J. L. Merz, H. Kukimoto, K. Nassau and J. W. Shiever, Phys. Rev. B6, 545 (1972).
47. K. Nassau, C. H. Henry and J. W. Shiever, Proc. Int. Conf. Phys. Semicond., Cambridge, 1970, ed. by S. P. Keller, J. C. Hensel and F. Stern (National Bureau of Standards, Va., 1970), p. 629.
48. H. Venghaus and P. J. Dean, Phys. Rev. B21, 1596 (1980).
49. P. J. Dean, Luminescence of Crystals, Molecules and Solutions, ed. by F. E. Williams (Plenum Press, New York, 1973), p. 538.
50. P. J. Dean and D. C. Herbert, Bound Excitons, in Excitons, ed. by K. Cho, Topics in Current Physics 14 (Springer, Berlin, 1979), p. 55.
51. D. C. Herbert and J. Inkson, J. Phys. C10, 695 (1977).
52. M. Rotenberg, J. Stein, Phys. Rev. 182, 1 (1969).
53. A. Baldereschi and N. O. Lipari, Proc. Int. Conf. Phys. Semicond., Rome, 1976 (Topografia Marves, Rome, 1976), p. 595.
54. J. R. Haynes, Phys. Rev. Lett. 4, 361 (1960).
55. P. J. Dean, R. A. Faulkner, S. Kimura and M. Ilegems, Phys. Rev. B4, 1926 (1971).
56. D. J. Ashen, P. J. Dean, D. T. J. Hurle, J. B. Mullin, A. M. White and P. D. Greene, J. Phys. Chem. Sol. 36, 1041 (1975).
57. A. M. White, P. J. Dean, K. M. Fairhurst, W. Bardsley, E. W. Williams and B. Day, Solid State Commun. 11, 1099 (1972).
58. A. M. White, I. Hinchcliffe, P. J. Dean and P. D. Greene, Solid State Commun. 10, 497 (1972).
59. N. Magnea, J. L. Pautrat, K. Saminadayar, B. Pajot, P. Martin and A. Bontemps, Rev. Phys. Appl. 15, 701 (1980).
60. D. C. Herbert, J. Phys. C10, 3327 (1977).
61. R. E. Halsted, Physics and Chemistry of II-VI Compounds, ed. by M. Aven and J. S. Prener (North Holland, Amsterdam, 1967), Ch. 8.
62. R. N. Bhargava, R. J. Seymour, B. J. Fitzpatrick and S. P. Herko, Phys. Rev. B20, 2407 (1979).
63. W. Schmid and P. J. Dean, unpublished data.
64. P. J. Dean, D. C. Herbert, C. J. Werkhoven, B. J. Fitzpatrick, and R. N. Bhargava, Phys. Rev. B23, 4888 (1981).
65. R. Romestain and N. Magnea, Solid State Commun. 32, 1201 (1979).
66. A. M. White, P. J. Dean and B. Day, J. Phys. C7, 1400 (1974).
67. E. C. Lightowlers and M. O. Henry, J. Phys. C10, L247 (1977).
68. K. R. Elliott, G. C. Osbourn, D. L. Smith and T. C. McGill, Phys. Rev. B17, 1808 (1978).

69. M. L. W. Thewalt, Phys. Rev. Lett. 38, 521 (1977).
70. D. G. Thomas, M. Gershenzon and J. J. Hopfield, Phys. Rev. 131, 2397 (1963).
71. T. N. Morgan, Phys. Rev. Lett. 21, 819 (1968).
72. P. J. Dean, R. A. Faulkner and S. Kimura, Phys. Rev. B2, 4062 (1970).
73. P. J. Dean, W. Schairer, M. Lorenz and T. N. Morgan, J. Lumin. 9, 343 (1974).
74. K. R. Elliott and T. C. McGill, Phys. Rev. B21, 2426 (1980).
75. H. Mathieu, J. Camassel and P. Merle, Phys. Rev. B21, 2466 (1980).
76. P. J. Dean, Phys. Rev. 157, 655 (1967).
77. P. J. Dean, J. Lumin. 1,2, 398 (1970).
78. P. Merle, B. Archilla, J. Camassel and H. Mathieu, Solid State Commun. 31, 205 (1979).
79. L. M. Roth, B. Lax and S. Zwerdling, Phys. Rev. 114, 90 (1959).
80. P. J. Dean and R. A. Faulkner, Phys. Rev. 185, 1064 (1969).
81. A. A. Bergh and P. J. Dean, Light Emitting Diodes (Clarendon Press, Oxford, 1976).
82. M. Gershenzon, F. A. Trumbore, R. M. Mikulyak and M, Kowalchik, J. Appl. Phys. 37, 483 (1966).
83. C. H. Henry, P. J. Dean and J. D. Cuthbert, Phys. Rev. 166, 754 (1968).
84. P. J. Dean, C. H. Henry and C. J. Frosch, Phys. Rev. 168, 812 (1968).
85. P. J. Dean and C. H. Henry, Phys. Rev. 176, 928 (1968).
86. Jane van W. Morgan and T. N. Morgan, Phys. Rev. B1, 739 (1970).
87. B. C. Cavanett, Proc. Int. Conf. Phys. Semicond., Kyoto, 1980, J. Phys. Soc. Japan 49, Suppl. A, 611 (1980).
88. M. Schmidt, T. N. Morgan and W. Schairer, Phys. Rev. B11, 5002 (1975).
89. J. L. Deiss, A. Daunois and S. Nikitine, Solid State Commun. 8, 521 (1970); B. S. Razbirin, I. N. Ural'tsev and A. A. Bogdanov, Fiz. Tverd. Tela. 15, 878 (1973). [Engl. Transl. Sov. Phys. Solid State 15, 604 (1973)].
90. D. F. Nelson, J. D. Cuthbert, P. J. Dean and D. G. Thomas, Phys. Rev. Lett. 17, 1262 (1966).
91. M. Lax, Phys. Rev. 119, 1502 (1960).
92. A. Nakamura and K. Morigaki, J. Phys. Soc. Japan 34, 672 (1973).
93. G. A. Osbourn and D. L. Smith, Phys. Rev. B16, 5426 (1977).
94. G. Benz and R. Conradt, Proc. Int. Conf. Phys. Semicond., Stuttgart, 1976, ed. by M. H. Bilkuhn (Teubner, Stuttgart, 1976), p. 1262.
95. E. I. Rashba and G. E. Gurgenishvili, Fiz. Tverd. Tela 4, 1029 (1962). [Engl. Transl. Sov. Phys. Solid State 4, 759 (1962)].
96. C. H. Henry and K. Nassau, Phys. Rev. B1, 1628 (1970).
97. D. G. Thomas and J. J. Hopfield, Phys. Rev. 116, 573 (1959).
98. J. D. Cuthbert and D. G. Thomas, Phys. Rev. 154, 33 (1967).
99. P. J. Dean, J. D. Cuthbert and R. T. Lynch, Phys. Rev. 179, 754 (1969).

100. K. K. Rebane, Impurity Spectra of Solids (Plenum, New York, 1970).
101. T. N. Morgan, B. Welbert and R. N. Bhargava, Phys. Rev. 166, 751 (1968).
102. V. Heine and C. H. Henry, Phys. Rev. B11, 3795 (1975).
103. C. H. Henry, P. J. Dean, D. G. Thomas and J. J. Hopfield, Localised Excitations in Solids, ed. by R. F. Wallis (Plenum, New York, 1968), p. 267.
104. A. M. Stoneham, Phil. Mag. 36, 983 (1977).
105. D. G. Thomas, Proc. Int. Conf. Phys. Semicond., Kyoto, 1966 (The Phys. Soc. Japan, 1966), p. 265.
106. P. J. Dean, Phys. Rev. B4, 2596 (1971).
107. P. J. Dean, unpublished data.
108. J. L. Merz, K. Nassau and J. W. Shiever, Phys. Rev. B8, 1444 (1973).
109. P. J. Dean, J. D. Cuthbert, D. G. Thomas and R. T. Lynch, Phys. Rev. Lett. 18, 122 (1967).
110. A. C. Carter, P. J. Dean, M. S. Skolnick and R. A. Stradling, J. Phys. C10, 5111 (1977).
111. P. J. Dean, J. R. Haynes and W. F. Flood, Phys. Rev. 161, 711 (1967).
112. C. H. Henry and K. Nassau, Phys. Rev. B2, 997 (1970).
113. P. J. Dean, W. J. Choyke and L. Patrick, J. Lumin. 15, 299 (1977).
114. P. J. Dean, D. C. Herbert and A. M. Lahee, J. Phys. C13, 5071 (1980).
115. A. Nakamura and C. Weisbuch, Solid State Electron. 21, 1331 (1978).
116. R. Ulbrich, Phys. Rev. Lett. 27, 1512 (1971).
117. Photoconductivity oscillatory structure is reviewed by P. G. Harper, J. W. Hodby and R. A. Stradling, Rep. Prog. Phys. 37, 1 (1973), while Photoluminescence effects are discussed by S. Permogorov, Phys. Status Solidi b68, 9 (1975).
118. P. J. Dean, Phys. Rev. 168, 889 (1968).
119. D. Munz and M. H. Pilkuhn, Solid State Commun. 36, 205 (1980).
120. H. Haug, J. Appl. Phys. 39, 4687 (1968).
121. C. Benoit a la Guillaume, J. M. Debever and F. Salvan, Phys. Rev. 177, 567 (1969).
122. H. Mahr, Excitons at High Density, ed. by H. Haken and S. Nikitine (Springer Verlag, New York, 1975), p. 265.
123. K. L. Shaklee, R. E. Nahory and R. F. Leheny, J. Lumin. 7, 284 (1973).
124. M. G. Craford and N. Holonyak, Jr., Optical Properties of Solids, new developments, ed. by B. Seraphin (North Holland, Amsterdam, 1976), p. 187.
125. D. G. Thomas and J. J. Hopfield, J. Appl. Phys. 33, 3243 (1962).
126. M. H. L. Thewalt, Solid State Commun. 25, 991 (1978).
127. G. M. Gale and A. Mysyrowicz, Phys. Rev. Lett. 32, 727 (1974).
128. R. C. Casella, J. Phys. Chem. Solids 24, 19 (1963).

129. L. L. Chase, N. Peyghambarian, G. Grynberg and A. Mysyrowicz, Phys. Rev. Lett. 42, 1231 (1979).
130. M. Ojima, T. Kushida, Y. Tanaka and S. Shionoya, Solid State Commun. 24, 841 (1977).
131. J. L. Merz, R. A. Faulkner and P. J. Dean, Phys. Rev. 188, 1228 (1969).
132. W. Czaja, L. Krausbauer, B. J. Curtis and P. J. Dean, Solid State Commun. 12, 807 (1973).
133. A. S. Kaminskii and Ya. E. Pokrovskii, Pisma Zh. ETF 11, 381 (1970). [Engl. Transl. JETP Lett. 11, 255 (1970)].
134. M. L. W. Thewalt, Bound Multiexciton Complexes, in Excitons, ed. by M. D. Sturge, Modern Problems in Solid State Physics (North Holland, Amsterdam, to be published).
135. R. Sauer, Proc. Int. Conf. Phys. Semicond., Stuttgart, 1974, ed. by M. H. Pilkhun (Teubner, Stuttgart, 1974), p. 42.
136. R. Sauer and J. Weber, Phys. Rev. Lett. 36, 48 (1976).
137. P. J. Dean, D. C. Herbert, D. Bimberg and W. J. Choyke, Phys. Rev. Lett. 37, 1635 (1976).
138. R. Sauer and J. Weber, Phys. Rev. Lett. 39, 770 (1977).
139. D. C. Herbert, P. J. Dean and W. J. Choyke, Solid State Commun. 24, 383 (1977).
140. M. L. W. Thewalt, Solid State Commun. 25, 513 (1978).
141. M. L. W. Thewalt, Can. J. Phys. 55, 1463 (1977).
142. G. Kirczenow, Can. J. Phys. 55, 1787 (1977).
143. A. E. Mayer and E. C. Lightowlers, J. Phys. C12, L539 (1979).
144. R. Sauer, W. Schmid, J. Weber and U. Rehbein, Phys. Rev. B19, 6502 (1979).
145. P. J. Dean and D. C. Herbert, J. Lumin. 14, 55 (1976).
146. C. Weisbuch, Ph.D. Thesis, cited by M. Voos, R. F. Leheny and J. Shah, Radiative Recombination, in Handbook on Semiconductors, Vol. 2, Optical Properties of Solids, ed. by M. Balkanski (North Holland, Amsterdam, 1980), p. 329.
147. Y. L. Yarnell, J. L. Warren, R. G. Wenzel and P. J. Dean, Neutron Inelastic Scattering 1, 301 (1968).
148. D. G. Thomas and J. J. Hopfield, Phys. Rev. 116, 573 (1959).
149. P. J. Dean, B. Monemar, H. P. Gislason and D. C. Herbert, to be published.
150. S. A. Lyon, D. L. Smith and T. C. McGill, Phys. Rev. B17, 2620 (1978).

EXCITONS IN INSULATORS

R. Grasser and A. Scharmann

I. Physikalisches Institut der Justus-Liebig-Universität
Heinrich-Buff-Ring 16, 6300 Giessen
Federal Republic of Germany

ABSTRACT

Our contribution to this course will begin with an introduction into the concepts underlying the Wannier-Mott and Frenkel exciton models, two extreme pictures of an exciton in crystalline materials. Excitons in wide gap insulators belong to the intermediate coupling case.

Our presentation will be essentially confined to the aspects of free and/or self-trapped (localized) excitons in alkali halides and rare gas solids. On that account, we will present and discuss absorption, reflectivity, and intrinsic luminescence spectra of these materials.

I. INTRODUCTION

An exciton is an elementary excitation of nonmetallic crystals. It represents a quantum of electronic polarization.

Within the one-electron approximation, nonmetallic or insulating crystalline materials are characterized by the existence of an energy gap E_g between ground and excited states. The energy band scheme of an insulating crystal is shown in Fig. 1. Depending on materials, the gap occurs at various energies. Solids which are insulators at temperature T = 0, but whose energy gaps are of such a size that thermal excitation of electrons can lead to measurable conductivity at temperatures below the melting point, are known as semiconductors. Evidently, the distinction between a semiconductor and an insulator is not a sharp one. The fraction of electrons ex-

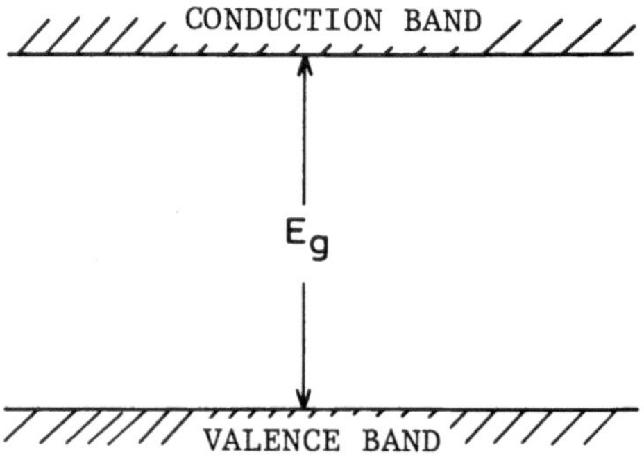

Fig. 1. Energy band scheme of a crystalline insulator.

cited across the gap E_g at temperature T is roughly of order

$$e^{-E_g/2kT},$$

where k is Boltzmann's constant. At room temperature ($kT \sim 0.025$ eV), with $E_g = 4$ eV, this factor is e^{-80}. Essentially no electrons are excited across the gap.

Our knowledge about excitons in semiconductors is very extensive and also very detailed [1]. That does not count for wide gap insula-show structure for photon energies below the band gap in an energy very limited results, only. Despite the fact that the optical properties of insulators have been studied for a long time in many cases very fundamental problems, such as the question about the character of the excitons, are still up for discussion. A notable breakthrough in this field of research has been achieved within the last years.

We have arbitrarily chosen to present only a limited number of exciton phenomena in wide gap insulators, where the latter are nearly entirely restricted to alkali halides and rare gas crystals. The most interesting feature of the exciton behaviour in these solids is the existence of free and/or self-trapped (localized) excitons. Therefore, an essential part of our discussion is devoted to the intrinsic luminescence of these materials.

II. THEORETICAL AND EXPERIMENTAL BACKGROUND

Especially at low temperature, absorption spectra of insulators show structure for photon energies below the band gap in an energy range where due to the one-electron approximation we should expect the crystal to be transparent. What one observes is the effect of the electron-hole Coulomb interaction on the fundamendal absorption edge.

In the case of an allowed band edge one finds a behaviour as illustrated schematically in Fig. 2, where the optical density α is drawn versus the energy of the incident photons $h \cdot \nu$. Due to the one-electron approximation we should observe an absorption pattern as indicated by the dotted line. Without final-state interaction, theory predicts a square-root absorption continuum for photon energies $h \cdot \nu > E_g$. In the case in point (Fig. 2), the result of the electron-hole Coulomb interaction on the fundamental absorption edge is above all a Rydberg-like series of discrete spectral lines at energies $E < E_g$. Furthermore, the final-state interaction drastically alters the one-electron continuum absorption, especially near the band gap E_g, as shown in Fig. 2. The higher members of the discrete line spectrum overlap with each other because of their finite widths. Thus, in the region of the band gap a quasi-continuum is produced which connects smoothly to the true continuum.

The resulting additional excited electronic crystal states are considered from two different points of view. The first model regards these states as essentially atomic or molecular excitations moving from site to site, the second is based on the energy band scheme of crystalline solids. While the first approximation, the

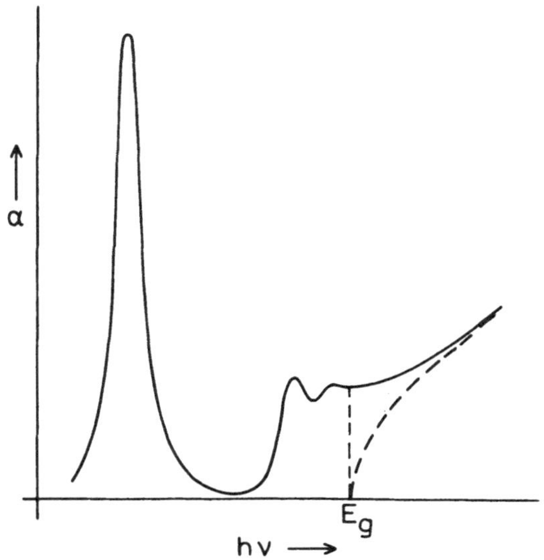

Fig. 2. Schematic representation of the effect of electron-hole Coulomb interaction on the fundamental absorption edge. The dotted line corresponds to the pure interband absorption edge.

Frenkel exciton, is appropriate for molecular crystals, the second, the Wannier-Mott exciton, applies to semiconductors and to most insulators. The Frenkel excitons have only translational degree of freedom. However, in the Wannier-Mott case, born out of the band model concept, where the excited state consists of an electron and a hole, the excitons are described in terms of a relative and a translational motion of this two-particle system electron and hole. In the Wannier-Mott approach, the limiting case of weak electron-hole interaction, the electron and hole orbits extend over many sites and the crystal can be exactly replaced by a medium with an effective mass and effective charge $e/\sqrt{\varepsilon}$. The problem can be solved within the effective mass approximation. This treatment leads to an equation formally identical to the Schrödinger equation for the free hydrogen atom.

The theory yields the following form:

$$E_n(\vec{K}) = E_g - \frac{\mu \cdot e^4/2\hbar^2 \cdot \varepsilon^2}{n^2} + \frac{\hbar^2 \cdot K^2}{2M} \quad ; \quad n = 1,2,\ldots.$$

$E_n(\vec{K})$ = exciton energy; μ = reduced mass of the electron and the hole, $\mu^{-1} = m_e^{-1} + m_h^{-1}$; e = elementary charge; $\hbar$ = Planck's constant ($= h/2\pi$); ε = dielectric constant; $\vec{K}$ = wave vector; $M = m_e + m_h$ = translational mass.

The Coulomb attraction between electron and hole causes a series of bound states below the continuum of nearly free electron-hole pair states. The exciton energy consists of the band gap energy less the binding energy according to the level spectrum corresponding to the hydrogen problem, and the kinetic energy of the centre of gravity of the exciton. Due to the translation symmetry, for every hydrogenic state we get an exciton band.

In the case of the Frenkel-type excitations when the exciton radius is of the order of the lattice constant a theoretical approach must be used which involves an explicit consideration of the lattice discreteness. For a definite crystal the number of distinct Frenkel excitons corresponds to the number of possible atomic or molecular excited states.

For some time past we investigate the optical properties of oxygen dominated phosphors, mainly of tungstates and molybdates which crystallize in the scheelite structure. The crystal structures of these ABO_4 compounds are ordered arrays of BO_4^{2-} oxyanions ionically bonded to A^{2+} cations. The covalently bonded BO_4^{2-} radicals are slightly compressed along the c-axis. As an example, Fig. 3 represents the reflection spectrum of a nominally pure $CaMoO_4$ crystal at room temperature (RT). At low temperatures the spectrum shows the same form. We have demonstrated that this spectrum can be discussed in the frame of a molecular orbital energy level diagram of the tetrahedral MoO_4^{2-} complex. In the following list an assignment is given of the measured reflectivity peaks to definite charge transfer transitions within the MoO_4^{2-} oxyanions.

Peak number	Transition	Peak energy (eV)
1	$t_1 \rightarrow 2e$ $(^1A_1 \rightarrow {}^1T_2)$	5.0
2	$t_1 \rightarrow 4t_2$ $(^1A_1 \rightarrow {}^1T_2)$	7.5
3	$3t_2 \rightarrow 4t_2$ $(^1A_1 \rightarrow {}^1T_2)$	8.4
4	$2t_2 \rightarrow 2e$ $(^1A_1 \rightarrow {}^1T_1)$	10.3
5	$2t_2 \rightarrow 2e$ $(^1A_1 \rightarrow {}^1T_2)$	11.7

The intrinsic luminescence of these materials is caused by a radiative decay of strongly relaxed excitons. All experiments clearly indicate that the band edge of these substances is an active arena for Frenkel excitons. Due to the present state of affairs it seems as if among inorganic wide gap insulators Frenkel excitons are realized only in oxygen dominated solids.

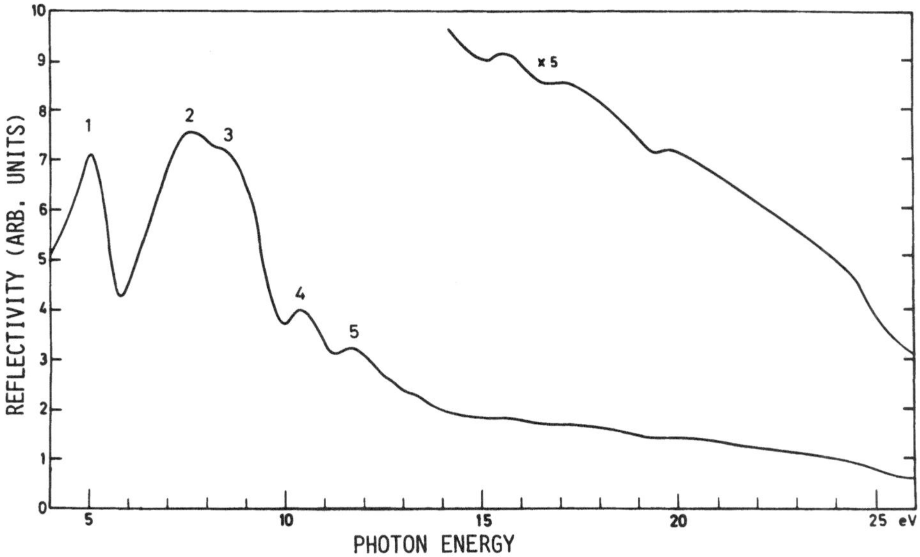

Fig. 3. Reflectivity of a $CaMoO_4$ crystal as function of photon energy at room temperature [2].

The Frenkel and the Wannier-Mott excitons are two extreme pictures of an exciton in crystalline materials with the most important common feature that the excitations propagate through the crystals.

Excitons in wide gap insulators which are the subject of our contribution belong to the intermediate coupling case. In this region the simple Wannier-Mott model fails because of a too small exciton radius in the n = 1 state. Theoretical efforts for exciton models that work in the intermediate coupling region have led to equations with bound exciton solutions which give rise to a modified Rydberg-Wannier-Mott series with corrections mainly for the lowest n = 1 exciton state. Resonant states fall into the continuum of uncoupled electron-hole pairs. The energy of exciton states in the intermediate coupling region has been calculated by Hermanson [3,4,5], Rössler and Schütz [6], and Altarelli and Bassani [7]. Calculations of the envelope functions F(R) of the lowest exciton states in solid argon within the formalism of the intermediate coupling model are shown in Fig. 4. The representation clearly demonstrates that the hydrogenic character of the exciton wave functions F(R) remains even for the intermediate case. Thus, it is reasonable to speak here also of 1s, 2s, ... excitons.

It does not make sense to reflect upon an "intermediate exciton". There is no smooth transition from the Frenkel exciton picture to

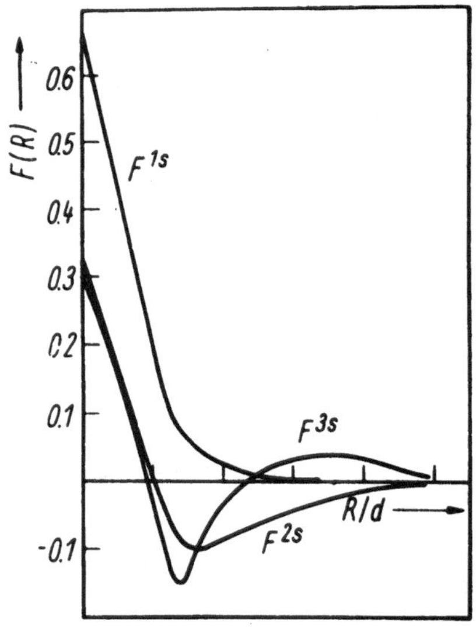

Fig. 4. Envelope functions of the lowest exciton states in solid argon due to the intermediate coupling approximation [6].

Table 1. Exciton binding energies in eV of the lowest state calculated within the Wannier-Mott $B_{1/WM}$ and the intermediate coupling approximation $B_{1/IC}$ for different rare gas solids. The Δ's are the relative hydrogenic defects, ε is the used dielectric constant.

	$B_{1/IC}$	$B_{1/WM}$	Δ	ε
Ar	1.97	2.17	+ 0.09	1.66
Kr	1.41	1.38	- 0.02	1.86
Xe	0.89	0.86	- 0.04	2.18

the Wannier-Mott model. Nature realizes only the Frenkel and the Wannier-Mott excitons, where in the case of wide gap insulators the n = 1 Wannier-Mott exciton state has to be modified to take account of the small exciton radius r_1.

In Table 1 the binding energies for the lowest exciton states in rare gas crystals as calculated in the Wannier-Mott $B_{1/WM}$ and intermediate coupling approximation $B_{1/IC}$ are compared. The binding energies $B_{1/WM}$ are taken from a paper of Rössler [8], the $B_{1/IC}$ from [6]. The central cell corrections of the Coulomb interaction within the intermediate coupling approximation lead to a change in the binding energies of the 1s excitons. For a comparison of the deviations of $B_{1/IC}$ from the corresponding values due to the Wannier-Mott model the so-called relative hydrogenic defect Δ has been defined [9]:

$$\Delta = \frac{B_{1/WM} - B_{1/IC}}{B_{1/WM}} .$$

The dielectric constants ε used to calculate $B_{1/WM}$ for the different rare gas solids are those of [10].

As shown in Table 1, Δ decreases with increasing atomic number, being positive for Ar and negative for Kr and Xe. Thus, according to the calculations done on rare gas solids, recent theoretical treatments yield that the exciton behaviour of wide gap insulators should be governed by modified Rydberg-Wannier-Mott sets.

The alkali halides just as the rare gas solids are closed shell systems. Both groups consist of ions or atoms with filled valence shells. Such configurations are especially stable. Their stability

Table 2. Calculated E_r^c and experimental E_r^e energies of the first resonant states for the rare gas solids (from ref. 6). E_i = ionization energy of the isolated atom. All energies are in eV.

	E_r^c	E_r^e	E_i
Ne	-	-	21.56
Ar	14.3	14.3	15.76
Kr	12.0	12.2; 12.22	14.00
Xe	9.4	9.6; 9.75; 9.56	12.13

corresponds to a large energy gap between the valence and the conduction band for these materials and, therefore, to insulating properties and a transparent appearance.

Neon and the elements directly below it in the periodic table, the normal inert gases, form the simplest closed shell systems. Rare gas crystals have face-centered cubic lattices. Their electronic structure is essentially that of the isolated atoms. The weak interatomic interaction is well described by the so-called Van der Waals interaction, an overlap interaction that includes a correlation energy contribution. The band gap of the rare gas solids may be correlated with the energy of the first resonant state within the optical spectra, respectively. The calculated (E_r^c) and experimental (E_r^e) resonant energies are listed in Table 2, together with the ionization energies (E_i) of the free atoms. As noticed by Pantelides [11], the gaps in rare gas crystals vary as $1/d^2$ with nearest neighbour distance d, a result found in covalent solids.

Ionic solids are the largest class of insulators. The alkali halides, the simplest ionic solids consist of equal numbers of positive and negative ions with charges of equal magnitude. They crystallize either in the rocksalt or the cesium chloride structure. The band gap in the electronic structure of ionic crystals is associated with the energy required to transfer an electron from a negative ion to a positive ion. The experimental values of the band gap energies in the alkali halides are summarized in Table 3. The values are from [12].

There is a strong trend in the gap energies with the atomic number of the halogen ion, while the dependence on the atomic number of the alkali ion is much weaker. The band gap decreases with increasing size of the halogen ion due to its decreased electronegativity. It is most interesting that these trends are independent of the crystal structure.

Table 3. Comparison of the experimental values of the band gap energies in eV of alkali halides.

	Li	Na	K	Rb	Cs
F	13.6	11.6	10.7	10.3	9.9
Cl	9.4	8.5	8.4	8.2	8.3
Br	7.6	7.5	7.4	7.4	7.3
I	-	-	6.0	6.1	6.2

Optical investigations on wide gap insulators are quite expensive. Tables 2 and 3 indicate that the electronic transitions already in the valence region of these materials occur in a spectral range which hardly can be covered without the use of synchrotron radiation.

III. EXCITONS IN ALKALI HALIDES

About 50 years ago, Hilsch and Pohl [13] started with a systematic investigation of the optical behaviour of alkali halides in the fundamental absorption region. They observed sharp absorption lines near the band gap of these crystals. The authors [14] assumed that this fine structure below the fundamental absorption edge is the consequence of charge transfer from the halogen ions to the alkali ions. In early discussions the sharp lines were interpreted as excitons of the Frenkel type.

III.A. Absorption and Reflectivity Measurements

Optical absorption measurements on all the alkali halides except LiF at 80K were performed by Eby et al. in 1959 [15]. The first theoretical discussion of these experimental results was given by Knox and Inchauspe [16]. The older picture of the Frenkel exciton was first questioned by Fischer and Hilsch [17,18]. The authors obtained from their experiments a relation between the ionization energy of the exciton and the optical dielectric constant, a result consistent with a hydrogenic model of the exciton. The measurements of Fischer and Hilsch have shown that effective mass states as well as relatively tightly bound states may participate in the electronic structure of the alkali halides. Later on, Huggett and Teegarden [19] strongly supported these ideas. They have found a photoconductivity threshold at an energy predicted by the effective mass model. Moreover, Baldini and Teegarden [20] have observed effective mass states in alkali halide alloys. Thereafter, Teegarden

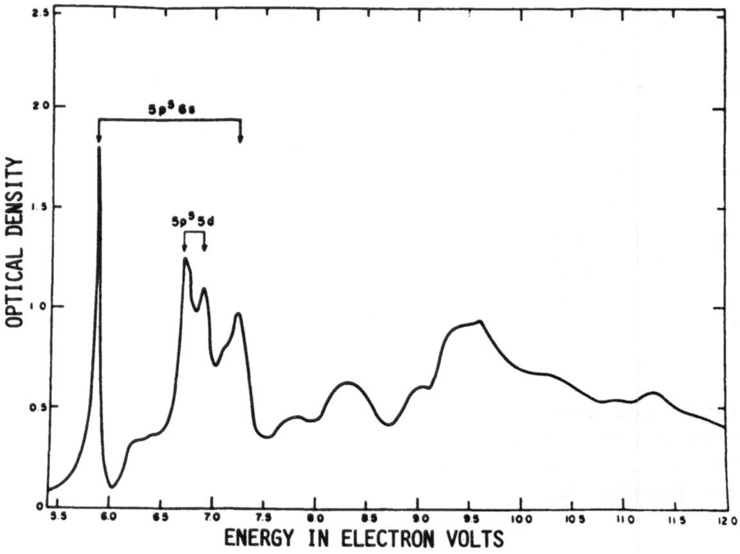

Fig. 5. Optical absorption spectrum of KI at 10K [21].

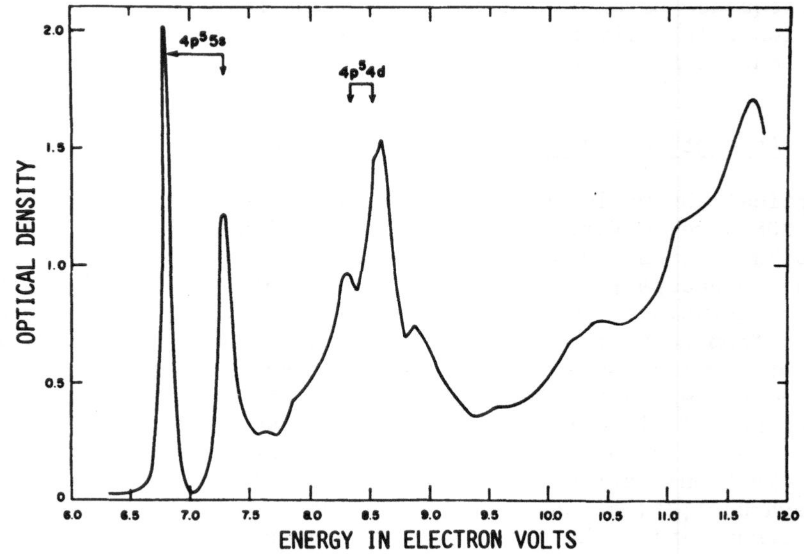

Fig. 6. Optical absorption spectrum of KBr at 10K [21].

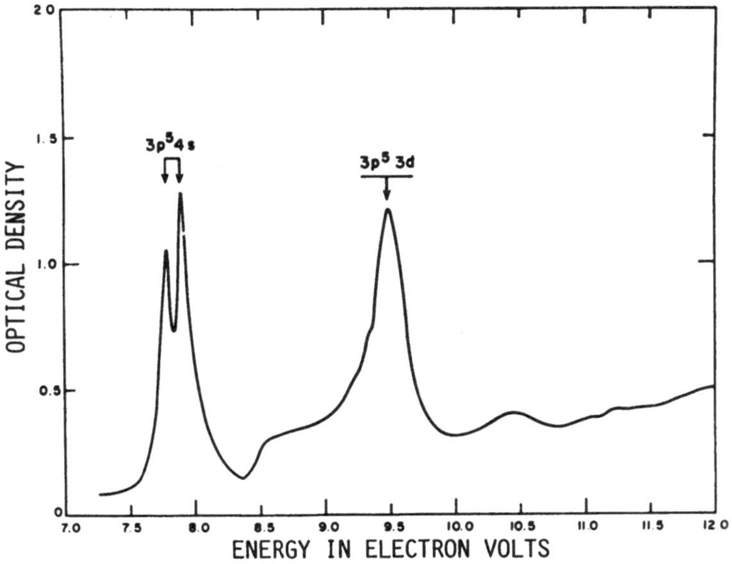

Fig. 7. Optical absorption spectrum of KCl at 10 K [21].

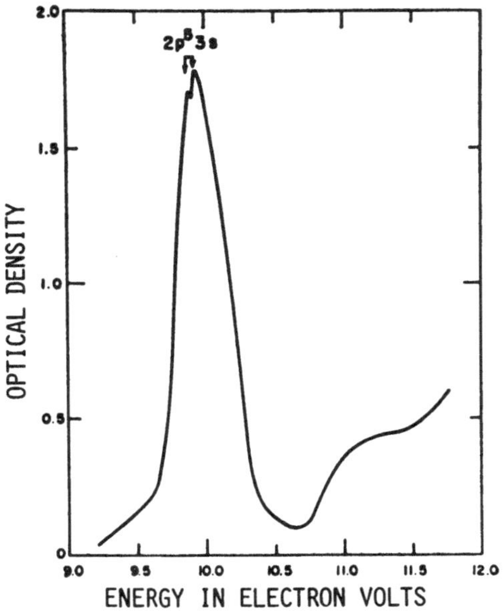

Fig. 8. Optical absorption spectrum of KF at 10 K [21].

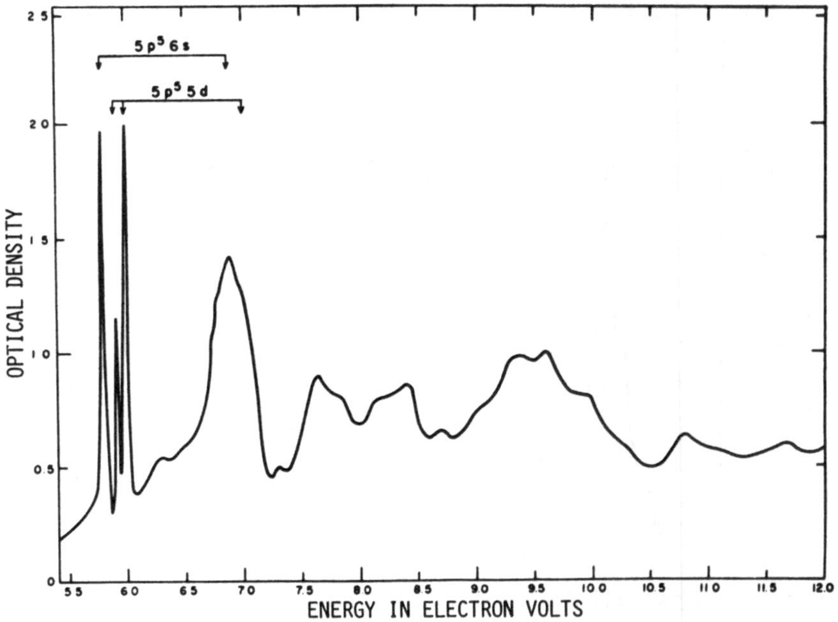

Fig. 9. Optical absorption spectrum of CsI at 10 K [21].

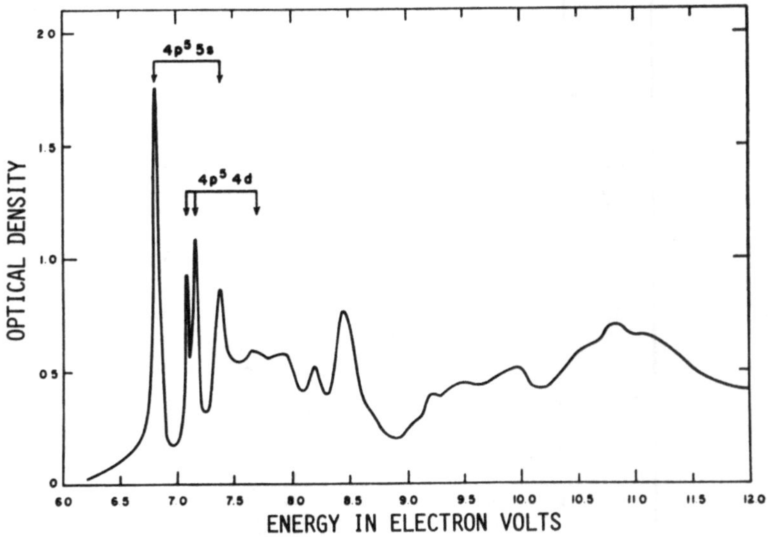

Fig. 10. Optical absorption spectrum of CsBr at 10 K [21].

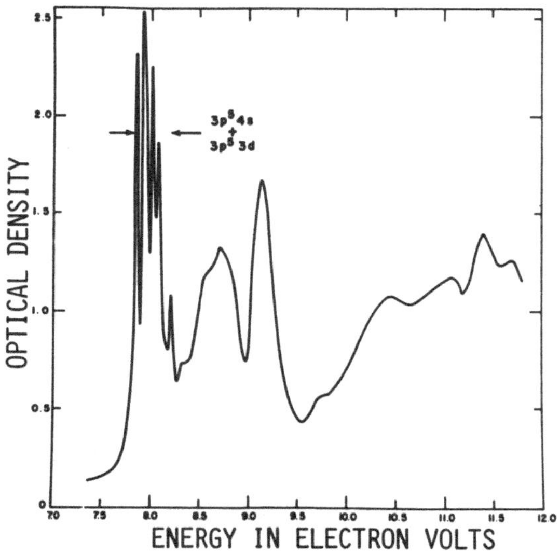

Fig. 11. Optical absorption spectrum of CsCl at 10 K [21].

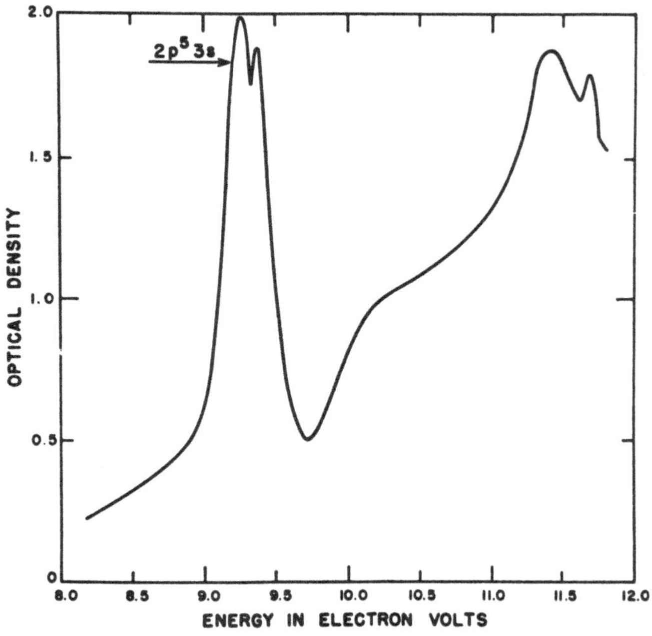

Fig. 12. Optical absorption spectrum of CsF at 10 K [21].

and Baldini [21] have reexamined the absorption spectra of the alkali halides with the exception of LiF and LiI. They measured in the spectral range from 5.0 to 12.0 eV at 10 K. A selection of these spectra are presented in Figs. 5-12. The absorption measurements were made on thin films of the alkali halides, prepared by vacuum evaporation onto a LiF substrate. The data are discussed in terms of an atomic-like model. The corresponding atomic configurations for some of the strong exciton lines are indicated. The transition energies to the lowest s-like and d-like conduction bands are given.

It is generally agreed that the first line in all the spectra shown is due to an exciton transition from the uppermost spin-orbit split halogen valence band, or in the atomic picture, from the closed p^6 shell of the halogen ion to an electron orbital of s-like symmetry. The second line in the spectra of the face-centered chlorides and bromides is generally assumed to represent the transition from the lower spin-orbit split halogen valence band to the same electron orbital of s-symmetry. This assignment is very reasonable, since the energy separation between these two lines is very close to the spin-orbit energy of the free halogen ion. For the iodides the identification of the second component of this spin-orbit doublet is very difficult because of overlap with other lines of different origin. In contrast to other authors [15,16,22], who also in the alkali iodides have assigned the line closest in energy to the first peak to the second component of the spin-orbit doublet, Teegarden and Baldini [21] suggest that the lines at 7.25 eV in the absorption spectrum of KI, and 6.9 eV in CsI, are the high energy component of the doublet, respectively. In Figs. 5 - 12 the spin-orbit doublets are denoted as belonging to the configuration p^5 s. In the absorption spectra of the fluorides the doublet is most difficult to resolve, since the free ion splitting is about 0.05 eV, only. Apparently, in the first peak in the KF spectrum the doublet character is indicated (Fig. 8).

Near the doublet associated with electron orbitals of s-symmetry, there are other sharp lines. Phillips [22] has required that these further lines are due to exciton transitions at critical points in the zone other than at the point Γ. Teegarden and Baldini [21], however, have identified these structures with exciton transitions, where the excited electron orbitals are of lower symmetry. From a comparison between the exciton spectra of the alkali halides and the absorption spectra of isoelectronic rare gas atoms, the latter authors propose that the additional lines in the halide spectra are the result of transitions in which the electron orbital has d-like symmetry. In Figs. 5 - 12 the analogous lines in the spectra are denoted by the configuration p^5d.

The absorption spectra of the simple cubic cesium halides in some cases show a lot of sharp lines in the spectral region of the

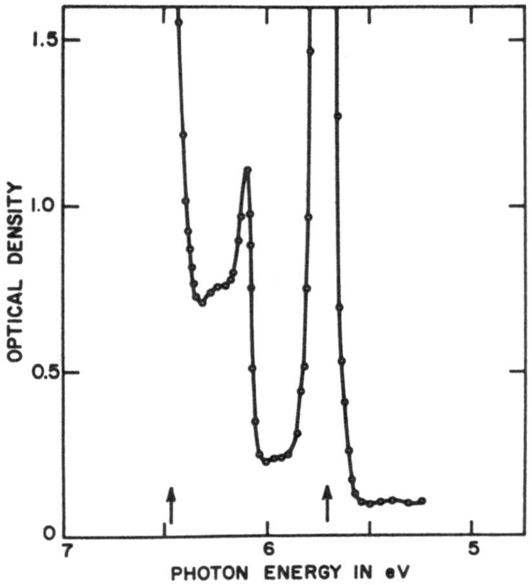

Fig. 13. Optical absorption of a RbI film at 10 K in the spectral region of the first spin-orbit doublet [21].

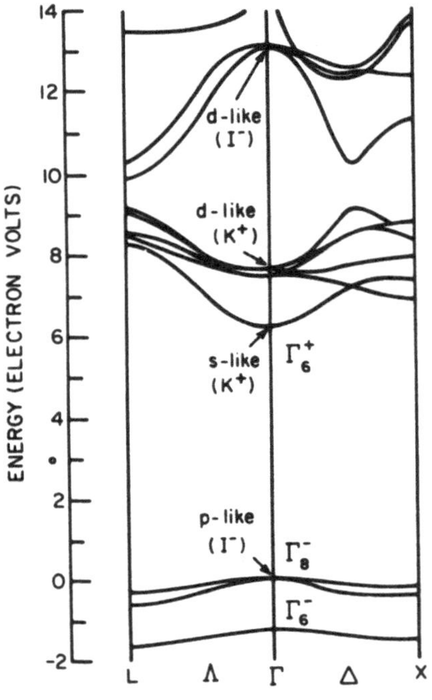

Fig. 14. Calculated band structure of KI [24].

spin-orbit doublet. This behaviour strongly complicates an interpretation and leads to even more controversial views. Teegarden and Baldini [21] have discussed the spectra of the simple cubic cesium halides in the same frame as that of the face-centered halides (Figs. 5 - 12).

In the case of the validity of the Wannier-Mott-model for the excitons in the alkali halides, we expect an additional weaker line structure associated with each of the exciton lines just discussed. As already shown by Hilsch and Fischer [17,18], who were the first to observe the higher members of an exciton series in the alkali halides, Teegarden and Baldini [21] have confirmed that the small shoulder on the high energy side of the first exciton peak in RbI consists of several overlapping weak lines. The experimental result is shown in Fig. 13. There is a clear structure in the step between the first two strong exciton lines which are indicated by arrows.

A band structure calculation of KI by Onodera et al., using a relativistic Green's function method, strongly supports the conclusions of Teegarden and Baldini taken from their measurements. The theoretical work indicates that a d-like conduction band exists just above the s-like conduction band at the Γ point (Fig. 14).

In the alkali halides the bottom of the conduction band is at $\vec{k} = 0$ with the orbitally nondegenerate Bloch function consisting mainly of the lowest unoccupied s state of the cation. The top of the valence band is also at $\vec{k} = 0$ with the triply degenerate Bloch function consisting of the outermost occupied p states of the anion, where the spin-orbit interaction leads to a splitting. The fundamental absorption spectra near the edge region of the alkali halides are of the allowed-edge type. Since in the case of an isotropic Wannier-Mott exciton the intensities of the discrete spectral lines are proportional to

$$\frac{1}{n^3} \qquad (n = 1,2,3,\ldots),$$

where n is the principal quantum number, it should be very difficult to detect the higher members of the modified Rydberg series. Furthermore, the discrete lines with large n overlap. They are unresolvable due to their finite widths. They create a quasi-continuum that connects smoothly to the true continuum of the interband absorption edge.

Shortcomings of the above discussed investigations result from the use of thin films during the measurements. Sydor [23] has studied the absorption spectra of RbI films of variable thickness with a photon-counting technique. The data show that the shpe of the critical step region following the first exciton peak is largely in-

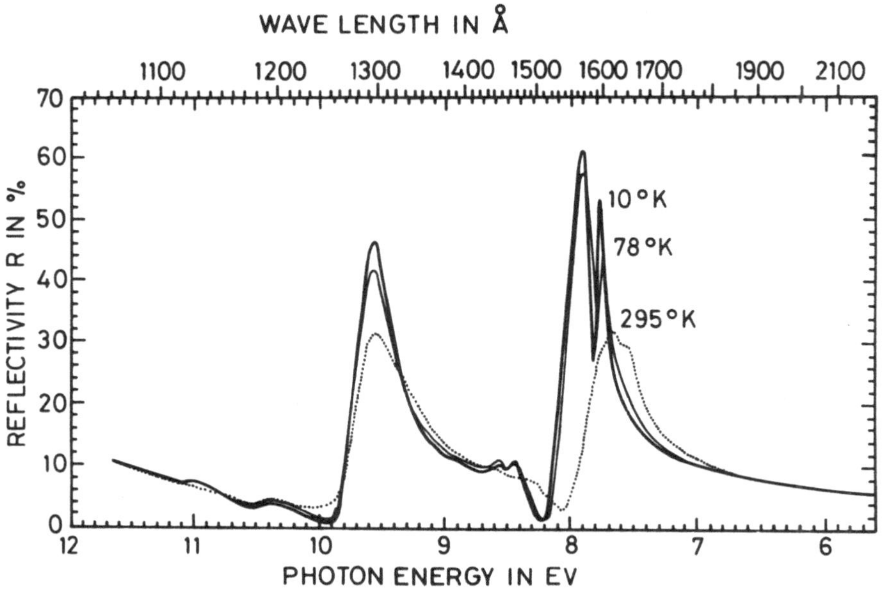

Fig. 15. Reflectivity spectra of KCl single crystals at temperatures T = 10, 78 and 295 K [25].

fluenced by surface states.

All these material specific difficulties have considerably delayed our insight into the properties of the excitons in alkali halides. Only very recent experimental investigations have led to a real breakthrough.

While the absorption measurements are restricted to thin films, investigations of the reflectivity can be made on thick crystalline samples. Due to this fact, the optical parameters obtained from reflectivity studies on single crystals may be more closely related to intrinsic effects and may be of higher quality.

The first very detailed investigation of the optical nature of the single crystalline state in the fundamental region of alkali halides was done by a Japanese research group [25 - 29].

Above all, the reflectivity spectra measured on alkali halide single crystals have supplied us with a clearer splitting of the halogen doublets as compared with the doublet spectra observed on thin film specimen. The spectral resolution is strongly enhanced in the case of single crystal spectroscopy.

The measured relectivity spectra [25] near the edge region of KCl single crystals observed at three different temperatures

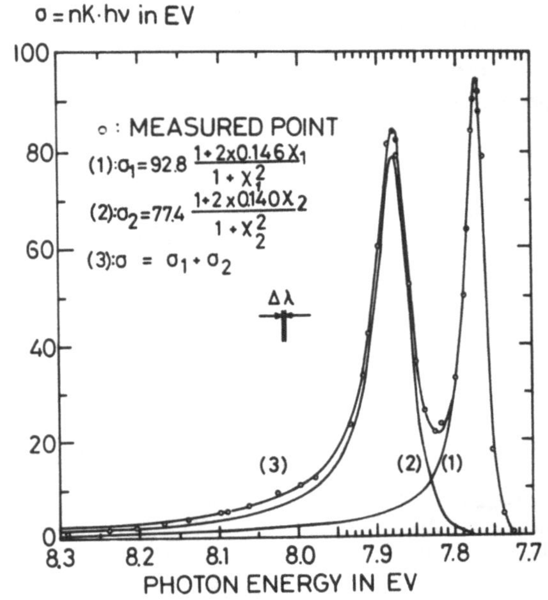

Fig. 16. Conductivity spectrum (3) of the first exciton band of KCl at 10 K. (1) and (2) are the two constituting asymmetric Lorentzian functions [25].

T = 10, 78 and 295 K are shown in Fig. 15. The spectra are, as already mentioned, of the allowed-edge type.

The first exciton band, made up of the chlorine doublet, indicates besides a remarkable narrowing of the lines a shift as a whole to higher energy with decreasing temperature of the sample. The step at the high energy side of the first exciton band also depends on temperature, while the peak position of the second prominent reflection band remains nearly constant. There is no possible splitting of this second exciton band, even at liquid helium temperature.

The author carried out a Kramers Kronig analysis of his reflectivity spectra. Fig. 16 describes the conductivity spectrum σ(E) in eV of the first exciton band of KCl at 10 K. It is possible to decompose the spectrum in two component functions of asymmetric Lorentzian type. Analytical expressions of those asymmetric Lorentzian functions σ(E) are given in a theory of Toyozawa [30] on exciton-phonon interaction. The conductivity σ(E) is related to the dielectric constant ε(E) as follows:

$$\varepsilon(E) = \varepsilon_1(E) - i\cdot\varepsilon_2(E)$$

$$\sigma(E) = \frac{\varepsilon_2(E) \cdot E}{2} = n \cdot K \cdot E \quad ,$$

where E is the photon energy in eV, n is the refractive index, and K the extinction coefficient.

Investigations of the conductivity $\sigma(E)$ of KCl single crystals as function of temperature for the range 10 K to 573 K in the intrinsic region up to 12 eV show that the $\sigma(E)$-spectra in the chlorine doublet region can be decomposed in the whole temperature range into the two components of asymmetric Lorentzian form with positive degree of asymmetry [27]. The conductivity spectra of n = 1 and n = 2 state exciton absorption bands with the temperature as parameter in the range 10 K to 573 K are shown in Fig. 17. The decomposition of the 295 K spectrum of KCl into two component bands of asymmetric Lorentzian shape is given in Fig. 18.

Figs. 19 and 20 demonstrate the conductivity spectra within the chlorine doublet region for NaCl at 10 K and 78 K, respectively [26].

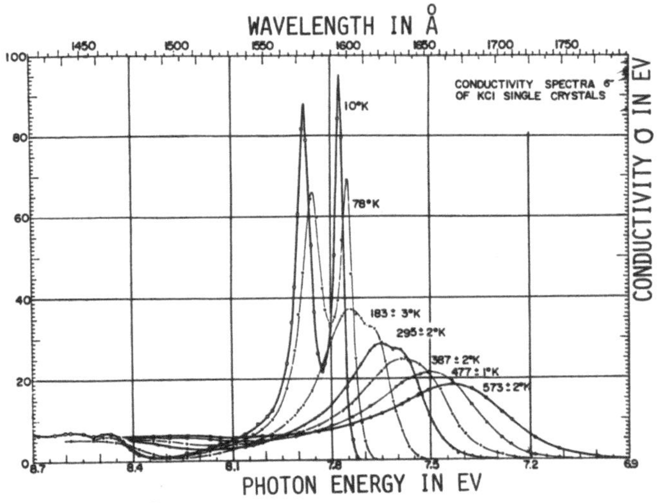

Fig. 17. Conductivity spectra of n = 1 and n = 2 state exciton absorption bands with the temperature as parameter in the range 10 K to 573 K [27].

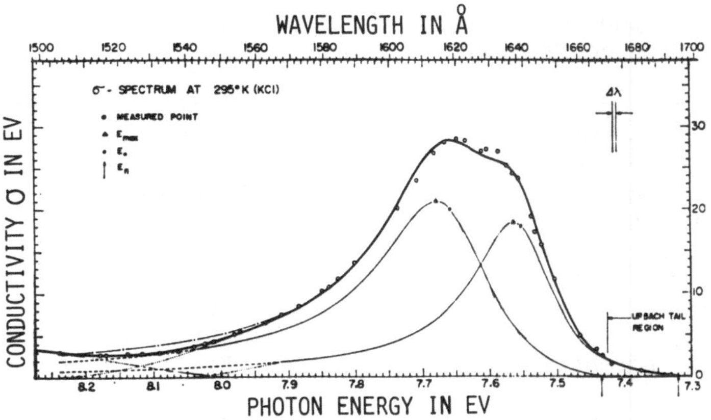

Fig. 18. Conductivity spectrum of KCl at 295 K decomposed into two component bands of asymmetric Lorentzian shape [27].

In the case of NaCl single crystals, too, a fit with a suitable combination of two asymmetric Lorentzian functions is excellent. The same holds for NaF. The reflectivity spectra of NaF single crystal at two temperatures and the decomposition of the 78 K conductivity spectrum in two component bands with asymmetric Lorentzian shapes are shown in Figs. 21 and 22, respectively [28].

The fluorine doublet in the reflectivity spectrum of NaF at 78 K exhibits a structure located at the higher energy side of the (Na $^2S_{1/2}$, F $^2P_{1/2}$)-peak. The separation nearly equals $2\hbar\omega_{LO}$, where $\hbar\omega_{LO}$ is the energy of the LO-phonon at the Γ-point. The side band results from a transition of a simultaneously created exciton and phonon. An investigation of the temperature dependence of the position of the first reflectivity peak ($^2S_{1/2}$, $^2P_{3/2}$) in the temperature range 10 K to 295 K clearly yields that the peak shift can be described within the frame of the Bose statistics [28].

This phonon side band effect of NaF single crystals can be very nicely observed in the spectra of NaBr and NaI crystals, here however on the Γ (3/2, 1/2)-exciton [29]. In Figs. 23 and 24 the temperature dependence of the reflectivity spectra in the Γ(3/2, 1/2)-exciton region of NaBr and NaI single crystals are represented. The shoulders on the high energy side of the main peaks develop into clear peaks with decreasing temperature of the samples. The

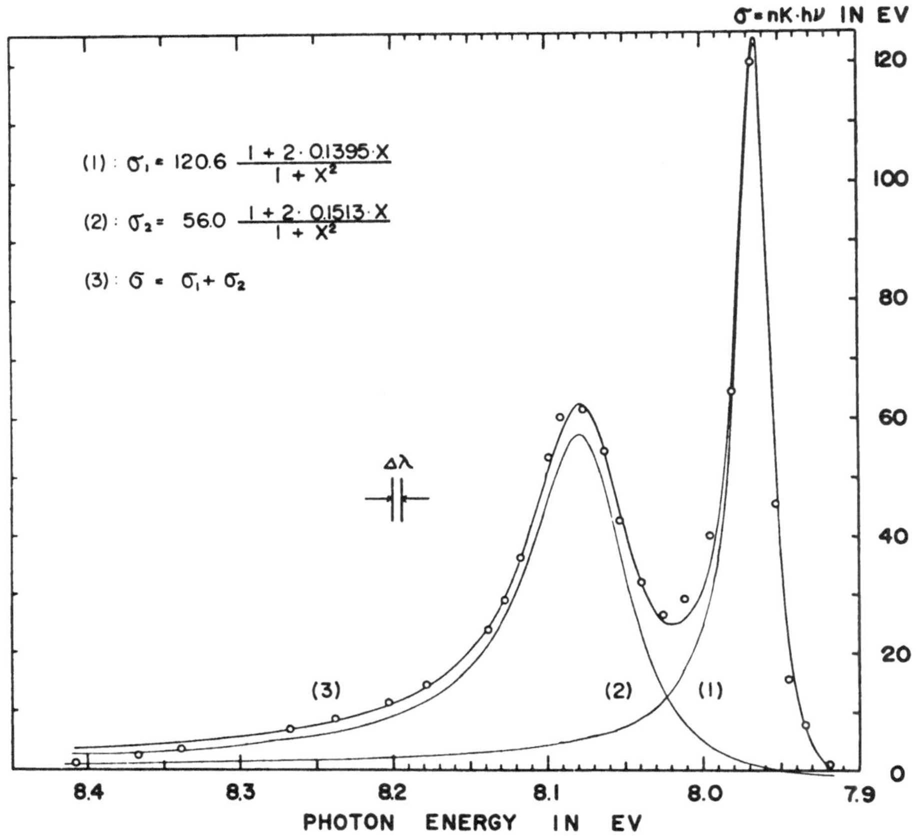

Fig. 19. Conductivity spectrum of the first exciton band of NaCl at 10 K built from two asymmetric Lorentzian functions [26].

energy difference is nearly equal to two LO-phonon energies. The phonon corresponds to the Γ point. NaBr shows at 27 K further LO-phonon structure (Fig. 23). The fine structure of NaI (Fig. 24) in the spectral region of the step seems to be due to an overlap of higher members of an effective mass state exciton series with a LO-phonon sideband related to the n = 2 exciton state.

The exciton-phonon interaction leads to an asymmetry of the optical absorption bands, has an influence of the band widths, and shifts the bands [30,31,32].

For the Γ-exciton in alkali halides the exciton energy band has its minimum at $\vec{K} = 0$. Excitons created in a direct transition

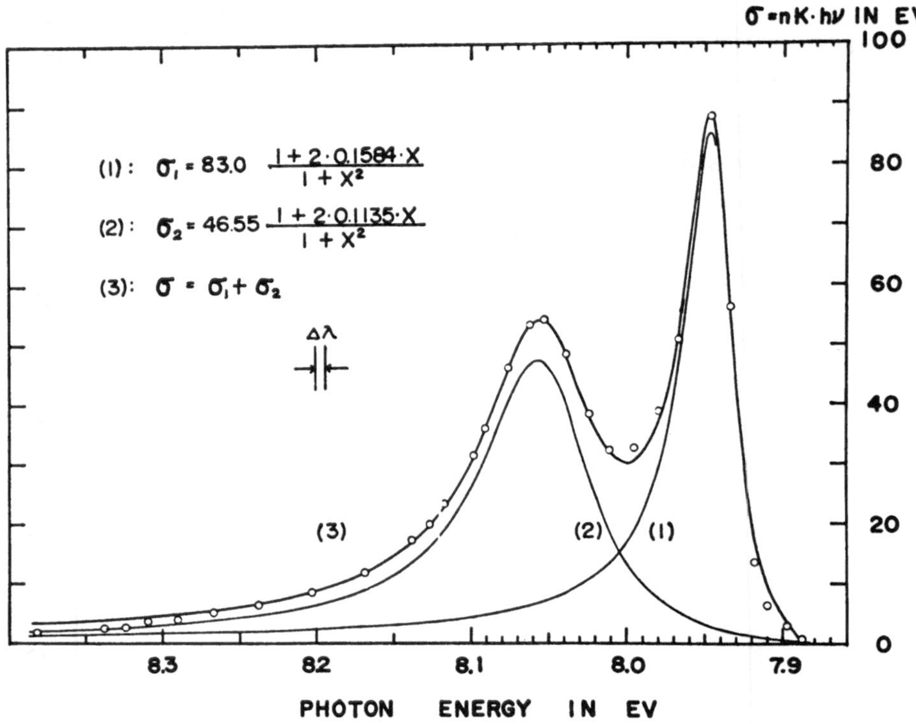

Fig. 20. Conductivity spectrum of the first exciton band of NaCl at 78 K decomposed in two asymmetric Lorentzian functions [26].

($\Delta\vec{k} = 0$) from the crystal ground state to the state $E_{\vec{k}=0,j}$ are characterized by a sharp absorption line with its peak at $E_{o,j}$. However, in the presence of a phonon field, the exciton produced by light excitation will be scattered to a state $E_{\vec{k},j'} = E_{o,j} \pm h\omega_{\vec{k},n}$ with the absorption or emission of a phonon. j denotes the exciton energy band and n the phonon mode. j = j' corresponds to intraband, $j \neq j'$ to interband scattering. In the case of intraband scattering within a distinct exciton band, the exciton will reach its final state by interaction with phonons of wave vectors $\vec{k}$ essentially limited to the range around $\vec{k} = 0$ due to energy momentum conservation. The resulting absorption band will be asymmetric. A band sharply truncated on the lower energy side of the peak has a tail on the higher energy side, since the minimum of the exciton band is at $\vec{k} = 0$. There is interband scattering, if the exciton excited in the state $E_{o,j}$ goes to the final state $E_{\vec{k},j'}$ with $j \neq j'$ and $E_{o,j} \sim E_{\vec{k},j'}$. This process enhances the absorption in the band tails on both sides of the peak. It leads to an increase in the half-width

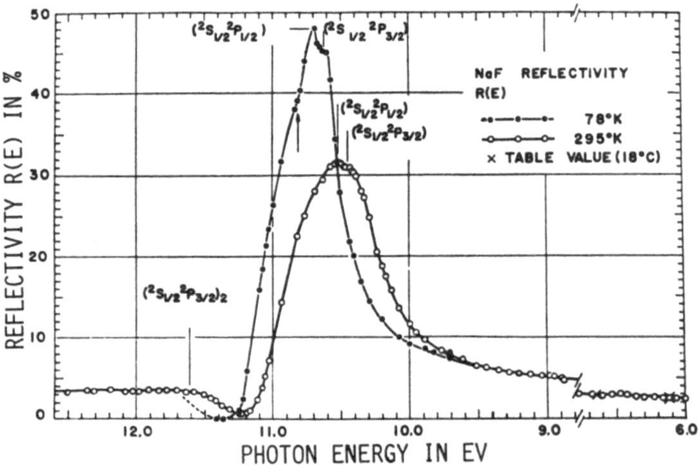

Fig. 21. Reflectivity spectra of a NaF single crystal at 78 K and 295 K [28].

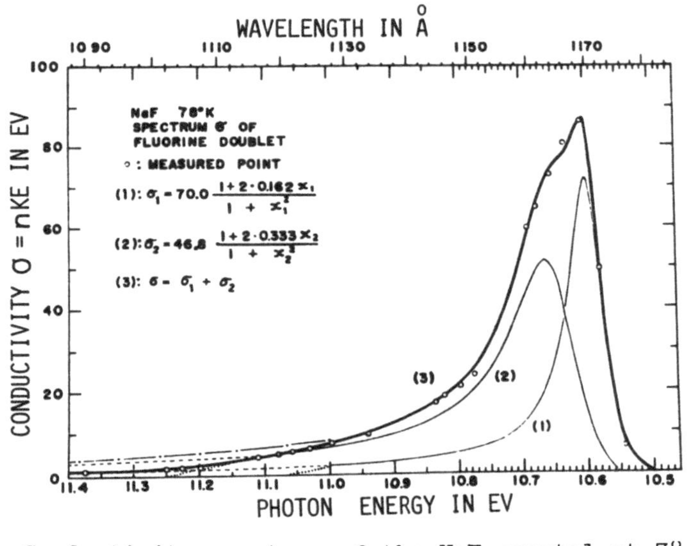

Fig. 22. Conductivity spectrum of the NaF crystal at 78 K and decomposition into two asymmetric Lorentzian lines [28].

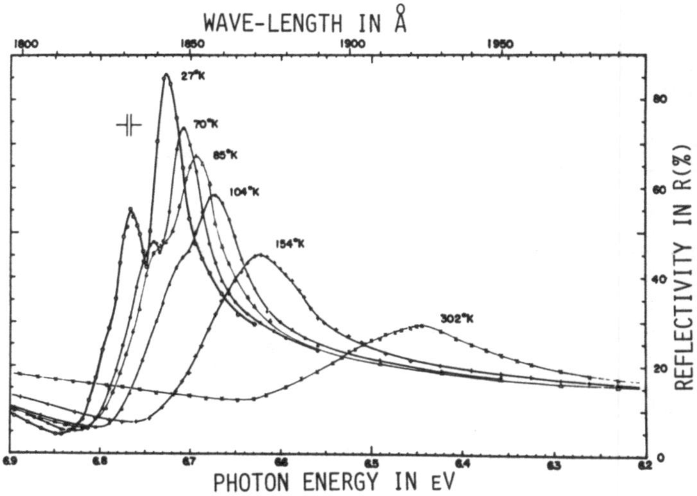

Fig. 23. Temperature dependence of the reflectivity spectra of NaBr in the spectral region of the Γ (3/2,1/2) exciton [29].

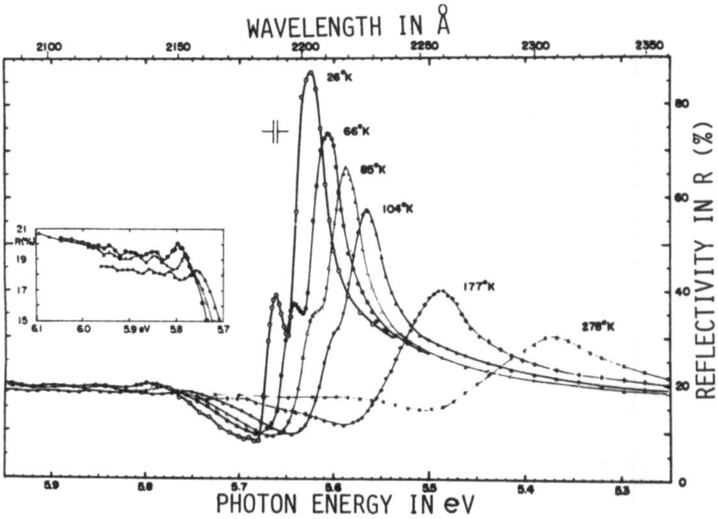

Fig. 24. Temperature dependence of the reflectivity spectra of NaI in the energy range of the Γ (3/2,1/2) exciton [29].

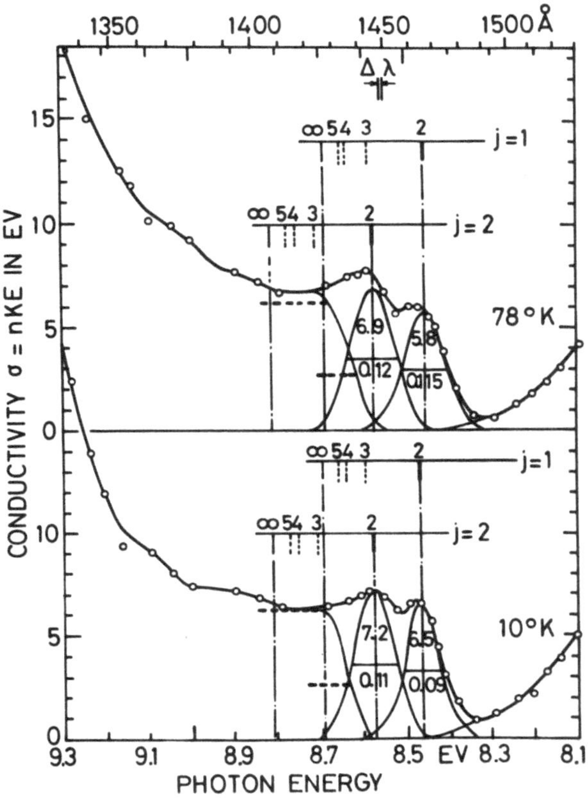

Fig. 25. Conductivity spectra of crystalline KCl at 78 K (upper curve) and 10 K (lower curve). The spectra are tentatively decomposed into 2s exciton bands, overlapping continuum and true-continuum. The two horizontal broken lines in each spectrum indicate the intensity level of the overlapping and true-continua for the two exciton series j = 1 and 2, respectively. j = 1 stands for Γ (1/2, 3/2) and j = 2 for Γ (1/2,1/2) [25].

of the band. The wave vectors of the acting phonons in interband scattering will not be limited to that near $\vec{k} = 0$ but to that with finite $\vec{k}$ values depending on the relative position of these exciton bands. Toyozawa [31] has shown that in the case of weak exciton-phonon interaction and an intraband scattering of excitons, only, the absorption band j jas an asymmetric line shape. For weak exciton-phonon interaction the absorption band j exhibits an asymmetric Lorentzian shape when the interband scattering is also allowed. Here, a part of the absorption strength of the band j will be transferred to the neighbouring band j'.

Thus, the asymmetric Lorentzian shape of the absorption bands

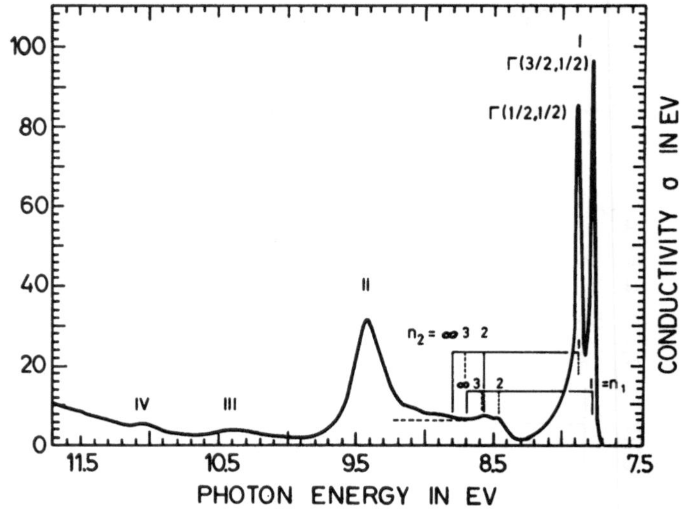

Fig. 26. Fundamental absorption spectrum of a KCl single crystal at 10 K [27].

in the range of the fundamental absorption edge of the alkali halide crystals proves that in this spectral region the optical behaviour is governed by several exciton energy bands.

A further very important experimental outcome obtained from the reflectivity measurements is the fact that the ratio of the absorption intensities of the component lines of the halogen doublet always differs from 2:1, a result that should be valid for jj-coupling of electron and hole as a reflection of multiplicity $2j_v + 1$. The spin-orbit split hole states are $j_v = 3/2$ and $j_v = 1/2$. The deviation of the intensity ratio of the spin-orbit split components Γ (1/2, 3/2) and Γ(1/2, 1/2) from 2:1 is a consequence of the exchange interaction [33]. The nonobservance of the 2:1 intensity ratio of the spin-orbit split components of the first halogen doublet in alkali halide crystals due to exchange interaction clearly demonstrates that the n = 1 excitons in these materials are beyond the effective mass approximation.

Now, let us look for possible higher members of exciton series in the reflectivity spectra of alkali halide crystals. For an identification of corresponding lines, we have to take a good look at the step behind the first doublet lines in the reflectivity spectra. As already shown in Fig. 15, the reflectivity spectra of the KCl single crystal at 10 K and 78 K present a fine structure in the spectral region of the step. A resolution of this substructure is possible with the aid of the conductivity spectra of Fig. 25.

The absorption near the step is resolved into two symmetric bands which are assigned to the n = 2 exciton of the j = 1 and j = 2 series, respectively. The attempts to represent the fine structure in the spectra of Fig. 25 by hydrogen-like series are indicated.

Figure 26 clearly shows the chlorine doublet lines followed by a weak doublet structure near the step. The association of the prominent lower energy doublet lines to the n = 1 state of a spin-orbit split exciton within a hydrogen-like series and of the weak higher energy doublet lines to the n = 2 exciton state is very reasonable. At the high energy side of the n = 2 exciton lines, the absorption spectrum consists of the unresolved lines. This overlap continuum finally terminates in the true continuum absorption [34].

Figures 27 and 28 yield that the n = 2 doublet lines near the absorption step are not confined to KCl crystals.

The asymmetric Lorentzian shape of the halogen doublet lines, the deviation of the intensity ratio of these two lines from the 2:1 value of the j-j coupling due to exchange interaction, and the hydrogen-like exciton behaviour in the fundamental absorption region strongly support the conclusion that the excitons in alkali halides belong to the intermediate coupling case. They have to be treated within a modified Wannier-Mott exciton model. The experi-

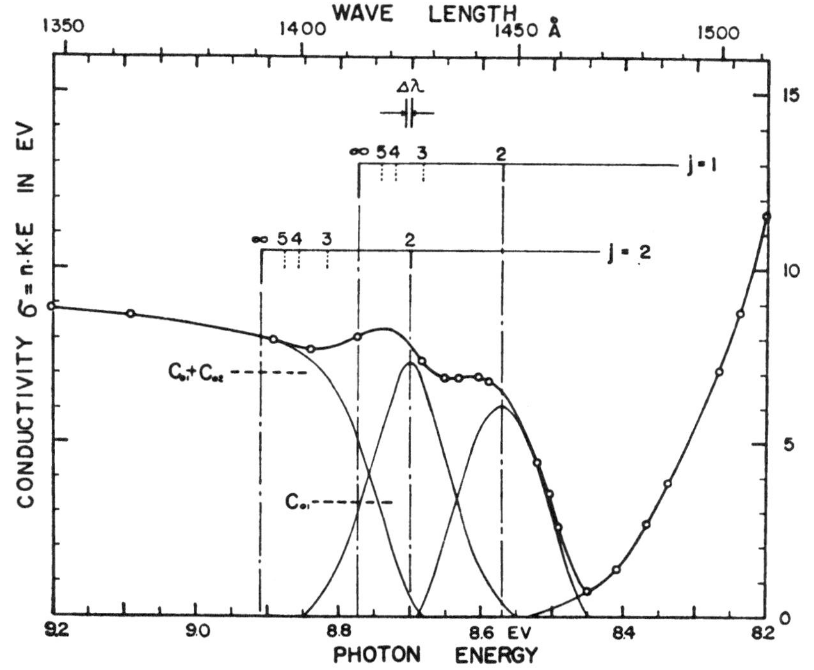

Fig. 27. Absorption spectrum of a NaCl single crystal at 10 K in the enrgy region of the step [26].

Table 4. Values characteristic for a hydrogenic approximation of the Γ-exciton in different alkali halides. $E_{n,j}$ = exciton energy; $E^{j}_{exc,n}$ = binding energy of the exciton; $r_{n,j}$ = exciton radius; T = temperature; Ref. = reference; j denotes the exciton series; n is a principal quantum number.

		NaBr		NaI	NaF	NaCl		KCl		KCl	
	n	j=1	j=2	j=1	j=1	j=1	j=2	j=1	j=2	j=1	j=2
$E_{n,j}$ (eV)	1	6.683	7.214	5.579		7.964	8.073	7.772	7.876	7.748	7.851
	2	7.020	7.62	5.790		8.570	8.700	8.465	8.575	8.455	8.575
	∞	7.132	7.756	5.860		8.774	8.91	8.696	8.808	8.691	8.816
$E^{j}_{exc,n}$ (eV)	1	0.449	0.542	0.281	1.5	0.81	0.837	0.924	0.932	0.943	0.965
	2					0.203	0.209	0.231	0.233	0.236	0.241
$r_{n,j}$ (Å)	1	5.77	4.8	8.31	2.62	3.51	3.40	3.508	3.477	3.44	3.36
	2					14.05	13.6	14.03	13.91	13.75	13.44
T(K)		85	85	85	25	10	10	10	10	78	78
Ref.		[29]	[29]	[29]	[28]	[26]	[26]	[25]	[25]	[25]	[25]

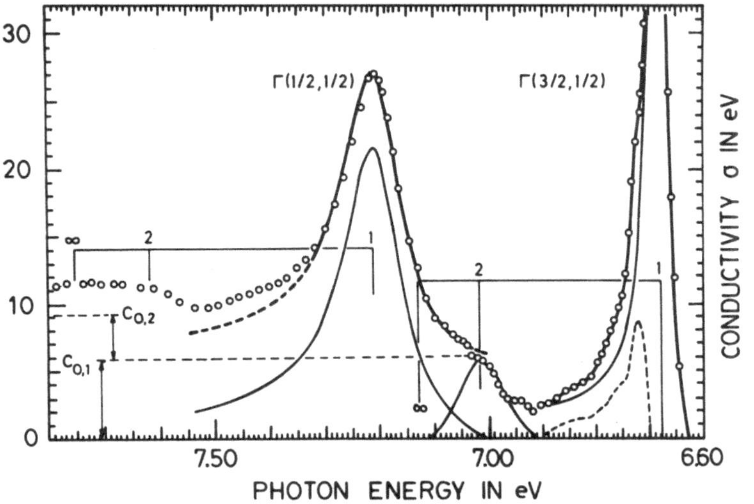

Fig. 28. Absorption spectrum of a NaBr single crystal at 85 K in the spectral region of the Γ (3/2,1/2) and Γ (1/2,1/2) excitons [29].

ments show that the simple hydrogenic model is only a crude approximation for the internal motion of excitons in alkali halides. In these substances the effective mass approximation is not appropriate to describe the n = 1 exciton lines since the n = 1 exciton radii become comparable with the interionic distances. Table 4 illuminates this problem.

III.B. Intrinsic Luminescence Measurements

Light emitted during the transition of an excited pure crystal to its ground state is called intrinsic luminescence. The excited electronic states of pure crystals may be either excitonic states or free electron and hole states. Free electron and hole pairs, which are not trapped elsewhere, may couple due to the long range Coulomb interaction to create an exciton, where the ensuing luminescence will have all the characteristics of direct exciton emission. This is indeed what one observes in the alkali halides. Thus, intrinsic luminescence is equivalent to excitonic luminescence.

At low temperatures, undoped alkali halides excited with ionizing radiation (electrons, X- and γ-rays) or UV-light of sufficient high energy show a typical luminescence [35,36]. Most alkali halides yield intense recombination luminescence at low temperatures. Figs. 29 and 30 indicate the X-ray luminescence spectra

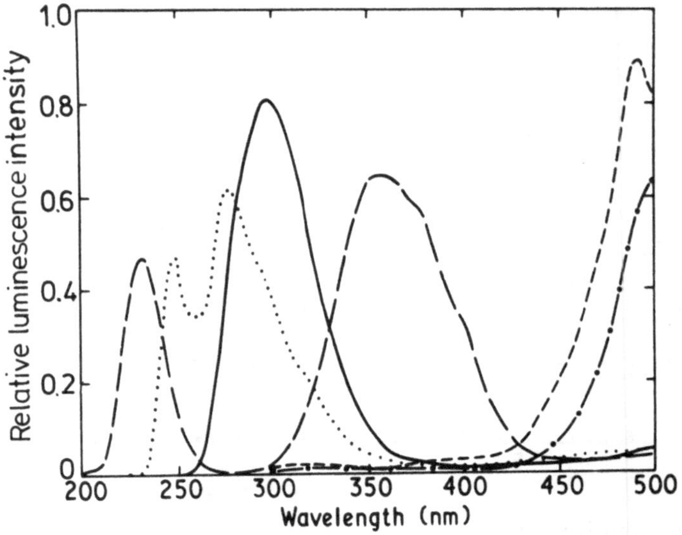

Fig. 29. X-ray luminescence spectra of chlorides at about 5 K. Full curve-LiCl; large broken curve-NaCl; small broken curve-KCl; chain curve-RbCl; dotted curve-CsCl [35].

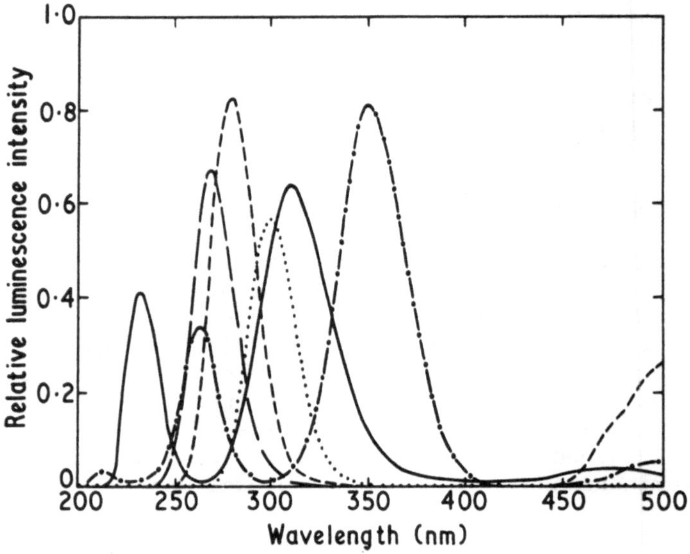

Fig. 30. X-ray luminescence spectra of bromides at about 5 K. Full curve-LiBr; large broken curve-NaBr; small broken curve-KBr; dotted curve-RbBr; chain curve-CsBr [35].

observed at about 5 K. The intensities given are measured relative to that of CsI at about 5 K which had the most intense band observed, with an efficiency of 1 photon/20 eV. The X-ray luminescence spectra of fluorides and iodides are analogous [35]. These broad emission bands with very large Stokes shifts have been known for a long time. Depending on the alkali halide, the broad luminescence may consist of one, two or even more emission bands. The broad luminescence is due to the recombination of the self-trapped excitons in alkali halides. In Table 5 values of peak energy and decay time of broad emission bands connected with the recombination of the self-trapped excitons in alkali halides at low temperature are summarized.

The self-trapped or relaxed exciton in alkali halides is a very thoroughly investigated and remarkably well documented entity. The electronic states of the relaxed exciton are of fundamental interest relating to intrinsic luminescence and also to damage processes. The simplest picture of the relaxed exciton is an electron trapped by a self-trapped hole. The self-trapped hole or V_K-centre can be considered as an X_2^- halogen molecular ion. The electron which it traps to form the exciton has Rydberg-like states split by the ax-

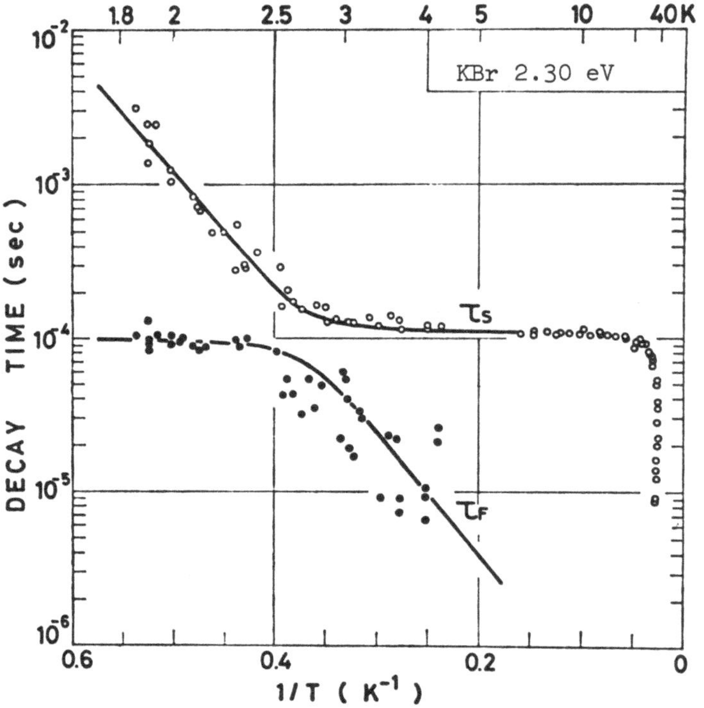

Fig. 31. Temperature dependence of the slow, τ_S, and the fast, τ_F, decay times of the 2.28 eV emission in KBr [48].

Table 5. Peak energy and decay time of luminescene bands of the self-trapped excitons in alkali halides at low temperature.

	High energy transition		Low energy transition		
	Peak energy (eV)	Decay time (10^{-9}s)	Peak energy (eV)	Decay time (10^{-6}s)	Ref.
LiF	5.80	-	3.53	-	[35]
NaCl	5.6	≤ 5	3.38	295	[35]
	5.47	-	3.47	310	[37,38]
KCl	-	-	2.32	5000	[35]
	-	-	2.54	5000	[37,38]
RbCl	-	-	2.27	5500	[35]
	-	-	2.41	8900	[37,38]
NaBr	-	-	4.60	0.49	[35]
	-	-	4.65	0.46	[37,38]
KBr	4.42	9	2.27	130	[35]
	4.40	-	2.44	100	[37,38]
RbBr	4.20	11	2.10	180	[35]
	4.13	-	2.36	150	[37,38]
NaI	-	-	4.20	0.1	[35]
	-	-	4.24	0.57	[39]
KI	4.15	≈ 9	3.34	4.4	[35]
	4.13	-	3.31	6	[37,38]
CsCl	5.07	-	4.52	-	[35]

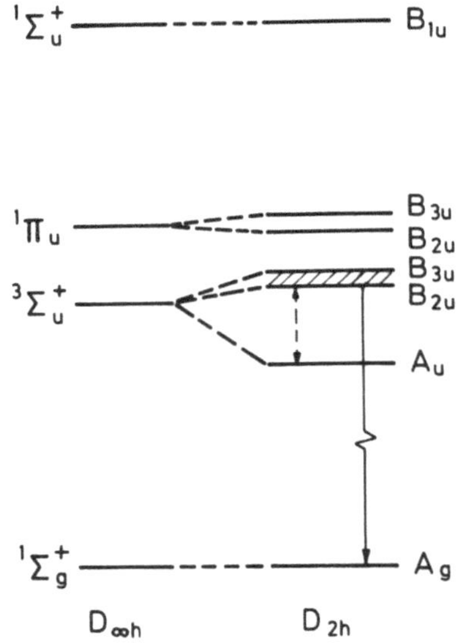

Fig. 32. Scheme of the lowest energy levels for the self-trapped exciton in the NaCl lattice.

ial symmetry of the X_2^--ion. After electron trapping the hole remains nearly unperturbed and the self-trapped configuration is conserved. The relaxed exciton is a $(V_K\text{-}e)$-centre.

The states of the relaxed exciton can be discussed in those of a diatomic rare gas molecule X_2^{2-} [40]. The unstable ground state has the electronic configuration

$$(\sigma_g np)^2\ (\pi_u np)^4\ (\pi_g np)^4\ (\sigma_u np)^2\ .$$

This is a $^1\Sigma_g^+$ state (closed shell configuration). The excited configuration of the lowest relaxed exciton states is

$$\ldots\ (\sigma_u np)\ (\sigma_g (n+1)s)\ .$$

The appropriate two states $^{1,3}\Sigma_u^+$ are bound states. The electronic configuration

$$(\sigma_g np)^2\ (\pi_u np)^3\ (\pi_g np)^4\ (\sigma_u np)^2\ (\sigma_g (n+1)s)$$

leads to the state $^1\Pi_u$.

Due to its long lifetime the triplet state of the relaxed exciton in alkali halides has been investigated in detail in recent years [41,42,43,44,45]. The results show that the hole is well localized and that the electron has a wave function very similar to that of the F-centre in its ground state. In the triplet state of the relaxed exciton, the electron and hole are substantially independent of each other. The optical transitions from the lowest triplet consist of hole excitations analogous to those of the V_K-centre and electron excitations analogous to those of the M-centre.

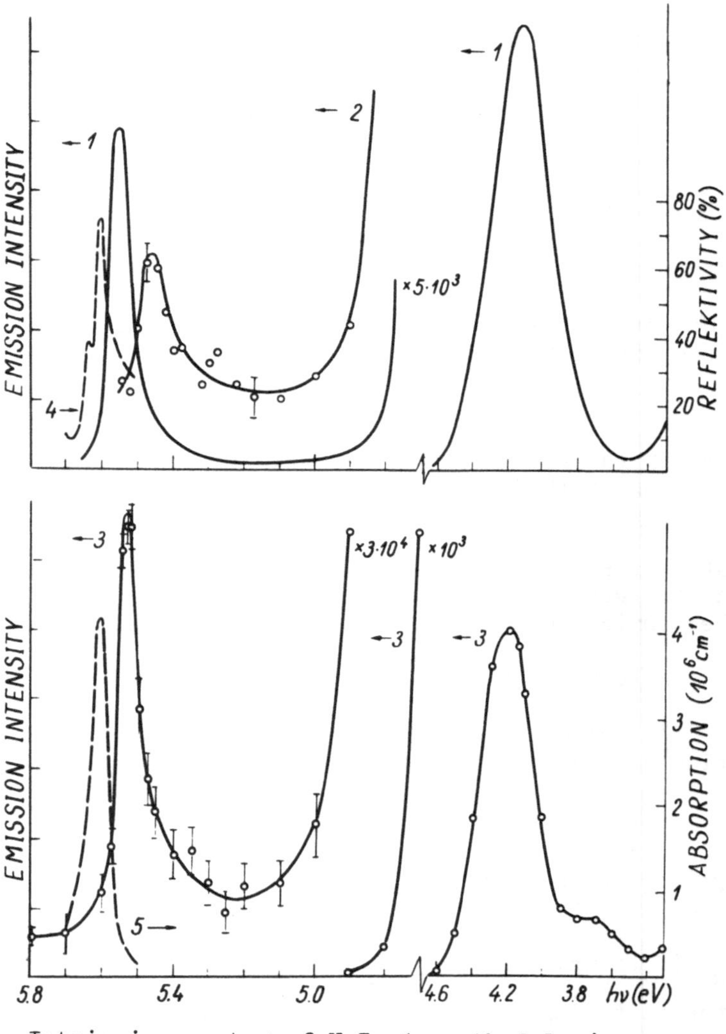

Fig. 33. Intrinsic spectra of NaI. 1: cathodoluminescence spectrum at 67 K; 2: photoluminescence spectrum at 80 K; 3: X-ray excited luminescence spectrum at 4.5 K; 4: reflectivity spectrum at 66 K; 5: absorption spectrum at 10 K [51].

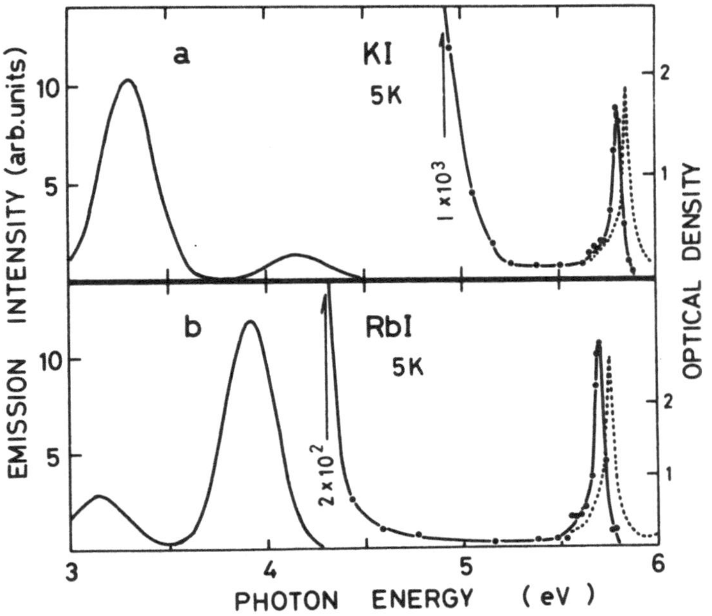

Fig. 34. Optical spectra of KI and RbI at 5 K. Solid curves: luminescence spectra under X-ray excitation; dashed curves: absorption spectra of thin films [52].

As shown in Table 5, in most cases the intrinsic luminescence due to the self-trapped exciton in alkali halides consists of two emission bands, a high energy and a low energy band. While the light emitted in the high energy band is σ-polarized, the emission in the low energy band is π-polarized. The decay times of the two bands are also different. The π-band, observed in all alkali halides, has a long lifetime of about 10^{-6} s which decreases with increasing atomic number of the anions. It is attributed to the spin-forbidden transition

$$^3\Sigma_u^+ \rightarrow {}^1\Sigma_g^+$$

made partially allowed by halogen spin-orbit coupling of the $^3\Sigma_u^+$ and $^1\Pi_u$ excited states. The σ-polarized high energy band must be due to a singlet-singlet transition. The assignment of the σ-band to a transition from the singlet $^1\Sigma_u^+$, corresponding to the above triplet $^3\Sigma_u^+$, to the ground state $^1\Sigma_g^+$ is still controversial.

The triplet-singlet transition of the relaxed exciton at low temperature is not a simple process. The decay time of the resulting luminescence band decreases exponentially with increasing

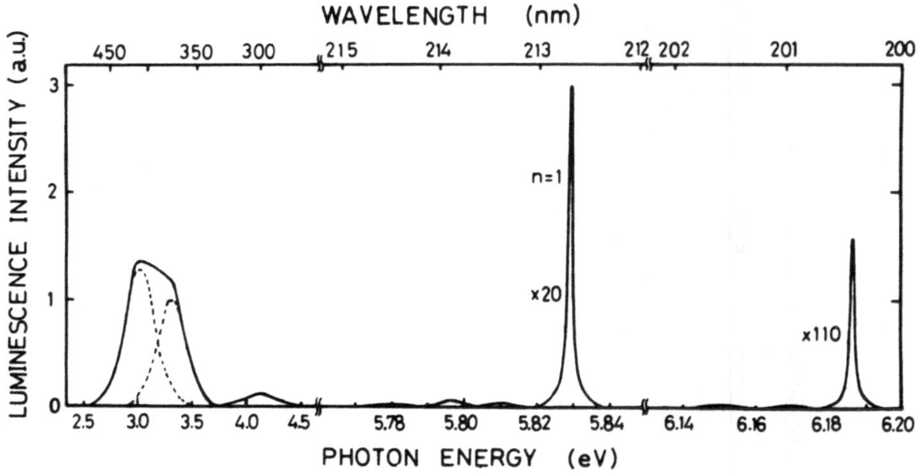

Fig. 35. Luminescence spectrum of KI under 6.186 eV excitation at 5 K [54].

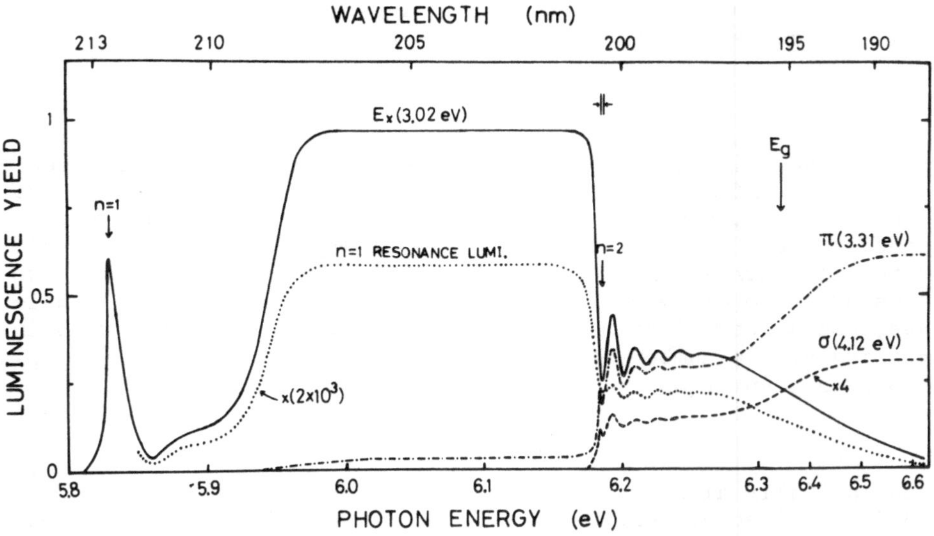

Fig. 36. Excitation spectra of KI at 5 K. Dotted curve: n = 1 resonance luminescence; broken curve: σ-emission band; chained curve: π-emission band; solid curve: E_X-luminescence band. The arrows n = 1 and n = 2 indicate the peaks of the n = 1 and n = 2 resonance luminescence lines, respectively [54].

temperature. This effect is probably due to a non-radiative transition between the triplet state and a singlet state lying slightly above it, for KCl approximately 15 meV. The singlet state is populated only by thermal excitation from the triplet [46]. Detailed analysis of the relaxed exciton decay yields two components at low temperatures [47,48]. In KBr, after excitation by a pulsed N_2-laser, the π-emission at 2.28 eV exhibits at 1.8 K two decay components:

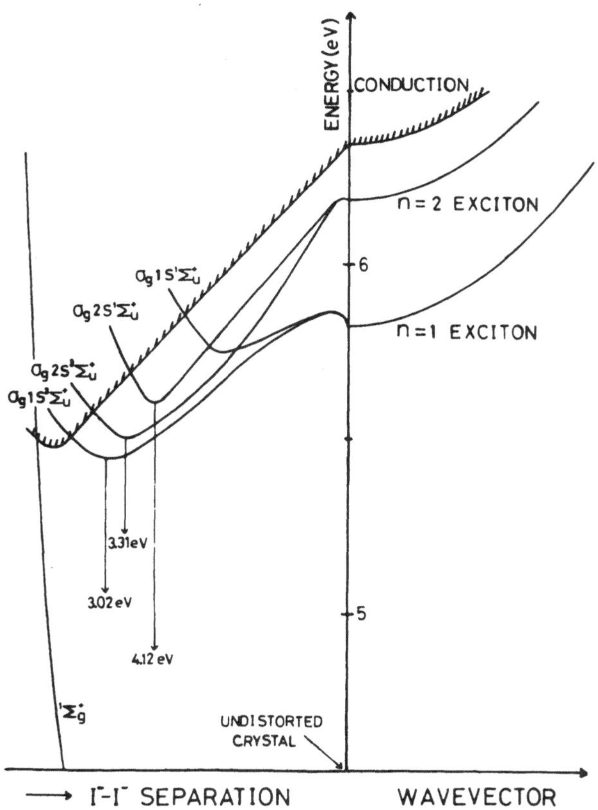

Fig. 37. Schematic energy diagram for the n = 1 and n = 2 free exciton states (on the right) and relaxed exciton states (on the left) [54].

a fast (100 μs) and a slow (3 ms) one [48]. The slow decay time of about ems at 1.8 K decreases to about 100 μs at 3 K and stayed at this value up to ~35 K, thereafter, it sharply decreases. The temperature dependence of the fast decay component shows normal behaviour (Fig. 31). KI indicates analogous properties. To understand these features of the π-emission of KI and KBr, a two-state model has been assumed. The self-trapped exciton due to its diatomic configuration X_2^{-*} has the symmetry $D_{\infty h}$, while in the NaCl lattice the actual site symmetry is D_{2h}. Fig. 32 shows schematically the ground and excited states of the relaxed exciton in $D_{\infty h}$ and D_{2h} symmetry.

In the crystalline environment spin-orbit and exchange interaction remove all degeneracy. Figure 32 demonstrates the corresponding levels. The $^3\Sigma_u^+$ state is split into the three levels B_{3u}, B_{2u}, and A_u. There is a mixing of the B_{3u} and B_{2u} states of the diatomic $^3\Sigma_u^+$ configuration with the B_{3u} and B_{2u} states of the molecular $^1\Pi_u$ level.

The π-emission with the fast decay time results from the transition (B_{2u}, B_{3u}) to the ground state A_g. The transition from A_u to A_g is strictly forbidden. The only way for an electron transition from the A_u state to the A_g ground state is via a thermal activation of the electron from A_u to (B_{2u}, B_{3u}). This mechanism leads to the component with the slow decay time.

All experimental and theoretical results clearly indicate that the broad intrinsic luminescence bands of the alkali halides are due to the transitions between states of the self-trapped exciton, an entity that may be described as a hole strongly localized on two neighbouring halogen ions in a <110> direction forming an X_2^- molecular ion with an electron loosely bound to it.

The belief that the intrinsic luminescence of the alkali halides is only governed by the Stokes-shifted broad emission bands from relaxed exciton states has been disturbed in the last years [49,50,51,52]. Figure 33 shows emission spectra of a NaI crystal observed at low temperatures. Besides the well-known intense and broad band at about 4.2 eV, a sharp and weak peak very close to the absorption edge is present. The authors suppose that the narrow band emission of NaI in the spectral region of 5.55 eV with a 0.07 eV halfwidth and a Stokes shift less than 0.06 eV is due to the radiative annihilation of the single halogen exciton of X^0-e type [51].

Emission spectra of KI and RbI at 5 K are shown by solid curves in Figs. 34 (a) and (b). The dashed curves are absorption spectra of the corresponding thin films for reference. Here also, additionally to the broad bands of the self-trapped exciton sharp but very weak emission is found close to the fundamental absorption edge of each crystal. The narrow band "edge emission" in KI and RbI is assumed to originate from a free exciton state [52].

Recently, Nishimura et al. [53] have observed additionally to the zero phonon resonance line of the free exciton emission from the lowest exciton energy band several very weak LO phonon assisted luminescence lines in KI and RbI crystals. The zero phonon resonance line and the LO phonon side bands within the free exciton luminescence have been investigated under UV light excitation in the first exciton absorption band in the temperature range between 3 and 80 K. The most interesting results have been achieved in the course of a study concerning luminescence lines due to the n = 2 free exciton state of the modified Wannier-Mott series in KI crystals [54]. The emission spectra of KI were measured in the temperature range between 3 and 40 K. The energy of the exciting light was 6.18 ~ 6.22 eV. Figure 35 shows the emission spectrum under 6.186 eV excitation at 5 K. Besides the luminescence bands due to self-trapped excitons (σ(4.12 eV), π(3.31 eV), and E_x(3.02 eV) bands) and the n = 1 free exciton lines, three further weak luminescence lines at 6.186, 6.169, and 6.151 eV are obtained. The energy splitting of these three lines is nearly equal to the LO phonon energy (18 meV) in KI. The excitation spectrum of the 6.186 eV luminescence line has its maximum at about 6.186 eV. The excitation spectrum of the n = 1 resonance luminescence at 5.829 eV and those of the σ(4.12 eV), π(3.31 eV), and E_x(3.02 eV) emission bands from self-trapped excitons are presented in Fig. 36.

The excitation spectra show a well-defined structure of four or five dips in the spectral region at 6.2 eV. Since an excitation spectrum has minima at reflectivity peaks near strong exciton absorption lines [55], a comparison with the reflectivity spectrum of KI yields that the minimum near 5.87 eV in Fig. 36 corresponds to the n = 1 exciton, while the dips near 6.2 eV belong to the n = 2 exciton peaks. As described in Fig. 36, the broad σ(4.12 eV) and π(3.31 eV) emission bands are excited above 6.186 eV, the peak energy of the zero phonon absorption line due to the n = 2 exciton. Thus, the above mentioned σ and π luminescence bands must be correlated with the n = 2 exciton state. These bands are due to transitions between self-trapped n = 2 exciton states. The excitation spectra of the E_x(3.02 eV) emission band and the n = 1 resonance luminescence line show the same behaviour (Fig. 36). Therefore, we have to conclude that only the E_x luminescence band is connected with the n = 1 free exciton state. The different relaxation processes of the optically created n = 2 free excitons can be understood within a schematic energy diagram for the n = 2 and n = 1 free exciton states and their corresponding self-trapped exciton states (Fig. 37). In Fig. 37 energy barriers between the free and self-trapped exciton states are indicated. From measurements of the quantum yield as a function of temperature the authors receive for the height of the energy barriers related to the n = 2 and n = 1 exciton states 3.0 meV and 33 meV, respectively. There is a strong decrease of the barrier energy with increasing principal quantum number n of the exciton state.

IV. EXCITON IN RARE GAS SOLIDS

Lectures on the subject of this paragraph were already given by G. Zimmerer and N. Schwentner during the course entitled "Luminescence of Inorganic Solids", held at Erice, 1977. This course was an activity of the International School of Atomic and Molecular Spectroscopy of the "Ettore Majorana" Centre for Scientific Culture [56]. For that reason, we will present the exciton behaviour of rare gas solids only in a short review.

The absorption spectrum of crystalline Xe in the vacuum-ultraviolet region provides us with detailed informations about excitons in rare gas solids. Exciton effects in rare gas solids were at first observed in Xe. As shown in Fig. 38, the fundamental absorption edge of solid Xe is characterized by two strong narrow peaks near 8.4 and 9.5 eV. The peak energies of the $\Gamma(3/2)$ and $\Gamma(1/2)$ lines are nearly equal to the energies of the transitions $^1S_0 \rightarrow {}^3P_1$ (8.44 eV) and $^1S_0 \rightarrow {}^1P_1$ (9.57 eV) in a free Xe atom. In the solid phase, the width of the absorption bands is about 0.15 eV. The close correspondence of the transition energies for the gaseous and solid phases strongly supports the idea that to a good approximation the optical spectra of Van der Waals crystals should reproduce those of the corresponding atoms. Hence, the n = 1 $\Gamma(3/2)$ and $\Gamma(1/2)$ excitons, considered separately, should be appropriate applicants for the Frenkel exciton

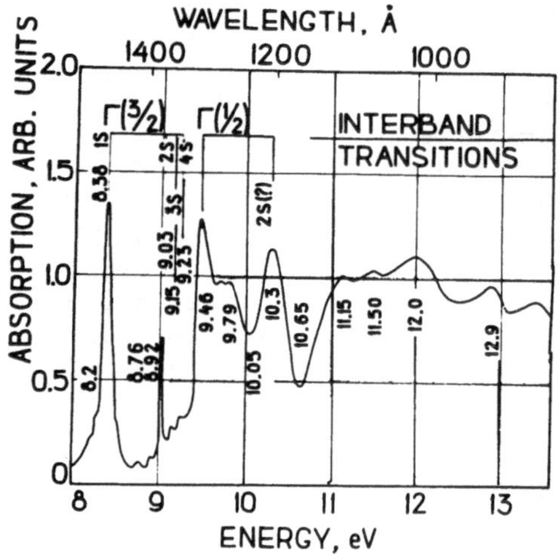

Fig. 38. Absorption spectrum of solid xenon [57].

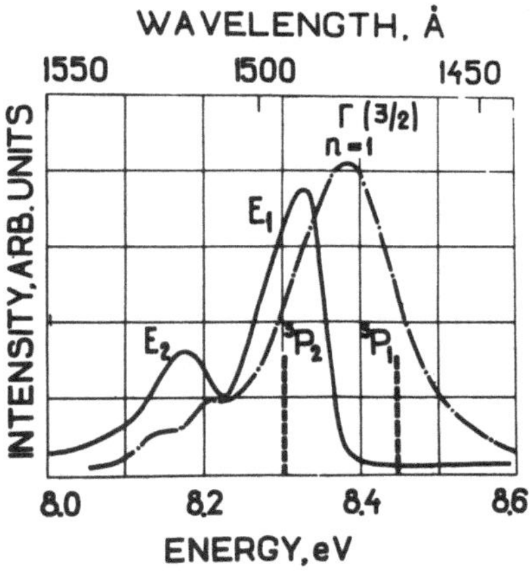

Fig. 39. Luminescence (solid curve) and absorption (chained curve) spectra of xenon crystals. $^3P_1$ and $^3P_2$ are resonant atomic lines [60].

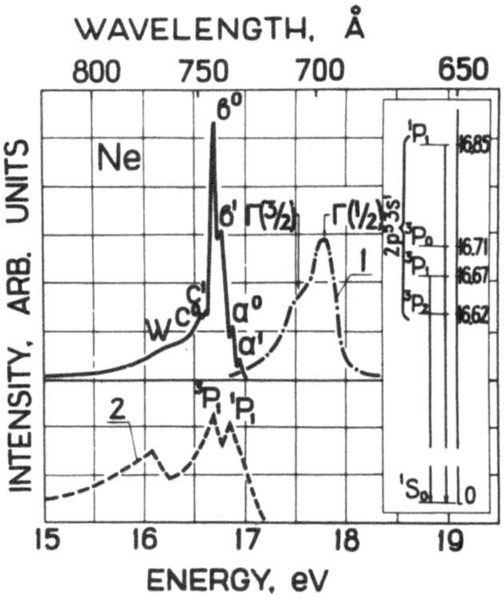

Fig. 40. Luminescence spectrum of solid neon at 5 K [61]. 1: absorption spectrum of solid neon [62]; 2: emission spectrum of gaseous neon under a pressure of 500 Torr [63]. The lowest atomic levels of neon are shown on the right [60].

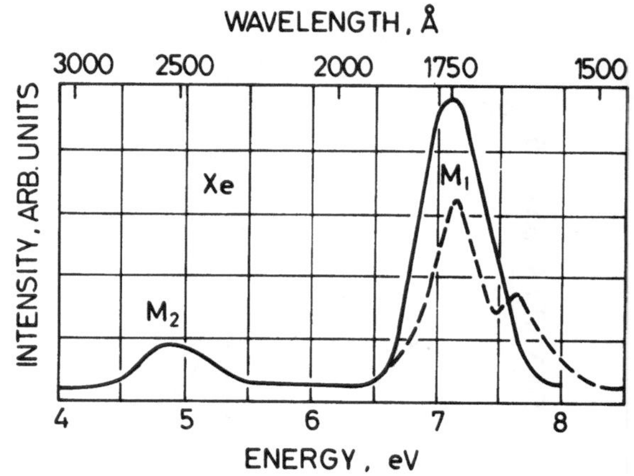

Fig. 41. Luminescence spectra of self-trapped excitons in xenon crystals at 5 K (solid curve) and at 60 K (broken curve) [64].

model. However, a view on Fig. 38 sends the Frenkel picture into the wilderness. On the higher energy side both n = 1 exciton lines are accompanied by a group of sharp bands which are not correlated with higher excited states of the free Xe atom but form hydrogen-like series, respectively [57,58,59]. The excitons in solid Xe are modified Wannier-Mott excitons corresponding to the Γ-point in the Brillouin zone ($\vec{k} = 0$). The two series Γ(3/2) and Γ(1/2) result from the spin-orbit split valence band. The more pronounced Γ(3/2) series in xenon shows exciton lines up to n = 4. The ionization limit of the Γ(3/2) series determines the minimum gap value E_g in the band structure and the exciton binding energy E_{exc}. The exciton (Γ(3/2) series) parameters for solid xenon are:

$$E_g = 9.3 \text{ eV}$$

$$E_{exc} = 1.0 \text{ eV}$$

$$r_1 = 3.2 \text{ Å}$$

r_1 is the exciton radius for n = 1. The absorption spectra of crystalline rare gases clearly indicate that the exciton behaviour of these substances can be discussed within the intermediate coupling case.

The intrinsic luminescence of rare gas crystals shows a remarkable diversity. Free exciton luminescence (E_1 band) in the VUV region from a Xe crystal is shown in Fig. 39. The E_1 emission band overlaps with the n = 1 absorption band of the Γ(3/2) exciton series.

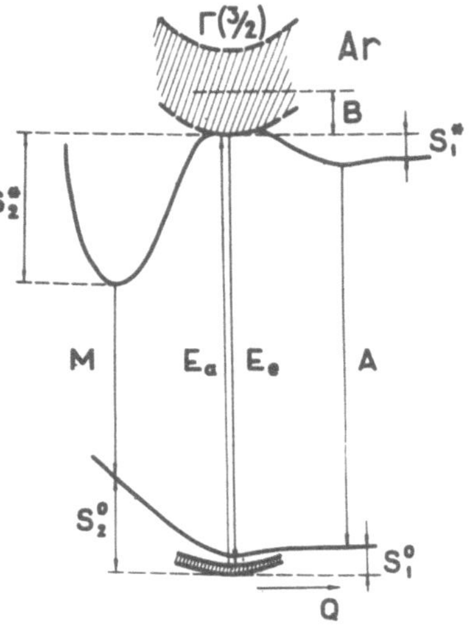

Fig. 42. Transition scheme for excitons in argon crystals. The hatched area corresponds to Γ (3/2) free excitons [60].

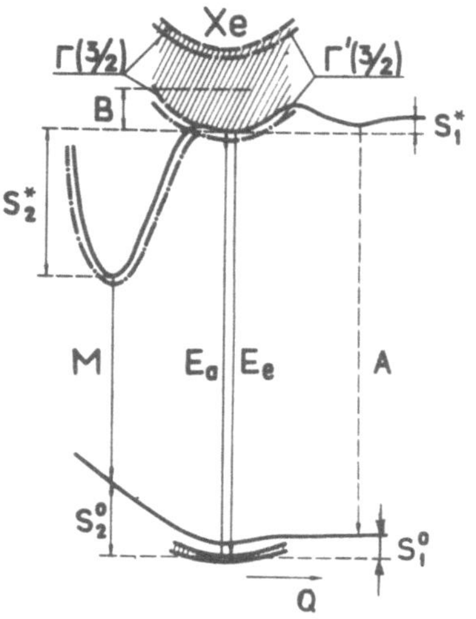

Fig. 43. Transition scheme for excitons in xenon crystals [60].

The maxima of the two bands are separated by 0.05 eV. The asymmetric shape of the E_1 band points to a partial re-absorption of the luminescence light by the n = 1, $\Gamma(3/2)$ exciton absorption band. A correction for the re-absorption would remove the Stokes shift. Therefore, it can be assumed that the E_1 band is due to the radiative decay of the n = 1 free excitons of the $\Gamma(3/2)$ series [60]. The E_2 band may be surface-exciton luminescence.

The luminescence spectrum of crystalline neon at 5 K is shown in Fig. 40 (solid curve). This fluorescence is called quasi-atomic luminescence. The name results from the experimental fact that this emission occurs near the resonance lines of the gaseous phase (see Fig. 40). Compared with the exciton absorption band, the emission peak is shifted towards lower energies. The Stokes shift is about 0.9 eV. The luminescence spectrum consists mainly of three doublets a, b, and c. These lines can easily be associated with the $^1P_1 \rightarrow {}^1S_0$, $^3P_1 \rightarrow {}^1S_0$, and $^3P_2 \rightarrow {}^1S_0$ transitions of the free neon atom. The singlet transition corresponds to the a bands, the other two to the b and c bands. The energy difference between the lines of a doublet is of the order of 10^{-2} eV. With respect to the experimental results obtained from very accurate measurements, it is reasonable to state that the luminescence bands a, b, and c of crystalline neon observed in the spectral range of atomic transitions are due to the radiative decay of collective excitations localized at atomic-like centres.

The extension of a rare gas atom in the excited state is considerable larger than that in the ground state. As a consequence, the electron wave functions of the excited atom in the undistorted lattice should overlap with those of the nearest-neighbour atoms. If the interaction between the normal and excited atoms is governed by the short-range repulsion forces, one should expect that a lattice distortion occurs in such a way as to reduce the overlap in the wave functions, i.e. to push aside neighbouring atoms. Thus, a deformation of this kind creates a microcavity around the localized excitation. As a result, the localized quasi-atomic excitations interact only slightly with the surrounding atoms and the transition energies in the crystal and the gas are nearly the same. The doublet structure of the a, b, and c bands is a crystal field effect. The splitting is produced by a non-spherical symmetry of the cavity surrounding the excitation [60].

Besides the free exciton luminescence, xenon crystals exhibit several broad and structureless emission bands. The most intense band is denoted M_1 in Fig. 41. The M_1 band has a slightly asymmetric shape with a tail extending in the short-wave region. Jortner et al. [65] have explained the M_1 luminescence band of solid xenon as due to the radiative decay of the Xe_2^* excimer. The excimers are molecular-like self-trapped excitons in the lattice of rare gas crystals. The M_1 luminescence band consists of two components with decay times of 3 ns and 900 ns, respectively [66].

A schematic transition diagram for excitons in crystalline argon is shown in Fig. 42. The figure illustrates the co-existence of free excitons and both, single (quasi-atomic) and double-centered localized excitons. The scheme of transitions from the lowest states of xenon is described in Fig. 43. The luminescence of xenon crystals is characterized by the simultaneous existence of free and molecular-like self-trapped excitons, only. The absence of the quasi-atomic fluorescence may be due to a higher electronic polarization in xenon.

V. CONCLUDING REMARKS

The intrinsic optical behaviour of wide gap insulators in the range of the fundamental absorption edge is controlled by modified Wannier-Mott excitons. While the alkali halides only show free and relaxed mulecular-like exciton emission, in rare gas crystals luminescence due to free, single and double centered localized excitons is observed. The simultaneous existence of free and self-trapped excitons in these solids requires an energy barrier for self-trapping.

REFERENCES

1. P.J. Dean and D.C. Herbert in Excitons, K. Cho, ed., Topics in Current Physics, Vol. 14, Springer-Verlag, Berlin-Heidelberg-New York, 1979.
2. R. Grasser, E. Pitt, A. Scharmann, and G. Zimmerer, phys. stat. sol. (b) 69, 359 (1975).
3. J. Hermanson, Phys. Rev. 150, 660 (1966).
4. J. Hermanson, Phys. Rev. Lett. 18, 170 (1967).
5. J. Hermanson, Phys. Rev. 166, 893 (1968).
6. U. Rössler and O. Schütz, phys. stat. sol. (b) 56, 483 (1973).
7. M. Altarelli and F. Bassani, Proc. 11th Internat. Conf. Phys. Semicond., Warsaw, 196 (1972).
8. U. Rössler, phys. stat. sol. 42, 345 (1970).
9. J. Hermanson and J.C. Phillips, Phys. Rev. 150, 652 (1966).
10. A.C. Sinnock and B.L. Smith, Phys. Rev. 181, 1297 (1969).
11. S.T. Pantelides, Phys. Rev. B11, 5082 (1975).
12. R.T. Poole, J.G. Jenkin, J. Liesegang, and R.C.G. Leckey, Phys. Rev. B11, 5179 (1975).
13. R. Hilsch and R.W. Pohl, Z. Physik 28, 384 (1928).
14. R. Hilsch and R.W. Pohl, Z. Physik 29, 812 (1929).
15. J.E. Eby, K.J. Teegarden, and D.B. Dutton, Phys. Rev. 116, 1099 (1959).
16. R.S. Knox and N. Inchauspe, Phys. Rev. 116, 1043 (1959).
17. F. Fischer and R. Hilsch, Nachr. Akad. Wiss. Göttingen IIa, Nr. 8 (1959).
18. F. Fischer, Z. Physik 160, 194 (1960).

19. R. Huggett and K. Teegarden, Phys. Rev. 141, 797 (1966).
20. G. Baldini and K. Teegarden, J. Phys. Chem. Solids 27, 943 (1966).
21. K. Teegarden and G. Baldini, Phys. Rev. 155, 896 (1967).
22. J.C. Phillips, Phys. Rev. 136, A1705 (1964).
23. M. Sydor, Phys. Rev. Lett. 27, 1286 (1971).
24. Y. Onodera, M. Okazaki, and T. Inui, J. Phys. Soc. Japan 21, 816 (1966).
25. T. Tomiki, J. Phys. Soc. Japan 22, 463 (1967).
26. T. Miyata and T. Tomiki, J. Phys. Soc. Japan 24, 1286 (1968).
27. T. Tomiki, J. Phys. Soc. Japan 26, 738 (1969).
28. R. Sano, J. Phys. Soc. Japan 27, 695 (1969).
29. T. Miyata, J. Phys. Soc. Japan 31, 529 (1971).
30. Y. Toyozawa, Progr. Theor. Phys. 20, 53 (1958).
31. Y. Toyozawa, Progr. Theor. Phys. 27, 89 (1962).
32. Y. Toyozawa, J. Phys. Chem. Solids 25, 59 (1964).
33. Y. Onodera and Y. Toyozawa, J. Phys. Soc. Japan 22, 833 (1967).
34. R.J. Elliot, Phys. Rev. 108, 1384 (1957).
35. D. Pooley and W.A. Runciman, J. Phys. C: Solid St. Phys. 3, 1815 (1970).
36. M.N. Kabler in Radiation Damage Processes in Materials, C.H.S. Dubuy, ed., Noordhoff International Publishing, Leyden, 1975.
37. M.N. Kabler and D.A. Patterson, Phys. Rev. 136A, 1296 (1964).
38. R.B. Murray and P.J. Keller, Phys. Rev. 137A, 942 (1965).
39. M.P. Fontana, H. Blume, and W.J. Van Sciever, phys. stat. sol. 29, 159 (1968).
40. M.N. Kabler and D.A. Patterson, Phys. Rev. Lett. 19, 652 (1967).
41. R.G. Fuller, R.T. Williams, and M.N. Kabler, Phys. Rev. Lett. 25, 446 (1970).
42. M.J. Marrone, F.W. Patten, and M.N. Kabler, Bull. Am. Phys. Soc. 18, 631 (1973).
43. R.T. Williams and M.N. Kabler, Phys. Rev. B9, 1897 (1974).
44. D. Block, A. Wasiela, and Y. Merle D'Aubigné, J. Phys. C: Solid St. Phys. 11, 4201 (1978).
45. M.J. Marrone and M.N. Kabler, Phys. Rev. Lett. 27, 1283 (1971).
46. A.E. Purdy, R.B. Murray, K.S. Song, and A.M. Stoneham, Phys. Rev. B15, 2170 (1977).
47. J.U. Fischbach, D. Fröhlich, and M.N. Kabler, J. Luminescence 6, 29 (1973).
48. T. Karasawa and M. Hirai, J. Phys. Soc. Japan 40, 128 (1976).
49. I.L. Kuusmann, P.K. Liblik, and C.B. Lushchik, JETP Lett. 21, 72 (1975).
50. I.L. Kuusmann, P.K. Liblik, G.G. Liid'ya, N.E. Lushchik, C.B. Lushchik, and T.A. Soovik, Sov. Phys. Solid State 17, 2312 (1976).
51. C.B. Lushchik, I. Kuusmann, P. Liblik, G. Liidja, N.E. Lushchik, V.G. Plekhanov, A. Ratas, and T. Soovik, J. Luminescence 11, 285 (1975/76).

52. T. Hayashi, T. Ohata, and S. Koshino, J. Phys. Soc. Japan 42, 1647 (1977).
53. H. Nishimura, Y. Tanaka, H. Miyazaki, C. Ohhigashi, and M. Tomura, J. Phys. Soc. Japan 46, 123 (1979).
54. H. Nishimura, H. Miyazaki, Y. Tanaka, K. Uchida, and M. Tomura, Phys. Soc. Japan 47, 1829 (1979).
55. J. Ramamurti and K. Teegarden, Phys. Rev. 145, 698 (1966).
56. B. Di Bartolo, ed., Luminescence of Inorganic Solids, Plenum Press, New York, 1978.
57. G. Baldini, Phys. Rev. 128. 1562 (1962).
58. R. Haensel, G. Keitel, E.E. Koch, M. Skibowski, and P. Schreiber, Phys. Rev. Lett. 25, 128 (1970).
59. I. Steinberger, C.A. Alturi, and O. Schnepp, J. Chem. Phys. 52, 2723 (1970).
60. I. Ya Fugol', Adv. Physics 27, 1 (1978).
61. I. Ya Fugol', E.V. Savchenko, and A.G. Belov, Sov. Phys. JETP Lett. 16, 245 (1972).
62. D. Pudewill, F.-J. Nimpsel, V. Saile, N. Schwentner, M. Skibowski, and E.E. Koch, DESY Report, SR-75/12, Hamburg, 1975.
63. T.E. Stewart, G.S. Hurst, T.E. Bortner, J.E. Parks, F.W. Martin, and H.L. Weidner, J. Opt. Soc. Am. 60, 1290 (1970).
64. N.G. Basov, E.M. Balashov, O.V. Bogdankiewich, V.A. Danilychev, G.N. Kashnikov, N.P. Lantsov, and D.D. Khodkievich, J. Luminescence 1, 834 (1970).
65. J. Jortner, L. Meyer, S.A. Rice, and E.G. Wilson, J. Chem. Phys. 42, 4250 (1965).
66. H. Hahn, N. Schwentner, and G. Zimmerer, DESY Report, SR-76/15, Hamburg, 1976.

INELASTIC SCATTERING OF FAST PARTICLES BY PLASMONS

A. A. Lucas

Département de Physique
Facultés Universitaire N.D. de la Paix
B-5000-Namur, Belgium

ABSTRACT

This article is divided into two main parts. In the first part the basic classical concepts of bulk and surface plasmons are reviewed for different surface geometries and model hamiltonians are constructed for the description of the approximate independent boson behaviour of these collective electronic degrees of freedom and for their interaction with charged particle probes. In the second part, a number of applications in various plasmon spectroscopies will be studied. Only cases where the semi-classical, high-energy approximation holds will be examined.

Energy loss spectra of fast charged probes will be explained for both bulk transmission and for surface reflection geometries in terms of coherent excitation of plasmons. A recent method in which the particle probe is internally excited by scattering off plasmons will also be studied.

I. INTRODUCTION

The development of spectroscopic methods for the investigation of surfaces in clean and adsorbed conditions has given access to the elementary excitations of these systems. The resolution and sensitivity of some of these methods have reached quite an impressive level.

Parallel to the experimental progress, the theoretical description of the interaction of probe particles with surfaces has considerably sharpened over the last twenty years. Broadly speaking, this interaction has been investigated in three energy regimes: i) for probe particles with low kinetic energy ($\lesssim 1$ eV) one deals with phonons and intraband excitations leading to vibrational losses, sticking, chemisorption etc...; ii) at intermediate energies ($\lesssim$ a few hundred eV) multiple scattering by interband and plasmon excitations dominate such as in ILEED, UPS, etc...; iii) at high energies ($\gtrsim 1$ keV) core level energies and plsmons are probed predominantly in such methods as XPS, ELS, etc.

In the present article, we will address ourselves to excitation phenomena of bulk and surface plasmons by fast charged particles. High energy probes are particularly well suited for investigating plasmons. This is because the theoretical description of the inelastic scattering process of, say, fast electrons by plasmons can be reduced, to a fairly good approximation, to an exactly soluble model of a classical force driving a boson field. This was realized immediately with the introduction of the concept of bulk plasmon in solid state physics but has since been extended to surface physics problems involving surface phonons, plasmons and polaritons. Whereas describing the fast probe as a classical force field is a handy simplification, the plasmons must be thought of as quantum mechanical quasi-particles and not simply through their classical electromagnetic fields. This is manifested in experimental inelastic scattering spectra which, as we shall see in this article, often show a multiple peak structure corresponding the discrete energy ladder of the boson oscillators.

It is desirable to begin with elementary considerations, using classical electrodynamics which is quite adequate for obtaining the basic plasmon concepts in the long wavelength limit. This is done in the first section of Part I. We then proceed with the construction of model Hamiltonians and necessary refinements are introduced to treat also the short wavelength bulk and surface plasmons. In Part III fast particle scattering by plasmons will be calculated on the basis of the model Hamiltonians, restricting ourselves to the high energy approximation which, as mentioned above, is the ideal condition to probe the collective modes.

The present treatment is theoretically oriented but frequent comparisons with experiments will be made.

II. BULK AND SURFACE PLASMON HAMILTONIANS

II.A. Classical Concepts

We begin in this section by introducing the classical concepts of bulk and surface plasma oscillations such as they can be derived from Maxwell's equations and associated boundary conditions.

Throughout these articles the Jellium model for the metal will be used : it consists of a rigid positive background of uniform charge density $\rho_o = |e|n$ compensated by valence electrons of average number density n.

1. Bulk Modes. If a local charge density imbalance $\delta\rho = \rho(r) - \rho_o$ iN the neutral Jellium plasma occurs, as a result of thermal or zero-point density fluctuations of the electron gas, its associated electric field will cause charge disturbances throughout the plasma which in turn will act back on the fluctuations at $\vec{r}$ by Coulomb interactions, thus setting up self sustained oscillations. The plasma oscillation "modes" are the eigenmodes of such charge density fluctuations. The "modes" are the eigenmodes of such charge density fluctuations. The work "plasmon" usually refers to the energy quantum of the modes.

To derive the plasma frequency, we shall be using the quasi-static approximation to Maxwell's equations ($c \to \infty$). Referring to the wave equation of electrodynamics

$$[\Delta + \varepsilon(\omega)\frac{\omega^2}{c^2}]\vec{E} = 0 \qquad (1)$$

We see that the second term is negligible if we restrict ourselves to oscillation frequency and wavevector such that $k \gg \sqrt{|\varepsilon|}\,\omega/c$, which defines the "nonretarded" limit of the $\omega(k)$ dispersion relation in the ω vs k plane. Other speakers in this school will study cases where consideration of retardation is important.

The equation of motion for a collective electron displacement field $\vec{x}(\vec{r})$ is

$$m\ddot{\vec{x}} = -e\vec{E} \qquad (2)$$

where $\vec{E}$ is the electric field of the density fluctuation. Both $\vec{E}$ and $\vec{x}$ are related to the charge density fluctuation $\delta\rho(\vec{r})$ through the Maxwell equation

$$\mathrm{div}\,\vec{E} = -4\pi\delta\rho \qquad (3)$$

and the continuity equation

$$\mathrm{div}(\rho\dot{\vec{x}}) + \frac{\partial\rho}{\partial t} = 0 \simeq \mathrm{div}(\rho_o\dot{\vec{x}}) + \frac{\partial\rho}{\partial t} \qquad (4)$$

where, in the right hand side, the second order term $\mathrm{div}(\delta\rho\dot{\vec{x}})$ is neglected. Substitution into Eq. (2) gives

$$\frac{\partial^2\delta\rho}{\partial t^2} + \frac{4\pi e\rho_o}{m}\delta\rho = 0 \qquad (5)$$

which shows that the charge fluctuation $\delta\rho$ freely oscillates at the classical plasma frequency

$$\omega_p = \sqrt{\frac{4\pi ne^2}{m}} \; . \qquad (6)$$

In this derivation, the spatial dependence of the fluctuation $\delta\rho(\vec{r})$ is immaterial [1] . The "eigenvectors" can be chosen as any complete orthonormal set of functions of $\vec{r}$ consistent with the translational invariance of the bulk Jellium model in all three dimensions of space. Such a set are the 3-D plane waves $\exp(i\vec{k}.\vec{r})$:

$$\delta\rho(\vec{r},t) = \sum_{\vec{k}} \rho_{\vec{k}}(t)\, e^{i\vec{k}.\vec{r}} \qquad (7)$$

The plasma frequency (6) gives the approximate zero of the classical Drude, longitudinal dielectric function of the electron gas

$$\varepsilon(\omega) = 1 - \frac{\omega_p^2}{\omega^2 - i\omega\gamma} = \lim_{\vec{k}\to 0} \varepsilon(\vec{k},\omega), \qquad (8)$$

valid as the longwavelength limit of the true frequency and wave-vector dependent dielectric function. We recall that this expression is easily arrived at by considering the response of an electron to a long-wavelength external field as prescribed by Newton's law

$$m\ddot{\vec{x}} + \gamma\dot{\vec{x}} = - e\vec{E}$$

where $\gamma = \tau^{-1}$ is the inverse relaxation time. Eq. (8) follows from the definition of the dielectric function $\varepsilon = 1 + 4\pi\alpha$ where $\alpha = ne|\vec{x}|/|\vec{E}|$ is the polarizability.

2. Planar Interface Modes. Suppose now we have two semi-infinite Jellium with dielectric functions $\varepsilon_1(\omega)$ and $\varepsilon_2(\omega)$ or bulk plasma frequencies ω_{p1}, ω_{p2}, separated by a planar interface at

z = 0 (Fig. 1). What are the eigenmodes of this system? In the longwavelength, nonretarded approxiamtion we must solve for

$$\operatorname{div} \vec{D} = \operatorname{div} \varepsilon(\omega)\, \vec{E} = 0 \tag{9}$$
$$\operatorname{rot} \vec{E} = 0 \tag{10}$$
$$\text{continuity of } \vec{E}_{\parallel} \text{ and } \varepsilon E_{\perp} \text{ across } z = 0. \tag{11}$$

This set of equations is satisfied in two ways :

i) Bulk modes :
$$\begin{cases} \varepsilon_1(\omega) = 0 \text{ for } z < 0 \\ \varepsilon_2(\omega) = 0 \text{ for } z > 0 \\ \vec{E} \text{ longitudinal but arbitrary.} \end{cases} \tag{12}$$

i.e. the solutions found before for the two infinite Jellium;

ii) Surface modes : $\varepsilon(\omega) \neq 0$, $\operatorname{div} \vec{E} = 0$ (13)

Combined with (10), (13) is equivalent to Laplace's equation

$$\Delta V = 0 \tag{14}$$

for the scalar potential $\vec{E} = -\vec{\nabla} V$.

The translational invariance parallel to the planar interface dictates to search for solutions of the form

$$V(\vec{r}) = \sum_{\vec{k}} V_k(z)\, e^{i\vec{k}.\vec{r}_{\parallel}} \tag{15}$$

where $\vec{k} = (k_x, k_y)$ is now a two-dimensional wavevector parallel to the interface. Eq. (14) is then satisfied by

$$V_k(z) = A_k\, e^{kz} + B_k\, e^{-kz} \quad . \tag{16}$$

The linear coefficients are determined by the regularity of V at $z = \pm\infty$ and the continuity conditions (11) :

$$\begin{cases} A_k = B_k \text{ (continuity of V),} \\ \varepsilon_1 A_k = -\, \varepsilon_2 B_k \text{ (continuity of } \varepsilon \frac{\partial V}{\partial z}\text{).} \end{cases} \tag{17}$$

Hence one must have for the surface modes

$$\varepsilon_1(\omega) + \varepsilon_2(\omega) = 0 \ . \tag{18}$$

Their amplitudes, according to (16) (Fig. 1), decay exponentially away from the interface which is one reason why they are called surface modes. Using Drude-like dielectric function, (18) gives the surface plasmon frequency

$$\omega_s = \left[\frac{\omega_{p1}^2 + \omega_{p2}^2}{2}\right]^{1/2} \tag{19}$$

which reduces to

$$\omega_s = \frac{\omega_p}{\sqrt{2}} \tag{20}$$

for the famous surface plasmons of the metal-vacuum interface.

The density fluctuation associated with surface modes is sharply localized at the interface. Indeed, we have Poisson's equation

$$\delta\rho = -\frac{1}{4\pi}\Delta V \tag{21}$$

where, from (16), (17)

$$V = \sum_{\vec{k}} A_k\, e^{-k|z| + i\vec{k}.\vec{r}_\parallel} \tag{22}$$

Substituting (22) into (21) and remembering that $\frac{\partial^2}{\partial z^2}|z| = 2\delta(z)$ we get

$$\delta\rho = \sum_{\vec{k}} \frac{kA_k}{2\pi}\, e^{i\vec{k}.\vec{r}_\parallel}\,\delta(z) \tag{23}$$

Although the charge fluctuation is purely superficial, one should note that the collective electron displacement $\vec{x}(\vec{r})$ extends into the metal. The electric field of the surface plasmon is, from (22)

$$\vec{E} = -\vec{\nabla}V = -\sum_{\vec{k}} A_k[i\vec{k}, -k\ sg(z)]\, e^{-k|z| + i\vec{k}.\vec{r}_\parallel} \tag{24}$$

and, hence, from (2)

$$\vec{x} = \frac{-e}{m\omega_s^2}\vec{E} = \frac{1}{2\pi ne}\sum_{\vec{k}} A_k(i\vec{k}, k)\, e^{kz + i\vec{k}.\vec{r}_\parallel} \qquad (z \leqslant 0) \tag{25}$$

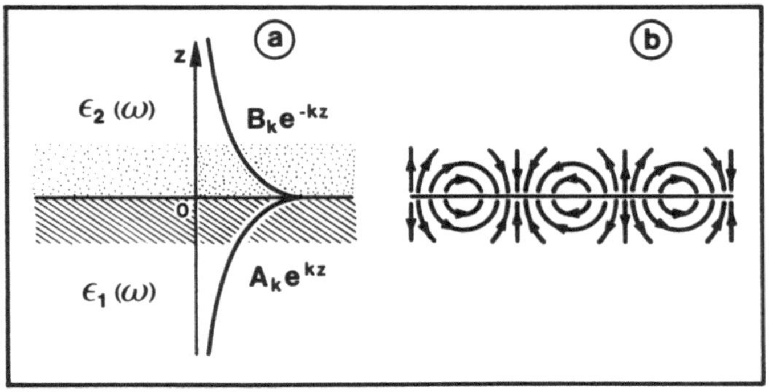

Fig. 1. a. Potential distribution of surface modes at a planar interface.
b. Electric field lines associated with the surface mode.

There is a $\frac{\pi}{2}$ phase shift between the parallel and perpendicular components of the field $\vec{E}_k$ or displacement $\vec{x}_k$ of the plasmon modes. The field lines are illustrated in Fig. 1.b.

3. Thin Film Modes. The surface mode potential is again given by Eqs. (15) and (16) but the coefficients A_k, B_k etc. must satisfy boundary conditions at the two interfaces (Fig. 2). The calculation is left to the student. The two frequencies are obtained for each k. The dispersion relation is

$$\frac{\varepsilon_1(\omega)}{\varepsilon_2(\omega)} = - \mathrm{th}^{\pm 1} ka \tag{26}$$

where the + 1 or - 1 exponents refer to the symmetric or antisymmetric mode, respectively. The amplitudes of the modes, as functions of z, are sketched in Fig. 2. Solved for ω in the case of a Drude metallic film in vacuum, (26) gives the dispersion relation illustrated in Fig. 2b.

$$\omega_{\pm}(k) = \frac{\omega_p}{\sqrt{2}} (1 \pm e^{-2ka})^{1/2} \tag{27}$$

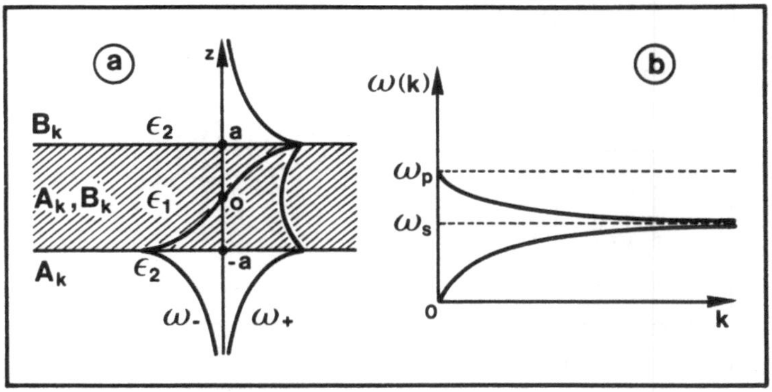

Fig. 2. a. Potential distribution of thin film modes.
b. Dispersion relation of thin film modes.

In addition to the pair of surface modes $\omega_\pm(k)$, for each $\vec{k}$, there are two discrete sets of longitudinal bulk modes in the film labelled by a positive integer m and parity index $p = \pm 1$. The quantity $k_z = m \frac{\pi}{2a}$ is the quantized third component of the wavevector for these bulk modes of degenerate frequency ω_p.

4. Sphere and Void Modes. The consideration of these geometries (Fig. 3a) is important for treating optical and other properties of powder and porous materials [2]. From the rotational symmetry, plasma oscillations in a sphere or around a spherical cavity must be proportional to the spherical harmonics Y_ℓ^m. The radial part of Laplace's equation is easy to solve in terms of power functions r^ℓ, $r^{-(\ell+1)}$ of the distance to center. Application of the boundary conditions results in the dispersion relation

$$\frac{\varepsilon_1(\omega)}{\varepsilon_2(\omega)} = -\frac{\ell}{\ell+1} \tag{28}$$

illustrated in Fig. 3 b for the cases $\varepsilon_1 = \varepsilon(\omega)$, $\varepsilon_2 = 1$ (sphere)

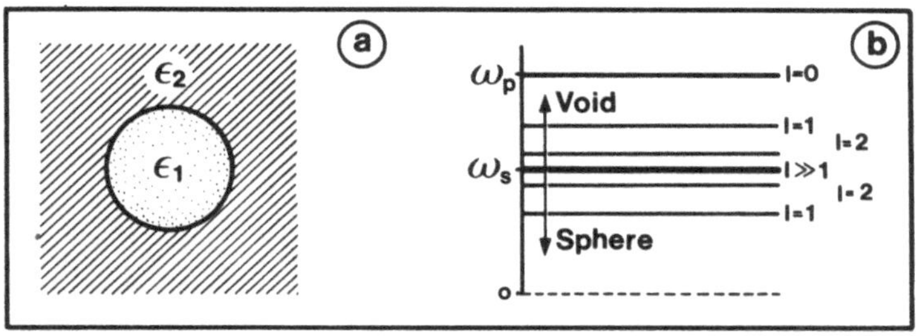

Fig. 3. a. Spherical dielectric inclusion.
b. Spherical surface plasmon spectrum.

and $\varepsilon_1 = 1$, $\varepsilon_2 = \varepsilon(\omega)$ (void).

Before proceeding to construct a Hamiltonian description of collective modes, one should make an important remark regarding the validity of the plasmon concept in real materials. A collective oscillation mode is genuinely defined if the number of oscillations is large before they are damped out. One measure of this is provided by the so-called loss function $\text{Im}\varepsilon^{-1}(\vec{k},\omega)$ which, as we shall see in Chapter III, is proportional to the energy loss spectrum of fast electrons going through the solid. From Eq. (8), this function

$$\text{Im}\ \varepsilon^{-1} \quad \div \quad \frac{\omega_p \gamma}{(\omega-\omega_p)^2 + \frac{1}{4}\gamma^2} \simeq \pi\omega_p\ \delta(\omega-\omega_p)$$

should be a sharp Lorentzian if $\gamma \ll \omega_p$. Fig. 4 illustrates the case of a free electron metal such as Al which indeed has the sharpest bulk plasmon and a d-band metal such as Cu which exhibits the more common type of broad, rather ill defined collective mode.

II.B. Model Hamiltonians

It is possible to develop classical Lagrangian and Hamiltonian descriptions of the longwavelength plasma oscillations in metals in terms of harmonic oscillator-like field variables. This will be done explicitly for the metal-vacuum interface, below in this section, as an illustrative example.

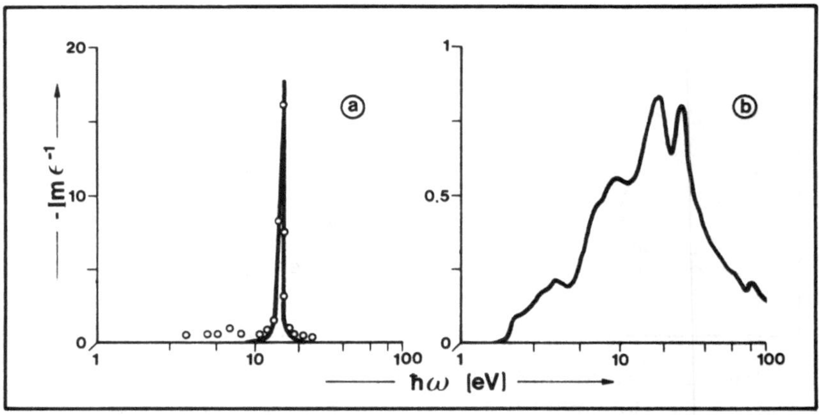

Fig. 4. a. Lorentzian loss function Im ε^{-1} for Al.
b. Loss function Im ε^{-1} for Cu.

For arbitrary wavelength, the particular difficulty met by the correct microscopic description of the electron gas is that the decoupling of the plasmon modes from the individual degrees of freedom is only approximate. There is no Born-Oppenheimer theorem here to strictly decouple the latter from the former because, as seen from the excitation spectrum of the electron gas in Fig. 5, the plasmon energies are degenerate with e-h excitation energies, at least at intermediate $k \sim k_F$. For surface plasmons, the degeneracy extends down to $k = 0$.

Nevertheless for bulk modes, the separation should be possible at least for small $k \ll k_F$ and for large $k \gg k_F$. Indeed, for small k, plasmons dominate the spectrum as can be seen by noting that the classical Drude expression (8) saturates the exact f-sum rule [3]

$$\int_0^\infty d\omega\ \omega\ \mathrm{Im}\ \varepsilon^{-1}(\vec{k},\omega) = -\frac{\pi}{2}\omega_p^2 \tag{29}$$

which is a statement that all degrees of freedom are accounted for. For large k on the other hand the electron gas behaves as free electrons of excitation energy $\hbar^2k^2/2m$. This is seen from Fig. 5 where the e-h continuum is bounded by the two parabolas

$$\omega(k) = \frac{\hbar}{2m}\left[(k \pm k_F)^2 - k_F^2\right] = \frac{\hbar k^2}{2m} \pm \frac{\hbar\ kk_F}{m} \tag{30}$$

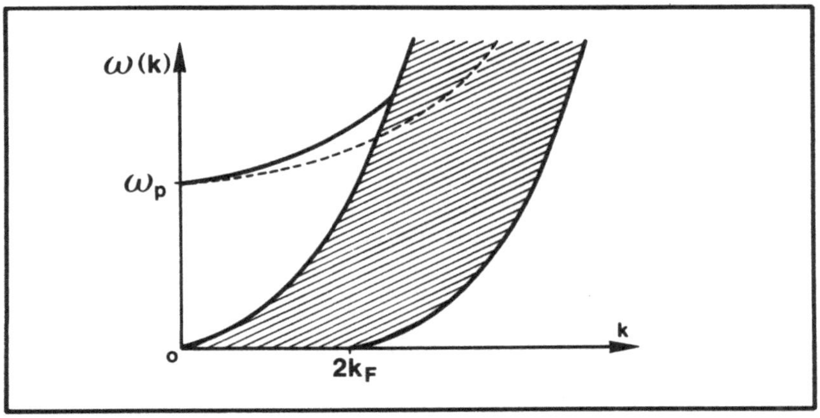

Fig. 5. Spectrum of elementary excitations in the bulk electron gas. The dashed line represents the "plasmon pole" model of Eq. (45).

and hence the spectral width $2\hbar\ kk_F/m$ becomes negligible as compared to $\omega(k)$ in the large-k limit.

Whether a particular region of the $\omega(k)$ spectrum or the whole of it is relevant in a given problem depends very much on the experimental conditons under which the electron system is explored. In some spectroscopies such as EELS (Electron Energy Loss Spectroscopy), it is possible to select one particular momentum transfer, e.g. the small ones, and thus interpret the inelastic scattering cross sections in terms of plasmon excitations only. For other properties, such as the cohesion energy of the electron gas, involving summations over momentum space, all parts of the $\omega(k)$ spectrum enter and thus the separation into plasmons and e-h excitations is not so useful.

It would be desirable however to have at one's disposal a model Hamiltonian for the electron gas which would adequately describe the small-k, boson-like plasmons and the large-k, electron-hole pairs (which are also approximate boson "particles") as well as provide a reasonable interpolation model for the intermediate region $k \sim k_F$. Such a model Hamiltonian has been introduced by Overhauser [4] for bulk modes and we will follow his derivation below.

1. Bulk Plasmon Hamiltonian. The effective bulk plasmon Hamiltonian is, in second quantized form,

$$H_o = \sum_{\vec{k}} \hbar\omega_k a_k^+ a_k \tag{31}$$

where the effective dispersion relation ω_k is to be determined. Since boson-like behavior is assumed throughout the spectrum, the creation and annihilation operators a_k^+, a_k satisfy

$$[a_k, a_\ell] = 0, \qquad [a_k, a_\ell^+] = \delta_{k,\ell} . \tag{32}$$

These operators create and annihilate 3-D plane-wave density fluctuations and must be linearly related to the Fourier components ρ_k of these density waves :

$$\rho_k = \Omega \lambda_k (a_k^+ + a_{-k}) \tag{33}$$

where Ω is the volume and λ_k is a second parameter to be determined.

The two model parameters ω_k, λ_k are chosen to satisfy the two following requirements :

i) the f-sum rule must be satisfied;
ii) a static test charge must be screened by the model plasmons in the same way as by the actual electron gas.

The first condition guarantees that the effective plasmon of a given $\vec{k}$ carries the same oscillator strength as the true electron gas excitations for that $\vec{k}$. It is shown in all textbooks [3] that the sum rule, of which Eq. (29) is a particular form, reads

$$\sum_i \hbar \omega_{oi} |(\rho_k)_{oi}|^2 = n e^2 \Omega \frac{\hbar^2 k^2}{2m} \tag{34}$$

where $\hbar \omega_{oi}$ are the exact excitation energies of the system between ground state $|o\rangle$ and excited states $|i\rangle$. Now for the model Hamiltonian (31), there is only one nonvanishing matrix element $(\rho_k)_{oi} = \Omega \lambda_k \delta_{i1}$ with excitation energy $\hbar\omega_k$ and thus (34) reduces to

$$\hbar \omega_k \lambda_k^2 = \Omega n e^2 \frac{\hbar^2 k^2}{2m} \tag{35}$$

The second condition concerns the interaction of a test charge at $\vec{r}_o$ with the model plasma. The interaction energy is

$$H_I = \int d\vec{r} \frac{e\delta\rho(\vec{r})}{|\vec{r}-\vec{r}_o|} = \sum_{\vec{k}} \frac{4\pi e}{k^2} \rho_k e^{-i\vec{k}.\vec{r}_o} \tag{36}$$

where the 3-D Fourier transform of the Coulomb potential $r^{-1} = (2\pi^2)^{-1} \int d\vec{q}\, q^{-2} \exp(-i\vec{q}.\vec{r})$ has been used. Hence the total hamiltonian is, combining (33), (36)

$$H = \sum_{\vec{k}} \hbar\, \omega_k a_k^+ a_k + \sum_{\vec{k}} M_k (a_k^+ e^{i\vec{k}.\vec{r}_o} + a_k e^{-i\vec{k}.\vec{r}_o}) \tag{37}$$

where

$$M_k = \left[\frac{2\pi e^2 \hbar \omega_p^2}{\Omega k^2 \omega_k}\right]^{1/2} . \tag{38}$$

The test charge interaction energy in the plasma ground state is the difference between the zero-point energies of (37) and (31) which is easily obtained with the usual displaced harmonic oscillator trick

$$a_k \to a_k + M_k/\hbar\omega_k . \tag{39}$$

One finds

$$W = - \sum_{\vec{k}} \frac{M_k{}^2}{\hbar\omega_k} . \tag{40}$$

This result must coïncide with the interaction energy of the test charge with the electron gas which is the energy content of the electrostatic field of the charge :

$$W = \frac{1}{8\pi} \int d\vec{r}\, (\vec{E}.\vec{D} - |\vec{E}|^2) \tag{41}$$

where the substracted term is the infinite self energy of the charge in vacuum. Fourier transforming, one finds

$$W = \frac{\Omega}{8\pi} \sum_{\vec{k}} (\vec{E}_k.\vec{D}_k^\star - |\vec{E}_k|^2) \tag{42}$$

and from Maxwell's equation div $\vec{D} = 4\pi e\delta(\vec{r}-\vec{r}_o)$,

$$D_k = \varepsilon_k E_k = \frac{4\pi e}{ik\Omega} e^{-i\vec{k}.\vec{r}_o} . \tag{43}$$

Substitution of (43) into (42) gives finally

$$W = \sum_{\vec{k}} \frac{2\pi e^2}{\Omega k^2} \left(\frac{1}{\varepsilon_k} - 1\right) . \tag{44}$$

Identification with (40) imposes the dispersion relation

$$\omega_k^2 = \omega_p^2 \frac{\varepsilon_k}{\varepsilon_k - 1} \tag{45}$$

in terms of the actual static dielectric function ε_k of the electron gas. The result (45) is schematically illustrated by a dashed line in Fig. 5. It has the required limits ω_p^2 and $(\hbar k^2/2m)^2$ for small and large k respectively since the function ε_k can be shown to have the following limiting behavior

$$\lim_{k\to 0} \varepsilon \sim \alpha k^{-2}, \qquad \lim_{k\to\infty} (\varepsilon_k - 1) \sim \beta k^{-4} \tag{46}$$

Thus the model plasmon system is completely specified by ε_k and M_k. The total interaction Hamiltonian (37) is then adequate for studying such problems as the screening of protons, the energy loss of fast electrons, the positron dynamics in metals etc., i.e. all problems where the test charge is distinguishable from the metal electrons making up the plasmons. If the "test charge", on the contrary, is one of the valence electrons, M_k must be modified to account for exchange and correlation effects [4].

2. Surface Plasmon Hamiltonian. Recently, effective, all-bosons Hamiltonians for the metal surface and its interaction with a test charge have been developed and justified on microscopic grounds [5,6].

The basic ideal is again to construct a "plasmon pole" approximation for the metal surfact in a way similar to the bulk procedure shown above. The effective boson degrees of freedom are required to satisfy an important f-sum rule for the surface and the matrix element for the coupling with a static test charge must be consistent with some appropriate surface dielectric constant. We refer to the work of Eguiluz [5] for a detailed study and references to earlier works in this area.

Surface plasmon and thin film plasmon Hamiltonian had been introduced before on semi-empirical grounds [7] and extensively used in a number of surface physics problems [7 - 10], some of which will be dealt with in later sections of this course.

In the present section, it is only feasible to give a simple derivation of the long wavelength surface plasmon Hamiltonian for the metal-vacuum interface (following ref. 11) and indicate the necessary modifications to make it suitable for shorter wavelengths.

Calling $\vec{x}$, as before, the collective electron motion and $\dot{\vec{x}}$ the associated velocity field, the classical Langrangian is

$$L \equiv T - V = \int d\vec{r}\ (\frac{1}{2}\, n\, m\, \dot{\vec{x}}^2 - \frac{1}{2}\, \delta\rho\ V) \tag{47}$$

where the first term is the kinetic energy and the second the potential energy of the charge density fluctuation $\delta\rho$ in its own electrostatic potential V (the factor $\frac{1}{2}$ takes account of the self-induced nature of this energy). Substituting in L the 2-D Fourier transforms of Eqs. (22), (23) and (25) and taking into account

$$\sum_{\vec{k}} = \frac{A}{(2\pi)^2} \int d^2k, \qquad \int d\vec{r}_\parallel\ e^{i\vec{k}.\vec{r}_\parallel} = (2\pi)^2\ \delta(\vec{k}), \tag{48}$$

we get

$$L = \sum_{\vec{k}} (\frac{Anm}{4k}\, \dot{\vec{x}}_k \cdot \dot{\vec{x}}_{-k} - \frac{Ak}{4\pi}\, A_k\ A_{-k}) \tag{49}$$

Eqs. (23), (25) actually allow us to express all Fourier amplitudes in terms of the z-component x_{kz} of $\vec{x}_k$ only. Hence L becomes

$$L = \sum_{\vec{k}} (\frac{Anm}{2k}\, \dot{x}_{kz}\, \dot{x}_{-kz} - \frac{A\pi n^2 e^2}{k}\, x_{kz}\, x_{-kz}) \tag{50}$$

The momentum variable π_{kz} canonically conjugated to x_{-kz} is now introduced by

$$\pi_{kz} = \frac{\partial L}{\partial \dot{x}_{-kz}} = \frac{An}{k}\, \dot{x}_{kz} \tag{51}$$

and allows us to write the Hamilton function

$$H_o = \sum_{\vec{k}} \pi_{kz}\, \dot{x}_{-kz} - L \tag{52}$$

$$H_o = \frac{1}{2} \sum_{\vec{k}} (\frac{k}{Anm}\, \pi_{kz}\, \pi_{-kz} + \frac{2\pi A n^2 e^2}{k}\, x_{kz}\, x_{-kz}) \tag{53}$$

This has the canonical form

$$H_o = \sum_{\vec{k}} \left(\frac{|p_k|^2}{2m} + \frac{1}{2} m \omega_s^2 |q_k|^2\right) \tag{54}$$

provided one makes the substitutions

$$q_k = \sqrt{\frac{nA}{k}} \; x_{kz}, \qquad p_k = \sqrt{\frac{k}{nA}} \; \pi_{kz} \tag{55}$$

$$\omega_s = \sqrt{\frac{2\pi ne^2}{m}} = \frac{\omega_p}{\sqrt{2}}$$

The interaction Hamiltonian between a test charge e at $\vec{r}_o$ and the surface plasmons is, according to (23), (25)

$$H_I = eV(\vec{r}_o) = e \sum_{\vec{k}} A_k \; e^{i\vec{k}.\vec{r}_o - k|z_o|} \tag{56}$$

or

$$H_I = \sum_{\vec{k}} \frac{4\pi^2 ne^4}{Ak} \cdot q_k \; e^{i\vec{k}.\vec{r}_o - k|z_o|} \tag{57}$$

First quantization is achieved by imposing the commutation relations

$$[p_k, q_\ell] = \frac{\hbar}{i} \; \delta_{k,\ell} \tag{58}$$

and the "second" quantization form of H and H_I is obtained by the linear substitutions

$$q_k = \sqrt{\frac{\hbar}{2m\omega_s}} \; (a_k^+ + a_{-k}) \; ; \quad p_k = i \sqrt{\frac{m\hbar\omega_s}{2}} \; (a_k^+ - a_{-k}) \tag{59}$$

giving

$$[a_k, a_\ell^+] = \delta_{k,\ell} \tag{60}$$

and the total interacting Hamiltonian $H = H_o + H_I$

$$H = \sum_{\vec{k}} \hbar\omega_s (a_k^+ a_k + \frac{1}{2}) + \sum_{\vec{k}} C_k \; e^{-k|z_0|} (a_k^+ e^{i\vec{k}.\vec{r}_0} + a_k e^{-i\vec{k}.\vec{r}_0}) \quad (61)$$

which is similar to (37) but with the matrix element

$$C_k = \left[\frac{\pi \hbar e^2 \omega_p^2}{2 A k \omega_s} \right]^{1/2} \quad (62)$$

To check the consistency of this result with what we started from, we consider the ground state interaction energy between the test charge and the surface plasmons which is the zero-point energy shift between H and H_0. This is easily obtained again by the displacement

$$a_k \to a_k + \frac{1}{\hbar\omega_s} C_k \; e^{i\vec{k}.\vec{r}_0 - k z_0} \qquad (z_0 > 0) \quad (63)$$

which yields

$$W = - \sum_{\vec{k}} \frac{C_k^2}{\hbar\omega_s} e^{-2kz_0} = - \frac{e^2}{2} \int_0^\infty dk \; e^{-2kz_0} \quad , \quad (64)$$

i.e. the classical image potential energy

$$W = - \frac{e^2}{4z_0} \quad (65)$$

This, at first surprising relationship between a purely static screening property and surface plasmons, which are usually thought of as giving the response to dynamical probes, was first pointed out by the author, in an application to Field Ion Spectroscopy [11-12] It is actually no miracle since small-k surface plasmons are nothing but quantized eigensolutions of electrodynamics in the quasistatic limit, as seen in Section II.A.

We should now like to insist that the matrix element (62) holds only for discussing problems where long wavelengths only are important (e.g. when $z_0 \gg 1$ Å), thus ignoring short range effects such as the k-dependence of the surface plasmon frequency $\omega_s(k)$, the diffuseness of the metal-vacuum interface, ion lattice effects, etc. To obtain an accurate Hamiltonian description valid in the surface region, several important corrections to both bulk and surface matrix elements (38) and (62) must be introduced. These have been obtained and discussed very clearly in Ref. 5 . For our future purpose, we mention here three major qualitative effects brought about by the proper treatment of the surface screening problem :

a) The bulk plasmon vertex M_k is modified by the presence of the surface so as to cancel the long range part (65) of the static screening by surface plasmons when the test charge penetrates into the metal.

b) For $|z_o| \leqslant v_F/\omega_p$, the coupling of the test charge to surface and bulk plasmons must be treated on an equal footing as dispersion effects make bulk and surface contributions to screening of comparable magnitude.

c) For $z_o \geqslant v_F/\omega_p$ surface plasmons only are important and Hamiltonian (61) is adequate provided the coupling vertex C_k be modified into

$$C_k = \sqrt{\frac{\pi \hbar e^2 \omega_p^2}{2 A k \omega_s(k)}} \cdot S(k) \tag{66}$$

where S(k) is a square-integrable function of k which cuts off the interaction for large k.

The most important effect of S(k) on the static screening property of surface plasmons is to remove the divergence of the classical image potential when $z_o \to 0$. We now have, instead of (64),

$$W = -\frac{e^2}{2}\int_0^\infty dk\, e^{-2kz_o}\, S^2(k) \cdot \frac{\omega_s^2}{\omega_s^2(k)} . \tag{67}$$

S(k) depends on which approximation to the surface density response function is adopted [5] and so does the dispersion relation $\omega_s(k)$ for surface plasmons. For practical calculations, it may be convenient to use simple forms such as a unit step function

$$S(k) = \theta(k - k_a), \tag{68}$$

an exponential function

$$S(k) = e^{-k/k_b} \tag{69}$$

for which the image potential has the approximate shifted form $-e^2/4(z_o + k_b^{-1})$ or an algebraic function (Thomas-Fermi form [14-15])

$$S(k) = (\sqrt{k^2 + k_c^2} - k) / \sqrt{k^2 + k_c^2}, \tag{70}$$

where $k_{a,b,c}$ are cutoff wavevectors of order 1 $Å^{-1}$. From the model leading to (70), k_c is found [5] to be

$$k_c = \frac{2\omega_s}{\beta} = \sqrt{\frac{10}{3}}\ \frac{\omega_p}{v_F} \tag{71}$$

where $\beta=\sqrt{\frac{3}{5}}\,v_F$ gives the k-dependence of the dispersion relations for bulk and surface plasmons according to

$$\omega_B^2(q) = \omega_p^2 + \beta^2 q^2 + \dots \tag{72}$$

$$\omega_s^2(k) = \omega_s^2 + \beta\omega_s k + \dots \tag{73}$$

k_c is the inverse width of the surface charge density fluctuation associated with surface plasmons $\delta\rho \sim e^{-k_c|z|}$ [instead of the δ-function of the classical result (23)].

In Fig. 6, are plotted the surface plasmon (V_s) and bulk plasmon (V_B) contributions to the total static "image" potential V as a function of distance z to the surface [5], clearly illustrating the three major effects mentionned above as a), b), c).

3. Multipole Hamiltonian. In problems such as chemisorption and physisorption and in a number of surface spectroscopies such as INS (Ion Neutralization Spectroscopy) BFS (Beam-Foil Spectroscopy), MBS (Molecular Beam Scattering) etc., the particle which probes the surface possesses internal degrees of freedom. In order to be able to describe processes of internal electron or vibrational excitations of the particle by the surface, we must generalize the model Hamiltonian (61) and (66). We can write

$$H = H_{particle} + H_{plasmons} + H_I \tag{74}$$

where H_I is the sum of interaction terms such as in (61) extending over all the point charges e_i composing the particle :

$$H_I = \sum_i \sum_{\vec{k}} C_{ki}\, e^{-kz_i}\, (a_k^+\, e^{i\vec{k}.\vec{r}_i} + a_k\, e^{-i\vec{k}.\vec{r}_i}), \tag{75}$$

$$C_{ki} = e_i \left[\frac{\pi\hbar\ \omega_p^2}{2A\ k\ \omega_s(k)}\right]^{1/2} S(k) \tag{76}$$

It will prove useful to develop (75) in multipolar terms, writing

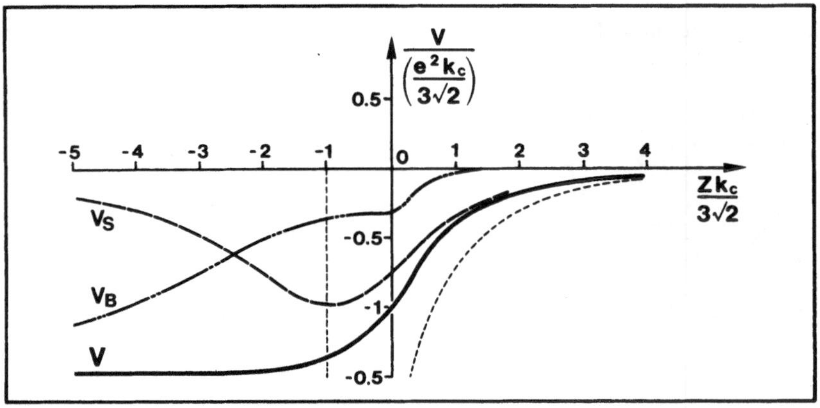

Fig. 6. z-dependence of the static potential of a test charge near a metal vacuum interface, after ref. 5.

$$\vec{r}_i = \vec{R} + \vec{x}_i \tag{77}$$

where $\vec{R}$ is the position of the nucleus of, say, an atom or an ion. Expanding the exponentials in (75) in powers of $\vec{x}_i$, we get

$$H_I = \sum_{m=0,1,2...} H_m \tag{78}$$

where m is the multipolar order and

$$H_m = \sum_{\vec{k}} C_k \, M_m(\vec{k}) \, e^{-kZ} \, (a_k^+ \, e^{i\vec{k}.\vec{R}} + a_k \, e^{-i\vec{k}.\vec{R}}), \tag{79}$$

$$M_m(\vec{k}) = \frac{1}{m!} \sum_i e_i \, (\vec{K}.\vec{x}_i)^m, \tag{80}$$

$$\vec{K} = (i\vec{k}, -k). \tag{81}$$

The first few terms are

$$M_o = \sum_i e_i \equiv Z_{ion} \qquad \text{(monopole)}, \tag{82}$$

$$M_1 = (\sum_i e_i \, \vec{x}_i) \, . \, \vec{K} \equiv \vec{D}.\vec{K} \qquad \text{(dipole)}, \tag{83}$$

$$M_2 = \frac{1}{2} \vec{K} \, . \, (\sum_i e_i \, \vec{x}_i\vec{x}_i) \, . \, \vec{K} \equiv \frac{1}{2} \vec{K}.\overleftrightarrow{Q}.\vec{K}, \qquad \text{(quadrupole)} \tag{84}$$

etc.

In what follows, we will consider a scattering problem for which the monopole and dipole terms will be treated simultaneously.

III. CHARGE PARTICLE SPECTROSCOPIES OF PLASMONS

In this chapter we exploit the Hamiltonians developped in II to treat a class of problems in which plasmons dominate the metal response.

There are two extreme cases for which a problem such as (61) can be solved exactly. One case has already been met with and consists in finding the metal response to a static or slowly moving charge or multiple, i.e. the quantum mechanical description of the image potential concept. Another example in this class which we will not examine here arises with the physisorption problem. The other extreme concerns the response of the metal to fast particles. Slow and fast refer here to particle velocities v much smaller or much larger than, say, v_F, the Fermi velocity.

III.A. High Energy Approximation

Let us assume that the probe, say a point charge of mass m, is moving at such high velocity that its energy $\frac{1}{2}mv^2$ is much larger than the plasmon quantum $\hbar\omega_p$ and that its momentum $m\vec{v}$ is much larger than plasmon momenta $\hbar\vec{k}$. The effect of the coupling is to induce real plasmon excitations (in contrast to the virtual plasmons making up the image potential). The basic assumption inherent in the high energy approximation developped below is that even multiple plasmon excitations cannot disturb appreciably the motion of the fast probe, neither in its energy nor in its momentum (i.e. direction of motion). For calculating the plasmon excitation state, we are then justified to neglect altogether the perturbation of the motion of the particle and assume that the latter follows whatever unperturbed classical trajectory it would have in the absence of coupling to plasmons.

The interest of this approach stems from the fact that it allows the final energy distribution of the plasmons to be calculated exactly. Now energy conservation must ultimately be satisfied so that any excitation energy of the plasmon system must be drawn out of the particle motion. Thus the calculated exact energy distribution of the final plasmon state coincides with the energy distribution of the scattered particles, i.e. the inelastic scattering spectrum which one actually measures, e.g. in EELS.

It is clear that self-consistency can be restored if this approach serves as a zeroth order approximation of an iterative procedure in which the resulting particle energy spectrum is used to construct a modified, average classical trajectory which in turn allows to obtain a new plasmon excitation spectrum, etc.

We are then to solve for Hamiltonians (37) or (61) in which the particle position $\vec{r}_o$ is a time-dependent c-number prescribed by the classical trajectory

$$\vec{r}_o = \vec{r}_o(t) \tag{85}$$

Substituting in (37) and (61), we obtain a plasmon Hamiltonian of the general form

$$H = \sum_{\vec{k}} \hbar\omega_k\, a_k^+ a_k + \sum_{\vec{k}} [\, f_k(t) a_k^+ + f_k^\star(t)\, a_k] \tag{86}$$

where

$$f_k(t) = M_k\, e^{i\vec{k}.\vec{r}_o(t)} , \tag{87}$$

or

$$f_k(t) = C_k\, e^{\vec{K}.\vec{r}_o(t)} . \tag{88}$$

Thus we have a collection of independent harmonic oscillators linearly driven by time-dependent "forces" $f_k(t)$. The exact time evolution of the oscillator wavefunctions is well known from elementary quantum mechanics, and best described in terms of the so-called coherent states which are eigenstates of the destruction operator a_k [16].

If $|\{o_k\}\rangle$ designates the plasmon ground state at $t = -\infty$, the exact Schrödinger state at t is

$$|\psi(t)\rangle = e^{-\frac{i}{\hbar} H_o t}\, D|\{o_k\}\,\rangle \equiv e^{-\frac{i}{\hbar} H_o t}\, |\{I_k(t)\}\,\rangle \tag{89}$$

where

$$D = \exp\ \{ \sum_{\vec{k}}\ [\, I_k(t)\ a_k^+ - I_k^\star(t)\ a_k] \tag{90}$$

is the plasmon displacement operator which stretches each plasmon half-amplitude a_k by the coherent displacement I_k according to

$$D^+ a_k\, D = a_k + I_k(t) \tag{91}$$

and where I_k is given by

$$I_k(t) = \frac{i}{\hbar} \int_{-\infty}^{t} d\tau \; f_k(\tau) \; e^{-i\omega_k \tau}. \tag{92}$$

Since $|\{I_k\}\rangle$ is an eigenstate of a_k (eigenvalue I_k) and a_k does not commute with H_o, the final state $|\psi\rangle$ must be a superposition of energy eigenstates. As indicated above, the particle energy spectrum $P(\omega)$, i.e. the probability for the particle to lose energy $\hbar\omega$ coincides, by energy conservation, with the density of energy states at $\hbar\omega$ to be found in $|\psi\rangle$. $P(\omega)$ can be expressed as the Fourier transform

$$P(\omega) = \frac{1}{2\pi} \int_{-\infty}^{+\infty} dt \; e^{i\omega t} \; P(t) \tag{92}$$

of the correlation function

$$P(t) = \langle \{I_k\}| \; e^{-\frac{i}{\hbar} H_o t} \; |\{I_k\}\rangle = \langle \{o_k\}| \; D^+ \; e^{-\frac{i}{\hbar} H_o t} \; D|\{o_k\}\rangle \tag{93}$$

The boson operator algebra necessary to calculate the expectation value is standard and the result is

$$P(t) = P(o)^{-1} \; \exp(\sum_{\vec{k}} |I_k(\infty)|^2 \; e^{-i\omega_k t}) \; . \tag{94}$$

Substitution of this result into (92) and expansion in powers of $|I_k|^2$ generates a Poisson distribution of δ-functions at integer multiples of ω_k. The strength of the distribution is given by

$$Q = \sum_{\vec{k}} |I_k|^2 \tag{95}$$

and is a measure of the ratio P_1/P_0 of the one-plasmon excitaiton line on the no loss line P_o in the Poisson spectrum.

III.B. Bulk, Transmission Spectrum

Due to the spatial homogeneity of the bulk Hamiltonian (37) we meet the difficulty that when $t \to +\infty$ the strength Q grows indefinitely. This is avoided by computing the accumulated coherent displacement I_k in (92) during a time interval 2 T over which the particle travels a distance a = 2 Tv (e.g. a film thickness) :

$$I_k(a) = \frac{i}{\hbar} \int_{-T}^{T} d\tau \; M_k \; e^{-i(\omega_k - k_z v)t} \tag{96}$$

$$= \frac{i}{\hbar} 2T \frac{\sin(\omega_k - k_z v)T}{(\omega_k - k_z v)T} \cdot M_k \tag{97}$$

Note that for large $T \gg \omega_k^{-1}$, this becomes singular :

$$I_k(\infty) = \frac{2\pi i}{\hbar} M_k \ \delta(\omega_k - k_z v) \ . \tag{98}$$

Now we obtain the strength Q in (95) for a travel distance a by using the trick

$$Q(a) = \frac{\Omega}{(2\pi)^3} \int d^3k \ I_k(a) \ I_k^\star(\infty) \tag{99}$$

$$= \frac{\Omega T}{\hbar^2 \ 2\pi^2} \int d^2k \int_{-\infty}^{+\infty} dk_z \ M_k^2 \ \delta(\omega_k - k_z v) \tag{100}$$

Substituting M_k from (38), we find

$$Q(a) = \frac{e^2 \omega_p^2 a}{2\pi \hbar v^2} \int d^2k \frac{1}{\omega_k \ (k^2 + \frac{\omega_k^2}{v^2})} \ . \tag{101}$$

The angular integration gives 2π whereas the remaining k-integral requires an explicit dispersion relation for ω_k, such as in Eq. (72) where $q^2 = k^2 + \omega_k^2/v^2$. Neglecting this dispersion, we get

$$Q(a) = \frac{e^2 \omega_p a}{\hbar v^2} \int_0^{k_c} \frac{kdk}{k^2 + \frac{\omega_p^2}{v^2}} \tag{102}$$

where k_c is either the necessary cutoff (having neglected dispersion) or the maximum transverse momentum transfer allowed by the spectrometer aperture $\theta = k_c/k_0$ (see Fig. 7), whichever is the smallest. The final result is the celebrated formula for the scattering probability

$$Q(a) = \frac{e^2 \omega_p a}{2\hbar v^2} \ln\left(1 + \frac{k_c^2 v^2}{\omega_p^2}\right) \ . \tag{103}$$

Q/a gives the probability per unit path length in a <u>bulk</u> metal

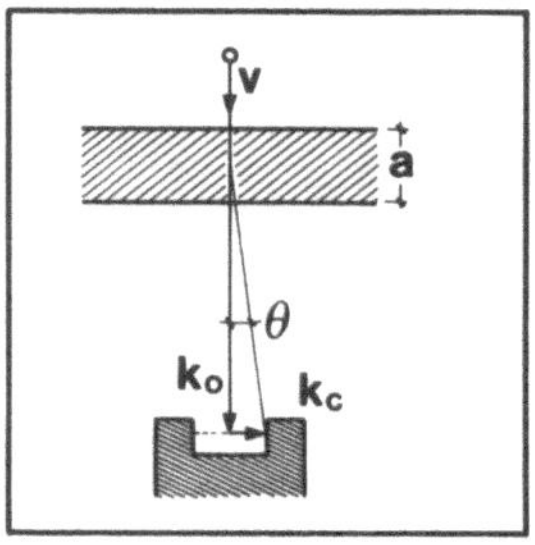

Fig. 7. Geometry defining the transverse cutoff wavevector k_c as determined by the spectrometer aperture.

for exciting one plasmon. Hence the inelastic mean free path for fast charged particles in a bulk plasma is

$$\Lambda(v) = \frac{a}{Q} = \frac{\hbar v^2}{e^2 \omega_p} \left[\ln \frac{k_c v}{\omega_p} \right]^{-1} \tag{104}$$

This result is plotted in Fig. 8 and compared with experiment [16] for Al and C, two materials in which bulk plasmons are fairly well defined. The Poisson nature of the EELS spectra in Al is clearly demonstrated in Fig. 9. [17, 18].

Note that the energy-dependent mean free path in (104) can also be derived from a classical dielectric treatment of the stopping power of the metal for charged particles. Such a treatment, based on Maxwell's equations, leads to the formula for the rate of energy loss

$$W = - \frac{e^2}{\pi^2} \int d^3k \; k^{-2} \int_0^{\infty} d\omega \; \omega \; \mathrm{Im}\, \varepsilon^{-1}(\vec{k},\omega) \; (\omega - \vec{k}.\vec{v}), \tag{105}$$

in terms of a loss-function $-\mathrm{Im}(1/\varepsilon)$. In (98) and (105) the δ-function expresses the fact that the energy dissipation process takes place through the excitation of collective modes which "ride" at a phase velocity ω/k_z equal to the particle velocity v.

It is important to mention here that for thin films the bulk plasmon excitation strength Q(a) in (103) is reduced by a pure

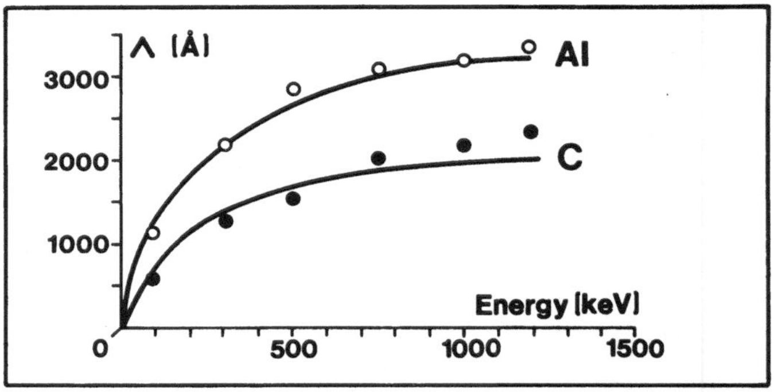

Fig. 8. Energy-dependent electron mean free path in Al and C (dots) compared with the bulk plasmon excitation mean free path obtained in Eq. (104).

surface term which, in the limit $a \ll \frac{v}{\omega_p}$ (~ 100 Å for 200 keV electrons and ~ 0.5 μ for 200 keV protons, in Al), cancels this bulk term. This important reduction has been given the name "Begrenzung" effect [19] by the German school when it was pointed out in the classical dielectric theory of energy losses. A similar compensation has already been mentionned to occur in the static image charge problem where, inside a metal, the long range part of the bulk screening is essentially cancelled by the surface image potential (see Fig. 6). The microscopic origin of the Begrenzung effect has been discussed in refs.[9-19] .

III.C. Surface, Reflection Spectrum

We will focus the attention on intrinsic surface effects by assuming that the charged particle trajectory $\vec{r}(t)$ does not penetrate into the metal sufficiently deep to excite bulk plasmons. This assumption is made here for the sake of simplicity but is probably not always justified, especially when penetration does occur, like with fast electrons which, to be reflected, require the coherent elastic scattering of a few surface layers of atoms. In such cases it is necessary to use the surface and bulk coupling functions C_k and M_k for the near surface region [5] rather than their asymptotic expressions (38) and (66) valid at large distances from the surface. This hasn't been done yet and awaits for further work.

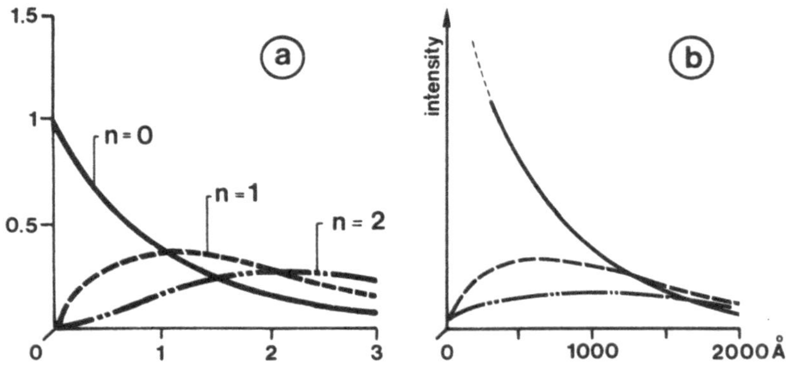

Fig. 9. a. Relative intensity of the n-th order peak in a theoretical Poisson distribution $e^{-\alpha}\,\alpha^n/n!$ as a function of the strength α.

b. Observed intensities of the first three loss peaks in EELS spectra of Al films as a function of film thickness.

Let $\vec{v} = (\vec{v}_\parallel, \pm v_\perp)$ be the constant velocity components of the particle before and after specular reflection [20]. The trajectory which we will use

$$\vec{r} = (\vec{v}_\parallel t,\ v_\perp |t| + z_o) \tag{106}$$

neglects the complicated multiple elastic scattering path near the surface and simply assumes a distance of closest approach z_o from the surface. Introducing (106) into (88) and (92), we get

$$I_k(t) = -\frac{i}{\hbar} C_k e^{-kz_o} \left[\frac{e^{-i\Omega_k t - kv_\perp |t|}}{i\Omega_k + kv_\perp |t|/t} - (1 + \frac{|t|}{t}) \mathrm{Re}(kv_\perp - i\Omega_k)^{-1} \right] \tag{107}$$

where

$$\Omega_k = \omega_k - \vec{k}.\vec{v}_\parallel \tag{108}$$

is the surface plasmon frequency, Doppler-shifted by the particle motion parallel to the surface. It is instructive to plot the time-evolution of $I_k(t)$ in the complex-I plane, as illustrated in Fig.10.

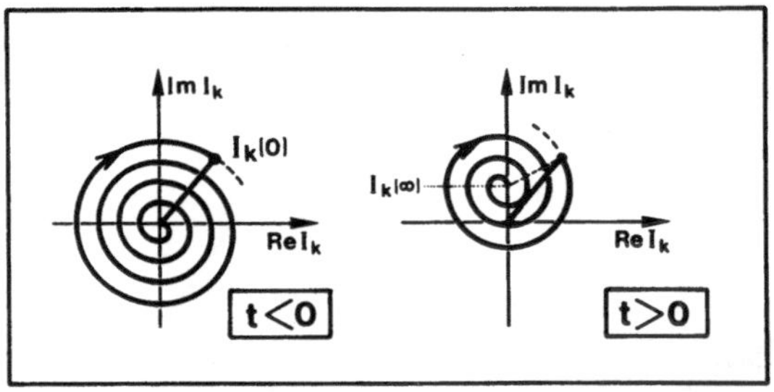

Fig. 10. Time evolution of the complex amplitude $I_k(t)$ for surface plasmons under the excitation by specular reflected charged particles.

I_k rapidly oscillates at the frequency Ω_k, growing from zero at $t = -\infty$ to a maximum value at the time of specular reflection $t = 0$ and then decreasing towards a finite value given by the second term in (107), namely

$$I_k(t) = 2\frac{i}{\hbar} C_k e^{-kz_0} \mathrm{Re} \frac{1}{kv_\perp - i\Omega_k} \tag{109}$$

$$= 2\frac{i}{\hbar} C_k e^{-kz_0} \mathrm{Re} \frac{kv_\perp}{k^2 v_\perp^2 + \Omega_k^2} \tag{110}$$

In this result, $kv_\perp$ is a measure of the inverse time during which the particle is strongly coupled to the surface plasmon of wave-vector k, before and after specular reflection. If this rate of approach is not small as compared to the Doppler shifted frequency Ω_k, then the asymptotic coherent amplitude $I_k(\infty)$ is not small either.

The total loss strength Q in (95) has been calculated for a general angle of incidence [19] but here, for illustration purpose, we will consider only two limiting cases, i.e. the normal and grazing incidence trajectories for which the $\vec{k}$-integration in (95) is particularly easy.

For normal incidence, we have $v_\parallel = 0$, $v_\perp = v$ and $\Omega_k = \omega_k$.

Hence

$$Q = \frac{A}{(2\pi)^2} \int d^2k \, C_k^2 \, e^{-2kz_o} \frac{k^2v^2}{(k^2v^2 + \omega_k^2)^2} . \tag{111}$$

Using the coupling function (66) with the exponential cutoff (69) and neglecting dispersion in ω_k, we find

$$Q = \frac{e^2}{\hbar v} I(\alpha) \tag{112}$$

where

$$I(\alpha) = 2 \int_0^\infty dx \, e^{-\alpha x} \frac{x^2}{(1 + x^2)^2}$$

$$= \text{Ci}\,\alpha(\sin\alpha + \alpha\cos\alpha) - \text{si}\,\alpha\,(\cos\alpha - \alpha\sin\alpha) \tag{113}$$

and

$$\alpha = 2(z_o + k_c^{-1}) \frac{\omega_s}{v} . \tag{114}$$

In this form, the loss strength Q appears as the product of a particle "fine-structure constant"

$$\frac{e^2}{\hbar v} = \frac{1}{137} \frac{c}{v} \tag{115}$$

and an integral whose value depends on the model parameter α as sketched in Fig. 11 a. Another transparent way to write Q is

$$Q = \frac{e^2}{(z_o + k_c^{-1})\,\hbar\omega_s} \frac{1}{2} \alpha I(\alpha) \tag{116}$$

where the function $\alpha I(\alpha)$ is plotted in Fig. 11 b where it is seen to have a broad maximum around $\alpha \simeq 1$.

<u>At grazing incidence</u>, we have $v_\perp \to 0$, $v_\parallel = v$ and the loss strength can be derived from the property

$$\lim_{v_\perp \to 0} \text{Re} \frac{1}{kv_\perp - i\Omega_k} = i \lim_{v_\perp \to 0} \text{Im} \frac{1}{\Omega_k - ikv_\perp} = i\pi\delta(\Omega_k) \tag{117}$$

This expresses, as before in the bulk problem (Eq. 98), the resonant excitation of surface plasmons whose phase velocity coïncides with the particle velocity. The δ-function allows the angular integral

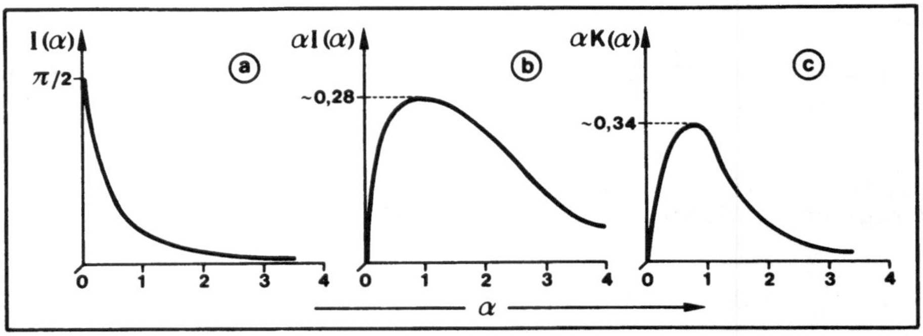

Fig. 11. Strengths of the Poisson distribution for surface plasmon excitation in the conditions of Eqs. (112), (116) and (120) respectively, as a function of the parameter α defined in Eq. (114).

in (95) to be done as

$$\int_0^2 d\phi\ \delta(\omega_k - kv\cos\phi) = \frac{2}{\omega_k\sqrt{x^2-1}}\ \Theta\ (x-1) \tag{118}$$

where $x = kv/\omega_s$ and Θ is the unit step function. Inserting (118) into (95), we find that the loss strength is again the probe fine structure constant times an integral depending on α :

$$Q = \frac{e^2}{\hbar v_\perp}\ 2 \int_1^\infty dx\ e^{-\alpha x}\ \frac{1}{x\sqrt{x^2-1}} \tag{119}$$

which can be rewritten an

$$Q = \frac{e^2}{(z_0 + k_c^{-1})\ \omega_s}\ \frac{1}{\Phi}\ \alpha K(\alpha) \tag{120}$$

where $\Phi = v\ /v$ is the grazing incidence angle and

$$K(\alpha) = \int_\alpha^\infty dz \int_1^\infty dx\ e^{-zx}\ \frac{1}{x\sqrt{x^2-1}} = \int_\alpha^\infty dz\ K_0(z) \tag{121}$$

$K_0(z)$ is the zeroth order modified Bessel function tabulated, as well as its integral in (121), in ref. [21]. The function $\alpha K(\alpha)$ is plotted in Fig. 11 c and exhibits a maximum around $\alpha \simeq 0.8$.

The $\frac{1}{\theta}$ dependence of Q in (120) has been confirmed in electron scattering experiments on liquid metal surfaces [21,22] which offer the flatness and smoothness required for a grazing incidence geometry. Fig. 12 shows a typical Poisson spectrum of multiple surface plasmon excitation on liquid In surface [23] . The incidence angle ($\sim 1°$ grazing) and electron beam energy (10 keV) are such that the penetration of the beam is very small so that bulk plasmon excitation, if any, is undetectably weak.

It is worthwhile to emphasize that, for the surface inelastic scattering, no meaningful concept of mean free path similar to the bulk mean free path can be defined. The usefulness of this quantity in the bulk stems from the fact that the multiple scattering spectrum such as in Fig. 9 can be interpreted as a random succession in space or time of single bulk plasmon excitation events, the probability of which could be calculated from first order perturbation theory or even from classical electrodynamics. No such picture holds for the surface plasmon excitation spectrum, such as in Fig. 12, because the excitation probability is a continuously varying function of time or space (see Fig. 10). The discreteness of the multiple peaks is then a direct manifestation of the quantized nature of the plasmon oscillations, not of the scattered particles [24] as has been wrongly argued sometimes in the litterature. The Poisson-nautre of the energy distribution is in fact a consequence of the classical and unperturbed nature of the probe since an harmonic oscillator driven by such a probe is always

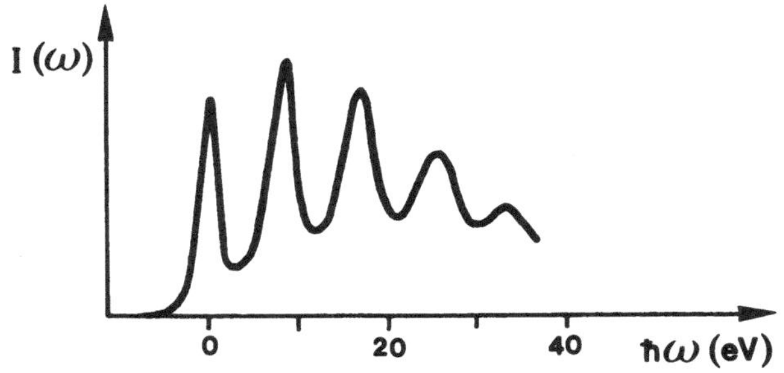

Fig. 12. Observed Poisson distribution of peak intensities for EELS spectra of fast electrons (10 keV) at liquid In surface at $\theta = 1.3°$ grazing incidence (after ref. 23).

prepared into a coherent state whose energy spectrum obeys Poisson statistics.

Altough the quantization of the collective modes is essential to understand the multiply peaked spectrum, the strength Q can nevertheless be obtained from an entirely classical approach, just as for the bulk loss rate in Eq. (105). One uses the classical trajectory of the charged particle (106) as a time dependent current source in Maxwell's equations and compute the power loss of the beam. The result of this dielectric approach is for the power loss spectrum

$$W_s = \frac{e^2}{v_\perp} f(\theta,k_c) \; \mathrm{Im} \frac{-1}{\varepsilon(\omega) + 1} \tag{122}$$

where f is a slowly varying function [19] of the incidence angle Φ and of k_c, the cutoff wavevector. The expression $-\mathrm{Im}(\varepsilon+1)^{-1}$ is known as the surface loss function. The argument has a pole when $\varepsilon(\omega)+1 = 0$ which determines the surface plasmon frequency, as in Eq.(18) for the metal-vacuum interface. Using the Drude ε of Eq. (8), one gets $\mathrm{Im}(\varepsilon+1)^{-1} \simeq \omega_s \delta(\omega-\omega_s)$ and the coefficient of the δ-function in (122) coïncides with Q as obtained from the quantum mechanical treatment.

III.D. Fluorescence

More complex inelastic scattering processes at surfaces involving internal excitation of the probe particle have been recently studied. One example is the radiation fluorescence of fast ions transmitted through thin foils or reflected at metal surfaces. In this section we will examine the fluorescence scattering of ions at grazing incidence as this case is particularly simple and instructive [25].

When a fast ion impinges on a smooth surface at grazing incidence, it sees the atomic rows as a near continuum. If the incidence angle with the surface is smaller than some critical value, the ion experiences a coherent sequence of two-body elastic or nearly elastic collisions in the near forward direction which deflect the ion in the specular direction, as shown in Fig. 13. Experimental evidences of this are shown in Figs. 14 and 15. In Fig. 14, the sputtering yield S of secondary particles emitted per primary ion hitting a smooth surface at incidence θ (with respect to normal) is seen to drop towards zero when θ approaches a critical angle close to 90°. In Fig. 15, the angular distribution of scattered primary ions is observed to peak sharply in the specular direction at grazing incidence [26]. This surface phenomenon corresponds, in the bulk, to planar channeling where fast ions are trapped between two atomic planes by successive specular reflections

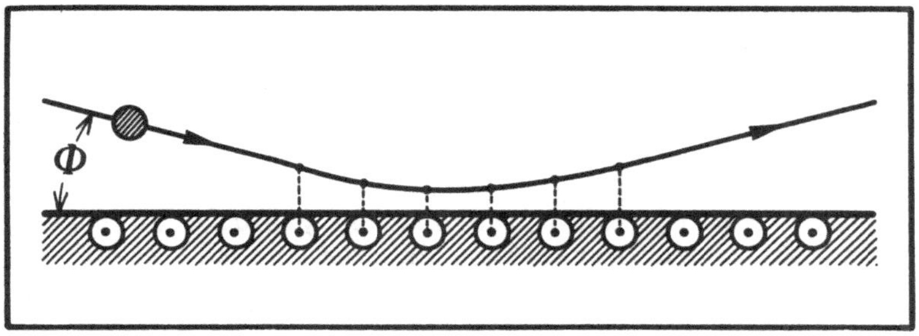

Fig. 13. Multiple forward scattering collisions of a fast ion with surface atoms resulting in specular reflection.

such as in Fig. 13. The ions are kept away from the plane of atomic nuclei by a distance of order of the sum of the core radii of the projectile and target ions. Here then is a case where there is no penetration at all (ignoring surface defects) and a description in terms of surface excitations only has a chance to be accurate.

In order to predict the spectral characteristics of the fluorescence, we need to calculate the final electronic state of the ion after scattering by the surface. Even in the absence of penetration, the surface scattering processes are still too complex to be described completely in a single theory. In particular, we intend to neglect :

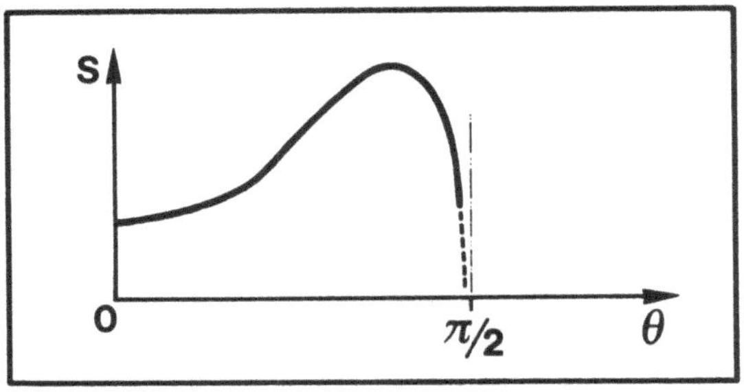

Fig. 14. General trend observed for the sputtering yield of fast ions at surfaces as a function of incidence angle Φ.

a) charge transfer processes, such as ion neutralization; they can, in principle, be discriminated against experimentally.

b) retardation effects in the ion-surface plasmon coupling. This is a rather severe approximation in view of the fact that radiation lifetime of exited atoms have been shown to be drastically altered with respect to the vacuum case) when the atoms are close to a surface [27,28] . Ideally, coupling to radiative surface plasmons (plasmons polaritons) should be considered but here we will calculate a spectral property, the Stoke parameters, that do not require the absolute fluorescence yield. A full scattering theory, including transverse photons, remains to be worked out.

We must then evaluate the transition probability from the ground state to excited states when the ions have been reflected along the specular direction. What can cause such transitions ? Referring to the multipolar Hamiltonian (78)-(79), we envisage that, due to the dime dependence of the ion trajectory $\vec{R}(t)$ in the coupling terms, excitation of plasmons and electron raising and lowering transitions will take place during the scattering. To be specific, let us consider the first two terms in (78), i.e. the monopolar and dipolar interactions, and restrict ourselves to one-electron ions such as He^+.
The total system Hamiltonian can be written, in second quantized form, as :

$$H \equiv H_e + H_p + H_m + H_d = \sum_{\nu=o,x,y,z} E_\nu c_\nu^+ c_\nu + \sum_{\vec{k}} \hbar\,\omega_k a_k^+ a_k$$

$$+ \sum_{\vec{k}} (f_k a_k^+ + f_k^\star a_k) + \sum_{\vec{k}} \sum_{\nu=x,y,z} (g_{k\nu} a_k^+ + g_{k\nu}^\star a_k)(c_o c_\nu^+ + c_\nu c_o^+) \qquad (123)$$

where the first two terms are the free-electron and free-plasmon Hamiltonians and where, from (82) and (83),

$$f_k(t) = C_k\, e^{\vec{K}.\vec{R}(t)} \quad , \quad g_{k\nu}(t) = Q_\nu K_\nu f_k(t) \qquad (124)$$

the vector $\vec{K}$ being defined in (81).
The c-numbers

$$Q_\nu = \langle 0|\nu|\nu\rangle \qquad (\nu = x,y,z) \qquad (125)$$

are the "dipole lengths" for the transitions from the ground state $|0\rangle$ to the excited states $|x\rangle$, $|y\rangle$, $|z\rangle$ of excitation energy $\hbar\omega_o = E_\nu - E_o$. These coefficients are related to the 1S → 2P oscillator strength f by $|Q_\nu|^2 = 3\,\hbar\, f/2m\,\omega_o$. Implicit in writing (123) is the further assumption of a 2-level model E_ν only. This is in keeping wiht our intention to use first order perturbation theory which allows these raising transitions only (dipole selection rule).

We already know from the last section [Hamiltonian (86)] that the monopole-surface plasmon coupling H_m is responsible for multiple excitation of plasmons. These excited plasmons are, in turn, able to excite the ion through their electric field which, as we know, extends into the vacuum. In addition, the fourth term H_d will also produce plasmons and induce dipole raising transitions. Hence, there are two channels through which the ion can be excited. A particularly illuminating, though quantitatively inaccurate picture of the total scattering process in terms of quasi-static images is sketched in Fig. 15 : the monopole charge e and dynamical dipole moment $\vec{Q}$ scatter on their images -e and $\vec{Q}_I$ in the metal surface. Since the ion moves rapidly ($v \sim v_F$) and the dipole fluctuation is fast ($\hbar\omega_0 \sim 40$ eV in He^+), this static picture is invalid and the plasmon response must be treated dynamically.

To achieve this we will exploit our knowledge of the exact wave function of the monopole part of the full Hamiltonian and we shall treat the dipole term H_m, which cannot be solved exactly, by time-dependent perturbation theory. Thus the unperturbed, exact evolution operator for $H_e + H_p + H_m$ is

$$U_0(t) = e^{-\frac{i}{\hbar}(H_e + H_p)t} D(t) \tag{126}$$

where D(t) is given in Eqs. (90), (92). From scattering theory [29] the exact evolution operator of the full system is then

$$U(t) = U_0(t) \, T \exp [- \frac{i}{\hbar} \int_{-\infty}^{t} dt' \, H_d^I(t')] \tag{127}$$

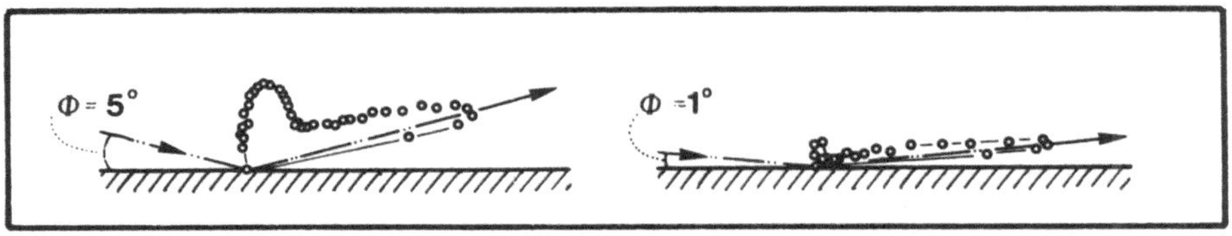

Fig. 15. Observed angular distributions of Ar^+ ions (300 keV) scattered by a Cu surface at decreasing grazing angles (after ref. 26). Angles are not to scale.

where H_d^I is the interaction representation of H_d :

$$H_d^I(t) = U_o^+(t)\ H_d(t)\ U_o(t) \ . \tag{128}$$

The canonical transformation is easily carried out by using the basic displacement property (91) :

$$H_d^I(t) = \sum_{\vec{k}} \sum_{\nu} [\, g_{k_\nu}^{\star} (a_k + I_k)\ e^{-i\omega_k t} + h.c.\ ]\ (c_o^+ c_\nu\ e^{-i\omega_o t} + h.c.) \tag{129}$$

If $|\{o_k\}, \{o_\nu\}>$ is the initial ground state of the system, the wavefunction, the first order in H_d, is

$$|\Psi(t)> \ = U(t)\ |\{o_k\}, \{o_\nu\}>$$

$$\simeq U_o(t)\ [\ 1 - \frac{i}{\hbar} \int_{-\infty}^{t} dt'\ H_d^I(t')\ ]\ |\{o_k\}, \{o_\nu\}>$$

$$= e^{-\frac{i}{\hbar}(H_e+H_p)t}\ D(t) \left\{ |\{o_k\}, \{o_\nu\}> - \sum_{\nu} [F_\nu(t) + \hat{G}_\nu(t)]\ |\{o_k\}, \nu> \right\} \tag{130}$$

where we have introduced the c-numbers

$$F_\nu(t) = \frac{i}{\hbar} \sum_{\vec{k}} \int_{-\infty}^{t} d\tau\, [\, g_{k\nu}^{\star}(\tau)\ I_k(\tau)\ e^{-i\omega_k \tau} + cc\ ] \tag{131}$$

and the plasmon field operators

$$\hat{G}_\nu(t) = \frac{i}{\hbar} \sum_{\vec{k}} \int_{-\infty}^{t} d\tau\ \ g_{k\nu}(\tau)\ a_k^+\ e^{i(\omega_k+\omega_o)\tau} \tag{132}$$

The first term in (130) describes the monopole scattering as studied in the previous section and the second term gives the coherent population of excited P states $|\nu>$ as well as further plasmon excitations by the dipole coupling operators.

With the wavefunction (130) we can now compute scattering probabilities. The probability amplitude of finding at $t = \infty$

the plasmons in a final energy state $|\{n_k\}>$ of excitation energy $E_f = \Sigma\, n_k\, \hbar\omega_k$ and the electron in the excited state $|\nu>$ is $<\Psi(t)|\{n_k\},\nu>$. Therefore the probability $P_\nu(\omega)$ for the ion to be excited to state $|\nu>$ and lose the kinetic energy $\hbar\omega$ is

$$P_\nu(\omega) = \sum_{\{n_k\}} |<\Psi(\infty)|\{n_k\},\nu>|^2\, \delta(\omega - \frac{E_f + \hbar\omega_o}{\hbar}) \quad (133)$$

The total probability

$$\sum_\nu \int_{-\infty}^{+\infty} d\omega\; P_\nu(\omega) = 1 \quad (134)$$

is normalized to unity due to completeness of the plasmon states

$$\sum_{\{n_k\}} |\{n_k\}><\{n_k\}| = 1 \quad (135)$$

and the assumed completeness of our two-level ion states :

$$\sum_\nu |\nu><\nu| = 1 \quad (136)$$

The property (135) can be exploited in the usual way be introducing the correlation function $P_\nu(t)$ which is the Fourier transform of the excitation loss spectrum :

$$P_\nu(t) = \int_{-\infty}^{+\infty} d\omega\; P_\nu(\omega)\; e^{-i(\omega-\omega_o)t} \quad (137)$$

i.e.

$$P_\nu(t) = <\Psi_\nu(\infty)|e^{-\frac{i}{\hbar}H_p t}|\Psi_\nu(\infty)> \quad (138)$$

where

$$|\Psi_\nu(\infty)> = <\nu|\Psi(\infty)> = D\,[\,F(\infty) + \hat{G}(\infty)\,]\,|\{o_k\}>. \quad (139)$$

It is not difficult to work out the expectation value (138). The final result is

$$P_\nu(t) = \langle o|\; (F_\nu^\star + G_\nu^\star)\; D^+\; e^{-\frac{iH_p t}{\hbar}}\; D(F_\nu + G_\nu)\; |o\rangle$$

$$P_\nu(t) = P_0(t) \{-[F_\nu + G_\nu(e^{-i\omega_s t}-1)] [F_\nu^\star + G_\nu^\star(e^{-i\omega_s t}-1)] + D_\nu e^{i\omega_s t}\} \quad (140)$$

where $P_0(t)$ is the no-excitation correlation function obtained in Eqs. (93)-(94) and where

$$G_\nu = \frac{i}{\hbar} \sum_{\vec{k}} \int_{-\infty}^{+\infty} d\tau \; g_{k\nu}(\tau) \; I_k^\star(\tau) \; e^{i(\omega_s+\omega_0)\tau} \quad (141)$$

$$D_\nu = \frac{1}{\hbar^2} \sum_{\vec{k}} \left| \int_{-\infty}^{+\infty} d\tau \; g_{k\nu}(\tau) \; e^{i(\omega_s+\omega_0)\tau} \right|^2 \quad (142)$$

$G_\nu^\star$ is the eigenvalue of $\hat{G}_\nu^+$ for the coherent eigenstate $|\{I_R\}\rangle$.

The overall correlation function $P_\nu(t)$ thus appears as a direct product of the no-excitation factor $P_0(t)$ and a time-dependent factor involving the square of the dipole oscillator strength f. Hence the excitation-loss spectrum $P_\nu(\omega)$ in (137) will be a convolution of the no-excitation spectrum (i.e. a Poisson distribution) with additional, ω-dependent excitation and loss factors. The processes of internal excitation and kinetic energy losses are therefore interfering and cannot be disentangled in the total spectrum. It will also not be easy to measure $P_\nu(\omega)$ as this requires discrimination of the excited and non-excited particles in the loss spectra. But the fluorescence intensity and its polarization state have been explored in detail with an experimental arrangement schematized in Fig. 17. The fluorescence yield measures the total internal excitation probability P_ν irrespective of the plasmon state left behind. This quantity can be obtained easily from the spectrum $P_\nu(\omega)$ by integrating out the kinetic energy losses :

$$P_\nu = \int_{-\infty}^{+\infty} d\omega \; P_\nu(\omega) = \int_{-\infty}^{+\infty} dt \; P_\nu(t) \; \delta(t) \quad (143)$$

Introducing (140) into (143) we finally obtain the very simple result

$$P_\nu = |F_\nu|^2 + D_\nu \; . \quad (144)$$

From the structure of F_ν in (131) which, through I_k and $g_{k\nu}$, involves both monopole and dipole couplings, the significance of the first term in (144) is clear : the ion monopole prepares the plasmon field into the coherent state $|\{I_k\}\rangle$ which, in turn, can 'decay' by inducing dipole raising transitions on the ion. The second term in (144) on the other hand accounts for direct excitation processes by the dipole-plasmon coupling. Thus the probabilities from the two excitation channels add up incoherently in P_ν even though the excitation-loss amplitudes interfere in the spectrum $P_\nu(\omega)$.

It is important to note that, in the absence of monopole coupling, i.e. when $I_k = 0$, the preceding results simplify to

$$P_\nu(t) = D_\nu \, e^{i\omega_s t} , \tag{145}$$

$$P_\nu(\omega) = D_\nu \, \delta[\omega - (\omega_o + \omega_s)] , \tag{146}$$

$$P_\nu = D_\nu , \tag{147}$$

i.e. the first order Born result. Hence fast neutral atoms or molecules scattering off metal surfaces at grazing incidence can also be excited by plasmons. To our knowledge, no experiment has attempted to verify this yet, although neutral Hydrogen fluorescence has been observed to follow proton scattering and attributed to neutralization via excited states [30].

When one evaluates the formal result (144) in practical cases of grazing incidence scattering [25], the indirect F_ν term is generally found to be several orders of magnitude smaller than the D_ν term. This is easily understood in view of the fact that the image monopole scatters off the ion at vanishingly small approach velocity $v_\perp$ (see Fig. 16) and hence causes quasi-adiabatic polarization of the ion and no strong real excitation. We will therefore forget about the F terms and evaluate only the pure dipole term D_ν. Introducing the specular trajectory into (124), we get

$$D_\nu = \frac{e^2 Q_\nu^2 \omega_s}{\pi\hbar} \int d^2k \, \frac{|K_\nu|^2}{k} \, S^2(k) \, e^{-2kz_o} \left(\mathrm{Im} \frac{1}{\omega_o + \omega_s - \vec{k}.\vec{v}_\parallel + ikv_\perp}\right)^2 \tag{148}$$

The total fluorescence yield is

$$Y = \sum_\nu D_\nu = 2 D_z \tag{149}$$

where D_z is the $|o\rangle \rightarrow |z\rangle$ excitation probability.

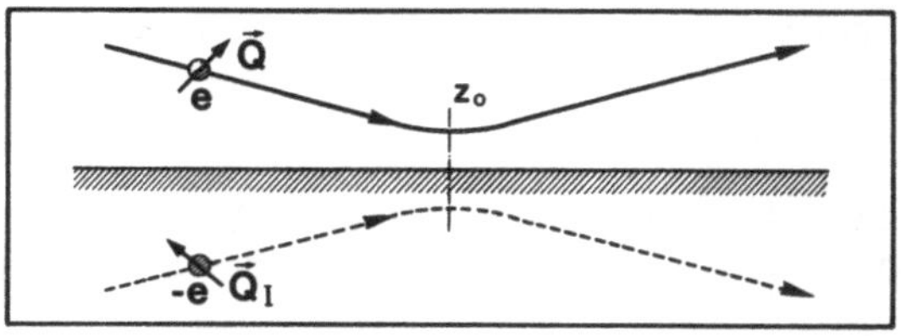

Fig. 16. Specular reflection of a fast ion seen as a scattering by its image monopole -e and image dynamical dipole $\vec{Q}_I$.

For grazing incidence, this is easily obtained with the help of (117) and (118). The result is

$$D_z = 2 \frac{e^2}{\hbar v_\perp} (k_c Q)^2 \frac{\omega_o + \omega_s}{\omega_o} \cdot F(\beta) \tag{150}$$

where

$$F(\beta) = y_c^{-2} \int_o^\infty dy\; e^{-\beta y}\; y/\sqrt{y^2-1} = y_c^{-2} K_1(\beta) \tag{151}$$

$$y_c = k_c v / (\omega_o + \omega_s) \tag{152}$$

$$\beta = \frac{2}{y_c} (1 + k_c z_o) \tag{153}$$

The dimensionless integral in (151) is the modified Bessel function of first order K_1 [21]. For convenience the function $F(\beta)$ is plotted in Fig. 18 for a few values of the reduced impact parameter $k_c z_o$. We note that this function peaks around $y_c \sim 2$, just as the other functions in Fig. 11 b,c.

Due to the dependence of (150) on the particle "fine structure constant" $e^2/\hbar v_\perp$, the total "probability" 2 D_z is easily made larger than 1 ($k_c Q \sim 1$ for He^+ 1S → 2P transition). This absurd result simply means breakdown of first-order perturbation theory to calculate excitation probabilities in such conditions. A non-perturbative approach suitable to deal with this strong coupling case remains to be developped. Meanwhile, the result (150) can

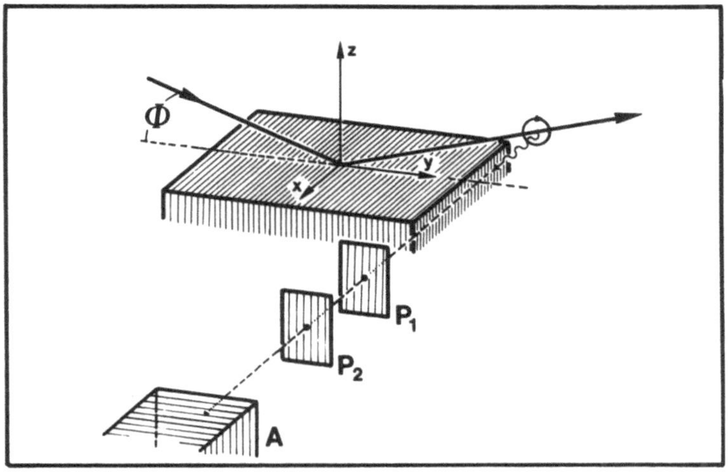

Fig. 17. Experimental arrangement to observe fluorescence of an ion beam scattered by a surface at incidence angle Φ. $P_1, P_2, \ldots$ are polarizer-analyser arrangements to observe the polarization state of the light with spectrometer A.

only be used for experimental conditions such that $Y < 1$.

The fluorescence emitted by particles scattered by thin films or reflected by surfaces at oblique incidence has been found to be elliptically polarized when viewed in direction perpendicular to the plane of incidence [26](see Fig. 17). The circularity of the polarization is observed to increase with increasing incidence angle and is maximum for grazing incidence. The experimental parameter which measures this observation is the ratio $(I_+ - I_-)/I_0$ of the difference between the righ- and left-handed circular polarization intensities and the total intensity I_0 in the viewing direction. Elliptic polarization for the 2P → 1S fluorescence in He^+ and isoelectronic ions implies and unequal probability of excitation for the three magnetic sublevels m = -1,0,+1 of the 2P state with the viewing direction as quantization axis. This anisotropic excitation is inherent in the surface plasmon mechanism studied here. Indeed, the probability amplitudes for exciting the P orbitals $|x\rangle$, $|y\rangle$, $|z\rangle$ are proportional to the components ik_x, ik_y, $-k$ of the vector $\vec{K}$ defined in (81). Note the $\frac{\pi}{2}$ phase shift (factor i) between the y component of $\vec{K}$ and its z component. This means that after scattering by the surface, the electron transition dipole moment rotates in a clockwise manner (Fig. 17) as viewed from the x direction. The amount of net circular polarization can be computed on the basis of the wavefunction (130). The relevant quantity here is the "orientation" parameter S/I defined by

$$\frac{S}{I} = \hbar \frac{\langle \Psi_{ex} | J_x | \Psi_{ex} \rangle}{\langle \Psi_{ex} | J^2 | \Psi_{ex} \rangle} = \Big(\sum_{m=-1,0,+1} m\sigma_m \Big) / \ell(\ell+1) \sum_m \sigma_m \qquad (154)$$

where

$$|\Psi_{ex}\rangle = \frac{i}{\hbar} \sum_i \sum_{\vec{k}} \int_{-\infty}^{+\infty} d\tau \, g_{ki}(\tau) \, e^{i(\omega_o+\omega_s)\tau} \, |\{0_\ell\}' 1_k\rangle |i\rangle \qquad (155)$$

is the final excited state of the scattered ion and plasmon system. In (154) the σ_m's are the production cross sections for the J_x angular momentum eigenstates $|m\rangle$ which are linearly related to our previous p-orbitals $|x\rangle$, $|y\rangle$, $|z\rangle$. It is not difficult to calculate the σ_m's from the wave-function (155). The final result [25] is

$$\frac{S}{I} = \frac{1}{2} K_o(\beta)/K_1(\beta) \qquad (156)$$

where K_o and K_1 are the modified Bessel function of zeroth and first orders which have been defined previously in Eqs. (121) and (151). This is plotted in Fig. 19 as a function of the parameter β^{-1} which is proportional, through Eqs. (152) and (153), to the ion velocity. The general trend of this theoretical velocity dependence is not inconsistent with observations for some scattering systems but a detailed test for the validity of Eq. (156) will require especially designed experiments. The rotational orientation effect obtained here from interaction with the inhomogeneous surface plasmon field is easily understood in classical terms : one can visualize the electronic cloud of the ion as a billard ball spinning, in the clockwise direction, after hitting a cushion at oblique incidence (see the geometry of Fig. 18).

REFERENCES

1. Alternative derivations in textbooks often use special geometries such as the planar, cylindrical or spherical capacitor configurations for the fluctuation. We leave it to the student to try their elementary electrostatics on such situations as an exercice. See e.g. C. Kittel, Introduction to Solid State Physics (Wiley, New York, 1971).

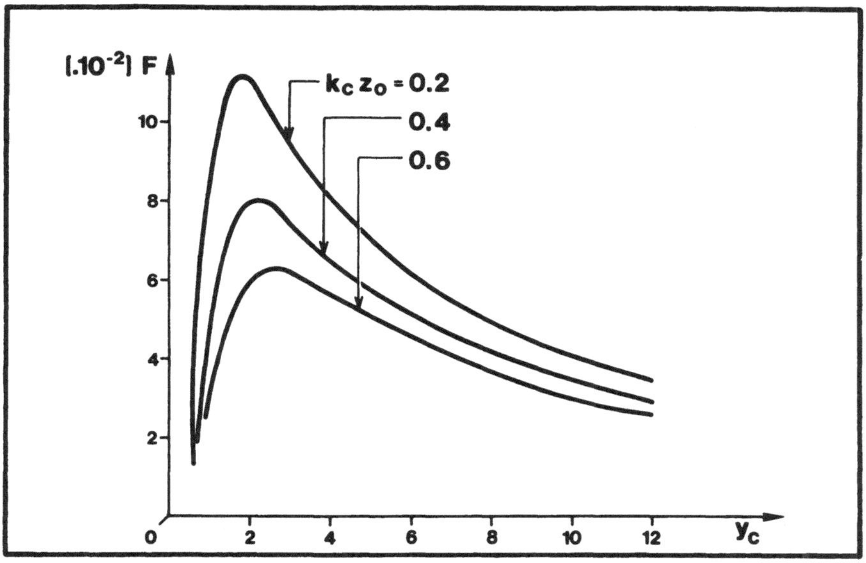

Fig. 18. Velocity dependence of the fluorescence yield as predicted by Eqs. (150), (152) for various impact parameters z_o.

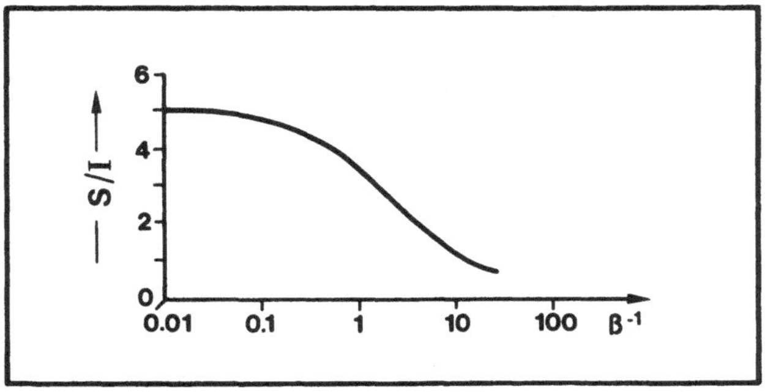

Fig. 19. Velocity dependence of the Stokes parameter obtained in Eq. (156).

S. Raimes, The Wave Mechanics of Electrons in Metals (North-Holland, Amsterdam 1967).
2. A number of physical systems in which spherical surface plasmons play a role have been studied recently in :
   a) A.A. Lucas, A. Ronveaux, M. Schmeits and F. Delanaye, Phys. Rev. B12, 5372 (1975);
   b) K. Ohtaka and A.A. Lucas, Phys. Rev. B18, 4643 (1978);
   c) K. Ohtaka, H. Miyazaki and A.A. Lucas, Phys. Rev. B21, 467 (1980);
   d) J.C. Rife, S.E. Donnelly, A.A. Lucas, J.M. Gilles and J.J. Ritsko, to be published.
   e) M. Schmeits and A.A. Lucas, Surface Sci. 64, 176 (1977); Surface Sci. 74, 524 (1978).
3. D. Pines, Elementary Excitations in Solids, W.A. Benjamin, N.Y. (1964).
4. A.W. Overhauser, Phys. Rev. B3, 1888 (1971).
5. A.G. Eguiluz, Phys. Rev. B, to be published.
6. J. Rogan, J.E. Inglesfield and T.B. Grimley, J. Phys. C, to be published.
7. A.A. Lucas, E. Kartheuser and R.G. Badro, Phys. Rev. B2, 2488 (1970).
8. G.D. Mahan, in Elementary Excitations in Solids, Molecules and Atoms, ed. by J.T. Devreese, A.B. Kunz and T.C. Collins (Plenum, London, 1974).
9. M. Sunjic and A.A. Lucas, Phys. Rev. B3, 719 (1971); Phys. Rev. Letters 26, 229 (1971).
10. A.A. Lucas, Phys. Rev. Letters 26, 813 (1971);Phys. Rev. B4, 2939 (1971).
11. E.N. Economou and K.L. Ngai, Adv. Chem. Phys. XXVII, 265 (1974).
12. A.A. Lucas and M. Sunjic, J. Vac. Science and Technology, 9, 725 (1971).
13. R. Gomer and L.W. Swanson, J. Chem. Phys. 38, 1613 (1963).
14. D. Newns, J. Chem. Phys. 50, 4572 (1969).
15. J. Heinrichs, Phys. Rev. B8, 1346 (1973).
16. A.A. Lucas,in "Elementary Excitations in Solids, Molecules and Atoms", ed. by J.T. Devreese, A.B. Kunz and T.C. Collins, NATO ASI Series B, Plenum (1974) Part A, p. 65.
17. H. Raether, Springer Tracts in Modern Physics, Vol. 88, Springer-Verlag, Berlin (1980).
18. J.O.H. Spence and A.E.C. Spargo, Phys. Rev. Letters 26, 895 (1971).
19. A.A. Lucas and M. Sunjic, Progress in Surface Sci.,2, 75 (1972), ed. by S.G. Davison, Pergamon Press.
20. Cases of Bragg reflections, charge exchange scattering and curved trajectories have been considered in A.A. Lucas and M. Sunjic, Surface Sci. 32, 439 (1972); A.A. Lucas, Phys. Rev. B20, 4990 (1979).
21. Handbook of Mathematical Functions, ed. by M. Abramowitz and I.A. Stegun, NBS Applied Math. Series 55 (G.P.O. Wash. D.C., 1964).

22. C.J. Powell, Phys. Rev. 175, 972 (1968).
23. J. Schilling, Z. Phys. B25, 61 (1976).
24. Classical particles such as heavy ions have been observed to exhibit Poisson-like energy distributions after interaction with a metal surface. See ref. 10.
25. A.A. Lucas, Phys. Rev. Letters 43, 1350 (1979); Phys. Rev. B20, 4990 (1979).
26. H.J. Andra, R. Fröhling, H.J. Plöhn and J.D. Silver, Phys. Rev. Lett. 37, 1212 (1976).
27. H. Morawitz and M.R. Philpott, Phys. Rev. 10, 4863 (1974).
28. K. Ohtaka and A.A. Lucas, Phys. Rev. B18, 4643 (1978).
29. A. Messiah, Quantum Mechanics (North Holland, Amsterdam, 1963) Vol. II.
30. N.H. Tolk, J.C. Tully, J.S. Kraus, W. Heiland and S.H. Neff, Phys. Rev. Lett. 42, 1475 (1979).

FROM MAGNONS TO SOLITONS

D. C. Mattis

Department of Physics
University of Utah
Salt Lake City, UT 84112, U.S.A.

ABSTRACT

This article concerns magnetism in nonmetallic solids. The analogy between magnons and phonons is established. Then, the exact 1- and 2-magnon spectra of ferromagnets are obtained, with bound states and scattering states being analyzed separately. The many-magnon states are obtained in a low-density approximation, and the effects of surfaces assessed.

In antiferromagnetism, the XY and XYZ (Heisenberg) antiferromagnets are solved exactly in one dimension by means of Bethe's ansatz.

Finally, the topological excitationk known as "solitons" are introduced, their nonlinear equations of motion are solved exactly, and various properties derived semi-rigorously.

I. INTRODUCTION

There exist striking similarities between the elementary excitations of a ferromagnet and those in an elastic solid. In the latter, we know that an atom displaced from equilibrium position will oscillate with the motion and frequencies of the normal modes of the cyrstal. The effect of quantum mechanics on this motion is to quantize the amplitudes of the individual normal modes, into units known as phonons. The analogous normal modes in the magnetic system are the spin waves; when the quantum mechanical nature of spins is taken into account, these also are quantized, with the basic unit being the magnon.

In the ferromagnet, for which the ground state has all spins parallel, the small amplitude oscillations are accurately described by harmonic-oscillator type dynamical variables. The interaction between neighboring oscillators is responsible for the dispersion, the dependence of frequency ω of a normal mode, on the wave-vector k (the wavelength $\lambda = 2\pi/k$). Because the ground state of a ferromagnet is known exactly, we are able to extend the analysis to obtain the exact one- and two-magnon eigenstates. The latter number bound states and scattering states, the interactions being caused by an inherent non-linearity in the spin dynamics, due to the coupling of the motion in various directions by the commutation relations of spin components. The analysis of 3- or more-magnon states requires the solution of a full many-body problem.

In the antiferromagnet (a material in which neighboring spins tend to be antiparallel), the inherent nonlinearities even prevent us from obtaining the exact ground state or elementary excitations, with the exception of one dimension (and there, only for spins 1/2 or spins $s \to \infty$, the classical limit). Nevertheless, spin waves are as useful for an approximate analysis as are the phonons in the study of the anharmonic solid. The ground state and elementary excited states, and the ground state correlation functions, can all be computed with spin wave theory. A check on the accuracy exists, in a variety of numerical experiments on representative systems.

II. MAGNONS

For concreteness, let us assume (for the present) a ferromagnetic Heisenberg Hamiltonian with nearest-neighbor interactions, in an external homogeneous magnetic field H:

$$H = -J \sum_{\substack{i>j=\text{nearest}\\ \text{neighbor}}}^{N} \sum \mathbf{S}_i \cdot \mathbf{S}_j - Hg\mu_B \sum_i^N S_i^z , \qquad J > 0, \tag{1}$$

where

$$\mathbf{S}_i \cdot \mathbf{S}_j \equiv S_i^x S_j^x + S_i^y S_j^y + S_i^z S_j^z = \frac{1}{2}(S_i^+ S_j^- + S_j^+ S_i^-) + S_i^z S_j^z$$

Beside the isotropic interactions which are included therein, we may also consider anisotropic interactions of various origins. Those of lowest order are of "dipolar" structure,

$$\frac{1}{2}\sum_i \sum_j D_{ij} \frac{\mathbf{S}_i \cdot \mathbf{S}_j r_{ij}^2 - 3\mathbf{S}_i \cdot \mathbf{r}_{ij} \mathbf{S}_j \cdot \mathbf{r}_{ij}}{r_{ij}^2} \tag{2}$$

which represents the magnetic dipole-dipole interaction (the magnetostatic potential of each spin in the field of the others) with

$$D_{ij} = g^2 \mu_B^2 r_{ij}^{-3} \tag{3}$$

But there may be additional contributions to D_{ij}, principally those arising from the spin-orbit coupling mechanism.

In Chapter 3 of our book on magnetisim [1] M1, we show that representations of spins may be constructed with the aid of harmonic-oscillator operators; and in particular, the Holstein-Primakoff representation, and approximations to it, were introduced. Let us review the latter for completeness. First, the spins are defined by the three nonvanishing matrix elements:

$$\langle n_i | S_i^x | n_i + 1 \rangle = \langle n_i + 1 | S_i^x | n_i \rangle^* = \frac{1}{2} \sqrt{(n_i + 1)(2s - n_i)}$$

$$\langle n_i | S_i^y | n_i + 1 \rangle = \langle n_i + 1 | S_i^y | n_i \rangle^* = - \frac{1}{2} i \sqrt{(n_i + 1)(2s - n_i)}$$

$$\langle n_i | S_i^z | n_i \rangle = s - n_i , \qquad 0 \leqslant n_i \leqslant 2s , \qquad \hbar = 1 . \tag{4}$$

The quantum numbers of other spins (n_j) have been suppressed, because the components of S_i cannot change these; but all occupation numers $n_1, n_2, \ldots, n_N$ must be specified in a definite state of the N spins, and therefore there are exactly $(2s + 1)^N$ such states if all spins have the same magnitude s.

Matrix elements of harmonic-oscillator operators have a structure very similar to the above. For example, if we symbolize the integrals

$$\int \psi_n^*(x_i)(op)\psi_m(x_i)dx_i$$

where $\psi_n(x)$ is the nth harmonic-oscillator function, by means of the Dirac bracket notation, $\langle n|(op)|m\rangle$, then for the very important operators x_i coordinate and p_i momentum, we have

$$\begin{aligned} \langle n_i | x_i | n_i + 1 \rangle &= \langle n_i + 1 | x_i | n_i \rangle^* = \sqrt{\frac{\hbar}{2m\omega} (n_i + 1)} \\ \langle n_i | p_i | n_i + 1 \rangle &= \langle n_i + 1 | p_i | n_i \rangle^* = - i \sqrt{\frac{\hbar m\omega}{2} (n_i + 1)} \end{aligned} \tag{5}$$

and the quantum number n_i may also be defined as the eigenvalue of a operator, now the energy operator:

$$\frac{P_i^2}{2m} + \frac{1}{2} m\omega^2 x_i^2 - \frac{1}{2}\hbar\omega|n_i\rangle = n_i\hbar\omega|n_i\rangle \tag{6}$$

The connection between harmonic oscillator and spin is now established, by use of the dimensionless canonical variables P, Q, defined by

$$Q_i = x_i\sqrt{\frac{m\omega}{\hbar}} \qquad P_i = p_i\sqrt{\frac{1}{\hbar m\omega}} \qquad [P_i,Q_j] = \delta_{ij}\frac{1}{i}, \tag{7}$$

by comparison of Eqs. (4) and (5). In terms of the P, Q, the matrix elements are:

$$\langle n_i|Q_i\sqrt{s}|n_i + 1\rangle = \langle n_i + 1|Q_i\sqrt{s}|n_i\rangle^* = \frac{1}{2}\sqrt{(n_i + 1)2s}$$

$$\langle n_i|P_i\sqrt{s}|n_i + 1\rangle = \langle n_i + 1|P_i\sqrt{s}|n_i\rangle^* = -\frac{1}{2}i\sqrt{(n_i + 1)2s} \tag{8}$$

$$\langle n_i|s - \frac{1}{2}(P_i^2 + Q_i^2 - 1)|n_i\rangle = s - n_i$$

quite similar to Eq. (4) <u>for small values of the</u> n_j. Note that for values of $n_i \approx s$, the harmonic-oscillator approximation becomes quantitatively incorrect, although the matrix structure is still qualitatively similar to the correct one. The error, however, becomes catastrophic only when some n_i equals or exceeds the value 2s, for although Eq. (8) continues to define matrix elements for the harmonic oscillators, there is no corresponding structure in angular-momentum space. So it must be understood that the whole theory which will be developed on the base of similarity between spins and harmonic oscillators will only be valid for low occupation numbers of every harmonic oscillator.

In the Hamiltonian of Eq. (1) we now make the substitutions,

$$S_i^x = Q_i\sqrt{s}, \qquad S_i^y = P_i\sqrt{s}, \qquad S_i^z = s - \frac{1}{2}(P_i^2 + Q_i^2 - 1) \tag{9}$$

and in accord with the above instructions, systematically discard cubic and quartic terms. One obtains the "linearized" Hamiltonian (i.e., the equations of motion are linearized, the Hamiltonian itself is of course quadratic):

$$H_{lin} = E_o + g\mu_B(H + \tilde{H}_o)\sum_i \frac{1}{2}(P_i^2 + Q_i^2 - 1)$$

$$- Js \sum_{\substack{\text{nearest}\\\text{neighbor}}} (P_i P_j + Q_i Q_j) \tag{10}$$

where

$$\tilde{H}_o = \frac{Jsz}{g\mu_B} \tag{11}$$

(z = number of nearest neighbors of any given spin) is precisely the molecular field constant which is introduced in the molecular field approximation, discussed in M2. The constant energy term

$$E_o = - NHg\mu_B s - \frac{1}{2} NzJs^2 \tag{12}$$

is the energy of the completely saturated state in which all spins are parallel to the applied field.

The linearized Hamiltonian bears an obvious resemblance to the Hamiltonian of lattice vibrations, which also involves interactions among neighboring harmonic oscillators. An important difference arises from the presence of "velocity-dependent forces" P_iP_j, to which may be attributed the difference between the spectra of spin waves and that of sound.

We now make the plane-wave transformation to the new set of generalized coordinates and momenta:

$$Q_i = \frac{1}{\sqrt{N}}\sum_k e^{ik\cdot R_i} Q_k \qquad P_i = \frac{1}{\sqrt{N}}\sum_k e^{ik\cdot R_i} P_k \tag{13}$$

note $Q_k^* = Q_{-k}$ etc., and

$$[P_k^*, Q_{k'}] = \frac{\delta_{kk'}}{i}$$

where $k_{x,y,z}$ = integer multiples of $2\pi/L$, are components of the <u>wavevector</u>. This transformation diagonalizes Eq. (10) for which

$$H_{lin} = \sum_k \frac{1}{2} (P_k^* P_k + Q_k^* Q_k - 1)\hbar\omega(k) + E_o = \sum_k n_k \hbar\omega(k) + E_o \qquad (14)$$

It may be verified that all harmonic operators, P_k, $Q_{k'}^*$, etc. which refer to different momenta, commute; and so we have N new oscillators, which are uncoupled in the above approximation. For each of these, the quadratic "number operator":

$$n_k \equiv \frac{1}{2} (P_k^* P_k + Q_k^* Q_k - 1) \qquad (15)$$

had integer eigenvalues = 0,1,2,.... Fortunately, it is not at all necessary that each of these eigenvalues be restricted to the range $\ll s$, and it is sufficient for the validity of the linearization procedure merely that $\sum n_k \ll Ns$.

As for the energies of the plane-wave "magnons,"

$$\hbar\omega(k) = Hg\mu_B + Js \sum_\delta (1 - \cos k\cdot\delta)$$

$$\simeq Hg\mu_B + Jsa^2k^2 + O(k^4) \qquad (16)$$

the sum runs over the vectors δ which connect a typical spin to the z nearest neighbors with which it interacts; $a \equiv |\delta|$ for the simple-cubic (sc) lattice, and the wave-vectors k are restricted to the range $|k_x|$, $|k_y|$, $|k_z|$ all $< \pi/a$, the "first Brillouin Zone."

So far, we have not taken advantage of the unique fact that the ferromagnetic ground state of H is known, where, reversing the sign of the applied field,

$$H = -J \sum_{\substack{i>j=\text{nearest}\\ \text{neighbor}}}\sum S_i \cdot S_j + Hg\mu_B \sum S_i^z \qquad (17)$$

and for this direction of applied field, the state in question consists of all spins "down" and will be hereinafter denoted

$$\Psi_o = |0) \qquad \text{energy } E_o = - NHg\mu_B s - \frac{1}{2} NzJs^2 \qquad (18)$$

the "vacuum" state. There are N different orthogonal and normalized states containing one spin deviate each:

$$\psi_i = (2s)^{-1/2} S_i^+ |0) \tag{19}$$

corresponding to all the choices of R_i. Now, amazingly enough, one finds that when H is applied to the state ψ_i, it generates other states of the same type but no states with two or more spin deviations are introduced. This would not be the case if we had included the anisotropic interactions in the Hamiltonian, and it is a simplifying property of the Heisenberg ferromagnetic exchange Hamiltonian. Thus H may be diagonalized within the N-dimensional subspace of the functions of Eq. (19).

If, as we shall assume, there is translation invariance (and periodic boundary conditions), the energy is immediately diagonalized by the introduction of plane waves:

$$\psi_k \equiv \frac{1}{\sqrt{N}} \sum_i e^{ik \cdot R_i} \psi_i \quad \text{and} \quad H \psi_k = E(k)\psi_k , \tag{20}$$

and the energy eigenvalues are

$$E(k) = E_o + \hbar\omega(k) \tag{21}$$

with E_o and $h\omega(k)$ exactly as defined in Eqs. (12) and (16). This is precisely the result we would have obtained in the harmonic-oscillator approximation to the isotropic Hamiltonian by setting $n_k = 1$ and all other $n_k' = 0$ in Eq. (14).

III. MAGNON-MAGNON FORCES

We have found that the ferromagnetic ground state, and the one-magnon eigenstates have precisely the energies predicted in the linearized harmonic-oscillator approximation. This is very encouraging, but not entirely unexpected in view of the agreement with the earlier semiclassical analysis. One might even be tempted to predict that states of two or more plane wave magnons are approximate eigenstates with an error of no more than O(1/N). But this prediction is only partly correct, as we shall see by a study of two-magnon states. For although most of the two-plane-wave magnon states undergo negligible scattering and energy shifts (which may be ascribed to the nonlinear corrections to the harmonic-oscillator approximation), a number of bound states appear. These are totally unexpected on the basis of our previous considerations, although they are of considerable importance in the demise of the spin-wave picture at or below the Curie temperature. Fortunately, the two-magnon problem can be solved exactly in any number of dimensions, and for any magnitude of the spins s, and so the relative importance of these bound states can be estimated under the various regimes.

Let us introduce a notation, so that we may develop qualitative arguments for why nonlinearities should be significant at long wavelengths. This will be followed by an exact analysis. Let the normalized, two-spin-deviate basis states be

$$\psi_{ij} = C_{ij} S_i^+ S_j^+ |0) = \psi_{ji} \tag{22}$$

For s > 1/2, there are 1/2 N(N + 1) states in this orthonormal set. For s = 1/2, the non-existence of states of type ψ_{ii} reduces the number of 1/2 N(N − 1). In any case, the energy and wavefunctions of N of these may be found without further calculation using only previously derived results. To see this, consider the two-plane-wave states

$$\psi_{kk'} = C \sum_{i,j} e^{i(k\cdot R_i + k'\cdot R_j)} \psi_{ij} = \psi_{k'k} \tag{23}$$

where C is an appropriate normalization constant. When k = 0, (23) is just a one-magnon eigenstate, to which the total spin raising operator $S^+ = \sum S_i^+$ has been applied:

$$\psi_{ok'} = C' S^+ \psi_{k'} \tag{24}$$

Now while S^+ (as well as S^- and S^z) commutes with the exchange part of the Hamiltonian, it is a raising operator for S^z. (The total spin is, of course, a good angular momentum.)

$$[S^z, S^+] = S^+ \qquad (\text{for } \hbar = 1) \tag{25}$$

that is, $S^z(S^+\psi) = S^+(S^z + 1)\psi$. Thus, for the Hamiltonian H of Eq. (17),

$$H\psi_{ok'} = C' S^+ (H + g\mu_B H)\psi_{k'} = \left(E(k') + g\mu_B H\right)\psi_{ok'} \tag{26}$$

and therefore,

$$E(Ok') = E_o + \hbar\omega(k') + g\mu_B H = E_o + \hbar\omega(k') + \hbar\omega(0) \tag{27}$$

In the absence of an external field H, $\omega(0) = 0$, and this particular two-magnon state is degenerate with the one-magnon state to which it is related by a spatial rotation. Since it is in any case an eigenstate, there is no "scattering," and there is no energy shift in a magnetic field beyond the simple change in Zeeman energy. By continuity, we may expect that if k and k' are reasonably small,

$$E(kk') = E_o + \hbar\omega(k) + \hbar\omega(k') + \frac{1}{N}\,\delta E(kk') \tag{28}$$

with $\delta E(kk') \to 0$ when either k or k' $\to 0$; and this is indeed the computed result, if the states Eq. (23) are taken as a set of variational states, and the energy is computed by the variational formula:

$$E(kk') = (\psi^*_{kk'}|H|\psi_{kk'})/(\psi^*_{kk'}|\psi_{kk'}) \tag{29}$$

But such a computation would not be exact--nor even adequate, for several important reasons, as follows:

1. The functions $\psi_{kk'}$ are not orthogonal. Their failure to be orthogonal, although it is only to O(1/N), leads to what Dyson has denoted the "kinematical interactions" which must be taken into account in solving for the energy eigenvalues. Note that for S = 1/2, the set of two-magnon states is overcomplete: there are 1/2 N(N + 1) functions to describe 1/2 N(N - 1) physical states, which shows that the functions in that case cannot all be orthogonalized, even in principle.

2. These functions do not diagonalize the Hamiltonian, except when k or k' vanish; to the extent that both wavevectors are finite, there is scattering, i.e., "dynamical interactions" in Dyson's language, which may even lead to bound states. To see this, it is necessary to go beyond the approximate treatment and diagonalize H exactly within the proper subspace of states ψ_{ij}. Fortunately, the Hamiltonian has no matrix elements connecting any of these states with a state outside the subspace, so that the problem is quite well defined, as it was in the case of the one-magnon eigenstates.

3. If the number of magnons $\sum n_k$ is not 1 or 2 but is O(Ns), the magnon-magnon interactions add up to a finite contribution to the energy of each magnon, and must be taken into account in a correct statistical mechanics even at low T. Eq. (28) can be generalized to this purpose, by writing:

$$E_{total} = E_o + \sum_k \hbar\omega(k)n_k + \frac{1}{2N}\sum\sum_{k,k'} \delta E^R(kk')n_k n_{k'} \tag{30}$$

an expression that must be valid at finite, but low, density of excitations, with δE^R the real part of the binary interaction energy (calculated in Eq. (76)). The imaginary part of this term also has a direct interpretation in terms of scattering lifetimes. We define the inverse scattering lifetime of a given magnon (k) in the presence of a finite, but low, density of other excitations as:

$$1/\tau(k) = \frac{2}{N}\sum_{k'} \delta E^I(kk')n_{k'} \quad . \tag{31}$$

Thus, all the important properties of the low-density many-body magnon fluid may be expressed in terms of the <u>complex</u> binary interaction function $\delta E(kk')$ that we shall calculate in this section. This is also known as the <u>t-matrix</u>.

The exact 2-magnon eigenstate is a linear combination of all possible configurations of 2 spin "flips," i.e., is of the form:

$$\psi = \sum_{i>j} f_{ij} S_i^+ S_j^+ |0) \qquad f_{ji} \equiv f_{ij} \tag{32}$$

Let us solve the eigenvalue equation,

$$H\psi = E\psi \tag{33}$$

for the amplitudes f_{ij}. There is no need to use the normalized configurations as the normalization constants may be considered absorbed in the f_{ij}. Also, eigenstates belonging to different energies are automatically orthogonal. One takes the inner product of both sides of the Schrödinger equations, (33), with every basis state:

$$(0|S_i^- S_j^-$$

to obtain the equations obeyed by the amplitudes f_{ij}, which are,

$$(E - E_o - 2g\mu_B H - 2sJz)f_{ij} + s\sum_n (J_{nj} f_{in} + J_{in} f_{nj})$$

$$= \frac{1}{2} J_{ij}(f_{ii} + f_{jj} - f_{ij} - f_{ji}) \tag{34}$$

where each of the bonds, J_{ij}, J_{nj}, J_{in} = J, when their respective scripts (ij), (nj) and (in) are nearest neighbor pairs, and vanish otherwise. This equation is correct for all values of s. Even when s = 1/2, the unphysical amplitudes f_{ii}, etc., cancel between both sides of Eq. (34). In that case, nevertheless, it is convenient to define fictitious f_{ii} in order to simplify the calculation. There are at least two manners of doing this: The way chosen by Bethe in his solution of the linear chain problem (see section on his one-dimensional solution, which follows) was to set $f_{ii} + f_{jj} = f_{ij} + f_{ji}$ when j,i are nearest neighbors, a sort of "boundary condition" ensuring that the homogeneous equation is obeyed at every pair of sites. But in general, it is more satisfactory to treat the special case s = 1/2 on the same footing as s > 1/2, and allow Eq. (34) with i = j to define the unphysical amplitude f_{ii}. Note that in any event, J_{ij} vanishes unless i,j are nearest neighbors, so that with this exception, the right-hand side of Eq. (34) vanishes.

Periodic boundary conditions require

$$f_{ij} = f_{ij} \qquad \text{where } R_j - R_j \equiv (L,0,0) \quad \text{or } (0,L,0) \tag{35}$$

etc., for a volume L^3 in three dimensions, a square of area L^2 in two dimensions, or a length L in one dimension. And if the right-hand side of Eq. (34) did in fact vanish everywhere (instead of nearly everywhere), then the above boundary conditions could be met by the plane-waves:

$$f_{ij} = e^{i(k\cdot R_i + k'\cdot R_j)} + e^{i(k'\cdot R_i + k\cdot R_j)}$$

with the Cartesian components of the wavevectors

$$k_x, \ldots \text{ and } k'_x, \ldots = \frac{2\pi}{L} \cdot \text{integer} \tag{36}$$

These are the eigenstates when either k or k' = 0. For nonzero k, k' it is necessary, because of the interaction, to take linear combinations of such plane waves to obtain the eigenstates. It is convenient to separate out the center of mass motion by introducing the center of mass wavevector and coordinate,

$$K = k + k' \qquad R = \frac{R_i + R_j}{2} \tag{37}$$

and a relative wavevector and coordinate,

$$q = \frac{k - k'}{2} \qquad r = R_i - R_j \tag{38}$$

and in terms of these, express f_{ij} in the form

$$f_{ij} = Ae^{iK\cdot R}\{\sum_q e^{iq\cdot r} f(q)\} \equiv Ae^{iK\cdot R} F(r) \tag{39}$$

where A is a normalization constant, required for $\Sigma|f_{ij}|^2 = 1$. Because spin operators on different sites commute,

$$F(r) = F(-r) \quad \text{or equivalently,} \quad f(-q) = f(q) \; . \tag{40}$$

In writing f_{ij} in the simple form above, one takes advantage of the fact that the interactions J_{ij} depend on the coordinates R_i and R_j only through the relative coordinate r and therefore the total momentum (wavevector K) is a constant of the motion. We shall solve for F(r), or rather, f(q), at fixed K. Once the solutions are obtained, they can be expressed in the original variables k,k' by use of Eqs. (37) and (38).

Inserting f_{ij} in the form above into the set of difference equations (34), we have:

$$(E - E_o - 2g\mu_B H - 2sJz)F(r) + 2sJ \sum_\delta \left(\cos \frac{K}{2} \cdot \delta\right) F(r + \delta)$$

$$= J(r)\left[\left(\cos \frac{K}{2} \cdot r\right) F(0) - \frac{F(r) + F(-r)}{2}\right] \tag{41}$$

in which the vectors δ connect a spin to its $\underline{z}$ nearest neighbors for which the interaction $J(\delta) = J \neq 0$. At all other distances J(r) vanishes. Next, we multiply the equation for F(r) by $e^{-iq\cdot r}$, and sum on all r to obtain the equation for f(q):

$$\left[E - E_o - 2g\mu_B H - 2sJ \sum_\delta \left(1 - \cos \frac{K}{2} \cdot \delta \cos q\cdot\delta\right)\right] f(q)$$

$$= \frac{J}{N} \sum_\delta \cos q\cdot\delta \sum_k \left(\cos \frac{K}{2} \cdot\delta - \cos k\cdot\delta\right) f(k) \tag{42}$$

We have made use of the usual identity,

$$\frac{1}{N} \sum_r e^{i(k-k')\cdot r} = \delta_{k,k'} \tag{43}$$

For brevity, define a symbol which will be used repeatedly:

$$\gamma_K(q) \equiv E_o + 2g\mu_B H + 2sJ \sum_\delta \left(1 - \cos \frac{K}{2} \cdot\delta \cos q\cdot\delta\right) \tag{44}$$

which is, basically, the "incoming" 2-magnon energy $E_0 + \hbar\omega(k) + \hbar\omega(k')$ of Eqs. (27) or (28). (We assume a simple-cubic (sc) lattice in 3D, a square (sq) one in 2D, and for either of these (as well as for the linear chain in 1D) we have the obvious symmetries $\gamma_K(q) = \gamma_{-K}(q) = \gamma_K(-q)$. In terms of the original k,k' we can write: (see Eqs. (37) and (38)),

$$\gamma_K(q) = \gamma(kk') = E_o + 2g\mu_B H + sJ \sum_\delta \left(2 - \cos k\cdot\delta - \cos k'\cdot\delta\right)$$

$$= E_o + \hbar\omega(k) + \hbar\omega(k') \text{ , QED.)} \tag{45}$$

Bound states, if they exist, must be outside the continuum of states lying between the maximum and minimum value of $\gamma_K(q)$ and so at fixed K must lie either lower than

$$E_o + 2g\mu_B H + 2sJ \sum_\delta \left(1 - \left|\cos \frac{K}{2}\cdot\delta\right|\right)$$

or higher than (46)

$$E_o + 2g\mu_B H + 2sJ \sum_\delta \left(1 + \left|\cos \frac{K}{2}\cdot\delta\right|\right)$$

Let us start by an analysis of the scattering solutions in the continuum. It is conventional to distinguish between "incoming" and "scattered" waves. This is easily done in Eq. (42) by singling out:

$$f(q_o) = f(-q_o) = \frac{1}{2} N \tag{47}$$

to be the amplitude of the "incoming" wave and solving for the remaining amplitudes f(k) [= O(1)] as a smooth function of the continuous variable k in the large N limit. Equation (42) can be put in a most compact form by defining D-dimensional vectors as follows:

$$C(q) = (\cos q_x\delta_x, \ldots) \tag{48}$$

and

$$V = (V_x, \ldots) \tag{49}$$

where

$$V_x = \frac{1}{N}\sum_{\substack{k \\ k \neq \pm q_o}} \left(\cos \frac{K_x}{2}\delta_x - \cos k_x\cdot\delta_x\right) f(k) \tag{50}$$

This yields:

$$f(q) = \frac{2J}{E - \gamma_K(q)} C(q)\cdot\left[C(\tfrac{1}{2}K) - C(q_o) + V\right] \tag{51}$$

for $q \neq \pm q_o$ and

$$E = \gamma_K(q_o) + 4(J/N)\, C(q_o)\cdot\left[C(\tfrac{1}{2}K) - C(q_o) + V\right] \tag{52}$$

at $q = \pm q_o$. The procedure is now to solve the integral (51) for $f(q)$ and V, then to obtain the $1/N$ correction to the energy, Eq. (52). In so doing, we run into expressions such as:

$$(2J/N) \sum_{q\neq\pm q_o} \frac{(\cos\frac{1}{2}K_\alpha\delta_\alpha - \cos q_\alpha\delta_\alpha)\cos q_\beta\delta_\beta}{E - \gamma_K(q)} \equiv M_{\alpha,\beta}(E) \tag{53}$$

with $\alpha,\beta = x, y,$ or z. In such a sum the value of E, accurate to $O(1/N)$ given Eq. (52), can be used, but the denominators can vanish at values of $q \neq \pm q_o$ (in fact, everywhere "on the energy shell"), therefore it is convenient to write $E = \gamma_K(q_o) - i\varepsilon$ with ε an infinitesimal positive quantity that we may allow to vanish at the conclusion of the calculation. It is now permissible to replace the sum in Eq. (53) by an integration.

The integrals are complex:

$$M_{\alpha,\beta}(q_o) = 2J\left(\frac{a}{2\pi}\right)^D \int d_Dq\left(\cos\tfrac{1}{2}K_\alpha\delta_\alpha - \cos q_\alpha\delta_\alpha\right)\cos q_\beta\delta_\beta$$

$$\times\left[\frac{P.P.}{\gamma_K(q_o) - \gamma_K(q)} + i\pi\delta\left(\gamma_K(q_o) - \gamma_K(q)\right)\right] \tag{54}$$

where we use the identity,

$$\lim \varepsilon \to 0\ (1/x - i\varepsilon) = P.P.(1/x) + i\pi\delta(x)$$

with $\delta(x)$ the D-dimensional Dirac delta function and $P.P.(1/x) \equiv x/(x^2 + \varepsilon^2)$. Thus the contribution to the integral on the energy shell is entirely imaginary, the real part coming from the entire Brillouin Zone.

We combine Eqs. (50) and (51), and with M the DxD matrix of which $M_{\alpha,\beta}$ is the matrix element, and C and V column vectors, we have

$$V = M\cdot\left[C(\tfrac{1}{2}K) - C(q_o) + V\right] \tag{55}$$

This is formally solved by multiplication by $[1 - M]^{-1}$, with 1 the D-dimensional unit matrix. Inserting the result into Eqs. (51) and (52)

$$f(q) = \frac{2J}{\gamma_K(q_o) - \gamma_K(q)}\, C(q)\cdot[1 - M]^{-1}\cdot\left[C(\tfrac{1}{2}K) - C(q_o)\right] \tag{56}$$

$$(q \neq \pm q_o)$$

and

$$E = \gamma_K(q_o) + 4(J/N)C(q_o)\cdot[1 - M]^{-1}\cdot\left[C(\tfrac{1}{2}K) - C(q_o)\right] \tag{57}$$

Thus,

$$\delta E(kk') = 4J\, C(q_o)\cdot[1 - M]^{-1}\cdot\left[C(\tfrac{1}{2}K) - C(q_o)\right] \tag{58}$$

We observe that this f(q) and the 1/N correction to the energy both vanish at $q_o = \pm 1/2\ K$; these are the special cases k or k' = 0, for which we already anticipated the absence of scattering.

The integrals (54) making up the 3x3 matrix M are difficult to express in terms of elementary functions (although they simplify in the long wavelength limit $ka, k'a \to 0$, precisely the limit studied by Dyson because of its applicability to low-temperature and low-energy behavior). In leading semiclassical approximation, we just ignore M which is O(1/s), and evaluate:

$$\delta E(kk') \approx 4JC(q_o)\cdot\left[C(\tfrac{1}{2}K) - C(q_o)\right]$$

$$= -(1/s)\left[\hbar\omega(k) + \hbar\omega(k') - \hbar\omega(k - k') - \hbar\omega(0)\right] \tag{59}$$

with $\hbar\omega(k)$ given in Eq. (16), as usual.

In 1D, M is just a number. Equation (54) yields:

$$M = \frac{2J}{2\pi}\int_{-\pi}^{+\pi} dq \frac{(\cos \tfrac{1}{2} K - \cos q) \cos q}{4sJ \cos \tfrac{1}{2} K (\cos q - \cos q_o) + i\varepsilon} \tag{60}$$

or, in terms of k and k':

$$M = \frac{-1}{s \cos \tfrac{1}{2} (k + k')} \sin\tfrac{1}{2} k \sin \tfrac{1}{2}k'\{1 = i \cot \tfrac{1}{2} (k - k')\} \tag{61}$$

Insertion into Eq. (56) yields the scattering amplitudes characteristic of the continuum of scattering states, expressed in terms of the unperturbed k values of Eq. (36). Later (see Eq. (105)) we shall reexamine this point as energy and momentum can be conserved with just two values of k, provided the k values are shifted somewhat. This is part of the so-called Bethe's ansatz; its applications to date have been limited to 1D, its relation to the above, somewhat obscure.

IV. TWO-MAGNON BOUND STATES

We now turn our attention to the two-magnon bound states, and look for E lying outside the range of values given in Eq. (46). As there is no corresponding (real) q_o there is no special amplitude (47). The eigenvalue equation (52) is now an homogeneous equation and f(q) is given by:

$$f(q) = \frac{2JC(q) \cdot V}{E - \gamma_K(q)} \tag{62}$$

with:

$$V = M \cdot V , \tag{63}$$

The matrix elements $M_{\alpha\beta}$ having been given earlier, as functions of E. The bound state energy E is adjusted until V is an eigenvector of M. This determines V to within a multiplicative constant required for normalization.

Before considering the region of small K, which because of the low spin-wave and bound-state energies will be of the greatest importance in thermodynamics, let us first examine the special case when all components of K are equal to π, the maximum possible value, in which case

$$\cos \frac{K_x}{2} = \cos \frac{K_y}{2} = \cos \frac{K_z}{2} = \cos \frac{\pi}{2} = 0$$

and all the integrals in the eigenvalue equation become trivial. Because the denominators are now all constant, the nondiagonal matrix elements $M_{x,y}$ defined in Eq. (54) all vanish, while the diagonal ones are all equal. The bound state will therefore be D-fold degenerate. Note that the continuum has collapsed to a single point at

$$\gamma_x \equiv \gamma_x(q) \equiv E_o + 2g\mu_B H + 2sJz \qquad (64)$$

and therefore the integral equation reduces to an algebraic one. The bound-state solutions of Eq. (63) are easily found to have energy

$$E = \gamma_x - J , \qquad (65)$$

by solving this algebraic equation. The bound state is identified as the complex formed by keeping the two spin deviates nearest neighbors.

The existence of at least one bound state for every value of K in two dimensions is guaranteed by the behavior of the various integrals near $q = 0$. Let us denote the energy of the lowest state in the continuum by E_{min}, then the integrands in the vicinity of $q = 0$ all contribute on the order of

$$\approx \int \frac{\text{numerator}}{E - E_{min} - D'q^2}\, q dq$$

for suitable positive constant D', and appropriate nonvanishing numerator. This contribution, as a function of E, ranges from zero to $-\infty$ as E varies from $-\infty$ to E_{min}, and in the limit $E_{min} - E \equiv \delta E \to 0$, varies as $\approx \log \delta E$. Thus one expects, even without the benefit of explicit calculation, that the "binding energy" δE will depend exponentially on the various parameters.

Turning to the more realistic case of three-dimensional structure, for which the integrals cannot be evaluated analytically, one may find it advantageous to represent them by Laplace transform methods. That is, the substitution

$$\frac{1}{D} = \int_0^\infty dt\, e^{-Dt}$$

combined with the definition of the Bessel functions of imaginary argument

$$I_n(z) = \frac{1}{\pi}\int_0^\pi e^{z\cos x} \cos nx\, dx$$

permits the replacement of three-dimensional integrals by a rapidly converting one-dimensional integral. But these substitutions are not required to determine the threshold for the existence of a bound-state solution.

For this, we may presume $K_x = K_y = K_z$, which yields the lowest edge of the continuum and is more likely to produce the bound-state than some less isotropic direction. In this case the three diagonal matrix elements $M_{i,i}$ are all equal, and the off-diagonal elements are also equal to each other. This results in $V_x = V_y = V_z$ and in an eigenvalue equation:

$$1 = M_{x,x} + 2\,M_{x,y}$$

$$= \frac{2J}{N}\sum_q \frac{\cos q_x [3\cos(K_x/2) - \cos q_x - 2\cos q_y]}{E-E_o-2g\mu_B H-12sJ+4sJ[\cos q_x+\cos q_y+\cos q_z]\cos(K_x/2)}$$

$$= \frac{E - E_o - 2g\mu_B H - 12sJ\sin^2(K_x/2)}{24s^2 J\cos^2(k_x/2)}$$

$$\times\left[1 - \frac{E - E_o - 2g\mu_B H - 12sJ}{N}\sum_q \frac{1}{E - \gamma_k(q)}\right] \tag{66}$$

Use has been made only of the cubic symmetry, and of the equality of the three components of K to arrive at the last, simplified expression. The bound-state threshold occurs for that K at which E just drops below the bottom of the continuum at $\gamma_K(0)$. This occurs for K_{xo}:

$$E = E_o + 2g\mu_B H + 24sJ\sin^2\frac{K_{xo}}{4} \tag{67}$$

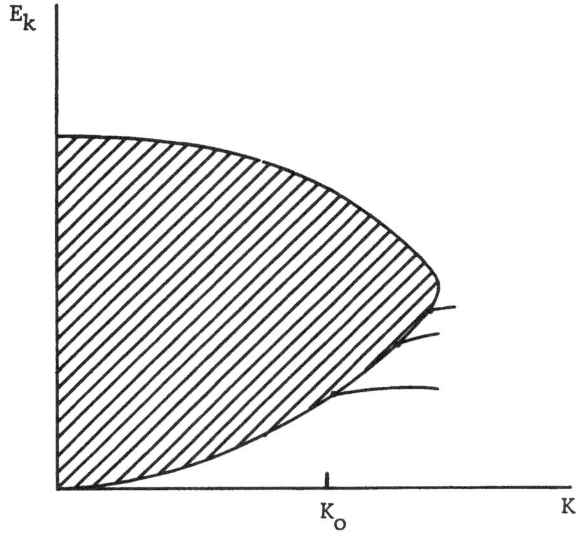

Fig. 1. Continuum of 2-particle scattering states lying in range given by Eq. (46). Bound states can form below the continuum, at momenta exceeding K_o. (Note: $K_o = 0$ for D = 1,2, but is finite at D = 3.) Thus there are no long wavelength bound states in three dimensions.

Replacing the sum in Eq. (66) by an integral in the usual manner, we use the first line to determine the threshold K_{xo}:

$$1 = \frac{-1 + \cos\frac{1}{2}K_{xo}}{2s\cos\frac{1}{2}K_{xo}}(1 - W), \quad \text{i.e.,} \quad \sin^2\frac{1}{4}K_{xo} = \frac{s}{2s + W - 1} \tag{68}$$

where W is Watson's integral:

$$W = \frac{1}{(2\pi)^3}\iiint_{-\pi}^{\pi} \frac{dq_x\, dq_y\, dq_z}{1 - \frac{1}{3}(\cos q_x + \cos q_y + \cos q_z)} = 1.516386 \tag{69}$$

Thus, the bound state exists in the range

$$0 < \cos\frac{K_x}{2} < \frac{.516386}{2s + .516386} \tag{70}$$

provided $K_x = K_y = K_z$. This results in $140° < K_x < 180°$ for $s = 1/2$, $157° < K_x < 180°$ for $s = 1$, $167° < K_x < 180°$ for $s = 2$, etc. As K is increased beyond the minimum, the other solutions make their appearance. For example, we have already found that there are three solutions at $K_x = K_y = K_z = 180°$, which is the largest total momentum wavevector in the positive direction. This small range of momentum space over which bound-state solutions are available, even in the most favorable case of spins one-half, is in sharp contrast with the results of the one- and two-dimensional calculations; and in the correspondence limit $s \to \infty$, the bound states simply disappear.

V. MANY-BODY SCATTERING THEORY

If it is desired to go beyond the harmonic oscillator approximation and calculate nonlinear corrections to the Heisenberg Hamiltonian, without the rigors of many-body scattering theory, a fairly direct manipulation of the spin operators themselves lends itself to this purpose. We start with the Holstein-Primakoff representation

$$S_i^+ = a_i^* \sqrt{2s}\sqrt{\left(1 - \frac{n_i}{2s}\right)} \qquad S_i^- = \sqrt{2s}\sqrt{\left(1 - \frac{n_i}{2s}\right)}\, a_i$$

$$S_i^z = n_i - s \tag{71}$$

The a's are boson operators, satisfying ordinary commutation relations:

$$a_i a_j^* - a_j^* a_i \equiv [a_i, a_j^*] = \delta_{i,j} \qquad [a_i, a_j] = [a_i^*, a_j^*] = 0 \tag{72}$$

with $n_i = a_i^* a_i$, the Boson number operator measuring the deviation from spin "down." The Heisenberg Hamiltonian takes on the form:

$$H = -\frac{1}{2}\sum_{i,j} J_{ij}\Big[\frac{1}{2} a_i^* \sqrt{\left(1 - \frac{n_i}{2s}\right)\left(1 - \frac{n_j}{2s}\right)}\, a_j + \text{H.C.}$$

$$+ (n_i - s)(n_j - s)\Big] + g\mu_B H \sum_i (n_i - s) \tag{73}$$

It is convenient to rationalize the square roots (M1, Chap. 3), by the similarity transformation:

$$a_i \to \left(1 - \frac{n_i}{2s}\right)^{-1} a_i \quad \text{and} \quad a_i^* \to a_i^*\left(1 - \frac{n_i}{2s}\right) \tag{74}$$

This transformation preserves the number operator as well as the commutation relations (72) within the physically relevant Hilbert space. Equation (73) now becomes:

$$H = E_o - s \sum_{i,j} J_{ij}(a_i^* - a_j^*)\left(1 - \frac{n_i}{2s}\right)a_j + g\mu_B H \sum n_i \tag{75}$$

the constant E_o being the usual ground-state energy first defined in Eq.(12). This non-Hermitean operator has all the eigenvalues of the Heisenberg Hamiltonian, plus an infinite number of unphysical ones. Whether an eigenstate is physically admissible or not can be tested by seeing whether it is an eigenfunction of the positive semidefinite operator

$$\left[\frac{1}{N} \sum_{i=1} (a_i^*)^{2s+1} (a_i)^{2s+1} \right] \tag{76}$$

with eigenvalue 0, or not.

For the purposes of spin-wave theory, one now transforms the above H to running waves. Let

$$a_i = \frac{1}{\sqrt{N}} \sum_k e^{ik\cdot R_i} a_k \qquad a_i^* = \frac{1}{\sqrt{N}} \sum_k e^{-ik\cdot R_i} a_k^*$$

$$k = \frac{2\pi}{L}(n,m,l) \tag{77}$$

where n,m,l = integers. It may be verified that this transformation is canonical, by showing that the a_k's obey a set of commutation relations identical to Eq. (72). For if one inverts the transformation, he obtains

$$a_k = \frac{1}{\sqrt{N}} \sum_i e^{-ik\cdot R_i} a_i \qquad a_i^* = \frac{1}{\sqrt{N}} \sum_i e^{ik\cdot R_i} a_i^* \tag{78}$$

and verifies they also satisfy the commutation relations (72). It is sufficient in Eq. (77) to use the set of smallest integers:

$n = 0, \pm 1, \pm 2, \ldots, \pm 1/2\ N_x$, $m = 0, \pm 1, \ldots, \mp 1/2\ N_y$, etc., for a solid $N_x X N_y X N_z$. This defines the first Brillouin Zone; an arbitrary k can always be reduced to the first BZ by subtracting an integer multiple of $(2\pi/a,0,0)$ or $(0,2\pi/a,0)$ or $(0,0,2\pi/a)$, where a = lattice spacing. This is because

$$e^{iK_n \cdot R_i} = 1 \qquad \text{hence } a_{k+K_n} = a_k \qquad \text{(by Eq. (78))},$$

denoting any integer multiple of the above vectors, K_n.

The Hamiltonian in terms of a_k's is

$$H = E_o - s \sum_{i,j} J_{ij} \sum_{k_1 k_2} \frac{a^*_{k_1}}{\sqrt{N}} \left(e^{ik_1 \cdot R_i} - e^{ik_1 \cdot R_j}\right)$$

$$\times \left(1 - \frac{1}{2s} \sum_{k_3 k_4} \frac{a^*_{k_3}}{\sqrt{N}} e^{ik_3 \cdot R_i} \frac{a_{k_4}}{\sqrt{N}} e^{-ik_4 \cdot R_i}\right) \frac{a_{k_2}}{\sqrt{N}} e^{-ik_2 \cdot R_j}$$

$$+ g\mu_B H \sum_i \sum_{k,k'} \frac{a^*_k}{\sqrt{N}} e^{ik \cdot R_i} \frac{a_{k'}}{\sqrt{N}} e^{-ik' \cdot R_i} \tag{79}$$

Let us isolate, in this large sum of terms, the contribution of operators diagonal in the occupation number representation (that is, of operators such as constants and functions of $n_k = a^*_k a_k$) from the nondiagonal terms. Terms bilinear in the a's, such as the last term in the magnetic field, are automatically diagonal by virtue of momentum conservation. Quartic terms are not, although a subset of them, for which

$$(k_1,k_3) = (k_2,k_4) \qquad \text{or} \qquad (k_4,k_2)$$

only involve n_k's and so will be included in the important diagonal part, which we shall call H_D. As the Fourier transform of J_{ij} consistently occurs, it is convenient to denote it J(k), defined as

$$J(k) = \frac{1}{2N} \sum_{i,j} e^{ik \cdot (R_i - R_j)} J_{ij} \tag{80}$$

In the cubic structure, $J^*(k) = J(-k) = J(k)$. For example, in the simple cubic structure nearest-neighbor

$$J(k) = 2J(\cos k_x a + \cos k_y a + \cos k_z a) \tag{81}$$

And in this manner one obtains the compact expression, which reproduces exactly the results of Eqs. (30) and (59) based on binary scattering:

$$H_D = E_o + \sum_k [2sJ(0) - 2sJ(k) + g\mu_B H]\, n_k$$

$$- \frac{1}{N} \sum_{k,k'} [J(0) + J(k - k') - J(k) - J(k')] n_k n_{k'} \qquad (82)$$

No bound states are found but, as already noted, none are expected at large values of s.

The study of H_D involves only minor generalizations of the theory of the linear Hamiltonian. The only important question is the following. Given a state occupied by n_1 magnons of wavevector k_1, n_2 of wavevector k_2, etc., how much energy is required to add one magnon of wavevector k to this state? Let $W(n_k)$ and $W(n_k + 1)$ indicate the eigenvalue of H_D in the two states $|n_k)$ and $a_k^*|n_k)$; then

$$H_D a_k^*|n_k) = a_k^* H_D |n_k) + [H_D, a_k^*]|n_k)$$

$$= W(n_k) a_k^*|n_k) + \Big\{2sJ(0) - 2sJ(k) + g\mu_B H$$

$$- \frac{2}{N} \sum_{k'} [J(0) + J(k - k') - J(k) - J(k')] \cdot n_{k'}\Big\}\, a_k^*|n_k)$$

$$\equiv W(n_k + 1) a_k^*|n_k) \qquad (83)$$

and therefore

$$W(n_k + 1) = W(n_k) + [2sJ(0) - 2sJ(k) + g\mu_B H]$$

$$- \frac{2}{N} \sum_{k'} [J(0) + J(k - k') - J(k) - J(k')] n_{k'} \qquad (84)$$

In terms of the magnon energy $\hbar\omega(k)$ as previously defined, for example, Eq. (16),

$$W(n_k + 1) - W(n_k) = \hbar\omega(k)$$

$$- \frac{1}{Ns} \sum_k [\hbar\omega(k) + \hbar\omega(k') - \hbar\omega(k - k') - \hbar\omega(0)] n_{k'} \equiv \varepsilon(k) \qquad (85)$$

with the sum representing the corrections to linear spin-wave theory.

VI. X-Y MODEL: A MODEL ANTIFERROMAGNET

The XY-model: a model antiferromagnet, exactly soluble in 1D for spins one-half.

$$H_{XY} = \frac{1}{2} \sum_{i=1} (S^+_i S^-_{i+1} + \text{H.c.}) \quad \left(S = \frac{1}{2} \quad \text{and} \quad S_{N+1} \equiv S_1\right) \tag{86}$$

The vacuum (all spins down) is an eigenstate, with energy 0. By translational invariance, one may guess that the one-particle states (N - 1 spins down, one spin up) are plane waves:

$$\psi_k = \frac{1}{\sqrt{N}} \sum_i e^{ik\cdot R_i} S^+_i |0) \tag{87}$$

with easily computed energy eigenvalues,

$$E_k = \cos ka \qquad \text{with } |ka| < \pi \tag{88}$$

Periodic boundary condition $e^{ikNa} = 1$ results in the discrete set,

$$k = \frac{2\pi}{Na} \times \text{integer} = \frac{\pi(2p)}{Na} \qquad p = 0, \pm 1, \ldots \tag{89}$$

Next, a product of two plane waves does not vanish in the configurations $S^+_i S^+_i$, as it should, but a determinant does. However, a determinant is antisymmetric under the interchange of the coordinates, whereas spins on different sites commute. The following choice thus imposes itself:

$$\psi_{k,k'} = \begin{cases} \dfrac{1}{\sqrt{N(N-1)}} \displaystyle\sum_{i,j>i} [e^{i(k\cdot R_i + k'\cdot R_j)} - e^{i(k\cdot R_j + k'\cdot R_i)}] S^+_i S^+_j |0) \\[2ex] \dfrac{-1}{\sqrt{N(N-1)}} \displaystyle\sum_{i,j<i} [e^{i(k\cdot R_i + k'\cdot R_j)} - e^{i(k\cdot R_j + k'\cdot R_i)}] S^+_i S^+_j |0) \end{cases} \tag{90}$$

When j is increased to N - 1, then to N, and finally to N + 1, the second spin becomes the first and the wavefunction changes discontinuously from the upper form to the lower unless the proper boundary condition is imposed. Because the position of the origin of the numbering system is completely arbitrary for the cyclic problem we are solving, such discontinuities at a particular site are inadmissible. The resolution of this difficulty is to take, instead of Eq. (89),

$$e^{ikNa} = -1, \qquad \text{or} \qquad k = \frac{\pi(2p+1)}{Na} \tag{91}$$

(and similarly for k'), an antiperiodic boundary condition. The generalization to any number of spins is straightforward. Let

$$\psi_{k_1, k_2, \ldots} = C \sum_{i_1, i_2, \ldots} F^{i_1, i_2, \ldots}_{k_1, k_2, \ldots} S^+_{i_1} S^+_{i_2} \ldots |0) \tag{92}$$

where C is the normalization constant, and F is the determinant

$$F^{i_1, \ldots}_{k_1, \ldots} = \varepsilon_p \begin{vmatrix} e^{ik_1R_{i_1}} & e^{ik_2R_{i_1}} & \ldots \\ e^{ik_1R_{i_2}} & e^{ik_2R_{i_2}} & \ldots \\ \vdots & \vdots & \end{vmatrix} \tag{93}$$

$\varepsilon_p = +1$ when the spins are in a natural order $i_1 < i_2 < i_3 < \ldots$, or an even permutation of this order, and $\varepsilon_p = -1$ when the spins are arranged in an odd permutation of the natural order. If i_n is the farthest spin, then the translation of i_n from N to N + 1 involves a reordering equivalent to an odd permutation for n + 1 odd, and an even permutation for n + 1 even. Thus

$$n + 1 = \text{odd} \rightarrow k = \frac{\pi}{Na}(2p + 1) \tag{94a}$$

and

$$n + 1 = \text{even} \rightarrow k = \frac{\pi}{Na}(2p) \tag{94b}$$

This is a very interesting situation. For although the many-spin wavefunctions at first appear to be essentially independent particle wavefunctions, yet when one "particle" is added (that is, when one more spin is turned up) all the other plane-wave states

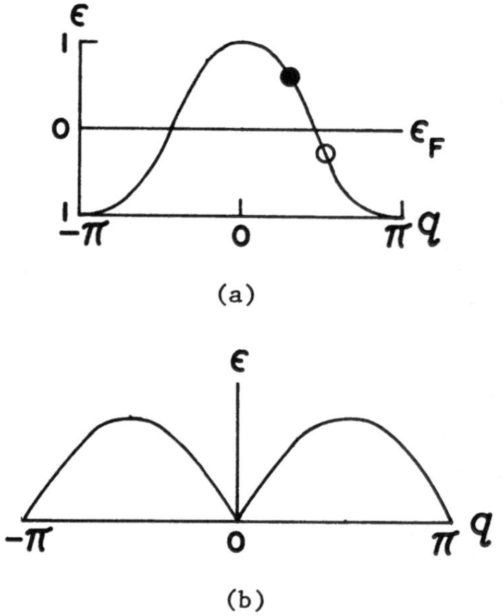

Fig. 2. D = 1, XY model (similar to Heisenberg antiferromagnet in one dimension). (a) Ground state, showing all negative energy states occupied (heavy line) and positive energy state empty. An elementary excitation is superposed: hole (empty circle below Fermi level) or particle (dot above Fermi level). (b) Energy of elementary excitation as function of wavevector, showing linear dispersion at long wavelengths characteristic of antiferromagnets.

are modified. This is in the nature of a cooperative effect, due to the effective hard core repulsion (form $(S_i^+)^2 = 0$) of two nearby spin deviations. Whether the k's are members of the even or of the odd set, the energy corresponding to the wavefunction above is

$$E_{k_1,k_2,\ldots} = \sum_{i=1}^{n} \cos k_i a = \sum_i \cos \frac{\pi}{N} \times \left\{ \begin{array}{l} 2p_i \\ \\ 2p_i + 1 \end{array} \right. \tag{95}$$

Note that all k's must be distinct, or else F ≡ 0. The ground state energy is achieved by allowing all the states of negative energy to be occupied, that is, all k's in the range

$$\frac{\pi}{2} < ka < \frac{3\pi}{2}, \quad \text{i.e.,} \quad \frac{N}{4} < p < \frac{3N}{4} \tag{96}$$

However, we must consider two ground states: the ground state for the even k's and the ground state for the odd ones. Denote these by E_0' and E_0'' respectively, with values

$$\left.\begin{aligned} E_o' &= \sum_p \cos \frac{\pi}{N}(2p+1) \\ E_o'' &= \sum_p \cos \frac{\pi}{N}(2p) \end{aligned}\right\} = -\frac{1}{\pi} N \quad (\text{when } N \to \infty) \tag{97}$$

with the sums in either set restricted to the range, Eq. (96). The separation of energies E' and E" depend on whether N is divisible by 4, or merely by 2, or whether it is odd. In any event, it is of order of magnitude of the smallest elementary excitations.

Either ground state may be represented by an energy-level diagram shown in Fig. 2. The Fermi level intersects the cosine curve at $ka = \pi/2$ and $3\pi/2$, and all allowed states below this are filled and all those above it are empty in either ground state. Excited states correspond to occupying states above the Fermi level, or emptying states below it. If k is a wavevector of either category in Eq. (94), as measured from $\pm \pi/2a$, the energy of an elementary excitation, $\varepsilon(q)$ is described by a doubly degenerate spectrum

$$\varepsilon(q) = |\sin q| \qquad \text{setting } q \equiv ka \tag{98}$$

in addition to any correction due to changes in parity of the state. Each elementary excitation represents an antiferromagnetic magnon, and it should be note that unlike the ferromagnetic magnons, their energies depend linearly on wavevector q, for small q. This appears to be a model-independent property of all antiferromagnetics.

VII. HEISENBERG ANTIFERROMAGNET: GROUND STATE

It is desired to calculate the eigenstates of

$$H = \sum_{i=1}^{N} \left[\frac{1}{2} \left(S_i^+ S_{i+1}^- + \text{H.c.} \right) + g S_i^z S_{i+1}^z \right] \qquad (S = \frac{1}{2},\ g = 1) \tag{99}$$

For $-\infty < g < -1$, this is similar to the Heisenberg ferromagnet. The region $-1 < g < +1$ has gapless excitations and includes the XY model ($g = 0$) as a typical case. For $g > 1$ the Ising antiferromagnet is a good model. Thus, $g = 1$ is a special case and must be treated with care.

For present purposes, it is most convenient to use a procedure valid only for spins one-half and assign to the unphysical amplitudes f_{ii} the value determined by

$$f_{i,i} + f_{i+1,i+1} = f_{i,i+1} + f_{i+1,i} = 2f_{i,i+1} \qquad (100)$$

where the physical amplitudes f_{ij} ($= f_{ji}$ for $j \neq i$) obey the equations

$$(E - E_f + 2)f_{ij} - \frac{1}{2}\left(f_{ij+1} + f_{ij-1} + f_{i+1j} + f_{i-1j}\right) = 0 \qquad (101)$$

This is the left-hand side of Eq. (34), with $s = 1/2$, $J = -1$, and $H = 0$, whereas Eq. (100), playing the role of a boundary condition, ensures that the right-hand side of that equation always vanishes. The homogeneous equation is solved by plane waves, <u>via Bethe's ansatz</u>:

$$f_{ij} = e^{(iki+ik'j+\frac{1}{2}\psi)} + e^{i(kj+k'i-\frac{1}{2}\psi)} \qquad (j > i) \qquad (102)$$

with f_{ij} in the range $j < i$ given by symmetry, $f_{ij} = f_{ji}$. The phase factor ψ ranges over the interval $-\pi$, $+\pi$. (In the XY model, Eq. (90), the effective ψ is a constant $\pm\pi$.) We make no attempt to normalize solutions here, as it is not required in the calculation of energy eigenvalues and represents some fairly involved analysis.

Inserting the above form of f_{ij} into the boundary condition and performing some algebra, we obtain:

$$2\cot\frac{1}{2}\psi = \cot\frac{1}{2}k - \cot\frac{1}{2}k' \qquad (103)$$

with the second boundary condition,

$$f_{iN} = f_{0i} \qquad (104)$$

determining the modified k,k':

$$k = \frac{\pi(2p) + \psi}{N} \quad \text{and} \quad k' = \frac{\pi(2p') - \psi}{N} \qquad (105)$$

with p,p' = integers. Thus, a two-magnon state is described by two wavevectors and a phase shift; there is refraction but no scattering. The sum k + k' is independent of ψ and is the center of mass wavevector K which parametrized the solution in an earlier section of this Chapter. The difference k - k' corresponds to the variable q. The bound state that appeared below the continuum threshold at q = 0 is now <u>above</u> the continuum, due to the change in sign of the interaction. Now, as k approaches k', Eq. (103) shows that ψ approaches $\pm\pi$, the upper sign applying if $k < k'$. From Eq. (105), as p,p' are integers, $k \leqslant k' \rightarrow p \leqslant p' - 2$, or $|p - p'| \geqslant 2$ (or else $f_{ij} = 0$ by Eq. (102).)

The energy, measured from the ferromagnetic level $E_f = 1/4\ N$, is given by the above eigenvalue equation as:

$$E - E_f = -(1 - \cos k) - (1 - \cos k') \tag{106}$$

just the energy of two noninteracting magnons; the interactions are included implicitly, <u>via</u> the phase shifts (103) and (105).

<u>Bethe's ansatz</u> consists partly in the statement that in the many-particle state, the amplitudes are subject to phase shifts that are simply given, as the sum of the two-particle phase shifts. Thus, for a given ordering $i < j \ldots < n$,

$$f_{ij\ldots mn} = \left\{ e^{i\left(k_1 i + k_2 j + \ldots + k_n n + \frac{1}{2}\sum\limits_{r<t}^{n}\sum^{n} \psi_{k_r k_t}\right)} + \left(\begin{array}{c}\text{all } n! - 1 \text{ remaining}\\ \text{permutations of the k's}\end{array}\right)\right\} \tag{107}$$

If we need $f_{i,j,\ldots}$ for any other ordering of the particles, we obtain it from the symmetry under permutations: $f_{i,j,\ldots,m,\ldots} = f_{i,m,\ldots,j,\ldots}$ so we need to know it only in the given interval. In this region, the amplitude f consists of a product over plane wave factors, summed over all permutations of the wavevectors with the ψ's antisymmetric in their subscripts and satisfying the equations:

$$2 \cot \frac{1}{2}\psi_{kk'} = \cot \frac{1}{2} k - \cot \frac{1}{2} k' \tag{108}$$

$$k = \frac{\pi(2p) + \sum\limits_{k'=1}^{n} \psi_{kk'}}{N} \qquad p = 0, \pm 1, \ldots \tag{109}$$

The sum in this equation, as well as in the next, is over the set of wavevectors present in Eq. (107). The energy, measured relative to the ferromagnetic reference energy, is

$$E - E_f = - \sum_{t=1}^{n} (1 - \cos k_t) \tag{110}$$

Note that if we translate the entire chain by one site, that is, let i,j, ... → i + 1, j + 1, ..., the wavefunction, Eq. (107), is multiplied by a phase factor

$$e^{i \sum_1^n k} = e^{\pi i \sum_1^n 2p/N} \tag{111}$$

The exponent is identified as the total momentum of the state <u>modulo</u> 2π, which is again independent of the phase shifts $\psi_{kk'}$. This is important, because the phase shifts themselves are now rather large. Each $\psi_{kk'}$ is O(1); there are a number of them contributing to each k, and the total shift in k is O(n/N) or a substantial fraction of k.

In the two-particle problem, there was no solution of Eqs. (103) and (105) for $k - k' \approx 0$. Similarly, in the many-particle state, one must choose the interval between k's such that no p = 0, and such that for all p and p',

$$|p - p'| \geq 2 \tag{112}$$

to ensure that a real solution exists. For N/2 particles, which is the ground state "population," the set {p} subject to the above restriction is

$$\{p\} = 1,3,\ldots,N - 1 \tag{113}$$

Notice an interesting effect of the $S_i^z S_{i+1}^z$ "interaction," which is to spread the set of 1/2 N integers p (restricted over the range $N/4 < p < 3N/4$ in the ground state of the XY model) to cover the entire range of phase space at present.

The regular spacing permits us to replace sums by integrals in the limit $N \to \infty$, and thus, with $x = p/N$ and $\Delta p = 1/2\ N\,dx$:

$$E - E_f = - \frac{N}{2} \int_0^1 [1 - \cos k(x)]dx = - N \int_0^1 \sin^2 \frac{k(x)}{2}\, dx\ . \tag{114}$$

where

$$2 \cot \frac{\psi(x,y)}{2} = \cot \frac{k(x)}{2} - \cot \frac{k(y)}{2} \tag{115}$$

and

$$k(x) = 2\pi x + \frac{1}{2} \int_0^1 \psi(x,y)\, dy \tag{116}$$

As $|\psi|$ cannot exceed π, Eq. (115) indicates that ψ has a jump discontinuity of -2π at $x = y$. More precisely, the derivative is:

$$\frac{\partial\psi(x,y)}{\partial x} = -\, 2\pi\delta(x - y) + \frac{1}{4}(\xi^2 + 1)\, \frac{dk(x)/dx}{1 + \frac{1}{4}(\xi - \eta)^2} \tag{117}$$

where we use the shorthand notation

$$\xi \equiv \cot \frac{1}{2} k(x) \qquad \text{and} \qquad \eta \equiv \cot \frac{1}{2} k(y) \tag{118}$$

We also find it convenient to introduce the functions,

$$f(\xi) = -(d\xi/dx)^{-1} \qquad \text{and} \qquad f(\eta) = -(d\eta/dy)^{-1} \tag{119}$$

and to differentiate Eq. (116) w.r. to x to obtain:

$$dk(x)/dx = \pi + \frac{1}{2} \int_{-\infty}^{+\infty} \frac{f(\eta)/f(\xi)}{1 + \frac{1}{4}(\xi - \eta)^2}\, d\eta \tag{120}$$

With the identity $f(\xi)dk/dx = 2(1 + \xi^2)^{-1}$ this becomes:

$$\frac{2}{1 + \xi^2} = \pi f(\xi) + 2 \int_{-\infty}^{+\infty} \frac{f(\eta)}{4 + (\xi - \eta)^2}\, d\eta \tag{121}$$

This type of equation is soluble because the kernel is a function of only the difference $(\xi - \eta)$. One may take Fourier transforms,

$$F_k \equiv \int_{-\infty}^{\infty} d\theta f(\theta) e^{ik\theta} \qquad \text{and} \qquad f(\theta) = \int_{-\infty}^{\infty} \frac{dk}{2\pi} F_k e^{-k\theta} \tag{122}$$

to obtain a simple solution of Eq. (121):

$$F_k = (2 \cosh k)^{-1} \tag{123}$$

The ground state energy (114) has the form

$$E - E_f = -N \int_{-\infty}^{\infty} d\xi \frac{f(\xi)}{1+\xi^2} = -N \int_{-\infty}^{\infty} \frac{dkF_k}{2\pi} \int_{-\infty}^{\infty} d\xi \frac{e^{-ik\xi}}{1+\xi^2}$$

$$= -2N \int_{0}^{\infty} dkF_k e^{-|k|} = -2N \int_{0}^{\infty} \frac{dke^{-2|k|}}{1+e^{-2|k|}} = -N\ell n2 \tag{124}$$

This is a very famous result, being one of the first exact solutions ever obtained of a nontrivial many-body problem in quantum mechanics.

VIII. SOME CONSEQUENCES

In the usual terms of energy per spin, the Bethe-Hulthén result for E/N represents $-\ell n2 + 1/4 = -0.443$. We can compare this to the ground state energy per spin in the XY model (Eq. (97) X1/N) which is $-\pi^{-1} = -0.318$. While this is a variational upper bound to the Heisenberg model ground state energy, it is evidently not a very close one. On the other hand, 3/2 × ground state energy of the XY model is a lower variational bound on the Heisenberg model (really, an "XYZ" model) and this yields $-3/2\pi = -0.477$ reasonably close to the exact answer. By way of contrast, we note that the Ising model ($S_i^z S_{i+1}^z$) has energy -0.250 per spin in these same units, which is a variational upper bound (although a poor one) to both XY and Heisenberg model. Moreover, 2X(-.25) = - 0.500 is a poor lower bound to XY model and 3X(-.25) = - 0.750 is an even worse lower bound to the Heisenberg energy per spin. (The reasons for these larger discrepancies may be attributed to the lack of spinwaves in the Ising model.)

Des Cloizeaux and Pearson, have extracted Bethe's analysis to obtain the energy of some of the lowest excited states, which are found to be triplet states. N/2 - 1 values of k are chosen by them, to obtain the eigenstates of lowest energy belonging to a finite total momentum Σk. This procedure is very delicate, however, and somewhat too complicated to reproduce or to justify here. There is an unanswered question of whether bound states participate among the low-lying excitations. If this is the case, then Bethe's ansatz formalism is incapable of describing them since the validity of the various equations above seems predicated on having real k's. Nevertheless, it is entirely possible that further mathematical study of Eqs. (102) and those following may allow

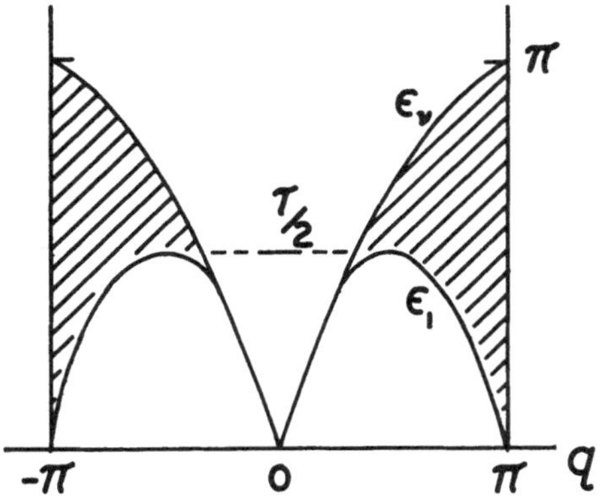

Fig. 3. Continuum of elementary excitations having S = 1 for linear chain Heisenberg antiferromagnet, lying between upper bound (126) and lower bound (125).

the extension of the known results to complex k plane and that a complete classification scheme will result. For example, following Orbach one might consider the $S_i^z S_{i+1}^z$ coupling terms with variable parameter g, and study the energy levels as g is adiabaticaly increased. At g = 0, we recover the XY model (of which the solutions are completely determined) and then increase it to g = 1 which is the present Heisenberg model. As g is further increased, the Ising model is approached. In fact, g = ±1 are two "critical points": for g >1 the properties approach those of an Ising antiferromagnet (e.g., the ground state energy approaches - 1/4 per spin at large g, the excitation spectrum has a gap), and for g < -1 they resemble those of an Ising ferromagnet. It is only in the range -1 < g < +1 that results are obtained similar to the gapless spectra we have seen here.

Des Cloizeaux and Person's results for the one-magnon triplet (S = 1) excitation spectrum is qualitatively the same as in Fig. 2 for the XY model. This comparison will have further significance when we study approximate theories of antiferromagnetism in the following sections. If we denote their triplet excitation spectrum ε_1, it is given by

$$\varepsilon_1 = \frac{1}{2}\pi \, |\sin q| \text{ vs. } |\sin q| \text{ for XY model .} \tag{125}$$

Yamada, and Bonner, Sutherland and Richards have pointed out that this spectrum is most likely the lower edge of a continuum of triplet excitations, with a curve ε_2 defining the upper edge:

$$\varepsilon_2 = \pi \left|\sin \frac{1}{2} q\right| \qquad (126)$$

This continuum is illustrated in Fig. 3. Undoubtedly, excitations belonging to S = 0,2,3,... lie in other continua that partly intersect this. However, in an experimental situation where one unit of angular momentum is transferred to the magnetic system, only the triplet continuum can contribute. For such processes, Hohenberg and Brinkman have computed that the matrix elements peak near the lower, des Cloizeaux-Pearson, threshold. This topic, and its nontrivial extension to the case of an applied magnetic field, is increasingly being studied theoretically and experimentally.

One important result follows immediately from Fig. 3, viz. the size of the Brillouin Zone remains 2π for the linear chain antiferromagnet. With only the spectrum of Fig. 2 one might suspect that the B.Z. could be folded in half, which is equivalent to the choice of 2 spins as the primitive unit cell. Physically, this is a very attractive idea: the couple of spin "up" and "down" nearest-neighbor make an ideal unit cell, and this is indeed the starting point of many approximate analyses. With the complete excitation spectrum, however, one sees that the translational invariance of the ground state is required and the larger B.Z. must be retained.

IX. EFFECTS OF SURFACES ON MAGNONS

The role of surfaces is dominant in the observed properties of magnetic substances; domains nucleate there, spinwaves are "pinned" there by impurities or stray fields, and phase transitions at a surface may foreshadow those in the bulk. To study these effects one appeals to a scattering theory which has some similarities with the two-magnon problem studied above. For definiteness, we consider only a single surface (in extremely thin films where the two surfaces interact, this treatment will have to be modified) perpendicular to one of the principal crystal axes in a simple cubic structure, ferromagnetic interactions, and an exchange parameter $J > 0$ which differs from the bulk value 1 only for the surface spins. A surface anisotropy anisotropy field $h_A > 0$ is also included, but the bulk ferromagnet is otherwise isotropic. With the surface located at $R_i = (X_i, Y_i, 0)$ and all interactions restricted to nearest-neighbors, we have the Hamiltonian:

$$H = -\frac{1}{2} \sum_{Z_i,} \sum_{Z_j \neq 0} S_i \cdot S_j - J \sum_{Z_i=0, Z_j \neq 0} \sum S_i \cdot S_j$$

$$- \frac{1}{2} J \sum_{Z_i=Z_j=0} \sum S_i \cdot S_j - h_A \sum_{Z_i=0} S_i^z \qquad (127)$$

The ground state $|0)$ of all spins "down" has energy:

$$E_o = E_o(\text{bulk}) - 3N_z Js^2 - h_A N_z s \tag{128}$$

in which we define the bulk energy to be that part which does not depend on the surface exchange parameters (J,h_A) and we note that each of the N_z surface spins has 4 n.n. in the surface (i.e., two bonds per spin), plus one in the neighboring plane, for a net total of 3 J-dependent bonds per surface spin. (This should not be confused with the obvious fact that each surface spin has 5 n.n. rather than 6 in the bulk.)

Absent the translational symmetry that enabled us to guess the plane wave form of the one-magnon states in the bulk, we construct the complete set of excited one-magnon eigenstates, in the form:

$$|\psi) = (2sN)^{-1/2} \sum_j f_j S_j^+ |0) \tag{129}$$

and extract from it a set of coupled equations for the amplitudes f_j by projecting as follows:

$$(2s)^{-1/2} N^{1/2} (0|S_i^- H|\psi) = (E_o + \varepsilon) f_i \tag{130}$$

ε is the excitation energy, to be compared with $\hbar\omega(k)$ for the translationally invariant case. Evaluating the l.h.s. of (130) we readily obtain:

$$(E_o + 6s) f(R_i) - s \sum_\delta f(R_i + \delta) \tag{131a}$$

for R_i in the bulk; for a spin in the plane adjacent to the surface, the left-hand side of Eq. (130) is, instead,

$$(E_o + 6s) f(R_i) - s \sum_\delta f(R_i + \delta) + (J - 1)s[f(R_i) - f(X_i, Y_i, 0)] \tag{131b}$$

and finally, for a surface spin:

$$(E_o + h_A + 5Js) f(X_i, Y_i, 0) - sJ[f(X_i + a, Y_i, 0) + f(X_i - a, Y_i, 0) + f(X_i, Y_i + a, 0) + f(X_i, Y_i - a, 0) + f(X_i, Y_i, a)] \tag{131c}$$

There remains translational invariance of sorts, i.e., in the X_i, Y_i plane. Therefore let us seek a solution of the form:

$$f_i = C\, e^{i(k_x X_i + k_y Y_i)} F(n) \tag{132}$$

where C is a normalization constant, Z_i = na and the surface plane is located at n = 0 (the first interior plane at n = 1, the bulk at n = 2,3,...). Without loss of generality, we write for the excitation energy of the continuum states:

$$\varepsilon = 2s(3 - \cos k_x a - \cos k_y a - \cos q) \tag{133}$$

and for the three regions considered above, the Schrödinger equation becomes,

$$2s\big(2 - \cos k_x a - \cos k_y a)F(n) + s(2F(n) - F(n + 1) - F(n - 1)\big) = \varepsilon F(n) \tag{134a}$$

for n > 2. At n = 1 we obtain:

$$2s\big(2 - \cos k_x a - \cos k_y a\big)F(1) + s\big(2F(1) - F(2) - F(0)\big) + (J - 1)s\big(F(1) - F(0)\big) = \varepsilon F(1) \tag{134b}$$

and at the surface,

$$\big(h_A + 2Js(2 - \cos k_x a - \cos k_y a)\big)F(0) + sJ\big(F(0) - F(1)\big) = \varepsilon F(0) \tag{134c}$$

As the bulk eigenstates take the form exp ± iqn, we try a linear combination: Let

$$F(n) = \cos(qn + \theta), \quad \text{for } n > 1 \tag{135}$$

in which $\theta(k_x, k_y, q)$ is a <u>phase shift</u> independent of <u>n</u>. However, this form is not suitable at the surface, n = 0. The substitution of Eq. (135) into Eq. (134a) solves that set of (bulk) equations exactly. We are left with (b) and (c), two equations in the two unknowns, θ and F(0):

$$F(0) = \frac{\cos\theta}{J} + \frac{(J - 1)}{J}\cos(q + \theta) \tag{136b}$$

and

$$F(0) = \frac{J}{D}\cos(q + \theta) \tag{136c}$$

where,

$$D(q) \equiv h_A/s + J + 2(J - 1)\left(2 - \cos k_x a - \cos k_y a\right) - 2(1 - \cos q) \tag{136d}$$

D(q) is a function of q alone, at fixed h_A, s, J, k_x and k_y but its variation, especially with k_x, k_y may be important. Solving the coupled equations for F(0), we find it to be a function of q and indicate this by a subscript:

$$F_q(0) = \frac{J \sin q}{\left[(J^2-DJ+D)^2 \sin^2 q + \left((J^2-DJ+D)\cos q - D\right)^2\right]^{1/2}} \tag{137}$$

Note that as $q \to 0$, this surface amplitude vanishes

$$F_q(0) \underset{(q\to 0)}{\to} \frac{\sin q}{|J - D(q)|} \to 0$$

and thus, long wavelength magnons have small amplitudes at the surface, unless $J = D(0)$. (For this special case, Eq. (137) yields:

$$\lim: (q \to 0)\ F_q(0) = 1.)$$

Bound surface states lying above the continuum of bulk magnon energies and below it are both possible. For bound states above we try $F(n) = C(-1)^n e^{-\lambda n}$ $(n \geq 1)$ and solve for λ, C, F(0) and ε_0 = bound state energy. For bound states below we try $F(n) = Ce^{-\lambda n}$ $(n \geq 1)$. When h_A is sufficiently negative, the bound state energy $\varepsilon_0 < 0$ and the surface is unstable against the emission of a large number of magnons. One then obtains a magnetization of the surface perpendicular to that of the bulk--i.e., half a domain.

X. MAGNONS vs. SOLITONS

It has been observed throughout this article that various problems in magnetism become simpler as s, the magnitude of the individual spin, becomes larger and the quantum fluctuations become correspondingly smaller. A century of progress has recently culminated in the exact solution of many problems in classical nonlinear dynamics in one spatial dimension, including those of greatest relevance to this article. A common thread is the identification of at least two different types of propagating excitations:

the ordinary wave (here, magnon) and the solitary wave, or soliton. The soliton is a particle-like manifestation which preserves its shape and identity after a long course of travel or period of time, and after collisions with stationary defects or other solitons. To quote its discoverer, J. Scott Russell:

> "I was observing the motion of a boat which was rapidly drawn along a narrow channel by a pair of horses, when the boat suddenly stopped--not so the mass of water in the channel which it had put in motion--it accumulated round the prow of the vessel in a state of violent agitation, then suddenly leaving it behind, rolled forward with a great velocity, assuming the form of a large solitary elevation, a rounded, smooth and well-defined heap of water, which continued its course along the channel apparently without change of form or diminution of speed. I followed it on horseback, and overtook it still folling on a rate of some eight or nine miles an hour, preserving its original figure some thirty feet long and a foot to a foot and a half in height. Its height gradually diminished and after a chase of one or two miles I lost it in the windings of the channel. Such, in the month of August 1834, was my first chance interview with that singular and beautiful phenomenon..."

To relate this remarkable phenomenon to magnetism, some concepts will have to be developed. Chief among these are Heisenberg's equations of motion for the spin dynamics, such as:

$$\frac{\partial}{\partial t} S_n = \frac{i}{\hbar}[H, S_n] \tag{138}$$

with H the Hamiltonian of Eq. (1). Abbreviating $\partial/\partial t$ by $(\cdot)$, we find:

$$\dot{S}_n = -\frac{i}{\hbar}\sum_m J_{mn}[S_m \cdot S_n, S_n] - \frac{i}{\hbar} g\mu_B [H \cdot S_n, S_n] \tag{139}$$

The Zeeman contribution is easy to compute by the familiar commutation relations, Chapter 3 in M1. The bracket coefficient of J_{mn} has also been previously calculated in the same Chapter. Combining the results, we find:

$$\dot{S}_n = S_n \times \left(\sum_m J_{mn} S_m + g\mu_B H\right) \tag{140}$$

Summing over all spins, we see that the cross terms cancel and the total spin merely precesses about the external field:

$$\dot{S}_T = S_T \times g\mu_B H \tag{141}$$

As the quantum unit of action $\hbar$ does not appear explicitly in the equation of motion (140), therefore they are valid without modification in the classical limit $s \to \infty$. (In proceeding to this limit, it is necessary to scale J and μ_B as 1/s to keep the results finite, while setting $\hbar = 1$ for dimensional convenience.) We can now represent a spin by classical unit vector: $S_n = s\, u_n$. Assuming a simple lattice (s.c., or sq., or n.n. l.c. with primitive unit vectors δ in the notation of Eq. (16), with $\delta = a$ the lattice parameter), the equations of motion take the form of difference equations:

$$\dot{u}(R_n) = u(R_n) \times \left(Js \sum_\delta (u(R_n + \delta) - u(R_n)) + h\right) \tag{142}$$

with $h \equiv g\mu_B H = (0,0,h)$. For all long wavelength phenomena, whether linear or not, a lattice may be approximated by a continuum. In Eq. (142) we have subtracted terms such as $u_n \times u_n$ which vanish in the classical limit, to obtain a form in which the continuum limit is easily taken. One expands in a Taylor series:

$$u(R_n + \delta) = u(R_n) + \delta\cdot\nabla\, u(R_n) + \frac{1}{2}(\delta\cdot\nabla)^2\, u(R_n) + \ldots \tag{143}$$

Terms in δ_α or in $\delta_\alpha\delta_\beta(\alpha \neq \beta)$ vanish in Eq. (142) by symmetry, leaving in leading order, only the following:

$$\dot{u}\,(R) = Jsa^2\, u(R) \times (\nabla^2\, u(R)) + u(R) \times h \tag{144}$$

This is better written in terms of $u^\pm = u_x \pm iu_y$ and u_z:

$$\dot{u}^+ = 9Jsa^2(u_z\, \nabla^2\, u^+ - u^+\, \nabla^2\, u_z) - iu^+ h \tag{145}$$

with u^- just the complex conjugate of u^+, and:

$$\dot{u}_z^+ = -\, i\, \frac{1}{2}\, Jsa^2(u^-\, \nabla^2\, u^+ - u^+\, \nabla^2\, u^-) \tag{146}$$

From our earlier studies we already know the form of a spinwave: it is a solution of these equations with constant amplitude and plane

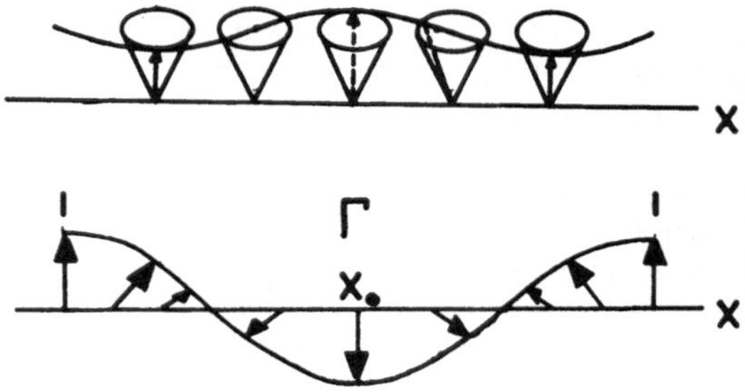

Fig. 4. Spinwave (S_z = p = constant) extending to x = $\pm\infty$. Soliton is moving pulse centered about x = vt.

wave character. Thus, $u^{\pm} = A \exp \pm i(k \cdot R - \omega t)$ and Eq. (145) yield:

$$\omega = Jsa^2 u_z k^2 + h \tag{147}$$

and

$$\dot{u}_z = 0 \,, \quad \text{hence} \quad u_z = p_A \,, \quad \text{a constant.} \tag{148}$$

As u is a unit vector, the amplitude A must be $\sqrt{(1 - p_A{}^2)}$. These results agree with Eq. (16) for A → 0, p_A → 1. At finite amplitude there is a "softening" of the frequency, in agreement with our expectations on the basis of nonlinear magnon interactions studied earlier in this Chapter, e.g., Eq. (85). But at finite amplitude, the excitation energy of the spinwave is extensive--i.e., it is infinite. A finite amplitude spinwave is the superposition of O(N) magnons. Thus the momentum and angular momentum are also infinite. This contrasts with the soliton, a localized excitation whose energy, momentum and angular momentum are all finite (1D) or proportional the cross-sectional area of the wave front (2 or 3D); see Fig. 4. To study it further, we shall need explicit expressions for the excitation energy, momentum and angular momentum. It will also be convenient to specialize to one dimension.

XI. SOLITON SOLUTIONS

The Hamiltonian (1) can be written:

$$H = \frac{1}{2} J \sum_{n=1}^{N} (S_{n+1} - S_n)^2 - h \sum_{n=1}^{N} (S_n^z - s) + E_o$$

$$+ \frac{1}{2} Js^2 a \int_0^{Na} dx \left(\frac{du(x)}{dx} \right)^2 +$$

$$+ \left(\frac{hs}{a} \right) \int_0^{Na} dx[1 - u_z(x)] + E_o \qquad (149)$$

proceeding to both classical and continuum limits. As unit vectors need only two parameters to specify their direction, we can introduce the two, canonically conjugate, dynamical variables p,q through the Villain representation (see Chap. 3) in the classical limit:

$$u^+ = e^{iq} \sqrt{1 - u_z^2} \; , \qquad u^- = (u^+)^+ \; , \qquad \text{and} \qquad u_z = p \qquad (150)$$

Thus, the excitation Hamiltonian, in the natural units ($Js^2a = 1$) is:

$$H = H_o - E_o = \frac{1}{2} \int dx \left[\frac{1}{1 - p^2} (dp/dx)^2 + (1 - p^2)(dq/dx)^2 \right]$$

$$+ h' \int dx(1 - p) \qquad (151)$$

writing hs/a as h'. The equations of motion (144) now take on the aspect:

$$-\dot{q}(x,t) = \frac{1}{1 - p^2} (d^2p/dx^2) + \frac{p}{(1 - p^2)^2} (dp/dx)^2 +$$

$$+ p\left(\frac{dq}{dx} \right)^2 + h' \qquad (152a)$$

and

$$p(x,t) = (1 - p^2)(d^2q/dx^2) - 2p\,\frac{dp}{dx}\frac{dq}{dx}$$

$$= \frac{d}{dx}\left[(1 - p^2)\frac{dq}{dx}\right] \tag{152b}$$

Along with the Hamiltonian, we identify two conserved quantities as, the total momentum,

$$P = \frac{s}{a}\int dx(1 - p)\frac{dq}{dx} \tag{153}$$

and the z-component of the angular momentum carried by the excitations,

$$M_z = \frac{s}{a}\int dx(p - 1) \tag{154}$$

respectively. The latter is naturally negative, as the maximum spin of a ferromagnet occurs in the ground state.

With time dependence arising only through x - vt and v an adjustable parameter, Eq. (152) simplifies

$$v\,\frac{dq}{dx} = \frac{1}{1 - p^2}\frac{d^2p}{dx^2} + \frac{p}{(1 - p^2)^2}(dp/dx)^2 + p(dq/dx)^2 + h' \tag{155a}$$

and

$$v(dp/dx) = \frac{d}{dx}\left\{(p^2 - 1)\frac{dq}{dx}\right\} \tag{155b}$$

(b) is integrated at once, yielding:

$$dq/dx = v\,\frac{p_o - p}{1 - p^2} \tag{156}$$

in which p_o is a constant of integration. Insertion into (a) now yields a nonlinear, second-order differential equation for p:

$$v^2 \frac{p_o - p}{1 - p^2} = \frac{1}{1 - p^2} \frac{d^2 p}{dx^2} + \frac{p}{(1 - p^2)^2} (dp/dx)^2$$

$$+ pv^2 \frac{(p_o - p)^2}{(1 - p^2)^2} + h' \tag{157}$$

This can nonetheless be solved, by the device of setting $(dp/dx)^2 \equiv F(p)$, so that $d^2p/dx^2 = 1/2\ dF/dp$. The above then turns into a linear, first-order differential equation in the new unknown F. This readily solved, yielding:

$$(dp/dx)^2 = F(p) = 2h'p(p^2 - 1) - v^2(1 + p_o^2 - 2p_o p) - p_1(p^2 - 1) \tag{158}$$

Here p_1 is the arbitrary constant of integration. It should be remarked that this equation is that of a classical particle of mass 1/2 and energy E, in a potential well V = E - F. There exists the additional constraint, that $|p| < 1$. Aside from this, the motion is confined to the positive region of F, which is a cubic polynomial with asymptotic behavior $2h'p^3$. The constants $p_{o,1}$ must thus be adjusted so that the solution is physically allowed. It is possible to integrate Eq. (158) and analyze the resulting elliptic functions. It is, however, much simpler to study F for the behavior between the turning points, from which one concludes that $p_{o,1}$ must be adjusted to allow one of three distinct patterns shown as (a), (b) and (c) in Fig. 5.

The first curve, (a), satisfies the conditions appropriate to a spinwave of amplitude p_A, previously analyzed. Only $p = p_A$ is allowed, hence Eq. (157) has the solution $q = k(x - vt) + q_o$.

The curve (b) describes the soliton superposed onto an otherwise perfect ferromagnetic background. The asymptotic value of p is $p_C = 1$, and the largest deviation from the asymptotic value is at p_A which can lie anywhere in the range $-1 < p_A < +1$, by suitable adjustment of the parameters $p_{o,1}$.

The final curve (c) leads to a periodic repetition of the soliton pulse, i.e., to a sort of wavetrain. With p_B and p_C very close and straddling 1, each pulse in the infinite train resembles the solution (b), although the period is finite rather than infinite. With p_A and p_B very close, the solution reduces to the spinwave. Spinwave and the soliton are thus rather opposite limiting cases of the general behavior exemplified by (c).

The soliton solution of this problem was first obtained by Nakamura and Sasada, later by Lakshmanan et al, and had its stability tested by Tjon and Wright. The construction of an F having the shape (b) in the Figure requires the right-hand side of Eq. (158) to have a double root at $p = 1$. $F(1) = 0$ implies $p_0 = 1$; then $dF/dp)_{p=1} = 0$ yields $p_1 = v^2 + 2h'$. Writing $p = 1 - 2\sin^2 1/2\,\theta$ and defining a new variable $y = (1 - v^2/4h')^{1/2} \sin 1/2\,\theta$ and a new independent variable $t = x\,(h' - 1/4\,v^2)^{1/2}$, we obtain

$$(dy/dt)^2 = y^2(1 - y)^2 \tag{159}$$

a well-known ("canonical") equation, known to have the solution $y = \operatorname{sech} t$. It is then a simple matter to substitute the original variables:

$$p = 1 - 2(1 - \frac{1}{4} v^2/h') \operatorname{sech}^2 \left(\frac{x - vt - x_o}{\Gamma} \right) \tag{160}$$

where $\Gamma \equiv (h' - 1/4\,v^2)^{-1/2}$, is the width of the pulse. This pulse is centered at x_o (an arbitrary constant, chosen to satisfy initial conditions) at $t = 0$, travels to the left ($v < 0$) or right ($v > 0$) with the given speed in the range $0 < |v| < 2(h')^{1/2}$. Integration of the equation for q proceeds by similar substitutions, to yield:

$$q = \emptyset_o + \frac{1}{2} v(x - vt - x_o)$$

$$+ \tan^{-1} \left\{ \left(\frac{4h'}{v^2} - 1 \right)^{1/2} \tanh\left(\frac{x - vt - x_o}{\Gamma} \right) \right\} \tag{161}$$

The distortions associated with the soliton are maximal at $v = 0$, yielding the narrowest possible pulse. Conversely, at $v^2 \to 4h'$, the pulse is infinitely broad and the local distortions infinitesimal.

Several quantities are of interest: the phase shift, the momentum, angular momentum, and energy. From Eq. (161) we find, for the total phase shift,

$$\Delta q = 2\tan^{-1} \left(\frac{4h'}{v^2} - 1 \right)^{1/2} \tag{162}$$

which varies from a maximum of π at $v = 0$ to zero as $v^2 \to 4h'$.

Using Eq. (151), we find the energy density to be

$$H(x,t) = \frac{4}{\Gamma^2} \operatorname{sech}^2\left(\frac{x - vt - x_o}{\Gamma}\right) \tag{163}$$

The total energy is time-independent, of course. Integrating the above, we find it to be:

$$\Delta E = 8/\Gamma$$

Restoring the original units:

$$\Delta E = 4Js^2a/\Gamma \tag{164}$$

Similarly, the momentum is:

$$P = \frac{4s}{a} \sin^{-1}\left(1 - \frac{1}{4} v^2/h'\right)^{1/2} \tag{165}$$

and

$$M_z = - 4s/h\Gamma a \tag{166}$$

Combining these gives a more perspicuous relation:

$$\Delta E = \frac{16Js^3}{|M_z|} \sin^2(Pa/4s) \tag{167}$$

For the spin one-half two-magnon bound state, $|M_z| = 2$, $s = 1/2$, this formula agrees precisely with the energy derived in M1 by quantum-mechanical analysis. The above also demonstrates the further lowering of the energy upon binding 3, 4, or more magnons into a bound state "soliton" of a given total momentum.

Tjon and Wright studied the collision of two solitons numerically, some of their results being displayed in M1. But the crucial progress has occured only recently. First Takhtajan and then Fogedby have come to the realization that the nonlinear equations of motion (145) or (146) can be replaced by a set of coupled linear eigenvalue equations. Difficult though the linear problem may be to bring to a closed form solution, there is the fact that the spectrum of eigenfunctions and eigenvalues is complete. This spectrum maps, in a one-to-one correspondence. It is therefore no accident that two solitons, having a collision described by Tjon and Wright, survive this collision with their identities umimpaired This is a necessary consequence of the large number of constraints--constants of motion--that the equations of motion imply and that a solution must satisfy.

The interested reader will have to seek details elsewhere [2] for the mathematics are beyond the scope of the present lecture. Recent applications of this theory include the XY model, modified somewhat to allow slight oscillations out of the XY plane as required for the experimental applications: future applications promise to explain the outstanding mysteries about one-dimensional magnets: their response to neutrons, to time-dependent forces and fields, etc.

Although there is no reason to exclude solitons from consideration in 2D and 3D systems, where indeed they must exist, their energy which is proportional to the area of the wave-front (N^{D-1}) will be too large for them to be included among those elementary excitations that are found spontaneously or easily created, although bound complexes of them may be.

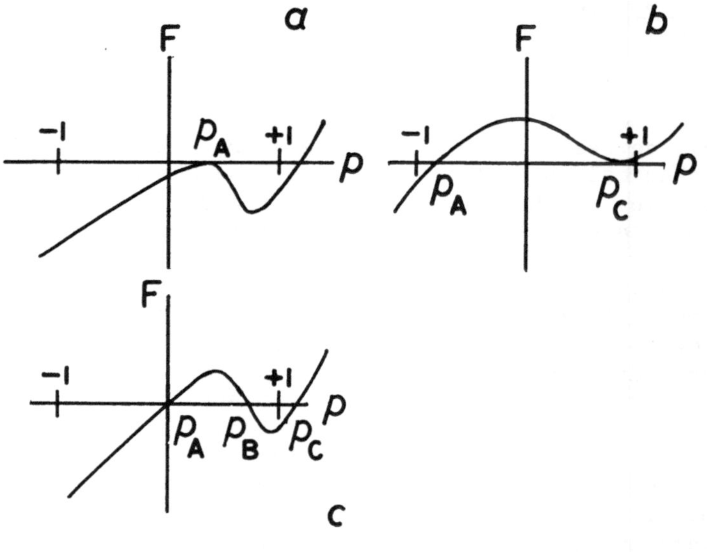

Fig. 5. Right-hand side of Eq. (158), which determines type of behavior: (a) spinwave, (b) soliton, (c) train of solitons.

REFERENCES

1. D. C. Mattis, The Theory of Magnetism, Vol. I (Springer, Berlin, New York, 1980). We shall refer to this book as M1 and to Vol. II (in preparation) as M2.

2. H. Fogedby, J. Phys. A13, 1467 (1980).

QUASIPARTICLES IN MAGNETIC METALS

D.C. Mattis

Department of Physics
University of Utah
Salt Lake City, UT 84112

ABSTRACT

The properties of a low-density electron gas are calculated, with all two-body scattering processes included in the t-matrix approximation. This yield a phase diagram indicating the possibility of itinerant electron ferromagnetism or antiferromagnetism. The results are confirmed at high density by an analysis of Nagaoka's model of a half-filled band. The collective excitations are electron- and hole-like quasiparticles having an interaction-dependent effective mass, and magons. The role of the atomic exchange forces ("Hund's rules") are assessed quantitatively in a two-band model. Without such exchange, no magnetic moments can exist. This chapter highlights some of the distinctions between magnetism in metals and in non-metals.

I. INTRODUCTION

The low density limit of the electron gas can be calculated systematically. First, one recognizes that the long-range part of the electron-electron Coulomb interaction (the troublesome part) is of no interest in the magnetic problem: quantized plasmons guarantee that charge neutrality is maintained throughout the metal. On the other hand, there is no requirement for charge neutrality on atomic scale, other than what is imposed by means of intra-atomic potentials. An Hamiltonian incorporating the kinetic energy of electrons in a single nondegenerate band, with intra-atomic Coulomb repulsion U between electrons of opposite spin (electrons of parallel spin cannot simultaneously occupy the same atom) has been extensively studied by Hubbard and a number of other workers, in the form:

$$H = H_o + U \sum_i n_{i\uparrow} n_{i\downarrow} \, , \text{ with } H_o = \sum_{k,m} c^*_{km} c_{km} E(k) \tag{1}$$

describing free (Bloch) states. For 2 electrons the exact triplet eigenstates and eigenvalues are: Eq. (2):

$$|k, -k';1) = c^*_{k\uparrow} c^*_{-k'\uparrow} |0) \quad \text{and } E_{kk'} \equiv E(k) + E(k') \tag{2}$$

for all k,k' in the first B.Z., with E(k) the Bloch energy. Two projections of

$$2^{-1/2}(c^*_{k\uparrow} c^*_{-k'\downarrow} + c^*_{k\downarrow} c^*_{-k'\uparrow}) \quad \text{and} \quad c^*_{k\downarrow} c^*_{-k'\downarrow} |0) \tag{3}$$

are also, trivially, eigenstates of H belonging to the same eigenvalue. However, for the same two electrons in a singlet state, the energy is different and must be obtained from scattering theory. For brevity define:

$$b_{kk'} \equiv 2^{-1/2}(c^*_{k\uparrow} c^*_{-k'\downarrow} - c^*_{k\downarrow} c^*_{-k'\uparrow})|0) \tag{4}$$

having band ("kinetic") energy $E_{kk'}$ (defined in Eq. (2)). Then let the eigenstate be of the form:

$$B_{kk'} \equiv |k, -k';0) = b_{kk'} + \frac{1}{N} \sum_{q\neq o} L_q b_{k+q,k'+q} \tag{5}$$

the usual form, an incoming wave and a scattered part. The Schrödinger equation consists of three parts:

$$H_o B_{kk'} = E_{kk'} b_{kk'} + \frac{1}{N} \sum_q L_q E_{k+qk'+q} b_{k+qk'+q} \tag{6a}$$

$$H' B_{kk'} = \frac{U}{N} B_{kk'} + \frac{U}{N} \Big\{ \sum_{q\neq 0} b_{k+qk'+q} + \frac{1}{N} \sum_{q',q\neq 0} \sum L_{q'} b_{k+qk'+q} \Big\} \tag{6b}$$

$$W_{kk'} B_{kk'} = W_{kk'} b_{kk'} + \frac{1}{N} \sum_q L_q W_{kk'} b_{k+qk'+q} \tag{6c}$$

Equating the coefficients of $b_{kk'}$ yields an expression for the energy eigenvalue:

$$W_{kk'} = E_{kk'} + \frac{U}{N}\left\{1 + \frac{1}{N}\sum_{q\neq 0} L_q\right\} \tag{7}$$

The scattering amplitudes L_q are obtained from the coefficients of the $b_{k+qk'+q}$. After minor algebra, the result is:

$$L_q = \frac{1}{W_{kk'} - E_{k+qk'+q}} \times \frac{U}{1 + UG_o} \tag{8}$$

where

$$G_o \equiv \frac{1}{N}\sum_q \frac{1}{E_{k+qk'+q} - W_{kk'}} \tag{9}$$

Consequently, the energy is:

$$W_{kk'} = E_{kk'} + \frac{1}{N}\left\{\frac{U}{1 + UG_o}\right\} \equiv E_{kk'} + \frac{1}{N}\{t(W_{kk'})\} \tag{10}$$

which also serves to define the scattering t-matrix, which is the expression of the interaction energy between the two particles with all multiple scattering included. For all practical purposes the $W_{kk'}$ can be replaced by the unperturbed $E_{kk'}$ from which it differs by only $O(1/N)$. (The exception is for bound states, i.e., the zeros of $1 + UG_o = 0$ which, if they exist, lie above the highest $E_{kk'}$ and and thus will not be relevant to the subsequent discussion. In passing, one notes that for negative U, no matter how weak, such a pole in the t-matrix can and will develop, below the lowest $E_{kk'}$; the lowest of these bound states belongs to $q = 0$, and is the zero-momentum Cooper pair, famous in the BCS theory of superconductivity. Our interest, however, is in fairly large, positive U.)

II. LOW DENSITY ELECTRON GAS

The modification of 2-body scattering required for the low density electron gas is straightforward: the range of q's in Eq. (6) *et seq*, especially G_o, Eq. (9), must be restricted to the unoccupied states: $k + q$, $-k' - q$ must lie outside the Fermi volume, i.e., $E(k + q)$ and $E(-k' - q)$ must exceed E_F. Denote the suitably modified G_o by a tilde and similarly for $t(E_{kk'})$ to obtain the effective Hamiltonian for the singlet pairs:

$$H_{eff} = \sum_{k,m} E(k)n_{km} + \frac{1}{N}\sum_{k,k'} \tilde{t}(E_{kk'})n_{k\uparrow}n_{-k'\downarrow} \tag{11}$$

To proceed we need evaluate $\tilde{t}(E)$. Our approach is greatly simplified, yet retains essential features: in a dilute electron gas, the Fermi level is near enough to the bottom of the conduction band(s) (or, for holes, the top) that the effective mass approximation $E(k) = \hbar^2k^2/2m^*$ is appropriate. With this, all the integrals that go into the calculation of the ground sate energy become simple.

First, we introduce the cutoff k_o to retain, properly, the volume of the first B.Z.

$$\frac{1}{N}\sum_{k<1st\ BZ} 1 = 1 = \frac{4\pi}{3}\left(\frac{k_o a_o}{2\pi}\right)^3 \tag{12}$$

The electron density ρ is defined as: the number of electrons per atom per band in a given spin direction $\uparrow$ or $\downarrow$. With a spherical Fermi surface at k_F, ρ is given as:

$$\frac{1}{N}\sum_{k<k_F} 1 \equiv \rho = (k_F/k_o)^3 \tag{13}$$

Similarly the kinetic energy (KE) per band per spin component is:

$$KE = \sum_{k<k_F} E(k) = \frac{N\sum_{k<k_F} E(k)}{\sum_{all} \cdot 1} = N\left[\frac{\int_o^{k_F} dkk^4\hbar^2/2m^*}{\int_o^{k_o} dkk^2}\right]$$

$$= \frac{9}{10} N \frac{\hbar^2 k_o^2}{3m^*} \cdot \rho^{5/3} \tag{14}$$

We start with the exact expression for $\tilde{G}_o$:

$$\tilde{G}_o = (a/2\pi)^3 \int_{\substack{|k + q| > k_F \\ |k' - q| > k_F \\ q < k_o}} d_3q \frac{1}{(\hbar^2/2m^*)[(k + q)^2 + (k' - q)^2 - k^2 - k'^2]}$$

and evaluate it, neglecting $k, k' << k_o$ in the dilute limit. An approximate but very convenient expression, independent of k and k', is what results:

$$\tilde{G}_o \approx (a/2\pi)^3 \int_{k_o > q > k_F} d_3q \frac{1}{(\hbar^2/2m^*)2q^2} = \frac{3m^*}{\hbar^2 k_o^2} \left(1 - \rho^{1/3}\right) \qquad (15)$$

using Eq. (13) to eliminate k_F. With this, we obtain a t-matrix which is independent of k, k' and depends explicitly on ρ. Kanamori has made a different simplification, with basically similar final results.

As the combination $h^2k_o^2/3m^*$ occurs in the interaction and in the KE, we eliminate it. It is recoginzed as a measure of the band-width, in units appropriate to the effective mass approximation. We therefore denote it W as usual:

$$W \equiv h^2k_o^2/3m^*$$

III. CRITERION FOR MAGNETIC GROUND STATE

With $\tilde{G}_o$ suitably reexpressed in terms of W and the $\tilde{t}$ then so written, and with the KE also simplified, we now have for the ground state energy of Eq. (11) the perspicuous expression:

$$E_o = NW\{2 \cdot \frac{9}{10} \rho^{5/3} + \frac{U\rho^2}{W + U(1 - \rho^{1/3})}\}$$

$$= NW \rho^{5/3} \{ \frac{9}{5} + \frac{U\rho^{1/3}}{W + U(1 - \rho^{1/3})} \} \qquad (17)$$

Both KE and interaction energies have been expressed in terms of the sole physical parameters of interest: the bandwidth W, the coupling constant U, and the electron density ρ. Writing a similar expression for the totally ferromagnetic state (at this stage, we do not bother with the added complications of partially magnetized states, so $\rho_{\uparrow} = 2\rho$, $\rho_{\downarrow} = 0$; the interaction energy disappears) we find:

$$E_{ferro} = NW(2\rho)^{5/3}\left(\frac{9}{5}\right) \tag{18}$$

A criterion for the occurrence of ferromagnetism can be obtained by comparing the two expressions. If Eq. (18) is lower, then the ground state has spin N 1/2 h. After a few most elementary algebraic manipulations, this criterion is seen to be:

$$\frac{U\rho^{1/3}}{U + W} > 0.5139 \tag{19}$$

or equivalently,

$$\frac{U}{W} > \frac{1}{\rho^{1/3} - (0.136)^{1/3}} \quad \text{for } \rho > 0.136 \,. \tag{20}$$

The phase diagram is plotted in Fig. 1, where the (uncertain) application to high-density (half-filled bands) is indicated by a dashed curve. Note the existence of a critical $\rho_c = 0.136$. For densities lower than this no magnetism is possible, even at infinite interaction strengths $U = \infty$. This general feature is in excellent accord with a theorem proved in M1, concerning the non-existence of ferromagnetism of 2 electrons, and extends it (for this particular interaction) to an infinite number of particles as long as the density does not exceed ρ_c. In Fig. 2 we indicate (without proof) the results of spatial ordering. As we have seen in the indirect exchange theory, the $\rho = 1/2$ density is always AF, and we may combine this knowledge with Eq. (20) to produce a phase diagram of the type shown in Fig. 2. Similar diagrams have been obtained from many other theories. The exact solution by Lieb and Wu of the one-dimensional Hubbard model for a half-filled band conclusively showed this to be an insulator for finite U (however weak) and to have a spectrum of low-lying excitations readily identifiable as antiferromagnons. This interesting work is the generalization to the interacting electrons of the Bethe solution for interacting spins, but involves mathematical techniques beyond the scope of these lectures. It should be noted that the replacement of G_o by its average value in Eq. (15) is not permissible in 1D

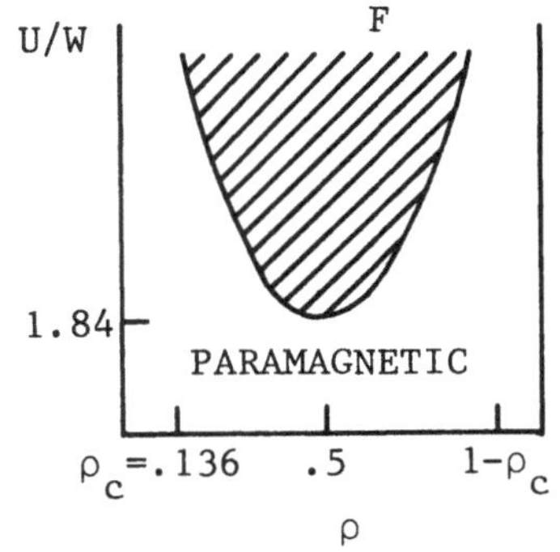

Fig. 1. Ground state phase diagram based on low-density approximation.

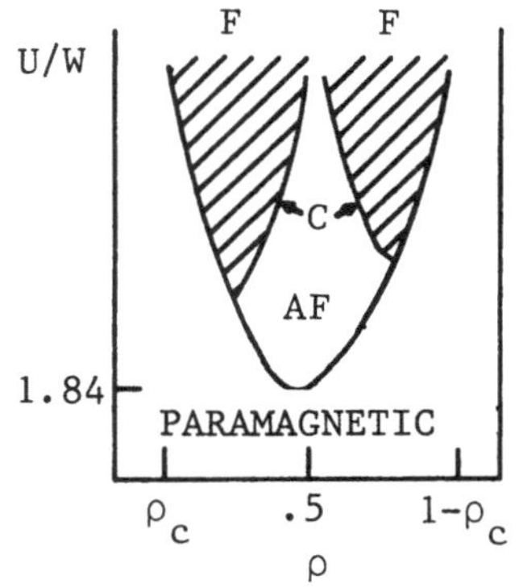

Fig. 2. Ground state phase diagram based on low-density approximation with magnetic ordering taken into account. Curve C is derived in Eq. (28) of text.

(nor perhaps in 2D), for the integral is very sensitive to the precise values of k,k' in this case. Indeed, for $k \sim -k' \sim \pm k_F$, $G_0 \to \infty$ in 1D and thus $t \to 0$. This has the interesting consequence that the well known instability of a 1D "metal" against perturbations of wave-vectors $2k_F$ is compensated by the vanishing, at precisely the same wavevectors, of the interactions!

IV. QUASIPARTICLES

Interactions affect the states of the system. At a given total momentum, only one state is absolutely stable; interactions cause higher lying states to decay. This will be examined by first taking the interacting Fermi sea, whether or not magnetic, to be the new "vacuum." Excited states are then described in terms of quasi-particles (fermions) or collective modes (bosons) in one-to-one correspondence with the familiar electrons (above the Fermi surface, FS) or holes (below the FS) of the noninteracting electron gas and the plasmons of the charged gas, as well as the magnons of the magnetic systems.

If one electron is added to those already present, the added energy is

$$E(k) + \frac{1}{N} \sum_{k'<k_F} \tilde{t}(E_{kk'}) \equiv E(k) + \Delta(k) \qquad (21a)$$

The quasi-particle energy must be measured from the FS, thus we define it to be:

$$\varepsilon(k) = |E(k) - E(k_F) + \Delta(k) - \Delta(k_F)| \approx \hbar^2 |k^2 - k_F^2| / 2\tilde{m} \qquad (21b)$$

which serves to define the renormalized quasi-particles mass $\tilde{m}$ in terms of the original band-structure mass m*. This quantity has been exhaustively studied for the low density electron gas with one of the results for the hard-core ($U = \infty$) gas being:

$$\tilde{m} = m^*\left\{1 + \frac{8}{15\pi^2} (7\ln 2 - 1)(k_F a)^2\right\}$$

in which a = hard sphere radius. We identify 2a as the distance of closest approach, i.e., a_o in the lattice analog. With Eq. (12) used to eliminate this, and Eq. (13) to bring the electron density, we obtain the simple result:

$$\tilde{m} = m^*\{1 + 0.79\ \rho^{2/3}\} \tag{22}$$

in which $\tilde{m}$ is the quasi-particle effective mass and m* the upperturbed, band-theoretic effective mass.

Along with a change in inertia, the excited particle above the FS can be scattered to lower energy while an electron within the Fermi sea is excited out of it. This process can conserve energy, and thus give an imaginary contribution to G_0. As the amount of phase space depends on how far above the FS we start, it is found that the lifetime for this scattering becomes infinite as $(k - k_F)^{-2}$. More precisely, with the above substitution for a, and u = ∞,

$$1/\tau_k = \frac{\hbar}{2m^*}\ 1.21\ \rho^{2/3}(k - k_F)^2 \tag{23}$$

This is not necessarily related to transport properties, as the interactions leading to this scattering do conserve momentum. An identical formula holds for hole quasi-particles.

At a density of 1/8, the increase in mass Eq. (22) is 20%; at a density of 1/2 (where the low-density theory is unreliable, however) the increase is 50%. The lifetime effects of quasi-particles near the FS are, according to Eq. (23), negligible.

V. NEARLY HALF-FILLED BAND: NAGAOKA'S THEORY

Figure 2 shows the possibility of ferromagnetism at extremely large U and an approximately half-filled band. In view of the usual tendency of a half-filled band toward antiferromagnetism, also shown in this figure but at lower values of the interaction parameter, Nagaoka's mechanism is worth of special note. It predicts what would happen if a sufficient number of electrons were removed from the prototypical antiferromagnet NiO, viz: a phase transition to the ferromagnetic metal. The following calculation will show why this occurs in 2D or 3D, and also why it can never occur in 1D, where the ferromagnetic phase is rigorously excluded.

Intially, we consider the U = ∞ limit. Some reflection shows that this effectively sets the tendency to antiferromagnetic ordering ~ W^2/U to zero. In this limit, N electrons, one at each site in the crystal, now have perfectly arbitrary spin configurations. The ground state is 2^N-fold degenerate. The introduction of even a single hole changes this dramatically in 2D or 3D (but not for nearest neighbor hopping in 1D, although n.n.n. hopping can give various results). To see this, let us calculate the eigenstates of the hole in a linear chain first.

Assume a matrix element-B for moving an electron from a site n to the two neighboring sites n ± 1 in the linear chain. As we know from the tight-binding theory, the eigenvalues are then $E(k) = -2B \cos ka$, with $B = 1/4\ W$. The infinite potential energy inhibits the motion of all the electrons except those in the vicinity of the hole. Let the initial configuration be:

$$\begin{array}{lcccccccc} m_n: & m_{-3} & m_{-2} & m_{-1} & 0 & m_1 & m_2 & m_3 & \cdots \\ \hline n: & -3 & -2 & -1 & 0 & +1 & +2 & +3 & \cdots \end{array} \qquad \text{(I)}$$

with the hole indicated by an open circle at $n = 0$ and the electrons being elsewhere. The spins $m_n = \pm 1/2$ of the electrons are a given, arbitrary, set. The initial configuration connects to a linear combination of two:

$$\frac{1}{\sqrt{2}} \left\{ \begin{array}{cccccccc} m_{-3} & m_{-2} & 0 & m_{-1} & m_1 & m_2 & m_3 & \cdots \\ \hline -3 & -2 & -1 & 0 & +1 & +2 & +3 & \cdots \\ \\ m_{-3} & m_{-2} & m_{-1} & m_1 & 0 & m_2 & m_3 & \cdots \\ \hline -3 & -2 & -1 & 0 & +1 & +2 & +3 & \cdots \end{array} \right. \qquad \text{(II)}$$

with matrix element $\sqrt{2}$ B. This, in turn, connects to I and to a new set of configurations denoted III (where the hole is either at -2 or at +2) with matrix element B. This connects back to II, and to a new set, IV, with matrix element B, and so forth. The resulting matrix has the appearance:

$$H = -B \begin{bmatrix} 0 & \sqrt{2} & 0 & \cdot & \cdot \\ \sqrt{2} & 0 & 1 & 0 & \\ 0 & 1 & 0 & 1 & \\ \cdot & 0 & 1 & 0 & \\ \cdot & 0 & & & \end{bmatrix} \qquad (24)$$

bearing some similarity to the Hamiltonian we diagonalized in

Chapter 5 for spin waves near the surface of ferromagnets. The eigenvectors .

$$V_q = A_q (v_o, v_1, \ldots, v_n \ldots) \tag{25}$$

are readily found:

$$v_o = 2^{-1/2}, \quad v_n = \cos(qn), \quad A_q = (2/N)^{1/2} \tag{26}$$

for $N \to \infty$ and $E(q) = -2B \cos q$. Thus, the local density of states at the origin of the motion:

$$N_o(E) = \sum_q |A_q v_o|^2 \, \delta\big(E - E(q)\big) \tag{27a}$$

$$= \frac{1}{N} \sum_q \delta(E + 2B \cos q) \tag{27b}$$

has precisely the usual form for free-particle states. As all the results obtained are for any, arbitrary, set of the $\{m_n\}$, we see that the extra hole does nothing to promote ferromagnetism. We can proceed to 2 or more holes by use of a determinantal function, and obtain that regardless of the number of holes, all the eigenstates in 1D are paramagnetic: their energy is independent of the spins in the absence of an applied magnetic field.

In two or higher dimensions the configurations which result from moving the hole to some point R_i depend on the path taken, except in the ferromagnetic case. Supposing for the ferromagnetic case a configuration ϕ_a connects to a configuration ϕ_b through two different paths: the matrix element will be -2 (in some units). If in the AF state the two final configurations are different, say $\phi_{b'}$ and $\phi_{b''}$, then the matrix element connecting ϕ_a to $2^{-1/2}(\phi_{b'} + \phi_{b''})$ in the same units will be $-\sqrt{2}$. This difference favors the ferromagnetic state over all other configurations.

For a small density n_h of holes, the energy favoring ferromagnetic order is An_hW, taking the bandwidth W to be proportional to B. On the other hand, at finite U the energy favoring AF near a half-filled band is $A'W^2/U$. With $n_h = 1/2 - \rho$, we set the opposing energies equal to obtain the phase boundary curve C, Fig. 2. By symmetry the equation for a more-than-half-filled band is similar, with the final result being:

$$|\rho - \frac{1}{2}| = \alpha W/U , \qquad (28)$$

α being some lumped constant.

Richmond and Rickayzen extended Nagaoka's work in a very interesting way, formulating a trial function in which all but one of the electron spins are aligned. The odd electron is localized about a particular site, whilst the remaining electrons adjust their motion to the presence of the repulsive potential that it represents. This is exactly soluble by the scattering theory that we have used in many connections throughout M1, and the bound state (if any!) energy is a variational upper bound to the true energy. If a bound state is found, the ferromagnetic state is unstable; thus their calculation also provides an equation for curve C. The results which they obtain are indistinguishable from Eq. (28).

VI. TWO OR MORE MAGNETIC SUB-BANDS

To assess the role of the intra-atomic exchange, we here turn to the special case of two degenerate d-like bands, with various Coulomb and exchange integrals.

The interaction of a singlet pair in the same band is parametrized by $U_{aa} = U_{bb}$ and by a t-matrix, which we approximate by the constant:

$$\tilde{t}_{aa}(\rho) = \frac{U_{aa}}{1 + \frac{U_{aa}}{W}(1 - \rho)^{1/3}} \qquad (29)$$

For singlet pair of which one member occupies one band and one the other, there is a similar expression with U_{ab} replacing U_{aa}. By a simple mathematical property of Coulomb integrals, $U_{aa} > U_{ab} > 0$. (Intuitively, electrons in two distinct orbitals overlap less, hence have a smaller Coulomb repulsion, than two electrons in identical orbitals.) Moreover, for triplet pairs (e.g., electrons of parallel spins) in the two different bands, this Coulomb integral is further reduced by an exchange correction, J_{ab} and becomes $U_{ab} - J_{ab} \equiv U_{ab}'$, which we also write as $U_{ab}(1 - j)$. The parameter j is thus the fractional interband exchange parameter, $0 < j < 1$. The ground state singlet energy is:

$$E_o = N\rho^{5/3}\{ \frac{18}{5} W + 2\rho^{1/3}(\tilde{t}_{aa}(\rho) + \tilde{t}_{ab}(\rho) + \tilde{t}'_{ab}(\rho))\} \qquad (30)$$

It is to be compared to the ferromagnetic state, with half the particles in band a and half in band b but all with spin "up." It has energy:

$$E_f = N\rho^{5/3}\{2^{2/3}\frac{18}{5}W + 4\rho^{1/3}\tilde{t}'_{ab}(2\rho)\} \tag{31}$$

The ferromagnetic state in which all the particles, of spin up, are also in a single band, say a (this represents spin and orbital magnetism, and should occur when the perturbations of the solid are too weak to quench the orbital moments of the individual atoms, as in f-shells of the rare earths), has energy:

$$E_{f,o} = N\rho^{5/3}\left(\frac{9}{5}4^{5/3}W\right) \tag{32}$$

The case of maximal interband exchange $j = 1$ is special. We compare the energies Eqs. (30)-(31) and conclude that Eq. (32), representing spin + orbital magnetism, can never lie lowest. The phase diagram showing the region when the spin-only magnetic moments form is remarkably similar to Eq. (19), with only the numerical value parameters $\rho_c = 0.04$ and $U_{min}/W = 1$ being different. Doubling the number of bands allows magnetic moment formation at much lower densities and interaction parameter U than previously.

In the physically more plausible cases of partial exchange $j = j_{ab}/U_{ab} < 1$ the situation may be quite different. We compared eqs. (30-32) by numerical calculation, assuming $U_{ab} = 1/2\ U_{aa} \equiv U$, and find: spin + orbital magnetism never occurs for $U/W < 20$ nor for densities far from 1/2, so this case may be eliminated from practical considerations even though it is good to know it exists (in principle) in the atomic limit $W \to 0$. Spin magnetism has itself a restricted range of stability, which depends strongly on j. For $j = 0.5$ or greater, the situation is qualitatively similar to $j = 1$, whereas for smaller values of j (0.2 or 0.1) the regions of stability shrink rapidly and become nonexistent at $j = 0$. Thus, regardless of the strength of the interaction parameter U, the existence of a magnetic moment ultimately depends on the Hund's rule exchange parameter, as a stabilizing factor. The situation is clearly summarized in Fig. 3.

VII. MAGNONS IN METALS

Once the Coulomb and exchange forces have created a magnetic solid, one queries the spectrum of elementary excitations. To the quasi-particles and collective modes which are common to all metals, one must add magnons for the magnetic metals. These are the quantized spinwaves which are the property of an ordered magnetic medium. Let us now examine the magnons for ferromagnetic metals.

With the exception of the strong-coupled one-band model, our knowledge of the itinerant ground state is imperfect even for ferromagnets. We cannot then proceed by computing an excited state, and subtracting the ground state energy therefrom, as even small errors of O(N) will obscure effects of O(1). A more direct method is required, that of the equation of motion. In this procedure, the scattering of a certain type excitation is treated exactly but other terms are neglected ("random phase approximation"). The neglected terms vanish at low density, and the procedure gives satisfactory answers at long wavelengths, therefore one can accept the results with the same confidence (or skepticism!) as for magnons in 3D Heisenberg antiferromagnets--as a semiquantitative, systematic approximation.

For simplicity, we treat the case of single band in some detail, including the cases--so far ignored--of partial magnetization. We then indicate the extensions of the theory to multi-band cases, and to questions of stability of the assumed ferromagnetic ground state, such as we already examined in connection with the indirect exchange mechanism.

So, let U and ρ be in the range shown in Fig. 2 or 3 for which the ground state is ferromagnetic. In the Heisenberg model, translational invariance was sufficient to construct the one-magnon states uniquely. Now, it is no longer enough. The degrees of freedom in a metal are sufficiently numerous, that if we only specify that the spin angular momentum must decrease by one unit and the total momentum wavevector must be q, there will be a very large number of excitations to satisfy this requirement. We therefore add to these requirements that of a very long or infinite lifetime, and find then that the magnon is bound state, a linear combination of all the states in which a hole is created in the majority-spin band (taken to be spins "down" henceforth) and a particle, with extra momentum hq, is added to minority spin ("up") subband.

The quasi-particle energies are,

$$E_m(k) = E(k) + \frac{1}{N}\sum_{k'} \tilde{t}(kk')f_{-m}(k') \equiv E(k) + \Delta_m(k) \tag{33}$$

where the Fermi function $f_m(k)$ is 1 if $E_m(k) < E_m(k_{Fm})$ and zero, otherwise, as discussed previously. Thus the elementary unit of magnetic excitation must be $a^+_{k+q\uparrow}\, a_{k\downarrow}|F)$, where $|F)$ is the ground state, and a_{km} destroys a particle (or creates a hole) in spin subband m, and a^+_{km} creates one. The one-magnon state must be $|q) \equiv \Omega_q|F)$, where Ω_q is the operator:

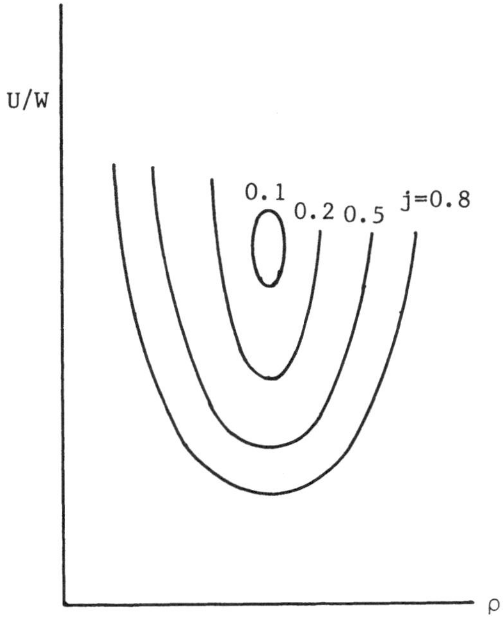

Fig. 3. Magnetic moment formation in 2-band model, assuming $U_{aa} = U_{bb} + U$, $U_{ab} = 1/2\ U$, and exchange parameter $j = J_{ab}/1/2\ U$. It is seen that small j precludes magnetic moment formation.

$$\Omega_q = N^{-1/2} \sum_k F_k a^+_{k+q\uparrow} a_{k\downarrow} \tag{34}$$

The energy of each unit must be,

$$E_{kq} \equiv \frac{\hbar^2}{2m^*}(2k\cdot q + q^2) + \Delta_\uparrow(k + q) - \Delta_\downarrow(k) \tag{35}$$

in the effective mass approximation. Each scatters into the others with a strength proportional to t(kk') and to the occupation-number factor. We determine this from the equations of motion: that is, Ω_q is a raising operator for H by an energy $\hbar\omega_q$ then

$$[H, \Omega_q] = \hbar\omega_q \Omega_q \tag{36}$$

the kinetic energy yields:

$$[H_o, \Omega_q] = N^{-1/2} \sum_q F_k(E_{kq}) a^+_{k+q\uparrow} a_{k\downarrow} \tag{37}$$

whereas the scattering part yields:

$$[H', \Omega_q] =$$

$$- N^{-3/2} \sum_{k,k'} F_{k'} \tilde{t}(kk')[f_\downarrow(k') - f_\uparrow(k + q)]a^+_{k+q\uparrow} a_{k\downarrow} \tag{38}$$

omitting terms such as $a^+_{k'+q'm}a_{k'm}a^+_{k+q-q'\uparrow}a_{k\downarrow}$ which are smaller by some power of the density ρ and are neglected in the random-phase approximation. We have identified $a^+_{km}a_{km} = n_{km}$ with its average $f_m(k)$, the Fermi function. Equating terms on both sides of the equation of motion (36) we obtain the equations of the amplitudes F_k:

$$F_k\left(E_{kq} - \hbar\omega_q\right) = N^{-1} \sum_{k'} \tilde{t}(kk')\left[f_\downarrow(k') - f_\uparrow(k' + q)\right] F_{k'} \tag{39}$$

This equation is exactly soluble at $q = 0$, by the choice F_k = const., and yields $\omega_o = 0$. This is an exact result, reflecting the rotational invariance of the magnetic ground state.

For $q \neq 0$, Eq. (39) is a transcendental equation that must generally be solved graphically or numerically. However, in the special case of $\tilde{t}(kk') = \tilde{t}$ = constant, there is a very simple solution $F_k = A(E_{kq} - \hbar\omega_q)^{-1}$ which leads to the secular equation for the eigenvalue:

$$1 = N^{-1} \sum_k \frac{\tilde{t}}{E_{kq} - \hbar\omega_q} \tag{40}$$

This yields an approximately parabolic magnon $\hbar\omega_q = Dq^2$ for $q < q_{max}$; the maximum q is not at the edge of the BZ but occurs when the denominator of Eq. (40) becomes complex. This signifies the onset of the scattering regime in which the bound state does not exist; an attempt to excite a magnon at $q > q_{max}$ will result only in a broad resonance. Rather than detail this calculation, we examine the case of two degenerate bands which we do calculate in detail. This case has an interesting feature: the elementary units in Eq. (34) can be $a^{+}_{k+q,a,\uparrow}a_{k,a\downarrow} \pm a^{+}_{k+q,b\uparrow}a_{k,b\downarrow}$, where a,b are the two bands in question. The (+) combination yields S^{+}_{tot} at q = 0, and thus commutes with the Hamiltonian. It is identified as the zero energy mode which expresses the rotational invariance of our approximation of the magnetic ground state; the (+) mode at q ≠ 0 is identical to that of a single band worked out above. These are denoted the "acoustic magnons" by analogy with lattice vibrations, where the low-lying spectrum are the acoustic modes. Correspondingly there exist the "optical" magnons" the (-) branch, which do not exist for a single band ferromagnet. The optical spectrum start at $2(\Delta_\uparrow - \Delta_\downarrow)$ at q = 0 in the two-band case, and increases slightly at higher wavevectors. The equations (with 2m* = 1, h = 1, and $k_F = 1$) are given below, and the calculated results plotted in Fig. 4. (The acoustic branch is identical to the solution of the one-band case, Eq. (40).) We replace the parameter $\Delta_\uparrow - \Delta_\downarrow$ by the single parameter Δ, and separately consider the cases when Δ < 1, Δ = 1, and Δ > 1.

When Δ > 1, we find the following transcendental equation:

$$\pm \frac{2}{3\Delta} = \frac{1}{2q} L(Q) \tag{44}$$

where

$$Q = \frac{2q}{\Delta + q^2 - \hbar\omega_q} \tag{45a}$$

and

$$L(Q) = \frac{1}{Q^2} \left[\frac{1}{2} (Q^2 - 1)\ln \left| \frac{Q+1}{Q-1} \right| + Q \right] \tag{45b}$$

For Δ < 1, $n_{k\uparrow} \neq 0$, and we obtain a slightly more complicated equation:

$$\pm \frac{2}{3} \left[\frac{1 - (1-\Delta)^{3/2}}{\Delta} \right] = \frac{1}{2q} \left[L(Q) - (1-\Delta)L(Q')\right] \tag{46}$$

where

$$Q' = \frac{2q(1-\Delta)^{1/2}}{\Delta - q^2 - \hbar\omega_q} \tag{47}$$

and Eqs. (45) define the other parameter and the function L(Q).

The optical magnon mode (-) starts at 2Δ for $q = 0$ and increases somewhat before merging with the continuum. The acoustic (+) branch starts at $\hbar\omega_0 = 0$ for $q = 0$ and increases approximately $\sim Dq^2 + O(q^4)$, with the parabolic approximation $\hbar\omega_q \sim Dq^2$ improving in relative accuracy as Δ is increased. Expansion of the equations leads to a formula for D:

$$D = \frac{1 + (1 - \Delta)^{3/2} - \frac{4}{5}[1 - (1 - \Delta)^{5/2}]/\Delta}{1 - (1 - \Delta)^{3/2}} \qquad \text{for } \Delta < 1 \tag{48a}$$

and

$$D = 1 - \frac{4}{5\Delta} \qquad \text{for } \Delta > 1 \tag{48b}$$

The acoustic branch enters the continuum and "dies" at $q_{max} = 0.75\Delta$ for $\Delta > 1$.

VIII. ANTIFERROMAGNETISM IN METALS

Antiferromagnetism occurs when spin-down electrons start to fill the Brillouin zone, as may be seen in the following demonstration for strong coupling. Assume $\Delta > 1$, and every state in the spin-down zone is filled, every state in the spin-up zone is empty. The eigenvalue equation can be expanded in powers of $E(k + q) - E(k) \equiv w(k,q)$, a procedure certainly valid for small q. We make use of the assumption $E(-k) = E(k)$ to prove,

$$\sum_k w^{2p+1} \equiv 0 \qquad p = 0,1,2,\ldots, \quad \text{all } q, \tag{49}$$

which together with the identity $\Sigma\, w^{2p} > 0$ readily established the desired result:

$$\hbar\omega_o = 0 > \hbar\omega_q \qquad \text{all } q \neq 0 \tag{50}$$

Not only does a collective magnon mode exist at every q in the Brillouin zone, but the q = 0 mode is a maximum, and the ferromagnetic state must be unstable against the emission of any number and any type of magnons. The new ground state must then be an antiferromagnetic, or a spiral spin configuration of the type previously discussed.

The antiferromagnetic behavior sets in even before the spin-down Brillouin zone is completely filled by electrons, in the neighborhood of a half-filled zone, although the precise point at which it occurs must be calculated numerically. It is akin to the antiferromagnetism of insulators. Whereas the exclusion principle

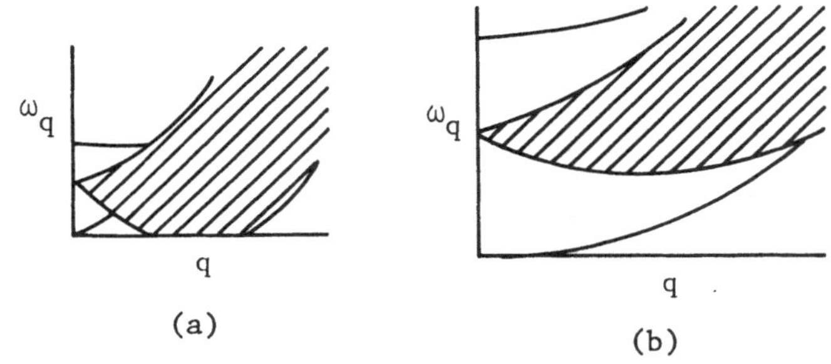

Fig. 4. Showing magnon branches emerging from continuum of electron spin-flip states in (a) weak-coupling, (b) strong-coupling limits.

has the effect in the ferromagnetic state of preventing the kinetic motion of electrons from atom to atom, such motion is not prohibited in the antiferromagnetic configurations. The difference with insulators, is that such "hopping" (band, or kinetic, energy) represents real transitions in the metallic state, but only virtual transitions in the insulator.

POLARITONS

R. Loudon

Physics Department, Essex University
Colchester CO4 3SQ, U.K.

ABSTRACT

Linear response theory for a bulk dielectric is used to obtain the electric-field fluctuation spectrum. The form of the spectrum provides the bulk polariton dispersion relation and damping. Measurements of bulk polaritons by light scattering and nonlinear optical processes are reviewed. The linear response theory is extended to a dielectric medium with a flat surface, thus providing the surface polariton dispersion relation and damping. Measurements by attenuated total reflection, light scattering and nonlinear optical processes are briefly reviewed. Both phonon-polaritons and exciton-polaritons are considered.

1. INTRODUCTION

Polaritons are coupled modes of the electromagnetic field and a crystal excitation [1]. They occur for all crystal excitations that have transverse electric-dipole coupling to the electromagnetic field. Bulk polaritons are the form that exists in infinite crystals where there is translational invariance in all three co-ordinate directions. Crystals with boundaries have additional coupled modes in the form of surface polaritons. The simplest case is that of a single flat boundary where translational invariance remains in two of the co-ordinate directions.

A typical dispersion relation for bulk polaritons is illustrated in Fig. 1. The dashed lines show the frequency ω as a function of the wavevector Q for the uncoupled electromagnetic wave and crystal excitation. For the electromagnetic wave

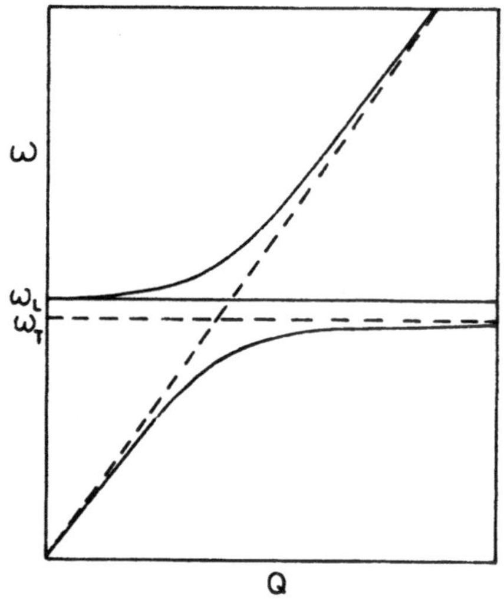

Fig. 1. Dispersion relations for uncoupled (dashed lines) and coupled (continuous lines) electromagnetic wave and crystal excitation.

$$\omega = cQ/\kappa_\infty, \qquad (1)$$

where c is the velocity of light and κ_∞ is the constant relative permittivity of the crystal in the absence of the excitation, which has itself been drawn with constant frequency ω_T.

Interaction between the electromagnetic field and the excitation dipole moment results in new dispersion curves shown by the continuous lines in Fig. 1. The crystal excitation considered is usually either an exciton, where typical co-ordinates of the cross-over of the dashed lines are $\omega_T \approx 6 \cdot 10^{14}$ s^{-1} and $Q \approx 2 \cdot 10^6$ m^{-1}, or a phonon, where typically $\omega_T \approx 1.5 \cdot 10^{13}$ s^{-1} and $Q \approx 5 \cdot 10^4$ m^{-1}. The wavevectors in both cases are very small compared to zone boundary values ($\approx 10^{10}$ m^{-1}), and the dependence of the bare exciton or phonon properties on the wavevector can often be ignored.

The polaritons represented by the continuous lines in Fig. 1 are called exciton-polaritons or phonon-polaritons depending on the parent excitation. The dispersion curve has two branches separated by a gap extending from the bare excitation frequency ω_T to a higher frequency ω_L which coincides with the frequency of the longitudinal crystal excitation. This excitation does not mix with the transverse photons but its frequency is raised from the value ω_T it would have for an excitation of zero dipole moment.

Ordinary linear optical processes are fully described by the frequency-dependent linear susceptibility $\chi^{(1)}$ or the relative permittivity

$$\kappa = 1 + \chi^{(1)}. \tag{2}$$

These functions show resonant effects for frequencies ω close to the frequency ω_T of an electric-dipole excitation. They do not explicitly depend on the wavevector, and there is generally no need to consider the coupled-mode properties of the excitations in the vicinity of ω_T.

Light scattering, and nonlinear optical processes in general, are described by higher-order components of the susceptibility, $\chi^{(2)}$, $\chi^{(3)}$, etc., which also show resonant effects associated with the electric-dipole frequencies ω_T of the crystal. However, in contrast to linear processes, the resonances in these higher-order processes depend upon both the wavevector Q and the frequency ω, to produce spectra centred on the polariton dispersion curve of Fig. 1. Thus light scattering and other nonlinear optical processes provide a means of studying the properties of polaritons, and the concept of polaritons makes a useful contribution to the interpretation of experiments.

The importance of polaritons in nonlinear processes stems from the basic mechanism by which the crystal produces interaction between different light beams. In many cases, the process is initiated by the coupling that occurs between monochromatic incident light and the dynamic variables of the dipole-active crystal excitations, to produce a frequency-shifted nonlinear polarization. The polarization in turn causes absorption or emission of light in frequency ranges determined by sums and differences of the incident and crystal excitation frequencies. For crystals in thermal equilibrium, the dynamic variables engage in random fluctuations whose statistical properties are governed by their thermal excitation probabilities. The spectrum of the light absorbed or emitted by the nonlinear polarization is controlled by the frequency spectra, or power spectra, of the relevant random fluctuations. The main aim of theory is the calculation and interpretation of the required power spectra.

The optical properties of polaritons are mainly determined by their associated electric-field fluctuations. The derivations that follow use linear response theory to obtain the electric-field fluctuation spectra. The simpler case of an infinite crystal is treated first; it is shown that the maximum fluctuation strength occurs for frequencies and wavevectors connected by the bulk polariton dispersion relation illustrated in Fig. 1. The more complicated case of a crystal with a single flat surface is treated subsequently; it is shown that additional maxima in the electric-field spectra now occur, and these determine the dispersion relation for the surface polaritons. The derivations of dispersion relations are followed by accounts of the spectroscopic influences of both bulk and surface polaritons and examples of relevant experiments.

II. BULK POLARITON LINEAR RESPONSE

Consider the dipole resonance associated with an exciton or phonon in a cubic insulating crystal of isotropic optical properties. Let W be one of the normal co-ordinates of the excitation, with the corresponding electric dipole moment parallel to the unit vector $\vec{\zeta}$. The equation of motion is

$$\ddot{W} + \Gamma\dot{W} + \omega_T^2 W = Z\vec{\zeta}.\vec{E}, \tag{3}$$

where ω_T is the resonance frequency, Γ is the damping, and Z is an effective charge (dimensions = charge/mass$^{\frac{1}{2}}$). For an optical vibration in a diatomic lattice, W is the relative displacement of the two types of atom multiplied by the square root of their reduced mass; for an exciton it is a similar quantity defined with respect to the electron and hole.

The field $\vec{E}$ is that which necessarily accompanies the excitation; conversely, an externally applied field excites an excitation amplitude W given by (3). If ω is the frequency of the motion,

$$(\omega_T^2 - \omega^2 - i\omega\Gamma)W = Z\vec{\zeta}.\vec{E}. \tag{4}$$

The dielectric polarization is

$$\vec{P} = (Z\vec{\zeta}W/\Omega) + \varepsilon_o(\kappa_\infty - 1)\vec{E}, \tag{5}$$

where Ω is the unit cell volume. The first term is the contribution of the resonance considered and the second term is the polarization in the absence of the dipole excitation.

The relative permittivity κ is defined by

$$\kappa\varepsilon_o\vec{E} = \varepsilon_o\vec{E} + \vec{P}. \tag{6}$$

In the case of a cubic crystal, where the mode polarization $\vec{\zeta}$ can be chosen parallel to the electric field, use of (4) and (5) gives

$$\kappa = \kappa_\infty + \frac{\omega_P^2}{\omega_T^2 - \omega^2 - i\omega\Gamma}, \tag{7}$$

where

$$\omega_P^2 = Z^2/\varepsilon_o\Omega \tag{8}$$

is the square of the plasma frequency associated with the excitation. The relative permittivity at zero frequency is thus

$$\kappa_o = \kappa_\infty + \omega_P^2/\omega_T^2. \tag{9}$$

The absorption of transverse electromagnetic waves in linear optics is entirely determined by Im κ and the absorption maximum occurs at the frequency ω_T. The frequency ω of the single electromagnetic wave is the sole variable in an experiment on an isotropic medium.

The electric field $\vec{E}$ is also related to the polarization $\vec{P}$ by Maxwell's equations, and the general form is

$$\nabla \times \nabla \times \vec{E} - (\omega/c)^2\vec{E} = (\omega^2/\varepsilon_o c^2)\vec{P}, \tag{10}$$

or equivalently

$$(\omega/c)^2\vec{P} = -\varepsilon_o\{\nabla^2 + (\omega/c)^2\}\vec{E} + \varepsilon_o\nabla(\nabla\cdot\vec{E}). \tag{11}$$

The wave equation (10) can also be cast into the forms

$$\nabla \times \nabla \times \vec{E} - \kappa_\infty(\omega/c)^2\vec{E} = (\omega^2/\varepsilon_o c^2)(Z\vec{\zeta}W/\Omega), \tag{12}$$

$$\nabla \times \nabla \times \vec{E} - \kappa(\omega/c)^2\vec{E} = 0 \tag{13}$$

with the help of (5) and (6). It follows from (13) that the wavevector q of free waves in the medium must satisfy

$$q^2 = \kappa\omega^2/c^2 \tag{14}$$

for transverse polarization.

Now consider the effect on a crystal of volume V of an applied polarization

$$\vec{P}^{ext} = \vec{P}_o^{ext} \exp(-i\omega t + i\vec{Q}.\vec{r}), \tag{15}$$

where the frequency and wavevector are independent real quantities. In all cases, real polarizations and electric fields are obtained from their complex representations by addition of the complex conjugates.

The applied polarization induces an electric field $\vec{E}$ in the crystal and an associated induced polarization $\vec{P}$ related to $\vec{E}$ by (6). Thus (11) becomes

$$(\omega/c)^2(\vec{P} + \vec{P}^{ext}) = \varepsilon_o\{Q^2 - (\omega/c)^2\}\vec{E} - \varepsilon_o\vec{Q}(\vec{Q}\cdot\vec{E}). \quad (16)$$

Take a Cartesian component and use (6) to remove $\vec{P}$:

$$(\omega/c)^2 P_j^{ext} = \varepsilon_o\{Q^2 - \kappa(\omega/c)^2\}E_j - \varepsilon_o Q_j(\vec{Q}\cdot\vec{E}). \quad (17)$$

This is a system of equations for the three components of $\vec{E}$, which can be solved without difficulty. It is convenient to express the solutions in terms of the linear response functions defined as

$$\langle\langle E_i(\vec{r});E_j(\vec{Q})^*\rangle\rangle_\omega \equiv \frac{E_i}{VP_{oj}^{ext}\exp(-i\omega t)} = -\frac{Q_iQ_j - q^2\delta_{ij}}{\varepsilon_o V\kappa(Q^2 - q^2)}\exp(i\vec{Q}\cdot\vec{r}),$$

$$(i,j = x,y,z) \quad (18)$$

where q^2 is given by (14). Alternatively, the linear response function, or Green function, can be Fourier transformed to a wavevector representation

$$\langle\langle E_i(\vec{Q}');E_j(\vec{Q})^*\rangle\rangle_\omega = (1/V)\int d\vec{r}\,\exp(-i\vec{Q}'\cdot\vec{r})\langle\langle E_i(\vec{r});E_j(\vec{Q})^*\rangle\rangle_\omega$$

$$= -\frac{(2\pi)^3}{V}\frac{Q_iQ_j - q^2\delta_{ij}}{\varepsilon_o V\kappa(Q^2 - q^2)}\delta(\vec{Q} - \vec{Q}'). \quad (19)$$

The response functions defined in this way determine the power spectra of the electric-field fluctuations in the crystal. The connection is made via the fluctuation-dissipation theorem, which can be stated in the form [2]

$$\langle E_i(\vec{Q}')E_j(\vec{Q})^*\rangle_\omega = (\hbar/\pi)\{n(\omega) + 1\}\mathrm{Im}\langle\langle E_i(\vec{Q}');E_j(\vec{Q})^*\rangle\rangle_\omega \quad (20)$$

where $n(\omega)$ is the Bose-Einstein thermal factor

$$n(\omega) = \{\exp[\hbar\omega/k_BT] - 1\}^{-1}. \quad (21)$$

The quantity on the left of (20), the power spectrum, is the contribution of frequency ω and wavevector $\vec{Q}$ to the time-dependent correlation function of the i and j field components

$$\langle E_i(\vec{Q}',0)E_j(\vec{Q},t)^*\rangle = \int d\omega\,\exp(i\omega t)\langle E_i(\vec{Q}')E_j(\vec{Q})^*\rangle_\omega, \quad (22)$$

where the angle brackets denote an ensemble average over the thermal probability distribution for the field components.

The denominator of the linear response function (18) or (19) can be factorized by introducing three orthogonal unit vectors $\vec{\lambda}$, $\vec{\mu}$ and $\vec{\nu}$, of which $\vec{\lambda}$ is parallel to $\vec{Q}$. Then if we use the property

$$\lambda_i\lambda_j + \mu_i\mu_j + \nu_i\nu_j = \delta_{ij}, \tag{23}$$

the main part of (18) or (19) can be written

$$-\frac{Q_iQ_j - q^2\delta_{ij}}{\varepsilon_o V\kappa(Q^2 - q^2)} = \frac{1}{\varepsilon_o V}\left\{-\frac{\lambda_i\lambda_j}{\kappa} + \frac{\mu_i\mu_j + \nu_i\nu_j}{(cQ/\omega)^2 - \kappa}\right\}, \tag{24}$$

where (14) has been used. This result is not restricted to the case where κ has a single resonance as in (7) but applies generally for cubic crystals.

Consider the form of the response function in the case of a single resonance frequency ω_T where (7) is appropriate. The first term in (24) contributes for field components polarized parallel to $\vec{Q}$ and it is thus the longitudinal term. With the use of (7) we obtain

$$-\text{Im}\,\frac{1}{\kappa} = -\frac{1}{\kappa_\infty}\,\text{Im}\,\frac{\omega_T^2 - \omega^2 - i\omega\Gamma}{\omega_L^2 - \omega^2 - i\omega\Gamma}, \tag{25}$$

where the longitudinal frequency ω_L is given by

$$\omega_L^2 = \omega_T^2 + \omega_P^2/\kappa_\infty = \kappa_o\omega_T^2/\kappa_\infty. \tag{26}$$

Provided that the separation between ω_L and ω_T is much larger than Γ, the longitudinal part of the response function is approximately

$$-\text{Im}\,\frac{1}{\kappa} \approx \frac{\omega\Gamma\omega_P^2/\kappa_\infty^2}{(\omega_L^2 - \omega^2)^2 + \omega^2\Gamma^2} \approx \frac{\Gamma\omega_P^2/4\omega_L\kappa_\infty^2}{(\omega_L - \omega)^2 + (\tfrac{1}{2}\Gamma)^2}\,. \tag{27}$$

The longitudinal contribution thus has a peak centred at frequency ω_L and with a frequency linewidth equal to Γ. It does not depend on the wavevector Q.

The second term in (24) contributes for field components polarized perpendicular to $\vec{Q}$ and it is thus the transverse term. Its magnitude depends on both Q and ω, and its form is illustrated in Fig. 2. The function was first analysed by Henry and Garrett [3] and later by Barker and Loudon [4]. For small Γ it can be approximated by

$$\text{Im}\,\frac{1}{(cQ/\omega)^2 - \kappa} \approx \frac{\omega_0(\omega_T^2 - \omega_0^2)^2}{4\{\kappa_\infty(\omega_T^2 - \omega_0^2)^2 + \omega_T^2\omega_P^2\}}\,\frac{\Gamma(\omega_0)}{(\omega_0 - \omega)^2 + [\frac{1}{2}\Gamma(\omega_0)]^2}, \tag{28}$$

where ω_0 is defined in terms of the wavevector Q by

$$\left(\frac{cQ}{\omega_0}\right)^2 = \kappa_\infty + \frac{\omega_P^2}{\omega_T^2 - \omega_0^2}, \tag{29}$$

and the linewidth parameter in the second factor is

$$\Gamma(\omega_0) = \frac{\omega_0^2\omega_P^2\Gamma}{\kappa_\infty(\omega_T^2 - \omega_0^2)^2 + \omega_T^2\omega_P^2}. \tag{30}$$

The transverse contribution thus has peaks at wavevectors Q and frequencies ω_0 that are related by (29), and this is taken to define the dispersion relation of the transverse polaritons. Fig. 1 shows the dispersion relation defined in this way, together with the longitudinal peak frequency ω_L from (26). Note that the function on the right of (29) is the same as the relative permittivity κ given by (7) but with the damping Γ removed; the damping plays no role in determining the peaks of the transverse response function and (29) is sometimes misleadingly called the "undamped" polariton dispersion

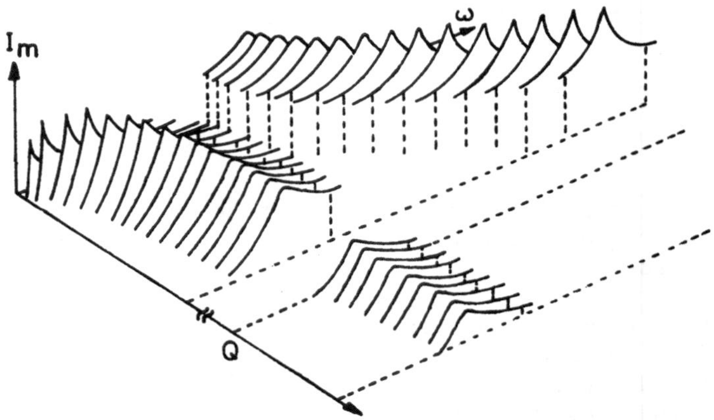

Fig. 2. Imaginary part (28) of the transverse polariton response as a function of Q and ω. Note that the vertical scale is logarithmic (after Barker and Loudon [4]).

relation. It is not to be confused with the dispersion relation (14) for propagation of free waves.

The linewidth of the transverse response at constant Q is given by (30). Note that the maximum linewidths are

$$\Gamma(\omega_T) = \Gamma(\omega_L) = \Gamma, \tag{31}$$

and the linewidth decreases as ω_O moves up the upper branch away from ω_L or down the lower branch away from ω_T. However, the basic damping parameter Γ varies with frequency in real crystals, and the polariton linewidth can vary with frequency in a complicated way (see, for example, [5,6]).

A similar calculation of the linear response can be carried out for crystal structures that are not optically isotropic [4,7] The results are similar in form to (18) and (19) but there is additional complication due to the existence of two or three independent components of the relative permittivity for uniaxial and biaxial crystals respectively.

III. EXPERIMENTS ON BULK POLARITONS

The simplest optical processes that determine the bulk polariton properties are light scattering and two-photon absorption. Consider first the light scattering process where incident monochromatic light of frequency ω_I and wavevector $\vec{q}_I$ inside the crystal produces light of frequency ω_S and wavevector $\vec{q}_S$ by scattering from a crystal excitation. The excitation frequency ω is determined by the energy conservation condition

$$\omega_I = \omega_S + \omega. \tag{32}$$

The optical wavevectors are real for a transparent crystal, and the excitation wavevector $\vec{Q}$ is then determined by the momentum conservation requirement

$$\vec{q}_I = \vec{q}_S + \vec{Q}. \tag{33}$$

The conditions of the experiment thus pick out the crystal spectrum at a given real frequency and a given real wavevector. The ω and $\vec{Q}$ can be varied to some extent by variation in the measured frequency ω_S and in the direction of the scattered light beam, but for crystals of constant optical refractive index the ω and Q available for study lie to the right of the oblique dashed line in Fig. 1 [8]. Thus light-scattering experiments are mainly concerned with the lower branch of the polariton dispersion curve. The excitations are usually phonon-polaritons since the frequency ω must be a difference of two optical frequencies.

An expression for the scattered intensity is obtained by considering the coupling of the incident and scattered field strengths E_I and E_S to the electric and vibrational amplitudes E and W of the polariton. The coupling energy has the form

$$U = -aE_I E_S^* W^* - bE_I E_S^* E^* + \text{comp. conj.}, \tag{34}$$

where a and b are derivatives of the linear susceptibility with respect to W^* and E^*. These parameters also determine the second and third-order nonlinear susceptibilities associated with the dipole resonance, according to [9]

$$\varepsilon_o \chi^{(2)} = \frac{aZ}{\omega_T^2 - \omega^2 - i\omega\Gamma} + b \tag{35}$$

$$\varepsilon_o \chi^{(3)} = \frac{|a|^2\Omega}{\omega_T^2 - \omega^2 - i\omega\Gamma} . \tag{36}$$

The polarization P_S of the crystal at the scattered frequency ω_S is obtained by differentiation of (34)

$$P_S = -\partial U/\partial E_S^* = aE_I W^* + bE_I E^*. \tag{37}$$

The normal co-ordinate W is related to the field E by (4) and we can write

$$W = \beta\vec{\zeta}.\vec{E} \qquad \text{where} \qquad \beta = Z/(\omega_T^2 - \omega^2), \tag{38}$$

provided that ω lies outside the linewidth of the resonance at ω_T. Thus (37) becomes

$$P_S = (a\beta\zeta + b)E_I E^*. \tag{39}$$

The light-scattering cross section of the crystal is proportional to the thermal average of the power radiated by the polarization P_S divided by the incident intensity [8]

$$d^2\sigma/d\Omega d\omega_S \propto \left\langle |P_S|^2 \right\rangle_{\omega_S} / |E_I|^2 = |a\beta\zeta + b|^2 \left\langle EE^* \right\rangle_\omega . \tag{40}$$

Thus with the use of (19), (20), (24) and (38), the cross section for the transverse part of the crystal excitation spectrum is

$$\frac{d^2\sigma}{d\Omega d\omega_S} \propto \{n(\omega) + 1\} \left| \frac{a\zeta Z}{\omega_T^2 - \omega^2} + b \right|^2 \text{Im} \frac{1}{(cQ/\omega)^2 - \kappa} . \tag{41}$$

The above derivation of this expression is somewhat skeletal and omits for example the tensor characters of the susceptibility derivatives and the vector natures of the mode polarization $\vec{\zeta}$ and electric field $\vec{E}$. The derivation is invalid within the linewidth of the resonance. However, the final result (41) includes the main features of the cross section derived by more rigorous methods [4,10].

It is seen that the wavevector dependence of the cross section (41) is determined by the same expression as treated above in (28) to (30) and illustrated in Fig. 2. Light scattering experiments thus measure the polariton electric-field fluctuation spectrum but with some distortion arising from the first two frequency-dependent factors on the right of (41). Fig. 3 shows the good agreement between the theoretical polariton dispersion relation for GaP obtained from (29) and experimental points obtained from the frequency and wavevector dependence of the cross section in the first light-scattering measurement of this kind [11].

Light scattering studies have since been extended to more complex crystals with several dipole-active phonons and/or noncubic structure. The dispersion relations depend on the direction of $\vec{Q}$ for noncubic crystals, and the simplest cases occur when $\vec{Q}$ is parallel

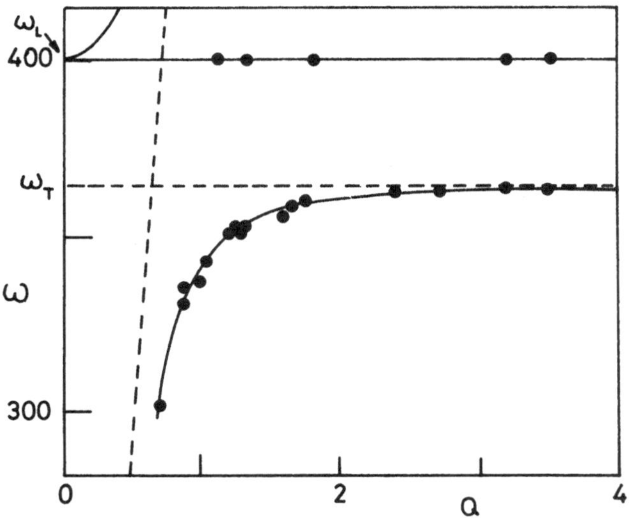

Fig. 3. Theoretical phonon-polariton dispersion relation for GaP with experimental points determined by light scattering (after Henry and Hopfield [11]). Units are $2\pi cm^{-1}$ for ω and $10^6 m^{-1}$ for Q.

to one of the principal axes. Fig. 4. compares theory and experiment for the dispersion relations of the polaritons that propagate perpendicular to the c-axis with polarization parallel to the c-axis in $LiIO_3$, a uniaxial crystal [12]. A representative measurement of the polariton dispersion relations for all three principal-axis directions in a biaxial crystal is that of Fukumoto et al. [13] on $KNbO_3$. The whole area of light-scattering work on phonon-polaritons is fully described by Claus et al. [14].

The theory of light scattering by polaritons outlined above needs only slight modifications to apply to two-photon absorption by polaritons. Suppose that two light beams with frequencies ω_1 and ω_2 are incident on a crystal. If $\vec{q}_1$ and $\vec{q}_2$ are the corresponding wavevectors inside the crystal, the frequency ω and wavevector $\vec{Q}$ of the crystal excitation created by absorption of one photon from each beam are given by

$$\omega_1 + \omega_2 = \omega \tag{42}$$

$$\vec{q}_1 + \vec{q}_2 = \vec{Q}. \tag{43}$$

Thus for a crystal that is transparent at frequencies ω_1 and ω_2, the two-photon absorption again probes the crystal spectrum at a real frequency and wavevector. A common experimental procedure keeps beam 1 fixed while the frequency and wavevector direction of beam 2 are

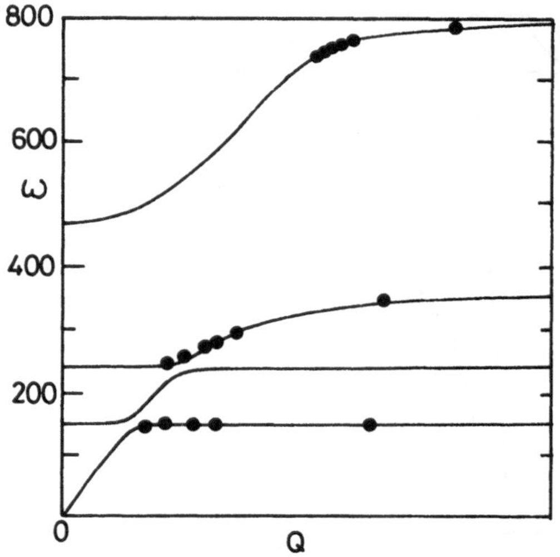

Fig. 4. Measured points and calculated dispersion relations for the extraordinary polaritons propagating perpendicular to the c-axis in $LiIO_3$ (after Otaguro et al. [12]). Units for ω are $2\pi cm^{-1}$.

varied. For a crystal of constant optical refractive index, the ω and Q available for study lie to the left of the oblique dashed line in Fig. 1. Two-photon absorption experiments are mainly concerned with the upper branches of exciton-polariton dispersion curves.

The processes of light scattering and two-photon absorption differ only in the replacement of emission by absorption for the behaviour of one of the photons. Two-photon absorption experiments commonly use one strong light beam ω_1, analogous to the incident beam ω_I in light scattering, and a weaker beam, analogous to the scattered beam ω_S in light scattering, whose absorption is measured as a function of its frequency ω_2. The theory [9,15] can be based on an energy density similar to (34) and the nonlinear susceptibilities continue to have forms similar to (35) and (36). The resulting absorption coefficient at frequency ω_2 has the same dependence on ω and Q as the scattering cross section (41) except that the thermal factor should be removed; the constant of proportionality includes the intensity of the beam of frequency ω_1. The experiments thus measure the polariton electric-field fluctuation spectrum and they provide a means of determining the polariton dispersion relation and linewidth via the analysis given in (28) to (30). The open circles in Fig. 5 show measurements [16] of the upper and longitudinal branches for excitons in CuCℓ obtained from the two-photon absorption spectrum.

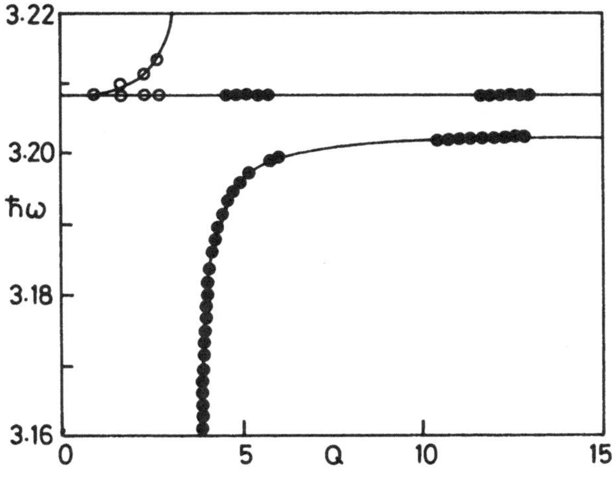

Fig. 5. Exciton-polariton dispersion relation in CuCℓ with measured points from two-photon absorption [16] (open circles) and two-photon light scattering [17] (solid circles) (after Hönerlage et al. [17]). Units are eV for $\hbar\omega$ and $10^7 m^{-1}$ for Q.

Both the phonon-polaritons generated in light scattering and the exciton-polaritons generated in two-photon absorption include a photon or electromagnetic-wave component that can be transmitted through the crystal boundaries. Thus polariton effects can be observed in various three-wave mixing processes where one or more of the frequencies lies close to a dipole resonance of the crystal. Experimental studies include difference-frequency generation of infrared radiation [18] in the region of the phonon-polaritons of GaP shown in Fig. 3, and sum-frequency generation of visible radiation [19] in the region of the exciton-polaritons of CuCℓ shown in Fig. 5.

The polariton dispersion relation and other properties can also be studied by higher-order nonlinear processes in which more than three quanta are involved in the interaction. There are for example several varieties of four-wave mixing processes in which two quanta of strong incident light combine to produce two different quanta, some or all of the frequencies being close to dipole resonances. One of these processes is two-photon light scattering where incident and scattered light couple to a crystal excitation whose frequency ω and wavevector $\vec{Q}$ satisfy

$$2\omega_I = \omega_S + \omega \tag{44}$$

$$2\vec{q}_I = \vec{q}_S + \vec{Q}. \tag{45}$$

All of the frequencies can now lie in the visible region and the additional quantum provides greater flexibility in the parts of the polariton dispersion curve that can be studied. The solid circles in Fig. 5 show the results of this kind of experiment performed on CuCℓ [17].

IV. SURFACE POLARITON LINEAR RESPONSE

We now generalize the infinite-crystal calculation of Section III to an isotropic crystal that has a single flat surface in the xy-plane. The crystal medium occupies the half-space $z < 0$ and free space occupies the half-space $z > 0$. The wavevectors of free waves in the medium continue to satisfy (14), but the wavevectors q in free space are given by

$$q = \omega/c. \tag{46}$$

There is translational invariance parallel to the surface plane and these components of the wavevectors are matched at the surface. It is convenient to consider wavevectors that lie in the zx-plane. The surface area A is assumed to have dimensions large compared to all other characteristic lengths of the system.

With the removal of translational invariance from the z-direction, it is better to consider the effect on the crystal of an applied polarization

$$\vec{P}^{ext} = \vec{P}_0^{ext} \exp(-i\omega t + iQ_x x)\delta(z-z') \qquad (47)$$

than to retain the three-dimensional wavevector forms of (15). Here z' is any co-ordinate within the medium (that is z' < 0). With this replacement, the principles of the calculation of the linear response remain the same as before. It is necessary to calculate the electric fields produced by the applied polarization using Maxwell's equations, but now taking account of the usual dielectric boundary conditions at the surface of the medium. The results of the calculation determine the linear response functions as in (18) where now the fields are evaluated at real-space z-co-ordinates but are Fourier components for the x and y co-ordinates. The linear response functions can be written in the form

$$\langle\langle E_i(Q_x',Q_y',z);\ E_j(Q_x,0,z')^*\rangle\rangle_\omega$$

$$= \{(2\pi)^2/A\}\delta(Q_x-Q_x')\delta(Q_y')\langle\langle E_i(z);\ E_j(z')^*\rangle\rangle_\omega, \qquad (48)$$

where [20,21]

$$\langle\langle E_i(z);\ E_j(z')^*\rangle\rangle_\omega$$

$$= \frac{i/q_z}{2\varepsilon_o A\kappa}\begin{bmatrix} -q_z^2 & Q_x q_z \\ -Q_x q_z & Q_x^2 \end{bmatrix}\frac{q_z - \kappa\mathcal{q}_z}{q_z + \kappa\mathcal{q}_z}\exp\{-iq_z(z+z')\}$$

$$+ \frac{i/q_z}{2\varepsilon_o A\kappa}\begin{bmatrix} q_z^2 & -Q_x q_z \operatorname{sgn}(z-z') \\ -Q_x q_z \operatorname{sgn}(z-z') & Q_x^2 \end{bmatrix}\exp\{iq_z|z-z'|\} \qquad (49)$$

for i,j = x,z, where

$$q_z^2 = \kappa(\omega/c)^2 - Q_x^2 \qquad (50)$$

$$\mathcal{q}_z^2 = (\omega/c)^2 - Q_x^2 \qquad (51)$$

$$\operatorname{sgn}(z-z') = \begin{cases} 1 \text{ for } z > z' \\ -1 \text{ for } z < z' \end{cases}, \qquad (52)$$

and

$$\langle\langle E_y(z);\ E_y(z')^*\rangle\rangle_\omega$$

$$= \frac{iq^2/q_z}{2\varepsilon_o A\kappa}\left\{\frac{q_z - \mathcal{q}_z}{q_z + \mathcal{q}_z}\exp\{-iq_z(z+z')\} + \exp\{iq_z|z-z'|\}\right\}. \quad (53)$$

Here z is also assumed to be a co-ordinate within the medium. The remaining components of the linear response function all vanish.

The above linear response functions all have similar structures. Their first terms are surface contributions that involve the round-trip distance z+z' from the stimulus point z' to the response point z via the surface. Their second terms are bulk contributions that involve only the direct distance $|z-z'|$ from z' to z. These bulk contributions are easily seen to be Fourier transforms, with respect to the wavevector z components, of the infinite-crystal functions (19). We thus need to consider only the new contributions produced by the surface. The power spectra of the electric-field fluctuations are determined by the imaginary parts of the linear response functions as in (20), and the surface polaritons correspond to the peaks that result from the surface contributions.

Suppose first that the damping Γ in the relative permittivity κ is negligibly small. Then q_z given by (50) is either purely real or purely imaginary. The additional peaks in the electric-field fluctuation spectrum occur only for imaginary wavevector z-components, obtained when

$$Q_x^2 > \begin{cases}\kappa(\omega/c)^2 \\ (\omega/c)^2.\end{cases} \quad (54)$$

These fluctuations are confined to the surface region if positive imaginary parts are taken, with the signs chosen in the z-dependences of the electric fields in the calculation of the linear response functions. It is seen that the surface contribution to (53) has no pole in this case, but the surface parts of the zx-plane response functions (49) do have additional poles for negative κ.

The resonant denominator of the response functions (49) in the small Γ limit can be analysed in a manner similar to the denominator in the bulk response functions (18) or (19). Analogous to (28), the imaginary part of the surface contribution is proportional to

$$\mathrm{Im}\,\frac{q_z - \kappa q_z}{q_z + \kappa q_z} \approx \frac{\kappa^2}{(\kappa+1)^2(\kappa-1)}$$
$$\times \frac{(\kappa_\infty+1)^2\omega_0(\omega_M^2-\omega_0^2)^2}{\kappa_\infty(\kappa_\infty+1)(\omega_M^2-\omega_0^2)^2 + (\kappa_0-\kappa_\infty)\omega_M^2\omega_T^2} \; \frac{\Gamma(\omega_0)}{(\omega_0-\omega)^2 + [\frac{1}{2}\Gamma(\omega_0)]^2}, \tag{55}$$

where κ is the relative permittivity (7) with the damping removed, ω_0 is now defined in terms of the surface wavevector Q_x according to

$$\left(\frac{cQ_x}{\omega_0}\right)^2 = \frac{\kappa}{\kappa+1}, \tag{56}$$

ω_M is the maximum frequency for which this equation gives a real Q_x, obtained with the help of (7) and (9) as

$$\omega_M = \left(\frac{\kappa_0+1}{\kappa_\infty+1}\right)^{\frac{1}{2}} \omega_T, \tag{57}$$

and the linewidth parameter in the final factor of (55) is [22]

$$\Gamma(\omega_0) = \frac{(\kappa_0-\kappa_\infty)\omega_0^2\omega_T^2\Gamma}{\kappa_\infty(\kappa_\infty+1)(\omega_M^2-\omega_0^2)^2 + (\kappa_0-\kappa_\infty)\omega_M^2\omega_T^2}. \tag{58}$$

The surface contribution thus has peaks at the wavevectors Q_x and frequencies ω_0 that are related by (56), and this defines the surface polariton dispersion relation. Fig. 6 shows the surface phonon-polariton dispersion relation of InSb calculated from (56). Note that these excitations exist only in the frequency interval

$$\omega_T \le \omega_0 \le \omega_M \tag{59}$$

and the corresponding range of values of the relative permittivity is

$$-\infty \le \kappa \le -1. \tag{60}$$

V. EXPERIMENTS ON SURFACE POLARITONS

Surface polaritons were first detected optically by Otto [24] using the method of attenuated total reflection. The method has been applied to the study of surface polaritons in a range of materials [25]. The existence of the technique relies upon an important qualitative difference between bulk and surface optical properties. An electromagnetic wave in a bulk material has a wavevector q that is

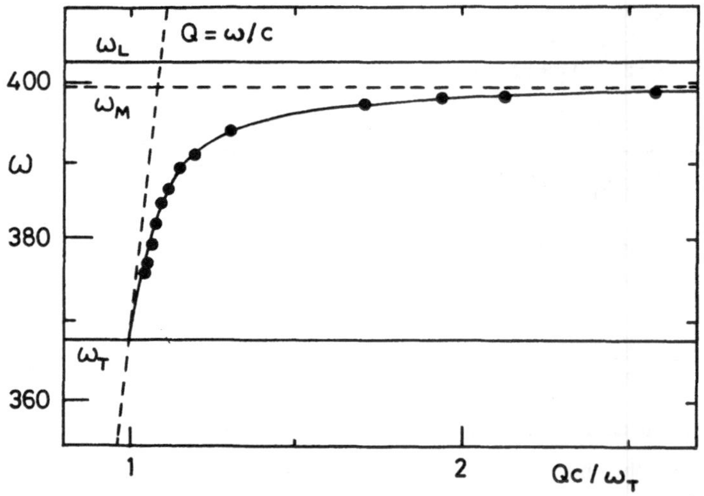

Fig. 6. Theoretical surface phonon-polariton dispersion relation in GaP with experimental points determined from attenuated total reflection (after Marschall and Fischer [23]). Units for ω are $2\pi cm^{-1}$.

determined by (14) for a given frequency ω; it is not possible to vary q and ω independently in a linear optical experiment and thereby produce a scan across the bulk polariton dispersion curve shown in Fig. 1. However, in the case of electromagnetic waves transmitted into a material through its surface, it is possible to vary the wavevector component Q_x parallel to the surface independently of the frequency ω to some extent by changing the angle of incidence. It is therefore possible to scan across the surface polariton dispersion curve shown in Fig. 6 in a linear optical experiment.

This feature of surface optical properties forms the basis of the method of attenuated total reflection. Since the inequalities (54) must be satisfied in the surface polariton region, the electromagnetic wave incident from outside the medium must produce a field inside the medium that decays exponentially with distance from the surface. This is achieved by totally reflecting the incident light from the surface of a prism adjacent to the surface of the material of interest. The incident light then produces an evanescent wave that leaks out of the prism, across the narrow space between the two surfaces, and into the material. When the wavevector and frequency of the evanescent wave match those of the surface polariton there is a loss of energy to the polariton accompanied by an attenuation of the 'totally reflected' wave. It is thus possible to determine the surface polariton dispersion curve by measuring the Q_x and ω of maximum attenuation. Fig. 6 shows measured points on the dispersion curve of surface phonon-polaritons in GaP determined in this way [23]. The same method can also be used for surface exciton-polaritons, as in the first measurement of this kind on ZnO [26].

The various two-photon and higher-order processes outlined in Section III can also be used to study surface polaritons. Thus, in the expression (40) for the light-scattering cross section, the electric-field power spectrum obtained by application of the fluctuation-dissipation theorem (20) to the response functions (49) shows peaks at frequencies ω and wavevector components Q_x that satisfy the surface dispersion relation (56). Consideration of energy and momentum conservation shows that most of the surface polariton dispersion curves generally lie within the accessible region for study by light scattering. A detailed theory of the cross section must include the effects of the boundary on the incident and scattered light beams [27-29] and the success of experiments depends upon the careful optimization of their geometrical arrangements. Most of the measurements have been made by Ushioda and his collaborators (see [21] for a review) and they include results on the richer variety of surface polaritons that exist in thin samples where the second surface affects the nature of the excitations.

Nonlinear optical experiments on surface polaritons were pioneered by De Martini and Shen [18,30]. Thus for example surface exciton-polaritons can be excited by absorption of two photons, and the dispersion relation in ZnO has been determined by detection of the polaritons generated in this way [31]. Four-wave mixing processes have been used to obtain the surface phonon-polariton dispersion relation in GaP [32].

VI. CONCLUSION

The dispersion relations of both bulk and surface polaritons have been measured by a variety of techniques and the results are in good agreement with the theory. There is little experimental information on polariton linewidths with which the theoretical predictions can be compared.

The dipole resonance frequencies ω_T have been assumed here to be independent of wavevector. Further polariton phenomena occur for resonance frequencies that vary with wavevector, when spatial dispersion effects must be included in the theory [33]. Resonant Brillouin scattering provides a means of measuring the polariton dispersion relations in this regime [34,35].

REFERENCES

1. D. L. Mills and E. Burstein, Rep. Prog. Phys. 37, 817 (1974).
2. R. Loudon, J. Raman Spect. 7, 10 (1978).
3. C. H. Henry and C. G. B. Garrett, Phys. Rev. 171, 1058 (1968).
4. A. S. Barker Jr. and R. Loudon, Rev. Mod. Phys. 44, 18 (1972).
5. S. Ushioda and J. D. McMullen, Solid State Commun. 11, 299 (1972).
6. D. Heiman, S. Ushioda and J. P. Remeika, Phys. Rev. Lett. 34, 886 (1975).
7. L. Merten, Phys. Stat. Sol. 30, 449 (1968).
8. W. Hayes and R. Loudon, "Scattering of Light by Crystals", Wiley, New York (1978).
9. R. Loudon, "Nonlinear Spectroscopy", N. Bloembergen, ed., North-Holland Pub. Co., Amsterdam (1977), p. 296.
10. H. J. Benson and D. L. Mills, Phys. Rev. B1, 4835 (1970).
11. C. H. Henry and J. J. Hopfield, Phys. Rev. Lett. 15, 964 (1965).
12. W. Otaguro, E. Wiener-Avnear, C. A. Aguello and S. P. S. Porto, Phys. Rev. B4, 4542 (1971).
13. T. Fukumoto, A. Okamoto, T. Hattori and A. Mitsuishi, Solid State Commun. 17, 427 (1975).
14. R. Claus, L. Merten and J. Brandmüller, "Light Scattering by Phonon-Polaritons", Springer, Berlin (1975).
15. D. Boggett and R. Loudon, J. Phys. C6, 1763 (1973).
16. D. Fröhlich, E. Mohler and P. Wiesner, Phys. Rev. Lett. 26, 554 (1971).
17. B. Hönerlage, A. Bivas and Vu Duy Phach, Phys. Rev. Lett. 41, 49 (1978).
18. F. De Martini, "Nonlinear Spectroscopy", N. Bloembergen, ed., North-Holland Pub. Co., Amsterdam (1977), p. 319.
19. D. C. Haueisen and H. Mahr, Phys. Rev. Lett. 26, 838 (1971) and Phys. Lett. 36A, 443 (1971).
20. A. A. Maradudin and D. L. Mills, Phys. Rev. B11, 1392 (1975).
21. S. Ushioda and R. Loudon, "Surface Polaritons", D. L. Mills and V. M. Agranovich, eds., North-Holland Pub. Co., Amsterdam, to be published.
22. J. S. Nkoma, R. Loudon and D. R. Tilley, J. Phys. C7, 3547 (1974).
23. N. Marschall and B. Fischer, Phys. Rev. Lett. 28, 811 (1972).
24. A. Otto, Z. Physik 216, 398 (1968).
25. A. Otto, Festkörperprobleme 14, 1 (1974).
26. J. Lagois and B. Fischer, Phys. Rev. Lett. 36, 680 (1976).
27. J. S. Nkoma and R. Loudon, J. Phys. C8, 1950 (1975).
28. D. L. Mills, Y. J. Chen and E. Burstein, Phys. Rev. B13, 4419 (1976).
29. V. M. Agranovich and T. M. Leskova, Sov. Phys. Solid State 19, 465 (1977).
30. F. De Martini and Y. R. Shen, Phys. Rev. Lett. 36, 216 (1976).
31. F. De Martini, M. Colocci, S. E. Kohn and Y. R. Shen, Phys. Rev. Lett. 38, 1223 (1977).

32. F. De Martini, G. Giuliani, P. Mataloni, E. Palange and Y. R. Shen, Phys. Rev. Lett. 37, 440 (1976).
33. F. Garcia-Moliner and F. Flores, J. Phys. (Paris) 38, 851 (1977).
34. W. Brenig, R. Zeyher and J. L. Birman, Phys. Rev. B6, 4617 (1972).
35. D. R. Tilley, J. Phys. C13, 781 (1980).

POLARONS

R. Evrard

Institut de Physique B5
Université de Liège, B-4000 Liège, Belgium

ABSTRACT

The concept of polarons and its evolution during the last decades are discussed. Fröhlich's Hamiltonian is derived and the main theoretical methods for large polarons are reviewed. The properties of polarons, as for example self-energy and effective mass are described. Finally small polarons and the difference between these latter and Fröhlich's polarons are discussed.

I. INTRODUCTION

Electrons or holes in an ionic crystal are the cause of the polarization of the lattice. The particle in interaction with the field of ionic polarization appears as a dressed particle. The situation is somewhat similar to that prevailing in quantum field theory. Therefore polarons as such are of a rather formal interest. They can serve as one of the simplest models of quantum-field theory. Another formal problem is the self-trapping of small polarons at strong coupling.

A second aspect has a more practical importance : the ionic polarization being related to lattice vibrations, the interaction between the charge carriers and the polarization is a form of electron-phonon interaction. This interaction plays an important role in many properties not only of ionic crystals but also of polar semiconductors like the II-VI and III-V compounds. For instance, it contributes to limit the electrical conductivity in the ohmic regime as well as at high electric field, it leads to a Stoke shift allowing laser emission in heterostructures, etc...

In these lectures, I intend to give a broad overview of the main properties of polarons. It seems to me that a more or less chronological account of the development of the ideas in the field suits this goal best.

II. THE LANDAU-PEKAR STRONG-COUPLING THEORY

In 1933, Landau [1] proposed that the F-centers are composed of electrons self-trapped in their polarization field. Of course, De Boer's alternative interpretation of F-centers is now universally accepted. It is true, however, that Landau was by far the first to study the effects of the interaction with the lattice polarization on the state of the conduction electrons in ionic crystals. Some ten years later, Pekar worked out this problem in more detail. He based his calculations on a method called crude-adiabatic approximation in molecular physics (*). An account of his work can be found in ref. [2]. The basic ideas are the following.

The polarization induced in the lattice gives rise to a force acting on the electron. As the ions are heavy, the lattice distortion which is associated with the polarization is assumed to remain at rest, so that the electron moves in a static potential well centered around a fixed point in the crystal, taken as the origin of the axes. The electron is far lighter than the ions; for this reason, Landau and Pekar made the assumption that the electron motion is much faster than the displacements of the ions, so that the source of the lattice polarization is the static average charge distribution $\rho(\vec{r})$ of the electronic particle. Obviously, this charge distribution is the product of the charge of the particle with the square of the wave function :

$$\rho(\vec{r}) = e|\varphi(\vec{r})|^2. \tag{1}$$

It gives rise to an electric displacement $\vec{D}(\vec{r})$ which is given by Gauss's law, so that

$$\vec{\nabla}.\vec{D}(\vec{r}) = 4\pi\rho(\vec{r}). \tag{2}$$

The total polarization $\vec{P}_t(\vec{r})$ induced in the crystal is given by

$$\vec{D}(\vec{r}) = \vec{E}(\vec{r}) + 4\pi\vec{P}_t(\vec{r}) \tag{3}$$

where $\vec{E}(\vec{r})$ is the electric field at the position $\vec{r}$. As

$$\vec{E}(\vec{r}) = \varepsilon_o^{-1}\,\vec{D}(\vec{r}) , \tag{4}$$

(*) For the definition of crude adiabatic see the glossary of terms (appendix A) in ref.[3].

one obtains

$$\vec{P}_t(\vec{r}) = (4\pi)^{-1}(1-\varepsilon_o^{-1})\vec{D}(\vec{r}). \qquad (5)$$

In these relations, ε_o denotes the static dielectric constant. In fact, the electronic polarization due to the distortion of the electron clouds of the ions follows instantaneously the displacements of the electron. Thus, the interaction with this part of the polarization is unable to perturb the motion of this latter and gives a constant contribution to the energy of the system. This contribution is already included in the energy of the band to which the particle belongs. Therefore, the electronic contribution $\vec{P}_e(\vec{r})$ must be substracted from the total polarization, leaving the ionic polarization $\vec{P}(\vec{r})$ in which we are interested here. Obviously

$$\vec{P}_e(\vec{r}) = (4\pi)^{-1}(1-\varepsilon_\infty^{-1})\vec{D}(\vec{r}) \qquad (6)$$

where ε_∞ denotes the high-frequency dielectric constant, which is the square of the refraction index, (i.e. $\varepsilon_\infty=n^2$). Therefore

$$\vec{P}(\vec{r}) = \vec{P}_t(\vec{r}) - \vec{P}_e(\vec{r}) \qquad (7)$$

$$= (4\pi\,\bar{\varepsilon})^{-1}\,\vec{D}(\vec{r}) \qquad (8)$$

where $\bar{\varepsilon}$ is an effective dielectric constant defined by the following relation

$$\bar{\varepsilon}^{-1} = \varepsilon_\infty^{-1} - \varepsilon_o^{-1}. \qquad (9)$$

The energy involved in the polarization of the lattice can easily be evaluated in the following way. The energy of interaction between the dipoles associated with the displacements of the ions and the external field $\vec{D}(\vec{r})$ is

$$E_{int} = -\int d^3r\ \vec{D}(\vec{r}).\vec{P}(\vec{r}). \qquad (10)$$

Now, producing the polarization $\vec{P}(\vec{r})$ requires an increase in energy due to two effects. The first part is an electrical contribution. It is due to the dipole-dipole interaction and is quadratic in the polarization $\vec{P}(\vec{r})$. The second is the elastic energy required to distort the lattice. For distortions small enough, Hooke's law applies and the elastic contribution is also quadratic in the polarization, so that the total polarization energy can be written as

$$E_p = c\int d^3r\ P^2(\vec{r}) - \int d^3r\ \vec{D}(\vec{r}).\vec{P}(\vec{r}) \qquad (11)$$

where c is a constant to be determined. The equilibrium value of $\vec{P}(\vec{r})$ corresponds to the minimum of (11), i.e.

$$\vec{P}(\vec{r}) = (2c)^{-1} \vec{D}(\vec{r}). \tag{12}$$

Comparing with (8) leads to

$$c = 2\pi\bar{\varepsilon} \tag{13}$$

and the energy associated with the polarization is equal to

$$E_p = -(4c)^{-1} \int D^2(\vec{r}) d^3r \tag{14}$$

$$= -(8\pi\bar{\varepsilon})^{-1} \int D^2(\vec{r}) d^3r. \tag{15}$$

Notice that the energy required to produce the polarization is equal, with the opposite sign, to half the energy of interaction between the polarization $\vec{P}(\vec{r})$ and the external field $\vec{D}(\vec{r})$. This result, due to the quadratic form of the polarization energy, is well known in electrostatics.

As for the electron wave function, it obeys the Schrödinger equation, which in a system of units such that $\hbar=2m=1$, can be written as

$$-\nabla^2\varphi(\vec{r}) + eV(\vec{r})\varphi(\vec{r}) = E\varphi(\vec{r}) \tag{16}$$

where $V(\vec{r})$ is the electric potential associated with the polarization field $\vec{P}(\vec{r})$. The polarization is itself related to the wave function through Eqs. (8) and (2). This latter (Gauss's law) requires an integration over r to obtain $\vec{D}(\vec{r})$. Therefore, if $\vec{P}(\vec{r})$ is replaced by its expression in terms of the wave function, the Schrödinger equation (16) becomes a non linear integro-differential equation, which looks impossible to solve. The interest of seeking an approximate solution is therefore obvious. The variational principle of least energy has been used by Pekar. A trial expression containing variational parameters is used for the wave function. The expectation value of the left-hand side of Eq.(16) is then evaluated. The first term gives the electron kinetic energy E_k :

$$E_K = \int |\nabla\varphi(\vec{r})|^2 d^3r . \tag{17}$$

The second term,

$$E_{int} = \int d^3\vec{r}\ V(\vec{r})\ e|\varphi(\vec{r})|^2 \tag{18}$$

$$= \int d^3r\ V(\vec{r})\rho(\vec{r}) \tag{19}$$

is simply the energy of interaction between the charge density $\rho(\vec{r})$ and the field of ionic polarization. The expression of this interaction energy has been obtained above (see Eq.(10)). It is

$$E_{int} = -\int \vec{D}(\vec{r}).\vec{P}(\vec{r})d^3r \ , \tag{20}$$

$$= -(4\pi\bar{\varepsilon})^{-1}\int D^2(\vec{r})d^3r \ . \tag{21}$$

To obtain the total energy of the system, it remains to add the potential energy E_{pot} required to produce the polarization. This contribution is the first term of Eq.(11), i.e.

$$E_{pot} = (8\pi\bar{\varepsilon})^{-1}\int D^2(\vec{r})d^3r \ . \tag{22}$$

Therefore, the total energy can be written as

$$E = \int|\vec{\nabla}\varphi(\vec{r})|^2 d^3r - (8\pi\bar{\varepsilon})^{-1}\int D^2(\vec{r})d^3r \tag{23}$$

where $\vec{D}(\vec{r})$ is given in terms of the wave function by Gauss's law

$$\vec{\nabla}.\vec{D}(\vec{r}) = 4\pi e|\varphi(\vec{r})|^2. \tag{24}$$

The variational procedure requires that this expression (23) be minimum, so that

$$\delta E = 0 \ . \tag{25}$$

Usually the calculations are easier when Fourier transforms are introduced. Let us define

$$\tilde{\rho}(\vec{k}) = (2\pi)^{-\frac{3}{2}}\int d^3r\rho(\vec{r})e^{-i\vec{k}.\vec{r}} \tag{26}$$

$$= (2\pi)^{-\frac{3}{2}} e\int d^3r|\varphi(\vec{r})|^2 e^{-i\vec{k}.\vec{r}} \ . \tag{27}$$

Since in electrostatics, $\vec{D}(\vec{r})$ has no transverse component, one can write

$$\vec{D}(\vec{r}) = (2\pi)^{-\frac{3}{2}}\int d^3k(\vec{k}/k)\tilde{D}(\vec{k})e^{i\vec{k}.\vec{r}} \tag{28}$$

which defines the Fourier transform $\tilde{D}(\vec{k})$ of the electric displacement. Gauss's law gives immediately

$$\tilde{D}(\vec{k}) = -4\ i\pi k^{-1}\tilde{\rho}(\vec{k}). \tag{29}$$

One of the simplest form that can be choosen as trial wave function is the ground-state wave function of the harmonic oscillator

$$\varphi(\vec{r}) = \kappa^{\frac{3}{2}}\pi^{-\frac{3}{4}}\ e^{-\kappa^2 r^2/2}, \tag{30}$$

where κ is the only variational parameter. With this particular choice of the trial function, one obtains

$$\tilde{\rho}(\vec{k}) = e(2\pi)^{-\frac{3}{2}}\ e^{-k^2/4\kappa^2}. \tag{31}$$

Parseval's theorem allows the immediate calculation of the average energy (23). Indeed,

$$\int D^2(\vec{r})d^3r = \int |\tilde{D}(\vec{k})|^2 d^3k \tag{32}$$

$$= 2\ e^2\pi^{-1}\int \frac{e^{-k^2/2\kappa^2}}{k^2}\ d^3k \tag{33}$$

$$= 4\ \kappa e^2\sqrt{2\pi}\ . \tag{34}$$

On the other hand,

$$\int |\nabla\varphi(\vec{r})|^2 d^3r = \frac{3}{2}\kappa^2, \tag{35}$$

so that

$$E = (\frac{3}{2})\kappa^2 - \frac{e^2\kappa}{\bar{\varepsilon}\sqrt{2\pi}}\ . \tag{36}$$

The minimum of (36) is reached when

$$\kappa = \frac{1}{3}\ \frac{e^2}{\bar{\varepsilon}\sqrt{2\pi}}\ . \tag{37}$$

The corresponding minimum value of the energy is

$$E = -\ \frac{1}{12\pi}\ \frac{e^4}{\bar{\varepsilon}^2}\ . \tag{38}$$

Reintroducing dimensions into this result (38) gives

$$E = -\ \frac{1}{6\pi}\ \frac{me^4}{\bar{\varepsilon}^2\ \hbar^2}\ . \tag{39}$$

For the sake of comparison, it is useful to introduce Fröhlich's coupling constant α, defined as

$$\alpha = \frac{e^2}{2\bar{\varepsilon}\hbar\omega}\ (\frac{2m\omega}{\hbar})^{\frac{1}{2}} \tag{40}$$

where ω is the frequency of the longitudinal optical phonons at the limit of infinite wavelength. Then

$$E = -\ \frac{1}{3\pi}\alpha^2\ \hbar\omega\ . \tag{41}$$

Other trial functions can of course be used ; for example, Pekar has tried an expression containing two variational parameters α and β,

$$\varphi(\vec{r}) = A(1+\alpha r+\beta r^2)e^{-\alpha r} \tag{42}$$

where A is a normalization constant. He then obtains for minimum energy

$$E = -0{,}0544 \frac{me^4}{\bar{\varepsilon}^2\hbar^2}$$

$$= -0{,}1088\ \alpha^2\hbar\omega\ , \tag{43}$$

which is to be compared with the previous result (41). This latter is approximately

$$E = -0{,}1061\ \alpha^2\hbar\omega\ , \tag{44}$$

so that the improvement brought by Pekar's trial function with two variational parameters is less than 3%.

In conclusion, in the framework of a crude adiabatic approximation, an electron or a hole in an ionic crystal has a self-energy of the order of

$$E = -\frac{1}{3\pi}\ \alpha^2\hbar\omega\ . \tag{45}$$

This self-energy is due to the interaction of the charged particle with the ionic polarization of the crystal.

This result concerns a particle at rest. If the particle is moving with a constant average velocity $\vec{v}$, the lattice distortion follows the motion of the particle with some delay. Due to the heavy mass of the ions, the distortion cloud contributes to the inertia of the system, replacing the electron (or hole) band mass m by a polaron effective mass m^*.

With the same crude adiabatic approximation, the change in the mass can be calculated in the following way. The wave function of the particle is now centered around a point in translation with a velocity $\vec{v}$, so that it can be written as $\varphi(\vec{r}-\vec{v}t)$ where φ has a spherical symmetry. The polarization is no longer static and we now need to know its dynamic response to the moving charge distribution

$$\rho(\vec{r},t) = e|\varphi(\vec{r}-\vec{v}t)|^2. \tag{46}$$

This requires the knowledge of the equation of motion for the polarization field. This equation of motion can, for instance, be derived from the Lagrangian density. We already know the polarization potential energy and the energy of interaction with the average charge for the static case (see Eq.(11)).The substitution of $\rho(\vec{r},t)$ given by (46) for the source of the displacement gives immediately the same quantities for the case of a moving particle. As for the kinetic energy associated with the motion of the ions responsible for the ionic polarization, it is obviously quadratic in the time derivative of the polarization. Thus the total Lagrangian density is

$$L(\vec{r},t) = \gamma\dot{\vec{P}}^2(\vec{r},t) - 2\pi\bar{\varepsilon}P^2(\vec{r},t) + \vec{D}(\vec{r},t).\vec{P}(\vec{r},t) \quad , \tag{47}$$

where $\vec{D}(\vec{r},t)$ is given by Gauss's law

$$\vec{\nabla}.\vec{D}(\vec{r},t) = 4\pi\rho(\vec{r},t) \tag{48}$$

and the dot denotes a derivative with respect to time. The constant γ in the Lagrangian density (47) remains to be determined. Lagrange's equation derived from (47) is

$$2\gamma\ddot{\vec{P}}(\vec{r},t) + 4\pi\bar{\varepsilon}\vec{P}(\vec{r},t) = \vec{D}(\vec{r},t) \tag{49}$$

where the double dot means a second derivative with respect to time. The solution for the static case is easily recovered. Indeed, in that case

$$\ddot{\vec{P}}(\vec{r},t) = 0 \tag{50}$$

and therefore

$$\vec{P}(\vec{r},t) = (4\pi\bar{\varepsilon})^{-1}\,\vec{D}(\vec{r},t) \tag{51}$$

which is the result (8) obtained above.

The constant γ is determined by considering the case of free ionic oscillations, i.e. when

$$\vec{D}(\vec{r},t) = 0 \; . \tag{52}$$

Dividing by 2γ, the equation of motion for the polarization field becomes

$$\ddot{\vec{P}}(\vec{r},t) + 2\pi\bar{\varepsilon}\gamma^{-1}\,\vec{P}(\vec{r},t) = 0 \; . \tag{53}$$

One usually assumes that the polarization is slowly varying in space, which means that the change of the polarization over the size of the elementary cell is negligible compared to the polarization itself. Therefore, the normal modes of vibration of the lattice involved in such oscillations of the polarization field are phonons with large wavelengths. In this range of wavelengths, the only phonons which give rise to a net polarization, i.e. the longitudinal optical phonons (L.O. phonons) have almost the same frequency ω. Thus the lattice polarization $\vec{P}(\vec{r},t)$ oscillates with a single frequency ω which is the LO-phonon frequency. Comparing with the result of the equation of motion (53) shows that

$$\omega^2 = 2\,\pi\bar{\varepsilon}\gamma^{-1} \tag{54}$$

or, if we now use units such that $\hbar=2m=\omega=1$,

$$\gamma = 2\pi\bar{\varepsilon} \ . \tag{55}$$

Let us now come to the case where the charge distribution

$$\rho(\vec{r},t) = e|\varphi(\vec{r}-\vec{v}t)|^2 \tag{56}$$

is present and is the source of the displacement field $\vec{D}(\vec{r})$. The equation (49) becomes

$$\ddot{\vec{P}}(\vec{r},t) + \vec{P}(\vec{r},t) = \frac{1}{4\pi\bar{\varepsilon}} \vec{D}(\vec{r},t). \tag{57}$$

Again, it is easier to use Fourier transforms. Define

$$\tilde{\rho}(\vec{k},t) = (2\pi)^{-\frac{3}{2}} e\int d^3r|\varphi(\vec{r}-\vec{v}t)|^2 e^{-i\vec{k}.\vec{r}} \ . \tag{58}$$

Taking $\vec{r}'=\vec{r}-\vec{v}t$ as variable of integration gives

$$\tilde{\rho}(\vec{k},t) = (2\pi)^{-\frac{3}{2}} e\int d^3r'|\varphi(\vec{r}')|^2 e^{-i\vec{k}.\vec{r}'} e^{-i\vec{k}.\vec{v}t} \tag{59}$$

$$= \tilde{\rho}(\vec{k}) e^{-i\vec{k}.\vec{v}t} \tag{60}$$

where $\tilde{\rho}(\vec{k})$ is the Fourier transform of the static charge distribution introduced previously.

Gauss's law gives as before

$$\tilde{D}(\vec{k},t) = -4\, i\pi k^{-1} \tilde{\rho}(k,t) \tag{61}$$

$$= -4\, i\pi k^{-1} \tilde{\rho}(k) e^{-i\vec{k}.\vec{v}t} \ . \tag{62}$$

The Fourier transform of the equation of motion (57) for the polarization field leads to

$$\ddot{\tilde{P}}(\vec{k},t) + \tilde{P}(\vec{k},t) = (4\pi\bar{\varepsilon})^{-1} \tilde{D}(\vec{k},t) \tag{63}$$

$$= -i \frac{1}{\bar{\varepsilon}k} \tilde{\rho}(k) e^{-i\vec{k}.\vec{v}t} \ . \tag{64}$$

In (64) use has been made of the fact that the polarization has only longitudinal components, allowing to write

$$\vec{P}(\vec{r},t) = (2\pi)^{-\frac{3}{2}} \int d^3k(\vec{k}/k)\tilde{P}(\vec{k},t)\, e^{i\vec{k}.\vec{r}} \ . \tag{65}$$

Consider the case of temperatures low enough for the population of thermal phonons to be negligible. This requires

$$KT << \hbar\omega \tag{66}$$

where K is the Boltzmann constant. As $\hbar\omega$ is usually of the order of 30 meV, this condition is satisfied at about the temperature of boiling nitrogen or below. Then no free oscillation of the polarization field exists and the solution of (64) is restricted to forced oscillations which gives

$$\tilde{P}(\vec{k},t) = -i \frac{1}{1-(\vec{k}.\vec{v})^2} \frac{\tilde{\rho}(k)}{\bar{\varepsilon}k} e^{-i\vec{k}.\vec{v}t} . \tag{67}$$

The total energy associated with the polarization is now easily evaluated. It has three contributions : the kinetic energy, the potential energy and the energy of interaction with the source. This gives

$$E_P = \frac{2\pi}{\bar{\varepsilon}} \int d^3k \frac{3(\vec{k}.\vec{v})^2-1}{|1-(\vec{k}.\vec{v})^2|^2} \frac{|\tilde{\rho}(\vec{k})|^2}{k^2} . \tag{68}$$

Here our aim is to determine the contribution of the lattice distortion to the inertia of the system. Therefore, we are interested in the energy term quadratic in the velocity v. For this reason, we expand the expression (68) of the energy of polarization up to the order v^2. This gives

$$E_P = -2\pi \,\bar{\varepsilon}^{-1} \int |\tilde{\rho}(\vec{k})|^2 k^{-2} d^3k +$$
$$+2\pi \,\bar{\varepsilon}^{-1} \omega^{-2} \int |\tilde{\rho}(\vec{k})|^2 k^{-2} (\vec{k}.\vec{v})^2 d^3k . \tag{69}$$

Obviously, the first term in this expression is the result obtained above for the case of polarons at rest and the second term gives the correction to the effective mass we are looking for. Using the wave function (30) of the ground state of the harmonic oscillator as trial function gives

$$E_c = \frac{e^8}{4.3^4\pi^2\bar{\varepsilon}^4} v^2 . \tag{70}$$

Reintroducing the dimensions leads to

$$E_c = \left[\frac{e^2}{2\bar{\varepsilon} \frac{\hbar}{2m\omega} \hbar\omega}\right]^4 \left(\frac{2}{3}\right)^4 \frac{1}{\pi^2} \frac{mv^2}{2} \tag{71}$$

$$= \left(\frac{2}{3}\right)^4 \frac{\alpha^4}{\pi^2} \frac{mv^2}{2} . \tag{72}$$

This means that, in the framework of the crude-adiabatic approximation (*), the lattice distortion contributes a quantity

(*) During a discussion with Prof. F.Williams, it appeared that crude adiabatic is probably not the most appropriate term to designate the Landau-Pekar approximation. Static is probably better suited.

$$m_p = (\frac{2}{3})^4 \frac{\alpha^4}{\pi^2} m \tag{73}$$

$$= 0{,}02001 \ \alpha^4 m \tag{74}$$

to the polaron mass.

In conclusion, the Landau-Pekar theory, with a harmonic approximation for the motion of the electron in its polarization field, predicts

$$E = -\frac{1}{3\pi} \alpha^2 \hbar\omega \tag{75}$$

and

$$m^* = \left[1 + (\frac{2}{3})^4 \frac{\alpha^4}{\pi^2}\right] m , \tag{76}$$

where E and m^* are respectively the polaron self-energy and effective mass. The largest corrections are expected to be found in the case of strongly ionic crystals such as the silver halides or the alkali halides. For conduction electrons, using the value for α and m given by E. Kartheuser in ref.[4], one finds the results of table 1.

TABLE 1. LANDAU-PEKAR RESULTS

	α	$\hbar\omega$ (meV)	m/m_e	$E/\hbar\omega$	E (meV)	m^*/m	m^*/m_e
KBr	3.05	21	0.37	-0.99	-21	2.73	1.01
KCl	3.44	27	0.43	-1.26	-34	3.80	1.65
AgBr	1.56	17	0.215	-0.26	-4.4	1.12	0.24
AgCl	1.94	23	0.30	-0.40	-9.2	1.28	0.38

α: coupling constant, $\hbar\omega$: L.O.phonon energy, m: band mass deduced from ref.[4]. E: Landau-Pekar results for the polaron self-energy, m^*: effective mass. The notation m_e is used for the electron mass in free space.

It can be concluded that, for conduction electrons, the interaction with the polarization is not strong enough to give rise to self-trapping. Indeed, even for the most ionic crystals, the polaron effective mass remains of the order of the mass of electrons in free space. Moreover, the correction to the energy is comparable to the excitation energy of the phonon field. This means that the oscillation frequency of the electron moving in its polarization field is not far larger than the frequency of the ion vibrations. Therefore, the Landau-Pekar approximation is probably not valid.

A question, particularly interesting from the theoretical point of view, concerns the existence of internal excited states, i.e. possible excitations of the electron in the potential well due to the polarization it has itself induced. This problem has been discussed by several authors (2,5,6). Two types of excitations seem possible at strong coupling : Fast Condon-like transitions between states in the unrelaxed polarization field which remains adapted to the initial state and transitions to relaxed states for which the polarization is adapted to the final state. For more information see ref.[5] and [6].

III. FIELD-THEORY FORMALISM, FRÖHLICH'S HAMILTONIAN AND WEAK-COUPLING THEORY.

In 1950, Fröhlich, Pelzer and Zienau [7] introduced a quantum field theory description of polarons. We can use the results of chapter I to derive the Hamiltonian obtained by these authors and usually called Fröhlich's Hamiltonian. Consider a particle of charge e located at $\vec{r}$. The charge density is now

$$\rho(\vec{x}) = e\delta(\vec{x}-\vec{r}) \tag{77}$$

with a Fourier-transform equal to

$$\tilde{\rho}(\vec{k}) = (2\pi)^{-\frac{3}{2}} e \; e^{-i\vec{k}.\vec{r}} . \tag{78}$$

The electrical displacement can be obtained by applying Gauss's law. This leads to the result (61) for the Fourier transform of this displacement, i.e.

$$\tilde{D}(\vec{k}) = -i(2/\pi)^{\frac{1}{2}} e \, k^{-1} e^{-i\vec{k}.\vec{r}} . \tag{79}$$

The Lagrangian for the polarization field in interaction with the particle is easily obtained by integrating the Lagrangian density (47) over the whole space. The value given by (57) is used for the constant γ. This leads to

$$L = \int d^3x \{ \frac{2\pi\bar{\varepsilon}}{\omega^2} \left[\dot{P}^2(\vec{x}) - \omega^2 P^2(\vec{x}) \right] + \vec{D}(\vec{x}) \, . \, \vec{P}(\vec{x}) \} , \tag{80}$$

where $\vec{P}(\vec{x})$ is the ionic polarization at the point $\vec{x}$. Notice that, by using the formalism of Section I, the electronic polarization directly induced by the electron is subtracted out. Again, Fourier transforms are used. Then, Parseval's theorem gives

$$L = \int d^3k \{ \frac{2\pi\bar{\varepsilon}}{\omega^2} \left| \, |\dot{\tilde{P}}(\vec{k})|^2 - \omega^2 |\tilde{P}(\vec{k})|^2 \right| + \tilde{P}(\vec{k}) \tilde{D}^*(\vec{k}) \} \tag{81}$$

where $\tilde{D}^*(\vec{k})$ is the complex conjugate of $\tilde{D}(\vec{k})$. As the polarization and the electric displacement are real, one has the following relations :

$$\tilde{D}^*(\vec{k}) = -\tilde{D}(-\vec{k}) , \quad (82a)$$

$$\tilde{P}^*(\vec{k}) = -\tilde{P}(-\vec{k}) , \quad (82b)$$

$$\dot{\tilde{P}}^*(\vec{k}) = -\dot{\tilde{P}}(-\vec{k}) . \quad (82c)$$

With these relations, the Lagrangian (81) becomes

$$L = -\int d^3k \left\{ \frac{2\pi\bar{\varepsilon}}{\omega^2} \left| \dot{\tilde{P}}(-\vec{k})\dot{\tilde{P}}(\vec{k}) - \omega^2 \tilde{P}(-\vec{k})\tilde{P}(\vec{k}) \right| + \tilde{P}(\vec{k})\tilde{D}(-\vec{k}) \right. . \quad (83)$$

Usually one prefers to work with discrete k-states. Therefore, consider a crystal with a finite parallelepipedal volume V and use Born-vonKarman periodic boundary conditions. This results in replacing the integral in (83) by a sum and introducting a factor $8\pi^3/V$. Moreover call

$$\tilde{P}(\vec{k}) = q_{\vec{k}} \sqrt{\frac{V}{8\pi^3}} . \quad (84)$$

The Lagrangian can then be written as

$$L = -\sum_{\vec{k}} \left[\frac{2\pi\bar{\varepsilon}}{\omega^2} (\dot{q}_{-\vec{k}} \dot{q}_{\vec{k}} - \omega^2 q_{-\vec{k}} q_{\vec{k}}) + \sqrt{\frac{8\pi^3}{V}} q_{\vec{k}} \tilde{D}(-\vec{k}) \right] . \quad (85)$$

The momentum $\pi_{\vec{k}}$ canonically conjugate to $q_{\vec{k}}$ is given by

$$\pi_{\vec{k}} = \frac{\partial L}{\partial \dot{q}_k} = -\frac{4\pi\bar{\varepsilon}}{\omega^2} \dot{q}_{-\vec{k}} . \quad (86)$$

Introducing the expression (79) of the electric displacement into (83) leads to the following form for the Hamiltonian describing the polarization field in interaction with a point charge located at $\vec{r}$:

$$H = \sum_{\vec{k}} \left[\frac{\omega^2}{8\pi\bar{\varepsilon}} |\pi_{\vec{k}}|^2 + 2\pi\bar{\varepsilon} |q_{\vec{k}}|^2 - 4i\pi \frac{e}{k\sqrt{V}} q_{\vec{k}} e^{i\vec{k}.\vec{r}} \right] . \quad (87)$$

The condition that the polarization field be real prescribes that

$$q_{-\vec{k}} = -q^*_{\vec{k}} , \quad (88a)$$

$$\pi_{-\vec{k}} = -\pi^*_{\vec{k}} . \quad (88b)$$

To quantize the polarization field, the variables $q_{\vec{k}}$ and $\pi_{\vec{k}}$ are now to be considered as operators obeying the commutation rules

$$\left| \pi_{\vec{k}} , q_{\vec{k}} \right| = -i\hbar , \quad (89)$$

all the other commutators being zero. For the sake of convenience, introduce now the annihilation and creation operators defined by the following relations

$$q_{\vec{k}} = \sqrt{\frac{\hbar\omega}{8\pi\bar{\varepsilon}}}\,(a_{\vec{k}} - a^{+}_{-\vec{k}}) \tag{90a}$$

$$\pi_{\vec{k}} = i\sqrt{\frac{2\pi\hbar\bar{\varepsilon}}{\omega}}\,(a^{+}_{\vec{k}} + a_{-\vec{k}}) \ . \tag{90b}$$

These operators comply with the commutation rules

$$[a_{\vec{k}}, a^{+}_{\vec{k}}] = 1 \ , \tag{91}$$

all the other commutators being zero again. Adding the kinetic energy of the electron, one obtains Fröhlich's Hamiltonian

$$H = \frac{p^2}{2m} + \sum_{\vec{k}} \hbar\omega a^{+}_{\vec{k}} a_{\vec{k}} + \sum_{\vec{k}} (V_{\vec{k}} a_{\vec{k}}\, e^{i\vec{k}.\vec{r}} + V^{*}_{\vec{k}} a^{+}_{\vec{k}}\, e^{-i\vec{k}.\vec{r}}) \tag{92}$$

with

$$V_k = -i\,\frac{\hbar\omega}{k}\left(\frac{\hbar}{2m\omega}\right)^{\frac{1}{4}} \sqrt{\frac{4\pi\alpha}{V}} \ . \tag{93}$$

Here α is the coupling constant introduced by Fröhlich, i.e.

$$\alpha = \frac{1}{2}\,\frac{e^2}{\hbar\omega(\hbar/2m\omega)^{\frac{1}{2}}}\left(\frac{1}{\varepsilon_\infty} - \frac{1}{\varepsilon_0}\right) \ . \tag{94}$$

As the phonon zero-point energy is a constant, it has been discarded from (92). Obviously, in (92), $\vec{r}$ now denotes the position of the electron.

Fröhlich, Pelzer and Zienau have solved the polaron problem for weak coupling strengths. Their method is essentially similar to a perturbation theory. Call

$$H = H_o + V \ , \tag{95}$$

where H_o is the Hamiltonian for the electron and the phonons in the absence of interaction, i.e.

$$H_o = \frac{p^2}{2m} + \sum_{\vec{k}} \hbar\omega\, a^{+}_{\vec{k}}\, a_{\vec{k}} \tag{96}$$

and V, the perturbation, denotes the electron-phonon interaction Hamiltonian which is

$$V = \sum_{\vec{k}} (V_k a_{\vec{k}}\, e^{i\vec{k}.\vec{r}} + V^{*}_k\, a^{+}_{\vec{k}}\, e^{-i\vec{k}.\vec{r}}) \ . \tag{97}$$

Again, use units such that $\hbar = 2m = \omega = 1$. The second-order perturbation energy of the state with total momentum K and no real phonon is

$$E(K) = K^2 + \langle\vec{K}|V(K^2-H_o)^{-1}V|\vec{K}\rangle \quad . \tag{98}$$

In this expression, $|\vec{K}\rangle$ denotes the eigenstate of H_o for which the electron has a momentum $\vec{K}$ and the phonon field is in the ground state, so that

$$p^2|\vec{K}\rangle = K^2|\vec{K}\rangle \tag{99}$$

and

$$a_{\vec{k}}|\vec{K}\rangle = 0 \; . \tag{100}$$

As the interaction Hamiltonian V is linear in the phonon operators, applying V to the unperturbed state $|\vec{K}\rangle$ in Eq.(98) leads to intermediate states with just one phonon which is to be annihilated by the second operator V to come back to the initial state. This leads to the following expression for the energy :

$$E(K) = K^2 + \sum_{\vec{k}} \frac{|V_k|^2}{K^2-(\vec{K}-\vec{k})^2-\hbar\omega} \quad . \tag{101}$$

Here, we restrict ourselves to the study of the polaron self-energy and effective mass. Therefore, to evaluate (101), a power expansion limited to order K^2 is used. The sum over $\vec{k}$ is replaced by $(V/8\pi^3)\int d^3k$. This gives

$$E(K) = -\alpha + K^2(1 - \frac{\alpha}{6}) . \tag{102}$$

Reintroducing dimensions we conclude that, in the framework of a second order perturbation theory, the polaron self-energy is

$$E = -\alpha h\omega \tag{103}$$

and its effective mass is

$$m^* = m/(1- \frac{\alpha}{6}) \; . \tag{104}$$

In the framework of perturbation theory, the electron can be thought of as undergoing quantum fluctuations about the polaron center itself in translation. These fluctuations are due to emission and absorption of virtual phonons. The polaron radius is easily estimated if one realizes that, using the expression (94) of the coupling constant α , the self-energy can be written as

$$E = -\frac{1}{2}\frac{e^2}{\sqrt{\hbar/2m\omega}}(\frac{1}{\varepsilon_\infty} - \frac{1}{\varepsilon_0}) \; . \tag{105}$$

This is the self-energy of an extended charge distribution in a polarizable medium with an effective dielectric constant $\bar{\varepsilon}$ given by

$$\frac{1}{\bar{\varepsilon}} = \frac{1}{\varepsilon_\infty} - \frac{1}{\varepsilon_0} . \tag{106}$$

The total charge is e and the radius of the charge distribution is of the order of

$$r_o = \sqrt{\hbar/2m\omega} , \tag{107}$$

which, therefore, is also the polaron radius. For KBr and AgBr, one respectively finds 21Å and 33Å. Though perturbation theory is somewhat questionable for the strongly ionic crystals such as KBr, this indicates that the electron wave function extends on many lattice cells and, therefore, that the continuum approximation used to describe the polarization is valid. It also gives a justification for using a single frequency for the phonons. Indeed, due to the relatively large polaron size, the interaction is restricted to phonons with a wave vector $\vec{k}$ close to the center of the Brillouin zone where the frequency dispersion is weak. Another consequence is the following : the potential acting on the electron is slowly varying and the band-mass approximation probably applies. For more detail on this point see ref.[8]. The fact that the interaction is linear in the phonon variables comes from the model of a polarizable medium used to describe the lattice. In this model, anharmonic effects and interaction due to higher-order multipoles are neglected. The study of these non-linear effects would be interesting, but, to my knowledge, has not yet been undertaken.

In fact, Fröhlich, Pelzer and Zienau obtained their results with a variational procedure, using a trial wave function of the type

$$|\psi\rangle = (1 + \sum_{\vec{k}} C_{\vec{k}} a^+_{\vec{k}} e^{-i\vec{k}.\vec{r}}) |\vec{k}\rangle \tag{108}$$

where the $C_{\vec{k}}$'s are the variational parameters. The minimum of

$$E(\vec{K}) = \frac{\langle\psi|H|\psi\rangle}{\langle\psi|\psi\rangle} \tag{109}$$

is then sought, leading to results similar to those of the perturbation theory described above. Since, with variational methods, the lowest energy obtained indicates the best result, we are now in a position to compare the predictions of the Landau-Pekar theory with that of the Fröhlich method. The comparison is given in table 2. Notice that the Fröhlich method is far better than the Landau-Pekar one, specially in the case of the slightly ionic crystals such as the III-V and II-VI compounds. This proves that Landau's ideas on localization and self-trapping by the polarization field are not applicable to the conduction electrons in common ionic crystals, even in the alkali halides.

Table 2. Comparison of Frohlich, Pelzer, Zienau and Landau-Pekar Results

	α	$E_{LP}/\hbar\omega$	$E_{FPZ}/\hbar\omega$	$(\frac{m^*}{m})_{LP}$	$(\frac{m^*}{m})_{FPZ}$
KBr	3.05	-0.99	-3.05	2.73	2.03
KCl	3.44	-1.26	-3.44	3.80	2.34
AgBr	1.56	-0.26	-1.56	1.12	1.35
AgCl	1.94	-0.40	-1.94	1.28	1.48
ZnS	0.65	-0.045	-0.65	1.004	1.12
GaAs	0.068	$-4.9\ 10^{-4}$	-0.068	1.000	1.011

α : coupling constant, E_{LP}: polaron self-energy as given by the Landau-Pekar method, E_{FPZ}: polaron energy obtained by the Fröhlich, Pelzer, Zienau approximation, m^*: polaron effective mass (in units of the band mass) given by Landau and Pekar (LP subscript) and by Fröhlich, Pelzer and Zienau (FPZ subscript).

IV. OTHER METHODS FOR LARGE (FROHLICH) POLARONS

Lee, Low and Pines [9] have developed a variational method for intermediate coupling strengths. A canonical transformation introduces a new frame of reference with the origin attached to the position of the electron. The trial wavefunction corresponds to a displacement of the equilibrium position of the ions in this new coordinate system, so that effects of correlations between the electron and the ion motions are taken into account, at least to a certain extent. The result for the polaron self-energy is the same as with the Fröhlich weak-coupling approximation, i.e.

$$E = -\alpha\hbar\omega \ . \tag{110}$$

As for the polaron effective mass, Lee, Low and Pines obtain

$$m^* = (1 + \frac{\alpha}{6})m \ , \tag{111}$$

which is close to Fröhlich's result

$$m^* = m/(1 - \frac{\alpha}{6}) \tag{112}$$

specially for small coupling strengths.

In 1955, Feynman [10] worked out what can still be considered as the best theory of Fröhlich's polarons. However, a detailed description of this theory would be too long to be given here. Let me just say that Feynman uses his path-integral formulation of quantum mechanics. It is then possible to eliminate the phonon variables from the path integral. One is left with the electron interacting with itself at different times. This interaction is approximated by a harmonic interaction. As a last step, a variational method is used, the variational parameters being the strength of the harmonic coupling and the time required for relaxing the memory effect.

Feynman's results for the polaron self-energy are given in table 3. They are better than the results given by the other methods for all coupling strengths. As for the effective mass, the results are given in table 4 where they are compared with the results of the other methods described in these lecture notes.

TABLE 3. COMPARISON OF THE POLARON SELF-ENERGIES PREDICTED BY DIFFERENT METHODS

α	3	5	7	9	11
$E_{LP}/h\omega$	-0.95	-2.65	-5.20	-8.59	-12.84
$E_{FPZ}/h\omega$	-3.00	-5.00	-7.00	-9.00	-11.00
$E_F/h\omega$	-3.13	-5.44	-8.11	-11.49	-15.71

E_{LP}: polaron self-energy as given by the Landau-Pekar theory, E_{FPZ}: result of the Fröhlich, Pelzer, Zienau approximation, E_F: result of the Feynman method.

TABLE 4. POLARON EFFECTIVE MASS OBTAINED WITH DIFFETEN DIFFERENT METHODS

α	3	5	7	9	11
Landau-Pekar	2.63	14.5	55.7	152	340
Lee,Low,Pines	1.50	1.83	2.17	2.50	2.83
Feynman	1.89	3.89	14.4	62.5	185

The table gives the ratio m^*/m (polaron effective mass divided by the band mass) as predicted by the Landau-Pekar theory, the Lee, Low, and Pines method and the Feynman theory.

It can be concluded that, for $\alpha \lesssim 2$, the results of the intermediate coupling theory due to Lee, Low and Pines, as well as those of the weak coupling method developed by Fröhlich, Pelzer and Zienau are quite satisfactory.

Another conclusion concerns the possibility of a phase transition between states describing polarons drifting in the lattice and self-trapped states. This phase transition would occur at a critical value of the coupling constant. Feynman's method, which predicts a smooth relation between the ground-state energy and the coupling constant, gives results lower than those obtained with the methods leading to phase transitions. The conclusions obtained with these methods are therefore most questionable. The situation is different in the presence of a magnetic field. In that case, as shown by Peeters and Devreese [11], Feynman's method at a critical value of α, predicts a transition between two different types of solutions. The paper by Peeters and Devreese gives an excellent discussion of the question of self-trapping as well as all the relevant references. Let me just point out that this question is very similar to the problem of self-trapping for small polarons that will be discussed next in these lecture notes.

Important applications of the theory of Fröhlich's polarons concern the energy levels of electrons (or holes) bound to impurities or defects in ionic crystals as well as the mobility of charge carriers in these ionic crystals. However, the discussion of these problems goes beyond the scope of the present lectures and will not be given here. The readers are referred to ref.[12], where they can find a rather complete bibliography.

V. SMALL POLARONS AND SELF-TRAPPING

The concept of small polarons has been introduced by Holstein [13] in 1959, in the framework of a model that he called the molecular-crystal model. Subsequent works have been devoted to small polarons by Holstein himself, Emin and others. A good review of the molecular-crystal model can be found in a paper that Emin wrote in 1973 [14]. In this model one considers a single excess electron in a lattice of identical molecular units. The electron is in interaction with one type of molecular distortion with which is associated a configurational coordinate. The distortion energy is quadratic in this coordinate and a linear coupling is assumed between the distortions of neighbouring molecules, leading to bulk phonons with well defined frequencies. For the study of the self-energy (and therefore of the possibility of self-trapping), the dispersion of the phonon frequency can be neglected. A tight-binding approximation is used to describe the electron states.(For description of the tight-binding approximation see the first article in this book).

The electron-phonon interaction is introduced in the following way : the energy of the electron when it sits on a molecule depends linearly on the configurational coordinate describing the distortion of this same molecule.

The delocalization of the electron through the formation of Bloch states is due to overlap integrals of the type

$$J = \int \psi^*(\vec{r}-\vec{R}_i) V(\vec{r}-\vec{R}_i) \psi(\vec{r}-\vec{R}_j) d^3r \tag{113}$$

where $\psi(\vec{r}-\vec{R}_i)$ is an atomic-like wave function describing the motion of the electron when it is on the molecule located at $\vec{R}_i$. As for $V(\vec{r}-\vec{R}_i)$, it is the potential acting on the electron and due to the undistorted molecule at $\vec{R}_i$. Due to the interaction between the electron and the molecular distortion, the wave functions to be used are now

$$|\psi_i\rangle = \psi(\vec{r}-\vec{R}_i)|\phi_i\rangle \tag{114}$$

where $|\phi_i\rangle$ is the phonon wave function appropriate to the electron sitting on the molecule i. Therefore the overlap integral becomes

$$J' = J\langle\phi_i|\phi_j\rangle \ . \tag{115}$$

For large bandwidth and therefore small band mass, the electron is delocalized so that the distortion extends over many cells and $|\phi_i\rangle$ is not much different from $|\phi_j\rangle$. Therefore

$$\langle\phi_i|\phi_j\rangle \sim 1 \tag{116}$$

and

$$J' \sim J \ . \tag{117}$$

Thus no self-trapping is expected under these conditions. On the contrary, when the bandwidth is small enough (large band mass), the particle is strongly localized. Then, with a short-range interaction as assumed here, the distortion corresponding to the electron on a site i has almost no overlap with the distortion induced by the electron on the neighbouring site j. Then

$$\langle\phi_i|\phi_j\rangle \sim 0 \tag{118}$$

and J' is far smaller than J. This leads to a kind of chain reaction with polaron states being still more localized than the bare-electron states would be. Therefore, when decreasing the bandwidth (by decreasing J), one can expect to come to a critical value for which there is a sudden evolution, and possibly a phase transition, from drifting states to self-trapped states.

To treat the problem in a more quantitative way, Emin suggests to apply the usual variational principle of minimum energy with, for trial wavefunction,

$$|\psi\rangle = \sum_i e^{i\vec{k}.\vec{R}_i}\,\psi(\vec{r}-\vec{R}_i)\,|\phi_i\rangle \tag{119}$$

where, for the phonons, states $|\phi_i\rangle$ corresponding to displaced equilibrium positions for the ions are chosen.

At weak coupling, a single solution is obtained, representing a delocalized (large) polaron. At strong coupling, there is again a single solution, representing this time a self-trapped (small) polaron. In the intermediate region of the coupling strength, two solutions are found, with the respective meaning of a drifting and a self-trapped polaron. The corresponding energies intersect at a critical value of the coupling constant. For coupling strengths weaker than the critical value, the large polaron solution has the lowest energy, whereas for coupling strengths larger than this cirtical value, the energy of the small polaron solution becomes the lowest one. The existence of two solutions in the intermediate range indicates the possibility of coexistence of large and self-trapped polarons.

However, it must be pointed out that, although a sudden localization probably occurs in a rather narrow range of coupling strengths, no definitive conclusions can be drawn from a variational method about the coexistence of two different polaron states and about the occurrence of a phase transition. Indeed, on the one hand, with variational methods, only the solution with the lowest energy can be retained doubtlessly. On the other hand, there is no proof that a better trial wavefunction cannot be found, giving a ground state energy which would connect smoothly the weak-coupling large polaron solution to the strong-coupling self-trapped solution. This is what happens with Fröhlich's polarons. Indeed, Feynman's theory gives a ground-state energy evolving smoothly with increasing coupling strengths from the weak coupling delocalized solution to the strong coupling localized solution. Other theories predict a kind of phase transition. However, these theories give a ground-state energy higher than that obtained with Feynman's method and therefore, their conclusion is questionable (for more details, ref.[11]). More work would be required to settle this question.

Self-trapping is well known to occur in some crystals. The V_k centers in alkali halides constitute probably the most remarkable case of self-trapping. In these colour centers, a hole is trapped in a region between two anions, say two Cl^- ions, which come closer and form a kind of Cl_2^- molecular ion embedded in the matrix constituted by the alkali halide itself. However, these centers are far from the idealized small-polaron model. Indeed, the interaction energy is probably not at all linear in the distance between the two anions. Moreover, the distortion is so strong that a harmonic approximation is certainly not valid. Also the overlap integrals cannot be considered as independent of the displacement of the cations.

REFERENCES

1. L.D. Landau, Z. Phys. Sowjet. 3, 664 (1933).
2. S.I. Pekar, Untersuchungen Ober die Elektronentheorie der Kristalle, Akademie-Verlag, Berlin, 1954.
3. F. Williams, D.E. Berry and J.E. Bernard in Radiationless Processes, B. Di Bartolo, ed., Plenum Press, New York and London, 1980, p.1.
4. E. Kartheuser in Polarons in Ionic Crystals and Polar Semiconductors, J.T. Devreese, ed., North-Holland, Amsterdam, 1972, p. 717.
5. J.T. Devreese, J. De Sitter and M. Goovaerts, Phys. Rev. B5, 2367 (1972).
6. E. Kartheuser, R. Evrard and J.T. Devreese, Phys. Rev. Lett. 22, 94 (1969).
7. H. Fröhlich, H. Pelzer and S. Zienau, Phil. Mag. 41, 221 (1950).
8. R. Evrard in Radiationless Processes, B. Di Bartolo, ed., Plenum Press, New York and London, 1980, p. 515.
9. T.D. Lee, F. Low and D. Pines, Phys. Rev. 90, 297 (1953).
10. R.P. Feynman, Phys. Rev. 97, 660 (1955).
11. F.M. Peeters and J.T. Devreese, Solid State Commun., in press.
12. E. Kartheuser, J.T. Devreese and R. Evrard, Phys. Rev. B19, 546 (1979).
13. T. Holstein, Ann. Phys. 8, 325 (1959); ibid, 8, 343 (1959).
14. E. Emin, Adv. Phys. 22, 57 (1973).

SURFACE COLLECTIVE EXCITATIONS

G. Benedek

Gruppo Nazionale di Struttura della Materia del Consiglio Nazionale delle Ricerche, Istituto di Fisica dell'Università, Via Celoria 16, Milano, Italy

ABSTRACT

The first part of this article is a basic introduction to the theory of surface collective excitations from the macroscopic point of view. The solution of the elastic wave equation and of the Maxwell equations combined with the appropriate boundary conditions yields the dispersion relations of Rayleigh waves and of polaritons associated with surface plasmons, excitons, optical phonons and magnons, respectively. The second part is devoted to the theory of surface phonons from the microscopic point of view. A short review of the various methods used in surface lattice dynamics serves as an introduction to the Green's function theory of surface vibrations in ionic crystals. Finally the recent progress in the spectroscopy of surface phonons by means of atom scattering is mentioned with an illustration of the late achievement in the theoretical interpretation of inelastic spectra.

I. INTRODUCTION

I.A. The Surface as a Perturbation

Collective small oscillations of nuclear coordinates, of valence and conduction electron coordinates, of spin orientations around the equilibrium configuration raise the ground state energy of a solid by quanta respectively termed as phonons, excitons, plasmons and magnons. In an indefinite ideal crystal with translational symmetry in three dimensions (3D), or equivalently in a finite solid with 3D periodic boundary conditions, the collective excitations of small amplitude, eigensolutions of the crystal hamiltonian, have harmonic character and are wavelike according to Bloch theorem. They are classified by the allowed values of the 3D momentum $\vec{q} = \hbar\vec{k}$ and, for any given $\vec{q}$, by a branch index j, which takes as many values as the number of degrees of freedom within the lattice unit cell. The quantum energy $\varepsilon_{kj} = \hbar\omega_j(\vec{k})$ depends on the quantum numbers ($\vec{k}$j) through the dispersion relations connecting the oscillation frequencies $\omega_j(\vec{k})$ of each branch to the wavevector $\vec{k}$.

The theory of collective excitations in solids seeks the determination of the dispersion relations. They are in turn the basic ingredient in the calculation of thermal and optical response functions, as well as of the equilibrium properties of the solid at finite temperature.

Local perturbations such as impurities, dislocations or surfaces modify the spectrum of collective excitations. Wavelike solutions are altered by a local perturbation only by infinitesimal amounts, due to their extension in space. Only a restricted number of exceptional solutions, comparable to the number of degrees of freedom involved in the symmetry breaking, may suffer a finite perturbation as a consequence of Ledermann theorem [1]. When the frequency of such an exceptional solution is removed outside the original excitation spectrum, no propagation into the crystal far from the perturbation region is allowed for such a frequency. Thus exceptional solutions lose their 3D-wave nature, their wavefunction being localized in space close to the perturbation. For instance, in a semiinfinite lattice formed by $N_xN_yN_z$ unit cells, with N_x, N_y, $N_z \to \infty$ and N_z enumerating the atomic layers parallel to the surface, the number of exceptional solutions localized at the free surface (and fastly decaying into the bulk) is of the order of N_xN_y, namely of the number of surface atoms. Since the wavevector component parallel to the surface, denoted by $\vec{K}$, is still a good quantum number owing to the preservation of the 2D translational symmetry, the surface solutions will be classified by the allowed values of $\vec{K}$. The respective frequencies $\omega_j(\vec{K})$ are organized in a set of surface dispersion relations ($j = 1,2,\ldots,n_s$), where n_s is comparable to the number of degrees of freedom in the surface unit cell.

I.B Macroscopic and Microscopic Surface Excitations

From a macroscopic point of view the solid surface is conceived as the boundary of a semiinfinite omogeneous continuum. The collective oscillations of microscopic variables, such as the positions of nuclear masses, of ionic and electronic charges, of spin orientations are regarded respectively as oscillations of the mass density (acoustic phonons), of the charge density (plasmons for conduction electrons, excitons for valence electrons, optical phonons for ions) or of the spin density (magnons). In this framework, the dynamical behaviour of the medium is fully accounted for by the macroscopic response functions of the crystal such as the elastic constants, the frequency-dependent dielectric susceptibility and magnetic permeability constants, respectively. Consequently, a surface acoustic phonon is seen as a vibrating strain field solving the elastic wave equation under the surface boundary conditions. In a similar way, charge and spin collective surface excitations are described by the associated oscillating electromagnetic field fulfilling Maxwell equations and the continuity conditions at the surface boundary.

As far as a solid is regarded as a continuum, the eigensolutions localized at the surface in the non-dispersive regime obey a simple scale rule, according to which the penetration length into the solid has to be proportional to the wavelength along the surface. This general property (only approximately valid when dispersion effects and polariton coupling are taken into account) characterizes macroscopic surface excitations.

At this point it is clear that the macroscopic point of view is valid only if the wavelength is much larger than lattice distances, and even in this case it is not at all exhaustive. In fact several microscopic surface excitations (of either phonon or plasmon or magnon nature) may appear at large $\vec{K}$ or exist even for $K \to 0$ as an effect of the lattice discreteness. In this case however the localisation at the surface is usually strong (few lattice distances) and almost independent of the surface wavevector.

The first part of these notes outlines the macroscopic theory of surface excitations. The surface elastic waves are explicitly derived for an isotropic medium (Sec. II) following the simple method reported in a review article by Farnell [2]. Surface plasmons, excitons, optical phonons and magnons are discussed in Sec. III with few elementary examples. They are obtained as solutions of Maxwell equations including retardation effects (polaritons). In this Section I have adopted the macroscopic approach illustrated in the recent book by Maradudin, Wallis and Dobrzynski [3] and in ref.[4]. The second part of the notes (Sec. IV) presents the microscopic theory of surface phonons in the framework of Green's function method. Conceptually the Green's function theory of surface lattice vibrations represents the natural extension of the macroscopic elastic wave theory, both treating the surface as a perturbation. Actually, by incorporating the differential elastic wave equation and the surface boundary conditions into a single integral equation, we have just the transcription of the ordinary dynamical problem in the Green's function language. The Green's function theory of surface excitations in the continuum limit is reported by Garcia-Moliner in a recent review article [5].

II. SURFACE ELASTIC WAVES

II.A Theory

The equation of motion for a continuous and homogeneous elastic medium reads

$$\rho \ddot{u}_\alpha = \sum_{\beta\gamma\delta} c_{\beta\alpha\gamma\delta} \nabla_\beta \nabla_\delta u_\gamma , \qquad (\alpha,\beta,\gamma,\delta = 1,2,3), \tag{1}$$

where ρ is the mass density, $\vec{u}(\vec{x})$ is the displacement vector field, $\vec{\nabla}$ the gradient operator and $c_{\alpha\beta\gamma\delta}$ the elastic tensor. For a semiinfinite medium occupying the half-space $x_3 < 0$, the displacement field at the free surface must fulfill the boundary condition

$$F_\alpha \equiv \sum_{\gamma\delta} c_{3\alpha\gamma\delta} \nabla_\delta u_\gamma = 0, \qquad \text{for} \quad x_3 = 0,\ \alpha = 1,2,3, \tag{2}$$

in order to have a vanishing external force F_α acting on the surface. Equation (1) admits travelling wave solutions like

$$\vec{u}^{(n)} = \vec{e}^{(n)} \exp[\ ik\,(\vec{\alpha}\cdot\vec{x} - vt\,)\,] , \tag{3}$$

where $\vec{e}^{(n)}$ is the unit polarization vector of the n-th eigensolution (n=1,2,3). The components of $\vec{\alpha}$ are the direction cosines of the wavevector $\vec{k} = k\vec{\alpha}$, and v is the phase velocity. Inserting (3) into (1), we have the secular equation for the polarization vector

$$\sum_\gamma C_{\alpha\gamma}\, e_\gamma^{(n)} = \rho\, v^2\, e_\alpha^{(n)} , \tag{4}$$

where

$$C_{\alpha\gamma} \equiv \sum_{\beta\delta} c_{\beta\alpha\gamma\delta}\, \alpha_\beta \alpha_\delta . \tag{5}$$

The three eigenvalues of v^2 are solutions of the 3x3 determinantal equation

$$\det \left| C_{\alpha\gamma} - \delta_{\alpha\gamma}\rho v^2 \right| = 0 . \tag{6}$$

So far we have found the usual wave solutions for the infinite medium, where all the components of $\vec{\alpha}$ must be real, and the eigenvalues $v = v_L$, v_{T1}, v_{T2} are the velocities of the longitudinal and transverse acoustical waves. In isotropic media or along certain symmetry directions of anisotropic media the two transverse velocities are degenerate: $v_{T1} = v_{T2} = v_T$.
The surface intervenes through the boundary condition. In general all stationary bulk solutions $\vec{u}^{(n)}$ having an antinode at the surface (namely $\nabla_\delta u_\gamma^{(n)} = 0$ at $x_3 = 0$) fulfill trivially Eq. (2).

There may be however other special solutions taking the linear form

$$\vec{u} = \sum_n a_n \vec{u}^{(n)} \tag{7}$$

which fulfill the boundary condition in a non-trivial way, provided that α_3 is allowed to be a complex constant assuming different values $\alpha_3^{(n)}$ for each wave component. When

$$\mathrm{Im}\,\alpha_3^{(n)} < 0 \tag{8}$$

we have physical solutions which are wavelike along the surface with parallel wavevector

$$\vec{K} = k\,(\alpha_1, \alpha_2)\ , \qquad \alpha_1^2 + \alpha_2^2 = 1\ , \tag{9}$$

and decay exponentially in the interior of the solid according to the factor $\exp(-x_3 k\,\mathrm{Im}\,\alpha_3)$. These special solutions, localized near the surface, are called surface waves.
When Eq. (7) is inserted into Eq. (2), we get a system of three equations for the coefficients a_n

$$\sum_n \left[\sum_{\gamma\delta} c_{3\alpha\gamma\delta}\, e_\gamma^{(n)} \alpha_\delta^{(n)} \right] a_n = 0, \quad (\alpha = 1,2,3), \tag{10}$$

where $\vec{\alpha}^{(n)} \equiv (\alpha_1, \alpha_2, \alpha_3^{(n)})$. A non-vanishing solution a_n is found if

$$\det \left| \sum_{\gamma\delta} c_{3\alpha\gamma\delta}\, e_\gamma^{(n)} \alpha_\delta^{(n)} \right| = 0. \tag{11}$$

Since $e_\gamma^{(n)}$ and $\alpha_\delta^{(n)}$ can be derived from Eqs. (4-6) and written as functions of v^2, Eq. (11) turns out to be an implicit equation in the unknown squared velocity v^2 for the surface waves. Except for special crystal symmetries combined with special surface and K-vector orientations, the set of simultaneous Eqs. (4), (6), (10) and (11) cannot be solved analytically. Surface wave velocities and polarizations are normally reported in a numerical form, even for cubic crystals. The reader can find several examples in Farnell's review article [2].

II.B Isotropic Crystals

For isotropic crystals, i.e. for cubic crystals whose elastic constants (hereafter given in Voigt notation) fulfill the isotropy condition

$$\eta \equiv 2\,c_{44} / (c_{11} - c_{12}) = 1, \tag{12}$$

the surface wave problem can be easily solved in an analytical way.

It is quite instructive to show how the solution is derived. We choose $\vec{K}$ along x_1, so that $\vec{\alpha} = (1,0,\alpha_3)$. Eq. (6) becomes (after the substitution $c_{12} = c_{11}-2c_{44}$)

$$\det \begin{vmatrix} c_{11}+c_{44}\alpha_3^2-\rho v^2 & 0 & (c_{11}-c_{44})\alpha_3 \\ 0 & c_{44}(1+\alpha_3^2)-\rho v^2 & 0 \\ (c_{11}-c_{44})\alpha_3 & 0 & c_{44}+c_{11}\alpha_3^2-\rho v^2 \end{vmatrix}$$

$$= \left[c_{44}(1+\alpha_3^2) - \rho v^2\right]^2 \left[c_{11}(1+\alpha_3^2) - \rho v^2\right] = 0. \tag{13}$$

Taking into account that $c_{11}/\rho = v_L^2$ and $c_{44}/\rho = v_T^2$, we can readily find the eigenvalues of α_3 as functions of v^2

$$\alpha_3^{(1)} = \alpha_3^{(2)} = -i\,(1 - v^2/v_T^2)^{\frac{1}{2}}$$
$$\alpha_3^{(3)} = -i\,(1 - v^2/v_L^2)^{\frac{1}{2}}, \tag{14}$$

and the eigenvectors

$$\vec{e}^{(1)} = \frac{v_T}{v}(-\alpha_3^{(1)},0,1); \quad \vec{e}^{(2)} = (0,1,0); \quad \vec{e}^{(3)} = \frac{v_L}{v}(1,0,\alpha_3^{(3)}). \tag{15}$$

Hence we write Eq. (11) as

$$\det \begin{vmatrix} \rho\frac{v_T^3}{v}\left(2-\frac{v^2}{v_T^2}\right) & 0 & 2\rho\frac{v_T^3}{v}\alpha_3^{(1)} \\ 0 & \rho\, v_T^2\alpha_3^{(2)} & 0 \\ 2\rho\frac{v_Lv_T^2}{v}\alpha_3^{(3)} & 0 & -\rho\frac{v_Lv_T^2}{v}\left(2-\frac{v^2}{v_T^2}\right) \end{vmatrix} = 0 \tag{16}$$

which is satisfied by either

$$\alpha_3^{(2)} = 0, \quad \text{i.e.} \quad v = v_T \tag{17}$$

or

$$(1 - \tfrac{1}{2} v^2/v_T^2)^4 = (1 - v^2/v_T^2)(1 - v^2/v_L^2). \tag{18}$$

The first condition, Eq. (17), tells us that the bulk transverse wave polarized parallel to the surface obeys the boundary condition in an essential way (since $\nabla_1 u_2 \neq 0$), and therefore has to be considered as a degenerate surface wave. The second condition, Eq. (18), is an equation in the unknown v^2 which admits (for $v_T < v_L$) a single solution below v_T^2. In this case we have a true surface wave polarized in the plane $(x_1 x_3)$ (sagittal plane) known as Rayleigh wave (RW). Two other solutions of Eq. (18) above v_L^2 do not yield any surface wave as both $\alpha_3^{(1)}$ and $\alpha_3^{(3)}$ become real. When the spurious solution $v^2 = 0$ is eliminated, we obtain the famous Rayleigh wave cubic equation ($x_T \equiv v^2/v_T^2$):

$$\frac{1}{16} x_T^3 - \frac{1}{2} x_T^2 + (\frac{3}{2} - v_T^2/v_L^2) x_T + v_T^2/v_L^2 - 1 = 0. \tag{19}$$

We can finally solve Eq. (10) and derive a_n. Then Eqs. (3),(7) and (15) we obtain the displacement field of Rayleigh waves

$$\begin{aligned}
u_1 &= -N\left[(1-v^2/v_T^2)^{1/4} e^{ik\vec{\alpha}^{(1)}\cdot\vec{x}} - (1-v^2/v_L^2)^{-1/4} e^{ik\vec{\alpha}^{(3)}\cdot\vec{x}} \right] e^{-ikvt} \\
u_2 &= 0 \\
u_3 &= iN\left[(1-v^2/v_T^2)^{-1/4} e^{ik\vec{\alpha}^{(1)}\cdot\vec{x}} - (1-v^2/v_L^2)^{1/4} e^{ik\vec{\alpha}^{(3)}\cdot\vec{x}} \right] e^{-ikvt}
\end{aligned} \tag{20}$$

where N is a normalization constant. The factor i in u_3 causes an elliptical polarization in the sagittal plane. At the surface ($x_3 = 0$) we have

$$(u_3/u_1)_{x_3=0} = i((1 - v^2/v_L^2)/(1 - v^2/v_T^2))^{1/4}. \tag{21}$$

Since $v_L > v_T$, the major axis is normal to the surface. For $v \to 0$ the polarization becomes circular, whereas for $v \to v_T$ the surface mode tends to be linearly polarized with a transverse polarization normal to the surface. The displacement field decays in the interior of the crystal with two components having different penetration lengths λ. The slowly decaying component penetrates a distance $\lambda = 1/k|\alpha_3^{(1)}|$; the fast decaying component has $\lambda = 1/k|\alpha_3^{(3)}|$. The inverse proportionality of λ to k descends also from a simple scaling argument. Moreover, the penetration tends to infinity as v tends to the transverse velocity v_T.

II.C Anisotropic Cubic Crystals

The above properties of Rayleigh waves, valid for the isotropic case, offer a guideline for a qualitative understanding of their behaviour in isotropic crystals with cubic symmetry. The bulk anisotropy implies anisotropy of Rayleigh waves with respect to the surface orientation and to the propagation direction on a given surface. If we consider the (001) surface of a cubic crystal, v varies as the propagation direction changes from (1,0) to (1,1), as shown in Fig. 1.

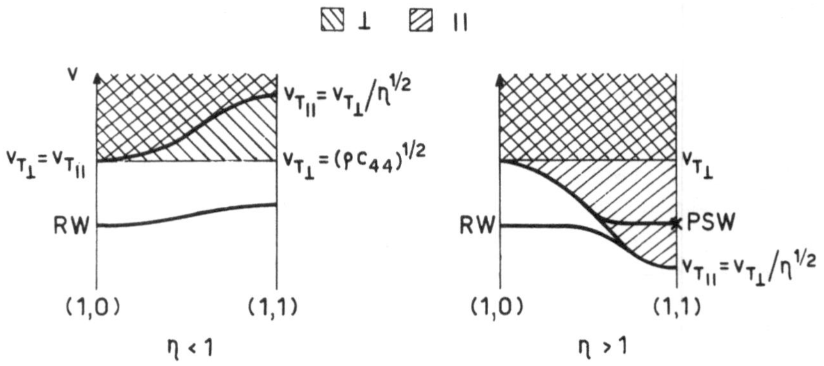

Fig. 1. Rayleigh wave velocity on the (001) surface of cubic crystals as function of the propagation direction and of the crystal anisotropy.

We note however an important difference between the cases $\eta < 1$ and $\eta > 1$. In the former the parallel transverse velocity $v_{T_\parallel}$ increases as direction is rotated from (1,0) to (1,1), and the RW exists everywhere. In the latter case v_T takes at (1,1) a value lower than at (1,0), and the RW dispersion curve (v as function of the propagation direction) may cross the dispersion of v_T and enter the continuum of bulk phase velocities. In this case the RW becomes a resonance and only at the symmetry direction (1,1) it recovers a pure sagittal surface wave character (pseudo-surface wave, PSW). Since the v_T dispersion corresponds to a degenerate surface wave, the crossing exhibits hybridization effects. Thus in the crossing region also the transverse parallel mode has a true surface character. However out of the symmetry directions the polarizations of the two hybridizing surface modes are no longer exactly parallel or sagittal.

III. SURFACE POLARITONS

III.A Introduction

From the macroscopic point of view, valid for a continuum, the collective oscillations of conduction electrons (plasmons), of valence electrons (excitons) and ionic charges in insulators (optical phonons), of spins (magnons) are accounted for by the oscillations of the associated electromagnetic field. Thus we have to solve Maxwell equations coupled to surface boundary conditions, the bulk behaviour being described by given frequency-dependent dielectric constant and magnetic permeability tensors, $\varepsilon(\omega)$ and $\mu(\omega)$, respectively. Electromagnetic waves propagating inside the medium obey the dispersion law

$$k^2c^2 = \omega^2\varepsilon(\omega)\mu(\omega). \tag{22}$$

In non-dispersive regimes, where ε and μ are constant, the natural excitations of the medium are just photons with a constant phase velocity $c/(\varepsilon\mu)^{\frac{1}{2}}$. In the highly dispersive regions where either $\varepsilon(\omega)$ or $\mu(\omega)$ has poles, k can vary over a wide range keeping ω almost constant; ω tends asymptotically to a pole ω_0 as $k \to \infty$. Such a flat branch of the disperion law (22), having vanishing group and phase velocities, is associated with matter oscillations of the medium, namely plasmons, excitons, etc., and loses completely any photon character. Since one can go continuously along the electromagnetic dispersion curve from the pure photon limit to the limit of pure matter oscillations, we shall use the general name of polariton for any electromagnetic wave travelling in a dispersive medium. Plasmons, excitons, optical phonons, magnons and photons are seen as limiting cases of polaritons.

III.B Surface Dielectric Polaritons

Consider a planar interface located at $z = 0$ between two non-magnetic isotropic media. They are characterized by their frequency dependent dielectric constant and conductivity $\varepsilon = \varepsilon_1(\omega)$ and $\sigma = \sigma_1(\omega)$ for $z > 0$, or $\varepsilon = \varepsilon_2(\omega)$ and $\sigma = \sigma_2(\omega)$ for $z < 0$, respectively. We start from the equation of motion for the electric field $\vec{E} = \vec{E}(x,y,z,t)$ as deduced from Maxwell equations

$$-c^2\,\vec{\nabla}\times(\vec{\nabla}\times\vec{E}) = \varepsilon\ddot{\vec{E}} + 4\pi\sigma\dot{\vec{E}} \tag{23}$$

with $\mu = 1$ and the assumption of charge neutrality

$$\vec{\nabla}\cdot\varepsilon\vec{E} = 0. \tag{24}$$

We look for special solutions, wavelike in two dimensions and localized at the surface:

$$\vec{E} = \vec{E}_1^o \exp\{-\kappa_1 z + i(\vec{K}\cdot\vec{R} + \omega t)\} \qquad \text{for } z > 0$$
$$= \vec{E}_2^o \exp\{+\kappa_2 z + i(\vec{K}\cdot\vec{R} + \omega t)\} \qquad \text{for } z < 0 , \tag{25}$$

with $\vec{R} \equiv (x,y)$, $\vec{K} = (k_x,k_y)$ and $\mathrm{Re}\,\kappa_1$, $\mathrm{Re}\,\kappa_2 > 0$. From eq. (24) and from the isotropy condition ($\varepsilon_{xx} = \varepsilon_{yy} = \varepsilon_{zz} \equiv \varepsilon$)

$$i\vec{E}\cdot\vec{K} = E_z \kappa \qquad (\kappa = \kappa_1, -\kappa_2). \tag{26}$$

Solutions with $\vec{E}$ orthogonal to $\vec{K}$ imply $\kappa = 0$. Thus surface localized solutions must be sagittal with vector amplitudes

$$\vec{E}_1^o = E_1^o(\vec{K}/K , -iK/\kappa_1), \tag{27'}$$
$$\vec{E}_2^o = E_2^o(\vec{K}/K , +iK/\kappa_2). \tag{27"}$$

Inserting the trial solution (25) with Eqs. (27) into (23) we recover the dispersion law, Eq. (22), in the form

$$(K^2 - \kappa_j^2)\, c^2 = \omega^2 \varepsilon_j(\omega) , \qquad (j = 1,2). \tag{28}$$

As usual the conductivity term has been incorporated into the dielectric term to give a complex-valued dielectric constant: $\varepsilon_j + 4\pi i\sigma_j/\omega \rightarrow \varepsilon_j$. However, in our examples the imaginary part of ε, yielding a damping of the polariton mode, can normally be neglected. The boundary conditions, consisting in the continuity through the interface of parallel $\vec{E}$ and normal $\vec{D}$ components, yield

$$E_1^o = E_2^o, \tag{29}$$

$$\kappa_1/\kappa_2 = -\varepsilon_1(\omega)/\varepsilon_2(\omega). \tag{29'}$$

Combining (29') with Eqs. (28) we get the surface polariton dispersion

$$K^2c^2 = \omega^2 \frac{\varepsilon_1(\omega)\varepsilon_2(\omega)}{\varepsilon_1(\omega) + \varepsilon_2(\omega)} . \tag{30}$$

We note that Eq. (30) is simply obtained by setting in Eq. (22) $\eta=1$ and replacing the inverse bulk dielectric constant $\varepsilon^{-1}(\omega)$ with the inverse surface dielectric constant

$$\varepsilon_s^{-1}(\omega) = \varepsilon_1^{-1}(\omega) + \varepsilon_2^{-1}(\omega). \tag{31}$$

Looking at Eq. (31) we note that the eventual poles of $\varepsilon_1(\omega)$ and $\varepsilon_2(\omega)$, corresponding to bulk transverse excitations in the two non-interacting media, are no longer poles of $\varepsilon_s(\omega)$. Rather, the poles of $\omega=\omega_s$ of $\varepsilon_s(\omega)$, giving interface phonons or plasmons, are solutions of the equation

$$\varepsilon_2(\omega) = -\varepsilon_1(\omega). \tag{32}$$

According to Eq. (29'), $\kappa_1 = \kappa_2$ in the limit $\omega \to \omega_s$. For a free surface, where one of the two media is the vacuum, say $\varepsilon_1 = 1$, surface phonons or plasmons occur when

$$\varepsilon_2(\omega) = -1. \tag{33}$$

Now I show two simple examples of calculation of surface polariton dispersion relations, referring the reader to the book of Maradudin, Wallis and Dobrzynski [3] for a systematic analysis of surface polaritons in more complicated systems.

III.C Surface Plasmon Polaritons

We write the complex dielectric constant of conduction electrons in the form

$$\varepsilon(\omega) = \varepsilon_\infty + \frac{\sigma_o}{\omega}\frac{4\pi i}{1 - i\omega\tau}, \tag{34}$$

where ε_∞ is the high-frequency dielectric constant, σ_o the dc conductivity, τ the conduction electron relaxation time. They are related to the plasma frequency ω_p by

$$4\pi\sigma_o/\tau\varepsilon_\infty = \omega_p^2 \equiv 4\pi ne^2/m^*\varepsilon_\infty \tag{35}$$

where n and m^* are conduction electron density and effective mass, respectively. For the vacuum-metal (semiconductor) interface [$\varepsilon_1=1$ $\varepsilon_2 = \varepsilon(\omega)$] teh dispersion relation for surface plasmon polariton can be written as

$$\frac{K^2c^2}{\omega^2} = \frac{\omega_p^2 - \omega^2\{1+i(\omega\tau)^{-1}\}}{\omega_p^2 - (1+\varepsilon_\infty^{-1})\omega^2\{1+i(\omega\tau)^{-1}\}}. \tag{36}$$

In this equation the term $i/\omega\tau$ can be neglected because in metals $\omega\tau$ is ~ 1 when $\omega \ll \omega_p$. Solving with respect to ω^2 one obtains

$$\omega^2 = \tfrac{1}{2}\left[\omega_p^2 + (1+\varepsilon_\infty^{-1})K^2c^2\right]\left[1-\sqrt{1-4\left(\frac{\omega_p Kc}{\omega_p^2+(1+\varepsilon_\infty^{-1})K^2c^2}\right)^2}\,\right] . \quad (36')$$

The plot of this dispersion relation is shown in fig. 2. Note that the solution with positive radical is spurious because for $\omega^2 > \omega_p^2$ $\varepsilon(\omega)$ becomes > 0, which is incompatible with Eq. (29') if Re κ_1 and Re κ_2 have to be positive.

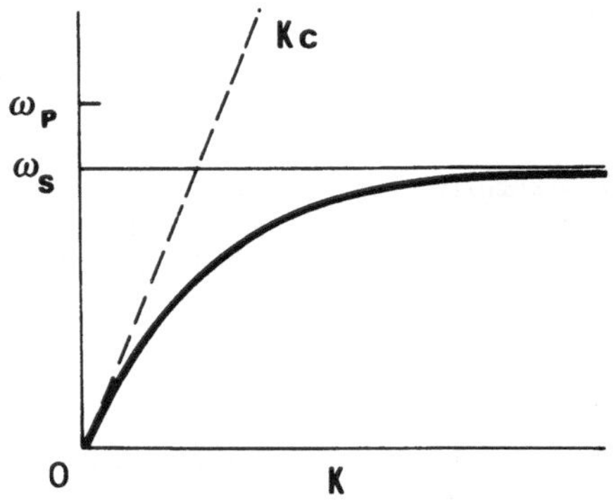

Fig. 2. Dispersion of surface plasmon polaritons.

For $K \to \infty$, ω tends to the surface plasmon frequency

$$\omega_s = \omega_p/(1+\varepsilon_\infty^{-1})^{\frac{1}{2}} , \quad (37)$$

whereas for $K \to 0$ the polariton displays a photon-like dispersion

$$\omega \simeq Kc . \quad (38)$$

The penetration length of the surface plasmon vanishes on both sides as $K \to \infty$, since $\kappa_1 \sim \kappa_2 \sim K$. The photon-like solution for $K \to 0$ has also a small penetration length inside the medium due to the screening of conduction electrons. Owing to the conductivity term, which cannot be neglected in this case, $\kappa_2 \equiv \kappa_2' - ik_z$ is a complex quantity. Solving Eq. (28) with respect to κ_2' for small K and ω, and keeping only the terms $\omega\tau$ (as τ may be very large), one finds

$$\kappa_2'^2 = \tfrac{1}{2}\varepsilon_\infty(\omega_p/c)^2 f(\omega\tau)\left[f(\omega\tau)+1\right] ; \quad f(x) \equiv x/(1+x^2)^{\frac{1}{2}} . \quad (39)$$

For $\omega\tau << 1$ (hydrodynamic limit) the penetration length becomes $\delta \equiv 1/\kappa_2' = c/(2\pi\sigma_0\omega)^{\frac{1}{2}}$ which is just the normal skin depth. However, if $\tau \to \infty$, $\varepsilon(\omega)$ tends to the Drude form $\varepsilon_\infty(1-\omega_p^2/\omega^2)$ without dissipation and δ takes the constant value $\delta_\infty = c/\varepsilon_\infty^{\frac{1}{2}}\omega_p$. For example, in potassium $\delta_\infty \sim 500$ Å.

III.D Polaritons Associated with Surface Phonons and Excitons

Another interesting example of surface polariton is that associated with the transverse infrared-active optical (TO) phonon of a cubic crystal with two atoms per unit cell. The contribution of the TO phonon branch to the dielectric constant can be written as

$$\varepsilon(\vec{k},\omega) = \varepsilon_\infty \left[1 + \frac{\omega_p^2}{\omega_T^2(\vec{k}) - \omega^2 - i\omega\tau^{-1}} \right], \tag{40}$$

where $\omega_T(\vec{k})$ is the frequency of the TO phonon of wavevector $\vec{k} \equiv (\vec{K},k_z)$, and the relaxation time τ is also, in principle, a function of $\vec{k}$. For optical phonons the plasma frequency ω_p is given by

$$\omega_p^2 \equiv \omega_L^2 - \omega_T^2 = (1 - \varepsilon_\infty/\varepsilon_s) \omega_L^2 , \tag{41}$$

where $\omega_T \equiv \omega_T(0)$; $\varepsilon_s = \varepsilon(0,0)$ is the static dielectric constant and ω_L is the $k = 0$ longitudinal optical (LO) phonon frequency, related to ω_T through the Lyddane-Sachs-Teller relation $\omega_L^2/\omega_T^2 = \varepsilon_s/\varepsilon_\infty$. Clearly for conduction-electron plasmons, where $\varepsilon_s = \infty$ and $\omega_T(\vec{k})=0$, Eq. (40) is identical to (34). The form of $\varepsilon(\vec{k},\omega)$ expressed by (40) is also valid for excitons, provided that ω_T is interpreted as the $k = 0$ excitation frequency, and $\omega_T(\vec{k})$ gives the exciton dispersion. In solving the surface polariton problem again we look for solutions where the third component of the wavevector $\vec{k}$ is imaginary, namely

$$\vec{k}_1 = (\vec{K},i\kappa_1) \quad \text{and} \quad \vec{k}_2 = (\vec{K},-i\kappa_2).$$

But now the conditions for κ_1 and κ_2, obtained by replacing the trial solution into Eqs. (23) and (24) are no longer explicity, since $\varepsilon_j(\vec{k}_j,\omega)$, with $j = 1,2$, depends on both $\vec{K}$ and κ_j. As shown by Maradudin and Mills [3], in the simpler case of a free surface ($\varepsilon_1 = 1$) there are three independent values of κ_2 whose corresponding trial solutions fulfill Eqs. (23) and (24). Like in the surface elastic wave problem, the boundary conditions are now satisfied only by a linear combination of the three trial solutions, and the resulting dispersion relation can be worked out only numerically. However, some relevant physical aspects of surface phonon (exciton) polaritons can be shown by assuming no dispersion and no damping of the optical branch, i.e., $\omega_T(\vec{k}) = \omega_T$ and $\tau^{-1} = 0$. Then, solving Eq.(30) with $\varepsilon_1 = 1$, the dispersion relation for surface polaritons turns out to be (Fig. 3)

$$\omega^2 = \tfrac{1}{2}[\ \omega_L^2 + (1+\varepsilon_\infty^{-1})K^2c^2\ ][1 - \sqrt{1-4\frac{(\omega_L^2+\varepsilon_\infty^{-1}\omega_T^2)\ K^2c^2}{(\omega_L^2+(1+\varepsilon_\infty^{-1})K^2c^2)^2}}\ ]. \quad (42)$$

The surface optical phonon frequency is obtained for $K \to \infty$:

$$\omega_s = \left(\frac{\omega_T^2 + \varepsilon_\infty\omega_L^2}{1 + \varepsilon_\infty}\right)^{\frac{1}{2}} = \omega_L\left(\frac{1 + \varepsilon_s^{-1}}{1 + \varepsilon_\infty^{-1}}\right)^{\frac{1}{2}} . \quad (43)$$

This is the famous Fuchs and Kliever (FK) mode [6] found for a non-dispersive semiinfinite dielectric medium. Note that $\omega_T < \omega_s < \omega_L$ and that no surface polariton exists below ω_T.

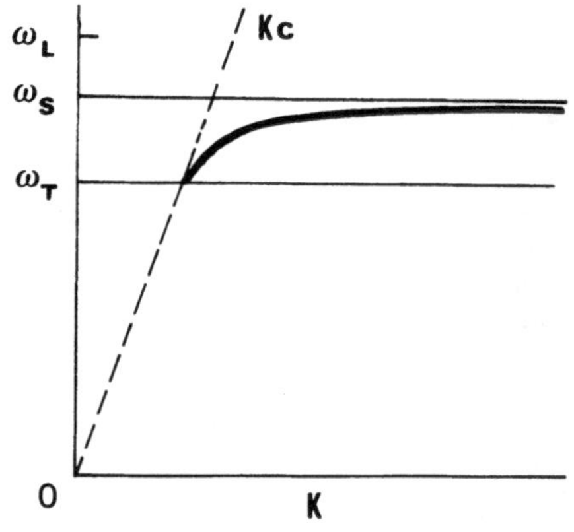

Fig. 3. Dispersion of surface optical phonon polaritons in a semi-infinite dielectric medium.

Setting $Kc = \omega_T$ in Eq. (41) one obtains $\omega^2 = \omega_T^2$. For $\omega < \omega_T$ the phase velocity of the photon-like wave is larger than c and tends to $(1+\varepsilon_s^{-1})c$ for $K \to 0$. Actually, κ_2 diverges at $\omega = \omega_T$ and becomes imaginary for $\omega < \omega_T$. Since κ_1 as well as ε_2 keep real, Eq. (29') is no longer filfilled. As an exercise, the reader could discuss the existence and find the dispersion relation of polaritons localized at the interface between two ionic solids whose optical phonon frequencies are $\omega_{T,1}$, $\omega_{L,1}$ and $\omega_{T,2}$, $\omega_{L,2}$.
An important remark: The above results, frequently used in discussing the optical properties of crystal surfaces, are not at all ri-

gorous, since the effects coming from LO and TO branch dispersion have been neglected. As a consequence, the frequency ω_s of the electromagnetic mode as given by Eq. (43), and corresponding to the root of $\varepsilon(0,\omega) = -1$, is expected to be in disagreement with the prediction of lattice dynamical calculations in the long wave limit. In real crystals the region between ω_T and ω_L is partially or even entirely filled by the continuous bands of TO and LO bulk frequencies $\omega_T(k)$ and $\omega_L(k)$, with $k = (0,k_z)$ and k_z running all over the allowed values in the 3D Brillouin zone. Thus the FK mode, falling in most cases into the LO bulk band, actually corresponds to a broad resonance, whose frequency may be quite apart from the prediction of Eq. (43). However in the special case when a sufficiently large gap exists between TO and LO bands, the FK mode might be a localized frequency in the gap. An example, discussed by Maradudin et al in ref. [3], is found in the lattice dynamical calculation of surface modes in NaCl(001) using a rigid-ion model. According to these authors, the FK mode occurs just below the bottom of the LO band, at a frequency which is quite lower than that predicted by Eq. (43). In several other crystals, where ω_s is sunk into the LO band [7], the FK mode loses much of its surface character, particularly when, at long wave, it is deeply penetrating into the bulk. In this case, however, the electric field is also penetrating deeply into the vacuum outside the crystal, since $\kappa_1 \sim \kappa_2$, allowing for a selective coupling with incident photons (e.g., in the attenuated total reflection method developed by Otto [8,9] and in Raman scattering [10]) or electrons (in high-energy [11] or low-energy [12] scattering experiments). On the other hand, neutral molecular beams, which interact exclusively with the topmost surface layers and are therefore the best probe for surface phonons, have no appreciable coupling with FK modes. Actually lattice dynamical calculations of surface phonon densities *projected* onto the first layer of the surface do not show any dynamical resonance associated with in-band FK modes [13].

III.E Surface Magnon Polaritons

We now consider a semiinfinite single domain ferromagnetic medium separated from the vacuum by a planar interface at z = 0. The spins are assumed to be oriented along x. In vacuum (z > 0) $\varepsilon = \mu = 1$; in the medium (z < 0) we assume $\varepsilon_2 = 1$ and a permeability tensor given by

$$\mu(\omega) = \begin{vmatrix} \mu_{xx} & 0 & 0 \\ 0 & \mu_{yy} & \mu_{yz} \\ 0 & \mu_{zy} & \mu_{zz} \end{vmatrix} \qquad (44)$$

with

$$\mu_{yy} = \mu_{zz} \equiv \mu ,$$

$$\mu_{yz} = -\mu_{zy} \equiv i\mu' , \qquad (\mu,\mu' \text{ real}). \qquad (45)$$

The equation of motion for the magnetic field as deduced from Maxwell equations reads

$$-c^2\vec{\nabla} \times (\vec{\nabla} \times \vec{H}) = \mu(\omega)\ddot{\vec{H}}, \qquad (46)$$

with the condition

$$\vec{\nabla} \cdot \mu(\omega)\vec{H} = 0. \qquad (47)$$

Surface-localized solutions with $\vec{K}$ oriented along y have the form

$$\begin{aligned} \vec{H} &= \vec{H}_1 \exp \{i(Ky - \omega t) -\kappa_1 z\} \qquad \text{for } z > 0 \\ &= \vec{H}_2 \exp \{i(Ky - \omega t) +\kappa_2 z\} \qquad \text{for } z < 0 \end{aligned} \qquad (48)$$

with Re κ_1, Re $\kappa_2 > 0$. It turns out from Eq. (47) that $\vec{H}$ is polarized in the sagittal plane (yz). Inserting Eq. (48) into (46) we have two conditions for κ_1 and κ_2:

$$\kappa_1^2 = K^2 - \omega^2/c^2 , \qquad (49')$$

$$\kappa_2^2 = K^2 - \mu_v\omega^2/c^2 , \qquad (49'')$$

with $\mu_v = \mu - \mu'^2/\mu$. We now apply the boundary conditions

$$H_{1y} = H_{2y} \qquad (50')$$

$$B_{1z} = B_{2z}, \qquad \text{i.e. } H_{1z} = \mu H_{2z} = -i\mu' H_{2y}. \qquad (50'')$$

Eqs. (50) and (47) form a homogeneous system in the components of the amplitude vectors $\vec{H}_1$ and $\vec{H}_2$. Nontrivial solutions exist for

$$\kappa_1\mu_v + \kappa_2 = K\mu'/\mu. \qquad (51)$$

Eliminating κ_1 and κ_2 by inserting Eqs. (49) into (51), we obtain the dispersion relation. Since μ and μ' are certain functions of ω, it is convenient to solve with respect to K:

$$K^2 = \frac{\omega^2}{c^2} \frac{\mu(\mu_v-1)(\mu\mu_v-1) - 2\mu'^2 \pm 2\mu' \{ \mu(2\mu-\mu\mu_v-1) \}^{\frac{1}{2}}}{(\mu\mu_v-1)^2 - 4\mu'^2} . \qquad (52)$$

The magnon contribution to the permeability tensor, in the absence of relaxation, is expressed by

$$\mu' = \mu_\infty\omega_m\omega/(\omega_0^2 - \omega^2) , \qquad (53')$$

$$\mu = \mu_\infty + (\omega_0/\omega)\mu' \ , \tag{53''}$$

where $\omega_0 = \gamma H_0$ and $\omega_m = 4\pi\gamma M_0$. H_0 and M_0 are the static field and magnetization, respectively, and γ is the gyromagnetic ratio. The resulting dispersion curves are schematically plotted in Fig. 4. Here we have two branches, since both signs of the radical, eq.(52), are acceptable. Indeed changing the radical sign is equivalent to change the sign of μ', and if the sign of μ' is changed Eq. (51) is still fulfilled by changing the sign of K. Therefore the sign of the radical in Eq. (52) determines also the sign of K. Thus surface magnon polaritons are not invariant for inversion in the momentum space ($\vec{K} \to -\vec{K}$). Such a lack of inversion symmetry is well understood if we consider that the magnetic field oscillation at the surface is, as usual, elliptically polarized. But, unlike elastic and dielectric cases, here clockwise and counterclockwise elliptical polarizations are not equivalent, as their axes would respectively be parallel or antiparallel to the spins.

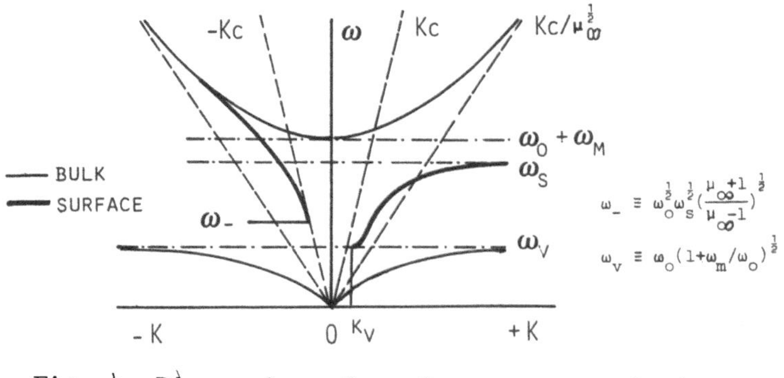

Fig. 4. Dispersion of surface magnon polaritons.

We see that only one branch extends to $|\vec{K}| = \infty$ yielding a surface magnon of frequency

$$\omega_s = \omega_0 + \omega_m/(1 + \mu_\infty^{-1}). \tag{54}$$

The theory here presented has been first developed by Hartstein, Burstein, Maradudin, Brewer and Wallis [4]. The reader should consult this important paper for an accurate analysis of the surface magnon problem.

IV. SURFACE LATTICE DYNAMICS

IV.A Introduction

When the wavelength of surface waves is so small to be comparable to the interatomic distances of the discrete crystal lattice, the previous continuum theory is no longer valid. This occurs for angular frequencies of the order or higher than 10^{11} s^{-1}. This limit is quite above the frequencies of surface waves used in devices of technological interest, but nearly the totality of the phonons involved in surface thermodynamic and response functions have higher frequencies, typically of the order of 10^{13} s^{-1}. To this respect the lattice dynamical approach is needed and a detailed knowledge of the interatomic force constants is required. It should be added that usually a crystal surface is non simply a step discontinuity in an otherwise perfectly homogeneous medium , but is accompanied by a local relaxation and sometimes by a reconstruction, which means local changes of elastic constants and density. In these cases, as well as in the presence of adsorbed layers of foreign atoms, the lattice dynamical approach is needed also in the long-wave limit.

Surface lattice dynamics has been studied from different points of view. A method, conceptually analogous to the elastic wave theory and frequently used in the past for simple crystal models with short range forces, is that based on the construction of a trial solution for the semiinfinite lattice [14]. The extension to crystals with long-range interactions of any kind has been given by Feutchwang [15], and successfully used for metal surfaces by Bortolani et al in Modena [16]. The hystorical importance is bound to the discovery of microscopic optical modes in diatomic crystals, frequently referred as to Lucas modes (LM) [17].

Another method, introduced and extensively used by the Austin group (Alldredge, Allen, de Wette, Chen et al) [18,19], consists in the direct calculation of eigenvalues and polarization vectors of a slab formed by a sufficiently small number of layers. This method yields all the variety of acoustic and optical surface modes all over the surface Brillouin zone (SBZ), provided that their penetration length is much smaller than the slab thickness. The above authors have shown that ten to twenty layers are normally sufficient to give all microscopic surface modes with excellent precision, which makes the direct method of practical use and quite convenient for most purposes.

Two limitations should be mentioned, however. One is represented by the poor statistics in the calculation of the momentum-selected surface phonon densities, due to the small number of layers. Such densities, entering several surface response functions, are more conveniently derived by a Green's function technique. Another difficulty in the slab calculation is the identification of resonances.

To some extent this can be done by examining the loci of hybridization of crossing bulk mode dispersion curves. In principle the Green's function method provides the most natural way to determine resonant surface modes.

The Green's function theory of surface vibrations, treating the surface as a perturbation which modifies the spectrum of bulk vibrations, was introduced by Lifschitz and Rosenzweig in the early forties [20]. Its formal advantage is the reduction of a big problem to a small one: it enables one to work in the perturbation subspace rather than in the large slab space. In practice the Green's function method has a drawback in the severe computational difficulties coming from the singular nature of the surface-projected Green's functions, which partly vanify the advantage of dealing with small matrices [13]. In this Section I present first the direct method in order to introduce standard nomenclature of surface lattice dynamics and to provide a basis to the main subject, the Green's function theory of surface lattice vibrations in crystals with general long-range interatomic forces.

IV.B Dynamics of a Thin Slab

We express the equilibrium positions of crystal atoms as

$$\vec{x}(\vec{\ell}\kappa) = \vec{x}(\vec{\ell}) + \vec{x}(\kappa), \tag{55}$$

where

$$\vec{x}(\vec{\ell}) = \ell_1\vec{a}_1 + \ell_2\vec{a}_2 + \ell_3\vec{a}_3 \tag{56}$$

denotes the positions of the unit cells of the periodic lattice; $\vec{a}_j$ are the basis vectors and ℓ_j are integer numbers. Each unit cell contains s atoms whose positions are $\vec{x}(\kappa)$, with $\kappa = 1,2,\ldots,s$. In the presence of a surface it is possible to select $\vec{a}_1$ and $\vec{a}_2$ parallel to the surface and to write

$$\vec{x}(\vec{\ell}\kappa) = \vec{x}(\vec{L}) + \vec{x}(\ell_3\kappa), \tag{57}$$

where

$$\vec{x}(\vec{L}) = \ell_1\vec{a}_1 + \ell_2\vec{a}_2, \qquad \vec{L} = (\ell_1,\ell_2). \tag{58}$$

Here ℓ_3 denotes the atomic layers parallel to the surface. In a slab-shaped lattice with N_L layers, $\ell_3 = 1,2,\ldots,N_L$. The two-dimensional reciprocal lattice associated with the periodic structure of the surface is

$$\vec{G} = g_1\vec{b}_1 + g_2\vec{b}_2, \tag{59}$$

with g_1 and g_2 integer numbers, and

$$\vec{b}_1 = 2\pi \frac{\vec{a}_2\times(\vec{a}_1\times\vec{a}_2)}{|\vec{a}_1\times\vec{a}_2|^2}, \qquad \vec{b}_2 = 2\pi \frac{\vec{a}_1\times(\vec{a}_2\times\vec{a}_1)}{|\vec{a}_2\times\vec{a}_1|^2}. \tag{60}$$

In the harmonic approximation [21] the interaction among the crystal atoms is represented by the set of force constants

$$\begin{aligned}\Phi_{\alpha\beta}(\vec{L},\vec{L}';\ell_3\kappa,\ell_3'\kappa') &= \Phi_{\alpha\beta}(\vec{L}-\vec{L}';\ell_3\kappa,\ell_3'\kappa') \\ &= \frac{\partial^2 U}{\partial u_\alpha(\vec{L},\ell_3\kappa)\partial u_\beta(\vec{L}',\ell_3'\kappa')},\end{aligned} \tag{61}$$

defined as the second-order derivative of the crystal potential energy with respect to the atomic displacement vectors $\vec{u}(\vec{L},\ell_3\kappa)$. Since the periodicity of the lattice with cyclic boundary conditions is preserved along the surface, the force constants depend only on the difference $\vec{L}-\vec{L}'$. According to eq.(61)

$$\Phi_{\alpha\beta}(\vec{L}-\vec{L}';\ell_3\kappa,\ell_3'\kappa') = \Phi_{\beta\alpha}(\vec{L}'-\vec{L};\ell_3'\kappa',\ell_3\kappa). \tag{62}$$

The force constants are usually derived from a model microscopic description of the two- and many-body interatomic potentials. For a two-body potential $\phi_{\kappa\kappa'}(r)$ the force constants are also symmetric with respect to the cartesian indices $\alpha\beta$ and the atomic labels, separately, since from eq.(61) it is found

$$\begin{aligned}\Phi_{\alpha\beta}(\vec{L}-\vec{L}';\ell_3\kappa,\ell_3'\kappa') &= \frac{r_\alpha r_\beta}{r^2}\left(\frac{\partial^2\phi_{\kappa\kappa'}}{\partial r^2} - \frac{\partial\phi_{\kappa\kappa'}}{r\partial r}\right) \\ &+ \delta_{\alpha\beta}\frac{\partial\phi_{\kappa\kappa'}}{r\partial r},\end{aligned} \tag{63}$$

with

$$\vec{r} = \vec{x}(\vec{L},\ell_3\kappa) - \vec{x}(\vec{L}',\ell_3'\kappa'), \qquad r = |\vec{r}|.$$

Since the crystal potential is invariant under a displacement field corresponding to an arbitrary rigid-body translation or rotation the force constants must satisfy the conditions [21]

$$\sum_{L'\ell_3'\kappa'} \Phi_{\alpha\beta}(\vec{L}-\vec{L}';\ell_3\kappa,\ell_3'\kappa') = 0 \qquad \forall\ \alpha,\beta \tag{64}$$

(translational invariance condition: TI)

$$\begin{aligned}\sum_{L'\ell_3'\kappa'} [&\Phi_{\alpha\beta}(\vec{L}-\vec{L}';\ell_3\kappa,\ell_3'\kappa')x_\gamma(\vec{L}'\ell_3'\kappa') \\ &- \Phi_{\alpha\gamma}(\vec{L}-\vec{L}';\ell_3\kappa,\ell_3'\kappa')x_\beta(\vec{L}'\ell_3'\kappa')] = 0,\end{aligned} \tag{65}$$

$\forall\ \alpha,\beta,\gamma$ (rotational invariance condition: RI)

Notice that the above conditions refer to rigid-body displacements and have nothing to do with the translational and point group symmetries of the lattice: they hold for any arbitrary array of atoms. The nature of the constraints due to the RI condition, Eq.(65), on the force constants near the surface has been illustrated in a classical example due to Lengeler and Ludwig [22]. In a model where the force constants are obtained by Eq.(61) from the crystal potential energy, whose phenomenological parameters fit the equilibrium condition of the crystal in the absence of external forces, the RI condition is authomatically satisfied.

The equation of motion for the harmonic lattice reads

$$M_\kappa(\ell_3)\ddot{u}_\alpha(\vec{L},\ell_3\kappa) = -\sum_{L'\ell_3'\kappa'\beta}\Phi_{\alpha\beta}(\vec{L}-\vec{L}';\ell_3\kappa,\ell_3'\kappa') \times u_\beta(\vec{L}',\ell_3'\kappa'), \tag{66}$$

where $M_\kappa(\ell_3)$ are the atomic masses. The reduction of Eq.(66) in the long wave limit, and the connection between the RI condition and Eq. (2) are shown in ref. [21]. Possible solutions of Eq. (66) are plane waves travelling along the surface of the form

$$\vec{u}(\vec{x},\ell_3\kappa) = [\hbar/2N_s\omega_j(\vec{K})M_\kappa(\ell_3)]^{\frac{1}{2}}\vec{e}(\ell_3\kappa|\vec{K}_j) \times \exp i[\vec{K}\cdot\vec{x}(\vec{L})-\omega_j(\vec{K})t], \tag{67}$$

where N_s is the number of surface unit cells, i.e., the number of values taken by the vector index $\vec{L}$. The polarization vectors $\vec{e}(\ell_3\kappa|\vec{K}_j)$ and the squared frequencies $\omega_j^2(\vec{K})$ are deduced from the secular equation

$$\sum_{\ell_3'\kappa'\beta}D_{\alpha\beta}(\ell_3\kappa,\ell_3'\kappa';\vec{K})e_\beta(\ell_3'\kappa'|\vec{K}_j) = \omega_j(\vec{K})e_\alpha(\ell_3\kappa|\vec{K}_j), \tag{68}$$

where

$$D_{\alpha\beta}(\ell_3\kappa,\ell_3'\kappa';\vec{K}) = [M_\kappa(\ell_3)M_{\kappa'}(\ell_3')]^{-\frac{1}{2}} \times \sum_{\vec{L}'}\Phi_{\alpha\beta}(\vec{L}-\vec{L}';\ell_3\kappa,\ell_3'\kappa')\exp[i\vec{K}\cdot(\vec{x}(\vec{L})-\vec{x}(\vec{L}'))] \tag{69}$$

is the dynamical matrix, whose dimensions are $3sN_L \times 3sN_L$. The branch index $j = 1,2,\ldots,3sN_L$ labels the eigensolutions. The dimensionless vectors $\vec{e}(\ell_3\kappa|\vec{K}_j)$ form a complete and orthogonal set, and are normalized to unity. They contain all the information about the spatial behaviour of the slab modes in the direction normal to the surfaces. For values of $\vec{K}$ along some symmetry directions of the SBZ the eigenvectors are perfectly polarized in the sagittal plane (sagittal polarization: $\perp$), or along the normal to the sagittal plane namely parallel to the surface (parallel polarization: $\parallel$). In this

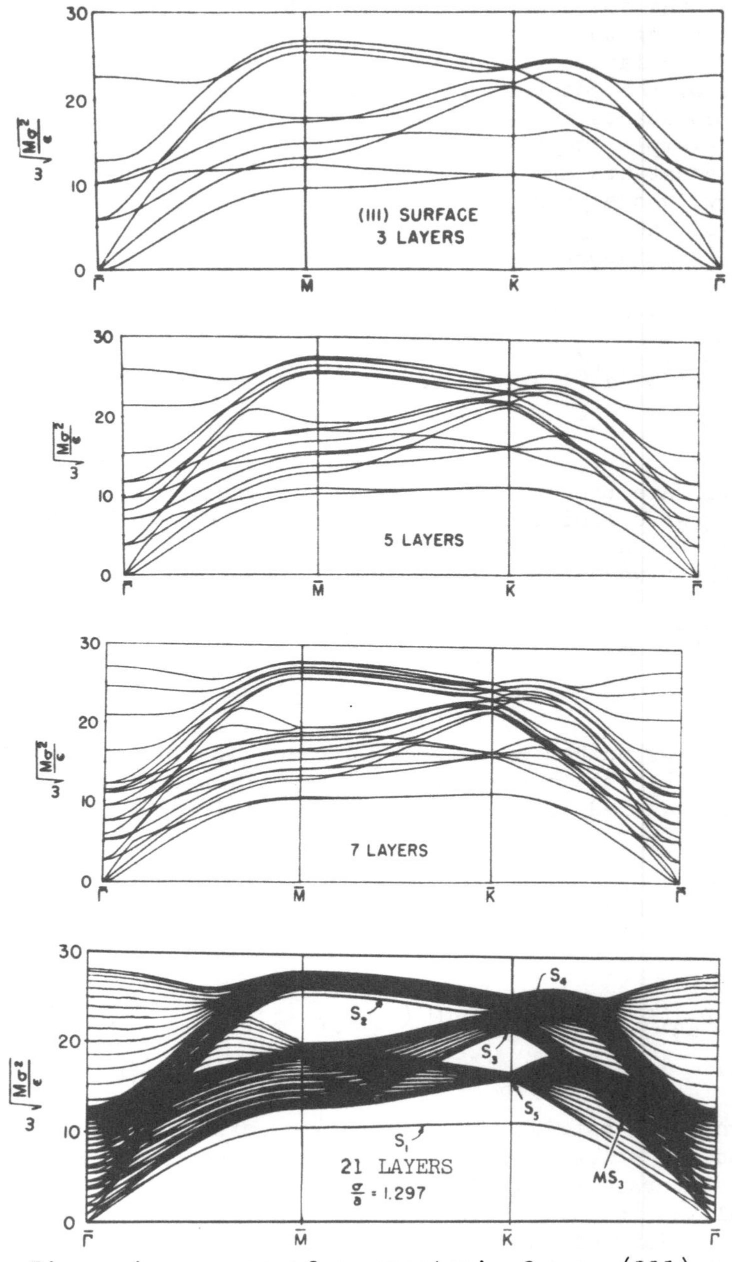

Fig. 5. Dispersion curves of a monoatomic f.c.c. (111)-oriented slab along the boundaries of the irreducible part of the SBZ, for different thickness (from Allen, Alldredge and de Wette, ref. [18]).

case the set of Eqs. (68) splits into two subsets for ⊥ and || modes, and the secular determinant of Eq. (68) factorized correspondingly. Fig. 5 shows an instructive example: the dispersion curves for a (111)-oriented monoatomic f.c.c. slab are plotted along the symmetry directions of the SBZ. The calculation, due to Allen, Alldredge and de Wette [18], are based on a Lennard-Jones two-body potential, and are reported for increasing number of layers (N_L = 3,5,7,21). It appears that, as N_L increases, the bulk modes thicken forming a certain number of bands. These bands become continuous in the limit of an infinitely thick slab: in this limit the band edges and the gaps appearing between different bands for each $\vec{K}$ are exactly those of the distribution of the frequencies $\omega_j(\vec{K})$ of the cyclic infinite lattice, where $j = 1,2,...,3s$, and $\vec{k} = (\vec{K},k_3)$ are the branch index and the 3D wavevector, when j and k_3 are allowed to assume all the possible values and $\vec{K}$ is kept constant. However, a finite number of slab modes, labeled S_1, S_2, etc., remain outside the bands. Since their eigenvectors are found to be large near the surface and rapidly decreasing with the increasing distance from the surface, they are interpreted as surface modes. Particularly the acoustic surface mode S_1 is localized below the acoustic band also in the long-wave limit. This mode is just the surface Rayleigh wave discussed in the previous Section. The other surface modes have no analogue in the continuum theory, since they appear in the dispersive region of the Brillouin zone. In the long-wave limit the monoatomic slab admits only acoustic surface waves, while the diatomic slab may have surface modes of optical character, whose frequencies do not vanish as $K \to 0$. In addition to the macroscopic FK modes, microscopic optical surface modes exist, such as the Lucas modes [17]. Fig. 6 shows the dispersion curves for a 15-layer slab of

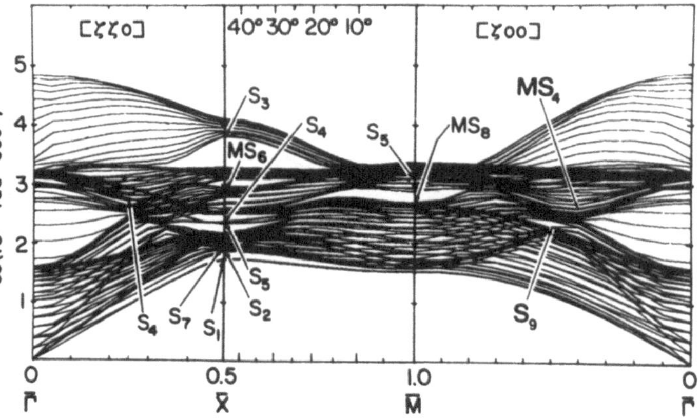

Fig. 6. Dispersion curves of a 15-layer (001)-oriented NaCl slab (shell-model calculation from ref. [19]).

NaCl with (001)-orientation. calculated from a 11-parameter shell model by de Wette et al [19]: the modes S_5 and S_4 are examples of Lucas modes with $\|$ and $\perp$ polarization, respectively. An interesting feature occurring for slabs with three-dimensional inversion symmetry (as in the case here considered) is that the surface modes occur in nearly degenerated pairs. The degeneracy occurs first in the short-wave region at the SBZ boundary, but, as N_L increases, it involves modes of larger and larger wavelength. This is due to the fact that each long-wave surface mode is deeply penetrating and involves the opposite surface. However, for $N_L = \infty$, all surface modes are degenerate everywhere. These modes should be regarded as single surface modes of a semiinfinite lattice. Actually, quite a precise information on the vibrations of the semiinfinite lattice can be extrapolated from slab calculations with reasonably large values of N_L.

IV.C The Green's Function Method

1. Free Surfaces as a Perturbation. The dynamical problem of a semiinfinite lattice can be approached by a perturbation method. We consider the infinitely thick slab to be originated by perturbing an infinite lattice with three-dimensional cyclic boundary conditions. The free surfaces are constructed by cutting the infinite lattice along an ideal plane Σ, as shown in Fig. 7, the resultant perturbation of the infinite lattice force constants

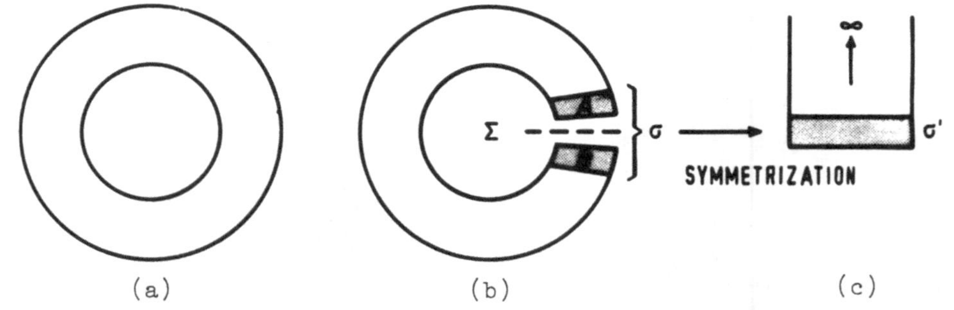

Fig. 7. Representation of an ideal lattice with cyclic boundary conditions (a), and of the same lattice after two free surfaces are created by a cut along the plane Σ(b); σ denotes the perturbation subspace. In the representation of symmetrized co-ordinates we have a semi-infinite lattice with a perturbation subspace σ', restricted to a single surface (c).

$\phi^{o}_{\alpha\beta}(\vec{\ell}\kappa,\vec{\ell}'\kappa') = \phi^{o}_{\alpha\beta}(\vec{L}-\vec{L}',\ell_3-\ell'_3,\kappa\kappa')$ is then described by setting to zero all the interatomic force constants crossing the plane Σ; more precisely, we define the perturbation matrix Λ according to

$$\Lambda_{\alpha\beta}(\vec{L}-\vec{L}';\ell_3\kappa,\ell'_3\kappa')=\phi_{\alpha\beta}(\vec{L}-\vec{L}';\ell_3\kappa,\ell'_3\kappa')-\phi^{o}_{\alpha\beta}(\vec{L}-\vec{L}';\ell_3-\ell'_3,\kappa\kappa'), \quad (70)$$

where both ϕ, the slab force constant matrix, and ϕ_o are assumed to satisfy the TI and RI conditions; thus also Λ must satisfy the TI and RI conditions. In Eq.(70) it is intended that the slab force constant elements connecting atoms across Σ are zero. In practice, Λ works as a perturbation only when its non-zero elements are restricted within a small pertubation subspace σ (Fig.7); for example if $\ell_3=\frac{1}{2}$ and $\ell_3=-\frac{1}{2}$ denote the two surface layers, the subspace σ is defined by ℓ_3 and $\ell'_3=\frac{1}{2}, \frac{3}{2}, \ldots, N_p-\frac{1}{2}$ or $-\frac{1}{2}, -\frac{3}{2}, \ldots, -N_p+\frac{1}{2}$, where N_p, the number of atomic layers on each side of the slab involved in the cutting procedure, is "small". This requirement seems to rule out the applicability of the perturbation method to lattices with long-range forces, like ionic crystals. However, we will see that in the limit $N\to\infty$ it is possible to work with a small perturbation matrix also for ionic crystals.

Let us rewrite Eq.(66) for stationary states in a self-explaining matrix notation

$$-M\omega^2\vec{u} = -\phi\vec{u} = -(\phi_o+\Lambda)\vec{u}, \quad (71)$$

where M is the mass matrix, and ω the variable frequency. We assume that Eq.(71) has been already Fourier-transformed, i.e. we work in the wave vector representation.

The eigenvalue spectrum of $D_o \equiv M^{-\frac{1}{2}}\phi_o M^{-\frac{1}{2}}$, the dynamical matrix of the infinite lattice, is represented for each $\vec{K}$ by a set of 3s continuous bands. First, we consider ω^2 outside the spectrum $\{D_o\}$ of D_o: in this case Eq. (71) can be written as

$$[I + (\phi_o-M\omega^2)^{-1}\Lambda]\vec{u} = 0 \qquad \omega^2 \notin \{D_o\} \quad (72)$$

The advantage of Eq. (72) with respect to Eq. (71) is that now we have a secular determinant having the same dimension of the perturbation Λ which in the K-representation is $6sN_p \times 6sN_p$. The values of ω^2 outside $\{D_o\}$ for which

$$\det|I+(\phi_o-M\omega^2)^{-1}\Lambda| = 0 \quad (73)$$

for any given $\vec{K}$ give the dispersion relations of the surface modes; they are localized at the surface since no frequency outside $\{D_o\}$ can propagate in the interior of the lattice.

When ω^2 falls into $\{D_o\}$, Eq. (71) is not reducible to the form (72) since $\phi_o-M\omega^2$ cannot be inverted. However, in this case we have a wave propagating in the bulk, which is distorted by the surfaces through scattering processes. Again the problem can be reduced to the perturbation subspace, according to the standard theory of scattering [23], by inverting ϕ_o - Mz, where $z = \omega^2 + i0^+$ is the squared frequency removed from the real axis by an infinitesimal positive quantity. In this case, Eq. (71) can be written as

$$\vec{u} = \alpha\vec{u}_o - (\phi_o - Mz)^{-1}\Lambda\vec{u}, \qquad \omega \in \{D_o\} \quad (74)$$

$$= \alpha\{I + (\phi_o - Mz)^{-1}\Lambda\}^{-1}\vec{u}_o, \quad (74')$$

$$= \alpha\vec{u}_o - \alpha(\phi_o - Mz)^{-1}\, T\vec{u}_o, \quad (74'')$$

where $\vec{u}^o$ is any eigenvector of the infinite lattice for the frequency ω, and α is a normalization constant. Each solution of Eq.(74) for $\omega^2 \in \{D_o\}$ can be seen as a superposition of an incoming unperturbed wave and a scattered wave, represented by $-(\phi_o - Mz)^{-1}T\vec{u}^o$ in Eq. (74"), where

$$T = \Lambda\{I + (\phi_o - Mz)^{-1}\Lambda\}^{-1} \quad (75)$$

is the transition matrix. T has the same dimensions as Λ; thus only the elements of $(\phi_o - Mz)^{-1}$ in the subspace σ are involved: they form a $6sN_p \times 6sN_p$ matrix

$$g \equiv (\phi_o - Mz)^{-1} \text{ projected onto } \sigma\,, \quad (76)$$

known as the unperturbed projected Green's function matrix. Resonant states, i.e. enhanced scattering waves, occur when

$$\mathrm{Re}\,\det|I + g\Lambda| = 0. \quad (77)$$

Since, for ω^2 outside $\{D_o\}$, g becomes just the real matrix $(\phi_o - M\omega^2)^{-1}$, Eq.(77) includes Eq.(73) as a particular case. Eq.(77) should be regarded as the general condition for the existence of surface modes. For each wave vector $\vec{K}$, Eq. (77) has a certain number of solutions $\omega^2 = \omega_i^2(\vec{K})$; the solutions outside $\{D_o\}$ give surface-localized modes, while the in-band solutions, inside $\{D_o\}$, correspond to surface resonant modes, or pseudosurface waves. For an infinite lattice, the perturbative effects of the free surfaces on the continuous frequency bands are understood in terms of phonon densities: useful concepts are the unperturbed and the perturbed projected phonon densities

$$\rho(\vec{K},\omega) = (2\omega/\pi)\, \mathit{Tr}\ \mathrm{Im}\ g(\vec{K},\omega^2) \quad (78)$$

$$\tilde{\rho}(\vec{K},\omega) = (2\omega/\pi)\, \mathit{Tr}\ \mathrm{Im}\ \tilde{g}(\vec{K},\omega^2), \quad (78')$$

respectively, where *Tr* denotes trace in the subspace σ of the perturbation, and

$$\tilde{g} \equiv (\phi - Mz)^{-1} \qquad \text{projected onto } \sigma \tag{79}$$

$$= (I + g\Lambda)^{-1} g \tag{79'}$$

is the perturbed Green's function matrix. Along symmetry directions all modes have either sagittal ($\perp$) or parallel ($\|$) polarization, yielding the factorization

$$\{D_o\} = \{D_o\}_{\perp} \{D_o\}_{\|}$$

and the block-diagonalization of all the above matrices. Having in mind Eq. (79'), the following cases occur for the values ω^2 which fulfill Eq. (77)

i) *Surface modes*: ω^2 is outside $\{D_o\}$. The imaginary part of g is infinitesimal and $\tilde{\rho}(\vec{K},\omega)$ exhibits a δ-peak.

ii) *Pseudo-surface modes*: ω^2 falls into a band of $\{D_o\}$, but the displacement vector $\vec{u}$ is orthogonal to all vectors $\vec{u}_o$ of that band; in eq. (74) we must set $\alpha = 0$ and again we have a local mode, crossing a transparent bulk band. This occurs, e.g., when ω^2 belongs to $\{D_o\}_{\perp}$ but is outside $\{D_o\}_{\|}$ (or viceversa). Clearly, pseudo-surface modes exist only along symmetry directions.

iii) *Surface resonances*: ω^2 falls into a band of $\{D_o\}$ whose $\vec{u}_o$ are not orthogonal to $\vec{u}$ ($\alpha \neq 0$). In this case we have a resonance: $\tilde{\rho}(\vec{K},\omega)$ displays a Lorentian-shaped peak, whose width is proportional to Im g , which is finite. When deviating from a symmetry direction any pseudo-surface mode transforms into a resonance. However, surface resonance may exist also along symmetry directions in addition to local modes.

2. The Semi-Infinite Lattice. In many cases the slab originated by cutting the cyclic lattice exhibits alike surfaces and a mirror symmetry with respect to the plane Σ (Fig. 7b). Sometimes, the mirror symmetry is recovered by a shear translation of the surface B with respect to A, as for the (001)-surfaces of a NaCl lattice. In these cases it is possible to replace the coordinate $|\ell_3)$ with symmetryzed coordinate $||\ell_3|,p)$ defined by the transformation

$$||\ell_3|,p) = 2^{-\frac{1}{2}}\{|\ell_3) + f(p)|-\ell_3)\}, \tag{80}$$

where $p = \pm$ is the parity index, and f(p) is a suitable function, having the properties $f(-p) = -f(p)$ and $|f(p)|^2 = 1$, choosen in such a way that the matrices Λ, g and $\tilde{g}$ are diagonalized with respect to p into blocks $3sN_p \times 3sN_p$. Call these blocks Λ_p, g_p and

$$\tilde{g}_p = \{I + g_p\Lambda_p\}^{-1} g_p , \qquad p = \pm. \tag{81}$$

This enables us to work in a reduced $3sN_p$-dimensional subsapce σ'. When $N_L \to \infty$, even- and odd-parity modes become degenerate: thus in subspace σ' we must have

$$\tilde{g}_+ = \tilde{g}_- \equiv \tilde{g} \,. \tag{82}$$

The $3sN_p$-dimensional matrix $\tilde{g}$ is interpreted as the perturbed projected Green's function for the single surface of the semi-infinite lattice. Eq. (81) can be written as

$$\tilde{g} = (I + \bar{g}\bar{\Lambda})^{-1}\bar{g}\,, \tag{83}$$

where

$$\bar{g}^{-1} = \tfrac{1}{2}(g_+^{-1} + g_-^{-1}), \tag{84}$$

and

$$\bar{\Lambda} = \tfrac{1}{2}(\Lambda_+ + \Lambda_-) \tag{83}$$

are the inverse of the unperturbed projected Green's function for the semiinfinite lattice, and the pertaining perturbation, respectively. Notice that all the inversions are performed in the subspace σ', The resonance condition, Eq. (77), in the subspace σ' becomes

$$\mathrm{Re}\,\det\left| I + \bar{g}\bar{\Lambda} \right| = 0. \tag{86}$$

Besides dimensional reduction, the symmetrization (80) has the advantage that the symmetrized perturbation $\bar{\Lambda}$ takes a very simple form. Indeed, the symmetrized components of Λ are

$$\begin{aligned}\Lambda_p &\equiv \Lambda_{\alpha\beta}(\vec{K};|\ell_3|\,\kappa p,|\ell_3'|\,\kappa' p) \\ &= \Lambda_{\alpha\beta}(\vec{K};|\ell_3|\,\kappa,|\ell_3'|\,\kappa') + f(p)\Lambda_{\alpha\beta}(\vec{K};|\ell_3|\kappa,-|\ell_3'|\kappa').\end{aligned} \tag{87}$$

According to Eq. (85) $\bar{\Lambda}$ is just the first term on the right-hand member of Eq. (87).

In the absence of surface elastic relaxation or reconstruction, the non-diagonal part of the perturbation concerns only pairs of atoms on opposite sides with respect to Σ. Thus all the elements of $\Lambda_{\alpha\beta}(\vec{K};|\ell_3|\kappa,|\ell_3'|\kappa')$ are zero, except those coming from the diagonal part of the force constant matrices (the Einstein part), which is obtined from the two-body force constants by imposing the TI condition, Eq. (64). We call Λ^T such diagonal part of $\bar{\Lambda}$.

Surface relaxation is caused by the unsaturated forces arising from the crystal fracture. They correspond to the perturbation in the first derivatives of the crystal potential, which are in turn related to the shear force constants (the terms $-r_\alpha r_\beta \phi'_{\kappa\kappa'}/r^3$ in Eq. (63)). Since the creation of a free surface destroys the inversion symmetry, the RI conditions relative to all the independent rotation axes parallel to the surface and lying within the reduced perturbation σ' yield non-trivial constraints on the shear force constants connecting atoms in σ'. Thus the shear force constants have to be perturbed within σ', and this contributes a non-diagonal part to $\bar{\Lambda}$. We argue that the independent non-diagonal elements of $\bar{\Lambda}$ are as many as the independent non-trivial RI conditions. These are in turn as many as the independent surface equilibrium conditions. We call Λ^R this non-diagonal part of $\bar{\Lambda}$. The role of Λ^R in surface dynamics is quite important, ensuring that the calculation refers to the surface in equilibrium. Maradudin et al show in their book [21] that the RI condition is equivalent to the boundary condition in the elastic theory and is therefore to be fulfilled in order to have Rayleigh waves with the appropriate velocity.

We write the perturbation matrix as

$$\bar{\Lambda} = \Lambda^T + \Lambda^R , \tag{88}$$

by which we mean that $\bar{\Lambda}$ is fully defined by TI and RI conditions. The procedure actually used to derive the elements of $\bar{\Lambda}$ is based on the equality

$$\Lambda_p = \bar{\Lambda} + \tfrac{1}{2} (g_{-p}^{-1} - g_p^{-1}) . \tag{89}$$

We apply now the sum rules associated with TI and RI conditions, Eqs. (64) and (65), which we formally re-write respectively as

$$T \phi = 0 , \qquad R \phi = 0 . \tag{90}$$

Since also $T \Lambda_p = 0$ and $R \Lambda_p = 0$, we have

$$\begin{aligned} T \bar{\Lambda} &= \tfrac{1}{2} T (g_p^{-1} - g_{-p}^{-1})_{\vec{K}=0} , \\ R \bar{\Lambda} &= \tfrac{1}{2} R (g_p^{-1} - g_{-p}^{-1})_{\vec{K}=0} , \end{aligned} \tag{91}$$

where the summation over all lattice sites selects, in the $\vec{K}$-space, only the $\vec{K} = 0$ matrix elements of the inverse Green's functions. Eqs. (91) form an inhomogeneos linear system whose unknowns are the independent elements of $\bar{\Lambda}$. Whatever is the assumed perturbation subspace σ', i.e., the number N_p of layers included in the perturbation, the unknown elements of $\bar{\Lambda}$ are as many as Eqs. (91). Thus such elements are uniquely (and self-consistently) defined in terms of the $\vec{K} = 0$ elements of the unperturbed Green's function.

Furthermore, in ionic crystals the summation over the ion index κ implicit in the operators T and R produces in the right-hand members of Eqs. (91) a cancellation of the long-range Coulomb contributions. Thus the diagonal elements of $\bar{\Lambda}$, as well as the non-diagonal ones, related to the surface elastic relaxation, decay in a very fast exponential way for increasing ℓ_3. It is therefore an excellent approximation to take only $\ell_3 = \frac{1}{2}$ ($N_P = 1$), namely to restrict the perturbation to the first layer [24]. This receives an *a posteriori* support from the very good agreement obtained between Green's function and slab calculations for alkali halide surfaces. Such a comparison is done in the next paragraph. As long as the perturbation can be cut at the first layer, the Green's function method applied to alkali halide (001) surfaces deals with 6 x 6 matrices, which makes this technique rather convenient and practical.

3. Examples and Comparison with Experimental Data. I show now a few examples of calculated dispersion curves of surface modes in alkali halide crystals, obtained using the Green's function method. The technical aspects of the calculation, based on the breathing shell model (BSM) for the bulk lattice dynamics and on a perturbation model restricted to the first layer, are illustrated in [13]. In fig. 8 I report the surface dispersion curves for the surface (001) of NaCl along the symmetry directions $\bar{\Gamma}\bar{X} = [\xi\xi 0]$, $\bar{X}\bar{M} = [1\xi 0]$

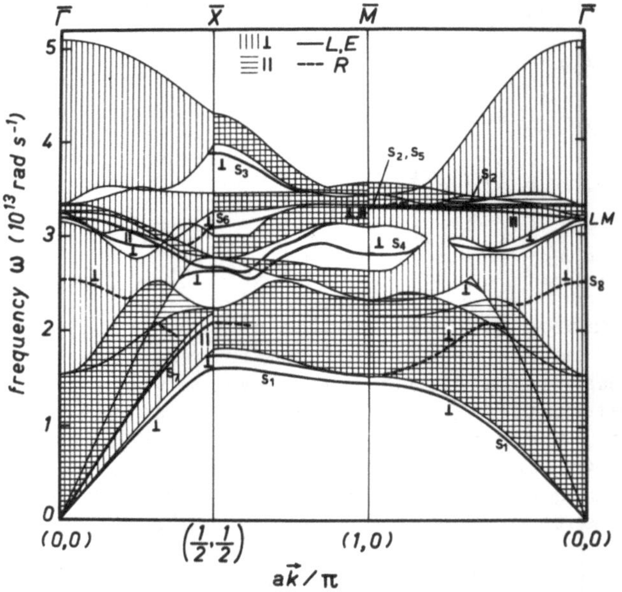

Fig. 8. Surface dispersion curves of NaCl(001) calculated by the Green s function method.

and $\bar{M}\bar{\Gamma}$ = [ξ00]. They should be compared with the slab calculations shown in Fig. 6, based on an 11-parameter shell model [19]. The main difference between the two calculations is that in the semiinfinite lattice the bulk bands become continuous (grey regions). Essentially all surface modes found in the slab calculation are also appearing in Fig. 8. There are of course small discrepancies due to the differences in the dynamical models and parameters, as well as to the fact the Green's function method is after all an approximation, but the overall agreement is very good. This method, however, has the advantage of giving the dispersion relations of resonances (broken lines): for instance, the resonance S_8 crossing the acoustic bands along both $\bar{\Gamma}\bar{X}$ and $\bar{\Gamma}\bar{M}$ directions. On the other hand no resonance corresponding to the Fuchs and Kliever macroscopic mode is found in the long wave limit. The deep penetration of this mode diverges as K → 0, giving a vanishing amplitude at the first surface layer. Actually no resonant structure associated with FK mode occurs in the K = 0 surface-projected sagittal phonon density (Fig. 9).

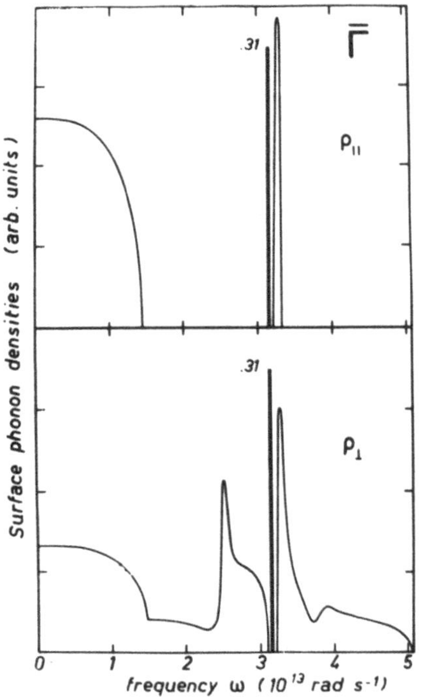

Fig. 9. Surface-projected phonon densities of NaCl(001) at the symmetry point $\bar{\Gamma}$.

Sagittal modes normally have elliptical polarization, but at the high-symmetry points $\bar{\Gamma}$ and $\bar{M}$ (C_{4v}) they become linearly polarized either along z or along x. But the x-mode is indistinguishable from the || mode polarized along y, so that at $\bar{\Gamma}$ and $\bar{M}$ all || surface modes must be degenerate with some x-polarized $\perp$ mode, each mode pair transforming like the two components of the irreducible representation E. At $\bar{\Gamma}$ the degenerate pairs are the Lucas modes S_4 and S_5, and the acoustic modes S_1 (Rayleigh wave) and S_7 (the so-called shear-horizontal mode discovered by Alldredge [25]). At $\bar{M}$ the $\parallel$ local mode S_5 meets the $\perp$ resonance S_2. Of course S_2 becomes a local mode at $\bar{M}$. This happens because the sagittal band which S_2 belongs to is polarized along z at the $\bar{M}$ point: thus we have a resonance which becomes a pseudo-surface mode not along a symmetry direction but only at isolated points of the SBZ! [26].

The second example is LiF(001). The bulk bands (Fig. 10), now reproduced separately for $\perp$ and $\parallel$ polarizations only along ΓM, are delimited by certain bulk dispersion curves (thin lines). The comparison of these curves with the available neutron data (black points) illustrates the quality of the BSM, here used with room temperature data. Very recently, thanks to the great progress in the nozzle beam technique, Brusdeylins, Doak and Toennies [27] succeded in measuring the dispersion curve of Rayleigh waves by means of inelastic scattering of He atoms. Rayleigh waves, more than any other surface modes, give sharp peaks in the time-of-flight spectra of the scattered atoms, which allows for a precise determination of phonon energy and momentum.

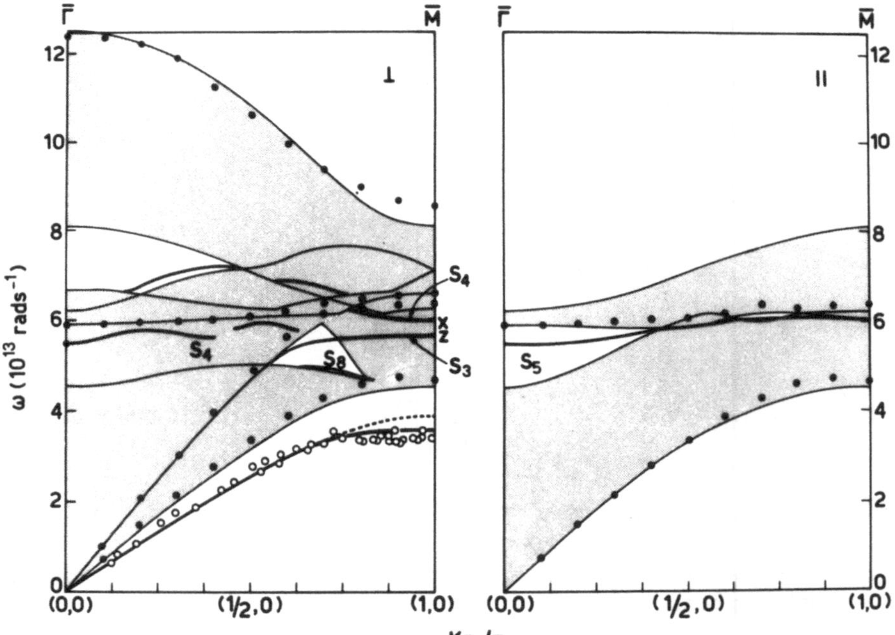

Fig. 10. Surface phonon dispersion curves of LiF(001) along (100) calculated by the Green's function method (room temperature data).

The experimental points (open circles) are quite in a good agreement with the dispersion curve calculated by either the Green's function method and BSM [26] (heavy line) or the slab method and an 11-parameter shell model [19] (broken line). The only small discrepancy, more pronounced for the slab calculation, occurs at the $\bar{M}$ point, and is probably removed if the effects of the surface change in ionic polarizabilities and/or anharmonicity are taken into account.

Besides RW dispersion curve, time-of-flight (TOF) spectra of the scattered atoms show something similar to the projected density of sagittal modes $\rho_{\perp}(K,\omega)$ with K and ω related through the experimental kinematical condition. Obviously, the projected density is modulated by the scattering amplitude (which is different for the various modes) as well as by the K-dependent Debye-Waller and ω-dependent Bose factors. When all these things are considered, the calculated phonon densities compare quite well with the observed spectra [28] . Fig.11 shows such a comparison with a spectrum taken

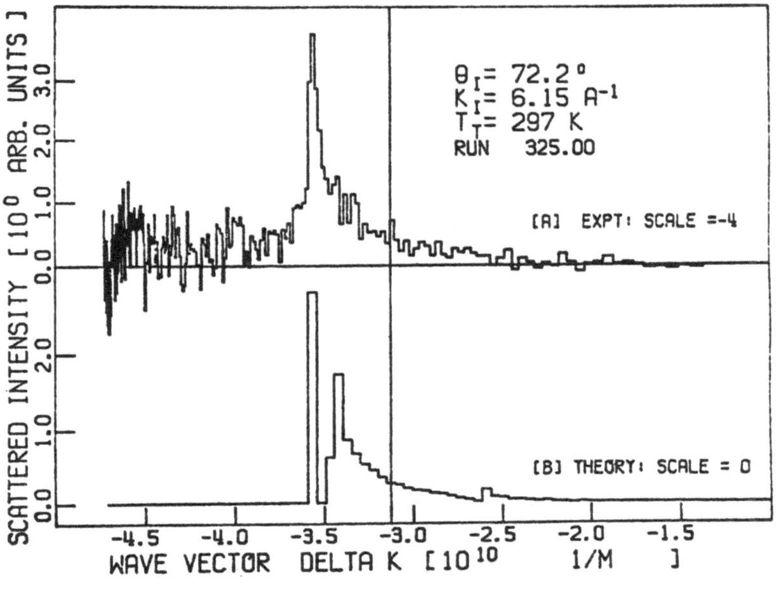

Fig. 11. Time-of-flight spectrum of He^4 atoms scattered from LiF(001) in the (xz)-plane with 90° geometry (incidence angle = 72.2°) (from refs. [27] and [29] compared with calculated energy-loss spectrum due to one-phonon inelastic processes (from ref. [28] in the framework of a hard corrugated surface model and breathing shell model for LiF bulk dynamics.

in the 90° scattering configuration for a small outgoing angle (17.8°). Above the sharp peak corresponding to RW, there are other features coming from the projected denstity of bulk modes which compare quite well with the experiment.

Unfortunately LiF has too high frequencies compared to the energy of the incident atoms and the optical surface modes are not detected. For this reason Doak et al. [29] have started experiments on KCl(001), whose maximum frequency is of the order of the incident atom energy and the RW intensity is weak compared to LiF. Indeed some weaker structures found in TOF spectra correspond to energy transfers above the RW frequencies. The experimental points (black points in Fig.12) fall all quite close to sagittal dispersion curves.

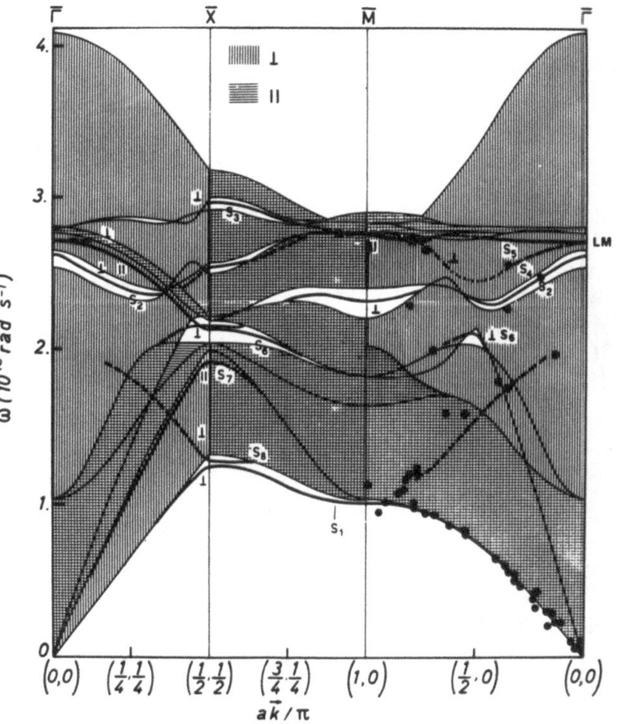

Fig. 12. Calculated surface phonon dispersion curves of KCl(001) by the Green function method using the breathing shell model and 0 K input data. Experimental points from [29].

Quite interesting are the few points along the dispersion curves of S_4 and S_8 modes, which could well be the first evidence of Lucas modes and of surface resonances, respectively. Indeed the resonance S_8 has quite a large intensity all over the SBZ. This appears from Fig.13 which shows the total projected phonon density at a point close to $\bar{X}$.

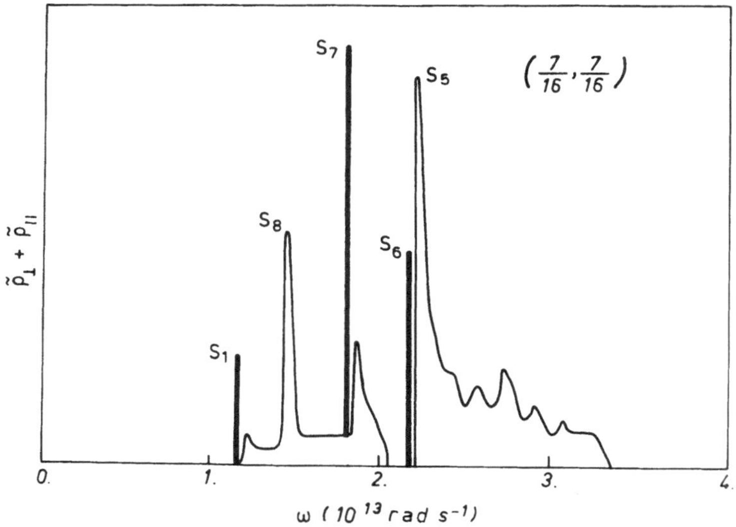

Fig. 13. Total surface projected phonon density for K=(7/16,7/16) in KCl(001) showing the sharp resonance S_8 of sagittal polarization.

ACKNOWLEDGEMENT

I am much indebted to Prof. Peter Toennies and to Drs. Bruce Doak and Guido Brusdeylins (Max-Planck-Institut für Strömungs-froschung, Gottingen) for several useful discussions.

REFERENCES

1. W. Ledermann, Proc. Roy. Soc. A182, 362 (1944).
2. G. W. Farnell, Physical Acoustics 6, 109 (1970).
3. A. A. Maradudin, R. F. Wallis and L. Dobrzinski, Handbook of Surfaces and Interfaces, Vol, 3: Surface Phonons and Polaritons (Garland STPM Press, New York, 1980).

4. A. Hartstein, E. Burstein, A.A. Maradudin, R. Brewer and R.F. Wallis, J. Phys. C: Solid State Physics 6, 1266 (1973).
5. F. Garcia-Moliner, Ann. Phys. (Paris) 2, 179 (1977).
6. R. Fuchs and K.L. Kliever, Phys. Rev. 140, A2076 (1965); and K.L. Kliever and R. Fuchs, Phys. Rev. 144, 495 (1966); 150, 573 (1966).
7. T.S. Chen, F.W. de Wette and G.P. Alldredge, Phys. Rev. B. 15, 1167 (1977).
8. A. Otto, Z. Phys. 216, 398 (1968).
9. V.V. Bryskin, Yu. M. Gerbshtein and D.N. Mirlin, Fiz. Tverd. Tela 13, 2125 (1972) (Sov. Phys. Sol. State 13, 1779 (1972)); and ibidem 14, 543, 3368 (1972) (Sov. Phys. Sol. State 14, 453, 2849 (1972).
10. D.J. Evans, S. Ushioda and J.D. McMullen, Phys. Rev. Letters 31, 369 (1973).
11. H. Boersch, J. Geiger and W. Stickel, Phys. Rev. Letters 17, 379 (1966).
12. H. IBach, Phys. Rev. Letters 24, 1416 (1970).
13. G. Benedek, Surface Scil 61, 603 (1976).
14. R.F. Wallis, Rendiconti SIF, Course LII (Academic Press, New York and London, 1972).
15. T. E. Feuchtwang, Phys. Rev. 155, 731 (1967).
16. V. Bortolani, F. Nizzoli and G. Santoro, Proc. Int. Conf. on Lattice Dynamics, M. Balkanski, ed. (Flammarion, 1978).
17. A.A. Lucas, J. Chem. Phys. 48, 3156 (1968).
18. R.E. Allen, G.P. Alldredge and F.W. de Wette, Phys. Rev. B4, 1648 (1971); 1661 (1971); 1682 (1971).
19. Ref. 7. See also T.S. Chen, G.P. Alldredge and F.W. de Wette Solid State Comm. 10, 941 (1972).
20. I.M. Lifshitz and Lm. Rozenzweig, Zh. Eksp. Teor. Fiz. 18, 1012 (1948); I.M. Lifshitz, Muovo Cim. Suppl. 3, 732 (1956).
21. A.A. Maradudin, E.W. Montroll, G.H. Weiss and L.P. Ipatova, Solid State Physics Suppl. 3 (2nd edition, 1971).
22. W. Ludwig and B. Lengeler, Solid State Comm. 2, 83, (1964).
23. S. Doniach and E.H. Sondheimer, Green's Functions for Solid State Physicists (Benjamin Inc., 1974).
24. G. Benedek, Phys. Stat. Solidi B. 58, 661 (1973).
25. G.P. Alldredge, Phys. Letters 41A, 281 (1972).
26. In my previous Green's function calculations (ref. 13) the $\bar{M}$ point degeneracy was not verified. Recently I could discover a computational error affecting the calculation in a small region around $\bar{M}$, which was causing the misfit. An erratum is to be published on Surface Science.
27. G. Brusdeylins, R.B. Doak and J.P. Toennies, Phys. Rev. Letters 46, 437 (1981).
28. G. Benedek and N. Garcia, Surf. Sci. 103, L143 (1981).
29. R.B. Doak, J.P. Toennies and G. Brusdeylins, private communication; R.B. Doak, Thesis (M.I.T., 1981; unpublished).

COLLECTIVE EXCITATIONS IN CONCENTRATED

Mn^{2+} SYSTEMS: SPECIAL PROPERTIES*

D. P. Pacheco

Department of Physics, Boston College
Chestnut Hill, MA 02167, U. S. A.

ABSTRACT

The spectroscopic properties of such Mn^{2+} systems as MnF_2, $RbMnF_3$, and $KMnF_3$ have been widely investigated over the past fifteen years or so. An important step in the understanding of these systems has been the realization of the role of collective excitations in determining these properties. In this article, the relationship between collective excitations and spectral features is explored. We begin with a description of the Mn^{2+} ion in a crystal and a discussion of its participation in collective excitations. Next, we briefly consider the magnetic properties of the crystals of interest here. Subsequent sections deal with excitons, magnons, and to a lesser extent phonons with particular attention to their effects on the optical properties of concentrated Mn^{2+} systems.

I. INTRODUCTION

In an earlier article in this volume, Di Bartolo discussed the general idea of collective excitations in solids and developed the formalism for describing their interactions with light and with each other. The aim of this article is to particularize these considerations to the special case of concentrated Mn^{2+} systems. The motivation for this lies in the fact that such compounds present the investigator with the opportunity to study several kinds of collective excitations in the same crystal: excitons, magnons, and phonons. The magnons enter the picture because these systems become antiferromagnetic at sufficiently low temperatures.

*Supported in part by NATO Research Grant Nol 1169.

In the past 15 years, considerable attention has been devoted to the optical properties of concentrated Mn^{2+} systems. MnF_2, $RbMnF_3$ and $KMnF_3$ are perhaps the best known in this regard. Two important spectral features have contributed to this interest. (1) Despite the high concentration, luminescence is observed from the Mn^{2+} system of ions. (Some of this Mn^{2+} emission is intrinsic, but most of it is due to the presence of crystal impurities.) Consequently, luminescence as well as absorption studies may be used to probe the characteristics of the collective excitations. The luminescence data provide important complementary information, as will be seen later. (2) Below certain temperatures, these compounds present "additional" spectral lines, both in absorption and in emission. Many of these lines can be related to electronic and magnetic collective excitations.

In this article, we would like to examine the extent to which the optical properties of concentrated Mn^{2+} systems can be explained by using a collective-excitation description. The organization of the paper is as follows. We will first discuss the basic properties of the Mn^{2+} ion in a crystal and its participation in collective excitations. This is intended to set the stage for the ensuing discussions. Next, we will briefly consider the magnetic properties of the crystals of interest here. Subsequent sections deal with excitons, magnons, and to a lesser extent phonons with particular attention to their effects on the optical properties of concentrated Mn^{2+} systems.

II. THE Mn^{2+} ION IN A CRYSTAL

II.A. Basic Properties

The Mn^{2+} ion has a configuration of the type (Ar core)^{18} $3d^5$, in which the half-filled 3d shell is the outermost. Due to the lack of shielding, the 3d electrons are sensitive to the environment of the ion. The crystal-field splittings of the free-ion levels are typically $\sim 10^4 cm^{-1}$. This may be contrasted with a value of $\sim 350\ cm^{-1}$ for the spin-orbit parameter. For the materials of interest here (e.g., MnF_2 and $RbMnF_3$) the crystal field is predominantly of octahedral symmetry, and so, as a first approximation, the energy levels may be labelled by the irreducible representations of the octahedral group. A schematic energy-level diagram for Mn^{2+} in such an environment is given in Figure 1. The noncubic components of the crystal field, the spin-orbit coupling, and the magnetic interactions produce a fine splitting of these levels. An interesting feature of Mn^{2+} is the isolated $^6A_1$ ground state, which is an orbital singlet. Because it is a pure spin state, it is unaffected by the crystal field. In addition, excitations within the $^6A_1$ multiplet are pure spin excitations. This makes the concept of a magnon as a spin wave a good one for concentrated Mn^{2+} systems. (In FeF_2 and CoF_2, for example, the free-ion ground

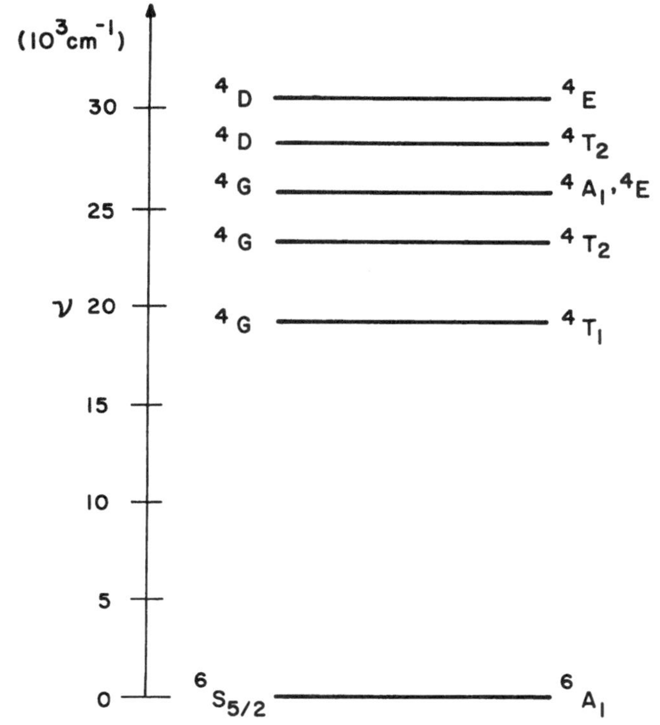

Fig. 1 Energy-level scheme of Mn^{2+} in MnF_2.

states are ^5D and ^4F, respectively. In a crystal field, these states split into a number of low-lying levels. As a result, the low-energy excitations contain an orbital as well as a spin character.)

Because the Mn^{2+} ions reside in an environment with inversion symmetry, parity is a good quantum number for the energy states. The parity of the $3d^5$ levels is even, since $P=(-1)^{\Sigma_i l_i}$, where l_i represents the orbital quantum number of the ith electron. This fact indicates that the purely electronic transitions within the $3d^5$ configuration are electric dipole-forbidden.

For our current purposes, we will focus attention on the $^6A_{1g} \leftrightarrow {}^6A_{1g}$ and $^6A_{1g} \leftrightarrow {}^4T_{1g}$ (^4G) transitions. (The "g" refers to the even parity.) This will allow for a discussion of the collective excitations without getting into unnecessary complications. Figure 2 gives a more detailed view of the energy levels involved. Note that the two lowest components of $^4T_{1g}$ are labelled E1 and E2. These sublevels will be of importance in our discussions of excitons. They are separated by about 17 cm^{-1} in energy and have been shown to be predominantly m_s = 3/2 in character [1].

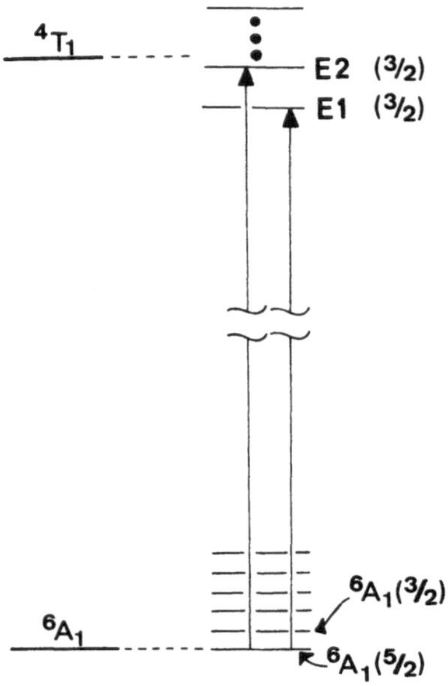

Fig. 2 Some of the components of the $^4T_{1g}$ and $^6A_{1g}$ states of the Mn^{2+} ion.

A few words about these transitions are in order at this point. As mentioned above, the materials of interest become antiferromagnetic at low temperatures. The resulting exchange field splits the $^6A_{1g}$ level into six components. Excitaitons within this multiplet give rise to the quantized spin waves, or magnons. The energies of these magnons are ∿ tens of cm^{-1}. The $^4T_{1g} \rightarrow {}^6A_{1g}$ transitions (luminescence) typically span the 14,000-18,000 cm^{-1} region, if we include all the sideband structure.

It is of note that the $^4T_{1g} \leftrightarrow {}^6A_{1g}$ transitions are both spin- and electric dipole-forbidden. The former is relaxed by the spin-orbit coupling between the two states. That is, $^6A_{1g}$ $(m_s=5/2)$ is not the true ground state of the Mn^{2+} ion. A better approximation is given by:[1]

$$|g> \cong |{}^6A_{1g}(5/2)> + \left(\frac{\sqrt{2}\xi}{\Delta E}\right)|\ {}^4T_{1g}\ (1,\ 3/2)>, \quad (1)$$

where ξ is the spin-orbit parameter (∿400 cm^{-1}) and ΔE is the energy separation between the two electronic levels (∿18,400cm^{-1}). For the current situation, the admixture of the $^4T_{1g}$ state into $|g>$ is only ∿0.03. The fact that this mixing is small is important in helping to explain why the purely electronic (exciton) lines are very weak, and in clarifying the nature of the mechanism responsible for the magnon sidebands (as will be discussed later).

II.B. Collective Excitations in Mn^{2+} Systems

For the current purposes, we are concerned with the optical properties of the tightly bound 3d electrons. In the previous section, several simplifications were made in discussing the Mn^{2+} energy-level scheme. Among these are (1.) that the neighboring ions are the source of static electric and magnetic fields at the ion sites, (2.) that the neighboring ions can be treated as point charges (monopoles), and (3.) that the resonance among the Mn^{2+} ions can be neglected. Such a treatment gives accurate values for the splitting of the free-ion levels, but it fails to account for the dynamical properties of the system. (It can readily be shown that a perturbation which produces a very small splitting of energy levels can nonetheless have a dramatic effect on energy delocalization [2].

Relaxation of the above conditions leads to a description in which the energy eigenstates span the entire Mn^{2+} system (i.e., are delocalized). The familiar excitons and magnons arise from such a description. The formalism for the collective-excitation picture has already been developed in Di Bartolo's article. We intend here to give a briefer treatment and indicate the application to concentrated Mn^{2+} compounds. In pursuing this, we will make use of the fact that for these materials the interionic interactions are much smaller than the intraionic ones. As a result, the Frenkel (tight-binding) description can be used for the excitons and magnons.

II.C. Antiferromagnetism in Mn^{2+} Systems

The Mn^{2+} compounds that have received the most attention are the manganese fluorides. These include MnF_2 (with the rutile structure), $RbMnF_3$ and $KMnF_3$ (perovskites) and $CsMnF_3$ (hexagonal system). These crystals become antiferromagnetic below ordering temperatures ∿50-90K. (See Table 1 for a summary of some important magnetic properties for these crystals.) The interionic interactions in manganese fluorides are well described by the Heisenberg exchange. The J values included in the table arise from a fitting process based on such a treatment [3]-[6].

The antiferromagnetic character of these compounds is attributed to covalency effects. The magnetic 3d electrons of a given Mn^{2+} ion bond covalently with a nonmagnetic ligand (in this case F^-); as a result, some of the magnetic moment resides on this ligand. In fact, several such magnetic ions surround the ligand and each produces a moment at that site. The interaction among these moments provides the antiferromagnetic superexchange. It may be noted

TABLE 1
MAGNETIC PROPERTIES OF SELECTED
MANGANESE FLUORIDES

CRYSTAL	T_N(K)	No. of sublattices	Exchange coupling parameters between nearest neighbors: 1 nn	2 nn	3 nn
MnF_2	67	2	N=2 d=3.31 Å J_1=0.22cm^{-1}	N=8 d=3.82 Å J_2=-1.22cm^{-1}	N=4 d=4.87 Å $\|J_3\|$<0.04cm^{-1}
$RbMnF_3$	82	2	N=6 d=4.24 Å J_1=-2.36cm^{-1}	N=12 d=6.02 Å J_2=0.0±0.1cm^{-1}	
$KMnF_3$	88 collinear 81 canted	2	N=6 d=4.19 Å J_1=-2.64cm^{-1}	N=12 d=5.93 Å J_2=0.08 cm^{-1}	
$CsMnF_3$	53.5	6	N=1 d=3.01 Å J_1=-3.31cm^{-1}	N=3 d =4.24 Å J_2=-4.47cm^{-1}	N=3 d=5.78 Å J_3=0.46 cm^{-1}

T_N = Néel (ordering) temperature
J>0 ferromagnetic ; J<0 antiferromagnetic
N= number of nearest neighbors

from the table that for MnF_2 both first and second nearest neighbors are needed to describe the magnetic behavior, while for $KMnF_3$ and $RbMnF_3$ only nearest neighbors contribute significantly. This is apparently due to the crystal structure. In the perovskites, the second-nearest neighbors have to couple through two nonmagnetic ions and this is expected to be relatively weak.

Except for $CsMnF_3$, the compounds listed in the table can be regarded as consisting of two interpenetrating sublattices in the ordered state. (These sublattices are often designated as "spin-up" and "spin-down.") Figure 3 shows the structure for MnF_2. $CsMnF_3$ is a more complicated system and will not be discussed here. Moncorgé and Jacquier consider some of the properties of this crystal in the following article.

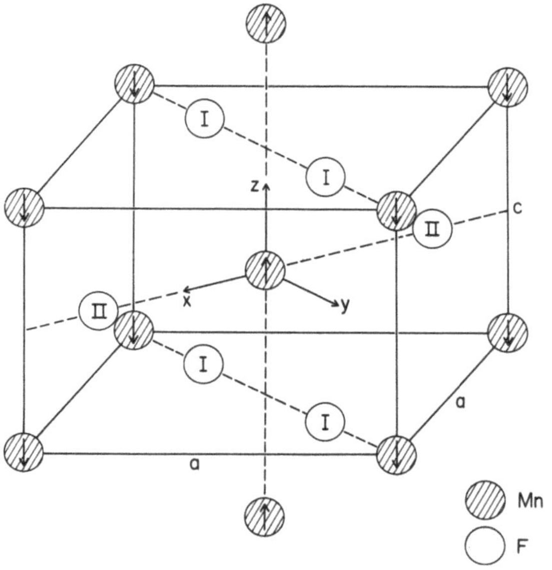

Fig. 3 Crystal structure of MnF_2 including spin arrangement.

III. EXCITONS IN CONCENTRATED Mn^{2+} SYSTEMS

In Di Bartolo's article, a simple Frenkel-exciton model was presented which considered only two ionic states: a ground and one excited state. This treatment is highly simplified in that there are in general many excited states for a given ion, and these may mix. For some states, the energy separations are large compared with the interionic coupling. In these instances the mixing is small and can be neglected. For the other situations, the mixing is often ignored to make the problem tractable.

For the Mn^{2+} ion, we can begin by constructing sublattice excitons in the usual fashion:

$$\psi_\alpha(\vec{k}) = N^{-\frac{1}{2}} \sum_{\vec{R}_h}^{(\alpha)} e^{i\vec{k}\cdot\vec{R}_h} \psi_\alpha(\vec{R}_h) \quad , \tag{2}$$

where the notation follows that of Di Barolo's article. The index α = 1,2 and refers to the two sublattices. In our subsequent discussions, $\Psi_\alpha(\vec{R}_h)$ will contain either E1 or E2 as the excited single-ion state (recall Fig. 2). The first-order energies of the excitons are given by:

$$\varepsilon_\alpha(\vec{k}) = \Delta E_o + \sum_{\vec{R}_h \neq \vec{0}}^{(\alpha)} e^{i\vec{k}\cdot\vec{R}_h} H'_{o\vec{R}_h} \tag{3}$$

where ΔE_o contains the k - independent terms.

Strictly speaking, the $\Psi_\alpha(\vec{k})$ in Eq. (2) are still not eigenstates of the Hamiltonian. This is because of nonvanishing off-diagonal terms of the form:

$$<\psi_1(\vec{k})|H'|\psi_2(\vec{k})> = H_{12}(\vec{k}). \tag{4}$$

That is, we have neglected the intersublattice coupling. This turns out to be a good approximation in concentrated Mn^{2+} systems (see below). In general, however, we must construct exciton wavefunctions that span both sublattices. Since $\psi_1(\vec{k})$ and $\psi_2(\vec{k})$ are degenerate, we may write:

$$\psi(\vec{k}) = a\,\psi_1(\vec{k}) + b\,\psi_2(\vec{k}) \tag{5}$$

with an energy given by

$$H\psi(\vec{k}) = \varepsilon(\vec{k})\,\psi(\vec{k}). \tag{6}$$

Since the interion interactions are known to be relatively small, we can use standard degenerate perturbation theory to obtain:

$$\psi_\pm(\vec{k}) = 2^{-\frac{1}{2}}\{\psi_1(\vec{k}) \pm \psi_2(\vec{k})\} \tag{7}$$

$$\varepsilon_\pm(\vec{k}) = \varepsilon_1(\vec{k}) \pm H_{12}(\vec{k}).$$

This result shows that there are in fact two $\vec{k}$-dependent exciton bands separated by an energy $2H_{12}(\vec{k})$. This is of course due to the presence of two inequivalent magnetic sites per unit cell. The energy separation between the two bands for $\vec{k} = \vec{0}$ is called the Davydov splitting.

As mentioned above, $H_{12}(\vec{k})$ is expected to be small for the systems of interset here. A simple argument may be used to demonstrate this. In the absence of spin-orbit coupling, the transition represented by Eq. (4) involves $|\Delta S_z| = 2$. Since H' (which is electrostatic in nature) does not operate on the spin, the matrix element is expected to be zerol One can obtain a nonvanishing contribution by appealing to spin-orbit coupling, but this reduces $H_{12}(\vec{k})$ by a factor $\simeq|\frac{\xi}{\Delta E_0}|^2$ ($\sim 10^{-3}$ or 10^{-4}) with respect to the intrasublattice term, which should not exceed several cm^{-1} itself [1]. The splitting between bands in concentrated Mn^{2+} compounds should therefore be very small, and is in fact not observed. To a good approximation, then, the excitons in these systems are degenerate sublattice excitons. This is an important feature, because it implies that there are no observable exciton effects for optical ($\vec{k} \simeq \vec{0}$) transitions involving E1 and E2. For example:

(1.) We have seen that there is no measureable Davydov splitting.

(2.) The selection rules are the same as for the single-ion transitions:

$$\langle\psi_1(\vec{k})|\sum_i \vec{\mu}_i|\psi_G\rangle = \frac{1}{\sqrt{N}} \langle\psi_e|\vec{\mu}_o|\psi_g\rangle \sum_{\vec{R}_h} e^{-i\vec{k}\cdot\vec{R}_h}$$

$$= \sqrt{N} \langle\psi_e|\vec{\mu}_o|\psi_g\rangle \delta_{\vec{k}\vec{0}}$$

$$\alpha \langle\psi_e|\vec{\mu}_o|\psi_g\rangle.$$

In the above equations, $|\psi_e\rangle$ and $|\psi_g\rangle$ represent the excited and ground states, respectively, of the single ion; the $\vec{\mu}_i$ are the single-site magnetic-dipole moments.

(3.) One can readily show that the response to an external magnetic field is the same as for the single-ion model.

In short, since we are dealing with sublattice excitons, then for $\vec{k} = \vec{0}$ it doesn't matter whether the excitation resides on one ion or many equivalent ions.

III.A. Exciton Dispersion and Energy Transport

Let us return now to the expression for the sublattice exciton energy (Eq. 3). If we apply this relation to MnF_2 and include interactions up to second-nearest neighbors, we find:

$$\varepsilon(\vec{k}) \simeq \Delta E_0 + 2H'_{01} \cos k_z c. \tag{8}$$

In this equation, c is the nearest-neighbor distance and H'_{01} the nearest-neighbor matrix element. The $\vec{k}$-dependence in equations (3) and (8) gives rise to a band of exciton energies in the crystal. The difference in energy between zone-boundary and zone-center excitons is often referred to as the exciton dispersion. In the somewhat simplified expression given above for MnF_2, the dispersion would be equal to $4H'_{01}$. The algebraic sign of H'_{01} determines whether the dispersion is positive or negative. The matrix element H'_{01} is of the form:

$$H'_{01} = \langle\psi_{e1}\psi_{go}|H'|\psi_{g1}\psi_{eo}\rangle$$

where, as before, ψ_e and ψ_g refer to excited and ground single-ion states. This is the same matrix element which is responsible for the transfer of excitation between nearest-neighbor ions. The size of the exciton despersion, therefore, is a measure of the rate of energy transfer between ions.

The question naturally arises as to the magnitude of the dispersion for E1 and E2 excitons in concentrated Mn^{2+} systems. The epxerimental evidence is contradictory on this point. Dietz et al [7,8] carried out stress studies on MnF_2 and obtained theoretical fits of the observed emission spectrum. They concluded that for both E1 and E2, the dispersion is less than 0.5 cm^{-1} and is probably $\simeq$0.02 cm^{-1}. On the other hand, Amer et al [9] investigated MnF_2 using two-magnon resonant Raman scattering and found evidence for a negative dispersion of 7cm^{-1} for E2. Wilson et al [10], in examining the temperature dependence of the E1 luminescence decay, also obtained results consistent with a negative dispersion for E2 ($\sim$6 cm^{-1}). Thus, although there is general agreement regarding E1, the status of the dispersion for E2 is still up in the air. But why should the dispersion for E1 or E2 be so small? There are two major reasons. (1) The intrasublattice exchange energy is itself small (e.g., 0.22 cm^{-1} for MnF_2). (2) The Franck-Condon overlap integral in H'_{01} reduces the electronic part of the transfer matrix element by a factor of about 100 [7,8].

III.B. Validity of the Wavevector Description

The theoretical development presented above is based on a description in terms of the wavevector $\vec{k}$, which in turn relies on the periodicity of the lattice for its form. In a real crystal, the presence of phonons and lattice imperfections breaks this translational symmetry. To what extent, then, is the wavevector description valid? The answer to this question appears to depend on the experimental conditions as well as the time scale of interest in the system. We can once again take MnF_2 as an example. Dietz et al [8] examined this question by monitoring the decay rates of $k=0$ and $k = 2\pi/3c$ excitons in the vicinity of liquid helium temperatures. In their experiments, they used low pump intensities and stressed samples. The results show unequal decay rates for excitons at these two points in the Brillouin zone over a time scale of several msec. This means that, after creation of an exciton population near $k = 0$, thermalization across the Brillouin zone would take at least several msec to occur. We can compare this with other times of interest. The rate of interion transfer of excitation (f) has been estimated to be $\sim 10^8$ sec^{-1} in MnF_2 [11], and the exciton trapping time for the stressed crystals is $\sim$200 μsec. We then have:

$$f^{-1} << \tau_{trapping} << \tau_{thermalization},$$

for the experimental conditions described by Dietz and co-workers. On the other hand, MacFarlane and Luntz[11] undertook a similar study on MnF_2 using high pump intensities and an unstressed crystal. For these conditions, thermalization time is $\sim$1 μsec. Since the trapping time is $\sim$250 usec for the unstressed sample, we can write:

$$f^{-1} << \tau_{thermalization} << \tau_{trapping}.$$

In both cases, the thermalization time is much longer than the interion transfer time. Whether or not the excitons have lost all phase memory by the time they are trapped depends on the experimental conditions.

IV. MAGNONS IN CONCENTRATED Mn^{2+} SYSTEMS

As mentioned earlier, the ground state for the Mn^{2+} ion in these systems is nearly a pure spin state ($^6A_{1g}$). This state is split into its six components by the exchange and anisotropy fields. The exchange interaction among the Mn^{2+} ions gives rise to a coupled spin system whose quantized excitations are called magnons. (These magnons are actually a type of Frenkel exciton. For clarity, however, the electronic excitations described earlier are referred to as excitons, and the spin excitations magnons.) The theoretical description of antiferromagnetic magnons can be found in many sources; only a brief account will be given here.

IV.A. Basic Theory

In what follows, we will assume that the system may be divided into two interpenetrating sublattices labelled 1 (spins up) and 2 (spins down). For simplicity, we will consider that only the nearest neighbors on the opposite sublattice are important. The number of such neighbors will be labelled Z. We will also assume $T<<T_N$ (to be discussed later) and allow for uniaxial anisotropy. The Hamiltonian for the spin system may then be written:

$$H_s = -2J\sum_{ij}^{nn} \vec{S}_i \cdot \vec{S}_j - \frac{g\mu_B H_A}{2S} \{\sum_i (S_i^z)^2 + \sum_j (S_j^z)^2\}, \quad (9)$$

where i and j refer to ions on sublattices 1 and 2, respectively. J is the appropriate exchange integral (which is negative for an antiferromagnet) and H_A is the effective anisotropy field. The first sum in Eq. (9) is the exchange term, which is electrostatic in origin and occurs as a direct result of the Pauli Exclusion Principle. The second and third terms represent the anisotropy energy. The procedure for diagonalizing H_S usually runs as follows. We first transform to spin-derivation operators by using the Holstein-Primakoff technique:

$$S_i^+ = S_i^x + i\, S_i^y = \sqrt{2S-a_i^+a_i}\; a_i \simeq \sqrt{2S}\; a_i$$

$$S_i^- = S_i^x - iS_i^y = \sqrt{2S-a_i^+a_i}\; a_i^+ \simeq \sqrt{2S}\; a_i^+ \quad (10)$$

$$S_i^z = S - a_i^+ a_i$$

for sublattice 1, with analogous expressions for sublattice 2. We note that a_i^+ creates and a_i destroys a spin deviation on the "up" sublattice. We have used the assumption $T \ll T_N$ in simplifing the square roots in Eq. (10). The next step is to transform to "magnon variables" to take into account the translational symmetry of the lattice:

$$a_i^+ = \frac{1}{\sqrt{N}} \sum_{\vec{k}} e^{-i\vec{k}\cdot\vec{R}_i} a_{\vec{k}}^+$$
$$a_i = \frac{1}{\sqrt{N}} \sum_{\vec{k}} e^{+i\vec{k}\cdot\vec{R}_i} a_{\vec{k}} \quad , \tag{11}$$

with similar relations for sublattice 2. At this point, H_s is still not diagonalized, because we have not included in the intersublattice coupling. (We cannot ignore this for the magnons; the intersublattice coupling is what gives rise to them in the first place). To do this we make the Weyl-Bogoliubov Transformation:

$$a_{\vec{k}} = u_{\vec{k}} \alpha_{\vec{k}} + v_{\vec{k}} \beta_{\vec{k}}^+$$
$$a_{\vec{k}}^+ = u_{\vec{k}} \alpha_{\vec{k}}^+ + v_{\vec{k}} \beta_{\vec{k}},$$

again with analogous expressions for sublattice 2. The specific forms for $u_{\vec{k}}$ and $v_{\vec{k}}$ are related to the crystal structure. They satisfy the equation $u_{\vec{k}}^2 - v_{\vec{k}}^2 = 1$, with $v_{\vec{k}} \to 0$ at the zone boundaries. The diagonalization of H_s is now complete:

$$H_s = \text{constant} + \sum_{\vec{k}} \hbar\omega_{\vec{k}} \left[(\alpha_{\vec{k}}^+ \alpha_{\vec{k}} + \tfrac{1}{2}) + (\beta_{\vec{k}}^+ \beta_{\vec{k}} + \tfrac{1}{2}) \right], \tag{13}$$

where

$$\hbar\omega_{\vec{k}} = 2ZS|J| \left[\left(1 + \frac{g\mu_B H_A}{2SZ|J|}\right)^2 - \gamma_{\vec{k}}^2 \right]^{1/2} \quad ; \tag{14}$$
$$\gamma_{\vec{k}} = \frac{1}{Z} \sum_{\vec{\delta}} e^{i\vec{k}\cdot\vec{\delta}} .$$

The vectors $\vec{\delta}$ join a given ion with its Z nearest neighbors on the other sublattice. At this point, several observations may be made.

(1.) There are two "types" of magnons, one created by the operator $\alpha_{\vec{k}}^+$ and the other by $\beta_{\vec{k}}^+$. The occupation numbers for these magnon states are generated by $\alpha_{\vec{k}}^+ \alpha_{\vec{k}}$ and $\beta_{\vec{k}}^+ \beta_{\vec{k}}$.

(2.) From Eq. (12), it can be seen that each magnon spans both sublattices in general. For zone-boundary magnons ($v_{\vec{k}} \to 0$), the sublattice magnons are normal modes of the system. For interior points of the zone, the α-magnon resides mainly on one sublattice, while the β-magnon is mainly on the other.

(4.) As $\vec{k} \to \vec{0}$, the magnon energy $\hbar\omega_{\vec{k}}$ does not vanish, but rather approaches some finite value ε_{min}. This is due to the anisotropy field H_A.

The above theoretical development is somewhat simplified, due in part to the restriction involving interactions with only one set of near neighbors. For such compounds as MnF_2, $KMnF_3$, and $RbMnF_3$, however, this treatment is quite good. As can be seen from Table 1 (presented earlier), one J value dominates in the description of these three compounds [3]-[5]. For $RbMnF_3$ in particular, only nearest neighbors have a measurable effect.

IV.B. Improvement to the Theory

A second important assumption in the foregoing analysis is that of low tempertures ($T \ll T_N$). Recall that in Eq. (10), we wrote:

$$S_i^+ = \sqrt{2S - a_i^+ a_i}\; a_i \simeq \sqrt{2S}\; a_i .$$

The last step above is strictly valid only for T = 0 K. A better approach would be to expand the square root:

$$S_i^+ = \sqrt{2S}\; \left(1 - \frac{a_i^+ a_i}{4S} - \ldots\right) a_i . \tag{15}$$

The correction terms introduce quartic and higher-order terms into our expression for H_s (magnon-magnon coupling). The result is that the energy of a given magnon mode depends on the thermal occupation of the other modes. As a measure of this effect, we may define a renormalization factor $R_{\vec{k}}(T)$:

$$R_{\vec{k}}(T) \equiv \frac{\omega_{\vec{k}}(T)}{\omega_{\vec{k}}(0)} \tag{16}$$

White [13] has investigated renormalization effects in MnF_2 for various assumptions. If we consider only intersublattice coupling and no anisotropy field, the renormalization factor is found to be:

$$R_{\vec{k}}(T) = 1 - (Z|J|S^2N)^{-1} \sum_{\vec{k}'} \hbar\omega_{\vec{k}'} \langle n_{\vec{k}'} \rangle , \tag{17}$$

where $\langle n_{\vec{k}'} \rangle$ represents the thermal occupation number and has the usual Bose-Einstein form. It is interesting to compare this result with the sublattice magnetization for an antiferromagnet:

$$\frac{M\ (T)}{M\ (0)} = 1 - (NS)^{-1} \Sigma_{\vec{k}'} \ (1 + 2v^2_{\vec{k}'}) \ \langle n_{\vec{k}'} \rangle . \tag{18}$$

Sell et al [1] plotted these two functions for T<30K and found a close correspondence. Therefore, if the temperature is not too high, one can think of the renormalized energies as roughly following the sublattice magnetization curve. This energy shift is of importance in the identification of magnon features (e.g., magnon sidebands) in optical spectra. In fact, this was one of the criteria used to identify the first magnon sideband reported in the literature [14].

V. SPECTRAL FEATURES OF Mn^{2+} SYSTEMS

In the optical region, there are two major features which are related to the presence of excitons and magnons: exciton lines and the associated magnon sidebands. For the sake of specificity, let us consider absorption at low temperatures. Here, the exciton lines represent simply the creation of a population of excitons near $\vec{k} = \vec{0}$ (Figure 4(a)). The magnon sidebands, however, involve the simul-

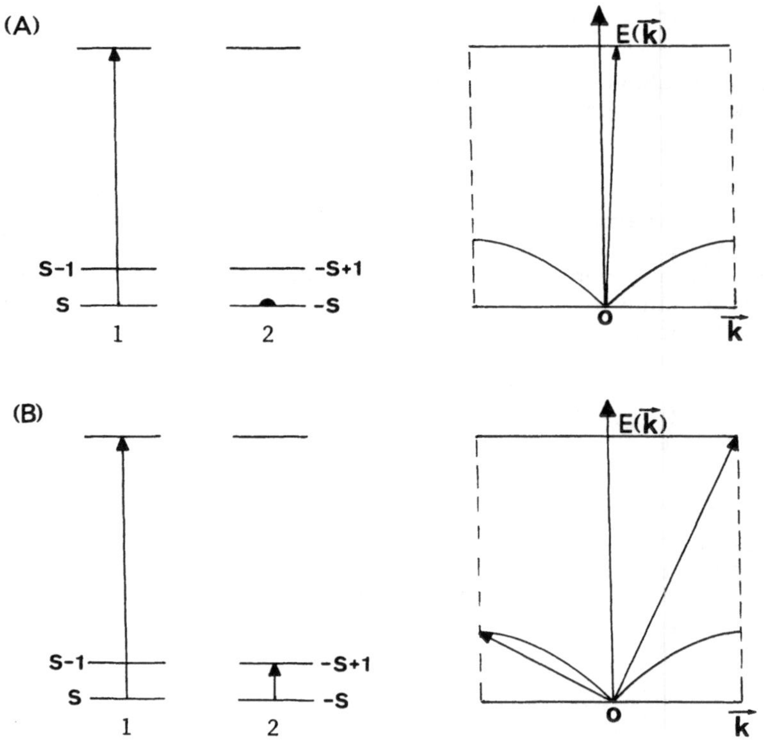

Fig. 4. Schematic representation of an exciton and exciton-magnon transition.

taneous creation of an exciton and a magnon (Figure 4(b)). Since the sideband transition is spin-allowed (see below), the exciton and magnon must appear on opposite sublattices in absorption.

A typical spectrum in absorption for these Mn^{2+} systems is shown in Fig. 5. The lines labelled E1 and E2 are the exciton lines, while σ1 and σ2 denote the magnon sidebands. These sidebands were first identified in MnF_2 by their positions (with respect to the excitons lines), shapes, and temperature behavior (due to renormalization) [14]. An important feature of this spectrum is that the exciton lines are relatively weak. This is because the transitions involved are spin- and electric dipole-forbidden (as discussed in Section II.A). The magnon sidebands, on the other hand, result from pair transitions which are spin- and electric dipole-allowed [1]. To understand this, we have to examine in some detail the basic mechanism responsible for the transition.

V.A. Appearance of Magnon Sidebands

A good discussion of the theory, with special emphasis on MnF_2, can be found in references [1] and [15]. We will present the most important points here. The interaction between the photon electric field $\vec{E}$ and the dipole moments of the electrons in the system can be written as:

$$H_{int} = \vec{E} \cdot \vec{P} , \tag{19}$$

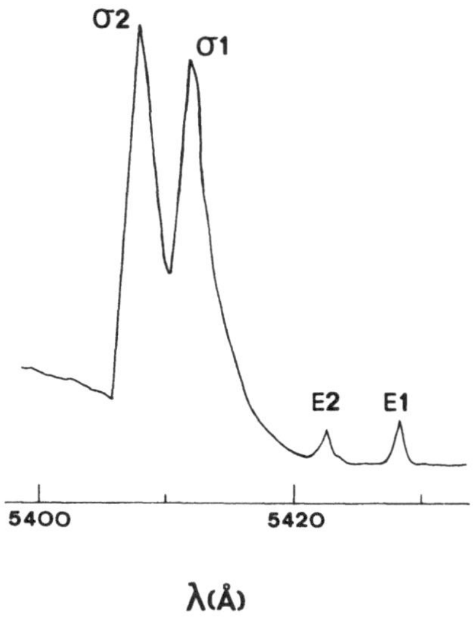

Fig. 5. Exciton lines and magnon sidebands in absorption.

where $\vec{P}$ is the sum of the electric-dipole operators for the electrons. The Hamiltonian of the crystal is then:

$$H = H_o + H' + \vec{E} \cdot \vec{P} \quad . \tag{20}$$

In the above relation, H_o is the sum of all the single-ion Hamiltonians, while H' is the ion-ion Coulomb interaction.

In first-order perturbation theory, the term H_{int} cannot be responsible for the relevant absorption, since it cannot induce the necessary pair transition. Nevertheless $\vec{E} \cdot \vec{P}$ and H', taken in second order, can produce the desired effect. The usual procedure is to construct an effective Hamiltonian which would connect the ground state to a two-particle excited state. The matrix element of the effective Hamiltonian for absorption may be written:

$$< \psi_F | H_{eff} | \psi_G > \quad =$$

$$\Sigma \; \vec{E} \cdot \left\{ \frac{<\psi_F|H'|\psi_I> \; <\psi_I|\vec{P}|\psi_G>}{E_G - E_I + \hbar\omega_{photon}} + \frac{<\psi_F|\vec{P}|\psi_{I'}> \; <\psi_{I'}|H'|\psi_G>}{E_G - E_{I'}} \right\} \tag{21}$$

where ψ_F is the two-particle excited state, ψ_I and $\psi_{I'}$ are intermediate states, and the sum is over the intermediate states. E_G, E_I, $E_{I'}$, are the energies associated with the states ψ_G, ψ_I, and $\psi_{I'}$, respectively.

Based on the form of the matrix element of Eq. (21), several interesting features are apparent. We first not that H' is a two-electron scalar operator and $\vec{P}$ is a sum of one-electron vector operators. Furthermore, let us recall that the initial and final electronic states have definite parity (as is the case in centro-symmetric systems) and belong to the same configuration.

(1) For these conditions, $|\psi_I>$ must be an opposite-parity exciton state, while $|\psi_{I'}>$ is a state consisting of an opposite-parity exciton and magnon.

(2) Conservation of wavevector implies that:

$$\vec{O} \simeq \vec{k}_{magnon} + \vec{k}_{exciton} \tag{22}$$

for these optical transitions.

(3) The interaction H' gives rise to a direct term and an exchange term. Each has its own properties and so must be considered separately. The direct term cannot contribute in the absence of spin-orbit coupling. This can be seen in the following way. The

interaction H' appears in Eq. (21) in two different combinations: $\langle\psi_F|H'|\psi_I\rangle$ and $\langle\psi_{I'}|H'|\psi_G\rangle$. In either case, H' connects a no-magnon state to a one-magnon state. Since these magnon states involve single-ion states with different m_s values and H' does not operate in spin space, these matrix elecments are equal to zero. Spin-orbit couping however, can provide a mixing of states so that the above matrix elements do in fact contribute. The exciton and magnon may be produced on the same or different sublattices with spin-orbit coupling.

(4) Let us now consider the exchange term (in the absence of spin-orbit coupling). The initial state of the system may be represented as:

$$|\psi_G\rangle = |\psi_G\,(\text{exciton});\ \psi_G(\text{magnon})\rangle$$

and the final state as

$$|\psi_F\rangle = |\psi_{\vec{k}_1}\,(\text{exciton});\ \psi_{\vec{k}_2}\,(\text{magnon})\rangle,$$

where $\vec{k}_1$ and $\vec{k}_2$ are the wavevectors of the exciton and magnon, respectively. Since H' and $\vec{P}$ are independent of spin, the total z-component of the spin must be conserved in the interaction. We can consider two cases: (a) exciton and magnon are created on the same sublattice, and (b) exciton and magnon are created on different sublattices.

In case (a), we consider that the interaction produces an optical excitation on one ion of sublattice 1 and a spin excitation on another ion of 1. In the ground state, the total z-component of the spin for these two ions is simply $m_s^{tot} = 2S$. For the final state, the spin-excited ion has $m_s = S-1$ and the value for the other ion may be designated S'. Therefore, the total z-component of the spin si S' +S-1. For the Mn^{2+} ion, we saw in Section II.A that $S' = 3/2$ and $S = 5/2$. Therefore, the required conservation of m_s^{tot} cannot be satisfied for this transition.

For case (b), the exciton is created on one sublattice (for example, 1) and the magnon resides mainly on the other sublattice (2). For the ground state:

$$m_s^{tot} = 0,$$

and for the final state:

$$m_s^{tot} = S' + (-S+1) = S' - S+1.$$

Here, m_s^{tot} is conserved for the transition. The exchange of an electron between sublattices can be used to "match" the states involved in the matrix elements.

In summary, we note that the direct term can contribute to the absorption process only if spin-orbit coupling is included. With such coupling, the exciton and magnon may appear on the same or different sublattices. In the case of exchange, no spin-orbit coupling is needed to produce the desired transition. The exciton and magnon, however, must be produced on different sublattices in order to conserve the total z-component of the spin. Because the spin-orbit coupling is weak in concentrated Mn^{2+} systems, it is the exchange mechanism which produces the sidebands observed in Fig. 5.

(5) From the treatment given in this section, we have seen how the sideband transition can be electri-dipole in nature. The fact that this may occur is a result of the cooperative behavior of the system. The relevant single-ion transition is parity-forbidden (as well as spin-forbidden); for pair transitions these problems can be avoided.

V.B. Sideband Profiles

The sideband shape may be calculated by applying Fermi's Golden Rule and using the matrix elements given in Eq. (21). The relations for absorption $A_p(\omega)$ and emission $E_p(\omega)$ strengths can be put in the form [1,16]:

$$
\begin{aligned}
A_p(\omega) &= \text{constant} \sum_{\vec{k}} |M_p^A(\vec{k})|^2 \, \delta(\omega - \omega_{\vec{k}}^{ex} - \omega_{\vec{k}}^{mag}), \\
E_p(\omega) &= \text{constant} \sum_{\vec{k}} |M_p^E(\vec{k})|^2 \, \delta(\omega - \omega_{\vec{k}}^{ex} + \omega_{\vec{k}}^{mag}),
\end{aligned}
\tag{23}
$$

where $M_p^{A,E}(\vec{k})$ is the total effective electric-dipole moment for the transition. Its dependence on wavevector $\vec{k}$ and polarization p is governed by the symmetry of the initial and final wavefunctions. The δ-function enforces conservation of energy for the absorption or emission process. From the form of Eqs. (23), it can be seen that the sideband profile is a convolution of the exciton and magnon density of states, weighted by $|M_p(\vec{k})|^2$. It is interesting to compare theory with experiment for the magnon sidebands. As a first approximation, we can neglect exciton dispersion as well as the exciton-magnon interaction. In emission, the fitting is excellent. Both Dietz et al [7,8] and Chiang et al [17] have performed a quantitative fitting for the magnon sideband of E1 in MnF_2. In absorption, however, the situation is quite different. A similar closeness of fitting is not possible with the two assumptions above. Sell et al [1] attempted to fit the absorption sidebands in MnF_2 by including exciton dispersion, but no exciton-magnon interaction. They were not, however, able to find a consistent set of exciton dispersion parameters for both σ and π polarizations. Parkinson, Loudon and co-workers [18,19] took the approach of neglecting exciton dispersion, but including exciton-magnon interaction. By adding an appropriate interaction term to H_{TOTAL} and using Green function methods, they were able to obtain a reasonable fit to the observed magnon sideband in $RbMnF_3$.

One can understand these results in the following way. In emission, only E1 is observed because of the Boltzmann population distribution. As described in Section III.A, there is no measured dispersion for this exciton, so it may be safely neglected. The emission sideband at low temperatures involves the destruction of an exciton and the creation of a magnon. They therefore, do not coexist in time and would not be expected to interact strongly. It is not surprising, then, that theory and experiment agree under assumptions of no dispersion or interaction. Since the magnon density of states is weighted towards zone-boundary $\vec{k}$ values, the separation between the exciton line and magnon sideband is approximately equal to a zone-boundary magnon energy. In absorption, the exciton and magnon are simultaneously created in close proximity. Their mutual interaction in this case may be significant and is apparently responsible for the problems encountered in fitting absorption data [1,17]. The overall effect of this interaction is to broaden the sideband and shift it to lower energy [18]-[20].

VI. PHONON SIDEBANDS

Because of their relative familiarity, phonons have not been mentioned much to this point. In Section III. A, it was pointed out that Franck-Condon factors reduce the excitation transfer matrix element by about two orders of magnitude. This is of course due to vibronic coupling. A well-known spectral feature resulting from this coupling is the phonon sideband, which is discussed at length in many sources. In the absorption spectrum of Figure 5, these sidebands would appear on the high-energy side of the magnon sidebands. Figure 6 shows a luminescence spectrum of MnF_2 dominated by broad phonon sidebands. The sharp lines lying immediately to the high-energy side (the "zero-phonon" lines) are not the exciton lines described earlier, but rather are impurity-perturbed emission lines characteristic of concentrated Mn^{2+} systems. The phonon sidebands appearing in these systems have been discussed at length in a previous paper [21], so no attempt will be made to duplicate it here. The interested reader is referred to this paper and also reference [22] for details.

VII. CONCLUDING REMARKS

In this article, we have examined various collective excitations and have shown how they may be used to gain a better understanding of the spectral properties of concentrated Mn^{2+} systems. Due to the

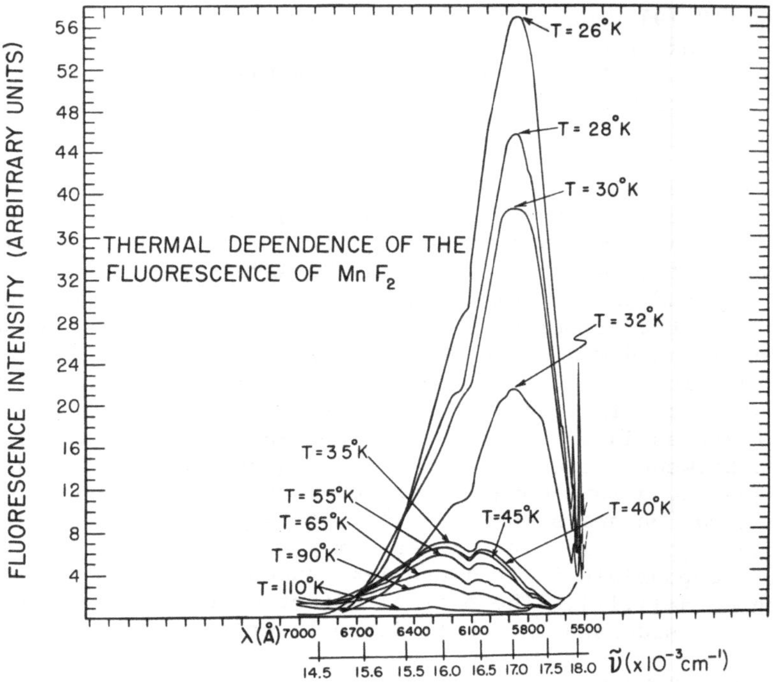

Fig. 6 Phonon sidebands in MnF_2

inevitable space limitations, however, many interesting topics have gone untouched or have been mentioned only in passing. These include two-magnon processes, exciton-exciton interactions and biexciton decay, effects of external perturbations (magnetic fields and stress), nonstoichiometric systems, and a number of others. We have attempted to give a basic introduction to the subject and not to be exhaustive. In the following paper, Moncorgé and Jacquier will take up some of the topics passed over here. We hope that the reader has gained some insight into the role of collective excitations in understanding these systems as well as a feeling for some of the questions that remain unanswered.

REFERENCES

1. D. D. Sell, R. L. Greene, and R. M. White, Phys. Rev. 158, 489 (1967).
2. D. P. Pacheco and B. Di Bartolo, in Luminescence of

Inorganic Solids (B. Di Bartolo, ed.), Plenum Press, New York, 1978, p. 295.
3. K. C. Turberfield, A. Okazaki, and R. W. H. Stevenson, Physics Letters 8, 9 (1964); Proc.Phys. Soc. 85, 743 (1965).
4. C. G. Windsor, and R. W. H. Stevenson, Proc. Phys. Soc. 87, 501 (1966).
5. S. J. Pickart, M. F. Collins, and C. G. Windsor, J. Appl. Phys. 37, 1054 (1966).
6. D. Khatamian and M. F. Collins. Can. J. Phys. 55, 773 (1977).
7. R. E. Dietz and A. Misetich, in Localized Excitations in Solids (Plenum Press, New York, 1968), p. 366.
8. R. E. Dietz, A. E. Meixner, H. J. Guggenheim, and A. Misetich, J. Luminescence 1,2, 279 (1970).
9. N. M. Amer, T. Chiang, and Y. R. Shen, Phys. Rev. Letters 34, 1454 (1975).
10. B. A. Wilson, W. M. Yen, J. Hegarty, and G. F. Imbusch, Phys. Rev. B19, 4238 (1979).
11. R. M. Macfarlane and A. C. Luntz, Phys. Rev. Letters 31, 832 (1973).
12. See, for example, reference 2 above. Other sources include: C. Kittel, Quantum Theory of Solids, John Wiley & Sons, Inc., New York, 1963, p. 58; V. Jaccarino, in Magnetism, Vol. II. A. (G. T. Rado and H. Suhl, eds.), Academic Press, New York, 1963.
13. R. M. White, Physics Letters 19, 453 (1965).
14. R. L. Greene, D. D. Sell, W. M. Yen, A. L. Schawlow and R. M. White, Phys. Rev. Letters 15, 656 (1965).
15. D. D. Sell, Ph.D. Thesis, Standord University, 1967 (unpublished)

OPTICAL DYNAMICS IN CONCENTRATED Mn^{2+} SYSTEMS *

R. Moncorgé and B. Jacquier

Université Lyon I (Bât. 205), E.R. N° 10 CNRS
69622 VILLEURBANNE, France

ABSTRACT

At very low temperatures, below the Neel point T_N, the optical spectra of the antiferromagnetic manganese fluoride systems exhibit fine structures which are related to both electronic and magnetic collective excitations. The experimental conditions, the optical pumping characteristics, the lattice temperature, the application of external factors such as an uniaxial stress or a magnetic field, have been used extensively in the way of identifying these transitions. More recently, these experimental conditions have been found to be of great interest in the study of dynamical properties.

I. INTRODUCTION

The luminescence of the concentrated magnetic insulators, in particular, that of the manganese fluoride antiferromagnets, shows that, in spite of an abundant literature on the subject, at least two questions of fundamental interest still remains :

1) - In a magnetic crystal such as MnF_2 and $AMnF_3$ - with A=K,Rb,Cs - numerous experimental results cannot be fully interpreted without including a non negligible effect of the exciton dispersion. However to our knowledge, no direct measurement of it has been achieved so far. As a consequence, the applicability of the exciton picture is still a matter open to discussion, specially when we are considering the fluorescence dynamics, since the rate of the excitation transfer between ions is proportionnal to the size of the exciton dispersion.

*Supported in part by NATO Research Grant No. 1169.

2) In the three dimensional systems, such as the above mentionned manganese fluorides, the same radiative lifetime is generally assumed for the unperturbed Mn^{2+} bulk ions, the impurity induced manganese traps and the broad phonon sidebands [1] ; usually, this radiative lifetime is of few tens of milliseconds for an effective exciton decay time constant of the order of one millisecond. On the other hand, if the dimensionality of the system is lower, the exciton localization is facilitated, as a result of which the intrinsic emission intensity and decay time should increase accordingly. This is what happens partially in the one dimensional compounds, manganese chlorides TMMC [2] and CMC [3] for example, in which due to the quasi complete exciton localization at low temperature, the observed fluorescence turns out to be essentially intrinsic. However, the lifetime of this intrinsic fluorescence is still of the order of one millisecond in good agreement with a value calculated from the oscillator strength of the transition as reported by D.S. McClure in TMMC [2]. Thus, one may ask again what is the actual radiative lifetime in the three dimensional systems. This question is to be related to recent findings according to which the nonexponential exciton decay mode at very low temperature could be due predominantly to an intrinsic de-excitation process [4] instead of the familiar fast diffusion and trapping mechanism [5].

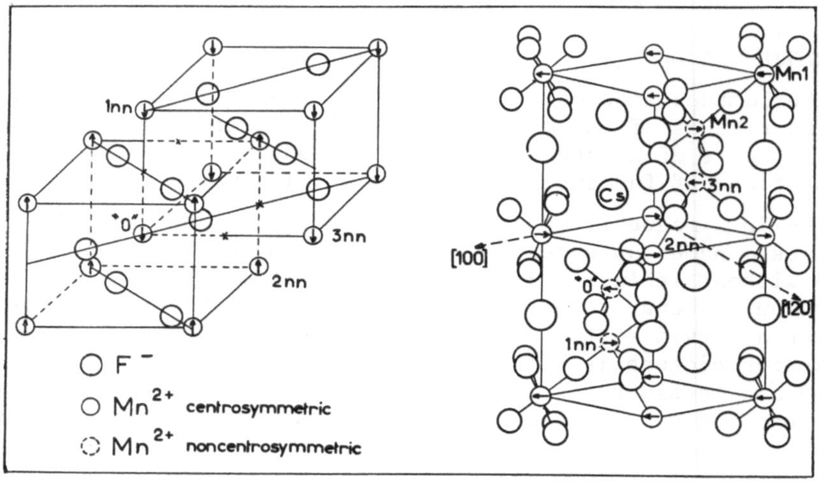

Fig. 1. Crystallomagnetic structures of MnF_2 (left) and $CsMnF_3$ (right).

The purpose of this article is threefold : 1) to give a general aspect of the optical properties of concentrated manganese fluorides at very low temperatures 2) to present some particular results in absorption taking into account directly the collective character of the electronic and magnetic excitations as well 3) to review the most important attempts performed in fluorescence in order to elucidate the above mentionned questions. A particular attention is addressed to $CsMnF_3$ which, due to its crystal structure, presents some interesting features not seen in the other systems.

II. GENERAL ASPECTS OF THE OPTICAL PROPERTIES OF MANGANESE FLUORIDE SYSTEMS

II.A. Crystal Structure

Among the manganese fluorides, the three dimensional systems MnF_2 and $AMnF_3$ (A=K,Rb,Cs) are very attractive for they are easily grown and they provide a wide range of crystallographic structures going from pure cubic site symmetry in the case of $RbMnF_3$ to tetragonal and orthorhombic distorted octaedral site symmetries in the case of $KMnF_3$ and MnF_2 respectively. These perovskite structures all lead to centrosymmetric Mn^{2+} surroundings of one type while in the hexagonal $CsMnF_3$ system two types of trigonal sites coexist with one third being centrosymmetric and two thirds being non centrosymmetric - see Fig. 1.

Due to the difficulty in getting good single crystals of reasonable size, single phase crystals of Cs_2MnF_4 have yet to be studied. On the other hand, the two-dimensional manganese fluorides K_2MnF_4, $RbMnF_4$ and $BaMnF_4$ are receiving particular attention due to their magnetic and optical properties. These crystals are mentioned for the sake of completeness but will not be specifically considered on the following study.

All these crystals are known to order antiferromagnetically below a reasonably high Neel temperature - see Table 1 - inducing additional structures in the optical spectra. In the three-dimensional fluorides the overall Heisenberg exchange character of the interionic interactions provides a good description of the magnetic properties.

II.B. Spectroscopic Properties

In addition to the "one-particle" excitons defined as such in the previous lectures because they are made of linear combinations of single ion excitations, "two-particle" transitions are also present. They are phased linear combinations of pair exciations. In the U.V.

Table 1. Magnetic Properties of Some Manganese Fluorides [6]-[10].

	T_N(K)	Nb of sublattices	Exchange coupling parameters J between nearest neighbour ions		
			1 nn	2 nn	3 nn
MnF_2	67	2	N=2 d=3.31Å J_1=.22cm^{-1}	N=8 d=3.82Å J_2=-1.22cm^{-1}	N=4 d=4.87Å J_3=.035cm^{-1}
$RbMnF_3$	82	2	N=6 d=4.26Å J_1=-2.35cm^{-1}	N=12 d=6.02Å J_2=.00±.02cm^{-1}	——
$KMnF_3$	88 collinear 81 canted	2	N=6 d=4.19Å J_1=-2.2cm^{-1}	N=12 d=5.93Å J_2=.08cm^{-1}	——
$CsMnF_3$	53.5	4 or 6	N=1 d=3.01Å J_1=-3.31cm^{-1}	N=3 d=4.24Å J_2=-4.47cm^{-1}	N=3 d=5.78Å J_3=.46cm^{-1}
Rb_2MnF_4	38.5	2	N=4 d=4.23Å J=-4.7cm^{-1}	N=4 d=5.98Å J=+. /	——
K_2MnF_4	45	2	N=4 d=4.2Å J=-5.87cm^{-1}	N=4 d=5.93Å J=+ /	——
$BaMnF_4$	26 2 dim.	2	N=2 d=3.92Å J=- /	N=2 d=4.2Å J=- /	N=4 d=5.74Å J=+ /

$J > 0$ ferromagnetic ; $J < 0$ antiferromagnetic
N = number of neighbors

and the visible regions, these one and two particle transitions give rise to the well-known exciton-magnon and exciton-phonon features, and the two exciton features. "Three particle" excitons may also occur in the spectra but they will not be considered here.

1. One Particle Optical Transition. Since we are dealing with U.V. and visible transitions in manganese compounds, in which the binding of the 3d electrons with a Mn^{2+} ion is much larger than the binding between the ions, the one particle optical excitations are well described in terms of electronic Frenkel excitons, in contrast with the Wannier excitons in the case of semiconductors and dielectrics with a high delectric constant.

Moreover, in manganese fluorides, all the single ion optical transitions of interest are spin forbidden ($\Delta S=1$). As a consequence, the intersublattice interactions (off diagonal terms which measure the transfer of excitation between ions on antiparallel spin sublattices) responsible for the splitting of the exciton bands, are vanishingly small and so no observable Davydov components are expected. On the other hand, the non negligible intrasublattice interactions (diagonal exchange terms measuring the transfer of excitation between ions on the same sublattice) which are typically of the order of few cm^{-1}, should contribute to excited state dispersions of comparable magnitudes. However, though there may be a sizeable exciton dispersion, only k=0 exciton lines can be observed in the absorption spectra since the wavelength of the incident light is much larger than the lattice constants, which allows us to take the wavevector of the photons equal to zero.

Finally one may ask how $^6A_{1g}(^6S) \rightarrow {}^4T_{1g}(^4G)$ transitions can occur since they are spin and parity forbidden in first order by the Laporte's Rule (3d - 3d transitions). The former interdiction is partially lifted due to a slight spin orbit coupling between the ground state and higher excited state components of the $^4T_{1g}(^4P)$. In MnF_2 such an admixture is found of the order of 3 % [11] and very weak exciton lines are to be expected. This leads indeed to the experimentally observed weak σ polarized magnetic dipole excitons E1 and E2. On the other hand, when the crystal is characterized, as in $CsMnF_3$, by the presence of sites without inversion center, strong electric dipole allowed optical lines appear in the spectra as a result of parity mixing with odd states of other configurations [12], $3d^44p$ for example, due to antisymmetric components of the crystal field or due to distortion of the crystal held during vibrations of the nuclei ; this leads to the observation of the π polarized electric dipole excitons A_2 and A_3 in $CsMnF_3$ [13].

2. Two Particle Transitions. Since the theory of two particle excitations, specially in the case of the exciton-magnon and exciton-

phonon transitions, has been developed in great details before, it looks more appropriate for the purpose of this paper to present here a very schematic momentum selection rules when necessary.

i) Two ion processes involving spin and electronic excitations

The exciton-magnon transitions appear in the optical spectra below the magnetic ordering temperature T_N as broad asymmetric satellites of the pure exciton lines with energy separations characteristic of points on the boundary of the first Brillouin zone where the magnon density of states is maximum. Due to the exchange interaction in a pair of adjacent ions they are of exchange induced electric dipole character and the selection rule $\Delta S=0$ is repspected. As we shall see, in addition to their particular lineshapes, the magnon sidebands can be identified very simply through their characteristic temperature induced shifts. Their stress and magnetic field studies are used to attribute them to their associated pure exciton lines and to determine on which particular sublattice the magnon is mainly localized.

The conditions of conservation of the wavevector $k_{ph} \sim 0$ of the incident photon and of the total spin momentum restrict the possibilities to the following four processes as illustrated on Fig. 2 in the case of a two sublattice antiferromagnet :

1) Cold absorption : simultaneous creation of an exciton +k and a magnon -k on antiparallel sublattices. Since they coexist in time the exciton and the magnon may strongly interact and the shape of the magnon sideband, which is a convolution of the exciton and the magnon densities of states, may be significantly distorted [14].

2) Hot absorption : an exciton k is created and a magnon k is annihilated on the same sublattice. In this case the magnon sideband is almost free of any exciton-magnon interaction (E.M.I).

3) Cold emission : an exciton k is annihilated as a magnon k is created on the same sublattice.

4) Hot emission : simultaneous annihilation of an exciton +k and a magnon -k on opposite sublattices.

Processes 3) and 4) have to be related respectively to processes 2) and 1).

The discussion of the exciton-phonon transitions which lead to the characteristic vibrational sidebands will not be developed here because of its strict similarity with the above description.

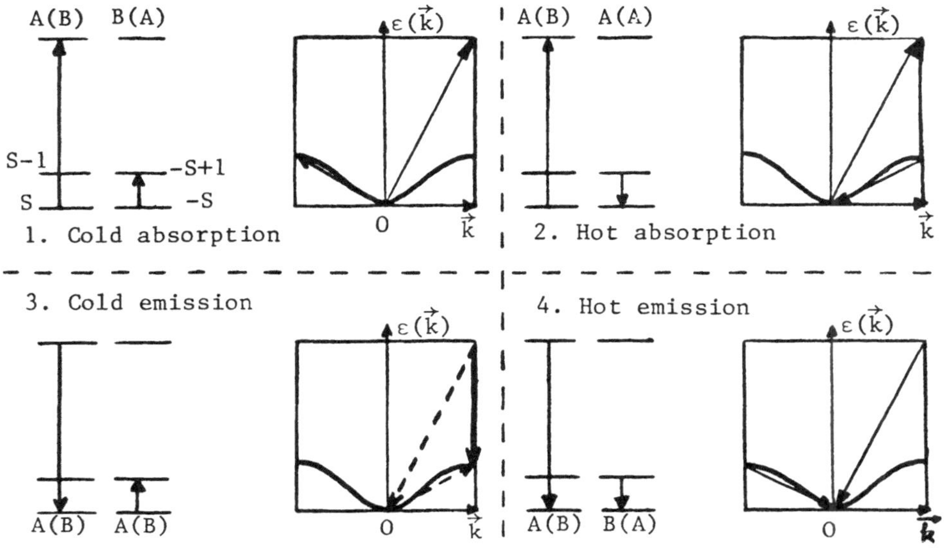

Fig. 2. Exciton magnon transitions

ii) Two ion processes involving optical transitions only

The following exciton pair processes are generally involved.

a) Absorption of one U.V. photon through the creation of two coupled visible excitons. This is the equivalent of cold absorption magnon/phonon sidebands. The incident U.V. photon directly creates single excitons at +k and -k on opposite sublattices, which interact in pairs to form an U.V. exciton of about twice their energy. Such an U.V. absorption has been reported in MnF_2 and $RbMnF_3$ [15] . As in the case of magnon/phonon sidebands, the transition is again of exchange induced electric dipole character ; moreover its energy may be strongly affected by the exciton-exciton interaction.

Cooperative non linear processes such as :

b) Absorption of one U.V. photon as a product of exciton-exciton annihilation. Under high optical pumping conditions, one may realize densities of visible k=0 excitons allowing efficient interactions (or collisions); the end products for such a process would be one ion in the ground state and the other in a higher exciton state at about twice the energy of the single excitons.

Such a mechanism is to be related to an elementary A.P.T.E. process (Addition de Photons par Transfert d'Energie) which is called "up-conversion" energy transfer [16].

c) Absorption of two visible photons through the creation of two visible excitons interacting in pairs. Such a process would occur as above under high pumping levels and result in the production of pairs of excitons with wavevectors +k and -k corresponding to maxima of the exciton density of states. As opposed to the previous case (b), the spin selection rule $\Delta S=0$ is now satisfied.

d) Emission of one U.V. photon as a de-excitation channel of process a. This has never been observed so far in manganese fluorides.

e) Emission of one visible photon as a decay channel of an excited state which has been fed by multiphonon relaxation from a higher excited state prepared itself through process a or b. Such a process has been tentativelly proposed to justify the non exponential behaviour of the exciton decay and its particular variation with the pump intensity in MnF_2 [4].

f) Emission of two visible $k \neq 0$ excitons as the reverse of process c.

g) Emission of $k = 0$ excitons as a result of a crossover relaxation from zone boundary emission (process c) to zone center emission.

3. Experimental Aspects. As in the case of dilute manganese systems, the U.V. and visible absorption of the MnF_2 and $AMnF_3$ crystals images the electronic structure of the Mn^{2} ion in a cubic environment. In a previous presentation [1] emphasis was placed on the luminescence study of the broad phonon sidebands of the lowest $^6A_{1g}$ (^6S) $\rightarrow$ $^4T_{1g}$ (^4G) transition. This present work is devoted to a deeper investigation of their parent fine structures which are the pure exciton and their associated sideband transitions discussed above. In particular, the fluorescence spectra are made of both intrinsic and impurity induced lines while the absorption spectra appear to be essentially intrinsic.

Depicted in Fig. 3 is the low temperature absorption structures appearing on the low energy side of the broad phonon sideband mentioned above in the case of MnF_2 and $CsMnF_3$. These absorption lines are attributed to intrinsic pure electronic transitions along with their magnon and phonon sidebands. As predicted, the σ polarized magnetic dipole E1 and E2 excitons in MnF_2 contrast strongly with the π polarized electric dipole A2 and A3 excitons in $CsMnF_3$

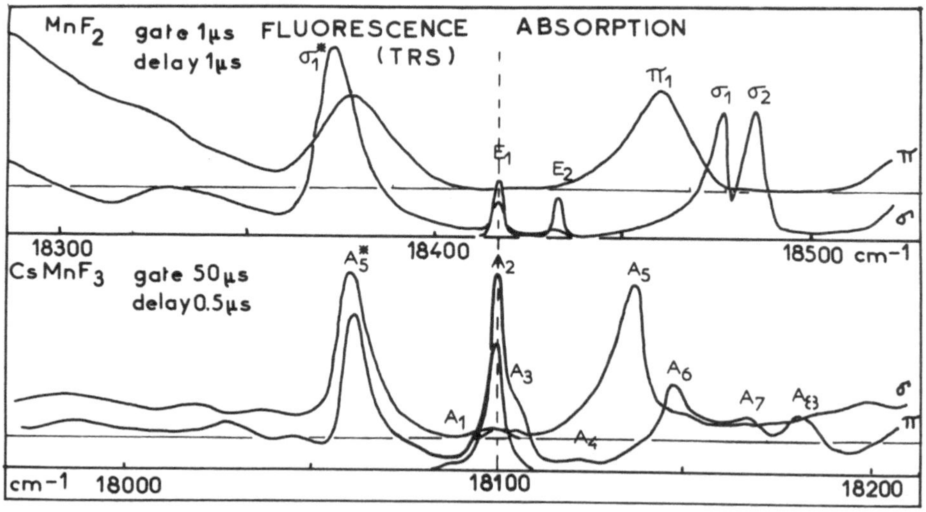

Fig. 3. Polarized absorption (right) and time resolved emission spectra (left) of MnF_2 [18] and $CsMnF_3$ [13].

while all the satellite lines appear to be electric dipole transitions in both cases.

The impurity induced lines which appear in the fluorescence spectra are attributed to the trapping of the excitons at some particular trap levels followed by their radiative deactivations.

Two types of traps have been considered in the visible and the infrared regions [5]:

- The shallow traps: they are Mn^{2+} ion sites perturbed by first, second, or third nearest neighbour non-luminescent impurities such as Mg^{2+}, Zn^{2+}, and Ca^{2+} (few parts per million) (see Fig. 4).

- The deeper traps: they are essentially fluorescing transition metal ions such as Ni^{2+} and Fe^{2+} (few parts per million) in nominally pure samples (infrared region), or trivalent rare earth ions in doped crystals, such as Er^{3+} and Eu^{3+}.

As we mentioned in the introduction, other undefined traps have to be introduced to account for the important fluorescence quenching characteristic of those materials.

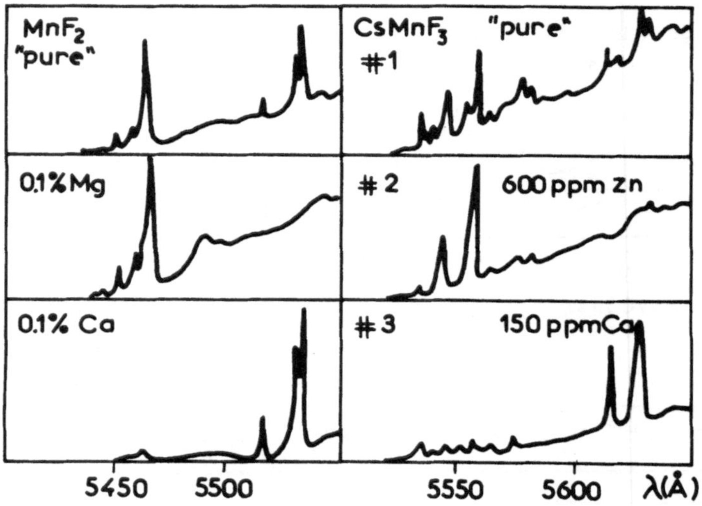

Fig. 4. Fluorescence structures in MnF_2 [19] and $CsMnF_3$ at different specific trap concentrations.

Due to the very efficient energy transfers to the traps or other unknown intrinsic deactivation processes, the lifetime of the intrinsic fluorescence is much shorter than that of the traps. Therefore, it is easy to wash out the spectra from any impurity-induced emission by looking at short time delay after the exciting laser pulse [17]; this is shown by comparing Figs. 3 and 4 in the case of MnF_2 and $CsMnF_3$ and it is found that these spectra image the lowest portion of the absorption fairly well.

III. OPTICAL COLLECTIVE EXCITATIONS IN CONCENTRATED MANGANESE FLUORIDES

III.A. Thermal Studies

The temperature dependence of the energy intervals of the exciton lines with their magnon/and phonon satellites provides a rather practical way of identifying each of the features.

As the temperature is raised, thermal disorder reduces the spin alignments and the magnon system changes according to that of the sublattice magnetization. Moreover, the exchange field on the ions is reduced which changes the electronic energies and causes a shift in the wavelength of the transitions. However, due to an effective deviation of the excited state interaction constant J' from its ground state value J and to non-negligible lattice expansion effects with increasing temperature, the shift of the electronic levels is appreciably reduced [20]. On the other hand, in the range of temperatures at which the shifts of the lines are observed, that is well below the Neel temperature, no noticeable phonon contribution is expected.

As a consequence, compared to the important lineshifts of the magnon sidebands, the changes in the pure excitons and their phonon satellites appear to be vanishingly small. In particular, if ν_0 is the frequency of the pure exciton line, which is assumed to be independent of temperature, the peak frequency of the magnon sidebands $\nu(T)$ transforms like

$$\nu(T) = \nu_0 + \nu_m(T) \tag{1}$$

where $\nu_m(T)$ stands for the magnon frequency alone. $\nu_m(T)$ is obtained by including in the spin hamiltonian quadratic and higher order terms in the creation and annihilation spin wave operators corresponding to an increase of the magnon scatterings with temperature (dynamical interaction). This results in a renormalization of the energy of the magnon at any given temperature. Within this frame, the above expression can be rewritten as :

$$\nu(T) = \nu_0 + \nu_m \times R(T) \tag{2}$$

where ν_m is the magnon frequency for T=0 and R(T) the so called renormalization factor [20].

Concurrently R(T) can be replaced in some cases by a modified Brillouin function $B_{5/2}$ (T/T_N) [21], which is a rough approximation compared to R(T), but has the advantage to give a rapid and sufficient information when one is interested in a qualitative interpretation of the immediate satellites of the pure exciton lines. The frequency shifts of the eight lowest absorption lines in $CsMnF_3$ with temperature are reported in Fig. 5. The dashed curves correspond to fittings with the above formula where R(T) has been replaced by $B_{5/2}(T/T_N)$.

The good agreement with the experimental data allowed us to attribute the lines A_2 and A_3 to pure exciton transitions and A_5, A_6 and A_7 to magnon sidebands with characteristic magnon frequency intervals of 35, 45 and 66 cm^{-1} respectively. The line A_8 does not shift with increasing temperature which would be expected for a phonon sideband; indeed its separation from the exciton A_2 matches very closely the frequency of a Raman active mode of about 78 cm^{-1}.

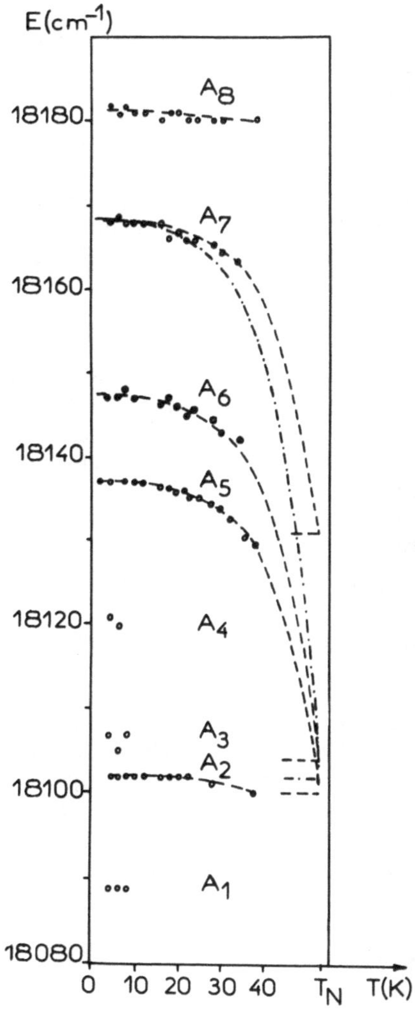

Fig. 5. Frequency shifts of the eight lowest absorption lines with temperature in $CsMnF_3$ [13].

The lines A_1 and A_4 could have been attributed later to hot and cold magnon sidebands associated with a maximum of the magnon density of states at about 15 cm^{-1}.

The interest in thermal studies of these optical lines is such that they give a rather good picture of the magnon density of states in the whole Brillouin zone and they are easily manageable.

III.B. Uniaxial Stress Effects

Since the strains induced by experiments are relatively small, less than 3000 kg/cm^2 in most cases, it is generally assumed that the magnon and phonon energies are independent of stress. Thus, stress studies may provide a very convenient method for assigning the sidebands to their specific exciton lines. In the case of $CsMnF_3$, for example, we have shown that whatever the stress is applied in the basal plane along a [100] or a [120] direction - see Fig. 6 -, the lines A_2, A_5, A_7 and A_8 move linearly by the same amount, 0.012 $cm^{-1}/kg/cm^2$, while A_3 and A_6 are shifted as a distinct group at a smaller rate, 0.006 $cm^{-1}/kg/cm^2$. Such a result enabled us to complete our description of the sidebands and to associate the magnon satellites A_5 and A_7 and the phonon sideband A_8 with the exciton A_2, the magnon satellite A_6 being assigned itself to the exciton A_3. No line splitting is observed which means that we are dealing with orbitally non degenerate excited states involving site symmetries lower than the reported trigonal environments. Similar results were pointed out in MnF_2 where stress applied along the z axis shifts the lines E1, π1 and σ1 as a group and E2 and σ2 as a distinct group [22].

For stress applied in other directions the results may be as much significant. For example, directions can be adequately choosen in order to distinguish between the different magnetic sublattices. Such studies may provide us with a very simple method for sorting out the respective sublattice excitons. An experiment was performed in that sense some years ago in the case of MnF_2 for a stress applied along the [110] direction [22]. Indeed a [110] stress affects the two Mn sublattices differently since it acts along an x direction for the one and along the y direction for the other. The result is that the magnetic sublattice degeneracy is lifted and that each exciton line splits into two. A similar observation was obtained in $CsMnF_3$ for a stress applied along an intermediate direction joining Mn1 (centrosymmetric) and Mn2 (non centrosymmetric) adjacent ions in order to distinguish between the adjacent Mn2 magnetic sublattices [23]. This is shown in Fig. 7 for stress up to about 2400 kg/cm^2. Each line A_2 and A_3 splits into two. Unlike MnF_2, the shifts are linear and practically the same for the highest and the lowest components respectively . The above results have been used to ascertain the Mn2 character of the doublet exciton A_2/A_3 and to confirm this doublet structure as low

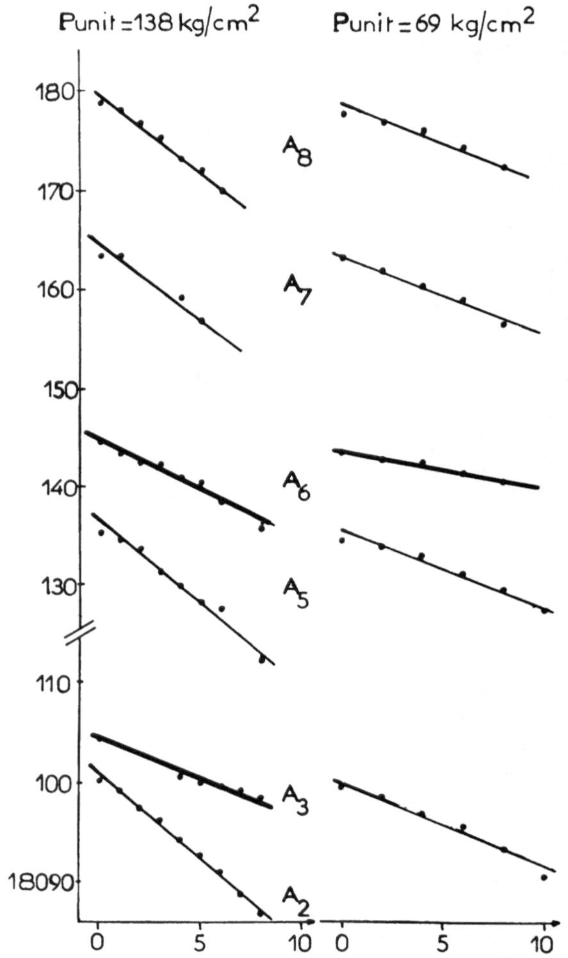

Fig. 6. The shifts of the $^6A_{1g} \rightarrow {}^4A_{1g}$ (^4G) lowest absorption lines under uniaxial stress (a) P// [120] (b) P// [100] in $CsMnF_3$.

crystal field components of an E-like sublevel of the cubic state $^4T_{1g}$, coming from an additional local site distortion. Such distortion is even more accentuated when the stress is applied in the basal plane which results in the effective separation of the lines A_2 and A_3 shown on Fig. 6.

III.C. Magnetic Field Effects

For fields lower than the spin-flop transition the following behaviours can be expected. When the magnetic field is applied along the direction of the magnetic moments we obwerve an increase in the transition energy for ions on one sublattice and a decrease for the other antoparallel ones by a quantity 2 $g\beta H$, where β is the Bohr magnetoc and g refers to the gyromagnetic factor for the Mn^{2+} ions (assumed to be the same in the excited and ground states and equal to 2). As they involve a unit change in their total spin momentum the pure exciton transitions then split and give rise, as in the presence of an uniaxial stress, to two distinct sublattice excitons. On the other hand, when we are considering the exchange induced electric dipole magnon sidebands, the total spin component is conserved and the increase in energy on one sublattice is compensated by a like decrease on the other; as a consequence no line splitting of the magnon sidebands occurs.

For fields greater than the critical field H_c (~ 90 kOe in MnF_2, ~ 1 kOe in $CsMnF_3$), the Zeeman components of the exciton lines coalesce into one, since the magnetic field energy is then identical for each sublattice, going back to their initial positions at H=0. However, some additional structures may appear because of new selection rules. As an illustration of the effect of high magnetic field we wish to present here our own observation of new electric dipole exciton transitions in $CsMnF_3$ induced in pairs of adjacent non centrosymmetric ions Mn2.

Ion pair process in non-centrosymmetric crystals:

In this mechanism a dipole moment is induced in a pair of magnetic ions coupled by exchange interaction, as a result of which an exciton is created in a single ion. This mechanism is remarkable since it is possible only in crystals containing magnetic ions displaced from a center of symmetry - this will be the case of the Mn2 ions in $CsMnF_3$ - and for non-collinear sublattice magnetic moments [24].

The transition matrix element $M^{f,o}_{\alpha,\beta}$ associated to the exchange interaction between adjacent ions α and β in the above conditions is given by [28]:

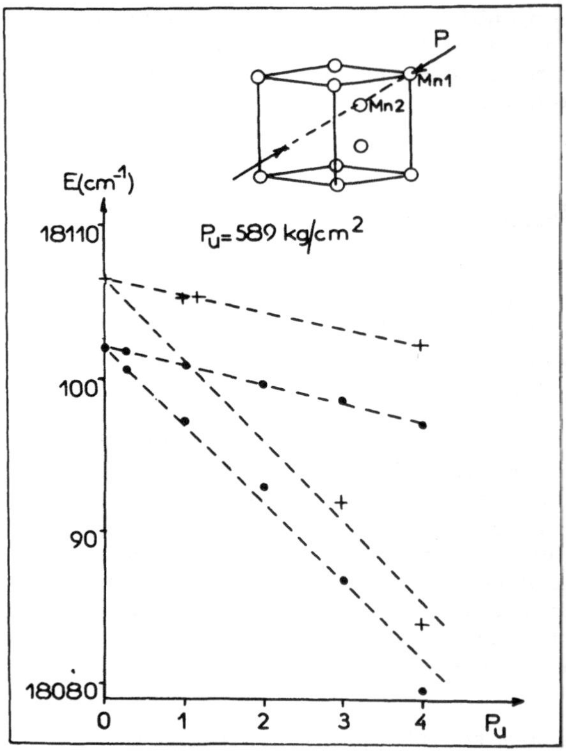

Fig. 7. Effect of a stress directed along an axis joining Mn1 and Mn2 adjacent ions in $CsMnF_3$ on the absorption exciton lines A_2 (.) and A_3 (o).

$$M^{f,o}_{\alpha,\beta} = <\phi^{f}_{\alpha}\ \phi^{o}_{\beta}/V^{ex}_{\alpha,\beta}/\phi^{o}_{\beta}\ \phi^{o}_{\alpha}> \tag{3}$$

where $V^{ex}_{\alpha,\beta}$ is the exchange part of interion interaction hamiltonian and f stands for the considered excited state.

Expressing $M^{f,o}_{\alpha,\beta}$ in terms of creation and annihilation operators for electrons in given spin orbitals it follows:

$$M^{f,o}_{\alpha,\beta} = \mathcal{M}^{f}_{\alpha,\beta}\ \sin\theta\ \cos\theta \tag{4}$$

where the amplitude value $\mathcal{M}^{f}_{\alpha,\beta}$ of the matrix element has been separated from its angular dependence θ.

θ is defined by: $\theta = \dfrac{\theta_\alpha - \theta_\beta}{2}$

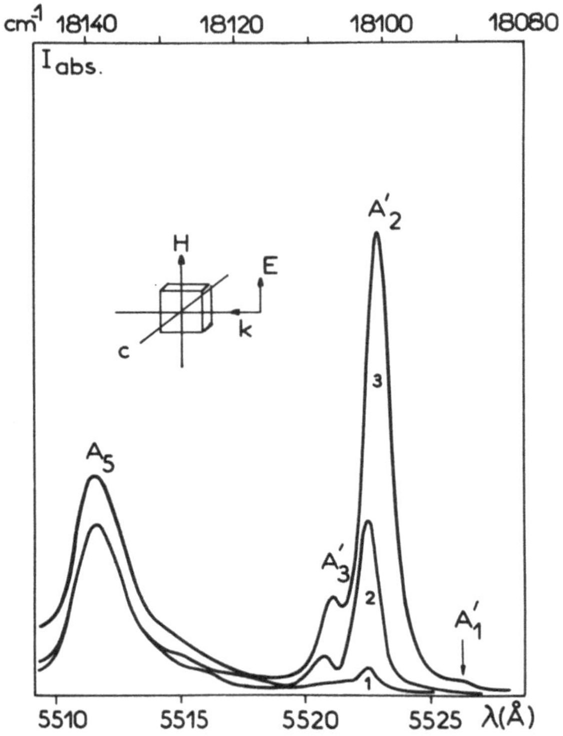

Fig. 8. Increase of the absorption lines A_1 to A_5 of $CsMnF_3$ in σ polarization at 2K with the applied magnetic field and for H//E; 1) H=0, 2) H-70 kOe, 3) H=125 kOe.

where θ_α and θ_β refer to the angles of the magnetic moments of the ions α and β with a given direction (that of the applied magnetic field for example).

At this point it is easily verified that:

1) If the magnetic ion is located in a centrosymmetric position then the functions ϕ^0 and ϕ^f are even and the operation of inversion with respect to α and β leads to $M^{f,o}_{\alpha,\beta} = -M^{f,o}_{\beta,\alpha}$. As a consequence, the total dipole moment

$$\pi^{f,o} \equiv \Sigma\, M^{f,o}_{\alpha,\beta}$$

characterizing the transition at the frequency E(o,f) of the exciton goes to zero and the proposed mechanism does not occur.

2) If $\theta_\alpha - \theta_\beta = {}_{f,o}^{0}$ (aligned) or π (antialigned sublattices), then $M_{\alpha,\beta}^{f,o} = 0 \ \forall\ \alpha,\beta$ and $\pi^{f,o} = 0$; the proposed mechanism will not appear either in that case.

In $CsMnF_3$, when the crystal is submitted to an external magnetic field one goes from conditions of collinear to non-collinear spin orientations. As shown on Fig. 8 this manifests itself in a pronounced intensification of new σ polarized absorption lines A_2' and A_3' appearing at the same frequencies of the lines A_2 and A_3. The intensities increase quadratically with the applied magnetic field in good agreement with the predictions of the above theory: for fields H much smaller than the exchange field on the ions Mn^{2+} $H_E \sim 350$ kOe, $|\sin\theta \cos\theta|^2$ is nearly proportional to H^2 [9,26].

These data have been used to ascertain the nature of the pure lines A_2 and A_3 in $CsMnF_3$ as excitonic transitions within the non-centrosymmetric Mn^{2+} ions labelled Mn2.

Moreover, the observed effects of high magnetic fields on the immediate magnons sidebands A_1 and A_5 allowed us to confirm them as hot and cold transitions with the involved magnon being mainly localized on an adjacent Mn1 or Mn2 sublattice respectively [23].

By way of conclusion of this section we like to point out a very interesting experiment, performed some years ago by J.F. Holrichter et al [22], which has the particular advantage to present a good illustration of both the uniaxial stress and magnetic effects. Transitory sublattice magnetizations are generated in MnF_2 by selective optical pumping of the distinct sublattice excitons which have been produced by applying [110] stress as discussed previously.

Indeed, the magnetic moment of a two sublattice antiferr0-magnet which is excited to one of its exciton (or magnon) levels $\Gamma_{A(B)}$ is :

$$M(t) = [n_A(t) - n_B(t)] \langle\Gamma_A|\vec{m}|\Gamma_A\rangle \quad (5)$$

where $n_A(t)$ and $n_B(t)$ are the numbers of excitons on sublattices A and B respectively at time t and $\vec{\mu}$ is the magnetic dipole operator. A pick up coil around the crystal detects the rate of change of magnetization after pulse tunable dye laser excitation.

In particular, the magnetization decay signals could have been attributed to an intersublattice coupling of the order of 10^{-5} cm^{-1} thus confirming the absence of Davydov splitting as asserted previously.

Up to this point, all the data were found to be consistent with a collective description of the excitation. The following development is intended to improve the validity of this picture through the dynamical aspect of fluorescence.

IV. VALIDITY OF THE EXCITON MODEL IN FLUORESCENCE

IV.A. Exciton Dispersion and Phase Memory

As mentioned previously, the exciton and the magnon involved in a cold exciton-magnon process may strongly interact ; thus the shape of the resulting magnon sidebands may be significantly affected. In addition, such a lineshape, which is a convolution of both the exciton and magnon densities of states, is quite sensitive to small amounts of exciton dispersion.

As a consequence, a more tractable problem is to consider a cold exciton-magnon emission process, for in this case the magnon sideband may be almost free of any exciton-magnon interaction. Thus, in a first step, one has to compute the shape of the magnon sideband, knowing the sideband matrix element and the effective magnon density of states, without including the exciton dispersion. Then the discrepancy, if any discrepancy is detectable, observed between the experimental and the calculated lineshapes may provide indirectly an estimate of the exciton dispersion $\mathcal{D}$ and an upper limit for the rate of interion transfer of excitation W_{max} to which it is related by the following expression

$$W_{max} \sim \frac{2\mathcal{D}}{\hbar} \qquad (6)$$

From the lineshape of the immediate σ_1^* intrinsic emission magnon sideband in MnF_2 (see Fig. 3), Dietz et al [28] have shown that the E1 exciton dispersion is certainly much less than .5 cm^{-1} which, according to the above relation, leads to a minimum transfer rate of the order of 10^{10} s^{-1}. These values have been used to reinforce the collective character of the exciton E1 and to represent it as being at least a moderate range exciton extending over a volume containing 100 Mn^{2+} ions as a minimum [29].

Some coherency (k vector memory) may be pointed out by selectively (optically) pumping the excitons by means of a high resolution tunable dye laser, thus preparing the system in a narrow band of k values and then, recording the fluorescence signal, at the shortest time delays, from excitons at other k values. For example, R.M. Mac Farlane et al [30] found in MnF_2, that the time it takes to the exciton to be scattered from the center to the edge of the Brillouin zone, and vice versa, i.e. the time it takes to lose any residual phase memory is of the other of 1μs at 2 K. This is found to be significantly larger that the hopping time of the exciton $\tau_h = W_{max}^{-1}$, as estimated before. Thus, at very low temperature, the excitons can be considered as somewhat delocalized excitations with a width in k, thus confirming, for example in the case of the magnon sidebands, the character of the wave vector selection rules. On the other hand, because the intraband relaxation of the excitons is fast compared with the time scale of the regular experiments - the deactivation process which limits the exciton decay is of the order of a few hundred of μs for both zone center and zone boundary excitons -, one may consider a localized phase incoherent picture by replacing, for example, the k dependent summations which may appear in the calculations by thermal averages.

In $CsMnF_3$ the A_5^* intrinsic emission magnon sideband is separated from its specific exciton line A_2 by about 38.7 cm^{-1}, thus, in perfect concordance with a magnon frequency interval of $\sim$ 38.6 cm^{-1} as obtained from neutron scattering and spin wave theory [10]. As a consequence, with the method used above an upper limit of about .1 cm^{-1} can be pointed out for the exciton dispersion.

On the other hand, first results at the shortest times after the laser pulse, using a fast transient recorder (Biomation) with a resolution of 10 ns, seems to indicate a zone center to zone boundary scattering time less than 100 ns, that is one order of magnitude faster than in MnF_2. If the same exciton-exciton scattering process is assumed to be responsible for the attainment of quasiequilibrum in k, this suggests that in $CsMnF_3$ the exciton dispersion is at least one order of magnitude smaller than in MnF_2, that is .05 cm^{-1} at most. This may be seen quantitatively by considering the crystallomagnetic structure of each compound knowing that the excitons reside essentially in parallel spin sublattices of same site symmetry. As shown on Fig. 1 and Table 1, each Mn ion in MnF_2 has two nearest neighbours at a separation of about 3.3 Å to which it is linked by only one fluorine ion. In $CsMnF_3$, the most probable channel for the transfer of the excitation are to the three third nearest neighbours at about 5.8 Å involving a bridge of two fluorine ions in each case.

IV.B. Exciton Lineshape and Impurity Effects

1. Inhomogeneous Broadening. By analogy with the broadening due to the Raman scattering of phonons, the homogeneous broadening of the magnetic excitons is attributed to a Raman scattering of magnons which, like the exciton-magnon interaction, is another consequence of the different J values in the excited and ground states of the Mn^{2+} ions [20]. This homogeneous broadening is too weak to be measured directly, and in general, the internal strains caused by the crystal defects produce inhomogeneously broadened Gaussian lines of widths at half maximum of the order of few cm^{-1} (~ 1 cm^{-1} in MnF_2, ~3 cm^{-1} in $CsMnF_3$ and ~ 4 cm^{-1} in $KMnF_3$).

On the other hand, such an inhomogeneous broadening is much larger than the underlying dispersion, which shows that the strain fields may be effective in the k mixing contributing also to the above mentioned loss of phase memory. However, as was shown recently in the case of the essentially localized exciton states of the shielded 4f electrons in $Tb(OH)_3$ the k vector delocalization may not be destroyed as a whole, thus confirming the scrambling effect of the strain fields to restricted range of k values due to experimentally separable microscopic regions [31]. In the end, the strain fields may impose energy barriers to the exciton mobility at low temperatures as a result of which the effective trapping probability may be significantly lowered ; this can be evidenced in some cases by observing the distinct temperature behaviours of the pure (k=0) and magnon assisted (k≠0) exciton decays [28].

At this point we wish to emphasize the role of crystal purity by showing the effects of the impurities on the spectral features and the energy transfer characteristics. In particular, J. Hegarty et al [32] have shown that for MnF_2 samples doped with important amounts of Zn^{2+} ions (.2 up to 1.2% Zn^{2+}) the E1 exciton line shifts to lower energies (up to 7 cm^{-1} for a 1.2 % Zn^{2+} sample) and significantly broadens (by a factor 5 for a 1.2 % Zn^{2+} sample) in absorption and emission as well. This is attributed to an increase of the inhomogeneous strains brought about by the random distribution of the Zn ions in the MnF_2 lattice. Such a data have been used tentatively to relate the various induced manganese trap fluorescences, currently observed in the spectra of nominally pure samples, to specific first, second and third nearest neighbours of a particular impurity, as mentioned in Section II [5].

2. Diffusion-Limited Energy Transfer. In self activated materials such as the concen rated manganese fluorides, one may adopt a localized phase-incoherent exciton description (which is legitimate in view of the time scale of our measurements). A rapid diffusion of the optical excitations among the donor system formed by the bulk active ions is generally assumed. In that case, at any time of the decay, all donors have an equal probability of being excited which considerably simplifies the configuration averaging problem arising in the calculation [33]; in particular, when starting with microscopic rate equations one needs an exact analytical description of the whole donor fluorescence versus time. Such a situation gives rise to an overall transfer rate which is time-independent and proportional to the acceptor concentration.

Writing $\Gamma = q + p_T$, where q is the sum of all the transfer rates except those which refer to traps T, i.e., p_T. The rate equations governing the populations n_e and n_T of the respective intrinsic and trapped excitons may be expressed as follows:

$$\overset{\circ}{n}_e(t) = - [\gamma_R + q + p_T]\, n_e(t) \quad (7)$$

and

$$\overset{\circ}{n}_T(t) = - \gamma_T\, n_T(t) + p_T n_e(t) \quad (8)$$

where γ_R and γ_T represent the respective radiative lifetimes.

The form of these equations leads us to the following remarks.

-The integration of Eq. (7) gives rise to an exponential exciton decay with time constant $[\gamma_R + q + P_T]^{-1}$.

-The feeding term of the trap fluorescence, i.e., $F(t) = p_T n_e(t)$- is directly proportional to the exciton population.

-In connection with the latter, the trap fluorescence is found to rise at short times with the same time constant as that for the exciton decay, reported above. This is determined by integration of Eq. (8), assuming $\gamma_R + q + P_T >> \gamma_T$; a situation which does prevail in this case.

In the opposite situation of dominant single step donor-trap energy transfer without diffusion, a situation which applies essentially to dilute compounds, the total transfer becomes time dependent thus reflecting for the excited donors a decreasing capability of sampling of new acceptor sites which are located at larger distances as time evolves. Such a situation refers to the so-called static-limit and has been worked out in detail by several authors [34,35].

In concentrated manganese fluorides containing abnormal impurity concentrations, direct donor-trap energy transfer may compete efficiently with diffusion. This is seen essentially at short times when excitation goes preferentially to nearest neighbor acceptor sites. This results in an increased deactivation of the donor system which is again perfectly describably within the above mentioned static scheme. At long times, the situation goes back asymptotically to a diffusion regime and the exciton decay becomes slower with the time constant $[\gamma_R + \Gamma]^{-1}$.

Diffusion limited energy transfers have been worked out in several specific cases [36-39]. There, the relative importance of diffusion and trapping was taken into account.

In view of these results, the macroscopic exciton rate equation can be written simply:

$$\mathring{n}_e(t) = -[\gamma_R + q + p_T + p_s(t)]\, n_e(t) \qquad (9)$$

where $p_s(t)$ stands for the multipolar or exchange static transfer rate given by:

$$p_s(t) = \gamma t^{-3/n} \qquad (10)$$

with n=6, 8 and 10 respectively corresponding to dipole-dipole, dipole-quadrupole, or quadrupole-quadrupole coupling and with γ as a given constant.

For the exchange case,

$$p_s(t) = \beta \sum_{k=0} \frac{(-x)^k}{k!\,(k+1)^4} \qquad (11)$$

where $x = \alpha\tau$ and where α and β are specific constant values.

Similarly the trap fluorescence rate equation transforms like:

$$\dot{n}_T(t) = -\gamma_T n_T(t) + [p_T + p_s(t)]n_e(t) \qquad (12)$$

It is immediately seen that the feeding term F(t) no longer tracks with the exciton population since $p_s(t)$ is now time dependent. In fact, this property is currently used to test the occurrence of diffusion limited energy transfer. This can be checked numerically without precising the form of $p_s(t)$ [29].

Indeed, using Eq. [12] and assuming that the populations n_e and n_T are proportional to the observed fluorescence intensities I_e and I_T, one can write:

$$p_D(t) = \frac{e^{-\gamma_T t}}{I_e(t)} \frac{d}{dt} [e^{\gamma_T t}\, I_T(t)] \qquad (13)$$

with $p_D(t) = p_T + p_s(t)$

or with $t = \dfrac{t_1+t_2}{2}$,

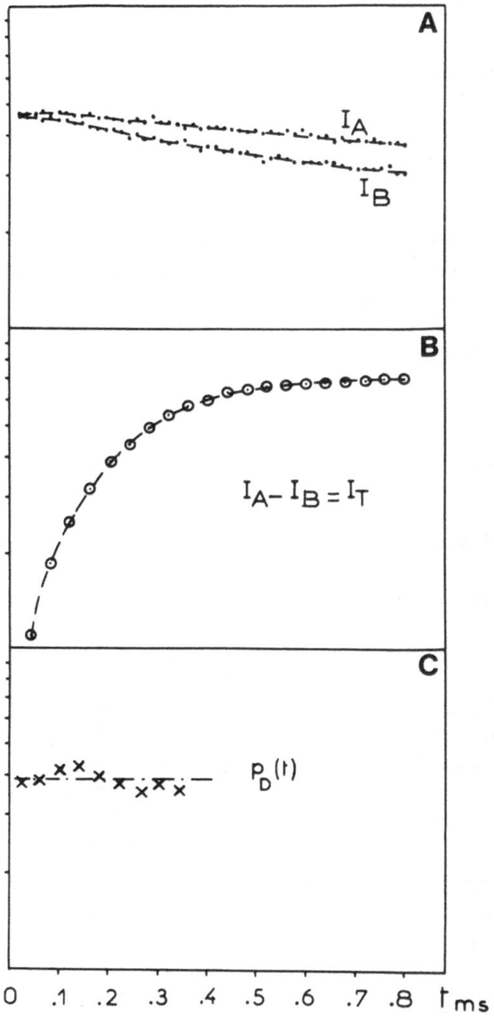

Fig. 9. (a) Short time fluorescence decay at two distinct positions A and B of the trap line located at 5618 Å
(b) Short time trap fluorescence rise
(c) Short time behaviour of the energy transfer rate in $CsMnF_3$ at T - 1.6 K.

$$p_D(t) = \frac{e^{\gamma_T t_2} I_T(t_2) - e^{\gamma_T t_1} I_T(t_1)}{e^{\gamma_T t}\ I_e(t)} \qquad (14)$$

Such a diffusion limited situation does predominate in samples which have been purposely overdoped with known impurities. The experiment was performed on MnF_2 doped with .5 and 1% Zn^{2+} impurities [29]. $P_D(T)$ decreases with time as expected from Eq. (11).

As a consequence, in nominally pure samples containing few parts per million of impurities, it is thought that the diffusion limited energy transfers, though they cannot be fully disregarded, are generally very weak. As shown on Fig. 9, this is verified in the case of $CsMnF_3$ (the intensity I_T of the trap fluorescence line located around 5618Å has been washed out of the contribution of the underlying intrinsic background emission by subtracting the data obtained at two distinct positions A and B of the line) : $p_D(t)$ is constant at short times.

IV.C. Fluorescence Decay Modes and De-excitation Models

1. Exciton Dynamics at Very Low Temperature. As an illustration of exciton dynamics at very low temperature we have reported on Fig. 10 the fluorescence decay of the intrinsic exciton A_2 in $CsMnF_3$ at T=1.6 K. Excitation is by laser pulses of moderate intensities i.e. few tens of μJ, into the absorption magnon sideband A_5. The semilogarithmic scale used here clearly shows the non-exponential behaviour of the exciton decay.

As we shall see in section IV.D.3. the short time variation (up to $\sim$1.5 ms) can be attributed to an intrinsic desexcitation process which depends non linearly on the excitation pump power. The theoretical decay curve predicted by this model has been reported on Fig. 10. In particular, the model gives rise to an asymptotic approach to a pure exponential with a decay rate characteristic of the regular diffusion and trapping mechanism mentionned before. However, as evidenced on Fig. 10, a slowing down of the exciton decay with respect to this prediction is taking place in this long time limit. B. Wilson already mentionned such a discrepancy in the case of MnF_2 [29]. This discrepancy was then tentatively assigned to "various back transfer mechanisms and limited diffusion rates across macroscopic faults". In fact, following recent findings from Sakun [40] on the hopping energy transfer mechanism, the exciton migration may be allowed to return more than once to the same donor site in the random walk process. This leads to additionnal terms in the transport equations which give rise to the observed slowing down effect of the intrinsic exciton decay at long times.

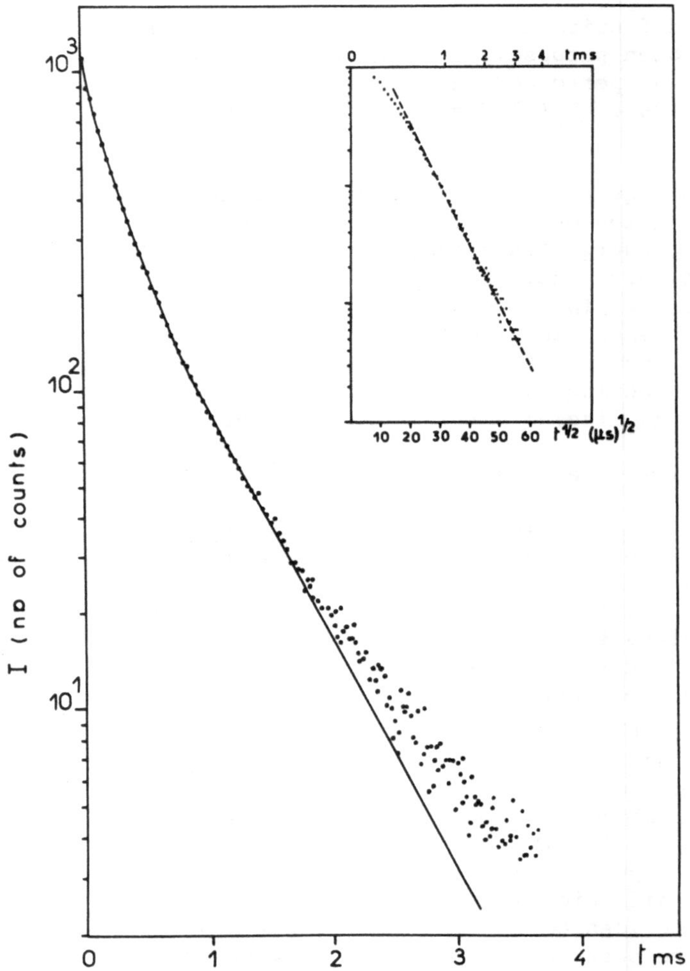

Fig. 10. Intrinsic exciton fluorescence decay in $CsMnF_3$ at T = 1.6 K versus t and $t^{1/2}$ (insert).

At this point we wish to say few words about the low temperature variation of the exciton decay in its short time portion.

In MnF_2, the E1 exciton decay time constant decreases very sharply with increasing temperature in the 2 to 10 K region. Such a result was attributed to a step up phonon deactivation of the exciton E1 to the exciton E2 located around 17 cm^{-1} above. Thus, the E1 exciton decay rate $p = \tau^{-1}$ satisfies the well known relation:

$$p = p_o + c[\exp(\Delta E/kT) - 1]^{-1}$$

where p_o is the residual rate due to the transfer and radiative decays observed at very low temperature (in the previous notations $p_o = \gamma_R + \Gamma$) and c measures the coupling strength.

In MnF_2 an activation energy gap ΔE of ~ 11.5 cm^{-1} was found, thus 5.5 cm^{-1} less than the energy separation of the excitons E1 and E2. Moreover, using the fluorescence intensity data instead of the fluorescence decay time constants, Dietz et al [28] reported an activation barrier ΔE of 6.3 cm^{-1}. While the latter [28] attributed the observed discrepancy with the spectral interval of the excitons E1 and E2 to the thermal activated trapping of an impurity exciton relative to E2, Wilson et al (41) advanced the possibility that the thermal activation process could occur at zone edge rather than zone center positions resulting in a negative dispersion of E2 of ~ 6 cm^{-1}, in agreement with a precedent estimate [11] and two magnon resonant scattering data [42].

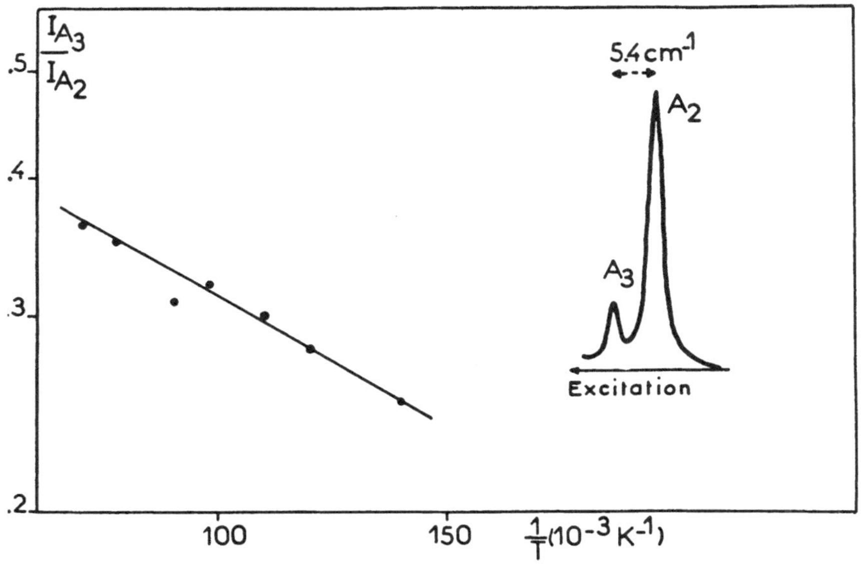

Fig. 11. Intensity ratio of the excitation lines A_3 and A_2 in $CsMnF_3$ as a function of the inverse temperature. $\Delta E = 3.9$ cm^{-1}.

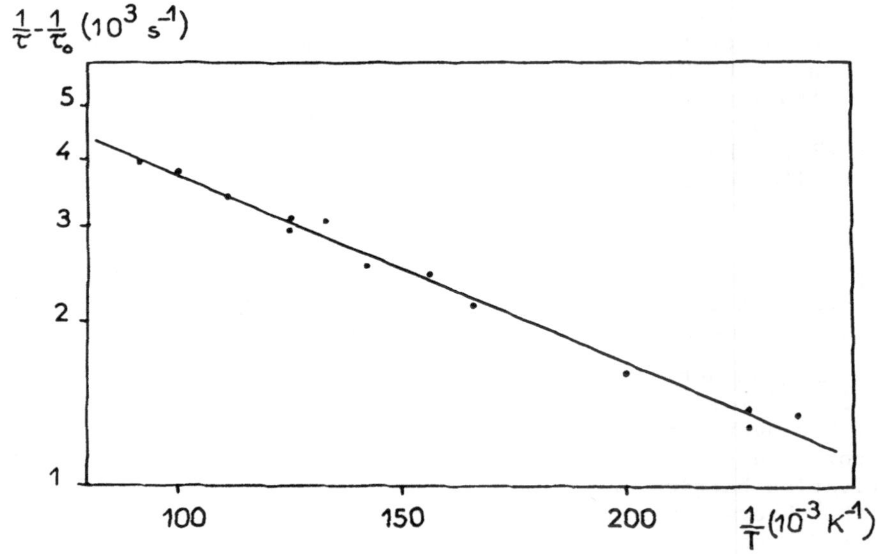

Fig. 12. Variation of the intrinsic decay rate τ^{-1} of the exciton fluorescence in $CsMnF_3$ as a function of inverse temperature with $\tau_o^{-1} \approx 1.4 \times 10^3 s^{-1}$ ΔE= $(5.4 \pm 5) cm^{-1}$.

We add that similar discrepancies were observed in $KMnF_3$ [43] and $CsMnF_3$ [23] using fluorescence intensity data, where it is found energy barriers of 6.4 and 3.9 cm^{-1} for zone center separations between E1 and E2 and between A_2 and A_3, of 9.3 and 4.9 cm^{-1} respectively (see Fig. 11).

Nevertheless, as it is shown on Fig. 12, the thermal variation of the A_2 fluorescence decay rate $p = \tau^{-1}$ in $CsMnF_3$ was also recorded taking a residual value $p_o = \tau_o^{-1}$ of $\sim 1.4 \times 10^3$ s^{-1} this allowed us to point out an activation barrier ΔE of $\sim 5.4 cm^{-1}$.

Though being highly sensitive to the presence of any trap level, our dye laser excitation spectra in the region of the excitons A_2 and A_3 never showed evidence for the existence of intermediate trapped excitons as would be expected from the intensity data. Also, as suggested before [41], it is thought that the intensity data may be spoiled by long lived phenomena such as thermal boil back from the shallower traps which in the other hand, are easily deconvoluted when essentially the early decay time constants are taken into account.

As a consequence, our measurements in $CsMnF_3$ are consistent with an activation process from A_2 to A_3 involving either a negligible or a positive exciton dispersion for A_3.

2. Boil-Back Process and Magnon Assisted Fluorescence at High Temperature. In general, for sufficiently high temperature the intensities and decay time constants of the various impurity induced fluorescences undergo similar behaviours, i.e., exponential decreases supported by a simple deactivation or boil-back process in which the energy barrier ΔE stands for the energy mismatch of the various trap levels with the intrinsic exciton band. Formally this is described by the same expression as above in Eq. (15).

This works fairly well in MnF_2; the energy barriers obtained from the intensity data compare favorably with the measured spectral intervals. In $CsMnF_3$ the agreement is as good - see Fig. 13 - expected for the shallower traps which deactivate differently. This discrepancy is attributed to the additional assistance of magnons. Indeed, in $CsMnF_3$ a number of lines are made of superpositions of a particular impurity induced transition, with a specific trap depth, and a magnon sideband of an upper exciton. As a consequence the deactivation of the latter may strongly affect the deactivation of the former. This is the case for example between the lines located at $\sim$ 5547 and 5535 Å and further, between the line at $\sim$ 5535 Å and the intrinsic exciton. In the end, the existence of hot magnon excitation [23] may also contribute significantly to the above discrepancy.

In any case, the high temperature regine, the mechanisms of fast diffusion trapping and thermal detrapping evidenced in the past, seems to apply satisfactorily again.

IV.D. Effect of External Perturbations

1. Effect of Magnetic Field on Exciton Delocalization. As mentioned previously, the optical electronic excitations are accompanied by a change of the ion spin by unity and, in collinear antiferromagnets, this leads to a rigourous spin exclusion on the transfer of excitation energy from one sublattice to another. On the other hand, in a material which presents a weak anisotropy, magnetic field induced spin-flop transitions can be easily achieved giving rise to the appearance of a weak ferromagnetic component along the applied field; the new spin configuration thus corresponds to a mixing of the different magnetic sublattices as a result of which the dispersion of the escitons may be greatly enhanced.

This has been observed in $CsMnF_3$ at very low temperature for continuous or pulsed laser excitation in the magnon sideband noted A_5 in magnetic fields up to about 120 kOe. As shown on Fig. 14

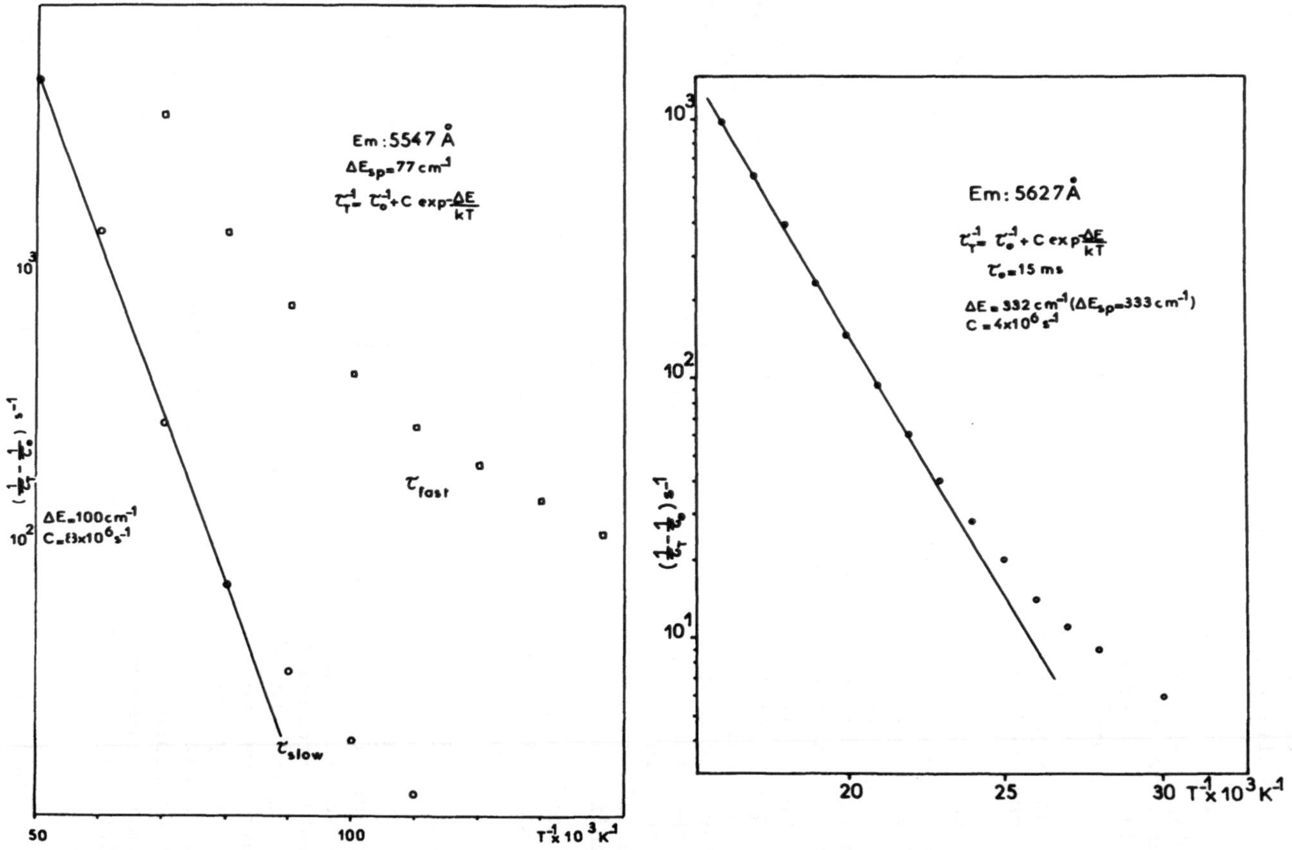

Fig. 13. Variation of the decay rates of the trap fluorescence lines located at 5547 (left) and 5618 Å (right) in $CsMnF_3$ as a function of inverse temperature.

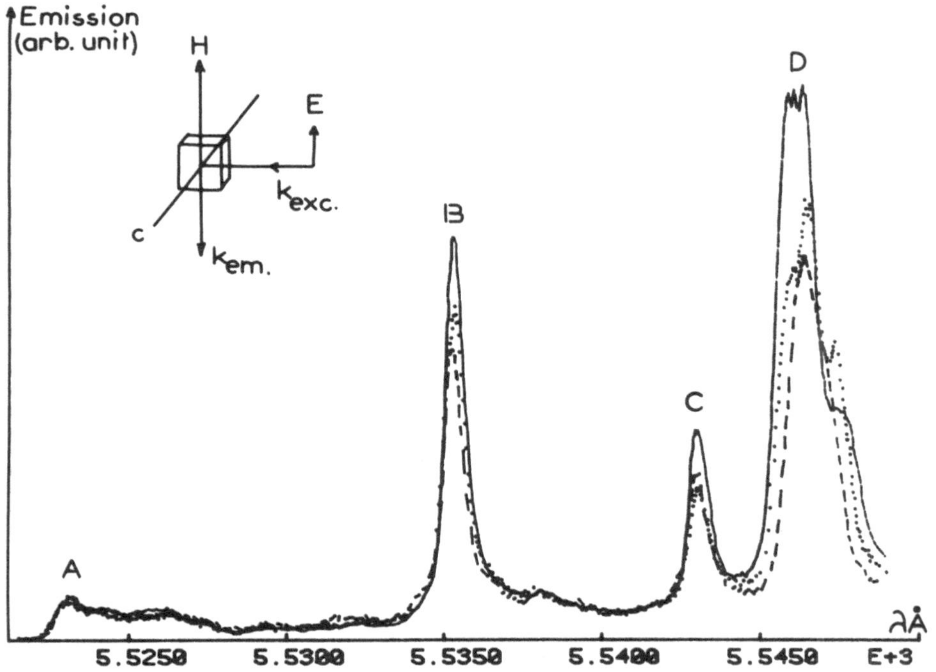

Fig. 14 Relative variations of the intensities of the trap fluorescences in $CsMnF_3$ as a function of magnetic field at T=2K for H c and for H=0 (---) H=70 kOe (...) H=120 kOe(——).

where the emission spectra have been calibrated in order to display the same exciton intensity, the impurity induced fluorescence increases as expected in a delocalization process. This overall increase in the exciton transfer rate to the traps is further verified through the decay time measurements and the time resolved spectra. In particular, both the intrinsic exciton decay and the trap fluorescence rise are shortened by a factor of ∿ 1.2 in a magnetic field of 70 kOe -.

Further, the dynamical results may be correlated to the magnetization data ; this is formulated as follows :

Within a fast diffusion model the exciton decay rate is directly proportionnal to an isotropic diffusion coefficient D. On the other hand, this diffusion coefficient, which is a measure of the exciton dispersion, is proportionnal to the square of the matrix element for the resonant excitation transfer between ions of different sublattices. This matrix element is given by an expression like [25]:

$$M^f_{\alpha,\beta} = \langle \varphi^f_\alpha \varphi^o_\beta / V^{ex}_{\alpha,\beta} / \varphi^f_\beta \varphi^o_\alpha \rangle \tag{16}$$

or taking into account the spin orientations of ions α and β :

$$M^f_{\alpha,\beta} = \mathcal{M}^f_{\alpha,\beta} \cos^2 \theta \text{ with } \theta = \frac{\theta_\alpha - \theta_\beta}{2} \tag{17}$$

Where, as defined in Section III.C, 2θ is the angle between the equilibrium positions of the spins on the ions α and β.

$\mathcal{M}^f_{\alpha,\beta}$ is an amplitude value independent of the orientations of the spins.

In fields greater than the spin-flop transition, one can write

$$\cos^2 (\theta) = (\frac{M}{M_o})^2 \tag{18}$$

where M is the magnetic field induced magnetization and M_o its saturation value.

As a consequence, we get the following relation:

$$\tau^{-1} \alpha (\frac{M}{M_o})^4 \tag{19}$$

The exciton decay time constant decreases as the fourth power of the induced magnetization. For example a change in the value of τ by a factor 1.2 will correspond to a change in magnetization of the order of 1.04. In order to test the validity of the model we have related this result to the magnetic field induced change in the exchange field H_{E2} on the ions Mn2 by assuming roughly H_{E2} given by A.A. Millner et al [26] we find that H_{E2} is increased by a factor of ~ 1.015 which compares favorably with the above data. In a field of ~ 120 kOe the predictions are 1.1 for H_{E2}, giving rise to an observed enhancement of the trap fluorescence by a factor 1.45 as shown on Fig. 14.

2. Increase of Exciton Density Under Uniaxial Stress and Intersublattice Relaxation. The increase of the exciton density observed when a crystal is submitted to an uniaxial stress applied along a particular direction was first pointed out in manganese fluorides by E. Strauss et al [43] in the case of $KMnF_3$. To explain this they proposed two models a) the stress increases the self trap depth of the exciton or b) the diffusion may become anisotropic and so the probability to find a trap is reduced. This effect manifests itself both in the emission spectra and the exciton decay mode. As a result of the increased exciton density

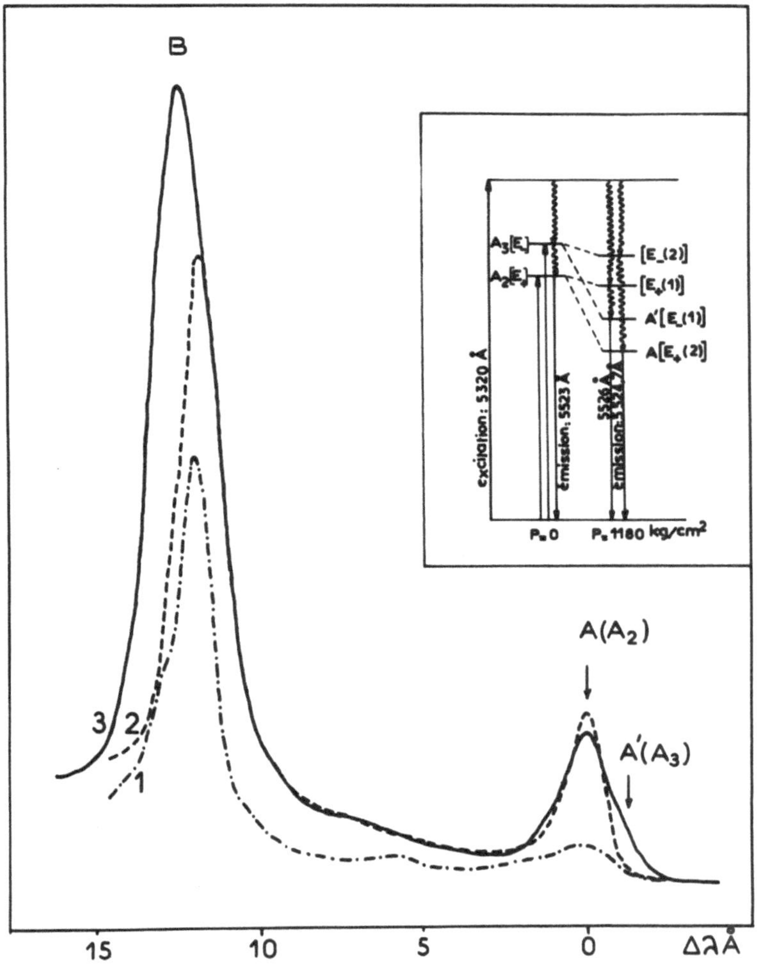

Fig. 15. Effect of a diagonal uniaxial stress on the emission spectra of $CsMnF_3$ at 2K. 1) continuous emission (-.-.) and 2) time resolved fluorescence (.5 s delay and 50 s gatewidth) at zero stress 3) continuous emission under a uniaxial stress P-1180Kg/cm^2. insert: excitation and emission energy level diagram at zero and non zero stresses.

and of the reduced trap fluorescence, the overall shape of the emission spectrum of the stressed crystal transforms progressively until it almost perfectly images the shape of the spectrum taken at short times after the laser pulse (T.R.S.) as it is observed in the unstressed crystal. Simultaneously the exciton decay shortens and its non exponentiality becomes more and more apparent.

In a previous section we have shown how the magnetic sublattice degeneracy of the absorption excitons can be lifted by applying a stress along a particular direction. Once the crystal is stressed and the degeneracy is lifted, one may test, in fluorescence, the validity of the statement that the excitations are on one or the other sublattice but not both simultaneous. In other words, the intersublattice energy transfers should be negligible, for a change of spin angular momentum of ± 2 would be needed. Such an observation has been reported by R.E. Dietz et al [28] in the case of MnF_2 by applying a [110] stress to lift the sublattice degeneracy of the exciton E1. The two sublattice excitons thus produced were separated by ~ 5 cm^{-1} (~ 1.5 Å) and equal emission intensities for both components were pointed out suggesting that a negligible relaxation occurs between them during the exciton decay.

In $CsMnF_3$ when the stress is applied along an intermediate direction joining Mn1 and Mn2 adjacent ions the lines A_2 and A_3 split into two components which were possible to attribute tentatively to distinct Mn_2 sublattice excitons. On Fig. 15 we have reported fluorescence spectra of $CsMnF_3$ at 2K in the range 5520-5540 A spectra 1) and 2) correspond to the case of zero applied stress and respectively represent continuous and time-resolved emission spectra, time-resolved spectra is used to separate the respective contributions of the intrinsic and trap fluorescences, the third spectrum again represents continuous emission but in the case of a crystal submitted to a stress of the order of 1180 kg/cm^2 i.e. a pressure at which the lower energy components of the splitted absorption excitons are separated by ~ 1.3 A (4.3cm^{-1})- see Fig. 7 and insert of Fig. 15.

It is seen immediately that the intensity ratio of the lines noted A (intrinsic exciton) and B (intrinsic magnon sideband + impurity induced exciton) compares favorably in the spectra 2 and 3 and strongly contrasts with that of spectrum 1. As expected, this agrees with a decreasing efficiency of the exciton to trap energy transfers and an increasing intrinsic fluorescence. We shall see in the following section that this uniaxial stress effect may be related to that of increasing exciton densities under high pumping conditions. In the end, Fig. 15 shows that the pure exciton line in the stressed crystal (spectrum 3) is made of two overlapping emission components coming from the splitting of the absorption lines A_2 and A_3 with approximately the same intensity ratio - see Fig. 11 -. Moreover, in the unstressed crystal the last lines do not appear together in the emission spectrum since they correspond to degenerate sublattice excitations and that the relaxation of A_3 to A_2 is very rapid. In the stressed crystal, it is thus suggested that the observation of the two overlapping emission components has to be related to the existence of specific antiparallel spin sublattice excitations with a negligible relaxation between each other within the time scale of the exciton decay, i.e. $\gtrsim$ 1 ms. However, supplementary data are neede to ascertain this point.

3. Effect of Intense Light Excitation. Fig. 16 indicates an other attractive point of the intrinsic exciton decay mode at very low temperature in $CsMnF_3$: the higher the intensity of the pump laser beam the faster the decay is at early times resulting in a more pronounced non exponentiality - see also Fig. 10 -. As in the case of stress experiments, eliminating all the other sources of non-exponential exciton decay such as a limited diffusion, back transfers from the shallow traps, radiative or phonon/magnon assisted energy transfers, the effect of pump power has to be related to increased exciton densities favoring one of the two exciton non linear de-excitation processes proposed in the beginning (section II.B,) The resulting exciton rate equation is given by:

$$\mathring{n}(t) = -p_1 n(t) - p_B n^2(t) \qquad (20)$$

where the quantities p_1 and p_B stand for the respective single and pair exciton decay rates averaged over the whole Brillouin zone - see Section IV.A-.

The solution of this equation is written as :

$$n(t) = \left[-\frac{p_B}{p_1} + \left(\frac{1}{n_o} + \frac{p_B}{p_1}\right) e^{-p_1 t}\right]^{-1} \qquad (21)$$

The exciton decay is obviously non exponential ; moreover it depends non linearly on the initial population n_o. The efficiency of radiative two exciton decays is further contained in the whole expression of the measured exciton emission intensity :

$$I(t) = \alpha p_1 n(t) + \beta p_B n^2(t) \qquad (22)$$

where α and β are the branching ratios of the one and two particle processes.

As shown in Figs. 10 and 16, the fittings are excellent. However, no n^2 dependence of the emission intensity was found and the best fits to the decays were obtained for a negligible two exciton branching ratio β, i.e. for I(t) proportional to n(t) and

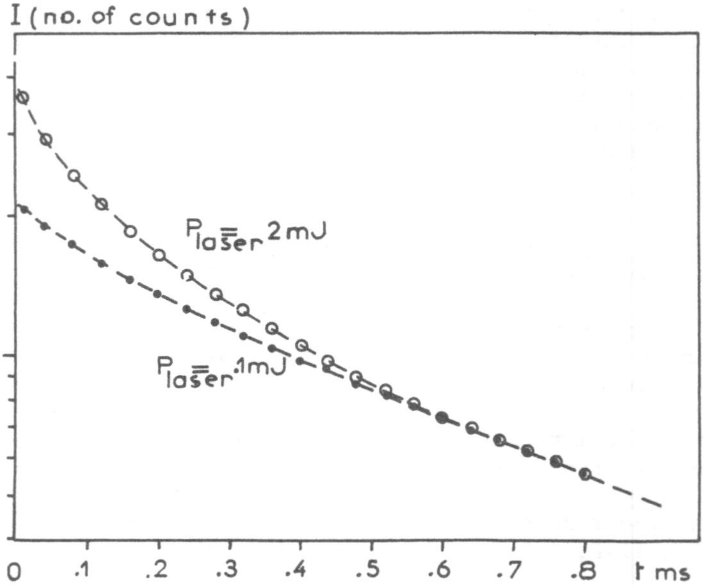

Fig. 16. Fluorescence decay of the intrinsic exciton in $CsMnF_3$ measured 1.6K with different excitation intensities. The dashed curves are the fits with Eq. (22).

$$I(t) = \left[-\frac{p_B'}{p_1} + \left(\frac{1}{I_o} + \frac{p_B'}{p_1}\right) e^{-p_1 t}\right]^{-1} \quad (23)$$

with

$$p_B \; 0 = p_B \; I_o . \quad (24)$$

The initial exciton population n_o is estimated from the laser pump power (10^{14} - 10^{16} photons), t-e absorption coefficient of the transition at the pump wavelength (~5 cm^{-1} at 532 nm) and the volume of the excited region of the crystal (~5 x 10^{-4} cm^3). In $CsMnF_3$, using Eq. [24] the data yield the following values:

$$p_1 = (1.4 \pm .1) \times 10^3 s^{-1}; \; p_B = (5.3 \pm 2.5) \times 10^{-15} \; cm^3 s^{-1}.$$

Similar results were reported in MnF_2 [4] and $KMnF_3$ [45] with p_B values of 5×10^{-13} and $4 \times 10^{-11} cm^3 s^{-1}$ respectively.

At this point, no particular assumption on the biexcitonic process was necessary. In fact no specific exciton-exciton mechanism could have been concluded so far. The most important attempts have been performed on $RbMnF_3$ and $KMnF_3$ [45,19] Under high pumping conditions, the time resolved emission of the intrinsic exciton exhibits a notable spectral shift, of the order of 1 cm^{-1}, during the exciton decay. An identical effect is found while observing the emission line at a fixed delay time after the laser pulse for different pump powers. Then the two following processes were advanced :

1) the annihilation of two excitons at +k and -k at short times and/or for high pumping levels, followed by a crossover from zone boundary emission to zone center k-0 emission as time evolves and/or for decreasing pump powers. We refer to process of Section II.B. This is plausible since all the exciton states with different k values are populated - see Section IV.A. - ; moreover a process involving the annihilation of excitons on opposite sublattices ($\Delta S=0$) with wavevectors $\pm k$ anywhere in Brillouin zone should predominate over the single exciton emission process ($\Delta S=1$) occurring at k=0 only. It results in a time and/or a pump power dependent displacement of the emission line which is directly related to the size of the exciton dispersion.

2) exciton-exciton collision process. The model has been developed in great details for the need of the atomic spectroscopy [6]; it depicts the migration of the excitons in terms of particle trajectories, which may apply here if the exciton gas is assumed to be sufficiently dilute to consider only "two particle" collisions at a time. This results in a reshifting as well as a broadening of the levels which is linearly dependent on the exciton density, in good agreement with the observed alteration of the E1 exciton line in $KMnF_3$ with time and/or excitation intensity.

In conclusion, though the above non-linear processes give a good description of the exciton dynamics (the exciton decay mode), their physical significance is still a matter open to discussion. Moreover, other experimental approaches are proved to be necessary for a better knowledge of the size (and sign) of the exciton dispersion.

ACKNOWLEDGEMENTS

Thanks are expressed to Dr. L. D. Brunel (CNRS, Grenoble, France), Prof. B. Di Bartolo and Dr. D. Pacheco (Boston College, USA) for their helpful contributions to this work.

REFERENCES

1. V. Goldberg, R. Moncorgé, D. Pacheco and B. Di Bartolo in Luminescence of Inorganic Solids, B. Di Bartolo, ed., Plenum Press, New York (1978).
2. H. Yamanoto, D. S. McClure, C. Marzaco and M.Waldman, Chem. Phys. 22, 79 (1977).
3. R. Ya. Bron, V. V. Eremenko and E. V. Matjushkin, Sov. J. Low Temp. Phys. 5, 314 (1979).
4. B. A. Wilson, J. Hegarty and W. M. Yen, Phys. Rev. Lett. 41, 268 (1978).
5. R. L. Greene, D. D. Sell, R. S. Fergelson, G. F. Imbusch and H. J. Guggenheim, Phys. Rev. 171, 600 (1968).
6. D. J. Breed, Physica 37, 35 (1967).
7. J. A. Van Jiujka, A. F. Arts and H. W. de Wijn, Phys. Rev. B21, 1963 (1979) and ref. therein.
8. G. A. Samara and J. F. Scott, Sol. St. Comm. 21, 167 (1977).
9. Y. Yanaguchi and T. Sakuraba, J. Phys. Soc. Jap. 38 1011 (1975).
10. D. Khatamian and M. F. Collins, Can. J. Phys. 55, 773 (1977).
11. D. D. Sell, R. L. Greene and R. M. White, Phys. Rev. 158, 489 (1967).
12. J. H. Van Vleck, Z. Phys. Chem. 41, 67 (1937).
13. B. Jacquier, R. Moncorgé and B. Di Bartolo, Sol. St. Comm. 31, 693 (1979) and ref. therein.
14. J. B. Parkinson and R. Loudon, J. Phys. C 1, 1568 (1968).
15. S. E. Stokowski, D. D. Sell and H. J. Guggenheim, Phys. Rev. B. 4, 3141 (1971).
16. F. Auzel in Radiationless Processes, B. Di Bartolo, ed., Plenum Press, New York, (1979).
17. R. Moncorgé, B. Jacquier and F. Gaume, J. Mag and Mag. Mat. 15-18, 817 (1980).
18. T. C. Chiang, P. Salvi, J. Davies and Y. R. Shen, Sol. St. Comm. 26, 217, 527 (1978).
19. V. Gerhardt, W. Gebhardt, U. Kellner and E. Strauss (D.P.C. Conference, Madison USA, Bull. Am. Phys. Soc. 24, 898 (1979).
20. W. M. Yen, G. F. Imbusch and D. L. Huber in Optical Properties of Ions in Crystals, H. Crosswhite and H. Moos, eds., Inter Science, N.Y. 301, (1967).
21. F. M. Johnson and A. H. Nethercot, Jr., Phys. Rev. 114, 705 (1959).
22. R. E. Dietz, A. Nisetch and H. J. Guggenheim, Phys. Rev. 16, (1966).
23. R. Moncorge, B. Jacquier and L. C. Brunel (to be published in J. Phys. Chem. Sol.)
24. E. G. Petrov and V. S. Oxtrovskii, Sov. J. Low Temp. Phys. 2, 713 (1976).
25. V. V. Eremenko and E. G. Petrov, Adv. in Phys. 26, 32 (1977).

26. A. A. Mil'ner and Yi. A. Popkov, Sov. J. Low Temp. Phys. 3, 92 (1977).
27. J. F. Holrichter, R. M. MacFarlane and A. L. Shawlow, Phys. Rev. Lett. 26, 652 (1971).
28. R. E. Dietz, A. E. Meixner and H. J. Guggenheim, J. of Lumin. 1, 2 279 (1970).
29. B. A. Wilson, Ph.D. Thesis (University of Wisconsin, Madison USA) 1978.
30. R. M. McFarlane and A. C. Luntz, Phys. Rev. Lett. 31, 832 (1973).
31. H. T. Chen and R. S. Meltzer, Phys. Rev. Lett. 44, 599 (1980).
32. J. Hegarty, B. A. Wilson, W. M. Yen, T. J. Glynn and G. F. Imbusch, Phys. Rev. B18, 5812 (1978).
33. D. L. Huber, Phys. Rev. B20, 2307 (1979).
34. Th. Förster, Disc. Faraday Soc. 27, 7 (1959).
35. M. Inokuti and F. Hirayama, J. Chem. Phys. 43, 1978 (1965).
36. A. I. Burshtein, Sov. Phys. JETP 35, 882 (1972).
37. M. Yokota and O. Tanimoto, J. Phys. Soc. Jap. 22, 779 (1967).
38. J. Heber, Phys. St. Sol. (b) 48, 319 (1971).
39. K. K. Gosh, J. Hegarty and D. L. Huber, Phys. Rev. B22, 2837 (1980).
40. V. P. Sakun, Sov. Phys. Sol. St. 21 (3), 390 (1979).
41. B. A. Wilson, W. M. Yen, J. Hegarty and G. F. Imbush, Phys. Rev. B19,4238 (1979).
42. N. M. AAmer, T. C. Chiang and Y. R. Shen, Phys. Rev. Lett. 34, 1454 (1975).
43. E. Strauss, Thesis (University of Regensburg, Germany) 1977.
44. E. Strauss, V. Gerhardt and W. Beghardt, J. Lumin. 18/19, 151 (1979).
45. E. Strauss, W. J. Miniscalco, W. M. Yen, U. C. Kellner and V. Gerhardt, Phys. Rev. Lett. 44, 824 (1980).
46. G. A. Thomas, A. Frova, J. C. Hensel, R. E. Miller and P. A. Lee, Phys. Rev. B13, 1692 (1976).

SPECTROSCOPY OF STOICHIOMETRIC LASER MATERIALS:

EXCITONS OR INCOHERENT TRANSFERS?

F. Auzel

Centre National d'Etudes des Télécommunications
196 rue de Paris
92220 BAGNEUX (France)

ABSTRACT

In the so-called stoichiometric laser materials, a high concentration of active ions is necessitated by the micrometric dimensions of the laser. Such a high concentration leads to strong interactions between ions giving rise to a self-quenching process explained either in an energy transfer model with fast diffusion among donors before transfer to a quenching center or in an excitonic model with exciton annihilation at the sample surface. Theoretical and experimental aspects of the spectroscopic and dynamical studies shall be presented with reference to usual laser materials along with a general discussion of self-quenching.

I. INTRODUCTION

In the search for micronsize lasers for telecommunications through optical fibers, new types of inorganic Lanthanide (Ln^{3+}) materials have recently emerged : namely stoichiometric or self-activated Ln^{3+} laser materials. In such materials what is looked for is essentially a high gain per unit length. To fulfil this requirement, a high concentration of active ions is necessary ($> 10^{21}$ cm^{-3}). The ideal case is the one where Ln^{3+} is no longer a dopant in the host crystal as in YAG, but a 100 % constituent of the matrix. The first example of such a stoichiometric laser material was given by Varsanyi who could obtain a laser action in $PrCl_3$ [1]. The second esample has been NdP_5O_{14} [2] the studies of which gave a new impulse to Ln^{3+} spectroscopy particularly in trying to solve the problem of self-quenching. The question is whi in such materails, this last process is so small as to permit the existence of fluorescence. In usual phosphors or in YAG for instance, the optimimum Ln^{3+}

centration amounts only to $10^{-17} \sim 10^{-19}$ cm^{-3}, whereas in $PrCl_3$ and in NdP_5O_{14} it reaches respectively 10^{22} and $4\ 10^{21}$ cm^{-3}. At this point it should be noted that in the literature dealing with such laser materials, the adjective "stoichiometric" is often taken as being synonymous with "small-quenching" since only this last case is of practical interest.

In a stoichiometric material the energy transfer situation can be though of as being different from that in the doped matrix. In fact it could be view exactly as a Frenkel exciton situation because of the existence of a translational symmetry for coupled active ions and because excited electrons remains on the same ion.

In each lattice cell there is a fixed number of Ln^{3+} ions, the electronic transitions of which can be coupled by the multipolar electrostatic interactions involved in usual energy transfers between Ln^{3+} .

In this article, we shall address ourselves the following questions :

- Does the excitonic situation provides new behaviour unexpected from the statistical doping case ?
- What could be the present understanding of the self-quenching problem in stoichiometric materials particularly the ones with Nd^{3+} ?

II. EXPERIMENTAL RESULTS [3]

Before discussing models, it is worthwhile to summerize the observations which they have to explain :

- When the concentration of active centers is increased, the pervading effect of self-quenching takes place. However, the probability of concentration quenching is found to vary linearly with concentration in small-quenching materials (stoichiometric materials), whereas it varies quadratically in strong-quenching ones.*
- There are negligible or very small spectral changes when the active ions are diluted with neutral ions of about same radius [4]. This last point has to be emphasized since for NdP_5O_{14} crystal dilution with Y^{3+} leads to crystal structural changes inducing spectral changes which could be misleading. Dilution of Nd^{3+} with La^{3+} does not produce such an artefact. Ultraphosphates of Ln^{3+} (LnUP) crystallize in two main classes : LaUP and TbUP are pseudo- orthorhombic ($\beta \simeq 90°$) ; TbUP to YbUP including YUP and LnUP are monoclinic ($\beta \simeq 91.5°$) with almost equal axes [5].
- Oscillator strengths are found to be concentration independent [4].

*concentration quenching rate W_Q is defined as :

$$W_Q = 1/\tau - 1/\tau_0$$

where τ_0 is the radiative lifetime and τ the actual lifetime with quenching.

- Fluorescence decay is exponential over several decades.
- Self-quenching is independent of temperature.

These are the basic gross features which have to be discussed and which have been presented in previous Erice Schools [3a,3b].

However some finer variations from these general lines could be found and shall be seen later.

III. INTERPRETATIONS OF RESULTS

III.A. The Exciton Concept

The strong self-quenching of Nd^{3+} in $(WO_4)_2$ NaGd:Nd [6], LaF_3:Nd [7], YAG:Nd [8] or glass [9] has been well-known for a long time to be due to the following cross-relaxation energy transfer (Fig. 1) :

$$Nd_A^{3+}(^4F_{3/2}) + Nd_B^{3+}(^4I_{9/2}) \rightarrow 2Nd_{A,B}^{3+}(^4I_{15/2}) \quad (1)$$

where A and B are two nearby sites.
such transfer is rather of the non-radiative resonant type. When energy matching for resonance is poor, then phonon-assisted energy transfer may take place in addition to Eq. (1) according to :

$$Nd_A^{3+}(^4F_{3/2}) + Nd_B^{3+}(^4I_{9/2}) \rightarrow 2Nd_{A,B}^{3+}(^4I_{13/2}) + N_1\hbar\omega_m \quad (2)$$

or

$$Nd_A^{3+}(^4F_{3/2}) + Nd_B^{3+}(^4I_{9/2}) \rightarrow Nd_A^{3+}(^4I_{13/2}) + Nd_B^{3+}(^4I_{15/2}) + N_2\hbar\omega_m \quad (3)$$

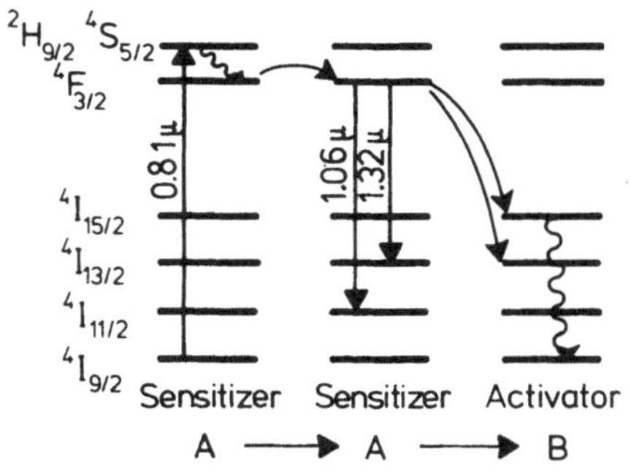

Fig. 1. Diffusion and cross-relaxation in self-quenching of Nd^{3+}.

All these cross-relaxation processes can be temperature independent since they are either resonant or with emission of phonons. Other phonon-assited transfers involving absorption of phonons would systematically involve a temperature dependence and for such reason are not considered.

However, cross-relaxations were first rejected in the case of NdP_5O_{14} because the linear variation of quenching with concentration was thought not to be compatible with a cross-relaxation process [10,11]. Excitonic behavior was then invoked to explain this variation [11,12]. In fact, very few papers have considered for itself the excitonic behavior of Ln^{3+} in stoichiometric material. Let us examine this hypothesis. Before considering the energy migration by excitons, the experimental proofs of the existence of excitons have to be found in their coherent behavior: namely the existence of some $\vec{k}$ dependence in the observed energy and the existence of Davidov splitting [13] when possible; k being the pseudo-momentum of the lattice.

For LnP_5O_{14}, the Ln^{3+} ions are in a regular array with four equivalent ions per unit cell [14]. For an electronic non-degenerate state of Ln^{3+}, Davidov splitting with four components is then likely to happen.

Recently, Shelby and MacFarlane, by a careful study at low temperature (≃ 2 K) of the $^5D_0 \rightarrow {}^7F_0$ transition of Eu^{3+} in EuP_5O_{14} [15] have shown that an inhomogeneous linewidth of 3.5 Ghz is found and that their photon-echo technics reveal that each inhomogeneous part of the line still vibrates with various longer lifetimes corresponding to varying smaller homogeneous linewidths as shown in Fig. 2.

Such line shape cannot be accounted for by a four component Davidov splitting mechanism. However, the observation of the frequency-dependent dephasing (homogeneous width) was attributed to a stronger Eu^{3+}-Eu^{3+} interaction in the center of the inhomogeneous line due to a larger effective concentration of resonant ions. Then it could be appropriate to characterize the excitation as a wave packet extending over several lattice sites [15]; a greater delocalization being found in the center of the inhomogeneous line. A completely localized model (hopping exciton) would require the dephasing to be fast with respect to the energy transfer between Eu^{3+} and hence would be independent of the position in the inhomogeneous line.

To the best of our knowledge, this is the only proof of a weak coherent excitonic behavior in the Ln^{3+} stoichiometric materials, not involving deconvolution of combined collective excitation [34].

This situation is confirmed when one looks at order of magnitudes for interaction strengths between Ln^{3+}.

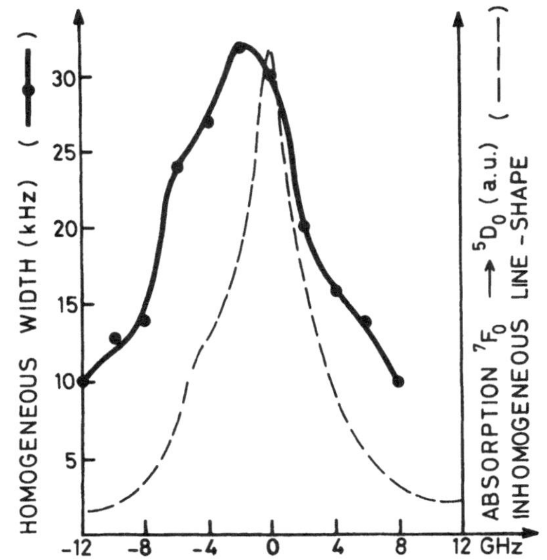

Fig. 2. Homogeneous line-widths in the inhomogeneous line shape for absorption of $^7E_0 \rightarrow {}^5D_0$ at 2 K (kT $\simeq$ 1.4cm^{-1}) in EuP_5O_{14} after [15].

Since in LnP_5O_{14} the distance between Ln^{3+} is smallest along the direction of the crystallographic a-axis [14], we are somewhat justified in considering that interactions occur preferentially along this axis and therefore in assuming a one-dimensional interaction with an excitonic energy of :

$$E(k) = E_o + 2\beta \cos ka$$

where E_0 is the ionic excited energy at one site, β the interaction between two ions separated by the lattice vector a and k the quasi-momentum vector of the crystal.

When the Brillouin zone is spanned the maximum excitonic width, $\Delta\nu_{excit.}$, is found to be 4β.

Taking a typical interaction between Nd^{3+} of $10^{10} s^{-1}$ [11,16] gives:

$$\Delta\nu_{excit} = 4\beta = 4\ 10^{10}/3\ 10^{10} = 1{,}3\ cm^{-1}$$

which is much larger than the observed homogeneous width and even the inhomogeneous one (respectively 10^{-6} cm^{-1} and 0.11 cm^{-1}) [15].

On the other hand, direct measurements of the diffusion distance of the $^4F_{3/2}$ excitation in NdP_5O_{14} along the a-axis of the crystal, making use of a standing wave grating of excited states, leads to a diffusion distance less than 320 Å [17]. This order of magnitude was quite recently confirmed by a four-wave mixing experiment [18] at room temperature.

From this value, we can try to see whether it is in accord with a hopping model where the exciton would have a diffusion length large compared with the region of coherence of the exciton. This situation being indicated by the small observed variation of bandwidth, and by the small oscillation strength ($\approx 10^{-6}$) yielding a nuclear relaxation before transfer.

Making use of the relation given by Wolf for diffusion of molecular Frenkel excitons [19] we have the following relations for the diffusion constant :

$$D = \frac{a^2 Z}{\tau} = \frac{\bar{r}^2}{\tau} \tag{4}$$

where $\bar{r}^2 = a^2 Z$ is the one-dimension diffusion length ; τ the ion measured lifetime ; Z the number of hops per lifetime ; a, the distance for one hop.

The hopping time is then given by :

$$t_H = \tau/Z \tag{5}$$

which can be linked to the interaction strength β by [20]:

$$\beta = \hbar/4t_H \tag{6}$$

From relations (4), (5) and (6) we can obtain an estimate of a, the hopping distance involved in such an exciton diffusion as:

$$a = \frac{\bar{r}}{2}\sqrt{\frac{\hbar}{\tau\beta}} \tag{7}$$

with $\bar{r} \leqslant 320$ Å ; $\tau = 120\ \mu s$ and $\beta = 6{,}6\ 10^{-17}$ erg ($10^{10}\ s^{-1}$)

One obtains : $a < 5.7$ Å indicating that this model is in accord with a model of exciton trapped at each Nd^{3+} site since the minimum distance between Nd^{3+} sites in NdP_5O_{14} is of same order (= 5.2 Å).

At this point it is worthwhile noting that Ln^{3+} interactions between excited states have also been considered as biexciton annihilation in stoichiometric materials such as $TbPO_4$ and TbP_5O_{14} [21, 22]. The argument for a biexciton model being the form of the phenomenological rate equation describing the kinetics of the excitation of the population (n) of the $^5D_4(Tb^{3+})$ state. At high excitation a bimolecular term has to be introduced giving an equation of the general form :

$$dn/dt = \alpha I - \beta n - \gamma n^2 . \tag{8}$$

Where αI is the excitation term, βn and γn^2 respectively the linear and bimolecular decays. This is analogous with biexcitonic singlet or triplet quenching of molecular crystal [23,24] described by the same phenomenological rate equation.

However, in the case of a molecular crystal such as anthracene, the existence of excitons as shown by the presence of Davidov splittings has been well documented [13]; such is not the case for Ln^{3+} crystals.

In rare-earth spectroscopy, equations of type (8) are now well known to describe radiative or non radiative up-conversion effects either of the sequential energy transfer (ATPE) or of the cooperative type [25,26] without having to rely on the exciton concept.

In the problem of self-quenching of stoichiometric materials, the concept of exciton annihilation was introduced essentially to explain the linear dependence of the quenching rate with the Ln^{3+} concentration [11].

In the following we shall see that it can be as well explained by the general energy transfer model.

III.B. The Energy Transfer Approach

In the absence of a typical Frenkel exciton behaviour for stoichiometric Lanthanide materials, we consider that they can be described by the usual Dexter theories for non radiative energy transfer just as in non-stoichiometric materials.

Probability calculations for such transfers take two aspects [3b]: i) a microscopic one ; it is the theoretical case of transfer from one sensitizer to one activator :

Probability is : $P_{SA}(R) = (R_o/R)^s/\tau_s$

with $R_o^s \propto S_{AB}$ where s = 6,8,10 respectively for dipole-dipole, dipole quadrupole and quadrupole-quadrupole interaction ; τ_s is the sensitizer lifetime ; R is the distance between the sensitizer (S) and the activator (A) S_{AB} is a generalized overlap integral between the line-shape functions of the sensitizer and the activator, taking into account eventually multi-phonon sidebands :

$$S_{SA} = \int g_s(\nu) g_A(\nu)\, d\nu, \quad \text{for the electronic resonant case.}$$

$$S_{SA} = S(0)\exp-\beta\Delta E \quad , \text{for the multiphonon}$$

assisted case ; ΔE being the electronic energy mismatch.

ii) The macroscopic one which is the experimental case of energy transfer from all sensitzer to all activators in the sample.

In a long time approximation, the overall emission of sensitizers is given by:

$$I_s(t) = \exp(-t/\tau_s - t/\tau_D)$$

and for a dipole-dipole interaction:

$$\tau_D^{-1} = VN_sN_A \text{ or } \tau_D^{-1} = U\ N_A \qquad (9)$$

where the diffusion among sensitizers is respectively slower or faster than the transfer between sensitizer to activator. These are the so-called diffusion limited and fast diffusion regime ; V and U are proportionality coefficients, the magnitudes of which depend on the strength of the microscopic probability and particularly on S_{SA}, as we shall see later.

In a stoichiometric material with only one type of Ln^{3+} ion, (cross-relaxation) equations (9) becomes:

$$\tau_D^{-1} = VN^2 \text{ or } \tau_D^{-1} = UN$$

showing that either a quadratic or a linear dependence can be expected according to the diffusion regime. Since diffusion involves identical multiplet levels (Fig.1) the diffusion is of same order for different materials [3b]; whereas energy transfer from S to A involving multiplets with different Stark crystal field splitting may be different. We have shown that, for Nd^{3+}, the energy matching needs a strong enough crystal field, see Fig. 3 [27]; for the free ion case (zero splitting), the energy transfer is not possible because of a defect in energy matching $(e_2 - e_1) < 0$. From this consideration, it is clearly understood that a small crystal field material (with small stark splitting) may have a small generalized overlap integral S for cross-relaxation which is then smaller than the constant overlap for diffusion S. This leads to a fast diffusion regime. Likewise, strong field materials may have larger S with

$$S_{SA} \simeq S_{SS}$$

leading to a limited diffusion regime.

Then the linear and the quadratic quenching are clearly found to belong to respectively small and strong crystal field materials when diffusion is independent of the material. If not, intermediate cases can be expected. So the different behaviour of $YAG{:}Nd^{3+}$ and NdP_5O_{14} can be explained without exciton consideration.

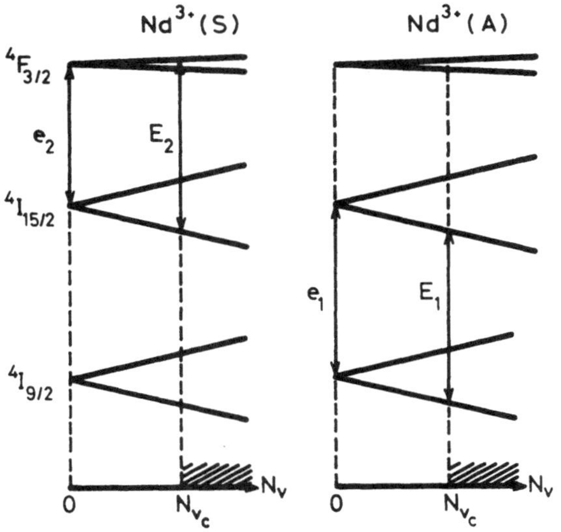

Fig.3 Maximum splitting of levels under crystal field (N_v) involved in energy-matching for cross-relaxation ; Hatched zone indicates strong-quenching region ($E_2-E_1 > 0$) ; free-ion case ($e_2-e_1<0$) and small field region ($E_2-E_1<0$) corresponds to small-quenching case ($N_v<N_{vc}$).

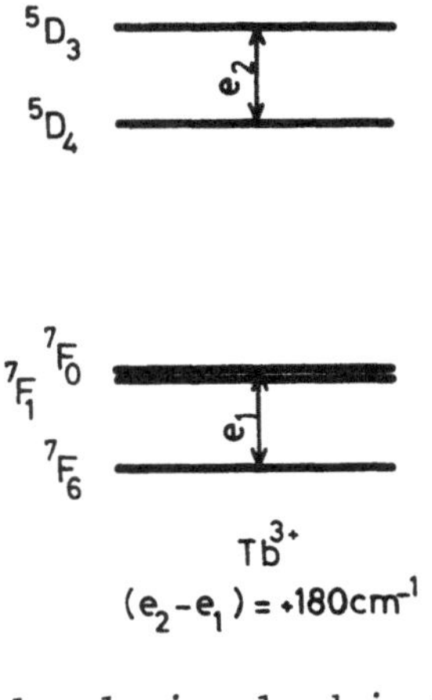

Fig. 4 Free ion energy levels involved in Tb^{3+} ($^5D_3$) cross-relaxation.

Such a crystal field consideration as the fundamental difference between small and strong quenching Nd^{3+} materials pushes somewhat aside the role of the maximum shortest distance between ions found in the literature :

A large shortest distance would favour small quenching [2,14, 28,30,31,32] and such is apparently the case in NdP_5O_{14} (Nd-Nd ≈ 5,2,Å in comparison with YAG (Nd-Nd ≈ 3,3 Å).

However, recent experimental results on TbP_5O_{14} and $TbNa(WO_4)_2$ [33] give clear proof that self-quenching can be important when energy matching conditions are fulfilled for transfers even in stoichiometric materials with large shortest distances. As predicted [36] it is not likely that weakly quenching hosts for $Tb^{3+}(^5D_3)$ could ever be found even for Tb^{3+} - Tb^{3+} distance of the order of 5.2 Å. This comes from a positive free ion energy difference (Fig. 4) These results are summarized in Table 1.

Table I shows in comparison with Nd^{3+} ($^4F_{3/2}$), that for $Tb^{3+}(^5D_3)$ quenching may even be smaller for materials with the shortest distance and that there is no direct correlation between this distance and the slope for the probability with respect to concentration (log-log).

Table 1. Opposite Self-Quenching Properties of Nd^{3+} ($^4F_{3/2}$) and Tb^{3+}($^5D_3$) in Stoichiometric Hosts with Large and Small Minimum Distance Between Active Ions at Concentration of 4.10^{21} cm^{-3}.

Ln^{3+} / Host	$Nd^{3+}(^4F_{3/2})$		$Tb^{3+}(^5D_3)$		d_{min} Ln^{3+} - Ln^{3+}
	W_Q	slope	W_Q	slope	
LnP_5O_{14}	$5\ 10^3 s^{-1}$	≃ 1	$2\ 10^5 s^{-1}$	≃2	5.2 Å
$NaLn(WO_4)_2$	$10^5\ s^{-1}$	≃ 2	$7\ 10^4 s^{-1}$	≃1	3.9 Å

IV. CONCLUSION

In stoichiometric Ln^{3+} materials, it is found that the hopping distance is of the same order as the lattice constant. A collective description of the excitation then leads to the consideration of an exciton trapped at each site. Also, due to the small interaction between ions with respect to their own excitation energy, a usually negligeable pseudomomentum dependence is found for the collective excitation. The consequence of this situation is that the same considerations of energy transfers as in random doped systems can explain the small-quenching behaviour of certain Nd^{3+} stoichiometric material. This peculiar behaviour is found to be essentially linked to the free ion energy matching conditions for cross-relaxation between Ln^{3+}. Generalization to another ion and host gives clear proof that there is no small-quenching material "per se"; the consideration of large smallest distance between like ions is not really important to insure small-quenching in stoichiometric materials as is generally proposed in the literature.

REFERENCES

1. F. Varsanyi, Appl. Phys. Lett. 19, 169 (1971).
2. H. G. Danielmeyer and H. P. Weber, IEEE Q.E.8, 805 (1972).
3. See, for instance:
   (a) A. A. Kaminskii, in Luminescence of Inorganic Solids, ed. by B. Di Bartolo (Plenum, New York, 1978), p. 511.
   (b) F. Auzel, in Radiationless Processes, ed. by B. Di Bartolo and V. Goldberg (Plenum, New York, 1978), p. 213.
   (c) G. Huber, Current Topics in Material Science, Vol. 4 (Springer, 1980), p. 2.
4. F. Auzel, IEEE Q.E.12, 258 (1976).
5. W. W. Krühler, G. Huber and H. G. Danielmeyer, Appl. Phys. 8, 261 (1975).
6. G. E. Peterson and P. M. Bridenbaugh, J. Optic. Soc. Am. 54, 644 (1975).
7. C. K. Asawa and M. Robinson, Phys. Rev. 1, 251 (1966).
8. H. G. Danielmeyer, in Advances in Lasers, Vol. II (Decker, 1975).
9. J. Chrysochoos, J. Chem. Phys. 6, 4596 (1974).
10. H. G. Danielmeyer and M. Blätte, Appl. Phys. 1, 2691 (1973).
11. J. M. Flaherty and R. C. Powell, Phys. Rev. B19, 32 (1979).
12. R. C. Powell, D. P. Neikirk, J. M. Flaherty and J. G. Gualtieri, J. Phys. Chem. Solids 41, 345 (1980).
13. A. S. Davidov, Theory of Molecular Excitons (Plenum, New York, 1971), p. 82.
14. H. Y. P. Hong, Acta Cryst. B30, 468 (1974).
15. R. M. Shelby and R. M. MacFarlane, Phys. Rev. Lett. 45, 1098 (1980).
16. M. Blätte, H. G. Danielmeyer and R. Ulrich, Appl. Phys. 1, 275 (1973).
17. P. F. Liao, H. P. Weber and B. C. Tofield, Solid State Commun. 16, 881 (1975).
18. C. M. Lawson, R. C. Powell and W. K. Zwicker, Phys. Rev. Lett., to be published.

19. H. C. Wolf, in Advances in Atomic and Molecular Physics, Vol. 3, ed. by D. R. Bates (Academic Press, 1967), p. 119.
20. M. D. Fayer and C. B. Harris, Phys. Rev. B9, 748 (1974).
21. P. C. Diggle, K. A. Gehring and R. M. MacFarlane, Solid State Commun. 18, 391 (1976).
22. T. Fukuzawa and S. Tanimizer, J. of Lumin. 16, 447 (1978).
23. A. Bergman, M. Levine and J. Jortner, Phys. Rev. Lett. 18, 112 (1972).
24. T. Kobayashi and S. Nagakura, Mol. Phys. 24, 695 (1972).
25. F. Auzel, Proc. IEEE 61, 758 (1973).
26. M. Hirano and S. Shionoya, J. Phys. Soc. Japan 33, 112 (1972).
27. F. Auzel, Mat. Res. Bull. 14, 2223 (1979).
28. S. R. Chinn, H. Y. P. Hong and J. W. Pierce, Laser Focus 64 (May 1976).
29. B. Blanzat, J. P. Denis and J. Loriers, Proc. 10th R. E. Conf. 2, 1170 (1973).
30. C. Brecher, J. Chem. Phys. 61, 2297 (1974).
31. Y. V. Denisov, Y. I. Krasilov, N. F. Perevoschikov, I. A. Rozanov and N. N. Chudinov, Izest. Akad. Nak. SSSR, Neorg. Mater. 12, 1061 (1976).
32. R. M. Brewer and M. Nicol, J. of Lumin. 21, 367 (1980).
33. F. Auzel and J. Dexpert, to be published.
34. First proof of the existence of W^{3+} excitons were obtained by deconvolution of magnon-exciton and exciton-exciton transitions: R. S. Meltzer and H. W. Moos, Phys. Rev. B6, 264 (1972); R. S. Meltzer, Solid State Commun. 20, 553 (1976).

ELECTRON-HOLE DROPLETS IN SEMICONDUCTORS

C. Benoit à la Guillaume

Groupe de Physique des Solides - Université de
Paris VII, Tour 23 - 2, Place Jussieu
75251 Paris, France

ABSTRACT

The main properties of electron-hole droplets (EHD) in semiconductors are presented, with particular emphasis on collective aspects.

1) The theoretical problem of EHD stability is outlined and a simple model is developped to show the effect of band structure on EHD properties. Next, the effect of coupling of electrons to Lo phonon is considered, and comparison with experimental data is given.

2) Modifications in EHD of the one electron properties are considered : mass renormalisation, effect of the electron-hole correlation on radiative or Auger lifetime.

3) The optical constants of EHD is considered in relation with FIR or IR absorption and light scattering experiments.

4) Problem of inhomogeneous electron hole liquid (EHL) is considered: surface energy, perturbation by shallow impurities or iso-electronic impurities. A critical review of the experiments available is given.

I. INTRODUCTION

The existence of a collective state analogous to a liquid metal in highly excited semiconductors was predicted by Keldysh in 1968, and was demonstrated and widely accepted in the period 1970-1974. A good historical introduction can be found in the review paper of T.M. RICE [1], as well as an excellent theoretical treatment of EHL. In this paper, we will recall the principle of EHL stability and develop a simple model to demonstrate the large role of the band structure on EHL stability. Then, the role on EHL stability of the coupling of electronic states to Lo phonons in polar semiconductor will be discussed. In part III, we will consider the modifications of one electron properties (which dominate for example the recombination radiation line shape) due to collective effects : electron hole correlation coefficient, mass renormalisation. In part IV, we will discuss the optical constant of EHL, which are dominated by plasmon resonance, as we shall guess. In part V, we will consider the problem of inhomogeneous EHL, with the help of the density functional formalism : surface energy, perturbation by shallow impurities, properties of EHL in doped material, effect of isoelectornic impurities.

II. EHL STABILITY

In non polar semiconductor, the mean energy E(n) of an electron hole pair in a plasma of density n appears as a sum of terms corresponding to increasing order in the perturbation theory.

$$E(n) = E_k + E_{ex} + E_{corr} \tag{1}$$

where E_k is the kinetic energy, E_{ex} the exchange energy and E_{corr} the correlation energy which contains all higher order terms. Only the first two terms can be calculated exactly, and the correlation energy requires sophisticated method of the many body theory [2-4].

We present here a simple "recipe" intended for practical utilisation. First, one should define appropriate units of length and enrgy which are respectively the exciton Bohr radius $a_x = \varepsilon/\mu e^2$ and the exciton Rhydberg $R_x = \mu e^4/2\varepsilon^2$, where ε is the dielectric constant and μ the exciton reduced mass.

Then, a dimensionless inter particle spacing r_s is defined as

$$r_s = (3/4\pi n)^{1/3}\, a_x^{-1}$$

Eq. (1) is now written, in the above defined unit, as:

$$E(r_s) = E_k(r_s) + E_c(r_s)$$

where $E_c(r_s)$ is now the Coulomb energy (that is the sum $E_{ex}+E_{corr}$). Our recipe is based on the following remark : By inspection of all the published theoretical calculations, it appears that $E_c(r_s)$ is a universal function, that is : it is nearly independant of the band structure. There is also an approximate theoretical justification to that statement : the solution of the many body problem is equivalent to the knowledge of $\varepsilon(q,\omega)$, the wave vector and frequency dependant dielectric constant. For a metallic system $\varepsilon(q,\omega)$ is dominated by the plasmon resonance. This is the so called "Plasmon pole approximation". Since the plasmon resonance is quite insensitive to details of the band structure, the stability of $E_c(r_s)$ against band structure change is thus justified.

It is easy to compute the Fermi energy $E_k(r_s)$.

$$E_k(r_s) = \frac{2,21}{r_s^2} \left(\frac{\mu}{\nu_e^{2/3} m_{de}} + \frac{\mu}{\nu_h^{2/3} m_{dh}} \right) \quad (3)$$

where ν_e and ν_h are the numbers of equivalent conduction or valence valley, m_{de} and m_{dh} are the density of state mass of electrons and holes. The main features of band structure which influence $E(r_s)$ can now be understood :

a) Increasing the number of valleys ν_e decreases E_k but leaves the units a_x and R_x unchanged.

b) An anisotropic electron valley lowers the ratio μ/m_{de} because $m_{de} = (m_{//e}\, m_{\perp e}^{2})^{1/3}$ and $\frac{1}{\mu} = \frac{1}{m_{oe}} + \frac{1}{m_{oh}}$ and $\frac{1}{m_{oe}} = \frac{1}{3}\left(\frac{1}{m_{//e}} + \frac{2}{m_{\perp e}}\right)$. Hence E_k is lowered.

c) A degerate band, like the top of valence band in Ge lowers also E_k because $\frac{1}{m_{oh}} = \frac{1}{m_{h\ell}} + \frac{1}{m_{hh}}$ and $m_{dh}^{3/2} = m_{\ell h}^{3/2} + m_{hh}^{3/2}$

Fig. 1 shows examples of the recipe. Three different band structures are considered : direct gap material with simple spherical band, Si and Ge. $E_c(r_s)$ is shown to be essentially unchanged ; on contrary $E_k(r_s)$ decreases dramatically in the sequence. The energy and density in the ground metallic state is thus easily determined.

We consider the role of the coupling of electronic states to Lo phonon on the energy of the metal liquid. This is of particular importance in direct band gap III-V or II-VI compound where the simple theory, according to Fig. 1 predict that EHD is not stable. One can understand the role of Lo phonons by noticing that the polarizability of the lattice enters now in $\varepsilon(q,\omega)$. If $\hbar\omega_{Lo} \gg \hbar\omega_p$, the contributions to ε of optical phonons and plasmons simply add. One can in this case simply use the $\varepsilon(o)$ dielectric constant and polaron masses for electron an hole. Clearly, this will not improve the binding of EHD with respect to free exciton. This is the case of GaAs. On the contrary, if $\hbar\omega_{Lo} \sim \hbar\omega_p$, we obtain coupled phonon-plasmon modes ω_+ and ω_- which causes a relative reduction of screening, thus a lowering of the energy of EHL. Of course, in that case, the ground state of the free excitons (FE) is also lowered, but in general, the effect is stronger for the EHL. This is for example the case of CdS. A much stronger effect is predicted in AgBr.

On the experimental side, the accuracy of the determination of the energy and density of the liquid depends primarily upon the lifetime of the excitations, FE and EHL. In direct gap material,

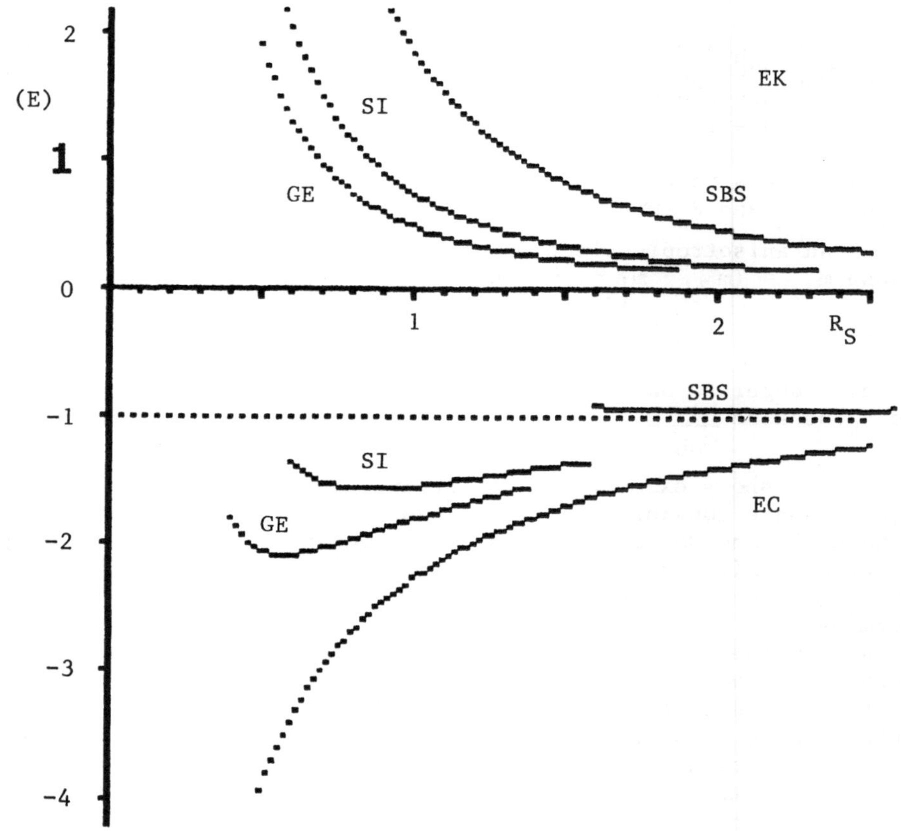

Fig. 1. The Coulomb energy has been taken as $E_c = -1.9\ r_s^{-.9} - .4$ in the region $.5 < r_s < 2.5$. $\langle E \rangle\ (r_s)$ is shown for Ge, Si and a semiconductor of simple band structure (SBS).

this lifetime is usually limited by the coupling to photons (n sec range). This explain why it is nearly impossible to observe EHL at equilibrium. In indirect gap material, the lifetime of FE is longer (μ sec range) and EHL lifetime is limited by an Auger process,($\tau \sim n^{-2}$). This is why experimental data are very good in Ge and Si, and quite good in GaP and SiC.Table I to VI, found in a recent paper of G. BENI and T. M. Rice, summarize the band parameter and FE and EHL parameters of 14 semiconductors.

III. MANY BODY CORRECTION TO ONE ELECTRON PROPERTIES IN EHL

III.A. Fluorescence Lineshape and Radiative Lifetime.

The analysis of the fluorescence lineshape of EHL was the main source of EHL parameters : energy, density. In indirect gap semiconductor, in the case of an allowed phonon assisted recombination, the lineshape results from a simple convolution of the densities of occupied electron and hole states. Implicitly in the lowest approximation, electron and hole states are assumed to be plane waves, and the wave function of EHL a Slater determinant of plane waves. In fact, the true wave function is modified by all the Coulomb interactions. The relevant correction here can be expressed in term of an electron-hole correlation parameter g_{eh} Itsmeaning is the following : the probability for an electron to found a hole at the same position in ng_{eh}, instead of the density n in the plane wave model.

Since g_{eh} depends on the wave function, it is much more difficult to compute than the energy. Theoretical values ranging from 2 to 10 have been obtained for a semiconductor of simple band structure [2,6].

From the experimental side, g_{eh} could be extracted from the radiative lifetime of EHL. More precisely, the radiative lifetimes $\tau_{R\ EHL}$ and $\tau_{R\ FE}$ of EHL and FE can be expressed as :

$$\tau_{R\ EHL}^{-1} = A\ g_{eh}\ n \qquad\qquad \tau_{R\ FE}^{-1} = A\ |\phi_{FE}(o)|^2$$

where A is a common factor and $|\phi_{FE}(r)$ the enveloppe wave function of FE. The ratio $R = g_{eh}\ n/|\phi_{FE}(o)|^2$ is obtained from the relative change in luminescence signal at constant excitation at two different temperatures such that the FE or EHL signal dominates,provided that the time decay of both states has been independently determined [7]. Unfortunately, the accuracy of such a procedure is poor.

III.B. Mass Renormalization

The change of the effective mass of carriers in EHL comes from a slight dependance of energy shift of one electron states upon k vector. On the experimental side, it as been obtained from the change of EHL luminescence lineshape in a strong magnetic field. The Landau quantization produces strong accidents in the density of state of electron and holes which are still apparent in the fluorescence lineshape which results from a convolution of these two. In Ge, an increase of the mass of 10% has been measured [8], in agreement with theoretical estimate.

IV. OPTICAL PROPERTIES OF EHL

The optical properties of EHL is described by the change $\delta\varepsilon$ of the dielectric constant with respect to the one of the pure semiconductor at equilibrium. $\delta\varepsilon$ is dominated by the plasmon resonance, in the whole ω range up to near band gap energy. The effective dielectric constant $\bar{\varepsilon}$ for a semiconductor of dielectric constant K with a density N_d or drops of radius R and dielectric constant ε is :

$$\bar{\varepsilon} = K \left| 1 + 4\pi N_d R^3 \frac{\varepsilon - K}{\varepsilon + 2K} \right| \qquad (4)$$

In the simplest form, ε can be taken according to the Drude formula

$$\varepsilon = -K \omega p^2/\omega (\omega + i\tau^{-1}) \qquad (5)$$

At very low frequency $\omega \ll \omega p$, the effect is an increase of the real part of $\bar{\varepsilon}$ which is proportional to the volume occupied by the liquid [9].

The absorption, which becomes important around the plasma resonance, is given in the Drude approximation by :

$$\alpha(\omega) = 4 \pi N_d R^3 C^{-1} K^{1/2} \omega_o^2 \omega^2\tau/[\omega^2 + (\omega^2 - \omega_o^2)\tau^2] \qquad (6)$$

where $\omega_o = \omega_p/\sqrt{3}$ and τ a scattering time in EHL. In fact, Eq. (6) is quite insufficient to explain the experimental data in Ge (see Fig. 2 from ref. 10). An additional damping comes from transitions between light and heavy hole bands. In addition, for large drops, the lowest order dipolar approximation is not sufficient, multipolar resonances contribute also.

In Ge, in the near infrared, the relevant experiments are light scattering at 3,39 μ and absorption in the region 0,3-0,4 eV corresponding to inter valence band transitions. It turns out that most of the index mismatch at 3,39 μ originate still from the plasma resonance [11]. The exact shape of the intervalence band absorption has to take into account the "bound character" of holes which induces non vertical transitions [12].

V. INHOMOGENEOUS ELECTRON HOLE LIQUID

We will consider two specific case of inhomogeneous EHL the surface of an EHD, and the effect of impurities. These problem can be treated with the help of the energy density formalism [13]

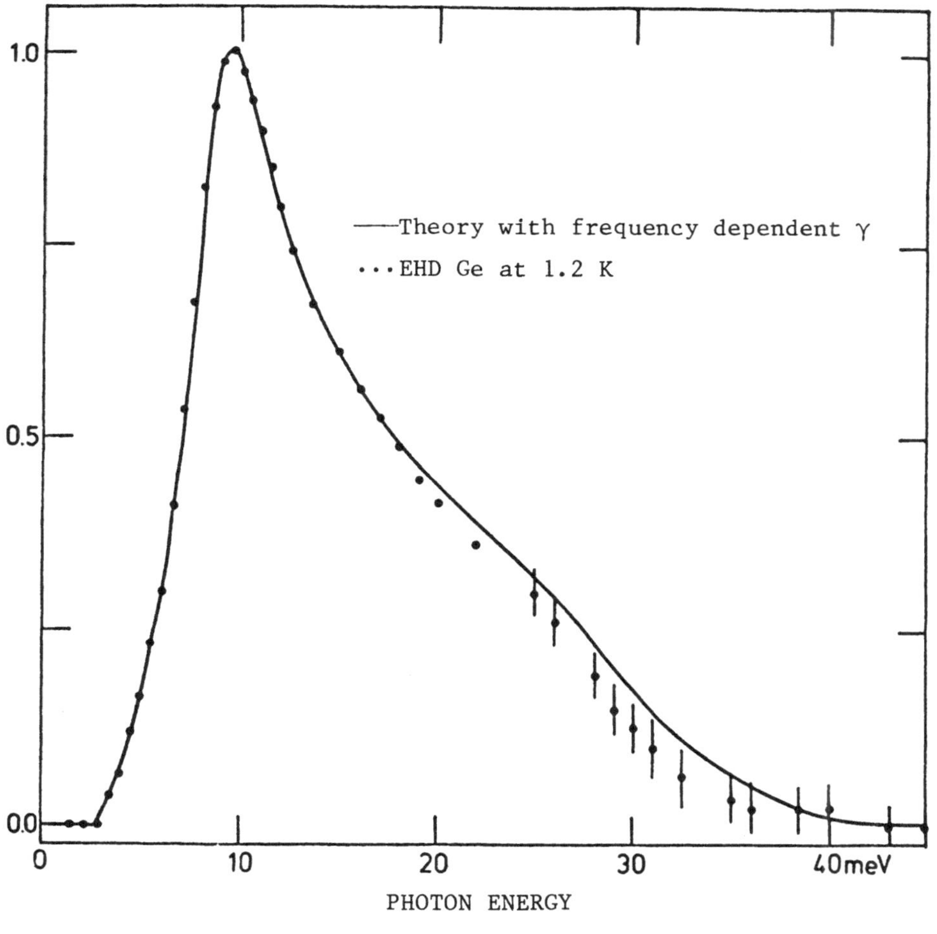

Fig. 2. Absorption of EHD at 1, 2° K. The radius of droplets estimated ∿ 7 μm [10].

where the total energy is written as an integral over the local energy and its derivatives.

$$E = \int d^3r \; [E\,(n_e,n_h) + \frac{1}{72m_e}\,\frac{(\nabla\, n_e)^2}{n_e} + \frac{1}{72\, m_h}\,\frac{(\nabla\, n_h)^2}{n_h} + E_{es}] \quad (7)$$

Here, the density of electron and holes n_e and n_h are function or r. The first term is the local energy, the second and third the gradient expansion of the kinetic energy and the fourth, the electrostatic energy which vanishes if $n_e = n_h$.

V.A. Surface Energy

In its crudest form, the problem of surface energy is solved assuming $n_e = n_h = n(r)$ minimizing (7) with a function n(r) depending mainly of one parameter, the width d of surface layer. The local term contribute a term like Ad and the gradient a term like B/d ; the minimum is obtained at $d = \sqrt{B/A}$. The result for the surface energy can be written as (14)

$$S \simeq 0{,}1\; E_x/a_x^{\;2} \qquad (8)$$

By adding an exchange correlation to the gradient terms [15], a substantial increase of S is found, giving for Ge a value $S = 3{,}5\;10^{-4}$ erg/cm^2. The best experimental value $S \sim 3\;10^{-4}$erg/cm^2 was obtained from the study of capillary resonance [16]. It has been proposed by Singwi and Tosi [17] to apply to the surface energy of EHL the method developped by Kohn and Yaniv [18] for the surface energy of metals. In this method, the surface energy S is obtained by integrating the work of the cleavage force F(z)

$$S = \frac{1}{2}\int_0^\infty F(z)\; dz \qquad (9)$$

F(z) is known at both limits, like Az for $z \to o$ and C z^{-3} for $z \to \infty$, where $A = \rho\,\mu^2/a$ (ρ = mass density, μ = sound velocity, a = interpaticle spacing) and C is the Lifshitz constant which depends on the dielectric constant, and is given for a jellium by $C = 1{,}8\cdot 10^{-3}\; \hbar\omega_p$.

Equation (9) is integrated assuming a universal shape for F(z). The result is $S = \alpha(AC)^{1/2}$. In metal α is found 0.476. In EHL in different semiconductors α is rather about 3.

V.B. Effect of Shallow Impurities of EHL

The perturbation of EHL by a charged donor impurity has been treated with the density functional approach [19]. The most striking feature is the following : around the donor, the electron density in EHL is quite analogous to the electron density around a neutral donor. This means that for Ge at the donor, n_e is 25 larger than the density in EHL. At an impurity density of 5.10^{16} cm^{-3}, the density in liquid in reduce by a factor 5 and the binding energy increased by about 1 meV. On the experimental side. such tendancy has been observed, particularly in Si [20]. The strong peak in $n_e(r)$ at the donor site can be proved by measuring the ratio of phonon assited to phonon less photoluminescence from EHL in doped Ge ; the recombination of free holes to neutral As donor in Ge gives a ration I_{NP}/I_{LA} = 10. Considering that $n_e(r)$ is just like around a neutral donor, one can predict for EHL in doped Ge a ratio :

$$\left(\frac{I_{NP}}{I_{LA}}\right)_{EHD} = \frac{10\, n_{As}}{n_{As} + n_h} \qquad (10)$$

where n_h is the hole density in EHL (= 2.10^{17} at low n_{As}). Eq. (10) fits closely the experimental finding for n_{As} up to 10^{17} cm^{-3} [21].

Does an EHL still exist at doping level above the Mott transition? This question was discussed for a long time theoretically and experimentally. It seems now that the answer is : no. It has been recognized [22] that in n type silicon the two series of luminescence lines which appear in order of increasing excitation are caused respectively by acceptor - donor band and valence band - donor band transitions.

V.C. Effect of Isoelectronic Impurities of EHL

We consider here the specific case of N in GaP. In contrast to charged impurities, the perturbation of N is much weaker, because the potential due to difference in electronegativity between N and P is strongly localized. The new feature here is the existence of a new collective state, in addition to EHL [23]. It is known that the correct picture to describe an exciton bound to N is the pseudo acceptor model : the hole is in an acceptor like wave function around a N^- ion. The new collective state here is just like a metal (with charge inverted) = a gas of delocalized holes around a lattice of N^-. It has been shown in [23] that at a nitrogen density of 10^{18} cm^3, such a state could be more stable than EHL. Some luminescence data by Schwabe et al [24] support this idea.

REFERENCES

1. T.M. Rice, Solid State Physics - Vol. 32 - page 1 (1977) - Edited by H.E. Ehrenreich, F. Seitz and D. Turnbull - Academic Press N.Y.

2. W.F. Brinkman and T.M. Rice, Phys. Rev. B7 1508 (1973)
3. M. Combescot and P. Nozieres, J. Phys. C 5, 2369 (1972)
4. P. Vashista , P. Bhattacharyya and K.S. Singwi, Phys. Rev. B 10 5108 (1974)
5. G. Beni and T.M. Rice, Phys. Rev. B 18 768 (1978)
6. P. Vashista, P. Bhattacharyya and K.S. Singwi, Phys. Rev. Let. 30 1248 (1975)
7. C. Benoit à la Guillaume and M. Voos, Phys. Rev. B 7 1723 (1973)
8. H.L. Störmer and R.W. Martin, Phys. Rev. B 20 4213 (1979)
9. J.C. Hensel and T.C. Phillips, Proc. Int. Conf. Phys. Semicond. 12th - Stuttgart, p. 51. Teubner - Stuttgart 1974.
10. H.G. Zarate and T. Timusk, Phys. Rev. B 19 5223 (1979)
11. J.M. Worlock, T.C. Damen, K.L. Shaklee and J.P. Gordon, Phys. Rev. Let. 33 771 (1974)
12. R.N. Silver and C.H. Aldrich, Phys. Rev. Let. 41 1249 (1978)
13. W. Kohn and L.J. Sham, Phys. Rev. 140 A 1133 (1965)
14. T.M. Rice, Phys. Rev. B 9 1540 (1974)
15. P. Vashita, R.K. Kalia and K.S. Singwi, Sol. St. Com. 19 935 (1976)
16. B. Etienne, L.M. Sander, C. Benoit à la Guillaume, M. Voos and Y. Prieur, Phys. Rev. Let. 37 1299 (1976)
17. K.S. Singwi and M.P. Tosi, Phys. Rev. B 23 1640 (1981)
18. W. Kohn and A. Yaniv, Phys. Rev. B 20 4948 (1979)
19. D.K. Fairobent and L.M. Sander, Phys. Rev. 23 4029 (1981)
20. R.W. Martin and R. Sauer, Phys. Stat. Sol. B 62 443 (1974)
21. C. Benoit à la Guillaume and M. Voos, Sol. St. Comm. 11 1585 (1972)
22. R.R. Parsons, Sol. St. Comm. 29 763 (1979)
23. M. Combescot and C. Benoit à la Guillaume, Phys. Rev. Let. 44 182 (1980)
24. R. Schwabe, F. Thuselt, M. Weinert and R. Bindeman, Phys. Stat. Sol. B, 95 571 (1979)

EXCITONS AND PLASMONS:

COLLECTIVE EXCITATIONS IN SEMICONDUCTORS

I. Egri

Institut für Theoretische Physik, RWTH Aachen
51 Aachen, F.R.G.

ABSTRACT

For a simple model consisting of a nondegenerate valence and conduction band, the spectrum of elementary excitations is calculated taking into account the Coulomb interaction between valence and conduction electrons. It contains, as collective excitations, excitons below and a valence plasmon above the continuum of single electron-hole pair excitations.

The importance of the so-called exchange interaction is pointed out as being responsible for the longitudinal-transverse-splitting of the excitonic solutions and for the existence of the plasmon. Using the equation of motion method not only the energy but also the wavefunction of the plasmonic solution is obtained.

Finally, the dielectric function of the model is derived and the exciton-polariton is discussed.

I. INTRODUCTION

Plasmons and excitons are both collective excitations of electrons in crystals. Traditionally, the plasmon is discussed for metallic or semiconducting crystals [1], whereas the exciton plays an important role for insulators or semiinsulators [2]. As a consequence, the theories describing them have developed along quite different lines and no simple relationship seems to exist between them. However, from a microscopic point of view excitons and plasmons are quite similar, both being coherent superpositions of electron-hole pairs.

To understand this, we must first realize that an elementary excitation from the ground state of the system consists of exciting one electron across the Fermi energy E_F leaving behind a hole below E_F. If the Coulomb interaction between the electrons was switched off, these single electron-hole pairs would constitute the only possible elementary excitations. Their energies form a continuum, which looks qualitatively different for intraband (metallic) and interband transitions. (In conventional free electron theory this continuum is misleadingly called single particle continuum.)

When the Coulomb interaction is turned on, bound states can emerge from the continuum, being linear combinations of single electron-hole pairs. These are the collective excitations of the system and we will classify them as excitons or plasmons, according to whether their energy lies below or above the single pair continuum. Figure 1 shows schematically the excitation spectrum of a metal (intraband transitions) and a semiconductor (interband transitions).

Presumably, the first authors who realized this connection were Quinn and Ferell [3] who considered the metallic plasmon as a "momentum exciton," i.e., a linear combination of electron-hole pairs in momentum space. The first treatment of excitons and plasmons in insulators was given by Horie [4].

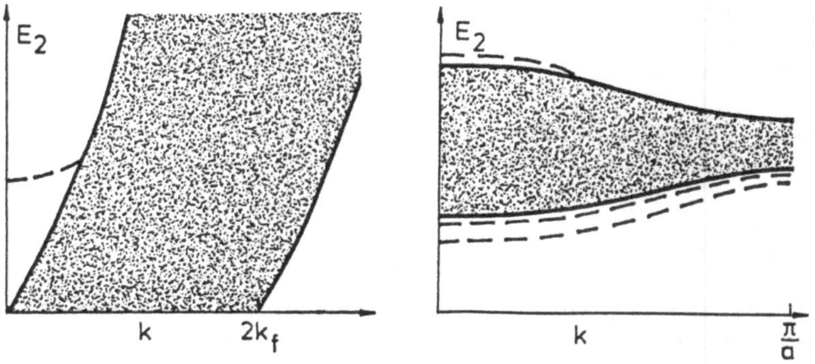

Fig. 1. Electronic excitation spectrum of a metal (left) and of an insulator (right). Dashed lines represent plasmon and exciton energies, shaded areas the respective single-pair continua.

It is the purpose of this contribution to present a unified treatment of excitons and plasmons in semiconductors using a simple two-band model. The emphasis is on general principles rather than on quantitative correctness. Some aspects of the present work have been published elsewhere [5].

Our simple model consists of a full valence band and an empty conduction band, separated by a direct gap at $\underline{k} = 0$. The transition between the bands is dipole-allowed and the spin of the electron is neglected. Additionally, we have included the Coulomb interaction between electrons and holes. It consists of two structurally different terms: the "direct" and the "exchange" term. Whereas the first is essential to the exciton, the second turns out to be essential for the plasmon.

We tackle the problem by setting up the equation of motion for electron-hole pairs which leads to an eigenvalue equation for the spectrum of pair excitations. Furthermore, we obtain the wavefunction of both collective excitations of our model, exciton and plasmon.

Due to the polarization carried by these excitations, the transverse ones interact with the light field forming a polariton [6]. We will derive the dielectric function of our system and discuss the resulting polariton dispersion briefly.

II. EIGENVALUE EQUATION

The Hamiltonian of our system is, in second quantization, given by:

$$H = H_o + H_c$$
$$H_o = \sum_{\underline{k}} \{E^c(\underline{k}) c^\dagger_{\underline{k}} c_{\underline{k}} + E^v(\underline{k}) v^\dagger_{\underline{k}} v_{\underline{k}}\} \tag{1}$$

$E^{c,v}(\underline{k})$ are the energies of conduction- and valence-band, $c^+_{\underline{k}}$, $v^+_{\underline{k}}$ create Bloch-electrons with wave-vector $\underline{k}$ in the conduction and valence band, respectively.

H_c represents the Coulomb interaction. Here, we consider only the largest matrix elements of the most relevant terms, namely the electron-hole interaction. Other contributions, like the hole-hole or electron-electron interaction as well as double creation or destruction of electron-hole pairs, are neglected. Then we have, in the Wannier-representation [7]:

$$H_c = \sum_{i,j} U(i-j) c_i^\dagger c_i v_j^\dagger v_j + \sum_{i \neq j} W(i-j) c_i^\dagger v_i v_j^\dagger c_j \quad , \tag{2}$$

$$U(i-j) = \int d\tau \int d\tau' \, |w^c(\underline{r}-\underline{R}_i)|^2 \frac{e^2}{4\pi\varepsilon|\underline{r}-\underline{r}'|} |w^v(\underline{r}'-\underline{R}_j)|^2$$

$$\cong \frac{e^2}{4\pi\varepsilon|\underline{R}_i-\underline{R}_j|} \tag{3}$$

$$W(i-j) = \int d\tau \int d\tau' w^c(\underline{r}-\underline{R}_i)^* w^v(\underline{r}-\underline{R}_i) \frac{e^2}{4\pi\varepsilon|\underline{r}-\underline{r}'|} w^v(\underline{r}'-\underline{R}_j)^* w^c(\underline{r}'-\underline{R}_j)$$

$$\cong - |\underline{\P}|^2 \frac{3\cos^2(\underline{\P},\underline{R}_i-\underline{R}_j) - 1}{4\pi\varepsilon|\underline{R}_i-\underline{R}_j|^3} \tag{4}$$

with $$\underline{\P} = \int d\tau w^c(\underline{r}) e\underline{r} w^v(\underline{r}) \quad . \tag{5}$$

The sum containing U is the interaction of the conduction electron density at lattice point $\underline{R}_i$ with the valence electron density at $\underline{R}_j$. The leading term in the multipole expansion of U is the monopole-monopole interaction. W can be interpreted as the interaction of an excited dipole at $\underline{R}_i$ with a deexcited dipole at $\underline{R}_j$. The Wannier-functions $w^c(\underline{r})$, $w^v(\underline{r})$ are orthogonal and hence the monopole term of W vanishes. Therefore, W is essentially a dipole-dipole interaction. In exciton physics, the U-term is called "direct" and the W-term "exchange," whereas the contrary is true in the plasmon literature. We have included a background dielectric constant in U and W to account crudely for screening due to other mechanisms than the interband transition under consideration. The Fourier-transform of H_c is given by:

$$H_c = -\frac{1}{N} \sum_{\underline{k}',\underline{k},\underline{q}} U_{\underline{q}} c_{\underline{k}}^\dagger v_{\underline{k}'+\underline{q}} v_{\underline{k}'}^\dagger c_{\underline{k}-\underline{q}} + \frac{1}{N} \sum_{\underline{k}',\underline{k},\underline{q}} W_{\underline{q}} c_{\underline{k}+\underline{q}}^\dagger v_{\underline{k}} v_{\underline{k}'}^\dagger c_{\underline{k}'+\underline{q}} \tag{6}$$

with $$U_{\underline{q}} = \sum_{\underline{R}_\ell} U(\underline{R}_\ell)\, e^{-i\underline{q}\underline{R}_\ell} \qquad W_{\underline{q}} = \sum_{\underline{R}_\ell \neq 0} W(\underline{R}_\ell) e^{-i\underline{q}\underline{R}_\ell} \quad . \tag{7}$$

W_g has been calculated [8]; the $q \to 0$ - limit is orientationally dependent (for an infinite crystal):

$$\lim_{\underline{q}\to 0} W_{\underline{q}} = \begin{cases} \frac{2}{3} p^2 & \underline{q} \parallel \underline{\P} \\ -\frac{1}{3} p^2 & \underline{q} \perp \underline{\P} \end{cases} \qquad p^2 = \frac{|\underline{\P}|^2}{\varepsilon v_w} \tag{8}$$

For this reason $W_{\underline{q}}$ is often called the non-analytic part of the exchange interaction. v_w is the volume of the unit cell.

This completes the discussion of the Hamiltonian and we can now proceed to construct the operator of our collective excitation. As pointed out in the introduction, it will be a linear coherent superposition of electron-hole pairs. Due to the translational invariance of the crystal, momentum is a good quantum number, and therefore only pairs with equal total momentum need to be considered. Hence, we define:

$$S_{\underline{q}} = \frac{1}{\sqrt{N}} \sum_{\underline{k}} \phi_{\underline{q}}(\underline{k}) v^{\dagger}_{\underline{k}} c_{\underline{k}+\underline{q}} \quad . \tag{9}$$

S_q has to satisfy the following equation of motion:

$$[S_{\underline{q}}, H] = E(\underline{q}) S_{\underline{q}} \quad . \tag{10}$$

$E(\underline{q})$ is the energy of the collective excitation and

$$\phi(\underline{R}_j) = \frac{1}{N} \sum_{\underline{k}} \phi_{\underline{q}}(\underline{k}) e^{i\underline{k}\underline{R}_j} \tag{11}$$

is the envelope wavefunction of the internal structure. It is a straightforward matter to commute $S_{\underline{q}}$ with H_0 of Eq. (1), but in commuting $S_{\underline{q}}$ with H_c quartic terms in the operators appear. In order to produce the right-hand side of Eq. (10) we therefore make the random phase approximation (RPA). Exploiting the fact that we have a full valence band and an empty conduction band in the ground state, we finally obtain [5]:

$$\left(E^c(\underline{k}+\underline{q}) - E^v(\underline{k}) - E(\underline{q})\right) \phi_{\underline{q}}(\underline{k}) = \frac{1}{N} \sum_{\underline{k}'} (U_{\underline{k}'-\underline{k}} - W_{\underline{q}}) \phi_{\underline{q}}(\underline{k}') \tag{12}$$

This is our eigenvalue equation which determines both energy and wavefunction of the collective excitations of our two-band model.

If we had applied our method to the intraband case, we would have reproduced the so-called "Ehrenreich-Cohen method" [9] and would have obtained the usual plasma frequency as given by the zero of Lindhard's dielectric function [10].

III. EXCITONS AND PLASMONS

In order to solve the eigenvalue equation (12) we will proceed in two steps: First, we neglect W_q and solve the remaining problem approximately. Then we include the effect of W_q on these approximate solutions. This can be done exactly.

With $W_q = 0$, the right-hand side of Eq. (12) reduces to a convolution integral, and therefore we Fourier-transform into real space. On the left-hand side of Eq. (12) we have to make two approximations: We expand the band energies E^c, E^v up to quadratic terms in $\underline{k}$, neglecting higher powers and we consider $\underline{R}_j$ as a continuous variable $\underline{r}$. Then, $k^2 \exp(i\underline{kr})$ can be replaced by $-\nabla_r^2 \exp(i\underline{kr})$ and introducing center-of-mass coordinates we arrive at:

$$\left(E_g + \frac{\hbar^2 \underline{q}^2}{2M} - \frac{\hbar^2}{2\mu} \nabla_{\underline{r}}^2 - \frac{e^2}{4\pi\varepsilon r}\right) \phi(\underline{r}) = E(\underline{q}) \phi(\underline{r}) \tag{13}$$

E_g is the direct gap at $\underline{k} = 0$, M and μ are total and reduced mass of the electron-hole pair, respectively. Thus we see that in this approximation we obtain the usual Schrödinger equation for the Wannier exciton [2].

Equation (13) has bound solutions below E_g, corresponding to the excitonic collective excitations, and scattering solutions with energy above E_g which represent the electron-hole continuum. We label these solutions with an index n which runs over all states, discrete and continuous. Within the above approximations they have the following property:

$$[S_{\underline{q}}^n , \tilde{H}] = E^n(\underline{q}) S_{\underline{q}}^n \tag{14}$$

where $\tilde{H}$ differs from H by the absence of the W_g-term.

We now want to include W_g. To this end we expand S_g, the solution of the full problem in terms of S_g^n. This can be done due to the completeness of the latter. Inserting such an expansion into Eq. (10) and using Eq. (14), we get the following eigenvalue for E(q):

$$1 = W_{\underline{q}} \sum_n \frac{|\phi_q^n(0)|^2}{E(\underline{q}) - E^n(\underline{q})} \tag{15}$$

This is an exact equation relating the energies $E(q)$ of the full problem to the exact solutions of the $\tilde{H}$-problem. It becomes an approximation, because we use the Wannier-model for $E^n(q)$ and $\phi_q^n(0)$ as approximate solutions.

As a first result, we may derive the longitudinal-transverse-splitting of the Wannier excitons. By neglecting all but one term in Eq. (15) we obtain in first order:

$$E(\underline{q}) = E^n(\underline{q}) + W_{\underline{q}}|\phi^n(0)|^2 \qquad (16)$$

Using Eq. (8) we find for the longitudinal-transverse-splitting:

$$E^{n,\parallel}(\underline{q}) - E^{n,\perp}(\underline{q}) = p^2|\phi^n(0)|^2 \qquad (17)$$

Returning to our main goal, namely finding a plasmonic solution of Eq. (15) we see that for the existence of such a solution $W_q > 0$ is a necessary condition. Thus our plasmon will be longitudinal, as it should be.

In calculating the plasmon energy we can safely neglect the contributions of the discrete exciton levels due to the large energy denominators. We are then left with the sum over the continuum states, where we can use $\underline{k}$, the momentum of the relative motion between electron and hole, as our label n. If we neglect the excitonic enhancement factor $|\phi^n(0)|^2$ in the numerator, which peaks at E_g and tends to 1 for $E \gg E_g$ [11] and insert Eq. (8) into Eq. (15) we obtain for $q = 0$:

$$1 = \frac{2}{3} p^2 \sum_{\underline{k}} \frac{1}{E(0) - (E^c(\underline{k}) - E^v(\underline{k}))} \qquad (18)$$

This equation determines the energy of the valence plasmon of our model. If the sum diverges as $E(0)$ approaches $\mathrm{Max}(E^c(\underline{k})-E^v(\underline{k}))$, we have a discrete line for arbitrarily small, non-vanishing p^2. This depends on the behavior of the (joint) density of states near the maximum energy; also, the excitonic enhancement factor may change the situation drastically. Preliminary results of our calculations indicate that the sum remains finite thus establishing a critical interaction strength for the existence of the plasmon. We can give a crude estimate of $E(0)$ if we replace $E^c(\underline{k})-E^v(\underline{k})$ by Δ, an average energy of the order of the order of the band gap E_g. Utilizing the f-sum rule for p^2 [4] which in our case reads as

$$2p^2 = \frac{\hbar^2\omega_p^2}{\Delta} \qquad (19)$$

where

$$\omega_p = \sqrt{\frac{ne^2}{m\varepsilon}} \tag{20}$$

is the usual plasma frequency, we obtain:

$$E(0) = \Delta\left(1 + \frac{1}{3}\left\{\frac{\hbar\omega_p}{\Delta}\right\}^2\right) \tag{21}$$

From the derivation of Eq. (18) it is clear that Eq. (21) is valid only for $\hbar\omega_p \ll \Delta$.

Equation (18) should be compared with the conventional definition of the valence plasmon being the zero of the dielectric function as, e.g., derived by Ehrenreich and Cohen [12]. Their result is:

$$1 = p^2 \sum_{\underline{k}} \left(\frac{1}{\hbar\omega_{p\ell} - (E^c(\underline{k}) - E^v(\underline{k}))} - \frac{1}{\hbar\omega_{p\ell} + (E^c(\underline{k}) - E^v(\underline{k}))}\right) . \tag{22}$$

We realize that it differs from Eq. (18) by the factor 2/3 in front of the sum and by the antiresonant term having the + sign in the denominator. Neither of these effects change the results qualitatively, especially if $\hbar\omega_{p\ell}$ is close to the upper bound of the continuum. We could have obtained the antiresonant term, if we had included double excitation of electron-hole pairs in the Hamiltonian. In fact, it was that term which was made responsible for the existence of the plasmon by Horie [4]. However, as we have shown, plasmons are already present in the absence of that term; they owe their existence to the dipole-dipole interaction W_q. We have no simple explanation for the factor 2/3, but we tend to attribute it to local field effects which transfer 1/3 of the oscillator strength to the transverse modes.

Making the same crude estimate as before, we get $\omega_{p\ell}$ from Eq. (22):

$$\omega_{p\ell} = \omega_p\left(1 + \frac{1}{2}\left(\frac{\Delta}{\hbar\omega_p}\right)^2\right) \tag{23}$$

with ω_p given by Eq. (20). In contrast to Eq. (21), this expression is valid for $\Delta \ll \hbar\omega_p$.

At this point, we have accomplished our main goal, namely the unified and simultaneous treatment of excitons and plasmons in semiconductors. They are both (approximate) solutions to the same equation, and there is no conceptual difference between them.

This unifying framework contains not only the above-mentioned exciton and valence plasmon, but also, as we have indicated, the intraband plasmon and the plasmon in inversion surface layers [13]; with little more effort we could even have included the magnons. Thus, we have introduced a certain order in the vast "zoo" of collective excitations. Apart from this general aspect, this treatment has also practical implications: just as for excitons, we can now calculate the wavefunction of the valence plasmon with the help of Eq. (12).

To be consistent with previous approximations, we have to neglect $U_{\underline{k}'-\underline{k}}$ in Eq. (12) and take the longitudinal value of W_q. Then, Fourier-transforming according to Eq. (11), we find:

$$\phi(\underline{R}_j) = \phi(0)\frac{2}{3}\, p^2 \sum_{\underline{k}} \frac{e^{i\underline{kR}_j}}{E(0) - (E^c(\underline{k})-E^v(\underline{k}))} \tag{24}$$

where E(0) is determined by Eq. (18).

$\phi(0)$ is obtained from the normalization condition:

$$1 = \sum_{\underline{R}_j} |\phi(\underline{R}_j)|^2 \tag{25}$$

In order to calculate the wavefunction, we must specify the bandstructures entering Eqs. (18) and (24). Since the main contributions to the sum come from the upper limit of the electron-hole continuum, we have made an effective mass approximation around the maximum of the conduction band and the minimum of the valence band. We have assumed isotropic effective masses and a spherical Brillouin-zone. Furthermore, the sums in Eqs. (18), (24) and (25) have been converted to integrals. The results are shown in Figs. 2 and 3.

It is interesting to note that with decreasing binding energy η, $|\phi(0)|^2$ tends to zero. This means that as the plasmon energy approaches the electron-hole continuum the "Bohr-radius" of the plasmon becomes very large. This is qualitatively the same behavior as that of excitons on the low energy side of the spectrum. There, the exciton radius varies as the inverse square root of the binding energy (in the Wannier model).

On the other hand, $|\phi(0)|^2$ approaches unity for large η. This means that electron and hole sit essentially on the same site just as they do in the Frenkel exciton, the excitonic counterpart. This case corresponds to a structureless homogeneous density oscillation.

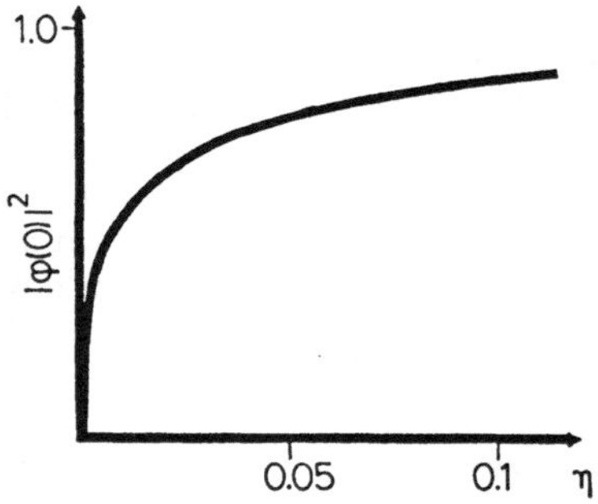

Fig. 2. Value of the plasmon wavefunction at the origin as function of η, the separation of the plasmon energy from the continuum in units of the combined bandwidth.

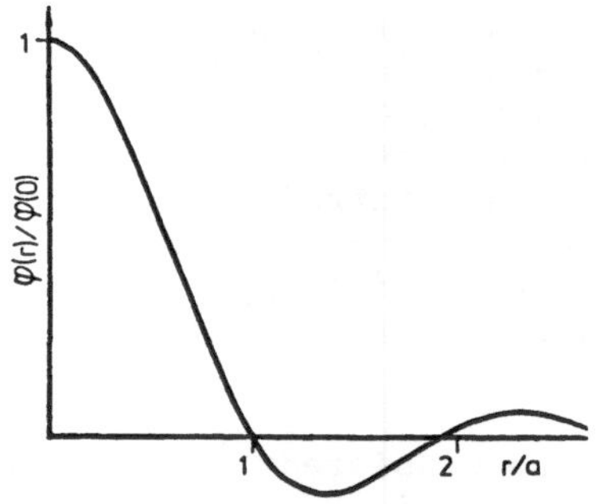

Fig. 3. Radial dependence of the plasmon wavefunction in units of the lattice constant for η=0.1.

Since the spatial extent of the plasmon enters into the scattering probability on, e.g., impurities [14], some information about it may be deduced from experiments.

IV. THE EXCITON-POLARITON

The general polariton problem has been already treated in this course [6], and we may therefore concentrate on special topics not covered so far. First, we will present a simple microscopic derivation of the dielectric function of an excitonic system and then we will discuss some problems specific to the Wannier-exciton-polariton.

To do so, we must add a term describing the interaction with the transverse light field $\underline{\varepsilon}^{\perp}(\underline{R}_i,t)$ to the Hamiltonian (1). This is given by:

$$H_1 = - v_W \sum_{\underline{R}_i} \underline{P}^{\perp}(\underline{R}_i)\, \underline{\varepsilon}^{\perp}(\underline{R}_i,t) \tag{26}$$

where the polarization operator is defined as

$$\underline{P}^{\perp}(\underline{R}_i) = \frac{\underline{\P}}{v_w}(c_i^{\dagger} v_i + v_i^{\dagger} c_i) \quad . \tag{27}$$

Using the solutions of the exciton problem having the property

$$[S_{\underline{q}}^{n,\perp}, H] = E^{n,\perp}(\underline{q}) S_{\underline{q}}^{n,\perp} \tag{28}$$

we can express $\underline{P}^{\perp}(q)$, the Fourier-transform of $\underline{P}^{\perp}(\underline{R}_i)$, in terms of those and obtain:

$$\underline{P}^{\perp}(\underline{q}) = \frac{\underline{\P}}{v_w}\sqrt{N}\sum_n \left((\phi^n(0))^* S_{\underline{q}}^{n,\perp} + \phi^n(0)(S_{-\underline{q}}^{n,\perp})^{\dagger}\right) \tag{29}$$

Since the exact solutions to Eq. (28) will have an l-t-splitting, we have used the superscript $\perp$ to indicate that we are interested in the transverse solutions.

Inserting Eq. (29) into Eq. (26) and using Eq. (28), we get:

$$i\hbar \dot{S}_{\underline{q}}^{n,\perp} = E^{n,\perp}(\underline{q}) S_{\underline{q}}^{n,\perp} - \phi^n(0)\frac{\underline{\P}}{\sqrt{N}}\,\underline{\varepsilon}\,(\underline{q},t) \tag{30}$$

where $\underline{\varepsilon}^{\perp}(q,t)$ is the Fourier transform of $\underline{\varepsilon}^{\perp}(\underline{R}_i,t)$. Taking the expectation value of Eqs. (29) and (30) we can perform a temporal Fourier transformation and arrive at:

$$\underline{P}^{\perp}(\underline{q},\omega) = \frac{|\underline{\P}|^2}{v_w}\sum_n |\phi^n(0)|^2 \left(\frac{1}{E^{n,\perp}(\underline{q})-\hbar\omega} + \frac{1}{E^{n,\perp}(-\underline{q})+\hbar\omega}\right)\underline{\varepsilon}^{\perp}(\underline{q},\omega) \tag{31}$$

and therefore

$$\varepsilon_{\perp}(\underline{q},\omega) = \varepsilon + \frac{|\underline{\P}|^2}{v_w}\sum_n |\phi^n(0)|^2 \left(\frac{1}{E^{n,\perp}(\underline{q})-\hbar\omega} + \frac{1}{E^{n,\perp}(-\underline{q})+\hbar\omega}\right) \tag{32}$$

is our expression for the transverse dielectric function of the excitonic system. The sum contains contributions from the discrete excitonic lines as well as from the continuum. $|\phi^n(0)|^2$ is the excitonic enhancement factor. ε is the same dielectric constant as that used in Eq. (4).

We obtain the polariton dispersion relation by inserting Eq. (32) into the polariton equation:

$$\omega^2 \varepsilon_\perp(\underline{q},\omega) = \varepsilon_0 c^2 q^2 \ . \tag{33}$$

For q = 0 we have one polariton branch starting at $\omega = 0$, but, in addition to that, every root of the equation

$$\varepsilon_\perp(0,\omega_\ell) = 0 \tag{34}$$

yields a polariton branch. To lowest order in $|\underline{\P}|^2$, these roots are:

$$\hbar\omega_\ell = E^{n,\perp}(0) + \frac{|\underline{\P}|^2}{\varepsilon v_W}|\phi^n(0)|^2 = E^{n,\parallel}(0) \tag{35}$$

where the second equality comes from comparison with Eq. (17). Thus, the zeros of the transverse dielectric function are given by the longitudinal exciton energies.

Finally, we discuss special features of the Wannier-exciton-polariton. In contrast to the phonon-polariton or even the Frenkel-exciton-polariton, the spectrum has more than two polariton branches; in fact, it has infinitely many, and in addition to that it has a continuum region where, due to intrinsic "Landau-damping," only heavily overdamped polariton modes exist. The influence of the continuum on the polariton spectrum has been discussed by Bendow [15] and Egri [16]. The presence of more than two spatially dispersive polariton branches complicates the ABC-problem [6]. As the frequency approaches the fundamental absorption edge, more than two exciton-polaritons can be excited thus requiring more than one ABC. Therefore, in this region, the conventional approach to ABC becomes questionable and new concepts may have to be developed [17]. The situation is summarized in Fig. 4.

V. SUMMARY

Using a simple two-band model augmented by the relevant parts of the Coulomb interaction, we were able to derive the spectrum of electron-hole excitations. Apart from the continuum it shows discrete excitonic and plasmonic lines. In addition, we also derived the wavefunction of the internal motion of the plasmon.

Taking into account the coupling of the excitons to the transverse photons, we also obtained the polariton spectrum. We have pointed out that the so-called "exchange" interaction is intimately related to the dipole moment of the transition. Therefore, we could define the energies of the longitudinal excitons either as roots of the eigenvalue equation (15) or as the zeros of the dielectric function (34), yielding identical results.

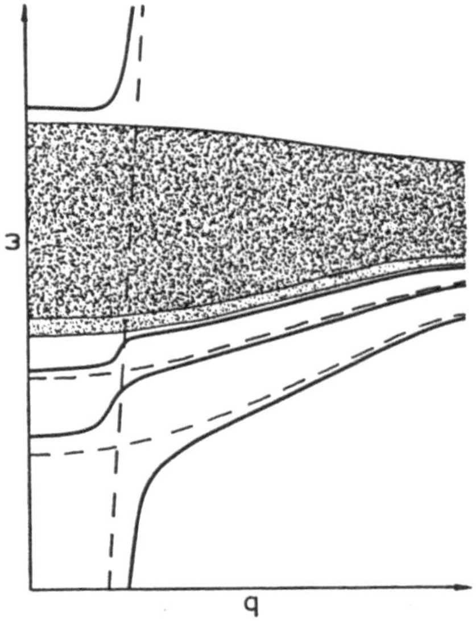

Fig. 4. Dispersion relation ω(q) of the Wannier - exciton-polariton in the cross-over region: full lines. Dashed lines represent photon and transverse exciton lines. Light and dark shaded areas represent the quasi-continuum of the n ⩾ 3 exciton lines and true single-pari continuum, resp.

REFERENCES

1. A. A. Lucas, this book.
2. R. S. Knox, this book.
3. R. A. Ferell and J. J. Quinn, Phys. Rev. 108, 570 (1957).
4. C. Horie, Prog. Theor. Phys. 21, 113 (1959).
5. I. Egri, Z. Physik B42, 99 (1981).
6. R. Loudon, this book.
7. B. Di Bartolo, this book.
8. M. H. Cohen and F. Keffer, Phys. Rev. 99, 1128 (1955).
9. H. Haken, Quantenfeldtheorie des Festkörpers (Stuttgart, Teubner, 1972).
10. See, e.g., J. M. Ziman, Principles of the Theory of Solids (Cambridge University Press, 1969).
11. R. J. Elliott, Phys. Rev. 108, 1384 (1957).
12. H. Ehrenreich and M. H. Cohen, Phys. Rev. 115, 786 (1959).
13. A. Tselis, this book.
14. A. Stahl, private communication.
15. B. Bendow, Springer Tracts in Modern Physics 82, 69 (1978).
16. I. Egri, J. Phys. C12, 5471 (1971).
17. A. Stahl, Phys. Stat. Sol. (b) 94, 221 (1979).

PICOSECOND EXCITON PHENOMENA IN CHLOROPHYLL COMPLEXES

R. S. Knox

Department of Physics
University of Rochester
Rochester, New York

Chlorophyll-portein complexes are now well established as the building blocks of the light-gathering system in all photosynthetic cells. Complexes containing N=6, 7, 13, and up to the order of 100 chlorophyll molecules have been characterized by many researchers with varying degrees of completeness. Homogeneous samples of these complexes are interesting from the point of view of exciton dynamics because the small value of N compared with 10^{23} in crystals means that exciton motion can be followed completely (in principle).

Experimental studies in Rochester have shown that aggregates of chlorophyll-a/b complexes have exciton behaviour different from that of simple complexes in the picosecond-to-nanosecond domain. This research, as well as some of the characterization studies and some work on excitons in chloroplasts, were reviewed.

BOSE-EINSTEIN STATISTICS IN EXCITON SYSTEMS

A. Mysyrowicz, D. Hulin and L.L. Chase*

Groupe de Physique des Solides de l'Ecole Normale
Supérieure, Université de Paris VII
Paris, France
*Physics Department, Indiana University
Bloomington, Indiana, U.S.A.

ABSTRACT

In the first part, the quantum-statistical properties of an ensemble of Bose particles are briefly reviewed. The ideal model of non-interacting bosons is introduced. Next, the effects due to the inclusion of residual forces are considered. The model is then applied to the case of excitonic particles in non-metallic crystals. The limitations of its validity are stressed.

In the second part, different methods of detecting quantum-statistical effects in an excitonic fluid are described. Experimental results obtained in Cu_2O and CuCl are then presented. In Cu_2O, free excitons show a gradual evolution, from a classical to a quantum degenerate regime with increasing particle densities. In CuCl, anomalies in the luminescence of excitonic molecules are attributed to the occurrence of Bose-Einstein condensation.

I. INTRODUCTION

Every elementary particle may be classified in two categories, according to the value of its associated spin quantum number. Particles with half-integer value (electron, proton, neutron, etc.) are fermions, those having integer value (pions, phonons, photons, etc., as well as composite particles with an even number of fermions, such as hydrogen atoms, ^{4}He atoms, excitons, etc.) are bosons. If one considers an ensemble of identical point particles in thermodynamical equilibrium, their energy distribution is expressed as:

$$N(E) = \rho(E) \cdot f(E) \qquad (1)$$

$\rho(E)$ is the density of states; $\rho(E)=AE^{\frac{1}{2}}$ for an ideal gas
$f(E)$ is the occupation probability of a level of energy E.

$$f(E) = \{\exp[(E-\mu)/kT] + 1\}^{-1} \quad \text{for fermions.} \qquad (2a)$$

$$f(E) = \{\exp[(E-\mu)/kT] - 1\}^{-1} \quad \text{for bosons.} \qquad (2b)$$

E is the energy, measured from the level with zero kinetic energy E=0, T is the temperature and μ is the chemical potential associated with the gas, determined by the condition $\sum_E N(E) = N_t$, the total number of particles. The negative sign in the denominator of Eq. (2b) for bosons imposes the constraint that μ be always negative or zero, since otherwise $f(E)<0$ for $|E|<|\mu|$, leading to the unphysical result of a negative number of particles. In the Fermi gas, μ can be either positive, negative or zero.

In usual situations the ratio $(E-\mu)/kT >> 1$ and both statistics are well approximated by the familiar Maxwell-Boltzmann distribution. In the reverse situation $(E-\mu) < kT$, marked differences occur in the behavior of these systems. Consider for instance the limit $T\to 0$. For a Fermi gas, the occupation probability cannot exceed unity (Pauli exclusion principle), leading to a high kinetic energy of the ground state of the system, even at T=0. By contrast, in the Bose gas, there is a tendency for the particles to accumulate in the levels of lowest energy, and at T=0 all particles occupy the same quantum oscillator K=0.

There is ample evidence for the degenerate quantum regime of a Fermi gas, most notably in free electrons in metals, in high density plasmas, in ^{3}He or in the electron-hole liquid in semiconductors. By contrast, the degenerate case of a Bose gas has proved to be much more elusive. In fact, the only system of massive particles so far (besides the case of excitonic particles discussed here) where Bose-Einstein statistics is believed to play an important role is ^{4}He below T=2.17 K [1]. In liquid ^{4}He, however, the interatomic forces, consisting both of an attractive part (responsible for the liquefaction) and a hard core repulsive part, are strong, and cannot be treated as a small perturbation on the ideal Bose gas. This makes the problem of superfluid ^{4}He a very difficult one to handle theoretically [2]. It is therefore of great importance to find new, much more dilute systems, in which purely quantum statistical effects are predominant so that the theory of the Bose gas can be tested under good conditions. In this paper, we present experimental evidence showing that excitonic particles may, to a certain extent, furnish such a case.

Before presenting these results, we briefly recall the main features predicted by the ideal Bose gas model.

II. THE IDEAL (OR NEARLY IDEAL) BOSE GAS

Consider an ideal ensemble of Bose particles (with zero potential energy). Hence the energy of each particle consists only of a kinetic term $\hbar^2K^2/2M$ where M is the mass of each particle.

The total number of particles $N_t=\sum_E N(E)$ may be evaluated, by converting the summation into an integral

$$N_t = N_o + \frac{V(2M)^{3/2}\cdot g}{4\pi^2\hbar^3}\int_0^\infty \frac{E^{\frac{1}{2}}\cdot dE}{\exp[(E-\mu)/kT]-1} \tag{3}$$

following the usual procedure. Here the term N_o, corresponding to the occupation of the oscillator with K=0, has been singled out. The vanishing of the denominator of the integrand for this E=0 state when $\mu\to 0$ makes it necessary to treat it separately from the others.

For $\mu=0$, the integral has a finite value

$$\int_0^\infty \frac{E^{\frac{1}{2}}\, dE}{\exp[(E-\mu)/kT]-1} = 2.315 \cdot T^{3/2} \tag{4}$$

Thus, the total number of atoms which can be accommodated in all excited states is limited to a maximum value

$$N_c = g(6.2\cdot 10^{15})\cdot V(M/m_o)^{3/2}T^{3/2} \tag{5}$$

where m_o is the free electron mass and g, the degeneracy factor indicates the number of states with same energy (and must not be confused with the degeneracy of the gas itself). V is the volume.

What happens if supplementary particles are added to the system, or if the temperature is lowered? Einstein [3] has given the following physical interpretation: all particles exceeding N_c will condense into the lowest state at K=0, which, for μ=0, can aquire a population comparable to the total number of particles. This phase transition occurs with considerable abruptness, since above T_c the occupation number N_0 is quite small (comparable with the occupation of low-lying states in the quasi continuum), whereas below T_c it is:

$$N_o/N_t = 1 - (T/T_c)^{3/2} \text{ with } T_c = \frac{2\pi\hbar^2}{Mk}\left(\frac{N_t}{V\cdot 2.612}\right)^{2/3} \quad (6)$$

A highly occupied single quantum state is expected to lead to unusual properties of the system, such as coherent properties (quantum mechanical phase relationship) over macroscopic distance, since all particles in the ground level are described by a single quantum mechanical wave function. The ideal Bose model has been successful in predicting a critical temperature T_c=3.14K for ^{4}He [4], close to the observed temperature for the onset of superfluidity. However, it is clear that a model which neglects all interparticle interactions is unrealistic [2].

The first attempt to include forces between particles has been performed by Bogoliubov [5]. Assuming the forces to be weakly repulsive, and taking into account only binary collisions between particles, Bogoliubov derives a Hamiltonian of the form

$$H = \sum_K (\hbar^2K^2/2M)a_K^+ a_K + (1/2V)\sum_K \sum_\ell \sum_m \phi_K a_{\ell-K}^+ a_{m+K}^+ a_m a_\ell \quad (7)$$

where a_K^+ and a_K are creation and annihilation operators for particles with wavevector K, and ϕ_K is the Fourier transform of the potential field [6]

$$\phi_K = \int e^{iKr}\phi(r)dr \quad (8)$$

It is assumed that the system is close to T=0, so that N_0 is large, $N_o \simeq N_t$. Then, the operators a_o^+ and a_o are treated as scalars of magnitude

$$a_o = a_o^+ \simeq N_o^{1/2}$$

Keeping only those terms quadratic in the particle operators, the Hamiltonian (7) may be reexpressed as:

$$H \simeq \frac{1}{2}\frac{N_t^2}{V}\phi_o + \sum_{K\neq 0}\left(\frac{\hbar^2K^2}{2M} + \frac{N_t}{V}\phi_K\right)a_K^+a_K + \sum_{K\neq 0}\frac{N_t}{2V}\phi_K(a_K^+a_{-K}^+ + a_Ka_{-K}) \quad (9)$$

In order to diagonalize (9) Bogoliubov introduces new Bose creation and annihilation operators, b^+ and b, which allow him to put the energy of the ensemble in the form

$$E = E_o + \sum_K E(K) b_K^+ b_K \tag{10}$$

Here b^+ and b represent creation operators of collective excitations in the coupled system, and E(K) is given by:

$$E(K) = \{2 \frac{N_t}{V} \phi_K \frac{\hbar^2 K^2}{2M} + \frac{\hbar^4 K^4}{4M^2}\}^{\frac{1}{2}} \tag{11}$$

Thus, the usual dispersion curve of non-interacting particles, $E_K = \hbar^2 K^2/2M$ is modified, because of the repulsive forces acting between particles, into a linear dispersion curve for small K values

$$E_K \simeq \hbar K (\frac{N_t \phi_o}{MV})^{\frac{1}{2}} = \hbar K v \tag{12}$$

where v is the sound velocity. This modification of the excitation curve at small K vectors reflects the fact that the system can entertain collective longitudinal modes of oscillation.

The non-vanishing slope of E(K) as $K \to 0$ has drastic consequences on the behavior of the condensed system. As shown by Landau [7], such a fluid with non-vanishing slope at $K \to 0$ will show superfluidity, i.e. a frictionless motion of a part of the fluid. This is a direct consequence of the laws of momentum and energy conservation, which cannot be satisfied simultaneously unless the system aquires a minimum velocity.

Another feature predicted by the Bogoliubov treatment is a depletion of the condensate. At T=0 the fraction of particles in the ground state is no more unity but is

$$N_o/N_t = 1 - (\frac{N_o}{N_t} M\phi_o)^{3/2} (\frac{N_t}{V})^{\frac{1}{2}} \frac{1}{3\pi^2} \tag{13}$$

Subsequent, more refined, microscopic treatments [8], have shown that the Bogoliubov model corresponds in fact to the first term in a perturbation expansion with respect to f_0 where f_0 is a characteristic scattering length describing the two-particle potential, which is typically of the order of the particle diameter.

III. THE CASE OF EXCITONS

The concept of the exciton as a quasi-particle has proved very useful to describe the properties of electronically excited non-metallic crystals [9]. The Coulomb attraction between an electron promoted to the conduction band and a hole in the valence band leads to bound electron-hole states called excitons with energies inside the forbidden gap E_g, which in simple cases take the form of an hydrogenic series of terms:

$$E_{n_p} = E_g - R/n_p^2 + \hbar^2K^2/2M \qquad (14)$$

where n_p=1.2,... is the principal quantum number, and

$$R = \frac{1}{2}\,\frac{e^4 m_r}{\varepsilon_o^2\hbar^2}$$

is the binding energy of the exciton, $M = m_e+m_h$ is the translational mass, $m_r = (m_e m_h)/(m_e+m_h)$ the reduced mass, m_e and m_h are the effective electron and hole mass and ε_o the dielectric constant.

In turn, each term of the series forms a quasi-continuum of energy levels, because of the translational motion of the particle (last term in (14)). Since an exciton consists of an even number of fermions, it falls in the category of the bosons and as such an ensemble of these particles is a candidate for the observation of Bose-Einstein condensation [10-12].

The most attractive feature here is the light mass of the excitons, of the order of the free electron mass or less. Assuming a density $n = N/V \sim 10^{19}\ cm^{-3}$, the Bose model predicts a very high critical temperature for the onset of the phase transition, of the order T = 100K.

Excitonic particles have two very useful properties which can be exploited to search for, and study, condensed states. The density of excitons may be easily varied, and the properties of the condensate can be probed directly by optical means. In particular, the luminescence emitted by the crystal can provide direct information about the distribution of the particles in K space.

Several authors [13,14] have examined in detail the possibility of Bose-Einstein condensation of excitons from a theoretical point of view. They obtain an expression for the dispersion of the exciton energy with momentum similar to that given by Bogoliubov:

$$E(K) = \left\{\left(\frac{\hbar^2K^2}{2M}\right)^2 + \frac{n\hbar^2K^2Rfa_o^3}{M}\right\}^{\frac{1}{2}} \qquad (15)$$

with f the effective scattering parameter of the order $f = 13\pi/3+\lambda$ where λ is a dimensionless constant of the order of unity [13].

Here $a_0 = \hbar^2 \varepsilon_0 / m_r e^2$ is the exciton radius.

The ground state energy per particle takes the form:

$$E_0/nV = - R(1-\tfrac{1}{2} f a_0^3 n) \tag{16}$$

Thus, as for ^{4}He, the condensate is expected to be stable at finite momentum values $K \neq 0$, and superfluid behavior should occur. Here, the lossless flow of the quasi-particles does not correspond to a displacement of mass in the system, (like in ^{4}He) or to an electric current, (as for Cooper pairs in metals), but it represents a "superconductor" for energy transport.

The model also predicts a decrease of the binding energy of excitons as the particle density is increased. The binding energy of the particles in the fluid is obtained as:

$$\mu = \frac{1}{V} \frac{dE_0}{dn} = f n a_0^3 R \tag{17}$$

Several assumptions are made in the above mentioned theoretical treatment of excitons. A first assumption is a repulsive exciton-exciton potential. Theoretical calculations, confirmed by experimental evidence, show that this is not true in crystals with indirect band gap structure, where it is energetically favorable for the system to separate into a low density exciton phase coexisting with regions of a high density electron-hole plasma [15]. An increase of injected excitons in this case merely increases the size of the electron-hole drops, without changing the free exciton density. Even in direct gap materials with simple band structure, where exciton liquefaction into plasma drops is not expected [15], the direct and exchange Coulomb interactions can lead to an overall attraction responsible for excitonic molecule formation. This attraction occurs for any mass ratio $\sigma = m_e/m_h$, according to a two-band calculation [16-18], leading to the conclusion that a Bose condensation of excitons is always impossible. However, a more complete treatment is necessary in the limit $m_e \simeq m_h$ where the binding is smallest. Such effects as the polar coupling of the carriers to the lattice and the electron hole exchange interaction must be considered. For example, Bassani and Rovere [19] have shown in the specific case of Cu_2O ($m_e = 0.73\ m_h$) that excitonic molecule formation is prevented if the electron-hole exchange interaction is taken into account. In the same limit $\sigma \to 1$, Keldysh and Kozlov [13] find that superfluidity could still occur even in the presence of a weak attraction between excitons, in the range of densities, n, such that

$$a_1^{-3} < n < a_o^{-3}$$

where a_1 is the radius of the excitonic molecule. In that case, the stabilization of the condensate is due to the Fermi repulsion between like particles in the two exciton (four fermions) collisions.

On the other hand, in the other limit $\sigma \to 0$, the formation of excitonic molecules is unavoidable and is experimentally well established. Biexcitons fall also in the category of bosons and therefore can undergo Bose condensation themselves. It is therefore necessary to examine whether the van der Waals attraction between molecules can lead to a first order phase transition into a dielectric liquid before BEC occurs in the gas phase. According to the quantum theory of corresponding states [20], this should not be the case if the reduced quantum parameter $\eta = \hbar^2/M\varepsilon\alpha^2$ is larger than 3.5, assuming a Lennard-Jones type potential. Here ε is the depth of the potential well and α the collision diameter. In actual excitonic systems η has a value much larger than 3.5 [21], because of the small particle mass (if compared to H_2) and therefore the system should remain a gas even at T=0. In that case, the results obtained for an exciton gas can be applied to the gas of excitonic molecules with little modification and Bose-Einstein condensation is expected for the biexcitons.

Another assumption in the theory is to treat excitons in first approximation as point particles. This condition is only fulfilled to the extent that the average distance between particles is large compared to their radius a_o, that is

$$na_o^3 << 1 \qquad (18)$$

Eq. (18) may be re-expressed at the critical density for BEC as: [12]

$$r_c/a_o = 1.6 \left(\frac{R}{kT}\right)^{\frac{1}{2}} \left(\frac{m_r}{M}\right)^{\frac{1}{2}} >> 1 \qquad (19)$$

which gives a measure of the deviation from ideality. Here r_c is the average distance between particles at the critical density. It is apparent from (19) that systems with large exciton (or biexciton) binding energy should be selected.

It is interesting to go to the other limit $na_o^3 > 1$. In that case, the starting point for a description of the system consists of degenerate electrons and holes, with the Coulomb attraction as a perturbation [22-24]. The treatment is similar to the BCS theory of superconductivity except that pairing of electrons and holes occurs, which, according to Keldysh and Kopaev [22], makes the system unstable against the formation of a dielectric superfluid at sufficiently low temperatures. Leggett [23], and Nozieres and Comte [24] come to similar conclusions. They treat the n-electron, n-hole system from a unified point of view, the Bose model of ideal particles being just the limit of the dilute system. Thus, it appears that the onset of superfluidity is not restricted by the condition $na_0^3 < 1$ but should be possible over a wide range of electron-hole densities, provided the temperature of the system can be kept sufficiently low. The temperature question, however, can present a serious difficulty due to the occurrence of Auger decay processes at high densities, which can lead to a substantial heating of the system. It must also be stressed that superfluidity in the high density limit necessitates a direct band gap, because pairing requires approximately equal but opposite momenta.

As a further assumption, the exciton lifetime is considered infinite, or at least long if compared with a characteristic relaxation time τ necessary to establish thermodynamical equilibrium between the excitonic particles and the lattice. If hot free electron-hole pairs are injected into the crystal, they rapidly lose their excess energy, forming excitons in the lowest kinetic energy band n=1. This relaxation process takes the order of 10^{-11}-10^{-12} sec, as long as the kinetic energy exceeds the energy of optical phonons of the crystal [25]. The further relaxation inside the n=1 kinetic energy band requires interaction with acoustic phonons, with a much longer interaction time, typically in the nanosecond range [25]. Thus it is necessary to have a particle population lifetime long compared to 10^{-9} sec. Hanamura and Inoue [26] have calculated the time evolution of excitons or excitonic molecules, with different initial conditions. They solve the Fokker-Planck equation for the kinetics of the system taking into account exciton-phonon collisions but neglecting interparticle collisions. The time required to establish a true equilibrium with the lattice is found to be very long, $t \gg \tau_{ac}$, although a quasi equilibrium, even in the absence of interparticle collisions, is brought about in a time comparable with the exciton-acoustic phonon relaxation time. Their results also indicate that Bose-Einstein condensation will occur in a time of the order 10^{-9} sec for an ensemble of excitonic molecules in CuCl. Since this time is comparable to the biexciton lifetime, it is unlikely that a condensate can develop if hot particles are initially injected. Recently, Yakhot and Levich [27] have extended this treatment to take into account the effect of interparticle collision. They come to the interesting conclusion that the time required for the onset of Bose-Einstein condensation can be considerably shortened, if an initial population of excitonic particles is present at low K vectors, acting as a nucleation center for the condensation. A similar suggestion had been made earlier by Hanamura [14], who pointed out that Bose condensation of excitonic molecules in CuCl should be realized if a high density of the particles is directly generated around K=0, by giant two-photon absorption of light.

Finally, a last assumption underlying the nearly ideal Bose gas treatment of excitons is a negligible interaction between excitons and the radiation field. If the exciton-photon coupling is strong, then the elementary electronic excitations of the system are better described in terms of polaritons [29]. It is immediately apparent that no accumulation of lowest energy polaritons near K=0 is possible, because of the predominant photon nature of the mixed mode in that region, and therefore Bose condensation is impossible [30]. Excitonic molecules, on the other hand, do not suffer from this complication even if they are formed from dipole active excitons, because they have a quadrupole moment and consequently do not interact in first order with the photon field.

To summarize, in order to have a Bose condensation of free excitons described in the framework of the nearly ideal model, it is necessary to consider a system with direct, simple band gap structure, nearly equal electron and hole effective masses, weak exciton-photon interaction, large exciton binding energy, and long exciton lifetime. If $m_e < m_h$, Bose-Einstein condensation is still possible, in principle, but will occur for long-lived excitonic molecules.

IV. METHODS OF DETECTION

The most obvious way to detect the occurrence of Bose quantum-statistical effects in a system of excitonic particles is through an analysis of the luminescence corresponding to their radiative recombination. However, in order to extract the distribution function, it is necessary to select a radiative decay process which involves the entire population of particles. This will necessarily correspond to a three-body process, in which a final particle is able to carry momentum, so that both momentum and energy conservation laws can be satisfied simultaneously. Examples of such transitions are the phonon-assisted decay of the free exciton, the conversion of an excitonic molecule into a photon and a free exciton, the Auger decay of two excitons into a photon and free carriers, etc.

The lineshape of the emission under consideration will be of the form:

$$I(\hbar\omega) = a\int_0^\infty \int_0^\infty d\bar{K}_i \; d\bar{K}_f \; f(\bar{K}_i) \left|M_{if}(\bar{K}_i,\bar{K}_f)\right|^2 \cdot \delta(E_i - E_f - \hbar\omega)\,\delta(\bar{K}_i - \bar{K}_0 - \bar{K}_f) \quad (20)$$

where $f(\overline{K}_i)$ is the distribution function of the decaying particles, $\overline{K}_i$, $\overline{K}_f$, $\overline{K}_o$ are the initial, final, and photon wavevectors respectively, and E_i, E_f, $\hbar\omega$ are the initial, final, and photon energies respectively. M_{if} is the matrix element for the transition.

The following features should become manifest in the lineshape analysis:
- At low particle density and high temperatures ($T \gtrsim T_c$) , f(E) should be given by a Maxwell Boltzmann function.
--Under a decrease of temperature, at fixed particle density, or alternatively at fixed T under increase of the particle density, a gradual deviation from classical statistics should occur, reflecting an increase of the value of the chemical potential.
- Above a critical density n_c, this emission band from thermally excited particles should approach a constant intensity. At the same time, a narrow peak should develop at the position corresponding to the K=0 state. This peak should increase linearly with the number of particles above n_c, at least in some range of density. The threshold value n_c should be a sensitive function of temperature, and the sharp line must vanish above the T_c corresponding to each value of the excitation.
- Further, according to (Eq. (17)), the binding energy of the excitonic particles in the condensate should decrease at very high densities. This can occur either through a high energy shift of the exciton resonance, or through a downshift of the free electron-hole continuum edge, so that the absolute value of the condensate emission energy may in fact vary very little [31].

There are other radiative processes in highly excited semiconductors which are known to give rise to sharp emission lines, and may therefore be a source of confusion. For instance, it is well known that residual impurities in a crystal can act as very efficient recombination centers leading to narrow lines in the exciton spectral region. Also, stimulated emission may distort and narrow the lineshape of an intrinsic, broader decay process. Finally, it is often necessary or desirable to minimize any heating of the distribution by injecting excitonic particles at the bottom of their kinetic energy band through resonant pumping. Under such conditions, the appearance of narrow lines in a phonon-assisted emission process may simply reflect an inelastic scattering of the incident light, with the excitonic particles only virtually excited. Such a Raman process becomes enhanced as the ingoing or outgoing photon energy approaches a resonance of the system.

In order to dispell any of those possibilities it is necessary to perform cross-check experiments. Usually, impurity-related effects may be recognized by the fact that they are sensitive to sample quality. In reasonably pure crystals with impurity concentrations $n_i < 10^{17}$ cm^{-3}, the bound exciton recombination lines should saturate in the carrier density range of interest so that intrinsic recombination should dominate here. Concerning the problem of stimulated emission, it is important to realize first that the observation of stimulated emission may be quite consistent with the presence of a Bose-Einstein condensation of excitonic particles. If a condensate is present in the system, it corresponds to a single (or a few) oscillation modes occupied by a macroscopic number $N \sim N_t \gtrsim 10^{18}$ cm^{-3}. It is then very likely that stimulated emission is observed in view of the large population inversion in a narrow spectral region. The pertinent question is therefore: whether stimulated emission alone can explain the onset of a sharp emission. In order to minimize the chances of forming accidental cavity modes in the sample by multiple reflections at the boundaries, it is advisable to have as small an excited volume as possible. An independent measure of the optical gain, by a pump and probe method, will also help in evaluating the importance of stimulated emission in the lineshape analysis. A simple test for the presence of resonantly enhanced spontaneous Raman scattering can be done by varying the intensity of the input light at a fixed input frequency: as can be seen from the solution of simple rate equation, the ratio of the Raman intensity to the luminescence component intensity should either be a constant, or it should decrease with increasing excitation if the intraband relaxation rate increases due to interparticle collisions.

A more serious problem, in the case of resonantly pumped excitons, is a stimulated Raman effect, since the Raman line will have many properties similar to those expected from a condensate formed at $\overline{K}_i$, where $\overline{K}_i$ is the incident photon wavevector. In this case, evidence is required from experiments other than simple luminescence or absorption measurements in order to establish conclusively that the sharp features are caused by condensation rather than by stimulated resonance Raman scattering.

A spectroscopic method for investigating a condensate, which may distinguish it from resonance Raman scattering, utilizes the so-called pump and probe technique [32]. A first intense pump beam generates enough particles that a substantial fraction is condensed. A second, weak probe beam, at another frequency and/or wavevector, adds a small amount of particles to the system, well below the critical density limit. Therefore, the probe beam alone gives rise to the emission characteristic of a classical gas. However, if the particles are added in the presence of the condensate, they cannot be accommodated in excited states anymore. Consequently, a quantum attraction must occur, with a preferential scattering of the probe excitons into the condensate, which can be observed by differential detection techniques.

As we have seen in Chapter III, the presence of a condensate leads to a distortion of the excitation spectrum of the system, from the usual quadratic law valid for isolated particles. It would be useful to detect such a density dependent dispersion as a function of $\bar{K}$, although it may be experimentally difficult to do so, in view of the small energies involved.

Finally, a striking feature expected from the condensate is superconductivity of energy current. This property might be verified by measurements of anomalous excitonic diffusion. Such experiments could prove difficult to realize however, since it must be shown unambigously that the energy migration is not simply due to an emission-reabsorption process. Recently, the so-called grating technique has been successfully applied to the measurement of ambipolar diffusion of carriers in crystals [33]. This method may be potentially useful here.

V. EXPERIMENTAL RESULTS IN Cu_2O

This compound is a good candidate for the observation of quantum-statistical effects in a free exciton gas. The band structure of Cu_2O consists of a non-degenerate upper valence band (symmetry Γ_7^+) and a lower conduction band (Γ_6^+), with extrema located at the center of the Brillouin zone at K=0 and separated by an energy gap of 2.175 eV [34]. This implies that electron-hole liquid formation will not occur here because of the simple, direct band structure [17]. Further, exciton states constructed from the Γ_7^+ and Γ_6^+ bands have a large binding energy and small radius $a_0 = 7$ Å. The n=1 exciton level is split, because of electron-hole exchange interaction, into a lower lying singly degenerate Γ_2^+ paraexciton (X_p=2.022 eV at 2K) and a triply degenerate Γ_{25}^+ orthoexciton (X_0=2.034 eV at 2K) [35]. The large electron-hole exchange energy ΔE = 12 meV and nearly equal electron to hole effective mass ratio ($\sigma = m_e/m_h = 0.73$) prevent excitonic molecule formation [19], so that free excitons remain the lowest electronic excited states of the crystal even at high exciton densities. Finally, because of the even parity of the valence and conduction bands, S-like excitons do not interact, in the dipole approximation, with the radiation field, so that the polariton effect can be neglected.

Luminescence spectra due to free exciton recombination obtained in high quality (natural growth) crystals under weak, cw optical excitation in the band to band absorption continuum are shown in Fig. 7 for three different lattice temperatures [36]. The line denoted X_0-Γ_{12}^- corresponds to the annihilation of free orthoexcitons in the n=1 state with the simultaneous emission of a parity conserving optical phonon of symmetry Γ_{12}^-, of energy 13.5 meV, and appropriate odd parity. Because of the non-polar character of the Γ_{12}^- phonon, the exciton-phonon coupling is independent of the

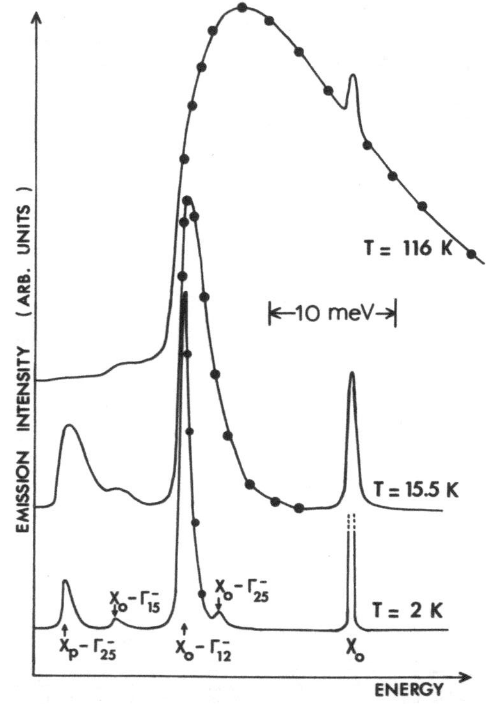

Fig. 1. Exciton luminescence of Cu_2O at different temperatures. Line $(x-\Gamma_{12}^-)$ is fitted with a Maxwell-Boltzmann formula (black points). Low excitation region $<10^{18}$ photons cm^{-3} sec^{-1}. Line X_0 corresponds to the direct recombination of orthoexcitons by quadrapole emission.

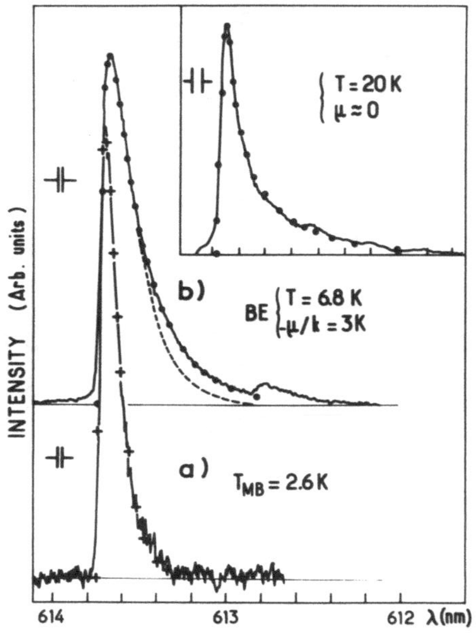

Fig. 2. Lineshapes of the phonon-assisted decay of orthoexcitons in Cu_2O held at 1.5K for different excitation intensities I_o. (a) $I_o \sim 10^2$ W/cm^2; (b) $I_o \sim 10^3$ W/cm^2; (c) $I_o \sim 10^7$ W/cm^2. In (a) the lineshape is fitted by a Maxwell-Boltzmann formula, in (b) and (c) by a Bose-Einstein distribution with μ and T as indicated. The spectral response of the detection is taken into account reolution as shown.

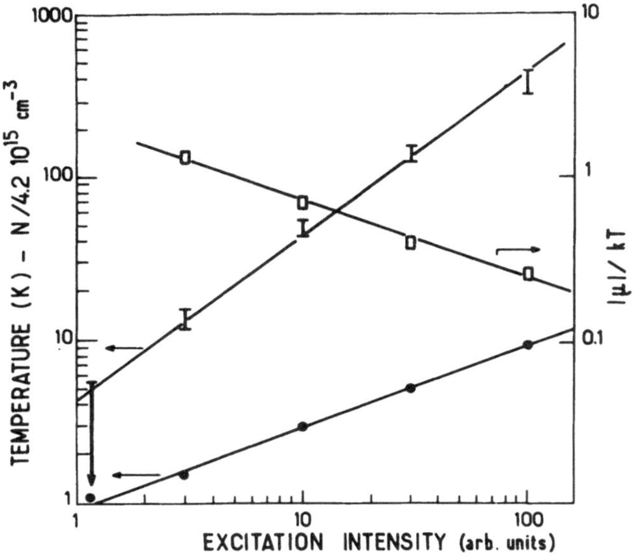

Fig. 3. Temperature increase ΔT (black dots), $|\mu|/KT$ (open rectangles), and density of orthoexcitons (vertical bars) as function of incident Ar^{++}-laser intensity, as deduced from best fits of X_0-Γ_{12} - lineshape analysis at T = 1.5 K. The density of particles at the lowest I_0 is undetermined, since it fits a Boltzmann distribution; the upper limit of N compatible with the line shape is shown.
f(E) the c

K vector. Further the Γ_{12}^- mode shows little dispersion, so that the exciton phonon-assisted emission lineshape is given by the simple formula

$$I(\hbar\omega) = g(E)\ f(E) \tag{21}$$

As can be seen in fig. 1, this relation reproduces well the experimental data over a wide range of temperatures if f(E) is taken as a classical Maxwell-Boltzmann function.

This is no longer the case at low temperatures, where deviations from Maxwell-Boltzmann statistics become apparent, if the excitation rate is increased gradually, keeping all other experimental conditions identical (see Fig. 2). On the other hand, the observed experimental lines are fitted adequately, simply by introducing for f(E) the correct Bose distribution function (Eq. (2b)) expected to apply for excitons [37]. Results are shown in fig. 2, and the values of μ and T_{eff} (the exciton effective temperature) obtained from a best fit are shown in Fig. 3. In turn, the knowledge of μ and T_{eff} permits a determination of an absolute value for the density of particles from the use of Eq. (3). Note that this is a very powerful method of extracting particle densities, since it only requires, besides μ and T_{eff}, the translational mass of the exciton as a material parameter. (A value $M=3\ m_o$ for Cu_2O is reported by Yu and Shen [38] from resonant Raman studies.) A plot of the density of orthoexcitons obtained from the lineshape analysis, versus input excitation intensity (i.e. generated particles number) yields a linear dependence, which is expected for a gas with weak interparticle forces (see Fig. 3). Volume expansion would occur in the presence of a strong repulsive potential. Under very intense excitation using a pulsed dye laser ($I_o \sim 10^7$ W/cm^2), with the laser wavelength tuned inside the phonon assisted n=1 exciton absorption continuum (so that heating of the excitons is minimized), it is possible to reach situations where the chemical potential of the gas is very close to zero [37] (see Fig. 2 inset). In this case, because of the low average power of the excitation laser $I \sim 10^{-3}$ W, it is not possible to change the excitation intensity from its maximum value by reducing the total number of photons since the signal to noise ratio would decrease too rapidly. However a decrease of input intensity obtained by defocusing the laser beam on the sample surface leads, as expected, to a decrease of the chemical potential to larger negative values.

In Fig. 4, densities of orthoexcitons extracted from a lineshape analysis are plotted against the gas effective temperature for various excitation conditions. Crosses are obtained if the sample is irradiated with a cw Ar^{++} laser, and triangles correspond to pulsed dye laser excitation. Also plotted in the same figure is the critical density for Bose condensation of orthoexcitons in Cu_2O. As can be seen, the rise of exciton temperature with increasing particle density prevents a large fraction of particles from condensing, although the critical density is reached symptotically at T_{eff}=10 K.

So far only the luminescence of the orthoexcitons has been examined. In principle, the lower lying n=1 paraexcitons should provide an even more favorable case, because the corresponding critical density for BEC is reduced by a factor of three, if compared to the orthoexciton population (degeneracy factor g=1 instead of g=3 in eq. 5). The paraexciton lifetime at low particle density is very long, of the order 10^{-5} sec., limited by non-radiative decay processes [36]. The orthoexciton lifetime, on the other hand, is limited by ortho-paraconversion to a value $\tau_o \sim 10^{-8}$ sec at 2 K so

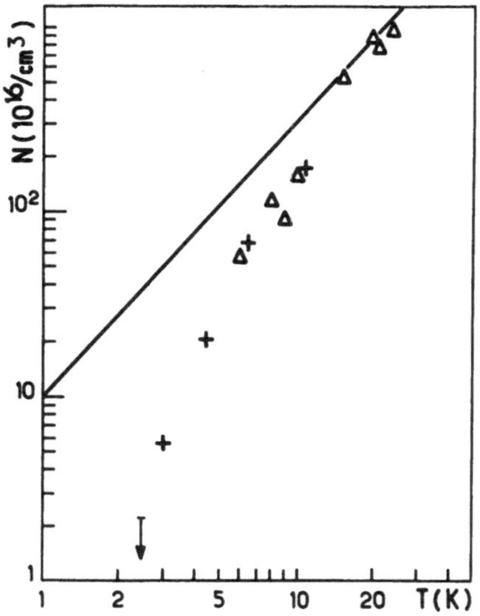

Fig. 4. Density of orthoexcitons as function of particle effective temperature. Crosses are from Ar^{++}-laser excitation, triangles from N_2-laser-pumped dye laser excitation at $\lambda = 598$ nm. The straight line of slope 1.5 gives the orthoexciton critical density for Bose-Einstein condensation, in the ideal Bose-gas model.

that an important accumulation of paraexcitons appears to be feasible [36]. The only possible radiative channel for paraexciton decay is by emission of a photon and a momentum and parity conserving phonon Γ_{25}^- of energy 10.8 meV. The corresponding transition probability is smaller by a factor 2 X 10^3 than the Γ_{12}^- phonon-assisted recombination of an orthoexciton [40]. Unfortunately, upon increase of excitation, the paraexciton emission intensity shows a marked sublinear behavior, which rapidly prevents any line shape analysis to be performed, due to the proximity of the Γ_{15}^- phonon-assisted orthoexciton emission, which still grows linearly with input intensity [37].

In order to understand the physical origin of the saturation of the paraexciton population and to obtain more information upon the parameters relevant to the problem of BEC in Cu_2O, Benoit à la Guillaume, D. Hulin and A. Mysyrowicz [41] have recently performed a detailed analysis of the exciton luminescence of Cu_2O at different temperatures and excitation rates: The experimental features are well reproduced by a set of coupled rate equations, for the ortho- and paraexcitons, in which an Auger-like recombination process from exciton-exciton inelastic collisions is introduced, which reduces the paraexciton lifetime at densities higher than $n \sim 10^{15}$ cm^{-3}. From an adjustment of the model to experimental results at low and intermediate excitation rates, it is possible to determine all parameters. Then, an extrapolation to the high intensity regime pertaining to the cases shown in Figs. 2, 3, 4 allows one to predict the densities of ortho- and paraexcitons present in the system. A good quantitative agreement is found for the orthoexciton densities evaluated in this way and those obtained from the line-shape analysis, giving additional support to the claim that strong quantum degeneracy is obtained. It also shows that the paraexciton population is reduced by Auger collisions to the point that it falls below the orthoexciton density when their respective decay times become comparable. Finally, an extension of the model to the problem of exciton heating shows that in the experimental conditions used, the critical density for Bose-Einstein condensation of orthoexcitons will only be reached asymptotically, in accordance with the results of Fig. 4.

To summarize, there is strong experimental evidence, supported by a model calculation, that free excitons in Cu_2O behave like a nearly ideal Bose gas. A very highly quantum degenerate regime with $\mu \simeq o$ can be reached. However, due to the temperature increase of the electronic system, it has not been possible so far to achieve a substantial condensation of particles in the K=0 state. To reach this goal remains a challenging task.

VI. EXPERIMENTAL RESULTS IN CuCl

This compound has a band structure similar to that of Cu_2O, except for two important features. First, the crystal point group, T_d, although cubic as in Cu_2O (O_h), does not have an inversion center, so that parity is not well defined. Secondly, the electron to hole mass ratio σ is small, of the order $\sigma=0.2$ or less [42]. The value of the forbidden gap is 3.43 eV at 4.2 K. The n=1 exciton states formed with the upper Γ_7 valence band and lower Γ_6 conduction band are split into a Γ_5 state at 3.208 eV with a large oscillator strength, $f \sim 10^{-3}$ and an optically inactive Γ_2 state at 3.202 eV. The simple band structure prevents the formation of an electron-hole liquid [15] as in Cu_2O. On the other hand, the small value of σ favors excitonic molecule (biexciton) formation at increasing exciton densities, and indeed, the evidence for the existence of this entity was first recognized in this material [43]. Since then, a considerable amount of work has been devoted to the investigation of the properties of the biexciton in CuCl [43]. Its energy of E_{xx}=6.372 eV, its binding energy of 29 meV and its symmetry Γ_1 are well established. Because of this relatively large binding energy, the biexciton is stable up to T~100°K. It is possible to directly generate excitonic molecules, from the ground state of the crystal, by a two-photon absorption process, using an intense laser beam tuned to the frequency $E_{xx}/2$ [44]. In view of the corresponding giant transition cross-section [14], [45], a very high density $n \sim 10^{19}$ cm^{-3} of particles may be readily obtained, even with moderate input beam intensities, in the MW/cm^2 range. The luminescence resulting from the radiative decay of the resonantly-pumped biexcitons, as obtained in high quality crystalline films, is shown in fig. 5 for different pump intensities. Here, a single, linearly polarized laser beam is used as an excitation source, so that the particles are created at $\bar{K}_{xx}=2\bar{K}_o$, where $\bar{K}_o=4.5\cdot10^5$ cm^{-1} is the wavevector of the incident photons inside the medium. At the lowest excitation rate ($I_o \sim 10^4$W/cm^2), the molecular luminescence spectrum is composed mainly of two inverted Maxwell-Boltzmann bands, denoted M_T and M_L, with a ratio of integrated intensities M_T:M_L~2:1. A similar luminescence is also obtained at low intensities if two counterpropagating laser beams of identical frequency $\hbar\omega/2$ are used (so that the biexciton population is created at K=0), or if the crystal is pumped in the band to band or exciton absorption region so that molecules are formed by subsequent exciton-exciton interactions. This indicates that the system has lost all memory of the particular excitation mechanism, due to a fast relaxation process which redistributes the particles into a quasi-equilibrium state. From the lineshape analysis, the quasi-equilibrium distribution function of the biexcitons may be inferred. The two bands M_T and M_L correspond to decay processes terminating on the transverse and longitudinal branches of the Γ_5 exciton respectively [46]. If a large fraction of the initial biexcitons have a K-vector $K_{xx} >> K_o$, the complication due to the polariton nature of the transverse exciton around $K=K_o$ may be ignored. The curvature of the biexciton dispersion is half as large as that of the exciton branches, because of the difference in effective masses [47]. For the decay of thermalized biexcitons to each exciton branch, equation (20) therefore reduces to an "inverse Maxwellian" intensity distribution of the form

$$I(\omega) \propto E^{\frac{1}{2}} \exp\{-E/kT_{eff}\} \quad . \tag{22}$$

The $E^{\frac{1}{2}}$ factor is a joint density of states, where $E = \{E_x(0)-E_B-\hbar\omega\}$ is the kinetic energy of a biexciton emitting at ω, $E_x(0)$ is the $\underline{K}=0$ exciton energy, E_B the biexciton binding energy, and $\hbar\omega$ is the energy of the emitted photon. The 2:1 ratio of M_T:M_L is due to the twofold degeneracy of the transverse exciton. Under those low excitation conditions the spectra are reasonably well fitted by taking a classical energy distribution for the gas of biexcitons, with an effective temperature, T_{eff}, between 15 K and 25 K [47].

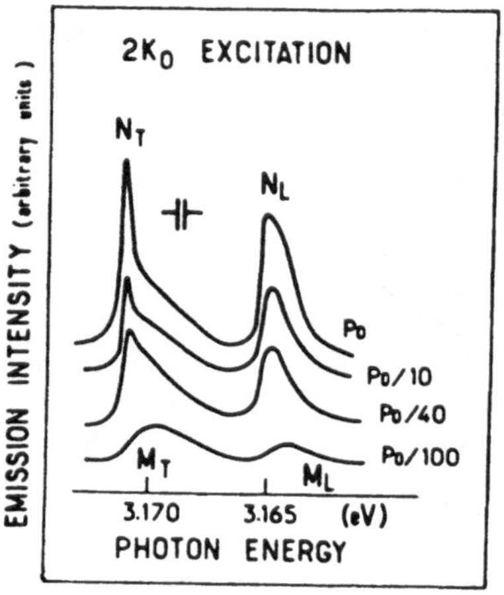

Fig. 5. Biexciton luminescence of a CuCl crystalline film held at T=4K, for different intensities P of the excitation laser ($\hbar\omega$=3.186 eV; $P_o \sim 10^7$ W/cm^2). The emission is detected at $\sim 90^\circ$ from the input laser propagation vector.

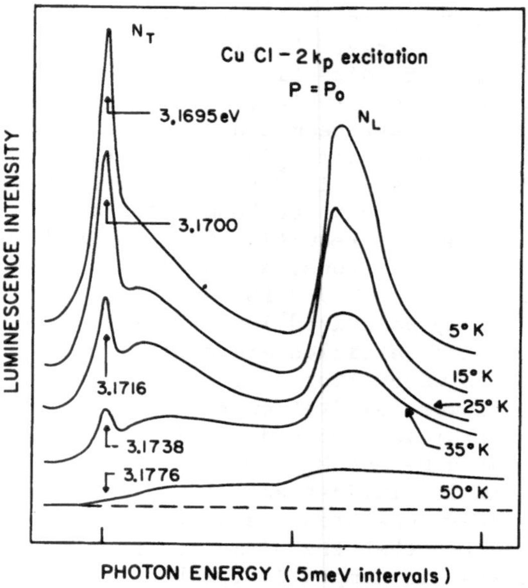

Fig. 6. Biexciton luminescence of CuCl film at different sample temperatures. Resonant two-photon excitation. Same detection geometry as in Figure 5.

At increasing excitation rates, large deviations from the Maxwell lineshapes are observed [48]. In particular, above a critical input intensity a sharp line, denoted N_T, grows on the high energy side of the M_T band if a single beam excitation is used. This line shows remarkable properties, such as a large spatial anisotropy in its intensity. It is intense in the backward detection geometry (with respect to the incident beam), but very weak in the forward geometry [48]. Further, the strength of the line shows a marked dependence upon sample temperature (see Fig. 6), with a threshold temperature for appearance varying with input excitation in the manner shown in the inset of Fig. 7.

The appearance of N_T has been interpreted as an evidence for the occurrence of a Bose-Einstein condensation in the biexciton gas [48], with the condensate taking place at the wavevector imposed by the incident excitation. Since the radiative decay of the condensate at $\underline{K}_{xx}=2\underline{K}_o$ will leave a final exciton with a wavevector somewhere between K_o and $3K_o$, the polariton aspect of the final exciton is now crucial. Inspection of the polariton dispersion curve of CuCl shows that in the backward detection geometry, the remaining transverse polariton at $3\overline{K}_o$ is essentially exciton like, with the complementary emitted photon lying on the high energy side of the M_T band. In the opposite limiting case (forward detection geometry), the final polariton is photon-like; in fact the whole decay process is better described in this latter case as a two-photon emission decay, with both emitted photons having the same energy $\hbar\omega = E_{xx}/2$. Further, in a collinear detection geometry, from symmetry arguments, the decay to the longitudinal exciton is forbidden, since the emitted photon and final exciton would have orthogonal polarization vectors. In a non-collinear geometry, however, this decay mode of the condensate is partially allowed, but only if the emitted photons have a polarization vector lying in the plane containing $\overline{K}_{xx}$. Fig. 7 shows these expectations to be verified experimentally.

Before discussing this interpretation of N_T as being due to BEC in more detail, it is necessary to consider alternative explanations [49-51]. All processes originating from a spatially isotropic biexciton distribution (such as spontaneous or stimulated emission from a cold, only partially thermalized gas [49-50], or processes related to the presence of impurities, may be disregarded because the N_T line shows a marked spatial anisotropy, a feature unusual in a cubic material. A spontaneous hyper-raman scattering (in which two incident photons are absorbed and one photon is emitted in a single step, with the biexciton acting as a virtually excited intermediate state) can also be eliminated from the fact that N_T nearly disappears at lower intensities but not M_T, as discussed in chapter IV. The only viable alternative would be a stimulated hyper-raman process, showing preferential gain over overlapping molecular luminescence. This possibility has been explored in a series of pump and probe experiments by Peyghambarian et al [52]. The optical gain experienced by a weak probe beam, with frequency tuned to N_T was measured under excitation of the crystal by an intense pump beam, with conditions similar to those of Fig. 7. The optical gain under these conditions is saturated at a value of about 1.2 through the 2μ film thickness. This result appears to be incompatible with an interpretation of N_T as a stimulated Raman process, if realistic material parameters are introduced. An even more definite conclusion is obtained from a series of pump and probe experiments which make use of two weak, counter-propagating probe beams, each of which is detuned by a small amount $\Delta E = 2$ meV to either side of the biexciton resonance [53]. The simultaneous absorption of photons from each of the probe beams leads to resonant excitation of biexcitons, initially near K=0. The resulting molecular luminescence shown in Fig. 9(b) shows the familiar inverted Maxwell-Boltzmann shape indicative of a classical distribution (note that either probe beam alone does

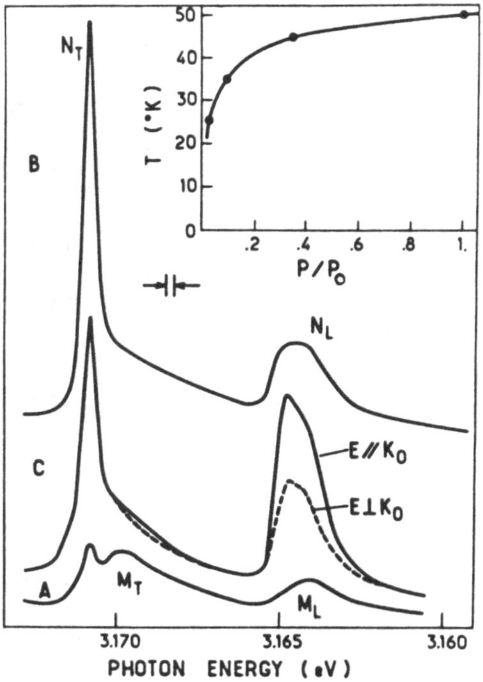

Fig. 7. Biexciton luminescence of CuCl films at T=4K, excited with a laser tuned at ħω=3.186, for different geometries, with respect to the input beam. a) forward detection, b) backward, c) non-collinear geometry. Inset: lattice temperature threshold for N_T vs laser intensity P/P_o.

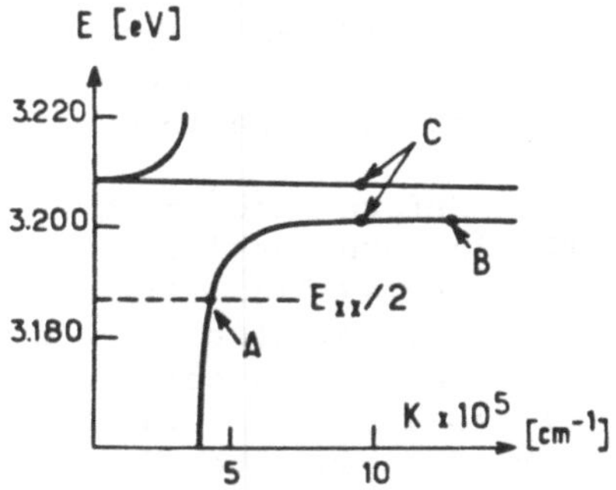

Fig. 8. Exciton-polariton dispersion curve of CuCl. A, B and C correspond to the final states reached in the decay of biexcitons with wavevector $2K_o$, for the three geometries of Figure 7.

not give rise to any luminescence). If the experiment is repeated in the presence of an intense pump beam, the recombination radiation from the two probe beams alone (as detected with differential techniques) is dramatically modified (see Fig. 9c) as long as the crystal is kept below a critical temperature T_c~40K. The observation of a sharp emission at the position of N_T instead of the M_T band, is indicative of a large compression of the phase volume occupied by the probe-excited biexcitons. A determination of the magnitude of this compression is not possible because of the inadequate energy resolution, but a reasonable lower limit is about a factor of 10^3. Such a feature is not expected to occur if N_T originated from a pure Raman effect, with biexcitons only virtually excited. The reason for this is that the biexcitons virtually excited at K~0 by the probe beams <u>cannot</u> produce Raman emission at the position of the N_T line because of the polariton dispersion around K=0. Furthermore, the small measured optical gain mentioned above is found to be insufficient to account for

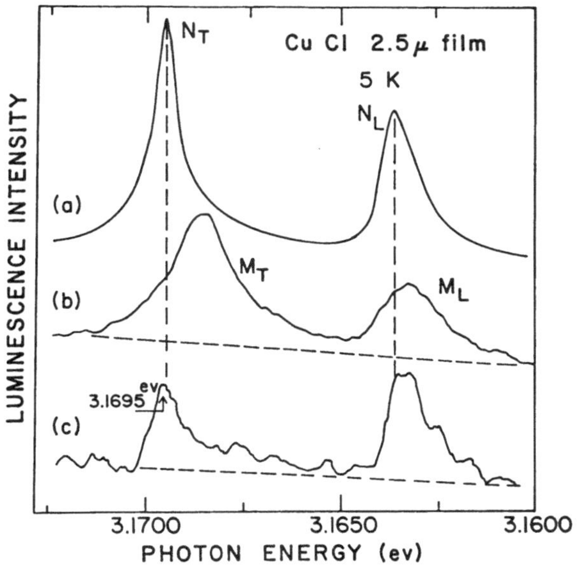

Figure 9. Biexciton luminescence of CuCl film at 5K.
a) from intense pump beam along ($h\omega_p$=3.186 eV)
b) from weak probe beams along ($h\omega_1$=3.184 eV), $h\omega_2$= 3.188eV). c) from weak probe beams alone, but in the presence of the pump beam.

these results in terms of an amplification of the emission induced by photons from the decaying probe biexcitons. On the other hand, this attraction in momentum space is precisely what is expected from the presence of a condensate of excitonic molecules in the system, produced by the pump beam.

There are several indications showing that the gas of biexcitons in CuCl does not behave like an ideal Bose gas. So far quantum statistical effects could only be observed if molecules were generated directly in their band with a low kinetic energy. This behavior is consistent with the model calculations of the biexciton dynamics mentioned in Section III. The lifetime of the particles is too short to allow the phase transition to develop if hot carriers are initially injected. On the other hand, if an initial population is prepared with a low $\bar{K}$-vector, it can act as a nucleation mechanism, thereby reducing the time for condensation [27]. In this respect Bose-Einstein condensation of excitonic molecules is a non-equilibrium situation. Further, there is evidence that the interactions among the particles are important. For instance, a study of the excitonic molecule resonance performed at densities comparable to those of Figs. 5-7 shows a marked broadening of the biexciton absorption line [54] [55], which has been interpreted as due to interparticle collisions in a dense, liquid-like fluid [54]. Finally, it must be also mentioned that several features of the M_L luminescence are not yet fully explained.

In summary, biexcitons in CuCl furnish a system where a Bose-Einstein condensation is believed to occur, although the useage of this terminology is somewhat loose, since it is a non-equilibrium phase transition.

VII. CONCLUSION

Despite a situation plagued by several premature claims, there is growing evidence that excitonic particles may furnish a new system where effects predicted by the Bose-Einstein model of integral spin particles can be studied. The luminescence from free excitons in Cu_2O and free biexcitons in CuCl shows anomalies at high densities which can be interpreted in a consistent way by the use of quantum statistics. In Cu_2O no deviations from the ideal gas have been observed so far, but Bose-Einstein condensation has not been achieved yet. In CuCl, biexcitons become stabilized at high densities in the quantum state where they are initially injected. This can be viewed as a non-equilibrium Bose-condensation.

REFERENCES

1. Cooper pairs of electrons in superconductors or pairs of ^{3}He atoms in superfluid ^{3}He below 3mK represent a special case of the degenerate Bose regime, which cannot be treated in the framework of the ideal Bose gas model (the average distance between different pairs is small compared to the radius of a pair).

2. Experimentally, it is found that the energy distribution of the He atoms is very different from the ideal Bose distribution of equation (2b). In fact, the uncondensed fraction appears to have an almost classical (Boltzmann) distribution characterized by an effective temperature of about 15°K even when the sample is well below the lambda point at 2.17K. The condensed fraction has been measured recently. It is the order of 10 percent well below T_λ , and it is found to have a temperature variation which is considerably different from that of an ideal Bose-Einstein condensate.

3. A. Einstein, Sitz, pr. Akad. Wiss. Phys. Math. Kl 22, 261 (1924), 23 3 (1925).
4. F. London, Nature, Lond 141, 643 (1938).
   F. London, Superfluids vol. 2 (Wiley 1954).
5. N. Bogoliubov, J. Phys. USSR 11, 23 (1947).
6. A detailed account may be found for instance in G.D. Mahan, Many Particle Physics, Plenum, New York, 1981.
7. See for instance L.D. Landau, E.M. Lifshitz, Statistical Physics (Pergamon Press, 1958).
8. T.D. Lee, C.N. Yang, Phys. Rev. 112, 1419 (1958).
   S.T. Beliaev, Z. Exp. Teor. Phys. 34, 433 (1958)(Sov. Phys. JETP 7, 299 (1958).
9. For a recent review, see for instance the articles by P.J. Dean and R.S. Knox in this volume.
10. S.A. Moskalenko, Fiz. Tverd. Tela 4 276 (1962) (Soviet Physics Solid State 4 199 (1962).
11. I.M. Blatt, K.W. Böer, W. Brandt, Phys. Rev. 126 1691 (1962).
12. R.C. Casella, J. Phys. Chem. Solids 24, 19 (1963).
13. L.V. Keldysh, A.N. Kozlov, Zh. Eksp. Teor. Fiz 54, 978 (1968) (Sov. Phys. JETP 27 521, 1968).
14. E. Hanamura, Solid State Comm. 12, 951 (1973).
15. See the article by C. Benoit a la Guillaume in this volume.
16. R.K. Wehner, Solid State Comm. 7 457 (1969).
17. O. Akimoto, E. Hanamura, J. Phys. Soc. Japan 33, 1537 (1972).
18. W.F. Brinkman, T.M. Rice. B. Bell, Phys. Rev. B8, 1570 (1973).
19. F. Bassani, M. Rovere, Solid State Comm. 19, 887 (1976).
20. L.H. Nosanow, J. de Physique 41 C7-1 (1980).
21. W.F. Brinkman, T.M. Rice, Phys. Rev. B7 1508 (1973).
22. L.V. Keldysh, Y.V. Kopeav, Fiz. Tver. Tela 6, 2791 (1964) Soviet Physics Solid State 6, 2219 (1965).
23. A.J. Leggett, J. de Physique 41 C7-19.
24. P. Nozieres, private communication.
25. Y. Toyozawa, Prog. Theor. Phys. Suppl 12 111 (1959).
26. M. Inoue E. Hanamura, J. Phys. Soc. Japan 41 771 (1976).
27. V. Yakhot, E. Levich, Physics Letters 80A, 301 (1980).
28. E. Levich, V. Yakhot, Phys. Rev. B15, 243 (1977).
29. For a recent review see for instance R. Loudon in this volume.
30. R.S. Knox, Theory of Excitons (Academic Press, New York, 1963).
31. J.G. Gay, Phys. Rev. B4, 2567 (1971).

32. N. Peyghambarian, L.L. Chase, A. Mysyrowicz, Optics Comm. 41, 178 (1982).
33. K. Jarasiunas, H.J. Gerritsen, Appl. Phys. Lett. 33 190 (1978).
34. For a review of excitonic properties of Cu_2O see for instance S. Nikitine in Optical Properties of Solids, Nudelman and Mitra editors, Plenum Press (1969).
35. E.F. Gross, I.I. Kreingold, V.L. Makarov, JETP Letters, 15, 383 (1972).
36. A. Mysyrowicz, D. Hulin, A. Antonetti, Phys. Rev. Lett. 43, 1123, 1275(E) (1979).
37. D. Hulin, A. Mysyrowicz, C. Benoit a la Guillaume, Phys. Rev. Lett. 45, 1970 (1980).
38. P. Yu, Y.R. Shen, Phys. Rev. Lett. 32, 939 (1974).
39. I.I. Kreingold, V.L. Makarov, Fiz. Tverd. Tela 15, 1307 (1973) [Soviet Physics Solid State 15, 890 (1973)].
40 P.D. Bloch, C. Schwab, Phys. Rev. Lett. 41, 514 (1978).
41. A. Mysyrowicz, D. Hulin, C. Benoit a la Guillaume, Proc. Int. Conf. Luminescence, Berlin, (1981) to be published in J. of Luminescence.
42. A. Goldmann, Phys. Stat. Sol. b 81 9 (1977).
43. For a review of the literature on biexcitons in CuCl see for instance C. Klingshirn, H. Haug, Physics Reports 70 315 (1981).
44. R.W. Svorec, L.L. Chase, Solid State Comm. 20, 353 (1976). N. Nagasawa, T. Mita, M. Ueta, J. Phys. Soc. Japan 41 929 (1976).
45. G.M. Gale, A. Mysyrowicz, Phys. Rev. Lett. 32, 727 (1974).
46. S. Suga, T. Koda, Phys. Stat. Sol. B61 291 (1974).
47. H. Souma, T. Goto, T. Ohta, M. Ueta, J. Phys. Soc. Japan 29, 697 (1970).
48. L.L. Chase, N. Peyghambarian, G. Grynberg, A. Mysyrowicz, Phys. Rev. Lett. 42, 1231 (1979).
49. T. Kushida, Solid State Comm. 32, 209 (1980).
50. R. Levy, C. Klingshirn, E. Ostertag, Vu Duy Phach, Y.B. Grun, Phys. Stat. Sol. b77, 381 (1976).
51. N. Nagasawa, T. Mita, M. Ueta, J. Phys. Soc. Japan 41, 929 (1976).
52. N. Peyghambarian, L.L. Chase, to be published.
53. N. Peyghambarian, L.L. Chase, A. Mysyrowicz, Optics Comm. 41, 178 (1982).
54. L.L. Chase, N. Peyghambarian, G. Grynberg, A. Mysyrowicz, Optics Comm. 28, 189 (1979).
55. Y. Masumoto, S. Shionoya, Solid State Comm. 38, 865 (1981).

SMALL POLARONS IN BIOLOGICAL SYSTEMS [1, 2]

J. Jortner

Department of Chemistry
Tel-Aviv University
Tel-Aviv 69978, Israel

The understanding of the primary events in bacterial photosynthesis is of central importance for the elucidation of the basic mechanisms of the acquisition, storage and disposal of energy in the photosynthetic process. During the last few years, extensive new information on the dynamics of charge separation in reaction centers has emerged from model experiments of picosecond spectroscopy. The theory of electron transfer in biological systems is advanced in terms of nonadiabatic multiphonon radiationless transitions. The conventional theory bears a close similarity to small polaron theory. Extension of the theoretical treatment to account for competition between vibrational relaxation and electronic processes will be presented to describe some ultrafast reactions. The dynamics of the primary events of charge separation are determined by the interplay between intramolecular reorganization and intermolecular engineering. These effects combine in determining the directionality, selectivity and efficiency of charge separation events. Apart from intrinsic biological interest, these systems provide a beautiful example for a micro-electronic gadget operating on a molecular level.

REFERENCES

[1] J. Jortner, Phil. Mag. B40: 317 (1979).
[2] J. Jortner, Amer. Chem. Soc. 102: 6676 (1980).

PERSPECTIVES OF FREE ELECTRON LASERS IN SOLIDS

G. Kurizki and J. McIver*

Max Planck Institut für Quantenoptic
Barching bei Munchen, F.R. Germany
*Institute for Modern Optics
Physics and Astronomy Department
Univeristy of New Mexico
Albuquerque, N.M. USA

Free-Electron Laser (FEL) is a device in which energy is transferred from an electronic beam to a monochromatic electromagnetic wave in the presence of a spatially periodic, and static electric or magnetic field. Present FELs use a magnet with a period of a few centimeters and radiate visible or infrared light. there is no fundamental reason why FELs that will emit x-rays or even γ-rays cannot be constructed.

The problem is in finding a static field of sufficiently small period and large amplitude. Electric fields of such short periods (of the order of Angstroms) and large amplitude (of the order of 10^5V/cm) occur naturally near the center of channels in crystals. In order to determine whether crystals can be used in FELs, the spontaneous radiation emitted by a relativistic positron channeled in a crystal was studied. It was shown that this radiation is emitted at two wavelengths with approximately equal intensities. One of these wavelengths, λ_1, is due to the interaction of a positron with that part of the potential that confines it to a channel. The other wavelength, λ_2, is proportional to the lattice period P along the channel axis. Typically, λ_2 is several orders of magnitude smaller than λ_1 and is less than P. Preliminary classical calculations showed the possibilities of obtaining laser action in such a system.

THE DISPERSION CURVES OF EXCITONIC POLARITONS AND THEIR DISTORTION WITH INCREASING DENSITY

C. Klingshirn

Institute für Angewandte Physik
D7500 Karlsruhe, F.R. Germany

Some common methods used to determine the eigenenergies and dispersion curves of excitonic polaritons in direct gap semiconductors were listed. They are the following:

1. one photon absorption measurements,
2. two photon absorption measurements,
3. reflection measurements,
4. thin prism method,
5. Fabry-Perot modes,
6. resonant Brillouin scattering, and
7. two-photon or Hyper-Raman scattering.

The methods, 2, 5, 6, allow direct spectroscopy in the $\underline{k}$-space. The combination of several of the above techniques allows one to determine complicated structures like multiple resonance (e.g. in case of wurtzite-type semiconductors) or light and heavy polariton branches (e.g. in case of blend-type materials). Distortions may occur due to the interaction of various quasi-particles if their density is increased. One example involving the biexciton level is given in the following: by an intense laser, a high density of polaritons is created at $\hbar\omega_i$ in the sample, then a strong one photon absorption becomes possible at $\hbar\omega_{an} = E_{biex} - \hbar\omega_i$. This leads, via the Kramers-Kronig relations, to a resonance-like anomaly of the dispersion curve at $\hbar\omega_{an}$. Furthermore, the distortion at $\hbar\omega_i$, itself, can be deduced from the experiment.

At the highest densities, excitons cease to exist and a new collective phase is formed, the so-called electron-hole-plasma. A strong renormalization of the gap occurs. In direct gap materials, high optical gain ($10^4 cm^{-1}$) occurs, mainly between Eg and the chemical potential μ of the EHP. For $\hbar\omega > \mu$ strong one photon absorption is observed, due to the band to band transitions. The shift of the Fabry-Perot modes indicates a decrease of the refractive index below μ and a resonance-like behaviour around μ.

MOTION OF A MAGNON SOLITON ABOUR A PHONON SOLITON IN A ONE-DIMENSIONAL FERROMAGNET

A. H. Nayyar

Physics Department
Quaid-i-Azam University
Islamabad, Pakistan

The nonlinear dynamics of a coupled magnon-phonon system in a one-dimensional ferromagnet, where both the spin and the lattice systems were taken to have their respective nonlinear interactions, was examined. The phonon soliton was shown to introduce spatial inhomogeneities in the propagation of the magnon soliton resulting in (a) trapping of the magnon soliton in the harmonically perturbing field of the phonon soliton, and (b) the amplitude and the width of the magnon soliton becoming space-time dependent.

INTERCONFIGURATION FLUCTUATIONS

C. Cordero-Montalvo

Universidad de Puerto Rico
Colegio Universitario de Cayey
Div. de Ciencias
Departmento de Matematica-Fisica
Cayey, Puerto Rico

A transition occurs in several rare-earth (RE) compounds which involves the promotion of an electron from a localized (F) to a delocalized state. The integral valence of the RE ion, however, changes to a non-integral value, intermediate between that of two configurations. A dynamic, rather than a static, model of an intermediate valence (I.V) state is required which involves a fluctuation in time between the two configurations. The I.V. state was found to exist at ambient conditions in several compounds or to be achieved by changes in parameters such as external pressure, concentration of an RE dopant and temperature. Magnetic susceptibility (χ) measurements served as the probe which revealed the existence of the I.V. state and have been useful for its characterization. The dependence of χ on pressure and concentration in $Sm_{1-x}RE_xS$ compounds at intermediate temperatures were reviewed in this seminar.

LUMINESCENCE OF Mn^{2+} IN $RbMn_xMg_{1-x}F_3$*

J. Danko, D. Pacheco, and B. Di Bartolo

Department of Physics
Boston College
Chestnut Hill, Massachusetts, U.S.A.

In this seminar, J. Danko reported on the concentration-dependent characteristics of Mn^{2+} luminescence in the non-stoichiometric crystals; $RbMn_xMg_{1-x}F_3$: x = 1.0, 0.8, and 0.4.

Mn ions, in the vicinity of Ca, Zn and Mg impurities, emit the crystal's dominant sharp-line luminescence. Since the result of this perturbation is a lowering of the ion's excited energy levels, then this strong luminescence is indicative of a rapid transfer of energy among the relatively unperturbed Mn ions. At low temperature, the perturbed ions trap the excitation and emit the resultant luminescence. The introduction of the perturbing impurity, Mg, into the system in appreciable quantity should hinder this effective transfer of excitation. In fact, a limit should eventually be met where such a transfer would be less probable than a downward radiative transition; in this case, one would be able to trace the demise of a collective excitation. The thermal behavior of each system's fluorescence spectra and corresponding lifetime data were presented and discussed. Previous assumptions concerning the origin of the broad-band emission of $RbMnF_3$ were questioned. A lower limit was placed on the transfer rate of excitation among the relatively unperturbed Mn ions in each system.

* Supported in part by NATO Research Grant 1169.

EFFECT OF HYDROSTATIC PRESSURE ON THE LUMINESCENCE SPECTRA OF THE S_2^- CENTRE AND THE EPR OF THIS S_2^- CENTRE FOR THE CRYSTAL SCAPOLITE

H. K. Liu,* D. Curie, P. Jaszczyn-Kopec, and
B. Canny

Laboratoire de Luminescence I
Université Pierre et Marie Curie
Paris, France

In this seminar, H. K. Liu reported on the effect of hydrostatic pressure on the vibronic spectrum of the molecular centre S_2^- for the crystal scapolite, $(Na,Ca)_4(Al,Si)_{12}O_{24}(Cl,CO_3,SO_4)$. The emission spectrum of the S_2^- centre was determined by using an excitation of 4100A. The measurements of the emission spectrum were performed at liquid nitrogen temperature with and without hydrostatic pressure. With pressure he reported: 1) a blue shift for all the zero-phonon lines of the vibronic spectrum for the S_2^- centre in scapolite, and 2) a redistribution of intensities in favour of the high energy vibronic lines. Under hydrostatic pressure of about six kilobars, he reported a mean value of $d(h\nu)/dp \doteq 1.50$ meV/Kbar for the blue shift.

In order to determine the orientation of the S_2^- centre in scapolite, an EPR study was made. The orientation can be found by measuring the angular variation of the g-factor. Finally it was emphasized that the observations of hyperfine and superfine structure lines proved: (1) the existence of the S_2^- centre in the crystal, and (2) that the S_2^- centre is substitute for a Cl^- ion in the unit cell of the scapolite crystal.

* Permanent address: Changchun Institute of Physics
Chinese Academy of Sciences
Changchun, P.R. of China

ON THE CALCULATION OF POLARON WAVE-FUNCTION IN THE STATIC ELECTRON-LATTICE COUPLING

I. Rojas-Hernandez

Departamento de Fisica
Universidad Michoacana
Morelia, Mich., Mexico

The wave function of the polaron has been calculated from a theoretical point of view. The microscopic model was developed by Stumpf [1] ; this model was later applied to calculate the wave function of the F-center by Schmid [2] .

In this seminar, the adiabatic approximation was used in order to separate the electronic from the ionic part in the crystal, as Dr. Rojas-Hernandez was only interested in treating the electronic problem. The Schrödinger equation for the many electrons was then solved. To obtain the expectation value, the Hartree-Fock method was used, taking into account the Slater determinant ansatz for the electronic wave functions. In the obtained expectation value, the microscopic electronic and ion polarizations in the crystal were taken into account. Finally, through the minimization principle, the numerical results from the parameters included in the analytical expression of the energy expectation value and, therefore, the wave function of the polaron were obtained.

REFERENCES

1. Stumpf, J. Phys. Kondens Matt. 13, 9 (1971); ibid. 13, 101 (1971).

2. Schmid, J. Phys. Kondens Matt. 15, 119 (1972).

SEMICONDUCTOR SURFACE INVERSION LAYERS AND THEIR COLLECTIVE MODES

A. Tselis

Department of Physics
Brown University
Providence, Rhode Island U.S.A.

An inversion layer may be formed at the surface of a semiconductor crystal by the application of an appropriate electric field normal to the surface. For a p-type semiconductor, the inversion layer consists of a quasi-two-dimensional electron liquid, in which the motion of the electrons parallel to the surface is free; the motion perpendicular to the surface is quantized. Some of the especially interesting features were pointed out and discussed in this seminar.

Since the electrons in the inversion layer form a strongly interacting system, they support a variety of collective oscillations. The momentum-exciton model of plasmons first introduced by Quinn and Ferrell was reviewed. This was then used to elucidate the types of electronic collective modes to be expected to an inversion layer. A calculation of plasmon relations using an energy functional perturbation method was briefly outlined and the relation in accordance to some experimental results was discussed. A derivation of some of the results using a Feynman diagram approach was also outlined.

Finally, possible extensions of the theory, taking into account effects of such things as valley structure, and retardation, and applications to exotica like superlattices were examined.

PRESENT TRENDS IN COLLECTIVE EXCITATIONS IN SOLIDS*

F. Williams

Physics Department
University of Delaware
Newark, Delaware 19711, U.S.A.**

and

Institut für Festkorperphysik II
Technische Universität
D 1000 Berlin 12, Germany ***

ABSTRACT

Excitations of solids are in general identified as collective because of the many-body interactions. Single particle descriptions are therefore approximations, and well-known quasi-particle excitations such as phonons, excitons and plasmons are themselves approximations. The exact many-body excitations are designated "collectons." The formulation is further generalized to include probes, sources and applied stresses in interpreting phenomena. Examples include the polariton which results from coupling of the electromagnetic probe with transverse polarization in solids, atomic beams coupled to surface phonons, and solid bodies coupled to high pressure apparatus. General trends in research on collective excitation are reviewed, including extremal conditions for experiments, applied research and sophisticated formulations such as quantum field analyses. Finally, trends specific to particular classes of collective excitations are reviewed.

* Supported in part by a grant from the Army Research Office and in part by a Humboldt Prize (Senior U.S. Scientist Award).
** Permanent address.
*** Address during the summer of 1981.

I. INTRODUCTION

In the opening articles of this book the main emphasis has been on simplifying the many-body problem of the solid state into separable single particle or quasi-particle problems. For example, the electronic band structure was obtained for a single electron moving in the periodic potential of perfect crystals; the quasi-particles, phonons, were obtained from the normal modes of the lattice of perfect crystals. In this way optical excitations in certain spectral regions were describable in terms of purely electronic transitions and excitations in other spectral regions were describable in terms of creation and annihilation of phonons. Similar separations led to other excitations describable in terms of excitons, magnons, and plasmons.

In this article we return to our original description of the solid state as a many-body problem and note that any experimental excitation is in general collective and to some extent involves all the particles in this many-body problem. Thus, only in some approximations are these excitations particle-like. Both from the conceptual and analytical viewpoint there are advantages to single particle methods in describing excitations. However, these are clearly approximations even in those cases where the quasi-particle includes part of the many-particle interactions. In the following we shall in part abandon the single particle viewpoint and retain as far as possible, at least conceptually, the full many-particle point of view. In our most general analyses we shall include in the formulation the probes, sources of excitation, and applied stresses - all used in studying collective excitations in the solid state.

II. THE SOLID STATE AS A MANY-BODY PROBLEM

The many particle system can be described by the many-particle wave function $\Psi(\underline{r},\underline{R},t)$ here chosen in the real space representation, where $\underline{r}$ and $\underline{R}$ denote, respectively, electron and nuclear spatial and spin coordinates as defined in the first chapter. However, $\Psi(\underline{r},\underline{R},t)$ is the time-dependent wave function and is the solution of the time-dependent Schroedinger equation. For stationary states $\Psi(\underline{r},\underline{R},t)$ separates into the $\psi(\underline{r},\underline{R})$ of the first chapter and a periodic time-dependent factor.

In formulating collective excitations we assume that the ground state described by $\Psi_g(\underline{r},\underline{R},t)$ has been fully determined. The excitation also creates a new state out of the ground state. The new state is described by $\Psi_e(\underline{r},\underline{R},t)$. Although the operator generating the new state may couple to only a few of the particles in the system, all particles, both electronic and nuclear, are coupled to each other and therefore the excitation is describable exactly only as collective and to some extent involves all the particles in the many-particle system.

In part as a consequence of our experiences with the macroscopic world we tend to approximate these excitations as single-particle excitations, either single real particles or single quasi-particles. The quasi-particle model accounts for some of the many-particle effects by clothing a single real particle with some of the effects of the many-body interactions. If we use the single quasi-particle point of view then the exact collective excitation can be designated as a "collecton." All the well-known quasi-particles such as excitons, magnons, phonons, plasmons and polarons are approximations to the collectons.

Most, but not all, of these quasi-particles are bosons, but only approximately. The approximation is illustrated by excitons which are bosons at low concentrations but at high concentrations - at the onset of Bose condensation - the structure of the exciton as interacting fermions intervenes and the condensation fails to occur.

There is an interesting problem in the separation of the ground state problem, $\Psi_g(\underline{r},\underline{R},t)$, and the collective excitation leading to the excited state $\Psi_e(\underline{r},\underline{R},t)$ for some excitations approximated in terms of bosons. This is the problem of the zero point energies of bosons. Thus the zero point fluctuations are part of $\Psi_g(\underline{r},\underline{R},t)$ and some of the effects of the excitation are already in the ground state. This is the case for example for phonons and plasmons but is not for excitons: for the phonons and plasmons the source excites the system to higher levels, whereas the exciton is created by the excitation source interacting with the ground state.

In general, better approximations to the collectons are being developed. Corrections are being applied to the simple single quasi-particle models. For example, magnon and phonon side bands are now well-known for exciton spectra. Also the co-existence of interacting excitons and plasmons in the same material appears possible, as discussed by Dr. Egri.

III. COMPLETE DESCRIPTION INCLUDING PROBES, SOURCES AND STRESSES

The description of phenomena involving collective excitations in terms of the solid state as a many-body problem dependent on the coordinates of the isolated system is itself an approximation. A more complete description includes the probes and sources, respectively, used to detect and to effect the excitations and also includes for solid bodies investigated with applied electrical, magnetic or mechanical stresses some properties of the apparatus responsible for these stresses. Especially at high excitation intensities where the magnitude of the coupling with the excitation source becomes comparable with the intra-system couplings are the limitations of the approximation in Section II evident. Even at low excitation intensities for those phenomena in which the source or probe particles couple with the quasi-particles of the system, the limitations of this approximation become significant.

A complete description of the solid state during measurements with sources of excitation (E), probes (P) of the excitation, and stresses (S) applied to the solid body, in the most general case, can be formulated in terms of a total wave function $\Omega(\underline{r},\underline{R},E,P,S,t)$. This description couples the solid body to the excitation source, to the probe and to the apparatus for application of stress. The total wave function $\Omega(\underline{r},\underline{R},E,P,S,t)$ describes the total system including these perturbations. We now consider some specific cases involving this more general description.

In the case of radiative excitation with high intensity lasers, for example free electron lasers, the radiation field in the solid body becomes comparable with inter-electron forces in the body. The usual quasi-particle viewpoint then becomes untenable and formulation in terms of quantum field operators is necessary as made clear by Doctor Mahler. The quantum field operators include contributions to the field from the excitation source and from charged particles in the solid body.

A well-established example of coupling between the particles of the excitation source and the quasi-particles of the solid body at low excitation intensities is the coupling between the electric field of electromagnetic radiation with polarization fields from transverse optical phonons or from excitons. The coupled photon and transverse optical phonon can be described as a quasi-particle, designated the vibrational polariton; the coupled photon and exciton constitute the excitonic polariton. In general, if the exact collective excitation of the solid body were known and found to contain transverse components of polarization, then the coupled system of an electromagnetic radiation source or probe and the generalized transverse polarization could be described as a "collectonic" polariton. This point of view focuses attention on the existence of other types of polaritons than the well-established excitonic and vibrational polaritons.

Another example where the inclusion of the excitation and probe are important to describe phenomena is the use of atomic beams to probe surface phonons. In this case the interaction of the incident atoms with the surface atoms of the solid body is comparable with the interatomic interactions in the solid body.

An example of a measurement which can be properly interpreted only by including some properties of the apparatus in the analysis is the change in optical spectra as a function of hydrostatic pressure. The effective applied force on the microscopic part of the solid body interacting with the radiation becomes comparable for high pressures with the interatomic forces within that microscopic component. For constant applied stress S from a pressure apparatus whose time constant is long compared to the time constants for all the internal modes of the solid body, both electronic and vibrational, then the effects of the pressure apparatus can be separated by an approximation similar to the adiabatic approximation used to separate internal modes: $\Omega(\underline{r},\underline{R},S,t) = \Psi_S(\underline{r},\underline{R},t)\mu(S)$, and $\Psi_S(\underline{r},\underline{R},t)$ smoothly adjusts to the pressure apparatus described by $\mu(S)$, shown in the parametric dependence s.

IV. GENERAL TRENDS

A continuing general trend in the experimental investigations of collective excitations in solids is measurements under an extremal condition: high intensity excitation, low temperature, high magnetic field or novel spectroscopic techniques including time-resolved, site-selection and high-resolution spectroscopy. There is also a trend to experiments utilizing several concurrent extremal conditions, for example, the study of electron-hole droplets with the combination of high intensity excitation, low temperature and high magnetic field.

There is a trend towards applied research utilizing collective excitations in solids, in part reflecting the world-wide trend in science support. The detailed spectroscopy of bound excitons in semiconductors, reviewed in detail by Dr. Dean, has contributed to the understanding of semiconductor devices, for example, the role of deep centers for non-radiative recombination in light emitting diodes. The investigations of surface phonons and surface plasmons by atomic beam spectroscopy is in part focussed at solving problems of heterogeneous catalysis, as emphasized by Doctor Benedek. Excitons and polarons are used in some experimental devices for the conversion of solar energy into electricity and in others for the generation of fuel from solar energy. There is extensive research based on applying solid state concepts, particularly those involving collective excitations, to biology, for example excitons participating in the energy transfer processes occurring during photosynthesis, as discussed by Professors Knox and Jortner.

The trend in the theoretical descriptions of collective excitations for solids is towards greater generality and sophistication. Second quantization techniques are well-established in describing large polarons as reviewed by Professor Evrard.

Quantum field analyses which avoid the concepts of particles are, as noted earlier, being used to describe the solid state under very high intensities of electromagnetic radiation.

V. TRENDS SPECIFIC TO PARTICULAR CLASSES OF COLLECTIVE EXCITATIONS

In this section we shall discuss briefly some trends which appear to be specific to particular classes of collective excitations. By dividing them into classes we are making, as discussed earlier, some well-established approximations. These specific trends reflect the status of the experimental and theoretical research on particular phenomena.

A trend in phonon research is to interface and local modes. Surface modes are now well-established and are a particular example of interface modes. The relevance of surface modes to heterogeneous catalysis has been noted earlier. Local modes have been extensively investigated for luminescent materials and to some extent for semiconductors. Professor Terzi has proposed some interesting theoretical questions on the coherent excitation of local modes.

The trends on exciton research are different for different classes of materials. For semiconductors there now exists a great diversity of bound Wannier excitons with well-defined characteristics; for insulators, fewer excitons have been well characterized. The availability of synchrotron radiation facilitates exciton spectroscopy on insulators. Dr. Grasser reviewed these studies on rare gas solids and also reported the observation of Frenkel excitons in tungstates. Opportunities are clearly evident for research on excitons in complex crystals and in biological materials.

Plasmon resonances are well-established and understood for simple metals and for degenerate semiconductors. Complex metals, such as transition metals, exhibit broad and complex resonances in the spectral region for plasmons, in part due to the presence of interband transitions. Valence band plasmons, observed with insulators, are even less well understood. The coexistence of excitons and plasmons is an active and somewhat controversial subject.

The Fröhlich polaron has been exhaustively investigated theoretically and is well-established experimentally for polar semiconductors. On the other hand, small polarons have been less extensively investigated.

The theoretical basis for polaritons was reviewed by Professor Loudon. Many beautiful experiments are being reported on polaritons, for example, the high resolution spectroscopy reported by Dr. Klingshirn.

Magnons, htat is, quantized spin waves in ordered magnetic materials, have become a well-developed, distinct discipline, as evident from the contributions of Professor Mattis. Magnons are evident not only in magnetic measurements but also in electrical, optical and thermal studies; for example, magnon side bands in optical spectra.

VI. CONCLUSIONS

From the many-particle point of view there is clearly evident a unity to all collective excitations. There are research trends which are general to all collective excitations and there are others which are specific to particular classes of excitations. The limitations of single particle descriptions and of conceptually isolating the solid body from the probes and sources of excitation are becoming evident. A more complete formulation of collective excitations includes the quantum fields (or particles) of the source of excitation and of the probe for detection, in addition to, and coupled with, those of the solid body itself. The interactions of collective excitations of different classes with each other and with applied stresses are becoming clear, both experimentally and theoretically.

ACKNOWLEDGEMENTS

The author is indebted to Professors Benedek, Di Bartolo, and Lucas for helpful suggestions during the preparation of this article.

CONTRIBUTORS

F. Auzel, Centre Nat'l d'Etudes des Télécommunications, Bagneux, France
G. Benedek, Istituto di Scienze Fisiche, Università degli Studi, Milano, Italy
C. Benoit à la Guillaume, Université de Paris VII, Paris, France
B. Canny, Université Pierre et Marie Curie, Paris, France
L. L. Chase, Indiana University, Bloomington, IN, USA
C. Cordero-Montalvo, Universidad de Puerto Rico, Cayey, Puerto Rico, USA
D. Curie, Université Pierre et Marie Curie, Paris France
J. Danko, Boston College, Chestnut Hill, MA, USA
P. J. Dean, RSRE Malvern, Worcestershire, U.K.
B. Di Bartolo, Boston College, Chestnut Hill, MA, USA
I. Egri, Institut für Theoretische Physik, Aachen, F. R. Germany
R. Evrard, Université de Liège, Liège, Belgium
R. Grasser, Justus-Liebig-Universität, Giessen, F. R. Germany
D. Hulin, Université de Paris VII, Paris, France
B. Jacquier, Université de Lyon I, Villeurbanne, France
P. Jaszczyn-Kopec, Université Pierre et Marie Curie, Paris, France
J. Jortner, Tel-Aviv University, Tel-Aviv, Israel
C. Klingshirn, Universität Karlsruhe, Karlsruhe, F. R. Germany
R. S. Knox, University of Rochester, Rochester, NY, USA
G. Kurizki, University of New Mexico, Albuquerque, NM, USA
H. K. Liu, Université Pierre et Marie Curie, Paris France (on leave from Changchun Institute of Physics, Changchun, China)
R. Loudon, University of Essex, Colchester, Essex, U.K.
A. A. Lucas, Facultés Universitaires Notre Dame de la Paix, Namur, Belgium
G. Mahler, Universität Stuttgart, Stuttgart, F. R. Germany
D. C. Mattis, University of Utah, Salt Lake City, UT, USA
J. K. McIver, University of New Mexico, Albuquerque, NM, USA
R. Moncorgé, Université de Lyon I, Villeurbanne, France
A. Mysyrowicz, Université de Paris VII, Paris, France
A. Nayyar, Quaid-i-Azam University, Islamabad, Pakistan
D. Pacheco, Boston College, Chestnut Hill, MA, USA

I. Rojas-Hernandez, Universidad Michoacana, Morelia, Michoacan, Mexico
A. Scharmann, Justus-Liebig-Universität, Giessen, F. R. Germany
N. Terzi, Istituto di Scienze Fisiche, Università degli Studi, Milano, Italy
A. Tselis, Brown University, Providence, RI, USA
F. Williams, University of Delaware, Newark, DE, USA

INDEX